HUMAN POLYOMAVIRUSES

HUMAN POLYOMAVIRUSES

MOLECULAR AND CLINICAL PERSPECTIVES

Edited by

Kamel Khalili
Director and Professor
Center for Neurovirology and Cancer Biology
Temple University

Gerald L. Stoner
Chief, Neurotoxicology Section
National Institute of Neurological Disorders and Stroke
National Institutes of Health

A JOHN WILEY & SONS, INC., PUBLICATION

This book is printed on acid-free paper. ♾

Published simultaneously in Canada.

For ordering and customer service, call 1-800-CALL-WILEY.

Library of Congress Cataloging-in-Publication Data:
Human polyomaviruses : molecular and clinical perspectives / edited by Kamel Khalili, Gerald L. Stoner.
p. cm.
Includes bibliographical references and index.
ISBN 0-471-39009-7 (cloth : alk. paper)
1. Polyoma virus. 2. Oncogenic DNA viruses. I. Khalili, Kamel, 1951– . II. Stoner, Gerald L., 1943–

QR372.O58 H85 2001
616′.0194—dc21

2001026948

Printed in the United States of America.

10 9 8 7 6 5 4 3 2 1

This book is dedicated to George Khoury (1944–1987) and Norman P. Salzman (1926–1997) for their pioneering studies on the molecular virology of the polyomaviruses.

CONTENTS

PREFACE

It has been 15 years since the late Norman P. Salzman edited the *The Papovaviridae*, Volume 1, *The Polyomaviruses*. Then single chapters covered JC virus (JCV) and BK virus (BKV), and the chapter on BKV was shared with monkey lymphotropic papovavirus. The explosion of knowledge in the last 15 years warrants an update on these viruses focusing on those of immediate consequence to humans. Two of them, BKV and JCV, were discovered in humans in 1971. The third, simian virus 40 (SV40), was discovered in monkey kidney cells 10 years earlier, but is now viewed as a potential human pathogen for reasons described herein.

We begin with some historical background, recognizing the generation of virologists and neuropathologists who first worked in this field and having them tell of the excitement of discovery in their own words, where possible. Dr. Hilleman, who discovered SV40, recently described that achievement (Hilleman, 1998). Later chapters cover all aspects of the human polyomaviruses, including the clinical issues involved in diagnosis and treatment of these difficult, persistent infections.

This book will be of interest to graduate students, medical students, and advanced undergraduates and to anyone engaged in the study of DNA viruses, their molecular biology, evolution, epidemiology, and pathologic potential. The latter includes their oncogenic properties and their roles in nephritis in renal allografts and in the fatal viral CNS demyelinating disease, progressive multifocal leukoencephalopathy (PML). PML was discovered in patients with leukemia and lymphoma, but is now primarily a disease of patients with AIDS. We have encouraged contributors set out areas of conflict, where these exist, which future research must resolve.

The polyomaviruses were originally considered to be a genus within the papovavirus family (the other genus including papillomaviruses), and sometimes the word *papovavirus* was applied to BKV, JCV, and SV40. The classification "papovavirus" has now been dropped by the International Committee on the Taxonomy of Viruses (ICTV). The polyomaviruses and the papillomaviruses are both small DNA viruses, but they do not share genomic organization or show homology of DNA sequence and are no longer placed in the same family. We have dropped the usage of *papovavirus* in this book in favor of *polyomavirus* except in its appropriate historical context. The mouse polyoma virus, which gave its name to this group of viruses, should be distinguished by the use of "polyoma virus" as two words.

We thank each of the contributors, without whom this book would not have been possible. The polyomaviruses are of international concern, and an international effort is underway to understand and defeat them. We thank our editor at Wiley, Luna Han and Danielle Lacourciere, for their expert assistance and dedication.

KAMEL KHALILI, PH.D.
Philadelphia, Pennsylvania

GERALD L. STONER, PH.D.
Bethesda, Maryland

REFERENCE

Hilleman MR (1998): Discovery of simian virus 40 (SV_{40}) and its relationship to poliomyelitis virus vaccines. Dev Biol Stand 94:183–190.

CONTRIBUTORS

Hansjürgen T. Agostini, M.D., Department of Ophthalmology, University of Freiburg, Freiburg, Germany

Amy S. Arrington, B.A., Department of Molecular Virology and Microbiology, Baylor College of Medicine, Houston, TX

Karl E. Åström, M.D., Karolinska Institute (retired Professor), Stockholm, Sweden

Walter J. Atwood, Ph.D., Department of Molecular Microbiology and Immunology, Brown University, Providence, RI

Giuseppe Barbanti-Brodano, Ph.D., Department of Experimental and Diagnostic Medicine, Section of Microbiology, and Center for Biotechnology, University of Ferrara, Ferrara, Italy

Joseph R. Berger, M.D., Department of Neurology, University of Kentucky College of Medicine, Lexington, KY

Janet S. Butel, Ph.D., Department of Molecular Virology and Microbiology, Baylor College of Medicine, Houson, TX

Alfredo Corallini, Ph.D., Department of Experimental and Diagnostic Medicine, Section of Microbiology, and Center for Biotechnology, University of Ferrara, Ferrara, Italy

Luis Del Valle, M.D., Center for Neurovirology and Cancer Biology, College of Science and Technology, Temple University, Philadelphia, PA

Kristina Dörries, M.D., Institute für Virologie und Immunbiologie, Julius-Maximilians-Universität Würzburg, Germany

Pasquale Ferrante, M.D., Laboratory of Biology, Don C. Gnocchi Foundation, Milan, Italy

Richard J. Frisque, Ph.D., Department of Biochemistry and Molecular Biology, Pennsylvania State University, University Park, PA

Jennifer Gordon, Ph.D., Center for Neurovirology and Cancer Biology, College of Science and Technology, Temple University, Philadelphia, PA

John W. Henson, M.D., Brain Tumor Center and Molecular Neuro-Oncology Laboratory, Massachusetts General Hospital, Boston, MA

Roland Hübner, Ph.D., Laboratory of Pathology, University of Antwerpen, Wilrijk, Belgium

Michael J. Imperiale, Ph.D., Department of Microbiology and Immunology, Comprehensive Cancer Center, University of Michigan Medical School, Ann Arbor, MI

David V. Jobes, Ph.D., Neurotoxicology Section, National Institute of Neurological Disorders and Stroke, National Institutes of Health, Bethesda, MD

Kamel Khalili, Ph.D., Center for Neurovirology and Cancer Biology, College of Science and Technology, Temple University, Philadelphia, PA

Hee-Sun Kim, Ph.D., Molecular Neuro-Oncology Laboratory, Massachusetts General Hospital, Charleston, MA

Wendy A. Knowles, Ph.D., Enteric and Respiratory Virus Laboratory, Virus Reference Division, Central Public Health Laboratory, London, UK

Dana E. M. Rollison, Sc.M., Department of Epidemiology, Johns Hopkins Bloomberg School of Public Health, Baltimore, MD

Mária Mázló, M.D., Ph.D., University of Pécs, Faculty of Medicine, Central Electron Microscopy Laboratory, Pécs, Hungary

Ugo Moens, Ph.D., University of Tromsø, Institute of Medical Biology, Department of Molecular Genetics, Tromsø, Norway

Kazuo Nagashima, M.D., Ph.D., Department of Molecular and Cellular Pathology, Hokkaido University School of Medicine, Sapporo, Japan

Avindra Nath, M.D., Department of Neurology and Department of Microbiology and Immunology, University of Kentucky, Lexington, KY

Massimo Negrini, Ph.D., Department of Experimental and Diagnostic Medicine, Section of Microbiology, and Center for Biotechnology, University of Ferrara, Ferrara, Italy

Ole Petter Rekvig, M.D., Ph.D., University of Tromsø, Institute of Medical Biology, Department of Molecular Genetics, Tromsø, Norway

Holly G. Ressetar, Ph.D., Department of Neurobiology and Anatomy, West Virginia University School of Medicine, Morgantown, WV

Todd D. Schell, Ph.D., Department of Microbiology and Immunology, Pennsylvania State University College of Medicine, Hershey, PA

Keerti V. Shah, M.D., Dr. P.H., Department of Molecular Microbiology and Immunology, Johns Hopkins Bloomberg School of Public Health, Baltimore, MD

Yukiko Shishido-Hara, M.D., Ph.D., Laboratory of Molecular Neurobiology, Human Gene Sciences Center, Tokyo Medical and Dental University, Tokyo, Japan

Gerald L. Stoner, Ph.D., Neurotoxicology Section, National Institute of Neurological Disorders and Stroke, National Institutes of Health, Bethesda, MD

Chie Sugimoto, Ph.D., Department of Microbiology and Immunology, The Institute of Medical Science, The University of Tokyo, Tokyo, Japan

Satvir S. Tevethia, Ph.D., Department of Microbiology and Immunology, Pennsylvania State University College of Medicine, Hershey, PA

Mauro Tognon, Ph.D., Department of Morphology and Embryology, Section of Histology and Embryology, and Center for Biotechnology, University of Ferrara, Ferrara, Italy

Duard L. Walker, Ph.D., Department of Medical Microbiology and Immunology, University of Wisconsin Medical School, Madison, WI

Thomas Weber, Dr. MED., Neurologische Klinik, Kath. Marienkrankenhaus GmbH, Hamburg, Germany

Yoshiaki Yogo, Ph.D., Department of Microbiology and Immunology, The Institute of Medical Science, The University of Tokyo, Tokyo, Japan

Gabriele M. Zu Rhein, M.D., Department of Pathology, University of Wisconsin Medical School, Madison, WI

1

PROGRESSIVE MULTIFOCAL LEUKOENCEPHALOPATHY: THE DISCOVERY OF A NEUROLOGIC DISEASE

KARL E. ÅSTRÖM, M.D.

1. INTRODUCTION

Progressive multifocal leukoencephalopathy (PML) was discovered and described in 1958 at the end of an era, approximately 100 years long, during which modern medicine, including neurology, acquired its form and content. Pioneered by the eminent French neurologist and pathologist Charcot and the renowned British neurologist Hughlings Jackson, the identification and description of diseases in the nervous system during this period was based on the method of clinicopathologic correlation in which gross and light microscopic changes, seen postmortem, are related to the clinical history, signs, and symptoms of an individual. The discovery of PML, by the same method, is part of this tradition. This historical essay is based on personal memories from 1956 and 1957 when, working in the Neurology Service and Neuropathology Laboratory of Massachusetts General Hospital (MGH) in Boston, I was participant in and witness to the discovery of PML and co-author of the first paper written about PML, which was published in 1958 (Åström et al., 1958). The discovery has also been described by Richardson et al. (1994) and Henson and Louis (1999).

Human Polyomaviruses: Molecular and Clinical Perspectives, Edited by Kamel Khalili and Gerald L. Stoner.
ISBN 0-471-39009-7

I went to MGH in April 1956 in order to study neuropathology, especially that of demyelinating diseases. I had earned my M.D. and Ph.D. at the Karolinska Institute in Stockholm, Sweden, spent 2 years as Research Fellow in its Department of Physiology and 1 year as Visiting Scientist at the Rockefeller Institute (later Rockefeller University) in New York, and completed a residency in neurology in Stockholm. My main interest since medical school had been the nervous system, especially its morphology.

2. MASSACHUSETTS GENERAL HOSPITAL

MGH has since its beginning been the main teaching hospital for students at Harvard Medical School (HMS). The first building, opened in 1821, and is named after its designer, the renowned New England architect of the U.S. Capitol, Charles Bulfinch. It has retained its original location near the Charles River (Fig. 1.1), but is now surrounded and dwarfed by new, towering constructions of glass and steel. It stands as a symbol of a dynamic institution,

Figure 1.1. The Bulfinch and White buildings of the Massachusetts General Hospital in the mid-1950s. The Bulfinch Building originally sat on the banks of Charles River. The land has since been filled in. Housed under a hemispherical roof is the "Ether Dome," where the first public demonstration of the use of ether took place in 1846 and where "weekly clinicopathologic exercises" are held, the proceedings of which are still published under this name in the *New England Journal of Medicine*.

one of the leading hospitals and centers for biomedical research in the world, the pride of Boston.

Dr. Raymond Adams, who was to become my mentor, teacher, and lifelong friend, was, when I first met him, concurrently Chief of the Neurology Service at MGH and Bullard Professor of Neuropathology at HMS. The importance given to pathology in relation to clinical neurology during his tenure at MGH is shown by the facts that the residents in neurology then had to spend 1 year out of 3 in the Neuropathology Laboratory and that three weekly conferences were devoted to cases in which the correlation of clinical and pathologic data was an important element. Adams' approach to teaching, research, the practice of clinical neurology, and the writing of textbooks was in the tradition of authors from the classic period such as Gowers, Hughlings Jackson, and Kinnier Wilson. This is the tradition in which we, who had the priviledge to work under and with him, were molded.

I spent my first months at MGH in the old Pathology Building (abandoned later in the year), where in 1927 Dr. Charles Kubik had established a neuropathologic laboratory (later given his name). Kubik, a neurologist, had learned neuropathology in the laboratory of Dr. J.G. Greenfield at National Hospital, Queen Square, London (Richardson et. al., 1994). After retirement he still came daily to the laboratory, where I met him and became his friend. There I also met the late Dr. E.P. Richardson, Jr., another lifelong friend. His association with the Neuropathology Laboratory lasted more than 50 years during which time he became a revered figure to many generations of students, residents, and visitors who passed through the laboratory.

The purpose of this introductory survey is to put the discovery of PML into its historical context by presenting the key figures in neurology and neuropathology at MGH in the mid-1950s and to show how closely pathology was integrated into neurology there at that time. The environment that then existed in the Neurology Department was conducive and indeed essential for this discovery.

3. AN ENIGMATIC CASE

Upon my arrival, Dr. Richardson suggested that I, as a Research Fellow, should study records and slides from three cases in which demyelination was a prominent feature. Demyelination, a destruction of myelin with relative preservation of the axis cylinders, is a nonspecific, nonpathognomonic phenomenon. The process was considered the common denominator of a group of diseases that, regarding etiology, were not known to be related (Adams and Kubik, 1952). Of my three cases, one proved to be an example of neuromyelitis optica and one was Schilder's disease, both of which were possibly variants of multiple sclerosis. It soon became clear, however, that the third one had no relation to either multiple sclerosis or any other disease known at that time. I focused my studies on this case, which would become the index case of a disease that we later described and gave the name *progressive multifocal leukoencephalopathy* (PML).

The patient was a 71-year-old woman who had developed neurologic symptoms after having had chronic lymphatic leukemia for 15 years. She was admitted to a cancer ward at MGH in November 1952, where Dr. Raymond Adams was consulted. Because he was the first neurologist at MGH to see her and to understand the unique character of her disease, I have asked him to describe his memories of this encounter. The following excerpts are from a letter that I recently received.

> A clinician at the Huntington Cancer Center at MGH . . . "had been following the patient and had concluded that her subacutely evolving left hemispheral lesion was a leukemic infiltrate. I thought it was not, based on lack of cerebral leukemic infiltrates in other chronic lymphatic leukemic patients. . . ." The clinician . . . "recommended radiation* which proved to be ineffective. . . ." "If my memory serves me correctly the hemispheral lesion continued to progress over a few weeks or months and led to the patient's death. Her lymphatic leukemia remained relatively inactive."

The patient died in March 1953. On pathologic examination, the most prominent feature was at first sight demyelination, which was, however, different from that in any then-known demyelinative disease. Glial reactions are to be expected with demyelinative lesions, but here a feature of special interest was the presence of giant, monstrous astrocytes with atypical mitoses of neoplastic type even though there were no other signs of brain tumor. From an early conversation with Dr. Adams about the case, I vividly remember his exclamation: "Have you ever seen astrocytes like these in multiple sclerosis?!" In his letter he wrote 44 years later:

> "You may recall my surprise at seeing both the inclusions and the multiple mitoses, often abnormal, in astrocytes. I had never seen the latter except in tumors (gliomas), never in demyelinative lesions. I asked J. G. Greenfield, when he was visiting Charles Kubik [in October 1956]. He said he may have but seemed uncertain."

Changes like these had never been seen before by members of the Neurology Group and Neuropathology Laboratory at MGH, and nobody understood their nature and cause. One hypothesis was that the lesions were side effects of radiation treatment, though clearly the lesions did not resemble our other cases of radiation necrosis. Alternatively, one could assume that the leukemia and the brain disease in some way were linked or that the latter was a condition *sui generis*, that is, unrelated to the leukemia. None of these questions could be answered on the basis of a single case, however. Its nature was an unsolved riddle as it had been during the 3 years since the death of the patient. An unexpected event then occurred.

---

*Radiation to the patient's head was given against Dr. Adams' advice.

Dr. Elliot Mancall, a clinical neurologist, was spending a year in the Neuropathology Department of MGH. Being on call one day in August 1956, he performed the autopsy of a 73-year-old woman, who, like the first patient, had had chronic lymphatic leukemia and a final episode of a progressive brain disease. The changes in the brain were also similar to those in the first case. An important point was that this woman had *not* received radiation. Similar histologic changes were also noted in a third case in which autopsy material had been provided by Dr. Herbert Karp of Emory University in Atlanta. The patient was a 42-year-old man with Hodgkin's disease who had had neurologic symptoms and signs during the last 3 months of his life.

4. CONCLUSIONS AND THE PAPER

This was the situation at the end of 1956: We—Richardson, Mancall, and myself—were confronted with three cases that were remarkably similar. All patients had had chronic diseases involving lymphoid tissue that had lasted from 2 to 15 years; all had developed terminal neurologic symptoms and signs that progressed and led to death within 4 months; and all had had the same unique type of lesions in their brains. Because the second patient had not received radiation of the head, we could eliminate such treatment as a cause of the cerebral lesions. We concluded that the neurologic signs and symptoms had been caused by the unique set of lesions in the brain, that the clinical features and changes in the brain were parts of an unknown disease, and that there was a linkage between this disease and chronic lymphatic leukemia and Hodgkin's lymphoma. This was our first formulation of PML, although the name came later. We decided to publish our findings and conclusions.

The work on the manuscript, which started in the early part of 1957, became an unusual, cooperative project. We met almost daily in Richardson's office for several months, looking at slides, discussing the findings, and writing the manuscript, word by word together. In this way we gradually formulated in our minds, and put on paper, the idea of a new disease, which eventually was called *progressive multifocal leukoencephalopathy*. Its fundamental elements were as follows.

Pathology

The pathology of PML is characterized by a mixture of small and larger demyelinative lesions, which are disseminated in the white and to a lesser degree gray matter of the brain. They create a pattern that is so typical of the disease that one can make a diagnosis by looking with the naked eye at the fixed brain, where the lesions appear as grayish spots, or microscopical slides thereof when stained for myelin. Especially noteworthy are disseminated small, round lesions that form groups (Fig. 1.2A). This arrangement of lesions had not been seen in our cases of multiple sclerosis. Very large lesions of PML also can be dis-

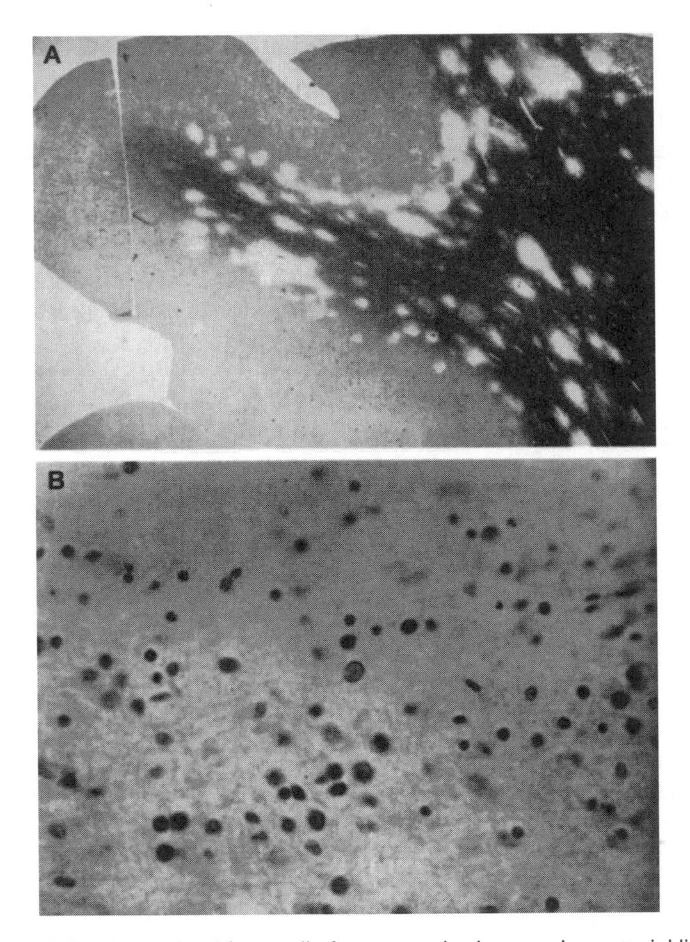

Figure 1.2. (**A**) Section stained for myelin from a cerebral gyrus shows sprinkling of small lesions with some tendency to confluence in the subcortical white and, to a small degree, gray matter. This pattern is typical of early phase PML. (**B**) Periphery of large lesion, stained for visualization of cells, shows normal and abnormal oligodendrocytes. The latter contain dark, enlarged, round or oval nuclei. An "inclusion body" fills the nucleus of an oligodendrocyte in the center of the picture except for a thin rim of chromatin at its margin. Taken, with permission of *Brain*, from Åström et al., 1958.

tinguished from those in Schilder's disease in that they are surrounded by small, so-called satellite lesions. The pathology reflects a progressive process: Small lesions appear to have merged to form larger ones. These continue to grow while new, small lesions successively appear in previously unaffected regions of the brain. The differing ages of the lesions are reflected in the variable microglial-histiocyte responses. Breakdown of myelin and gliotic reactions are of the same type as those in other demyelinative diseases (e.g., multiple sclerosis), except for the presence in larger lesions of gigantic astrocytes of a type that usually are seen only in gliomas.

While reviewing all the slides we observed a previously neglected feature. At the periphery of some lesions we noted round or oval, dark, basophilic cells that were similar to, but larger than, normal oligodendrocytes. Some of them contained nuclear inclusions of eosinophilic material (Fig. 1.2B). After further study and discussions we concluded that they were abnormal oligodendrocytes. Alterations of oligodendrocytes like these, never before described in the pathologic literature, are pathognomonic for PML. Cavanagh et al. (1959) and Waksman (quoted by Richardson, 1961) later suggested a viral origin of the disease.

A survey of the literature showed that cerebral lesions like those in our three cases had been seen by other authors and described in four papers under various diagnoses or declared to be idiopathic. Hallervorden (1930) found them in two patients, one of whom had died of tuberculosis. He understood that the pathologic changes were unique and called them *non-classifiable*, whereas the other authors tried to fit the lesions into molds of conventional types. Thus, Winkelman and Moore (1941) described them as a "lymphogranulomatous invasion of the brain" in a case of Hodgkin's disease. In another case of Hodgkin's disease, Bateman et al. (1945) believed the lesions to be those of Schilder's disease. Finally, Christensen and Fog (1955) reported extensive demyelination in the brain of a 59-year-old man with sarcoidosis as "Schilder's Disease in an Adult." Reviewing slides from three of these five cases, which we were allowed to borrow, and studying descriptions and photographs in relation to the others, we concluded that the pathologic changes in the brains were virtually identical to those in our three cases.

Some of the quoted authors had noted the monstrous astrocytes (although they were mistakenly called Reed-Sternberg cells in one paper), but none had seen the abnormal oligodendrocytes with eosinophilic inclusions, and none had appreciated the progressive nature of the lesions. Hence, ours was the first complete description of PML's pathology.

Clinical Picture

The clinical picture was remarkably similar in all eight cases. Multiple neurologic symptoms localized to one region of the brain had appeared successively, progressed without remissions, and terminated in the death of the patient, usually within 6 months or less. The fact that the clinical and pathologic features of the brain disease were similar in all eight cases, and that they had in at least seven of them developed against a background of a lymphproliferative or granulomatous condition, strengthened our conviction that the former were a complication of the latter.

Background Disease

The background or underlying disease had been lymphatic leukemia in two cases, Hodgkin's disease in three, and sarcoidosis and tuberculosis in one case each. No mention was made of a background disease or of a general autopsy in Hallervorden's second case. In the subtitle of our article we called PML "a hitherto unrecognized complication of chronic lymphatic leukemia and Hodgkin's disease." It was known in the mid-1950s that patients with these diseases could have neurologic symptoms due to accumulations of leukemic cells around and near blood vessels, nerve roots, and meninges, as well as growths of granulomatous tissue in the epidural space and, in at least one case, within the brain (along with infarcts and hemorrhages). PML was obviously a different kind of complication, however. It was a disease in itself with characteristic pathology and clinical features.

5. THE DISCOVERY

After the publication of our paper in 1958 our observations were substantiated, the name PML accepted, and its unique nature affirmed by others. The discovery was a product of preparedness and luck. As pointed out in the Introduction, the pervailing atmosphere in the Neurology Department at MGH in the 1950s was conducive to discoveries of this kind. Furthermore, in connection with my arrival at MGH in April 1956, the first case was brought out from the files and became a focus for studies by me and renewed interest on the part of Richardson and Adams. Thus, our minds were prepared when, a few months later, a similar case came to autopsy and when we had access to material from a third one. At this moment the concept of PML was born; a set of enigmatic pathologic findings had found its disease. This was our "Eureka" experience.

The essence of our discovery consisted of three parts, each of which has been of seminal importance for subsequent research on and understanding of PML. (1) PML is a disease *sui generis*, which has its own characteristic pathology and clinical features and, once started, follows its own progressive course, independent of the background disease. The discovery of PML as a well-defined entity in 1958 and its acceptance by the medical community opened the door to research in many fields, which continues. (2) The discovery of nuclear inclusions and enlargements of oligodendrocytes in PML lesions and the observation of abnormal astrocytes formed a starting point for research by others on the etiology and pathogenesis of the disease. (3) PML appears in the course of a number of different underlying conditions. In the paper we focused on the connection between PML and leukemia/Hodgkin's disease, which had existed in our three cases, although a review of the literature showed that it could appear also during the course of sarcoidosis and tuberculosis. The latter observation gave an important clue: PML is not linked to a specific disease. We know now that it can occur in the course of many conditions, the common denominator of which is impaired immunity, and in other situations where the immune system has been compromised.

The discovery of PML has given rise to various opinions regarding priorities, none, however, voiced by any of the three authors or anyone who was present when the discovery was made. For example, Berger and Major (1999) claimed that Hallervorden described PML in 1930. These authors have confused the concepts of pathology and disease. A disease in a living being is by definition an aberration of normal bodily functions that involves pathology (changes in structures and functions), etiology, clinical signs and symptoms, relation to other diseases, prognosis, and so forth. Our paper in 1958 dealt with these features except the etiology, although our discovery of the pathology of oligodendrocytes provided a starting point for prospective work in this field. Hallervorden was the first to describe the *pathology* (although incompletely) of a disease that 28 years later was discovered and named PML, but he did not describe or discover the *disease* PML.

In the mid-1950s, due to increasing specialization and the ever-expanding mass of new information, clinical neurology and neuropathology gradually became separated. Neurosciences were about to enter a new era. New methods for morphologic investigations allowed for higher resolution of structural details, identification of chemical products, and correlation between structure and function in cells. These methods coupled with progress in virology, immunology and molecular biology have made possible more recent additions to our knowledge of PML. For example, in the 1990s it was possible to apply methods to and obtain new information from material that had been collected and saved in the early 1960s, when these methods were not available (Åström and Stoner, 1994).

With today's knowledge in mind, I will claim that data in our paper support if not actually forecast the following ideas. PML is a slow viral disease (Sigurdsson, 1954), one of the first to be known in humans. It is an immune deficiency disease, in today's language an *opportunistic infection*. It is a gliopathy because it affects macroglial cells selectively. Demyelination, at first sight the most striking pathologic change, is secondary to degeneration of oligodendrocytes. Virus can cause demyelination and gliomatous changes in astrocytes. The discovery also supports the concept that oligodendrocytes are essential for synthesis of myelin.

6. CONCLUDING REMARKS

The PML paper was written in the same year as the eighth and last volume in the German series on the pathology of the nervous system appeared (Lubarsch et al., 1957). A reviewer in the *Journal of Pathology and Bacteriology* (74: 469, 1957) declared about one of the volumes in this monumental series, which is a summary of a century's accumulated knowledge of classic neuropathology: "The present volume is a much needed reminder of the existence of vast unexplored tracts of pathological morphology without which the psychiatrist, neurochemist, and others would be in danger of living in a fool's paradise." PML was discovered at the end of an era of research that is embodied in this Handbook, but it came also at the threshold of a new one.

In conclusion, I feel a deep gratitude to the institution where the work was done; to my collaborators and teachers at MGH; and to fate for letting me be part of and witness to a piece of medical history.

ACKNOWLEDGMENTS

I thank Drs. R.D. Adams and E.L. Mancall for reading the manuscript and making valuable suggestions.

REFERENCES

Adams RD, Kubik CS (1952): The morbid anatomy of the demyelinative diseases. Am J Med 12:510–546.

Åström KE, Mancall EL, Richardson EP Jr (1958): Progressive multifocal leukoencephalopathy. A hitherto unrecognized complication of chronic lymphatic leukemia and Hodgkin's disease. Brain 81:93–111.

Åström KE, Stoner GL (1994): Early pathological changes in progressive multifocal leukoencephalopathy: A report of two asymptomatic cases occurring prior to the AIDS epidemic. Acta Neuropathol 88:93–105.

Bateman OJ Jr, Squires G, Thannhauser SJ (1945): Hodgkin's disease associated with Schilder's disease. Ann Intern Med 22:426–431.

Berger JR, Major EO (1999): Progressive multifocal leukoencephalopathy. Semin Neurol 19:193–200.

Cavanagh JB, Greenbaum D, Marshall AHE, Rubinstein LJ (1959): Cerebral demyelination associated with disorders of the reticuloendothelial system. Lancet 2:524–529.

Christensen E, Fog M (1955): A case of Schilder's disease in an adult with remarks to the etiology and pathogenesis. Acta Psychiatry 30:141–154.

Hallervorden J (1930): Eigenartige und nicht rubrizierbare Prozesse. In Handbuch der Geisteskrankheiten; O. Bumke, Ed. Verlag von Julius Springer: Berlin, Vol. 11, part 7, pp 1063–1107.

Henson JW, Louis DN (1999): Edward Peirson Richardson Jr (1918–1998) and the discovery of PML. J Neurovirol 5:325–326.

Lubarsch O, Henke F, Rössle R, Eds (1957): Handbuch der speziellen pathologischen Anatomie und Histologie, Vol. 13, Nervensystem, Part 8. Springer-Verlag: Berlin.

Richardson EP Jr (1961): Progressive multifocal leukocencephalopathy. N Engl J Med 265:815–823.

Richardson EP Jr, Åström KE, Kleihues P (1994): The development of neuropathology at the Massachusetts General Hospital and Harvard Medical School. Brain Pathol 4: 181–195.

Sigurdsson B (1954): Rida: A chronic encephalitis of sheep. With general remarks of infections which develop slowly and some of their characteristics. Br J Vet 110:341.

Winkelman NW, Moore, MT (1941): Lymphogranulomatosis (Hodgkin's Disease) of the nervous system. Arch Neurol 45:304–318.

2

PAPOVA VIRIONS IN PROGRESSIVE MULTIFOCAL LEUKOENCEPHALOPATHY: A DISCOVERY AT THE INTERFACE OF NEUROPATHOLOGY, VIROLOGY, AND ONCOLOGY

GABRIELE M. ZU RHEIN, M.D.

1. INTRODUCTION

Personal memory has a dubious reputation, and great caution is indicated when bringing recollections to paper. This report, however, has been chiefly synthesized from work records, from publications, from professional documents, from my correspondence, and from my personal diaries. Friends, such as Dr. Sam Chou and Dr. Richard Johnson, were kind enough to contribute details.

Human Polyomaviruses: Molecular and Clinical Perspectives, Edited by Kamel Khalili and Gerald L. Stoner.
ISBN 0-471-39009-7

2. FROM PATHOLOGIST TO NEUROPATHOLOGIST— A PROFESSIONAL JOURNEY

Before starting my academic career in the United States in 1954, I lived in Munich, Germany, and graduated there from the Medical School of the Ludwig-Maximilian University in 1945. Already as a student my chief attention was devoted to pathology. As an old Sherlock Holmes addict, I enjoyed the required detective work, and, as a visually oriented person, I enjoyed the pleasures of microscopy, offering such a richness of colors and patterns. With little effort I was able to convince my father to buy me a microscope so that I could extend my slide studies into the weekends.

My favorite pathology professor, Dr. Ludwig Singer, hired me right after graduation. A few weeks later, however, units of the 6th U.S. Army marched into Munich and occupied the undamaged Municipal Hospital, where our laboratory was located. The new Chief of Laboratory of the 98th General Hospital, Dr. Maurice Lev, invited me to join his group, and this affiliation, as a pathologist, persisted for 8 years. My position was one of many given to German physicians, by the German government, to assist U.S. Medical Units. I was assigned to do autopsies and surgicals and to give clinicopathologic conferences. My training proceeded "on the go." I was extremely fortunate that many of the drafted pathologists, back home, had been university professors or directors of laboratories, and I received from them personal instruction and guidance. The medical library contained the latest editions of textbooks and a fine selection of journals.

Just outside the walls of the 98th General Hospital was located the Max Planck Institute of Psychiatry, where Professor Willibald Scholz was the Director of the Neuropathology Division. A person of international renown, he conducted biannually an introductory course in neuropathology for army officers, to which I was also invited. Untold times in later years did I think back in gratitude to this exceptional experience, which gave me my first understanding of the complexities of nervous system diseases. An additional contact with neuropathology was offered to me by Dr. E. Manuelidis, from Yale University, who worked on porcine encephalomyelitis (Teschen's disease). He was a member of an epidemiology team, led by Drs. John Paul and Dorothy Horstmann, that was hosted—including their well hidden pigs!—by our laboratory.

Dr. Alfred Evans, of the Yale University team, a strong advocate for my emigration to the United States, became Professor of Epidemiology at the University of Wisconsin, in Madison, after his return to civilian life. He scouted out for me a position in the Department of Pathology, and I gladly accepted the challenge. In late December 1953, I sailed in the company of my newly acquired Leitz Ortholux microscope, the best on the German market, purchased with the shares of my mother's inheritance.

Dr. D. Murray Angevine ran a department well known for its experimental pathology. Diagnostic human pathology was of secondary interest because not much scientific progress was expected from it. Whoever did not work with rats, guinea pigs, or dogs was a member of a lower caste. I was assigned to the autopsy service and to teaching medical students. A few months after my arrival, a professor, who had cut the brains, departed. With this came my greatest challenge. Very casually, Dr. Angevine said to me "The Germans have a good reputation for neuropathology; why don't you do it? The others know much less than you." He threw me into the water, and I had to swim. My anguish about this situation was communicated in a May letter to Professor Scholz (in translation): "My feelings are very mixed. On one hand I am glad that I can now put to use what I learned from you, but on the other hand it is difficult to find the courage to start as an embryo in this field while having to give lectures to medical students already in the fall."

Neuropathology as a subspecialty of pathology was still a novelty in 1954. Where would I find appropriate texts to read? In English, there was one small volume by Dr. Ben Lichtenstein, published in 1949. A German book on inflammatory diseases of the nervous system, of 1942, had been in my immigration luggage. Eventually, my "life jacket" proved to be the seven volumes on diseases of the nervous system edited by Professor Scholz, and published by Springer Verlag from 1955 to 1958, with 95% of the contributions written in German. Our medical school librarian permitted me to keep these on my desk until another reader would request them. This never happened. It was only in 1958, and 1959, respectively, that the first comprehensive textbooks of neuropathology written in English became available: Greenfield's *Neuropathology* and Russell and Rubinstein's *Pathology of Tumours of the Nervous System.*

A benefactor, without whose support and friendship I could have hardly succeeded, was Dr. Hans Reese, Chairman of Neurology and Psychiatry at the University of Wisconsin. A German by birth, he was sympathetic to my plight. He supplied me with helpful books and journals, made his tissue technician available to me, and arranged for personal contacts with neurologists and pathologists at national and international meetings and congresses. Later, I became a member of the Neurology Department, in a double appointment, conducting clinicopathologic conferences and teaching residents on a neuropathology rotation.

3. AN INTEREST IN VIRAL DISEASES SHARED WITH DR. CHOU; ARRIVAL OF PML CASES AND AN ELECTRON MICROSCOPE

Another interdisciplinary contact, which became of considerable importance later on, developed with Veterinary Science. In this Department, Dr. Carl Olson had created a research unit for the study of papilloma viruses, especially of bovine and canine types. A viral oncologist who appreciated the contributions of morphology, he early had acquired an electron microscope. My affiliation with Veterinary Science began in 1957. All such students had to do course work in pathology, and I became their instructor in neuropathology. Over the years I served on at least nine examination committees for M.S. or Ph.D.

candidates, most of them from Dr. Olson's group and others from Dr. Robert Hanson's group which focused on scrapie and transmissible mink encephalopathy.

Meanwhile, in clinical autopsy work, one disease that attracted my particular interest was an acute necrotizing encephalitis with intranuclear inclusion bodies. We observed three such cases from 1957 to 1960 and published the data in 1962. The first case had occurred after head trauma, and I had difficulties with the diagnosis. However, Dr. Stanley Inhorn, a resident engaged in virus research, exhorted me to persist in looking for inclusion bodies. The slow and tedious work was eventually successful, and a herpetic infection could be suggested. With trained eyes the search became much easier in the two following cases. This was a time when immunocytochemistry was not yet part of the diagnostic arsenal.

In 1959, Dr. Sam (Shi-Ming) Chou joined the Department of Pathology as a postdoctoral student with the aim of completing a Ph.D. program in Zoology and Pathology. Supported by the National Multiple Sclerosis Society, he engaged in research in neurolathyrism. By 1962, he had decided that he would choose neuropathology as his career. He opted to take the course that I gave for the neurology residents.

In the fall of 1962, a particularly stimulating consultation case was presented to me by the pathologist of a downtown Madison hospital. The patient, a 33-year-old woman with lupus erythematosus, had died after several weeks of progressive cerebellar disease. The slides showed a multifocal demyelinating disease with a most striking combination of giant tumor-like astrocytes and large numbers of oligodendrocytes with greatly enlarged nuclei deeply stained with hematoxylin. There were no distinct inclusion bodies as one sees with herpes viruses. I was fascinated and knew I had never seen this disease before. I showed the slides to a visiting neuropathologist and he, too, was at a loss. At that time I was in the midst of a very time-consuming experiment with a group of sophomore students. It involved the induction of brain tumors in chicken with Rous sarcoma virus. There was no time for a library search. However, I did show the slides to Dr. Chou, and to my utter surprise, and delight, he brought from his desk a folder with reprints on demyelinating diseases from which he extracted the paper entitled "Progressive Multifocal Leukoencephalopathy" by Åström, Mancall, and Richardson, Jr. (1958) and Richardson's follow-up paper of 1961. We had no doubt that our consultation case was one of the less than 30 cases of this disease known at that time. Only 2 months later, a 67-year-old woman came to autopsy in our department (A 62-393). She had suffered from chronic sinusitis and bronchitis and had developed a left hemiparesis and mental changes during a 7 month period. The clinical diagnosis was multiple infarcts. In the formalin-fixed brain I noticed extensive myelin destruction, and the cytopathology, without doubt, was that of progressive multifocal leukoencephalopathy (PML). We used our sudden wealth of two PML cases for local conferences and teaching exercises.

In 1963, the Pathology Department faculty insisted on the acquisition of an electron microscope to aid various research projects. Dr. Chou, who had learned the technique from Dr. Hans Ris in Zoology, became one of the first users of our RCA EMU 3G instrument (Fig. 2.1). I realized my chance to do some acceptable research at a raised level of morphology. Dr. Angevine had agreed to a sabbatical leave, and Dr. Reese had secured for me a position in the Neuropathology Laboratory of Dr. Harry Zimmerman, at Montefiore Hospital, Bronx, New York. Before thinking of leaving, however, I had to find a colleague who would pitch in for me at home. Dr. Chou graciously consented to help. Thus, in a reciprocal arrangement, I taught him more diagnostic neuropathology and he taught me how to run the electron microscope.

In May 1964, another of Dr. Olson's students took his Ph.D. examination, and I was a thesis reader. Among the illustrations for "The Cytology of Canine Oral Papilloma" were electron micrographs of cell nuclei with dispersed or aggregated virions. Listed among the references was Dr. Melnick's paper in *Science* (1962) entitled "Papova Virus Group." In it, he combined the papilloma and polyoma viruses into one group of oncogenic DNA viruses, capable of producing latent infections. The morphology of virions of this group, in thin sections, had been characterized just within thc last few years.

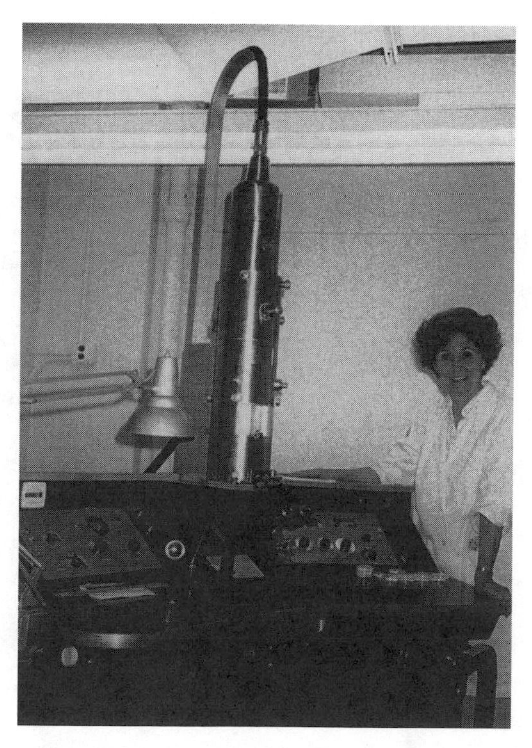

Figure 2.1. The electron microscope, an RCA EMU 3G, in which we first saw the intranuclear virions in brain tissue with progressive multifocal leukoencephalopathy (PML).

In August, one of our autopsy cases showed an extensive esophagitis with distinct intranuclear inclusion bodies. I planned to make up for my lost chance with the encephalitis cases and to search for herpes group virions.

4. A LUCKY DAY FOR TWO NEUROPATHOLOGISTS AND MORE LUCK ON A SABBATICAL

In discussions with Dr. Chou, we decided to give another project priority, namely, the clarification of the mystery of the abnormal oligodendrocytes in PML. We felt unencumbered by the strict rules for optimal ultrastructural images as worked out by cell biologists. The structural proteins of viruses were perhaps still preserved despite postmortem delays in fixation and immersion in 10% formalin for 2 years. What was there to lose in trying? On August 7, 1964, Dr. Chou and myself attended a celebration for a student friend who had passed her preliminary examination. My diary shows the following entry, in mixed English and German: "Pink Bacardi mit cherry!! Dann erste EM session mit Sam, 62-393. CRYSTAL INTRANUCLEAR VIRUS!!!! POLYOMA?? Bis 5 h." We found the crystalloid aggregate (Fig. 2.2) in the first ultrathin section,

Figure 2.2. Electron micrograph obtained on the day of the discovery of the papova virions, here arranged in a crystalloid pattern. ×65,000 orig. mag.

and we could not believe our luck and kidded about the influence of the Bacardi. After Dr. Chou developed the first photographic plates that also revealed scattered virions and filamentous forms, we hugged in the dark room; what an exciting day in our lives it was. Whereas the tissue in general showed moderate autolysis, the virions were so well preserved that their images compared favorably with those in the literature. The virions belonged to an identifiable group, with oncogenic potential, perhaps explaining the bizarre shapes of astrocytes, and this virus had penetrated deep into the brain, unlike papilloma viruses. Yet no human polyoma virus was known at that time. When we showed the plates to Dr. Angevine, his dry comment was: "This is the way discoveries are being made."

In early September, I arrived at Montefiore Hospital equipped with Epon blocks of the esophagitis case and PML case 62-393, with the first prints of the latter and with Ludwik Gross's book titled *Oncogenic Viruses* (Gross, 1961), which became my "bible." When Dr. Zimmerman saw the prints, he found them exciting enough to add a presentation, belatedly, to the program of the forthcoming ARNMD (Association for Research in Nervous and Mental Diseases) symposium entitled "Infections of the Nervous System." Dr. Zimmerman, at that time, was President of the ARNMD and the program coordinator.

In addition to continuing electron microscopy now also on our second case of PML, I was forced to rapidly pursue library studies. Which journals and books was I to read in a field at the crossroads of virus and cancer research and cell biology? It required a fast reorientation for a diagnostic neuropathologist. In September, I was able to resolve the nature of the nuclear inclusion bodies of the esophagitis case: They consisted of herpes-type virions, well preserved in this autopsy tissue. In mid-November, I got a phone call from Dr. Lucien Rubinstein, of Stanford University. He had just received the ARNMD program with the listing of "Papova Virus in Progressive Multifocal Leukoencephalopathy" by Zu Rhein and Chou and wanted to inform me that he and Dr. Silverman had found similar virus particles in a recent case of PML (Silverman and Rubinstein, 1965). He offered to back us up in the meeting, with Dr. Zimmerman's consent.

5. THE "COMING-OUT PARTY"

On December 5, 1964, with several illustrious virologists presiding on a dais and Dr. Chou watching intently from the balcony, I presented our two cases, followed by Dr. Rubinstein. His case had never been in formalin. At once Dr. Sabin took over and tore into us with vigor. "This is deplorable, everybody thinks everything is a virus" (quote from my diary), and "One cannot say one has morphologic evidence of a virus," and, with regard to the electron micrographs, "That is a good way not to get a virologist interested" (Sabin, 1968). Dr. Rubinstein countered that all electron microscopes should be thrown into a garbage can if Dr. Sabin's attitude were correct. None of the panel members commented in our favor. I remember wishing that a helicopter would come and lift me out of there, as I had read in the papers they did with beleaguered white people in the jungles of the war-torn Congo. After the session, Dr. Richard Johnson (Cleveland) gave me a big hug and consoled me with the idea that this attack might bring some prestige later on. Dr. Sabin had said to him in disdain: "She thinks there are warts in the brain!" At lunch, I ended up next to Dr. Sabin at the table in the Rough Rider room of the Roosevelt Hotel. He leaned over and gave this parable: "You know, Doctor, if you take a piece of wood, and you carve it, and you paint it, and you polish it, it will eventually look like an apple."

6. THE "SELLING OF THE APPLES": ACCEPTANCE, COLLABORATIONS, AND INVITATIONS

Well, our "apples" sold fast and widely. *Science* published our first paper (Zu Rhein and Chou, 1965). Dr. Zimmerman had agreed that we should not wait for the publication of the ARNMD volume, which indeed became delayed for 4 years. A neighbor in the Bronx was Dr. Ludwik Gross, Director of the Cancer Research Unit of the VA Medical Center. A pioneer in virus research, he had stood at the cradle of the polyoma virus. Some of his work had also been met with disbelief early on. He was a virologist well versed in viral morphology. When he reviewed our PML electron micrographs, he shared our interpretation with enthusiasm. He invited me subsequently to join him in a research seminar on "Viruses in Disease." We stayed in close contact for the later editions of the *Oncogenic Viruses*. Dr. Richard Shope (Rockefeller University), discoverer of the rabbit papilloma virus, also became an immediate supporter during a laboratory visit. "You are in business" he told me with confidence and true joy.

The first international support, early in 1965, came from Dr. Allan F. Howatson, of the Ontario Cancer Institute, who had extensive experience with the ultrastructure of wart and polyoma viruses. He requested PML tissue in order to apply the negative staining method developed in 1959 by Brenner and Horne for the visualization of viral capsid details in spray preparations. Despite the original formalin fixation of the tissues, it was possible to classify the PML virions as polyoma rather than papilloma virions (Howatson, et al., 1965). Previously, it had been the smaller particle size alone that made us favor polyoma virions. In 1966, Dr. Howatson collaborated with Dr. Chou and myself in a poster exhibit entitled "Polyoma-like Virions in a Human Demyelinating Disease," which was shown during the annual meetings of the American Academy of Neurology (Fig. 2.3) and the Electron Microscopy Society of America. In it, the PML virions were compared with polyoma virions in infected mouse kidney.

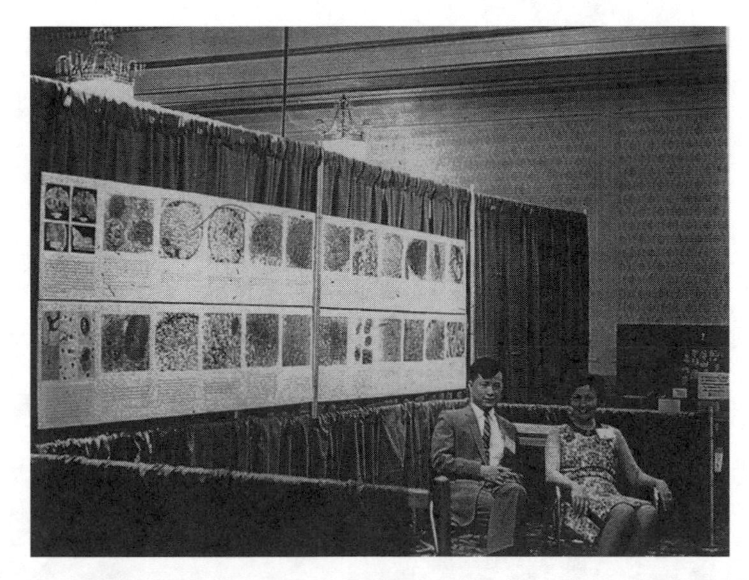

Figure 2.3. Dr. Sam Chou and myself with our poster exhibit "Polyoma-like Virions in a Human Demyelinating Disease" at the annual meeting of the American Academy of Neurology, in Philadelphia, in 1966.

The first European recognition came from Dr. W. Bernhard, Director of the Cancer Research Institute in Villejuif, France. He was a pioneer of viral research and electron microscopy with a special interest in murine polyoma virus and SV40. Following earlier correspondence, he invited me in the summer of 1965 to visit his laboratory—and also to enjoy French cuisine! He was convinced that the glial nuclei in PML were infected by a polyoma group virus.

The acceptance of our work by Dr. Joseph L. Melnick (Baylor College, Houston, Texas) was especially gratifying. As the Series Editor of *Progress in Medical Virology*, he solicited from me a review article (Zu Rhein, 1969) for which I could collect 27 cases of PML with ultrastructurally proven papova virions. Of these, 19 were confirmatory studies by other investigators. No negative findings came to my attention. Data on biologic studies were scarce chiefly due to the rarity of PML cases before the advent of AIDS. The results obtained by four laboratories, some related as personal communications, were negative in their in vivo and in vitro aspects. A variety of routinely used cell lines had been employed in these attempts.

Invitations to symposia, workshops, and meetings were gladly accepted to spread the knowledge of PML to a wider range of physicians in the hope for eventual tissue retrieval. In 1966, the National Multiple Sclerosis Society assembled a workshop at the USPHS Rocky Mountain Laboratory, in Hamilton, Montana (Fig. 2.4). Representatives of virology, pathology, epidemiology, and neurology discussed "slow virus" diseases such as subacute sclerosing panencephalitis, PML, scrapie, and Kuru at a time when the concept of "neurovi-

Figure 2.4. Workshop at the USPHS Rocky Mountain Laboratory, in 1966, dealing with "slow virus" diseases. Clockwise from left: Carl M. Eklund, Hamilton, Montana; John Seal, NIAID; Gabriele Zu Rhein, Madison, Wisconsin; Ellsworth C. Alvord, Jr., Seattle, Washington; Richard T. Johnson, Cleveland, Ohio; Jacob A. Brody, NINDB, Bethesda, Maryland; Hilary Koprowski, Philadelphia, Pennsylvania; William J. Hadlow, Hamilton, Montana; John Hotchin, Albany, New York; Clarence J. Gibbs, Jr., and D. Carleton Gajdusek, NINDB, Bethesda, Maryland.

rology" had not yet been developed. In 1967, an international symposium on the "Pathogenesis and Etiology of Demyelinating Diseases" was convened in Locarno, Switzerland, with the support of a German multiple sclerosis foundation. It brought to light several Japanese cases of PML that were collected and studied, with the demonstration of virions, by Dr. F. Ikuta at Niigata University.

7. A NEW CULTURE SYSTEM FOR VIRUSES PRESENTED AT A NEUROPATHOLOGY MEETING

A remarkable, lucky coincidence in viral research happened in June 1965, when, during the annual meeting of the American Association of Neuropathologists, in Atlantic City, Dr. Harvey Shein (Boston) presented a paper entitled "Interaction of a Tumor Virus (Simian Virus 40) with Human Fetal Spongioblasts and Astrocytes in Dispersed Cell Cultures." In this study, Shein, for the first time, had put to use a culture system that he had recently developed.

I was fascinated by the dual cytopathic effects of necrosis of spongioblasts and transformation of astrocytes so similar to the lysis of oligodendrocytes and "transformation" of astrocytes in PML. In a discussion of Shein's paper (Zu Rhein, 1966), I proposed the use of this culture system for the isolation of the postulated PML polyoma virus. Dr. Shein expressed to me personally, and in later letters, a strong desire for collaboration if fresh PML tissue became available.

8. HARVESTS OF FRESH PML TISSUES AND THE BIRTH OF JC VIRUS

A first ray of hope came from a resident in hematology at the VA Hospital in East Orange, New Jersey, in March 1967. In her letter, Dr. Aurea R. del Rosario referred to our paper in *Science* and asked for diagnostic help for a patient with Hodgkin's disease who had developed subacute neurologic symptoms consistent with PML. I offered to read a biopsy specimen. However, the patient declined the procedure. It became a long wait. Dr. del Rosario was able to obtain from the patient's mother an autopsy permit limited to the brain, with my name given as the recipient. The patient died on January 21, 1968, a Sunday, which I had spent with friends. As I learned the following day, Dr. Sidney Trubowitz, Chief of Hematology, who was anxious for a diagnosis but was unable to reach me, had already contacted another pathologist for help. After settling the issue amiably with this colleague in New York, I flew to East Orange, cut the brain, found the lesions compatible with PML, sampled for virology and pathology, and happily arrived with my harvest back in Madison around midnight, to find Dr. Duard Walker waiting with open arms in the autopsy room.

Dr. Walker, a Professor of Virology in the Department of Medical Microbiology, had been known to me since 1956, when he had invited me to participate in a paper—my very first one—on Coxsackie virus infection. He later followed the PML story with great interest and always received news from me, including that of Shein's culture system. With the tissue in hand, soon proved to contain the familiar virions, he wanted to have a first try at isolation before sharing material with Dr. Shein.

In June 1970, I received the histologic slide of a brain biopsy of John F. Cunningham, a patient with Hodgkin's disease and rather rapidly progressing neurologic deficits. The patient was under the care of Dr. Bertram Dessel, Chief of Hematology at the VA Hospital in Wood, Wisconsin. Dr. Dessel had known since 1967 of my interest in PML, when I had done ultrastructural studies on autopsy tissue of a previous PML case at his hospital. When Dr. Dessel conveyed my diagnosis of PML to Mr. Cunningham, he expressed the wish that his brain should aid research into this fatal disease. Dr. Dessel's call came on July 12, again on a Sunday. He was able to reach me in the laboratory. The alarm went promptly to Dr. Walker, and we joined forces for tissue retrieval in the Wood VA Hospital autopsy room. It was Mr. Cunningham's brain out of which the new human polyoma virus, named JC virus after him, was born, thanks to Dr. Walker and his research associate Dr. Billie Padgett, and to Dr. Harvey Shein. The "birthday party" took place on March 24, 1971. We rejoiced and thought of Dr. Sabin.

9. A POST-ISOLATION LIFE WITH HAMSTERS AND—AGAIN—THE ELECTRON MICROSCOPE

In the following months and years many strategies were designed to determine the characteristics, and the biologic effects, of the new virus. My share of these investigations consisted of ultrastructural studies of the virus in spray preparations, in fetal glial cell cultures, and in immune reactions with sera of other papova viruses and, predominantly, of experiments with Syrian hamsters. My collaborators during a more than 10 year period were Robert J. Eckroade, D.V.M.; Albertina E. Albert, Ph.D.; John N. Varakis, M.D.; and Takeo Ohashi, M.D. Our support came initially from an NIH grant instigated by Dr. Walker and later from a grant in my name.

The first sick hamster was autopsied 4 months after subcutaneous and intracerebral inoculation. A large cerebellar medulloblastoma (Fig. 2.5) demonstrated not only the oncogenic potential of JC virus but also a difference in tumor phenotype from neoplasms induced in the same host by mouse polyoma virus and SV40. Overall, in numerous experiments, medulloblastomas remained the most frequent tumor type. However, JC virus behaved uniquely polyoncogenic for the nervous system (Zu Rhein, 1983), a joy and challenge for a neuropathologist. Tumors of the pineal gland had never before been experi-

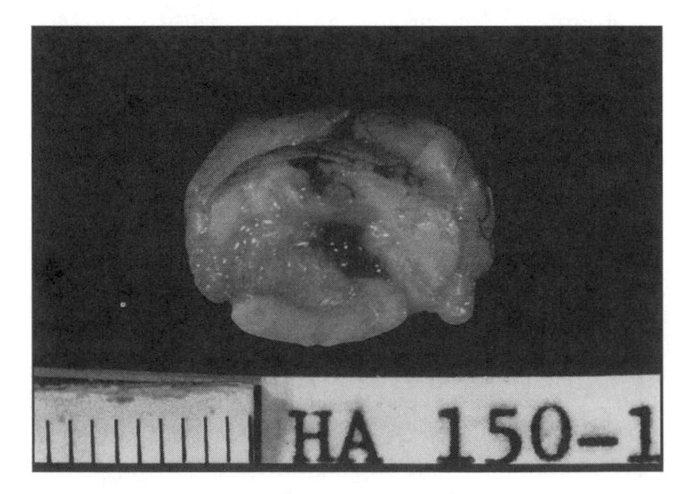

Figure 2.5. The first grossly noted JCV-induced hamster tumor, 4 months after inoculation: a medulloblastoma that has subtotally destroyed the cerebellum.

mentally induced. JC virus also became the first human virus to induce solid tumors, namely, glioblastomas, in primates. Neither in Syrian hamsters nor in owl monkeys did we observe a PML-like demyelinating disease. My sincerest wish is that JC virus would be found to be a causative factor in certain human brain tumors and that specific methods could be designed to prevent or treat them.

REFERENCES

Gross L (1961): Oncogenic Viruses. Pergamon Press: New York.

Howatson AF, Nagai M, Zu Rhein GM (1965): Polyoma-like virions in human demyelinating brain disease. Can Med Assoc J 93:379–386.

Sabin AB (1968): Discussion following presentation by G.M. Zu Rhein and S.M. Chou of "Papova virus in progressive multifocal leukoencephalopathy." Res Publ Assoc Nerv Ment Dis 44:362.

Silverman L, Rubinstein LJ (1965): Electron microscopic observations on a case of progressive multifocal leukocencephalopathy. Acta Neuropathol (Berl) 5:215–224.

Zu Rhein GM (1966): Discussion following presentation by H.M. Shein of "Interaction of a tumor virus (simian virus 40) with human fetal spongioblasts and astrocytes in dispersed cell cultures." J Neuropathol Exp Neurol 25:160.

Zu Rhein GM (1969): Association of papova-virions with a human demyelinating disease (progressive multifocal leukoencephalopathy). Prog Med Virol 11:185–247.

Zu Rhein GM (1983): Studies of JC virus–induced nervous system tumors in the Syrian hamster. Prog Clin Biol Res 105:205–221.

Zu Rhein GM, Chou SM (1965): Particles resembling papova viruses in human cerebral demyelinating disease. Science 148:1477–1479.

3

PROGRESSIVE MULTIFOCAL LEUKOENCEPHALOPATHY: CULTIVATION AND CHARACTERIZATION OF THE ETIOLOGIC AGENT

DUARD L. WALKER, M.D.

1. INTRODUCTION

Following the 1965 demonstration by Gabriele Zu Rhein and Shi-Ming Chou (Zu Rhein and Chou, 1965) and independently by Lloyd Silverman and Lucien Rubenstein that oligodendrocytes in the brain lesions of patients with progressive multifocal leukoencephalopathy (PML) contained numerous particles that looked like papovavirus virions, it was inevitable that many virologists would try to isolate and cultivate the virus in cell cultures or experimental animals. In electron micrographs the particles looked very much like the virions of the polyomavirus genus of the papovavirus family of viruses. But in 1965 no polyomavirus was known to infect people. The particles were smaller than virions of papillomaviruses, which make up the other genus of the papovavirus family. However, some of the papillomaviruses were known to infect people, producing skin and mucous membrane tumors, so they had to be considered. The two major polyomaviruses were simian virus 40 (SV40), a virus found in wild

Human Polyomaviruses: Molecular and Clinical Perspectives, Edited by Kamel Khalili and Gerald L. Stoner.
ISBN 0-471-39009-7

rhesus monkeys, and mouse polyoma virus found in mice. Those two viruses had been studied extensively and were not considered difficult to cultivate. Mouse polyoma virus is easily cultivated in many mouse cell types, but particularly in mouse kidney cells, and SV40 grows very well in primary monkey kidney cell cultures or kidney cell lines of several monkey species. Thus it was quite reasonable to expect that if the virus in PML was a polyomavirus, the cultural techniques that work with mouse polyoma virus and SV40 would work well with this human polyoma-like virus. If the virus should happen to be an unusual papillomavirus, then the chances of successful cultivation were poor because no one had succeeded in cultivating human papillomaviruses in cell cultures despite many attempts.

The simple way to start would be to put the virus into human fetal kidney cell cultures. I am quite certain that many virologists had that plan in mind—provided that they could get virus-containing tissue from a case of PML. That matter of obtaining diseased tissue for an inoculum was going to be a limiting factor. I do not know how many virologists were successful in obtaining appropriate tissue, but there were a few. There may have been others who did not report their trials, but a few did so. By mid-1966 we learned through a publication that Lucien Rubinstein, who had reported his electron microscopic observations of PML, had provided unfixed tissue to his Stanford virologist colleagues, P.R. Schwerdt and C.E. Schwerdt. The Schwerdts did some concentration, purification, and physical characterizations of the virus and concluded that it was a papovavirus and one that had the characteristics of the polyomavirus genus. They inoculated cultures of human diploid cell lines of embryonic lung, embryonic skin and muscle, and an established line of human glial cells but found no evidence of viral multiplication even after multiple subcultures. They inoculated newborn hamsters subcutaneously with brain tissue extract but found no evidence of infection or tumors during 10 months of observation.

In 1967, C.L. Dolman, in Vancouver, British Columbia, reported extensive work that obviously had been in progress for quite a long time. Dolman worked with tissue from two cases of PML and inoculated cultures of multiple human, monkey, and hamster cell types and made blind passages but saw no evidence of cytopathic effect or cell transformation. Two monkeys, newborn mice, and adult mice were inoculated intracerebrally and intraperitoneally, but no disease or tumor developed in 6 to 11 months of observation. In addition, Dolman sent tissue from one case to John Enders in Boston. There is no information available as to what Enders did with the tissue. However, Enders' main interest for years had been the propagation of viruses in cell culture, so it seems safe to assume that he inoculated cell cultures of some kind. His results were reported to Dolman simply as "negative."

Thus, by the end of 1966 it was beginning to appear that this virus that looked so abundant and enticing in Gabriele Zu Rhein's photographs was not going to be an easy one to cultivate. Little did we suspect at that time just how hard the task was going to be.

2. THE START OF A COLLABORATION

Gabriele Zu Rhein clearly wanted to see this virus cultivated and characterized. She had demonstrated its presence in oligodendrocytes in PML brain tissue. The logical next step was to cultivate and identify it in order to study its role in nervous system disease. She wanted to have a part in that process, but she did not have the virology laboratory or virology background to do this, so she began prodding me to take it on in a collaborative project. I had a virology laboratory in the Department of Medical Microbiology in the same building but one floor below the Department of Pathology and Gabriele's electron microscope suite. Gabriele and I had co-authored a paper in 1956 concerned with a virus infection, but had not worked together in the interim. I had been studying persistent, chronic viral infections in cell cultures and animals and also was in the middle of a study of myxoma and fibroma virus infections. I had no experience with polyomaviruses, but I had some experience with papillomaviruses, the other genus of the papovavirus family. In that project, like many other virologists, I had tried and failed to cultivate human papillomavirus in cell cultures, so I had experienced the frustrations of trying to cultivate a very fastidious and difficult virus.

In my consideration of an effort to try to cultivate the virus of PML, an important person was Billie Padgett (Fig. 3.1). Billie was an experienced virologist who had done her Ph.D. thesis research in my laboratory working with influenza virus. She continued in my laboratory as a postdoctoral fellow studying myxoma and fibroma viruses in both cell cultures and rabbits before going

Figure 3.1. Billie Padgett, Duard Walker and Gabriele Zu Rhein at Dr. Walker's retirement party in 1988 (Left to right). Courtesy of Gabriele Zu Rhein.

to Canberra, Australia, to work with Frank Fenner for 2 years. She returned to Madison, Wisconsin, in January 1967 to continue work with myxoma and fibroma viruses. Fortunately, she was willing to make a switch to work on PML, and, as the project developed, it was her patience and tenacity in learning how to get good cultures of fetal spongioblasts that led to the successful cultivation of JC virus. Since neither Billie nor I had experience with polyomaviruses, we were starting out as novices in that area. It would have been easy to obtain mouse polyoma virus and SV40 to study their cultivation, but we did not want to contaminate our laboratory with them. These are hardy viruses, and SV40, in particular, is notorious for contaminating a laboratory and persisting and appearing in subsequent experiments. Although our lack of experience with polyomaviruses might seem to have been a disadvantage, we did not view it as a serious problem, and we knew we would have one major advantage in our work. That was because this was to be a collaborative project with Gabriele, who would provide essential expertise in neuropathology and electron microscopy, and her prominence in the PML field would make it reasonably likely that we could obtain needed virus-containing tissue. Certainly, a serious problem facing anyone setting out to isolate the virus of PML was the paucity of virus-containing tissue available for study. PML was an uncommon disease, and locating terminal patients under circumstances where an autopsy could be performed and fresh or frozen tissue obtained was not likely to be achieved easily or quickly. As it turned out, Gabriele's recognition among pathologists and neuropathologists was essential to our obtaining good tissue.

Gabriele alerted pathologists and neurologists about our need for virus-containing tissue from cases of PML in sufficient quantity to allow significant work. We expected to have to wait a while for tissue, but we certainly hoped that the wait would not be as long as it turned out to be. It was nearly a year before a case appeared.

On March 9, 1967, Gabriele received a letter from Dr. Aurea del Rosario, a hematology resident at the Veterans Administration Hospital in East Orange, NJ. There was a patient in the VA Hospital who had Hodgkin's disease and the clinical features of PML. She had read Gabriele's paper on polyoma-like particles in PML. She was seeking help in confirming the diagnosis, but also offered any specimens that could be useful to Gabriele's research. Gabriele kept in very close touch with Dr. del Rosario by mail and telephone during the surprisingly long course of this patient's terminal illness. It was not until January 21, 1968, that Gabriele received word that the patient had died and that an autopsy would be performed the next day. Gabriele gathered appropriate containers, dry ice, and tissue culture medium and flew to New Jersey on the morning of January 22. She and the pathology resident removed the brain and sliced it immediately. Characteristic lesions were found in both hemispheres. Tissue was selected to go into culture medium, some was frozen, and some went into appropriate fixatives. Gabriele was soon on her way back to the airport and she was back in Madison very late that evening.

There is a later interesting addendum to this tissue procurement expedition. Three or four years later I was asked to discuss JC virus at a conference in New Jersey. During the discussion I mentioned, rather casually, that our first supply of fresh PML tissue had come from New Jersey. The Director of Public Health for the State of New Jersey happened to be in attendance and became quite agitated and incensed that he had not been informed about the case and that the tissue had been allowed to leave New Jersey rather than going to his laboratory. He had wanted to try cultivating the virus.

After obtaining the New Jersey tissue, we had enough virus-containing tissue to begin serious work. We began what turned into a long and frustrating series of efforts to cultivate the virus from the diseased tissue. We tried explants of the fresh brain tissue, but after many weeks of coaxing and coddling only a few fibroblasts and astrocytes grew. We made 10% extracts of diseased brain tissue and clarified it by centrifugation and then used the supernatant fluid to inoculate cell cultures. A similar extract of normal human brain tissue was used on control cultures. Monolayer cultures of primary and secondary human embryonic kidney cells, Hep-2 cells, and human fibroblasts were inoculated. None of these cell cultures showed any evidence of being infected even after weeks of cultivation and medium changes and subculturing. Mice, newborn mice, and hamsters were inoculated by the intracerebral and peritoneal routes and were observed for many months without showing signs of disease or tumor. Guinea pigs and rabbits were inoculated, guinea pigs intraperitoneally and rabbits intravenously to raise antibody against the virus. In this last effort we had a small measure of success. Using frozen sections of New Jersey brain tissue and the indirect method of immunofluorescent staining we found some guinea pig and rabbit sera that reacted at a low level with cell nuclei in the PML tissue. However, antibodies against human tissue were also present, and these interfered greatly. We had to absorb the antihuman antibodies out of the serum with dried human brain powder. This helped, but the antisera were very weak.

Billie began trying to cultivate primary human fetal glial (PHFG) cells. At the 1965 meeting of the American Association of Neuropathologists Gabriele had heard H.M. Shein describe a method for cultivating human fetal astrocytes and spongioblasts in dispersed cell cultures. He also described his use of these cultures to study the effects of SV40 infection on astrocytes and spongioblasts. Such cultures sounded very promising, but there were serious problems in producing them. Human fetal brain tissue was not easily obtained, and good cell cultures derived from the tissue were not easily produced. There was no commercial source, so they had to be produced in our laboratory from tissue obtained from local hospitals. We had gotten our fetal tissue supply problems worked out and Billie was making progress, but the cultures were still mainly what Shein considered to be astrocytes. Although we inoculated PHFG cell cultures with New Jersey brain tissue extract, they gave no indication of infection that we could recognize at that time.

Ten percent extracts of degenerating PML brain tissue tended to have a toxic effect on most cell cultures, and diluting away from that toxic effect obviously reduced the inoculum size. Therefore we began trying to purify and concentrate the virus to escape the toxicity and to obtain antigen of sufficient potency to develop antisera in rabbits and guinea pigs. The lipids in white matter presented a real challenge. How could we free the virus from that mass of myelin lipid and cell membranes? It became a matter of trial and error and using any clues available from the work of other investigators. Many techniques were tried, including genetron extraction and centrifuging to a pellet through 5–20% sucrose, but we eventually settled on one that started with homogenizing the tissue in a mortar or a blender. It was then sonicated and treated with sodium deoxycholate and trypsin and subjected to differential centrifugation. Supernatant fluid from a final low-speed centrifugation was diluted to the equivalent of a 10% tissue extract in buffered saline. Extracts of diseased tissue and normal human brain tissue prepared in this way were used in subsequent animal inoculations and for inoculating cell cultures. We referred to these extracts as semipurified because the virus certainly was far from really purified.

During the early phases of our PML project our financial support was from two National Institutes of Health (NIH) grants that were for research on persistent viral infections but were not directly aimed at PML. We needed additional funds and ones designated for research on the viral agent of PML. Fortunately, at about this time the concept of "slow virus infections" and particularly "slow virus infections of the central nervous system" was attracting attention. The Icelandic virologist Björn Sigurdsson had introduced the terms in his research on rida, scrapie, visna, and maedi of sheep because of the long incubation periods and slow progression of these diseases. Carleton Gajdusek used the terms in his work on kuru and Creutzfeldt-Jakob disease, and Gajdusek and others began including PML among the slow virus infections of the central nervous system. This gave PML some prominence and "pizzazz" and made it more likely that we could gain support.

However, in 1968 we were already beyond those halcyon days of NIH when research funds were easily obtained. Funding had already tightened up, so it seemed rather unlikely that a proposal to study just the viral etiology of PML would be well received, particularly because we did not yet have even a suggestion of success in cultivating the PML virus. NIH study sections had developed a preference for supporting "a sure thing," that is, a project already nearly accomplished, rather than risking money on an uncertainty. And we were a real uncertainty.

3. WIDENING OUR COLLABORATION

We thought that broadening our proposal might improve our chances, so we turned to colleagues in the Department of Veterinary Science. The Departments of Medical Microbiology and Pathology in the Medical School and the Department of Veterinary Science in the College of Agriculture (this was before the University had a School of Veterinary Medicine) had for many years maintained a cordial working relationship in both research and graduate teaching. Within the Department of Veterinary Science there were some very good virologists who were already working in the area of slow virus infections. One of these, Robert Hanson, was a prominent virologist whose eminence was based mainly on his extensive studies of Newcastle disease in poultry, but Bob had also maintained a smaller program studying transmissible mink encephalopathy (TME) and scrapie. Although the TME program had remained small with usually one or two graduate students working on it, enough work had been done on the epidemiology and pathology to provide a strong suggestion that the disease in mink was due to the same agent that caused scrapie in sheep. The pathology of TME in mink was very similar to that of scrapie in sheep. It appeared quite likely that in their pursuit of inexpensive high protein food mink ranchers were sometimes feeding meat of scrapie-infected sheep to their mink, thereby causing outbreaks of TME.

Gabriele had been working with Hanson's group on the neuropathology of TME and was currently mentoring Robert Eckroade, a graduate student who was working toward a Ph.D. with a double major in Veterinary Science and Pathology. Bob Hanson's program on TME had been supported by a mink grower's organization but was making enough progress to justify expansion, so Bob was interested in joining a proposal to NIH. In addition, June Osborn, who was a virologist in the Department of Medical Microbiology, had been developing a project to study subacute sclerosing panencephalitis. With her participation we had a three-pronged attack on slow virus infections of the central nervous system, and with this group approach we applied to NIH for support of the three laboratories. Today's NIH study sections would probably disapprove of our forming such a group and would tell us we were "not focused," but in 1968 NIH looked favorably on such programs as a way to support more research on less money.

Meanwhile, we began meeting as a slow virus disease group in 1968 to plan, analyze, and discuss experiments and to discuss reports from other laboratories. Originally the group consisted of Bob Hanson and Robert Eckroade from Veterinary Science, Gabriele Zu Rhein from Pathology, and Billie Padgett, June Osborn, and Duard Walker from Medical Microbiology, but as time went on many others participated. Roland Rueckert from Biochemistry joined in efforts to purify the agent of TME. Cornelius Hopper from Neurology participated. An important addition was Richard Marsh, who returned to Madison from NIH where he had been fulfilling his military obligation working in Carleton Gajdusek's laboratory as a Public Health Service officer. The doctor's draft was still in force and Dick who had a D.V.M. was therefore draftable. Before going to NIH Dick had obtained his Ph.D. in Veterinary Science working on TME. Both Dick and Bob Eckroade were veterinarians before obtaining a Ph.D., and both became major contributors to the TME and the PML projects.

Our slow virus disease group met every second Friday morning for about 2 hours. I usually prepared an agenda, but the discussions ranged far and wide. Because we were working with pathogenic agents of uncertain risk and potential, it was important that graduate students, technicians, and all associated persons be fully aware of what we were doing, so they attended and participated as well. The discussions were often intense. In my view, these meetings were very valuable. Each subgroup would plan experiments and then present them to the group where they were likely to be taken apart and reassembled, often in altered form. We were all "slow virologists" planning experiments to be measured in months and years rather than minutes, hours, or days. We had to think everything through very carefully, particularly with animal experiments, because once we started an experiment we were committed for a long time, often for several years. These meetings were of such value that we kept them going for about 10 years.

Our NIH grant application for the "Study of Chronic Viral Infections of the Nervous System" was approved and funded for 5 years to begin January 1, 1969. Although, of course, the grant was not funded for all the money we had applied for and thought we really needed, it did provide money for one very expensive piece of equipment, a new Philips electron microscope. All of Gabriele's electron microscopy work had been done using an RCA EMV 3D microscope, which by 1968 was considered obsolete by most microscopists. Unfortunately, due to one delay after another, the new microscope was not installed until October 1971.

The year 1968–1969 was disappointing for those of us working on PML. Although we had obtained the New Jersey tissue, we had not yet had success in cultivating virus. However, we had many inoculated animals to observe, and we remained optimistic. In November 1968 we received a very small piece of frozen diseased tissue from a PML case in West Virginia. The tissue was sent by Sam Chou. This was the Shi-Ming Chou (usually known as Sam) who had worked with Gabriele and was co-author of the 1965 paper describing the virions in PML lesions. Sam had moved to West Virginia University and autopsied a PML patient in Morgantown. He gave us only 1 g of tissue, though, and this was so small that we did not use it in our trials at that time. Later, after we had succeeded in isolating and cultivating JC virus, we got it out of our freezer and had no difficulty isolating JC virus from it.

In 1968–1969 our colleagues working on TME were making great progress infecting multiple species, including monkeys, with TME and scrapie agents and comparing the neuropathology in these species. The possibility that Creutzfeldt-Jakob disease in people is caused by the agent of scrapie or a related agent that is transmitted from food animals was viewed as a distinct possibility as early as 1968–1969, and Hanson, Marsh, and Eckroade were pursuing this possibility.

However, progress was slow with PML. One problem was that we were not receiving new PML tissue. In November 1969, George Ellison at Yale offered us some tissue from a patient who had died in 1965. The tissue had been stored in 50% glycerine-saline at 4°C since 1965. We certainly accepted it, but the specimen consisted only of two very small pieces, and, considering its storage history, we did not view it as first-rate material. We moved it to −70°C storage and continued working with our New Jersey tissue. Much later, after we had worked out our production of PHFG cultures and improved isolation techniques, we tried cultivating virus from this glycerin-stored tissue and succeeded in isolating JC virus from it. Storage of tissue containing hardy viruses in 50% glycerine-saline at 4°C was a technique used by virologists before freezers were available. Richard Shope stored his rabbit papillomas and fibroma tumors this way. Because JC virus is a relatively hardy virus, it remained viable during multiple years of storage by this method. After receiving this tissue from Ellison we learned that Ellison had also sent some of the tissue to Gajdusek, but we never learned its fate.

On May 20, 1970, our quest for tissue suddenly improved. Gabriele received word from Dr. Bert Dessel, Chief of Hematology at the VA Hospital in Wood, Wisconsin, a suburb of Milwaukee, that he had a patient with an 8 year history of Hodgkin's disease and neurologic signs strongly suggestive of PML. A brain biopsy had been done on May 20. Dessel sent the biopsy slides to Gabriele, who confirmed the diagnosis of PML. The patient, whose initials were J.C., died at 4:00 AM on July 12, and a brain autopsy was permitted.

Gabriele and I drove the 70 miles from Madison to Milwaukee armed with tubes of tissue culture medium, dry ice, and containers suitable for transporting tissue. When the brain was removed Gabriele sliced it immediately. Demyelination was found in all lobes. Extensive disease with cavitation was present in the right temporal and parietal lobes and the left occipital lobe. Many small samples were taken and placed in tissue culture medium in the hope that some brain cells would grow. Many areas of early lesions and the borders of large lesions were collected and frozen for later extraction of virus. The VA Hospital pathologists and hematologists made this very easy for us, and Gabriele and I were back in Madison in a few hours with a much improved reservoir of virus-containing tissue.

We did not, however, get all of the tissue destined for laboratory study. Dr. Dessel told us that Carleton Gajdusek had also been advertising his interest in PML tissue, so Dessel collected several grams to be sent frozen to Gajdusek. We never learned to what use this tissue was put or the outcome of its use, but it is very likely that it was used to inoculate one or more primates.

The tissue that we got from patient J.C. was much larger in quantity and proved to be much richer in virus than that from the New Jersey case. We renewed our efforts to concentrate and purify virus for use in inoculating cell cultures and animals and in producing antiserum. The only way to follow the virus through various purification procedures was to use negatively stained preparations examined by electron microscopy. Bob Eckroade did most of this microscopy.

In our previous efforts to concentrate and purify virus from brain tissue of the New Jersey case all our extractions and differential centrifugations had not seemed to accomplish much. However, when brain tissue from patient J.C. was treated in the same way, the semipurified extract contained virions in sufficient numbers that they could be found by electron microscopy in negatively stained preparations. This was probably because we started from a much richer source of virus. The New Jersey patient had a very slowly progressive clinical course of PML. From the first signs of PML to death was about 1 year, whereas patient J.C. had rapidly progressive disease that led to death in 2 months and had very extensive brain involvement.

Groups of animals were inoculated by the cerebral and peritoneal routes. These included mice, newborn mice, guinea pigs, rabbits, hamsters, ferrets, and mink. The mink were used because they were available from the associated TME project and because they were a species not likely to have been tested or to be tested by other investigators. Most of the inoculation of animals and much of the long-term observation of the animals was done by Dick Marsh and Bob Eckroade. The animals were housed in animal quarters belonging to and managed by the Department of Veterinary Science. They had the only animal quarters with what could be considered isolation rooms. These animal quarters were located on a university-owned experimental farm (Charmany Farm) on the outskirts of Madison. Although they were primitive by today's standards, the Charmany Farm quarters were considered adequate for that time, and we were very fortunate to have them available.

In an effort to develop antisera against the virus, rabbits and guinea pigs were injected by various routes, including intravenous, intramuscular, and intramuscular injection of extract mixed with adjuvant. These animals were bled for their serum over a period of 10 days to 6 weeks.

Inoculated animals were observed for many months, often for the lifetime of the smaller animals. We were able to hold animals for long periods only because the Department of Veterinary Science bore most of the cost. We contributed only the salary of one animal caretaker each year. Our work was done before the requirements for animal quarters and the costs for animal care escalated to levels that would have been prohibitive for us.

Even though the animals were observed for long periods and Gabriele autopsied every animal, we found no indication of disease attributable to inoculation with brain tissue extracts.

At the time all these animals were being inoculated and observed, work with cell cultures was also progressing—slowly. Many cell types were planted in Leighton tubes on top of cover glasses. The cell types included primary human embryonic kidney cells, human embryonic fibroblasts, Vero cells (a line of African green monkey kidney cells), primary African green monkey kidney cells, Hep-2, WI-38, mouse embryo cells, and PHFG cells. Each tube was inoculated with 0.1 ml of J.C. brain extract or a normal control extract. These cultures were incubated and observed for cytopathic effects for several weeks. Periodically the cover glasses were removed, and the cells were fixed and stained with hematoxylin and eosin (H&E) and searched microscopically for cells containing inclusion bodies. Cells were also removed from cover glasses, centrifuged into a pellet, and frozen and thawed to disrupt them. The cell debris was negatively stained and searched for virions in the electron microscope.

In general, no cytopathic effect or other evidence of infection was found in these cultures, even after several subcultures. However, we began to think something interesting was happening in some of the inoculated cultures of

PHFG cells. Subtle changes were slowly developing that were not found in control cultures.

4. FINALLY, SOME GLIMMERING OF SUCCESS

Billie had been working hard for months trying to improve cultures of PHFG cells. It was slow going. Initially the cultures consisted of astrocytes that would form a monolayer, but spongioblasts were either sparse or absent. We thought it important to have good cultures of spongioblasts because neurobiologists considered spongioblasts to be precursors of oligodendrocytes, the cells in which the virus was found in human brain.

Through trial and error over many months, Billie developed techniques that produced cultures with increasing numbers of spongioblasts. The spongioblasts were usually in clumps sitting on top of a layer of astrocytes. The procedure was complicated and required multiple medium changes and manipulations. It required periodic adjustments in fetal calf serum and high levels of glucose. In particular, it required much patience and tenacity. Fortunately, Billie had those attributes, and it was beginning to look like this effort was paying off.

Three or four weeks after inoculation with J.C. brain extract, PHFG cell cultures stained with H&E were found to contain a scattering of enlarged cells with dark-staining, bizarre nuclei. We judged these cells to be enlarged astrocytes. This effect could be transmitted to new cultures with frozen, thawed, and sonicated cell debris from affected cultures. The smaller number of spongioblasts that had been in the cultures were usually gone by the time we recognized enlarged astrocytes. Later, when we had cultures richer in spongioblasts, we came to realize that the spongioblasts had already been released from the astrocyte layer and were floating in the medium by the time the astrocytes showed abnormalities.

In addition to staining cells with H&E, we were periodically removing cells from tubes and freezing, thawing, sonicating, and negatively staining them for examination in the electron microscope. Bob Eckroade did the negative staining and electron microscopy in an electron microscopy suite that was one floor above my office. On the afternoon of March 24, 1971, Bob had been examining material from inoculated PHFG cell cultures when he came to my door with a big grin on his face and said, "Come see what we have." I ran back up the stairs with him and looked at the microscope screen. There was no question about what was there. Polyoma-like virions were evident in abundance, many times what could be carried over from the inoculum. This had to be a successful culture. Bob and I looked for Billie and Gabriele. Gabriele was away that afternoon, but Billie and the technicians got to see what they had been working toward for more than 3 years. Bob took pictures of the virions, and the next morning we showed them to Gabriele. That day the laboratory crew had a little champagne to mark the occasion. Gabriele chose to call this occasion the birthday party for JC virus.

We then began doing all the things necessary to identify and characterize the virus and to confirm our ability to cultivate it. We also began tightening our safety precautions. Up to this time we had been using techniques that were standard in microbiology laboratories working with pathogens. Our cell culture work and inoculations were done in cubicles under negative pressure, but without additional protection for the worker other than good aseptic technique. However, we now believed that we had the etiologic agent of PML in hand and in quantity. This caused some nervousness because at that time we knew nothing about its epidemiology or anything else about it except that when it caused disease it was lethal.

The concept of and term *biohazard* had fairly recently been introduced, popularized, and discussed. This had led to increasing pressure on laboratories and NIH to improve safety. Commercial companies responded with production of biosafety cabinets. We received the first one in the department and added others as they became available. All future handling of the virus was done in biosafety cabinets.

To demonstrate our ability to cultivate the virus, it was isolated twice from each of two extracts of J.C. brain tissue. By electron microscopy the isolated virions had the same size and shape as those in brain tissue. As PHFG cell cultures richer in spongioblasts became available we could demonstrate a 1000-fold increase in infectious virus in PHFG cell cultures during a 3 week period. Wallace Rowe, Werner Henle, Hilary Koprowski, and Kenneth Takemoto generously provided antisera and fluorescent conjugates that helped us establish that the virus was not SV40, mouse polyoma virus, or mouse K virus. At this point we had to bring SV40 and mouse polyoma viruses into our laboratory in order to compare what we now called JC virus (JCV) with those viruses. PHFG cell cultures that showed good evidence that JCV infection did not react with virus-specific antiserum against SV40, mouse polyoma virus, or K virus when tested by the indirect immunofluorescence method or with fluorescein-conjugated antiserum, nor did these antisera react with the virus in frozen sections of J.C. brain tissue. Antiserum that we had recently produced in rabbits by intravenous injection of J.C. brain extracts did react with oligodendrocytes in frozen sections of J.C. brain tissue and with enlarged spongioblasts in infected PHFG cell cultures. We produced antiserum in rabbits by injecting them intravenously with semipurified, concentrated virus from PHFG cell cultures.

This antiserum did react with both enlarged spongioblasts in infected PHFG cell cultures and enlarged oligodendrocytes in frozen sections of J.C. brain tissue, thus identifying the virus in cultures as the virus in J.C. brain tissue. The antiserum did not react with SV40 or mouse polyoma virus in cell cultures. JCV from PHFG cell cultures did not produce cytopathic effects in cultures of primary mouse embryo, BSC-1, Vero, or primary African green monkey kidney cells, and no cells in these cultures developed antigen reactive with antiserum against SV40 or mouse polyoma virus. JCV did not produce cytopathic effects in cell cultures of primary human embryonic kidney, primary human amnion, or an established human embryonic cell line. All evidence to this point indicated that JCV was a newly recognized virus and not one of the previously known polyomaviruses. It was also becoming clear that JCV had a remarkably restricted host cell range.

5. SURPRISES

About a month after our first successful isolation of virus from J.C. brain tissue, Gabriele received a telephone call from Richard Johnson at Johns Hopkins Medical School. He invited her to attend a workshop on the epidemiology of multiple sclerosis at Easton, Maryland, and to visit his laboratory where she "would be able to see the virus of PML." After Gabriele conveyed this news to me I talked by phone with Johnson and learned that Leslie Weiner and he had a papova-like agent growing in primary African green monkey kidney cells. They had made isolations from two cases of PML using cell fusion and co-cultivation techniques. Just the fact that their isolates grew well in green monkey kidney cells made it unlikely that their virus was the same as ours. I told him we had an isolate also, and he and I agreed that Gabriele and Billie both should attend the workshop so their isolates and ours could be compared.

At the workshop Weiner presented the data on the Baltimore isolates (Weiner et al., 1972), and Billie and Gabriele presented information on JC virus. One of the Baltimore isolates came from tissue sent to Johnson by Lucien Rubinstein from Stanford, while the other isolation was made from a brain biopsy of a case of PML in Baltimore. All data were preliminary at this time, but the Baltimore isolates were acting very much like SV40 while, at that time, we had found no antigenic relationship to SV40 for our isolate and in cultural characteristics it looked to be very different from SV40.

We were not through hearing about new papovavirus isolates. A month after learning about the Baltimore isolates we heard a rumor that investigators in London, England, had isolated a papovavirus from the urine of a patient with a kidney transplant and that the investigators thought it could be the virus of PML.

We learned no more about the London isolate at that time, but in the meantime we had assembled a manuscript describing our isolation and cultivation of what we now called JCV. We submitted the manuscript to *Lancet* for publication (Padgett et al., 1971). To our surprise, the editor told us that *Lancet* had received another manuscript describing the isolation of a papovavirus from a human patient and that the two papers were accepted and would be published together.

The two manuscripts going to the same publisher at approximately the same time was entirely a coincidence. The other paper was by Sylvia Gardner and Anne Field at the Virus Reference Laboratory in London and Dulcie Coleman and B. Hulme at St. Mary's Hospital, London (Gardner et al., 1971). We had not communicated and knew nothing of each other's work except what we in Madison had heard as a rumor.

The London isolate was initiated in the continuous line of African green monkey kidney cells called Vero cells and could be serially passaged in Vero cells. It had a minor antigenic relationship to SV40. Like mouse polyoma virus, it agglutinated human and guinea pig erythrocytes at 4°C, but was antigenically unrelated to mouse polyoma virus. Gardner and her associates had done as we did and used the patient's initials as a label in work with the virus, so they called it BK virus (BKV).

6. PROGRESS AT A SOMEWHAT FASTER PACE

We had not previously tested JCV for its hemagglutinating capacity, but if BKV caused hemagglutination, it seemed likely that JCV would. It was a great relief to find that, like BKV and mouse polyoma virus, JCV agglutinated human, guinea pig, and chicken erythrocytes at 4°C. This was very fortunate because finally we had a technique that offered some speed. Everything else about JCV was slow, and slower, but with hemagglutination and hemagglutination inhibition we could measure virus and antibody in hours rather than weeks or months.

After determining that antibody measured by hemagglutination inhibition correlated well with neutralizing antibody in human serum, we were able to perform a serum antibody survey to determine the prevalence of antibodies in human and animal populations. We were surprised to find that even though PML disease is uncommon, antibodies against the virus are very common in the human population. The presence of antibodies as evidence of past or present infection indicated that infection begins in early childhood and that by middle age 75% of persons have been infected. On the other hand, no antibodies were found in the sera of 12 animal species, including five primates. This suggested that JCV had a very narrow host range and that animals were not the source of infection for people.

Almost all of the persons working in our laboratories had good levels of serum antibody against JCV. To a considerable extent this relieved anxiety about working with the virus. Curiously, Billie and I were among the few without serum antibody, and we never did develop it during years of working with the virus. Billie and I have chosen to attribute our antibody-negative state to clean living and good laboratory technique. My reward for being JCV antibody-negative and having type O erythrocytes was to become the designated blood donor every Monday morning when the week's supply of red blood cells was collected for hemagglutination tests.

7. TUMOR PRODUCTION

Because mouse polyoma virus and SV40 were known to be highly oncogenic viruses, it was important to test the oncogenicity of JCV. Our previous injections of multiple animal species with New Jersey and J. C. brain extracts had not resulted in recognized tumors, but it seemed likely that the amount of virus in those inocula was quite small. Therefore experiments were planned as a direct test of oncogenicity.

Virus from the third passage in PHFG cells was concentrated and partially purified by differential centrifugation and tested to be certain that there was no contaminating SV40 or mouse polyoma virus in it. Newborn hamsters were inoculated both intracerebrally and subcutaneously with 10^6 infectious doses in each site. The control inoculum was a concentrated and partially purified extract of uninfected PHFG cells.

There was no acute disease caused by the inocula, but approximately 4 months after inoculation the virus-inoculated hamsters began showing overt signs of central nervous system disease. The first hamster to show clear signs and increasing illness did so while I was on vacation in San Francisco. An autopsy by Gabriele revealed a sizable brain tumor, so I received a telephone call from an excited group of people in Madison telling me about the tumor and the increasing number of neurologically sick hamsters. At first, one or two hamsters per week showed signs but soon the rate increased to five or six per week. Gabriele autopsied every hamster and examined each very closely for both gross and microscopic tumors. At 6 months after inoculation, the surviving animals were killed and autopsied.

The only significant pathogenic findings were tumors in the JCV-inoculated group. Eighty-three percent of 63 JCV-inoculated animals had developed brain tumors within the 6-month period, a remarkably high level of oncogenicity. None of 39 controls had tumors. The tumors ranged from 2 to 7 mm in diameter. On microscopy, Gabriele found them to be malignant and glial in origin. Most were glioblastomas, medulloblastomas, or unclassified primitive tumors (Walker et al., 1973).

JCV was isolated from some of these tumors by serially subculturing the tumor cells. Tumor cells did not contain JCV coat protein when tested by indirect immunofluorescence, nor did they contain coat protein of SV40 or mouse polyoma virus. They did, however, contain an intranuclear antigen that reacted with serum from certain of the tumor-bearing hamsters. This intranuclear antigen had the characteristics of the tumor antigens (T antigens) described for other polyomaviruses. When tumor cells were reacted with antiserum against SV40 and mouse polyoma virus it became clear that there was a strong cross reaction between the T antigens of JCV and SV40. Later it was found that the T antigen of BKV also cross reacted with JCV and SV40.

The demonstration of the high degree of oncogenicity of JCV in hamsters opened up a whole new area for additional research. Inoculation of newborn hamsters by various routes resulted in many varieties of tumors. Gabriele made an extensive and detailed study of the tumors over the next several years. Interest in JCV-induced tumors was further enhanced when, in a collaborative project with William London, John Sever, and associates at NIH it was found that intracerebral inoculation of JCV in adult owl monkeys resulted in malignant brain tumors in about 25% of the animals after an incubation period of 16 to 25 months. Another experiment showed that JCV also induced brain tumors in adult squirrel monkeys.

Soon after our paper appeared in *Lancet* we began receiving requests for JCV from federally supported juggernaut laboratories. These requests increased after our reports on the oncogenicity of JCV. This was at a time when the National Cancer Institute was pouring many millions of dollars into a frontal assault on viruses as a major cause of cancer and had established multiple strongly funded satellite laboratories to study the role of viruses in cancer. It was clear from most of these requests that those asking for the virus were confident that they could get this obstinate virus to multiply in cells other than PHFG cells. Some mentioned human glial cell lines that they had developed. As we produced new virus we shipped starter samples to these laboratories. Almost invariably we never heard from them again. I inquired about the results of their efforts from a few of them and was usually told that they did not have time to spend on a virus like JCV so they had given up trying to cultivate it. They had greater interest in fast viruses than in slow viruses.

8. COMPARISON OF THE NEW VIRUSES

After our papers appeared together in *Lancet* we exchanged viruses and information with the London group so that the viruses could be compared in each laboratory. As these comparisons progressed it was clear that JCV and BKV were distinct virus species, but they had many similarities. Both viruses circulated widely in the human population. Infection begins to occur in early childhood and persists, probably for life, as a seemingly harmless latent infection of the kidneys unless host immunity is impaired. During periods of impaired immunity virus is frequently released in urine, and overt disease can develop. The overt disease produced by the two viruses is, however, very different. JCV causes the lethal brain infection PML, while disease caused by BKV is usually limited to the urinary tract. JCV is highly oncogenic in hamsters, owl monkeys, and squirrel monkeys, while BKV is much less oncogenic.

It was important to clarify the antigenic relationships of JCV, BKV, and SV40, and this was worked out in several laboratories including those of Sylvia Gardner and associates, Keerti Shah, Opendra Narayan and associates in Baltimore, Kenneth Takemoto and M.F. Mullarkey at NIH, as well as ours. It was gradually determined that there is at least one antigen on the surface of JCV, BKV, and SV40 virions that is specific for each virus and allows them to be distinguished from each other by serologic methods. Antiserum against the species-specific antigen can be produced in rabbits by one or two intravenous injections of purified, intact virions. Multiple injections of virus or use of adjuvants leads to the appearance of cross-reacting antibodies. JCV, BKV, and SV40 share a minor surface antigen that sets them apart as a subgroup from other polyomaviruses. This was the minor relatedness first recognized by Gardner as a weak cross reaction between BKV and SV40. Use of intact virions in immunizing rabbits will produce antibodies against this antigen, but hyperimmunization is required. Keerti Shah and associates demonstrated that all members of the polyomavirus genus share a genus-specific cross-reacting antigen that is best demonstrated by the immunofluorescent technique in the nuclei of infected cells. Antibody can be produced in rabbits by immunizing with virions disrupted by detergent or alkali. This antigen is particularly useful because, unlike the other antigens, it is resistant to formalin and can be detected in the nuclei of infected cells in formalin-fixed, paraffin-embedded tissue using the peroxidase–antiperoxidase immunostaining technique.

Like SV40, JCV and BKV were shown to induce nonstructural T antigen in the nuclei of lytically infected cells, transformed cells, and tumor cells. The T antigens of JCV, BKV, and SV40 are antigenically very similar. Antiserum developed in tumor-bearing hamsters against the T antigen of one virus reacts strongly with the T antigens of all three viruses.

It gradually became clear that correct identification of one of the polyomaviruses by serologic methods requires recognition of these specific and cross-reacting antigens and requires reliable information about how an antiserum was produced plus appropriate tests and controls to certify the specificity or breadth of reactivity of an antiserum. Use of antisera of uncertain specificity has sometimes led to erroneous conclusions about the identity of viruses of this group.

After we had improved our isolation and cultivation techniques with JCV and had developed good virus-specific antisera for JCV, BKV, and SV40, we set about identifying the virus in more cases of PML. We went back to the New Jersey tissue and, using virus-specific antiserum and fluorescence microscopy, we identified the virus in the tissues as a JCV type of virus and were able to isolate and cultivate it. We also succeeded in cultivating virus from the small pieces of tissue we had received from Sam Chou and found it to be antigenically identical to JCV. Using frozen sections of the diseased tissue and virus-specific antisera, we found the virus in the brain tissue to be a JCV type of virus, and we were even able to do the same with the tissue that George Ellison had given us. This was the PML tissue that had been stored in 50% glycerin-saline at 4°C since 1965. This gave a good indication that JCV was a very hardy virus.

With virus-specific antisera, frozen sections, and immunofluorescence techniques it became possible for us to confirm a diagnosis of PML in biopsy and autopsy tissue in a few hours to a few days. When pathologists, neurologists, and other physicians learned this, we began receiving specimens of PML tissue at an increasing pace. We chose to cultivate viruses from about 15 cases so that we could compare isolates. We sometimes went through the work of isolating and cultivating the virus because the case was of special interest, such as the cases of two children who developed PML. All isolates were antigenically similar.

In the 30 years since JCV was isolated and cultivated, the identity of the virus in PML lesions has been determined many times. Most of these cases have gone unreported because JCV in PML is no longer news, but the number must be approaching 200 because it was over 100 in 1988 when I retired and stopped receiving reports. Identification has been accomplished by a variety of methods in laboratories in many countries on several continents. In no case, thus far, has BKV been found in the brain of a patient with PML, and all available evidence indicates that BKV has no part in PML. With the exception of a handful of cases, JCV has always been the virus in the brain lesions of PML patients.

However, in six cases the virus has been reported to be SV40. The SV40 identifications have been from laboratories working with SV40 or with monkey cells that frequently harbor latent SV40 and where cross contamination is a risk. Furthermore, antiserum used in these identifications usually has not been shown with certainty to be virus specific. The cross-reacting antibodies common in antiserum developed against any one of SV40, BKV, or JCV can lead to mistaken identification. When Gerald Stoner and Caroline Ryschkewitsch (Stoner and Ryschkewitsch, 1998) re-examined diseased brain tissue from three of the six supposed SV40 cases, they found that these three cases were indeed mistakenly identified as due to SV40. Stoner and Ryschkewitsch used virus-specific monoclonal antibodies, virus-specific DNA probes for in situ hybridization, and virus-specific primers in the polymerase chain reaction. These are techniques that were not available when the cases were originally studied. In each of these cases, one of which was Rubinstein's case from Stanford reported in 1972, the virus in the brain lesions was unequivocally shown to be JCV rather than SV40. Tissues from the other three cases were not available for examination by Stoner and Ryschkewitsch, but it is quite likely that had they been re-examined they, too, would have been found to be mistaken identifications and that JCV is the sole etiologic agent of PML.

ACKNOWLEDGMENTS

I gratefully acknowledge the help of Dr. Billie Padgett, who read the manuscript and made helpful suggestions.

REFERENCES

Gardner SD, Field AM, Coleman DV, Hulme B (1971): New human papovavirus (B.K.) isolated from urine after renal transplantation. Lancet 1:1253–1257.

Padgett BL, Walker DL, Zu Rhein GM, Eckroade RJ, Dessel BH (1971): Cultivation of papova-like virus from human brain with progressive multifocal leukoencephalopathy. Lancet 1:1257–1260.

Stoner GL, Ryschkewitsch CF (1998): Reappraisal of progressive multifocal leukoencephalopathy due to simian virus 40. Acta Neuropathol 96:271–278.

Walker DL, Padgett BL, Zu Rhein GM, Albert AE, Marsh RF (1973): Human papovavirus (JC): Induction of brain tumors in hamsters. Science 181:674–676.

Weiner LP, Herndon RM, Narayan O, Johnson RT, Shah K, Rubinstein LJ, Pregiosi TJ, Conley FK (1972): Isolation of virus related to SV40 from patients with progressive multifocal leukoencephalopathy. N Engl J Med 286:385–390.

Zu Rhein GM, Chou SM (1965): Particles resembling papova viruses in human cerebral demyelinating disease. Science 148:1477–1479.

4

SERENDIPITY—THE FORTUITOUS DISCOVERY OF BK VIRUS

WENDY A. KNOWLES, PH.D.

1. INTRODUCTION

It is a remarkable fact that the first isolations of the two human polyomaviruses, BK virus (BKV) and JC virus (JCV), were reported simultaneously in *The Lancet* (Gardner et al., 1971; Padgett et al., 1971), the work of each group unbeknown to the other, and even more interesting to contrast the route by which each virus was discovered. Whereas the history of JCV can be followed from disease to virus over a protracted course (see Chapters 1 to 3), the discovery of BKV arose from observations made during work on an unrelated virus and was followed by a search for a disease. This is an account of the discovery of a virus taken from notes, specimen forms, and laboratory notebooks of the time, with the personal recollections of Dr. Sylvia Gardner. I was then working on cytomegalovirus for my Master of Philosophy degree with Sylvia as my supervisor, and was not directly involved in the initial polyomavirus studies, but switched to BK virus later when I began work on my doctorate.

Human Polyomaviruses: Molecular and Clinical Perspectives, Edited by Kamel Khalili and Gerald L. Stoner.
ISBN 0-471-39009-7

2. THE INSTITUTE AND PEOPLE INVOLVED

The Public Health Laboratory Service (PHLS) began as a network of bacteriology laboratories in England and Wales, the Emergency Public Health Laboratory Service (EPHLS), brought together in 1939 to combat the threat of epidemics during the Second World War (Williams, 1985). In 1946 a permanent service was established and subsequently enlarged to include 63 laboratories by 1969 to monitor and control the spread of infectious disease in peacetime. In 1946 a collection of reference laboratories was assembled on the site of the Government Lymph Establishment at Colindale in North West London, where previously smallpox vaccine was produced. This formed the Central Public Health Laboratory (CPHL) and included the Virus Reference Laboratory (VRL), the initial function of which was to set up diagnostic facilities for smallpox. The work of CPHL expanded, and in 1951 the building of the "tower block" (Fig. 4.1), was begun. VRL was housed on the third floor of this building, and much of the early work on BKV was done in laboratory 316 (Fig. 4.2). The building was later demolished when CPHL relocated in 1985.

By the early 1970s the work of VRL was dominated by cell culture for virus isolation and virus neutralization tests, electron microscopy, and complement fixation tests. Hot rooms and incubators contained rack upon rack or rollers of cell culture tubes that had to be examined and medium laboriously changed once or twice a week, maybe for several months. There was a large tissue

Figure 4.1. The 1951 "tower block" at CPHL. BKV was first isolated in a laboratory on the third floor of this building. Courtesy of Mr. J. Gibson.

Figure 4.2. The laboratory (room 316) in the "tower block" where much of the early work on BKV was carried out. Courtesy of Dr. W.A. Knowles.

culture preparation laboratory and sections devoted to enteroviruses, respiratory viruses, smallpox, and rabies diagnosis.

Consultant Virologist Dr. Sylvia Gardner (Fig. 4.3) joined VRL in 1962, having studied medicine at Birmingham University. Her early interests were in respiratory viruses, and she ran a small virology unit in the hospital of a large childrens' home organized as a village. There was always a strong clinical aspect to her work, and she later became involved with investigations in VRL on transplant patients and pregnant women. This brought her into contact with Professor K.A. Porter, a pathologist at St. Mary's Hospital, Paddington, who, with colleagues in the United States was researching cystic duct obstruction in the first liver transplant recipients, and who was anxious for the role of viruses to be studied. Dr. Gardner retired from VRL in 1989.

Dr. Anne Field (Fig. 4.4) studied Microbiology at Bristol University and came to CPHL in 1959. She originally worked in the enterovirus laboratory, but in 1968 was appointed the first electron microscopist at CPHL and was responsible for setting up and running the electron microscope unit (EMU) until retirement in 1994. The first electron microscope in use at CPHL was an AEI 6B.

3. IDENTIFICATION OF A NEW VIRUS

In the late 1960s and early 1970s much work was being done to determine the epidemiology of cytomegalovirus (CMV), first isolated a decade earlier, and

Figure 4.3. Dr. Sylvia Gardner. Courtesy of Mr. J. Gibson.

its transmission in donated blood and organs. A collaboration was set up between Dr. Dulcie Coleman in the Department of Histopathology and Cytology at St. Mary's Hospital and Sylvia Gardner at Colindale to investigate, both cytologically and virologically, the presence of CMV infection in renal transplant patients.

On October 7, 1970, a midstream urine sample, collected a day earlier, arrived in the laboratory from a Sudanese patient who had received a renal transplant from his brother on June 24, 1970. A phone call the same day from St. Mary's Hospital informed us that this urine contained many inclusion-bearing cells, and so electron microscopy examination would be well worthwhile. Two days later, Anne Field saw very large numbers of papovavirus particles in the high-speed urine pellet and suggested that they were particles of common wart virus (the only human papovavirus known at the time). It was queried whether the patient had genital warts; a report earlier that year (Spencer and Andersen, 1970) had described a high incidence of warts after renal transplantation. Ultrathin sectioning of the cells in a subsequent urine sample collected on October 12 from the same patient showed many virus particles within enlarged cell nuclei.

On the day of receipt, the first urine was inoculated into tubes of both secondary rhesus monkey kidney (MK) cells and human embryo lung fibro-

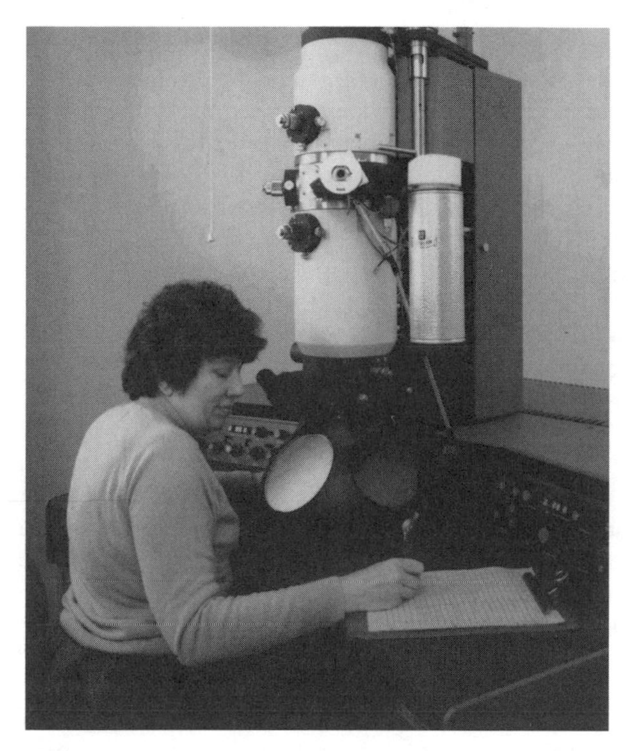

Figure 4.4. Dr. Anne Field at a Philips 420 transmission electron microscope. Courtesy of Mr. J. Gibson.

blasts (HEL) and observed every 3 or 4 days for a cytopathic effect (CPE). On day 19 a CPE appeared in the inoculated MK cell cultures, and the culture fluid was found to contain papovavirus particles. The fluid from uninoculated cells of the same batch was also examined as it was known that SV40, a monkey polyomavirus, was sometimes a contaminant in these cells; however, the controls were negative. In contrast, no CPE was seen in the HEL cells up to 27 days after inoculation. Vero cells were also inoculated on October 12, and on day 37 a CPE consisting of rounded cells was clearly visible.

The isolate was subsequently passaged in MK and Vero cells but appeared to die out in the MK cultures, and by the third passage CPE was no longer detectable even after 47 days in culture. Papovavirus was subsequently isolated from many more urine samples from this patient and others, but was always slow growing in culture, requiring immense patience. Virus harvested from the second passage in Vero cells was used as an antigen in complement fixation tests, and a seroconversion to the agent was demonstrated in the patient at the time the virus was isolated. The agent was named BK, the patient's initials, and the original isolate is known as the Gardner strain.

Measurement of the virus particles (45 nm diameter) showed the isolate to be a polyomavirus, as distinct from a papillomavirus, and work was undertaken to distinguish it from the only known polyomaviruses at the time: mouse polyoma, K virus of mice, monkey SV40, and rabbit kidney vacuolating agent. In early 1971 a range of biologic properties was investigated to characterize the new agent. It was found to agglutinate erythrocytes from different animal species, including humans, which immediately distinguished it from SV40, and enabled a hemagglutination inhibition (HI) test to be developed to add to the standard complement fixation test. Seroprevalence studies were later undertaken in various populations. The tumorigenic property of BKV in hamsters was also established early on.

After the virus was isolated and shown not to be human papillomavirus or SV40, it was thought that it might be the virus that had been described in association with the rare neurologic disease progressive multifocal leucoencephalopathy (PML) but that had thus far not been grown. Sylvia Gardner, therefore, wrote to laboratories in the United States that had published cases of PML and requested some PML brain tissue from them (hence the rumors heard by Duard Walker in Wisconsin!). No brain tissue was forthcoming, but we were later able to study patients in the United Kingdom with PML.

After our manuscript was submitted to *The Lancet*, the editors informed us that they had received another paper from the United States on the isolation of a human polyomavirus. Both papers had been accepted and would be published together.

Following the publication of the two original papers, it was obviously important to find out as soon as possible whether BK and JC viruses were related or even identical, and immediate collaboration was set up across "the pond" between Sylvia Gardner, Duard Walker, Billie Padgett, and Gabriele Zu Rhein. Viruses were exchanged, and it was established that, indeed, the isolates were two new human polyomaviruses, related to each other and also to SV40.

The original BKV paper was identified in 1989 as a "Citation Classic" by the Institute for Scientific Information.

4. SUBSEQUENT WORK ON THE HUMAN POLYOMAVIRUSES AT COLINDALE

Diagnostic and epidemiologic work on both BKV and JCV continued in VRL over the next 30 years, alongside developmental work that reflected the advances in new virologic techniques. BKV was always easier to grow than JCV, the latter requiring fetal brain cells. We became expert at growing these cells, which are no longer available. Large studies involving the original methods of cytology, electron microscopy, virus isolation, and serology were undertaken on renal transplant and PML patients, pregnant women, and children. BKV- and JCV-specific IgM detection methods were developed, and the molecular detection of BKV and JCV DNA was done first by hybridot and later by polymerase chain reaction. Antigenic subtypes of BKV were described, and subsequent work demonstrated the molecular basis for the observed variations.

Very recently collaborative studies have been undertaken on BKV and JCV VP1 expression in a yeast cell system.

ACKNOWLEDGMENTS

I am indebted to Dr. Sylvia Gardner for her recollections of events 30 years ago, which are unobtainable from any records or publications.

REFERENCES

Gardner SD, Field AM, Coleman DV, Hulme B (1971): New human papovavirus (B.K.) isolated from urine after renal transplantation. Lancet i:1253–1257.

Padgett BL, Walker DL, Zu Rhein GM, Eckroade RJ, Dessel BH (1971): Cultivation of papova-like virus from human brain with progressive multifocal leucoencephalopathy. Lancet i:1257–1260.

Spencer ES, Andersen HK (1970): Clinically evident, non-terminal infections with herpesviruses and the wart virus in immunosuppressed renal allograft recipients. BMJ iii:251–254.

Williams REO (1985): Microbiology for the Public Health. Public Health Laboratory Service: London.

5

THE HUMAN POLYOMAVIRUSES: AN OVERVIEW

MICHAEL J. IMPERIALE, PH.D.

1. INTRODUCTION

The subfamily of viruses named *Polyomavirinae* are small, naked viruses with icosahedral capsids and circular, double-stranded DNA genomes. These viruses have been isolated from a number of species, including human, monkey, rabbit, rodents, and birds. Their host range is rather restricted, and they generally do not infect other species productively. The first of these viruses to be identified was mouse polyoma virus (PyV), which, along with simian virus 40 (SV40), have been two of the most intensely studied viruses since their discovery 40 years ago (Sweet and Hilleman, 1960; Stewart et al., 1958). The study of these two viruses has led to major advances in our understanding of a variety of eukaryotic cell processes, including transcription, DNA replication, translation, signal transduction, cell growth, and oncogenic transformation. Indeed, the name *polyomavirus* is derived from the fact that the mouse virus causes a spectrum of tumor types when inoculated into newborn mice. Two polyomaviruses that solely infect humans have been identified, BKV and JCV. Both of these viruses were first isolated in 1971 (Padgett et al., 1971; Gardner et al., 1971), BKV from the urine of a renal transplant patient and JCV from brain tissue derived from a patient with progressive multifocal leukoencephalopathy

Human Polyomaviruses: Molecular and Clinical Perspectives, Edited by Kamel Khalili and Gerald L. Stoner.
ISBN 0-471-39009-7

(PML), of which JCV is clearly identified as the causative agent. In this chapter, the biology of the polyomaviruses is reviewed, with an emphasis on these two human viruses and on SV40, which historically has not been thought to be a human pathogen but, according to accumulating recent evidence, may well be. For a detailed treatment of the other polyomaviruses, the reader is referred to the chapter by Cole (1996) in *Fields Virology*.

2. POLYOMAVIRUS STRUCTURE

The icosahedral capsids of polyomaviruses are composed of three structural proteins, VP1, VP2, and VP3. VP1 is the major component, with 360 molecules per capsid, and VP2 and VP3 contribute 30–60 molecules each to the capsid. The VP1 protein is the most highly homologous among the three viruses (Table 5.1). The significant degree of homology among these viruses has complicated somewhat the development of highly specific serologic assays, an issue that is relevant to ongoing epidemiologic studies relating to the involvement of these viruses in human cancer and the possible spread of SV40 in the human population. These viruses have a somewhat unique structure in that the icosahedron is composed of 72 pentamers, with no apparent hexamers (Rayment et al., 1982; Liddington et al., 1991). Each pentamer consists of five VP1 molecules and one molecule of VP2 or VP3. The capsid surrounds the viral DNA, which is a supercoiled, circular, double-stranded molecule of approximately 5200 base pairs in the case of the primate viruses. Histones H2A, H2B, H3, and H4 are associated with the DNA in the virion. In the cell, histone H1 is also bound to the viral DNA, and in fact the viral chromosome in the cell is structurally indistinguishable from host cell chromatin (Louie, 1975; Keller et al., 1978; Zentgraf et al., 1978; Muller et al., 1978). The DNA sequence identity among the primate viruses is approximately 70% (Table 5.1).

The genomic organization of these viruses is also conserved (Figure 5.1). The early and late promoters, along with the origin of DNA replication, are

Table 5.1. Sequence Homologies Among BKV, JCV, and SV40

	BKV–JCV	BKV–SV40	JCV–SV40
Proteins[a]			
VP1	87	88	85
VP2	88	86	83
VP3	84	81	78
T antigen	90	84	82
t antigen	86	78	76
DNA[b]	72	69	68

[a]Percent amino acid homology.
[b]Percent sequence identity.

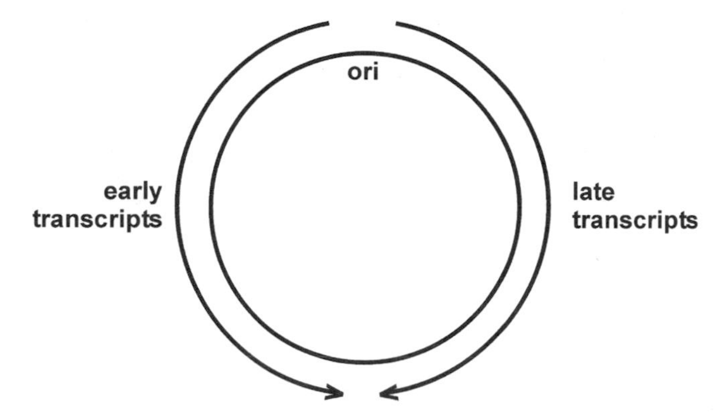

Figure 5.1. *Genomic organization of polyomaviruses.* The double-stranded, circular DNA genome is depicted with the origin of DNA replication (ori, which overlaps the transcriptional control region) at the top. The direction of transcription for the early and late transcription units is indicated by the arrows.

clustered together into a so-called viral regulatory or transcriptional control region. The early genes are transcribed from one strand of the genome, and the late genes are transcribed in the opposite direction from the complementary strand. In effect, then, the genetic map of the viruses has two halves, early and late, that are divided by the promoter/origin region on one side and the mRNA polyadenylation sites on the other.

3. THE POLYOMAVIRUS LIFE CYCLE

While the polyomaviruses can bind to and enter cells of multiple species, the outcome of the infection is species specific (Fig. 5.2). In permissive hosts, one generally obtains a lytic infection, while in nonpermissive hosts (e.g., rodents in the case of the primate viruses), there is a block to viral replication, leading to abortive infection or oncogenesis. For example, SV40 replication is restricted in mouse cells due to an inability to use the host DNA polymerase for viral DNA synthesis (Murakami et al., 1986). Infection of human cells by the human viruses is generally limited to epithelial cells, fibroblasts, lymphocytes, or cells derived from the nervous system. The outcome of that infection is also cell type specific. For example, while JCV persists in the urinary tract, infection of glial cells can be cytolytic. Most recently, it has been reported that cell type differences may dictate the outcome of SV40 infection. Specifically, it has been shown that while SV40 infection of human fibroblasts leads to a productive outcome, which has been known for some time, in mesothelial cells it established a persistent, nonlytic infection (Bocchetta et al., 2000). This result has important implications for the possible role of SV40 in human mesothelioma, as is discussed below. It should be noted that the concept of differences in the progression of the infection in different cell types within a given organism is

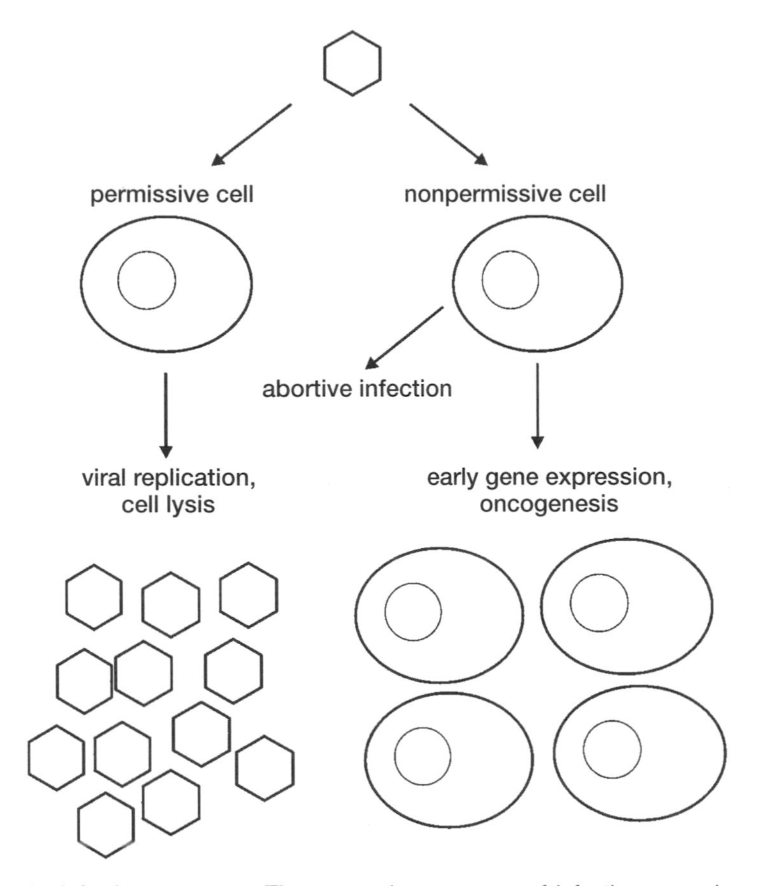

Figure 5.2. *Infection outcomes.* The two major outcomes of infections are shown. In a permissive cell (left side), the viral life cycle is completed and progeny virions are produced. In nonpermissive cells (right side), replication is blocked and abortive infection or transformation may ensue.

not limited to tissue specificity, as it has been known for some time that polyomaviruses require differentiated cells within a given tissue to provide a fertile environment for replication (Villarreal, 1991).

Infection of cells is initiated by binding of the virion to specific cell surface receptors. It is believed that SV40 binds MHC molecules (Norkin, 1999), although a differential ability to infect polarized epithelial cells via the apical or basolateral surfaces has been reported, even though MHC expression on the two membranes is the same (Basak et al., 1992). This may be a reflection of expression, or lack thereof, of additional molecules that are required for entry once the virus has bound. The receptors for BKV and JCV are not known, although it has been demonstrated that JCV binds specifically to glial cells, tonsilar stromal cells, and B lymphocytes, closely mirroring the observed cell tropism with respect to persistent and lytic infection (Wei et al., 2000). The viral capsids are then internalized by endocytosis and are transported to the nucleus, where the DNA is uncoated. The mechanism by which the virions escape endocytic vesicles is not clear. Once the DNA reaches the nucleus, however, transcription of the early genes begins, driven by promoter and enhancer sequences that overlap the origin of DNA replication. These promoter elements bind a large variety of ubiquitous and cell type-specific transcription factors (see Chapter 6). Early gene transcription has been demonstrated to play a regulatory role in infection in the replication of JCV in glial cells, as discussed below.

The early genes of the human polyomaviruses encode two major proteins, the large T and small t antigens, which are expressed from two mRNAs that are derived by alternative splicing of a single primary transcript (Fig. 5.3A). These proteins are so named because of the early finding that they were the dominant tumor antigens recognized by the immune systems of animals harboring virally induced tumors (Black et al., 1963). Overall, the degree of amino acid homology among the T antigens is similar to that observed for the capsid

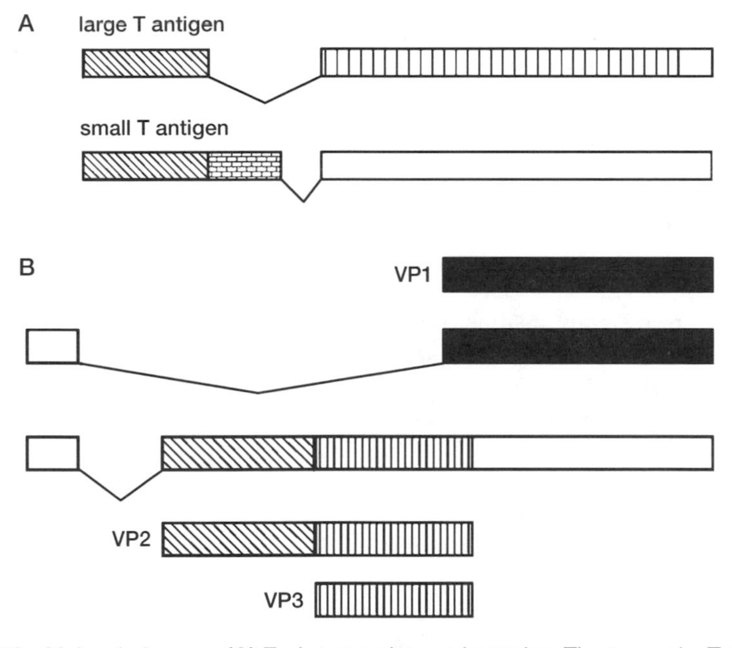

Figure 5.3. *Major viral genes.* **(A)** Early transcripts and proteins. The two major T antigens are expressed from two differentially spliced mRNAs. The angled lines indicate introns. The angled lines indicate introns that are removed by splicing. Hatched boxes, sequences shared in common by large T antigen and small t antigen; bricked box, small t-unique sequences; lined box, large T-unique sequences. **(B)** Late transcripts and proteins. The three late proteins VP1, VP2, and VP3 are expressed from two mRNAs, the shorter of which encodes VP1 and the longer of which encodes the other two proteins. The angled lines indicate introns. Black box, VP1; hatched box, VP2-unique sequence; lined box, sequences shared by VP2 and VP3.

proteins (Table 5.1), but if one examines the protein domains that have been shown to be functionally important for viral DNA replication and oncogenic transformation, the homology is much higher, closer to 100%. Mouse polyoma virus encodes a third T antigen molecule, middle T antigen, distinguishing it from the primate viruses. Most recently, an additional early mRNA has been identified in SV40 that would encode a T antigen with a predicted molecular weight of 17 kDa, sharing the first 131 amino acids of TAg and then terminating after four unique amino acids (Zerrahn et al., 1993). The role of this T antigen, sometimes referred to as "tiny t," has not yet been investigated in detail. For additional details on early mRNAs of JCV see Chapter 6.

Large T antigen (TAg), a nuclear phosphoprotein of approximately 700 amino acids in the primate viruses, is a critical player in the life cycle of the virus. One of its major roles during the early phase of the infection is to induce the cell to enter S phase. The virus must accomplish this because it does not encode any of the DNA synthetic machinery required for replication of its genome and therefore relies on the cell's machinery, which is expressed at high levels in S phase. Because the virus generally infects cells that are not actively cycling, it must provide them with a means to do so. The major mechanism by which the virus accomplishes this is by binding to and inactivating the retinoblastoma susceptibility protein, pRb, and its two related family members, p107 and p130 (DeCaprio et al., 1988; Ewen et al., 1989; Hannon et al., 1993; Dyson et al., 1989, 1990; Harris et al., 1996; Bollag et al., 1989; Trowbridge and Frisque, 1993). Inactivation results in the release of E2F transcription factors, which stimulate transcription of genes involved in entry into S phase and DNA synthesis (for review, see Cress and Nevins, 1996). The release of E2F is mediated by binding of TAg to pRb family members resulting in direct physical displacement of E2F, as well as TAg-mediated induction of the degradation of pRb family member proteins (Campbell et al., 1997; Stubdal et al., 1997; Zalvide et al., 1998; Sock et al., 1999; Harris et al., 1998). Both of these functions require a conserved LXCXE motif that is involved in pRb family protein binding, and degradation also requires the J domain, so named due to its homology to the DnaJ family of molecular chaperones (Rutila et al., 1986; Kalderon and Smith, 1984; Ewen et al., 1989; Christensen and Imperiale, 1995; Zalvide and DeCaprio, 1995; Brodsky and Pipas, 1998; Campbell et al., 1997; Stubdal et al., 1997; Sheng et al., 1997; Zalvide et al., 1998; Harris et al., 1998). The exact mechanism by which the J domain targets these tumor suppressors for degradation is currently unclear. TAg also plays a second, more direct, role in DNA replication, acting as an important initiation factor for DNA synthesis. First, TAg binds to specific sequences in the viral DNA at the origin of replication and recruits the DNA synthetic machinery to the origin of replication by directly binding to the DNA polymerase α/primase complex and replication protein A (RPA), a single-stranded DNA-binding protein that is required for efficient replication (Dornreiter et al., 1990; Melendy and Stillman, 1993). Second, TAg has helicase activity, unwinding the DNA at the origin and setting the stage for the initiation of DNA synthesis (Dean et al., 1987; Stahl et al., 1986). Indeed, much of what we know about eukaryotic DNA replication comes from in vitro studies of SV40 replication, and the entire system has been reconstituted from purified factors in recent years (Waga et al., 1994).

Small t antigen is a cysteine-rich protein of 172 amino acids in BKV and JCV (174 in the case of SV40), the first 80 of which (82 in SV40) are shared with TAg. It appears to be dispensable for lytic infection in cultured cells. However, whether it plays a role in vivo is not known (Shenk et al., 1976; Staneloni et al., 1977). It has the ability to bind the cellular phosphatase PP2A, inhibiting its function. This results in a signal transduction cascade that stimulates cell growth, implying that, like TAg, small t antigen may be important in setting up the cellular environment for efficient replication under certain conditions (Sontag et al., 1993; Pallas et al., 1990). Middle T antigen of mouse PyV, as its name implies, is intermediate in size to the large and small T antigens. This cytoplasmic protein is localized to the plasma membrane and is identical to small t antigen in its amino terminus but has a unique carboxy terminus, and it is also required for viral replication in mice (Freund et al., 1992b).

Once DNA replication begins, the infection enters the late phase. At the same time, TAg acts to stimulate transcription of the late genes and to repress transcription of the early genes. While the mechanism of repression is poorly understood, activation of late gene expression is accomplished through the interaction of TAg and components of the cellular basal transcription machinery (TAFs; Gruda et al., 1993; Berger et al., 1996). The late genes encode the three structural proteins described above, VP1, VP2, and VP3, which are also expressed from alternatively spliced mRNAs whose transcription initiation and polyadenylation sites are identical (Fig. 5.3B). Based on studies in vitro and using baculovirus-expressed capsid proteins, it is believed that these proteins assemble into capsid structures and the DNA is then inserted into the capsids (Sandalon et al., 1997; Forstova et al., 1993; Salunke et al., 1986). It has been speculated that the infected cell is killed through an apoptotic mechanism. It has been argued, for example, that the reason the virus inhibits p53 function (see below) is to keep the cell alive longer to maximize the yield of progeny virions. There is direct evidence from analyses of adenovirus, whose E1A and E1B proteins share many functions with TAg, to support this idea (Rao et al., 1992). Studies have shown SV40 virions being released directly from the plasma membranes of intact epithelial cells, however (Clayson et al., 1989). Therefore, it is not clear that cell lysis is required for release of progeny virions.

4. CLASSIC INFECTIOUS DISEASE

The major clear-cut associations between the human polyomaviruses and classically defined infectious diseases are those of JCV with PML and BKV with hemorrhagic cystitis (Weber and Major, 1997; Arthur et al., 1988; Barbanti-Brodano et al., 1998). BKV and JCV are ubiquitous in most human populations, infecting most individuals in early and late childhood, respectively (Gardner, 1973; Padgett and Walker, 1973; Taguchi et al., 1982). BKV is thought to cause mild respiratory illness, which would correlate with its appearance in the population at early ages (Goudsmit et al., 1981, 1982; Hashida et al., 1976; Mininberg et al., 1982; Padgett et al., 1983). Indeed, both viruses can be detected in tonsillar tissue, and JCV has been shown to infect tonsillar lymphocytes and stromal cells (Goudsmit et al., 1982; Monaco et al., 1998a,b). The viruses then disseminate and establish a subclinical, persistent infection in the urinary tract and lymphocytes (and, for JCV, the brain). To obtain overt disease, the individual usually must be immunocompromised. Most commonly affected are patients with AIDS or transplant recipients taking immunosuppressive drugs. The pathology of the disease is thought to be due to destruction of the infected tissue due to viral replication. The molecular mechanisms of disease have been described in a fair amount of detail for JCV (Chapter 12). For a detailed discussion of BKV and clinical disease see Chapter 14.

The pathology of PML is due to viral replication in oligodendrocytes, which are the cells in the brain that produce myelin. Numerous studies have demonstrated that the JCV promoter and enhancer contain binding sites for a variety of transcription factors that are expressed in glial cells, although only one of these appears to be truly glial cell specific (Frisque and White, 1992; Raj and Khalili, 1995). Disease is also often accompanied by rearrangements in the transcriptional control region that appear to enhance TAg expression and, subsequently, replication (Dörries, 1997, 1998; Weber and Major, 1997). The BKV promoter and enhancer have not been dissected in nearly as much detail, making the molecular pathogenesis of hemorrhagic cystitis less well understood.

5. ONCOGENIC TRANSFORMATION

Large T antigen is the major transforming protein of the primate viruses. Its expression is both necessary and sufficient for the oncogenic transformation of most cell types (Martin and Chou, 1975; Brockman, 1978; Tegtmeyer, 1975; Brugge and Butel, 1975). The TAg of PyV differs from its primate cousins in that it is not required for maintenance of the transformed state: that function falls upon middle T antigen (Di Mayorca et al., 1969; Eckhart, 1969; Fried, 1965; Basilico et al., 1980; Lania et al., 1980). Small t antigen is required for transformation under certain conditions, working through its interaction with PP2A. The ability of small t to bind PP2A correlates with its ability to induce cell growth (Mungre et al., 1994).

The capacity of TAg to transform cells maps to at least three domains of the molecule: the pRb binding and J domains discussed above and a more carboxy-terminal domain that allows TAg to bind to the tumor suppressor protein p53 (for review, see Pipas, 1992). It is worth noting here that p53 was originally identified through its ability to coprecipitate with TAg (Lane and Crawford, 1979; McCormick and Harlow, 1980; Linzer and Levine, 1979). Our current understanding of the mechanism of transformation is as follows. TAg expression leads to the induction of S phase through the pRb-binding and J domain functions described above. The cell responds to this abnormal signal to divide by turning on a p53 response, which would normally lead to cell cycle arrest or apoptosis. TAg overcomes this roadblock, however, by binding to p53 and inhibiting its function. Thus, the cell is induced to proliferate, and p53-mediated check point control is abrogated. The purpose of the interaction with pRb and p53 pathways during lytic infection is therefore to push the cell into a maximal environment for the support of viral DNA replication and to keep it alive as long as possible to optimize viral yields before the cell is killed. Oncogenic transformation occurs when there is a block to productive infection, but continued expression of TAg. This leads to cell proliferation without killing of the cell, ultimately resulting in tumor formation in the nonpermissive host due to the restrictive nature of the virus–cell interaction relating to DNA replication. One can also obtain oncogenic transformation in a permissive host if viral replication is blocked by other means, such as mutations in the origin of replication (Gluzman, 1981). Finally, one must keep in mind that TAg itself can induce DNA damage, making its inactivation of p53 doubly dangerous to the cell (Theile and Grabowski, 1990; Lazutka et al., 1996; Tognon et al., 1996; Trabanelli et al., 1998; Ray et al., 1990).

In addition to the large and small T antigens, the third PyV T antigen, middle T antigen, is required for transformation by the mouse virus. While PyV large T antigen does bind to and deregulate pRb family members, working through the same functional domains as the primate TAg (Larose et al., 1991; Freund et al., 1992a; Sheng et al., 1997), it does not bind to p53. PyV small t antigen does affect PP2A function and appears to serve an identical function to SV40 small t antigen (Pallas et al., 1988). Middle T antigen acts through a distinctly different mechanism than the other two oncoproteins. Its membrane-bound location puts it in a position to bind to and activate Src family kinases, thereby signaling the cell to grow and divide via a kinase cascade pathway (Kaplan et al., 1989).

6. HUMAN CANCER

Until recently, evidence for the involvement of JCV, BKV, and SV40 in human cancer was virtually nonexistent. The advent of polymerase chain reaction (PCR) technology, however, has sparked renewed interest in the issue and provided provocative evidence that these viruses may indeed play a role in certain neoplasms. The fact that these viruses can indeed cause tumors is not disputed, based on the in vitro studies described above as well as experiments in animals. For example, SV40 induces osteosarcomas and mesotheliomas in infected hamsters and causes brain tumors in transgenic mice; JCV also causes brain tumors in transgenic mice and infected hamsters; and BKV transgenic mice get kidney tumors and hepatocellular carcinomas (Cicala et al., 1993; Diamandopoulos,

1972; Brinster et al., 1984; Walker et al., 1973; Zu Rhein, 1983; Franks et al., 1996; Krynska et al., 1999; Small et al., 1986; Dalrymple and Beemon, 1990). The crucial questions with respect to their potential role in human cancer are (1) Is the detection of viral sequences in tumor samples reliable? (2) Is the association of viral sequences with human tumor samples a reflection of a causal role, or is the virus an innocent bystander?

The use of PCR to detect sequences in biopsy samples is somewhat tricky in that often only small amounts of tissue are available, requiring large numbers of cycles in the reaction. Such large numbers increase the possibility of detecting contaminants that might be inadvertently introduced into the reactions. While this may have called into question some early reports, most laboratories now take great care in segregating the various sample manipulations so as to prevent contamination problems. In addition, there have now been a number of cases in which viral sequences isolated from tissues have been sequenced and shown to be distinct from laboratory strains of the viruses, thereby ruling out simple contamination with cloned DNAs.

BKV sequences have been detected in a variety of human tumors, including pancreas, neuroblastoma, and urinary tract, as well as samples isolated from tissues that the virus has not been previously thought to infect (for review, see Imperiale, 2000). Two of these studies are notable as the conclusions were not solely based on PCR analysis. First is an analysis of urinary tract tumors by Monini et al. (1995) at the University of Ferrara, in which they were able to detect BK viral sequences in some tumors using Southern blotting. Analysis of these sequences indicated that the viral DNA had integrated into the host chromosome and also that rearrangements in the origin of replication were present that would be predicted to interfere with productive infection. As discussed above, a block to viral replication concomitant with TAg expression could lead the infection down the path to oncogenic transformation. Flaegstad et al. (1999), in Norway, detected BKV DNA sequences in 17 of 18 neuroblastomas by in situ PCR, BKV TAg by immunohistochemistry, and TAg–p53 complexes by immunoprecipitation.

JCV sequences have been reported in colorectal tumors (Laghi et al., 1999; Ricciardiello, et al., 2001), although this location, like that of some of the sites in which BKV has been detected, does not represent a known site of infection by the virus. JCV has also been detected in various neuronal tumors, and T antigen expression can be demonstrated by immunohistochemistry in some of these samples (see Khalili et al., 1999, and references therein). With both viruses, the analysis of whether they are truly involved in cancer is complicated by the fact that they are also detected in a large percentage of normal tissue: For example, in the urinary tract study of BKV discussed above, approximately 60% of the normal tissues were positive for the viral early region by PCR. Thus, although viral sequences can be detected in tumors derived from tissues in which the virus is known to persist, expression can be detected in some cases, and the state of the viral chromosome appears to be conducive to transformation in some cases, proving a role for these viruses in human cancer will require further study.

A possible role of SV40 in human cancer has also been postulated recently (Levine et al., 1998; Barbanti-Brodano et al., 1998; Lednicky and Butel, 2001). Viral sequences have been reported in three types of human cancer, brain tumors, osteosarcomas, and mesotheliomas, but have not been found at various other sites (Bergsagel et al., 1992; Lednicky et al., 1995; Carbone et al., 1994, 1996; Pass et al., 1996; Testa et al., 1998). Interestingly, brain tumors and mesotheliomas are the two most common tumors induced by SV40 in transgenic mice and infected hamsters, respectively (Cicala et al., 1993; Brinster et al., 1984). T antigen has been shown to be expressed in these human samples (Carbone et al., 1997, 1999; Butel and Lednicky, 1999). The argument for a role of SV40 in mesothelioma has received a strong boost recently from the finding that the virus can establish what appears to be a persistent infection in primary human mesothelial cell cultures and that these infected cells score positive in transformation assays in culture (Bocchetta et al., 2000). Again, this raises the possibility that nonproductive infection may predispose the cell to travel down an oncogenic path. In addition, viral DNA has been cloned from a number of these human tumors, and two findings have been described. First, the viruses isolated from humans do not have a characteristic duplication of the enhancer sequences that is found in laboratory strains after passage in tissue culture (Lednicky et al., 1995; Lednicky and Butel, 2001). Second, there is quite a bit of variability in the extreme carboxy-terminus of the T antigen gene (Stewart et al., 1996). These findings indicate that the samples have not been contaminated with laboratory stocks of the virus or plasmids containing viral sequences. It is clear that a large number of people may have been exposed to SV40 through administration of contaminated poliovirus vaccines in the late 1950s and early 1960s, but perhaps of more concern is the fact that the virus is being isolated from individuals who are too young to have received these contaminated lots of vaccine. Serologic studies also indicate the presence of SV40 infection in nonvaccinated populations (Butel et al., 1999). Thus, it is quite possible that SV40 is being transmitted horizontally throughout the population, and even vertical transmission cannot be ruled out. With the evidence for SV40 as a human pathogen growing, attempts to develop an SV40 vaccine may take on added importance (Xie et al., 1999; Imperiale et al., 2001).

7. CONCLUSIONS

The polyomaviruses have a long history of providing unique insights into how viruses interact with their host cells and organisms. Early investigations lent tremendous insight into basic molecular events that are critical to both viral replication and host cell biology. More recently, the mechanisms behind the role of JCV in PML have been teased apart. Finally, the evidence implicating BKV, JCV, and SV40 in human cancer has reached a point where the possibility of their contributing to neoplastic disease cannot be ignored. Continued studies into the molecular pathogenesis of these cancers will no doubt allow a more definitive conclusion in the near future. In addition, an increased understanding of the biology of these viruses will assist in the development of antiviral drugs and vaccines, holding promise that the human diseases associated with these viruses can be treated or prevented.

REFERENCES

Arthur RR, Shah KV, Charache P, Saral R (1988): BK and JC virus infections in recipients of bone marrow transplants. J Infect Dis 158:563–569.

Barbanti-Brodano G, Martini F, De Mattei M, Lazzarin L, Corallini A, Tognon M (1998): BK and JC human polyomaviruses and simian virus 40: Natural history of infection in humans, experimental oncogenicity, and association with human tumors. Adv Virus Res 50:69–99.

Basak S, Turner H, Compans RW (1992): Expression of SV40 receptors on apical surfaces of polarized epithelial cells. Virology 190:393–402.

Basilico C, Zouzias D, Della-Valle G, Gattoni S, Colantuoni V, Fenton R, Dailey L (1980): Integration and excision of polyoma virus genomes. Cold Spring Harb Symp Quant Biol 44(Pt 1):611–620.

Berger LC, Smith DB, Davidson I, Hwang JJ, Fanning E, Wildeman AG (1996): Interaction between T antigen and TEA domain of the factor TEF-1 derepresses simian virus 40 late promoter in vitro: Identification of T-antigen domains important for transcription control. J Virol 70:1203–1212.

Bergsagel DJ, Finegold MJ, Butel JS, Kupsky WJ, Garcea RL (1992): DNA sequences similar to those of simian virus 40 in ependymomas and choroid plexus tumors of childhood. N Engl J Med 326:988–993.

Black PH, Rowe WP, Turner HC, Huebner RJ (1963): A specific complement-fixing antigen present in SV40 tumor and transformed cells. Proc Natl Acad Sci USA 50: 1148–1156.

Bocchetta M, Di Resta I, Powers A, Fresco R, Tosolini A, Testa JR, Pass HI, Rizzo P, Carbone M (2000): Human mesothelial cells are unusually susceptible to simian virus 40–mediated transformation and asbestos cocarcinogenicity. Proc Natl Acad Sci USA 97:10214–10219.

Bollag B, Chuke WF, Frisque RJ (1989): Hybrid genomes of the polyomaviruses JC virus, BK virus, and simian virus 40: Identification of sequences important for efficient transformation. J Virol 63:863–872.

Brinster RL, Chen HY, Messing A, Van Dyke T, Levine AJ, Palmiter RD (1984): Transgenic mice harboring SV40 T-antigen genes develop characteristic brain tumors. Cell 37:367–379.

Brockman WW (1978): Transformation of BALB/c-3T3 cells by tsA mutants of simian virus 40: Temperature sensitivity of the transformed phenotype and retransformation by wild-type virus. J Virol 25:860–870.

Brodsky JL, Pipas JM (1998): Polyomavirus T antigens: Molecular chaperones for multiprotein complexes. J Virol 72:5329–5334.

Brugge JS, Butel JS (1975): Role of simian virus 40 gene A function in maintenance of transformation. J Virol 15:619–635.

Butel JS, Jafar S, Wong C, Arrington AS, Opekun AR, Finegold MJ, Adam E (1999): Evidence of SV40 infections in hospitalized children. Hum Pathol 30:1496–1502.

Butel JS, Lednicky JA (1999): Cell and molecular biology of simian virus 40: Implications for human infections and disease. J Natl Cancer Inst 91:119–134.

Campbell KS, Mullane KP, Aksoy IA, Stubdal H, Zalvide J, Pipas JM, Silver PA, Roberts TM, Schaffhausen BS, DeCaprio JA (1997): DnaJ/hsp40 chaperone domain of SV40 large T antigen promotes efficient viral DNA replication. Genes Dev 11: 1098–1110.

Carbone M, Fisher S, Powers A, Pass HI, Rizzo P (1999): New molecular and epidemiological issues in mesothelioma: Role of SV40. J Cell Physiol 180:167–172.

Carbone M, Pass HI, Rizzo P, Marinetti M, Di Muzio M, Mew DJ, Levine AS, Procopio A (1994): Simian virus 40–like DNA sequences in human pleural mesothelioma. Oncogene 9:1781–1790.

Carbone M, Rizzo P, Grimley PM, Procopio A, Mew DJ, Shridhar V, de Bartolomeis A, Esposito V, Giuliano MT, Steinberg SM, Levine AS, Giordano A, Pass HI (1997): Simian virus–40 large-T antigen binds p53 in human mesotheliomas. Nat Med 3: 908–912.

Carbone M, Rizzo P, Procopio A, Giuliano M, Pass HI, Gebhardt MC, Mangham C, Hansen M, Malkin DF, Bushart G, Pompetti F, Picci P, Levine AS, Bergsagel JD, Garcea RL (1996): SV40-like sequences in human bone tumors. Oncogene 13:527–535.

Christensen JB, Imperiale MJ (1995): Inactivation of the retinoblastoma susceptibility protein is not sufficient for the transforming function of the conserved region 2-like domain of simian virus 40 large T antigen. J Virol 69:3945–3948.

Cicala C, Pompetti F, Carbone M (1993): SV40 induces mesotheliomas in hamsters. Am J Pathol 142:1524–1533.

Clayson ET, Brando LV, Compans RW (1989): Release of simian virus 40 virions from epithelial cells is polarized and occurs without cell lysis. J Virol 63:2278–2288.

Cole CN (1996): Polyomavirinae: The viruses and their replication. In Fields Virology; Fields BN, Knipe DM, Howley PM, Eds. Lippincott-Raven: Philadelphia, pp. 1997–2026.

Cress WD, Nevins JR (1996): Use of the E2F transcription factor by DNA tumor virus regulatory proteins. Curr Top Microbiol Immunol 208:63–78.

Dalrymple SA, Beemon KL (1990): BK virus T antigens induce kidney carcinomas and thymoproliferative disorders in transgenic mice. J Virol 64:1182–1191.

Dean FB, Bullock P, Murakami Y, Wobbe CR, Weissbach L, Hurwitz J (1987): Simian virus 40 (SV40) DNA replication: SV40 large T antigen unwinds DNA containing the SV40 origin of replication. Proc Natl Acad Sci USA 84:16–20.

DeCaprio JA, Ludlow JW, Figge J, Shew JY, Huang CM, Lee WH, Marsilio E, Paucha E, Livingston DM (1988): SV40 large tumor antigen forms a specific complex with the product of the retinoblastoma susceptibility gene. Cell 54:275–283.

Diamandopoulos GT (1972): Leukemia, lymphoma, and osteosarcoma induced in the Syrian golden hamster by simian virus 40. Science 176:173–175.

Di Mayorca G, Callender J, Marin G, Giordano R (1969): Temperature-sensitive mutants of polyoma virus. Virology 38:126–133.

Dornreiter I, Hoss A, Arthur AK, Fanning E (1990): SV40 T antigen binds directly to the large subunit of purified DNA polymerase alpha. EMBO J 9:3329–3336.

Dörries K (1997): New aspects in the pathogenesis of polyomavirus-induced disease. Adv Virus Res 48:205–261.

Dörries K (1998): Molecular biology and pathogenesis of human polyomavirus infections. Dev Biol Stand 94:71–79.

Dyson N, Bernards R, Friend SH, Gooding LR, Hassell JA, Major EO, Pipas JM, Van Dyke T, Harlow E (1990): Large T antigens of many polyomaviruses are able to form complexes with the retinoblastoma protein. J Virol 64:1353–1356.

Dyson N, Buchkovich K, Whyte P, Harlow E (1989): The cellular 107K protein that binds to adenovirus E1A also associates with the large T antigens of SV40 and JC virus. Cell 58:249–255.

Eckhart W (1969): Complementation and transformation by temperature-sensitive mutants of polyoma virus. Virology 38:120–125.

Ewen ME, Ludlow JW, Marsilio E, DeCaprio JA, Millikan RC, Cheng SH, Paucha E, Livingston DM (1989): An N-terminal transformation-governing sequence of SV40 large T antigen contributes to the binding of both p110Rb and a second cellular protein, p120. Cell 58:257–267.

Flaegstad T, Andresen PA, Johnsen JI, Asomani SK, Jorgensen GE, Vignarajan S, Kjuul A, Kogner P, Traavik T (1999): A possible contributory role of BK virus infection in neuroblastoma development. Cancer Res 59:1160–1163.

Forstova J, Krauzewicz N, Wallace S, Street AJ, Dilworth SM, Beard S, Griffin BE (1993): Cooperation of structural proteins during late events in the life cycle of polyomavirus. J Virol 67:1405–1413.

Franks RR, Rencic A, Gordon J, Zoltick PW, Curtis M, Knobler RL, Khalili K (1996): Formation of undifferentiated mesenteric tumors in transgenic mice expressing human neurotropic polyomavirus early protein. Oncogene 12:2573–2578.

Freund R, Bronson RT, Benjamin TL (1992a): Separation of immortalization from tumor induction with polyoma large T mutants that fail to bind the retinoblastoma gene product. Oncogene 7:1979–1987.

Freund R, Sotnikov A, Bronson RT, Benjamin TL (1992b): Polyoma virus middle T is essential for virus replication and persistence as well as for tumor induction in mice. Virology 191:716–723.

Fried M (1965): Cell-transforming ability of a temperature-sensitive mutant of polyoma virus. Proc Natl Acad Sci USA 53:486–491.

Frisque RJ, White FA, III (1992): The molecular biology of JC virus, causative agent of progressive multifocal leukoencephalopathy. In Molecular Neurovirology; Ross R, Ed. Humana Press: Clifton, NJ, pp 25–158.

Gardner SD (1973): Prevalence in England of antibody to human polyomavirus (B.K.) BMJ 1:77–78.

Gardner SD, Field AM, Coleman DV, Hulme B (1971): New human papovavirus (B.K.) isolated from urine after renal transplantation. Lancet 1:1253–1257.

Gluzman Y (1981): SV40-transformed simian cells support the replication of early SV40 mutants. Cell 23:175–182.

Goudsmit J, Baak ML, Sleterus KW, van der Noordaa J (1981): Human papovavirus isolated from urine of a child with acute tonsillitis. Br Med J (Clin Res Ed) 283: 1363–1364.

Goudsmit J, Wertheim-van Dillen P, van Strien A, van der Noordaa J (1982): The role of BK virus in acute respiratory tract disease and the presence of BKV DNA in tonsils. J Med Virol 10:91–99.

Gruda MC, Zabolotny JM, Xiao JH, Davidson I, Alwine JC (1993): Transcriptional activation by simian virus 40 large T antigen: Interactions with multiple components of the transcription complex. Mol Cell Biol 13:961–969.

Hannon GJ, Demetrick D, Beach D (1993): Isolation of the Rb-related p130 through its interaction with CDK2 and cyclins. Genes Dev 7:2378–2391.

Harris KF, Christensen JB, Imperiale MJ (1996): BK virus large T antigen: Interactions with the retinoblastoma family of tumor suppressor proteins and effects on cellular growth control. J Virol 70:2378–2386.

Harris KF, Christensen JB, Radany EH, Imperiale MJ (1998): Novel mechanisms of E2F induction by BK virus large T antigen: Requirement of both the pRb binding and J domains. Mol Cell Biol 18:1746–1756.

Hashida Y, Gaffney PC, Yunis EJ (1976): Acute hemorrhagic cystitis of childhood and papovavirus-like particles. J Pediatr 89:85–87.

Imperiale MJ (2000): The human polyomaviruses, BKV and JCV: Molecular pathogenesis of acute disease and potential role in cancer. Virology 267:1–7.

Imperiale MJ, Pass HI, Sanda MG (2001): Prospects for an SV40 vaccine. Semin Cancer Biol 11:81–85.

Kalderon D, Smith AE (1984): In vitro mutagenesis of a putative DNA binding domain of SV40 large-T. Virology 139:109–137.

Kaplan DR, Pallas DC, Morgan W, Schaffhausen B, Roberts TM (1989): Mechanisms of transformation by polyoma virus middle T antigen. Biochim Biophys Acta 948: 345–364.

Keller W, Muller U, Eicken I, Wendel I, Zentgraf H (1978): Biochemical and ultrastructural analysis of SV40 chromatin. Cold Spring Harb Symp Quant Biol 42(Pt 1):227–244.

Khalili K, Krynska B, Del Valle L, Katsetos CD, Croul S (1999): Medulloblastomas and the human neurotropic polyomavirus JC virus. Lancet 353:1152–1153.

Krynska B, Otte J, Franks R, Khalili K, Croul S (1999): Human ubiquitous JCV(CY) T-antigen gene induces brain tumors in experimental animals. Oncogene 18:39–46.

Laghi L, Randolph AE, Chauhan DP, Marra G, Major EO, Neel JV, Boland CR (1999): JC virus DNA is present in the mucosa of the human colon and in colorectral cancers. Proc Natl Acad Sci USA 96:7484–7489.

Lane DP, Crawford LV (1979): T antigen is bound to a host protein in SV40-transformed cells. Nature 278:261–263.

Lania L, Hayday A, Bjursell G, Gandini-Attardi D, Fried M (1980): Organization and expression of integrated polyoma virus DNA sequences in transformed rodent cells. Cold Spring Harbor Symp Quant Biol 44(Pt 1):597–603.

Larose A, Dyson N, Sullivan M, Harlow E, Bastin M (1991): Polyomavirus large T mutants affected in retinoblastoma protein binding are defective in immortalization. J Virol 65:2308–2313.

Lazutka JR, Neel JV, Major EO, Dedonyte V, Mierauskine J, Slapsyte G, Kesminiene A (1996): High titers of antibodies to two human polyomaviruses, JCV and BKV, correlate with increased frequency of chromosomal damage in human lymphocytes. Cancer Lett 109:177–183.

Lednicky JA, Butel JS (2001): Simian virus 40 regulatory region structural diversity and the association of viral archetypal regulatory regions with human brain tumors. Sem Cancer Biol 11:39–47.

Lednicky JA, Garcea RL, Bergsagel DJ, Butel JS (1995): Natural simian virus 40 strains are present in human choroid plexus and ependymoma tumors. Virology 212:710–717.

Levine A, Butel J, Dörries K, Goedert J, Frisque R, Garcea R, Morris A, O'Neill F, Shah K (1998): SV40 as a putative human commensal. Dev Biol Stand 94:245–269.

Liddington RC, Yan Y, Moulai J, Sahli R, Benjamin TL, Harrison SC (1991): Structure of simian virus 40 at 3.8-Å resolution. Nature 354:278–284.

Linzer DI, Levine AJ (1979): Characterization of a 54K dalton cellular SV40 tumor antigen present in SV40-transformed cells and uninfected embryonal carcinoma cells. Cell 17:43–52.

Louie AJ (1975): The organization of proteins in polyoma and cellular chromatin. Cold Spring Harbor Symp Quant Biol 39(Pt 1):259–266.

Martin RG, Chou JY (1975): Simian virus 40 functions required for the establishment and maintenance of malignant transformation. J Virol 15:599–612.

McCormick F, Harlow E (1980): Association of a murine 53,000-dalton phosphoprotein with simian virus 40 large-T antigen in transformed cells. J Virol 34:213–224.

Melendy T, Stillman B (1993): An interaction between replication protein A and SV40 T antigen appears essential for primosome assembly during SV40 DNA replication. J Biol Chem 268:3389–3395.

Mininberg DT, Watson C, Desquitado M (1982): Viral cystitis with transient secondary vesicoureteral reflux. J Urol 127:983–985.

Monaco MC, Jensen PN, Hou J, Durham LC, Major EO (1998a): Detection of JC virus DNA in human tonsil tissue: Evidence for site of initial viral infection. J Virol 72: 9918–9923.

Monaco MC, Shin J, Major EO (1998b): JC virus infection in cells from lymphoid tissue. Dev Biol Stand 94:115–122.

Monini P, Rotola A, Di Luca D, De Lellis L, Chiari E, Corallini A, Cassai E (1995): DNA rearrangements impairing BK virus productive infection in urinary tract tumors. Virology 214:273–279.

Muller U, Zentgraf H, Eicken I, Keller W (1978): Higher order structure of simian virus 40 chromatin. Science 20:406–415.

Mungre S, Enderle K, Turk B, Porras A, Wu YQ, Mumby MC, Rundell K (1994): Mutations which affect the inhibition of protein phosphatase 2A by simian virus 40 small-t antigen in vitro decrease viral transformation. J Virol 68:1675–1681.

Murakami Y, Wobbe CR, Weissbach L, Dean FB, Hurwitz J (1986): Role of DNA polymerase alpha and DNA primase in simian virus 40 DNA replication in vitro. Proc Natl Acad Sci USA 83:2869–2873.

Norkin LC (1999): Simian virus 40 infection via MHC class I molecules and caveolae. Immunol Rev 168:13–22.

Padgett BL, Walker DL (1973): Prevalence of antibodies in human sera against JC virus, an isolate from a case of progressive multifocal leukoencephalopathy. J Infect Dis 127:467–470.

Padgett BL, Walker DL, Desquitado MM, Kim DU (1983): BK virus and nonhaemorrhagic cystitis in a child. Lancet 1:770.

Padgett BL, Walker DL, Zu Rhein GM, Eckroade RJ, Dessel BH (1971): Cultivation of papova-like virus from human brain with progressive multifocal leukoencephalopathy. Lancet 1:1257–1260.

Pallas DC, Cherington V, Morgan W, DeAnda J, Kaplan D, Schaffhausen B, Roberts TM (1988): Cellular proteins that associate with the middle and small T antigens of polyomavirus. J Viol 62:3934–3940.

Pallas DC, Shahrik LK, Martin BL, Jaspers S, Miller TB, Brautigan DL, Roberts TM (1990): Polyoma small and middle T antigens and SV40 small t antigen form stable complexes with protein phosphatase 2A. Cell 60:167–176.

Pass HI, Kennedy RC, Carbone M (1996): Evidence for and implications of SV40-like sequences in human mesotheliomas. Important Adv Oncol 89–108.

Pipas JM (1992): Common and unique features of T antigens encoded by the polyomavirus group. J Virol 66:3979–3985.

Raj GV, Khalili K (1995): Transcriptional regulation: Lessons from the human neurotropic polyomavirus, JCV. Virology 213:283–291.

Rao L, Debbas M, Sabbatini P, Hockenbery D, Korsmeyer S, White E (1992): The adenovirus E1A proteins induce apoptosis, which is inhibited by the E1B 19-kDa and Bcl-2 proteins. Proc Natl Acad Sci USA 89:7742–7746.

Ray FA, Peabody DS, Copper JL, Cram LS, Kraemer PM (1990): SV40 T antigen alone drives karyotype instability that precedes neoplastic transformation of human diploid fibroblasts. J Cell Biochem 42:13–31.

Rayment I, Baker TS, Caspar DL, Murakami WT (1982): Polyoma virus capsid structure at 22.5 Å resolution. Nature 295:110–115.

Ricciardiello L, Chang DK, Laghi L, Goel A, Chang CL, Boland CR (2001): Mad-1 is the exclusive JC virus strain present in the human colon, and its transcriptional control region has a deleted 98-base-pair sequence in colon cancer tissues. J Virol 75:1996–2001.

Rutila JE, Imperiale MJ, Brockman WW (1986): Replication and transformation functions of in vitro–generated simian virus 40 large T antigen mutants. J Virol 58:526–535.

Salunke DM, Caspar DL, Garcea RL (1986): Self-assembly of purified polyomavirus capsid protein VP1. Cell 46:895–904.

Sandalon Z, Dalyot-Herman N, Oppenheim AB, Oppenheim A (1997): In vitro assembly of SV40 virions and pseudovirions: Vector development for gene therapy. Hum Gene Ther 8:843–849.

Sheng Q, Denis D, Ratnofsky M, Roberts TM, DeCaprio JA, Schaffhausen B (1997): The DnaJ domain of polyomavirus large T antigen is required to regulate Rb family tumor suppressor function. J Virol 71:9410–9416.

Shenk TE, Carbon J, Berg P (1976): Construction and analysis of viable deletion mutants of simian virus 40. J Virol 18:664–671.

Small JA, Khoury G, Jay G, Howley PM, Scangos GA (1986): Early regions of JC virus and BK virus induce distinct and tissue-specific tumors in transgenic mice. Proc Natl Acad Sci USA 83:8288–8292.

Sock E, Enderich J, Wegner M (1999): The J domain of papovaviral large tumor antigen is required for synergistic interaction with the POU-domain protein Tst-1/Oct6/SCIP. Mol Cell Biol 19:2455–2464.

Sontag E, Fedorov S, Kamibayashi C, Robbins D, Cobb M, Mumby M (1993): The interaction of SV40 small tumor antigen with protein phosphatase 2A stimulates the map kinase pathway and induces cell proliferation. Cell 75:887–897.

Stahl H, Droge P, Knippers R (1986): DNA helicase activity of SV40 large tumor antigen. EMBO J 5:1939–1944.

Staneloni RJ, Fluck MM, Benjamin TL (1977): Host range selection of transformation-defective hr-t mutants of polyoma virus. Virology 77:598–609.

Stewart AR, Lednicky JA, Benzick US, Tevethia MJ, Butel JS (1996): Identification of a variable region at the carboxy terminus of SV40 large T-antigen. Virology 221: 355–361.

Stewart SE, Eddy BE, Borgese NG (1958): Neoplasms in mice inoculated with a tumor agent carried in tissue culture. J Natl Cancer Inst 20:1223–1243.

Stubdal H, Zalvide J, Campbell KS, Schweitzer C, Roberts TM, DeCaprio JA (1997): Inactivation of pRB-related proteins p130 and p107 mediated by the J domain of simian virus 40 large T antigen. Mol Cell Biol 17:4979–4990.

Sweet BH, Hilleman MR (1960): The vacuolating virus, SV40. Proc Soc Exp Biol Med 105:420–427.

Taguchi F, Kajioka J, Miyamura T (1982): Prevalence rate and age of acquisition of antibodies against JC virus and BK virus in human sera. Microbiol Immunol 26: 1057–1064.

Tegtmeyer P (1975): Function of simian virus 40 gene A in transforming infection. J Virol 15:613–618.

Testa JR, Carbone M, Hirvonen A, Khalili K, Krynska B, Linnainmaa K, Pooley FD, Rizzo P, Rusch V, Xiao GH (1998): A multi-institutional study confirms the presence and expression of simian virus 40 in human malignant mesotheliomas. Cancer Res 58:4505–4509.

Theile M, Grabowski G (1990): Mutagenic activity of BKV and JCV in human and other mammalian cells. Arch Virol 113:221–233.

Tognon M, Casalone R, Martini F, De Mattei M, Granata P, Minelli E, Arcuri C, Collini P, Bocchini V (1996): Large T antigen coding sequences of two DNA tumor viruses, BK and SV40, and nonrandom chromosome changes in two glioblastoma cell lines. Cancer Genet Cytogenet 90:17–23.

Trabanelli C, Corallini A, Gruppioni R, Sensi A, Bonfatti A, Campioni D, Merlin M, Calza N, Possati L, Barbanti-Brodano G (1998): Chromosomal aberrations induced by BK virus T antigen in human fibroblasts. Virology 243:492–496.

Trowbridge PW, Frisque RJ (1993): Analysis of G418-selected Rat2 cells containing prototype, variant, mutant, and chimeric JC virus and SV40 genomes. Virology 196: 458–474.

Villarreal LP (1991): Relationship of eukaryotic DNA replication to committed gene expression: General theory for gene control. Microbiol Rev 55:512–542.

Waga S, Bauer G, Stillman B (1994): Reconstitution of complete SV40 DNA replication with purified replication factors. J Biol Chem 269:10923–10934.

Walker DL, Padgett BL, Zu Rhein GM, Albert AE, Marsh RF (1973): Human papovavirus (JC): Induction of brain tumors in hamsters. Science 181:674–676.

Weber T, Major EO (1997): Progressive multifocal leukoencephalopathy: Molecular biology, pathogenesis and clinical impact. Intervirology 40:98–111.

Wei G, Liu CK, Atwood WJ (2000): JC virus binds to primary human glial cells, tonsillar stromal cells, and B-lymphocytes, but not to T lymphocytes. J Neurovirol 6:127–136.

Xie YC, Hwang C, Overwijk W, Zeng Z, Eng MH, Mule JJ, Imperiale MJ, Restifo NP, Sanda MG (1999): Induction of tumor antigen-specific immunity in vivo by a novel vaccinia vector encoding safety-modified simian virus 40 T antigen. J Natl Cancer Inst 91:169–175.

Zalvide J, DeCaprio JA (1995): Role of pRb-related proteins in simian virus 40 large-T-antigen–mediated transformation. Mol Cell Biol 15:5800–5810.

Zalvide J, Stubdal H, DeCaprio JA (1998): The J domain of simian virus 40 large T antigen is required to functionally inactivate RB family proteins. Mol Cell Biol 18: 1408–1415.

Zentgraf H, Keller W, Muller U (1978): The structure of SV40 chromatin. Philos Trans R Soc Lond B Biol Sci 283:299–303.

Zerrahn J, Knippschild U, Winkler T, Deppert W (1993): Independent expression of the transforming amino-terminal domain of SV40 large T antigen from an alternatively spliced third SV40 early mRNA. EMBO J 12:4739–4746.

Zu Rhein GM (1983): Studies of JC virus-induced tumors in the Syrian hamster. In Polyomaviruses and Human Neurological Disease; Sever JL, Madden DM, Eds. Alan R. Liss: New York, pp 205–221.

6

TRANSCRIPTION AND REPLICATION IN THE HUMAN POLYOMAVIRUSES

HEE-SUN KIM, PH.D., JOHN W. HENSON, M.D.,
and RICHARD J. FRISQUE, PH.D.

1. INTRODUCTION

JC virus (JCV) is a 5 kb circular double-stranded DNA virus that causes the fatal human disease progressive multifocal leukoencephalopathy (PML) by selective destruction of oligodendrocytes, leading to multiple areas of demyelination and attendant loss of brain function (Åström et al., 1958; Zu Rhein and Chou, 1965). The viral genome is divided into early and late gene coding regions, between which lies a regulatory region containing a bidirectional promoter and the viral origin of replication (Fig. 6.1). The JCV early promoter directs glial-specific expression of large T antigen (TAg). TAg is the major viral protein, containing several activities, through which the virus commandeers the cellular metabolic machinery for production of virions (Henson, 1996). Thus, glial-specific transcriptional regulation of TAg expression is believed to constitute a major mechanism of neural tropism in PML.

The promoters of numerous brain-specific genes have been partially characterized with respect to the ability of the 5′ untranslated region (UTR) to drive expression of a reporter gene in transgenic mice or in transient transfection assays. Studies of a number of viral and cellular genes have indicated that the

Human Polyomaviruses: Molecular and Clinical Perspectives, Edited by Kamel Khalili and Gerald L. Stoner.
ISBN 0-471-39009-7

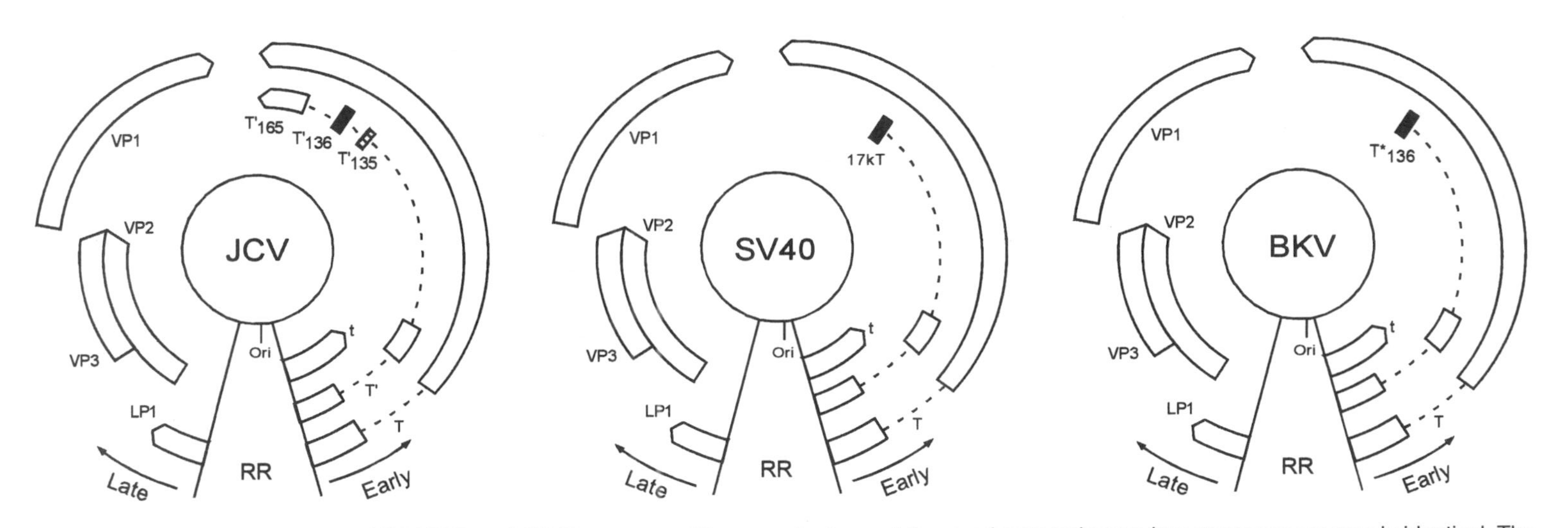

Figure 6.1. Comparisons of the JCV, BKV, and SV40 genomes. The organizations of these primate polyomavirus genomes are nearly identical. The inner circles represent the double-stranded DNA genomes (5130 bp, JCV[Mad-1]; 5098 bp, BKV[AS]; 5243 SV40[776]), and the outer arcs denote the encoded viral proteins. The genomes are divided into three regions. The early region specifies five (JCV) or three (BKV, SV40) regulatory proteins produced by translation of alternatively spliced early mRNAs. T′, T*, and 17kT on the JCV, BKV, and SV40 maps, respectively, represent proteins encoded by mRNAs composed of three exons that may or may not be shared with the TAg transcript. Each genome may encode an early leader peptide (JELP, BELP, or SELP) of unknown function. The late region specifies four proteins required for capsid assembly, VP1–3 and LP1. The regulatory region (RR) contains the *cis*-acting elements that control viral DNA replication (Ori) and transcription. The promoter/enhancer signals for transcription are the least conserved sequences in the three genomes.

temporal and cell-specific activation of eukaryotic genes by enhancer/promoter elements requires association of *cis*-acting DNA elements with *trans*-acting cellular proteins that recognize such DNA sequences. Specificity may arise from the expression of a specific transcriptional regulator in a restricted fashion or by virtue of a cell-specific combination of transcription factors and transcription factor binding sites within the promoter. For instance, a combination of NF-1 and AP1 transcription factor binding sites appears in numerous brain-specific promoters (Amemiya et al., 1992).

In the first half of this chapter, the transcriptional regulation of JCV is reviewed. First, the glial-specific nature of the JCV early promoter is described. Second, regulation by transcription factors, including TAg, are described. Third, cross-regulation by cytokines and HIV protein is reviewed. Finally, the possibility of regulation by epigenetic factors such as DNA methylation or chromatin structure is discussed.

In the second half of this chapter the regulation of primate polyomavirus DNA replication is reviewed. Our present understanding of JCV and BK virus (BKV) DNA replication relies in large part on studies involving the closely related monkey polyomavirus, simian virus 40 (SV40). SV40 has served as an important model with which to unravel the process by which small DNA tumor viruses duplicate their genomes; it has also served as a model to identify the mechanisms by which the DNA of eukaryotic cells is copied. A variety of molecular genetic and biochemical approaches have been employed to dissect the DNA replication process of this small, circular, double-stranded DNA genome. These approaches are now being redirected toward an investigation of human polyomavirus replication, and, while it is expected that the major events affecting replication of the human viruses will closely parallel those events already shown to influence SV40 replication, specific details unique to JCV and BKV have already been uncovered.

2. GLIAL-SPECIFIC NATURE OF JCV EARLY PROMOTER

A large number of observations have demonstrated the glial-specific nature of the JCV early promoter. Although oligodendrocytes are the only cells in the body known to be lytically infected by JCV, astrocytes also express viral early genes (i.e., large and small TAgs) in PML, whereas neurons do not (Itoyama et al., 1981). Successful JCV isolation followed the discovery that primary human fetal glial (PHFG) cells were permissive for infection in vivo, further demonstrating the restricted nature of the virus infection (Padgett et al., 1971). The expression of the TAg early gene was abolished following fusion of JCV-transformed hamster glial cells with mouse fibroblasts, and the degree of reduction of TAg expression correlated with the number of fibroblast nuclei in the heterokaryon fusion (Beggs et al., 1988). As discussed in more detail below, this was early experimental evidence in favor of JCV early promoter repression by a nonglial cell protein. In 1984 transcriptional analysis of the early promoter, using a transient transfection assay, demonstrated that the JCV early promoter directed higher levels of expression of a reporter gene in glial cells than in nonglial cells (Kenny et al., 1984). This latter observation has been confirmed in a wide range of glial and nonglial cells types and has led to detailed analyses of the promoter in vitro. Mice carrying a JCV early promoter–TAg transgene selectively expressed TAg in oligodendrocytes, giving rise to a phenotype of demyelination (Trapp et al., 1988). Additional transgenic experiments, in which the early promoters and TAg genes of JCV and SV40 were exchanged, demonstrated that the JCV early promoter was responsible for glial-specific expression of TAg (Feigenbaum et al., 1992).

Thus, it is now established that the JCV early promoter directs glial-selective gene expression. A large number of transcription factors are known to regulate the early promoter, some of which activate expression in a cell-specific way.

3. STRUCTURE OF THE JCV PROMOTER

A promoter region of 293 bp was isolated from Mad-1 JCV obtained from PML brain tissue (Kenny et al., 1984; Tada et al., 1991). The Mad-1 promoter contains two characteristic tandem 98 bp repeats and functions in opposite orientations to regulate the early and late gene expression (Figs. 6.2 and 6.3). Distinct transcription initiation start sites have been detected from the viral promoter in the early and late phases of infection. Additional promoters cloned directly from brain (MH1, GS/B, Mad-11.3) contain an identical 222 bp 5′ UTR region from the first codon of TAg (Loeber and Dörries, 1988; Yogo et al., 1990). Thus, each isolate is identical to the end of the tandem repeat that is adjacent to the TATA box (Fig. 6.2), but several isolates diverge somewhat from the original Mad-1 promoter (Henson et al., 1992). Most isolates contain an Sp1 binding site upstream of the TATA box in a location identical to the first of six Sp1 binding sites in the SV40 promoter. By comparison, the original Mad-1 JCV promoter had sequence identity only through 13 bp upstream of the TATA box, and it then diverged from the direct isolates due to a 23 bp deletion (lower case letters in the Mad-1 sequence in Fig. 6.2). It is important to note that while the structure of the Mad-1 promoter does not alter its glial specificity, it does serve to complicate functional analysis of the Mad-1 promoter since the TATA region is duplicated in the upstream repeat. On the other hand, the other promoters can be easily divided into proximal and upstream regions in a manner similar to that employed in earlier studies of the SV40 promoter (Ondek et al., 1988).

4. CELL-SPECIFIC TRANSCRIPTION ANALYSIS

Introduction of Mad-1 and MH1 (Mad-11) JCV early promoter reporter gene plasmids into glial (U87MG glioma cells) and nonglial (HeLa) cells demon-

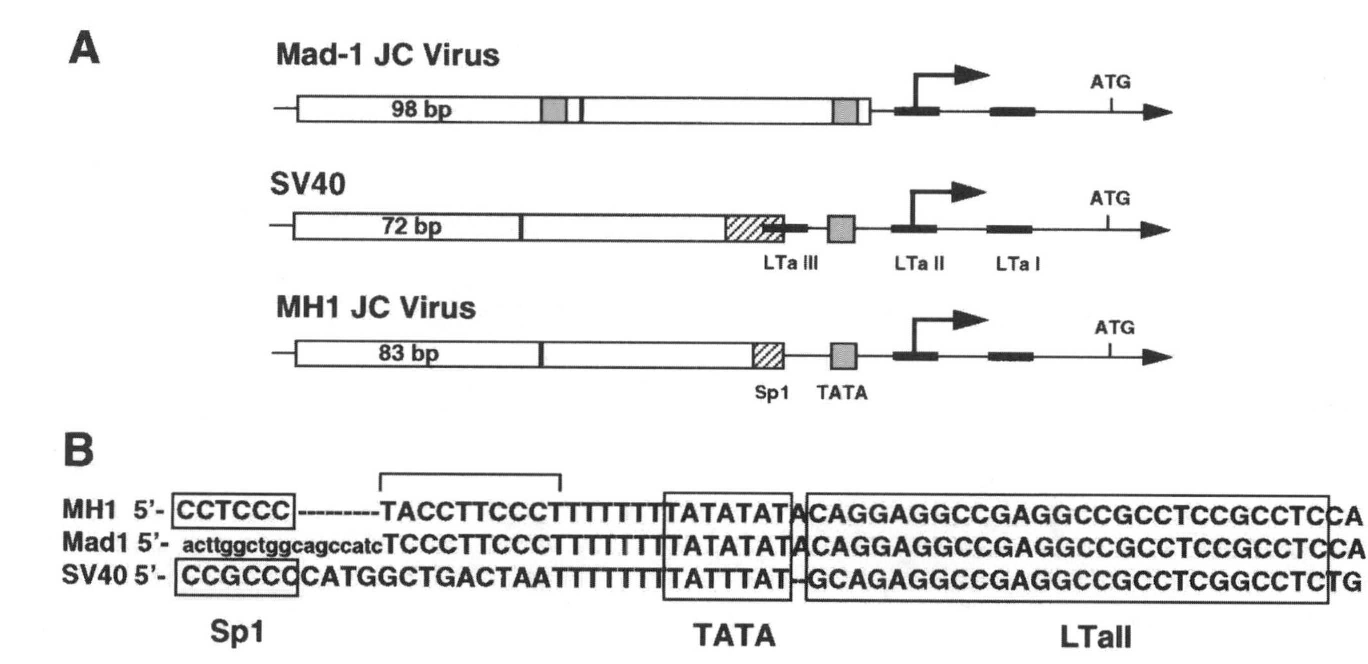

Figure 6.2. (A) Schematics of the Mad-1 JCV early promoter, the MH-1 JCV early promoter, and the SV40 promoter. Open boxes indicate direct tandem repeats of the indicated number of base pairs, dotted boxes represent TATA homologies, and the striped boxes represent Sp1 binding sites. The SV40 promoter contains three sequences that constitute binding sites for the viral protein large TAg (black boxes), whereas the JC virus early promoter contains only sites I and II. **(B)** Comparison of the basal promoter region DNA sequence reveals differences between JCV and SV40 in the region between the Sp1 binding sites and the second TAg binding site. MH1 and Mad-1 diverge upstream of a 5 bp TC(A)CCT repeat (horizontal bracket, B).

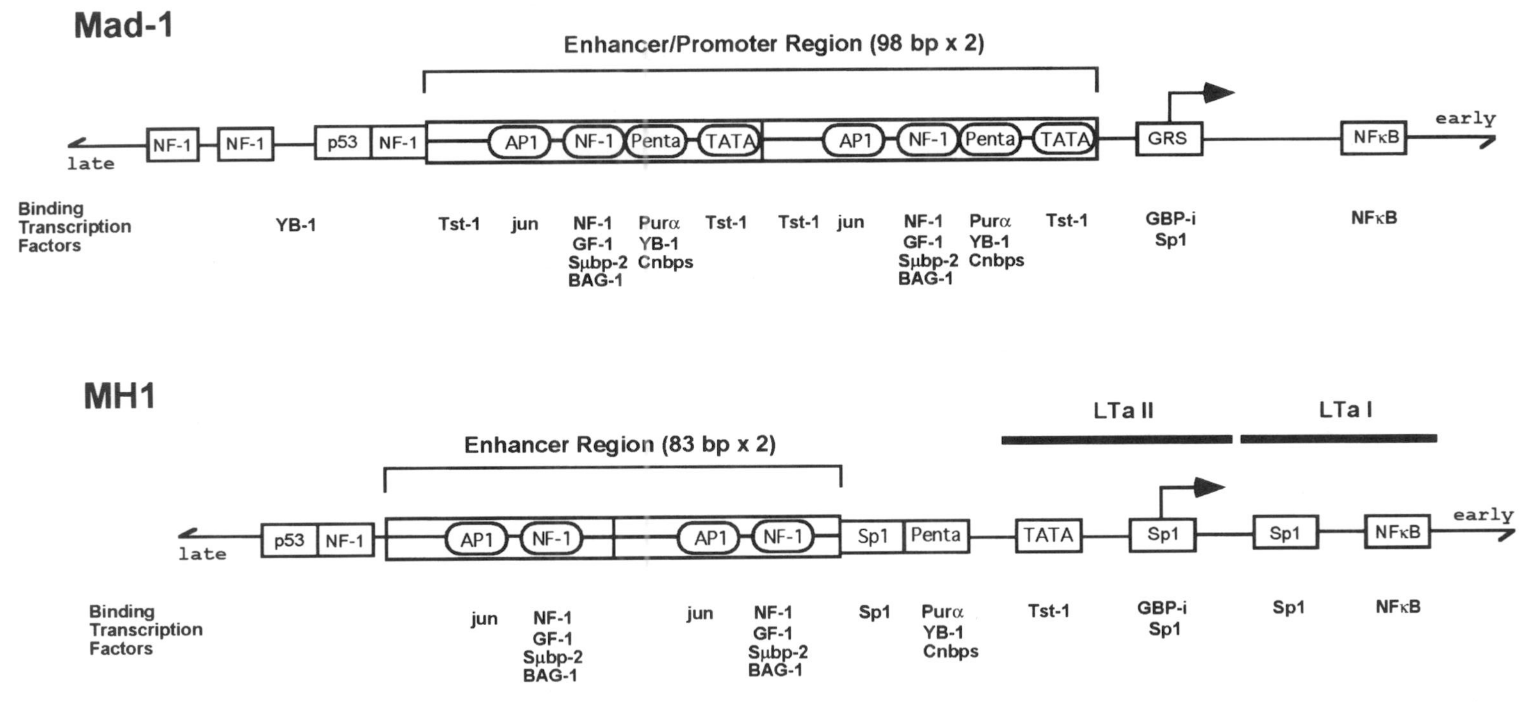

Figure 6.3. Multiple transcription factor binding sites in the enhancer/promoter regions of Mad-1 and MH1 JCV. Transcription factors that have been shown to bind the promoters are illustrated at the bottom of each schematic. Potential transcription factor binding sites derived from computer-aided analyses are depicted in the open boxes or circles in the schematics of two promoters. Two TAg binding sites (LTa I, LTa II) identified by DNase I footprinting analysis are exhibited on the MH1 promoter. Three Sp1 binding sites are also exhibited in the schematic of the MH1 promoter. The arrow indicates the transcription start site.

strated that the promoters directed 30- to 40-fold stronger expression in glial cells than in nonglial cells (Henson, 1994). More importantly, this analysis demonstrated that the MH-1 tandem repeats activated expression in both glial and nonglial cells, thus suggesting that the repeats are not the major determinant of cell specificity. Consistent with this, the proximal promoter region alone directed 19-fold more activity in glial cells than in nonglial cells (Henson et al., 1995; Krebs et al., 1995). Thus, the JCV basal promoter region is able to direct glial-specific gene expression, whereas the major role of the tandem repeats may be to increase promoter strength.

JCV early promoter function has been tested by in vitro transcription assays using whole cell extracts. A strong run-off transcript was identified across increasing concentrations of HJC (hamster glial cell) extract, whereas increasing HeLa whole cell extract concentration produced loss of the run-off transcript (Ahmed et al., 1990a,b). This not only suggested that the in vitro assay was glial specific but also provided additional evidence for a repressor in HeLa cells.

The expression of JCV is regulated by transcription factors, including activators and silencers. Many studies have been done to identify glial-specific transcription factors, as is discussed in the following section.

5. REGULATION OF JCV PROMOTER BY KNOWN TRANSCRIPTION FACTORS

Sp1, NF1, AP1, GF1, Sμbp-2

Sp1. Three Sp1 sites have been identified on the MH1 JCV early promoter region by gel-retardation and DNase I footprinting analysis, using recombinant protein and nuclear extracts (Sp1-I, Sp1-II, Sp1-III) (Henson et al., 1992; Henson, 1994; Kim et al., 2000). Sp1-I is located immediately upstream of the TATA box and appears to activate the JCV early promoter in both glial and nonglial cells (Henson, 1994) (Fig. 6.3). Sp1-II and Sp1-III sites are located downstream of the TATA box. Sp1-II was shown to be important for TAg-mediated transactivation of the JCV early promoter, but it is not involved in the regulation of basal expression (Kim et al., 2000).

Recently, Sp1 was shown to associate with Purα and regulate myelin basic protein (MBP) expression during brain development (Tretiakova et al., 1999a,b). TAg also appears to downregulate MBP transcription by interacting with Purα, leading to hypomyelination (Tretiakova et al., 1999a,b).

Based on its frequent occurrence in CpG islands, one possible mechanism underlying transcriptional control by Sp1 may be its role in maintaining methylation-free CpG islands in active genes (Graff et al., 1997; Macleod et al., 1994). In vitro methylation of the MH1 JCV early promoter leads to very strong repression of transcription after transfection into glial cells. Thus, cell-specific methylation is another potential mechanism regulating JCV early gene expression (see below).

NF1, AP1. The Mad-1 promoter has been subjected to extensive molecular analysis. Several investigators have demonstrated the lack of a glial-specific DNase I footprint on the enhancer region of the Mad-1 JCV early promoter (Amemiya et al., 1989, 1992; Tada et al., 1989). A possible explanation for the lack of a glial-specific DNA binding activity would be that there is a glial-specific form of a well-known factor, such as NF1. The presence of closely apposed NF1 and AP1 binding sites in many brain-specific promoters has led to the search for such gene families (Amemiya et al., 1992; Sumner et al., 1996). Specifically, NF1/AT1(NF1-A1) was shown to be expressed at a higher level in human fetal brain cells compared with the ubiquitous NF1/CTF1(NF1-C1) in HeLa cells (Sumner et al., 1996). In addition, the cerebellum-enriched form of NF1-A1 transactivated two gliotropic JCV early promoters to a greater extent than NF1-C1 in U87MG and HeLa cells (Krebs et al., 1996).

Recently, BAG-1, a novel Bcl-2–interacting protein, was cloned from retinoic acid–differentiated P19 embryonic carcinoma (EC) cells using the JCV NF1 binding site as a probe sequence (Devireddy et al., 2000). This protein was shown to bind to the NF1 site and activate JCV early and late promoters. However, BAG-1 is ubiquitously expressed in neuronal and non-neuronal cells, arguing that it is not a cell-specific transactivator.

Next to the NF-1 binding site are cAMP response elements (overlapping with the AP1 site) that appear to regulate the JCV early promoter and enhancer in a tissue-specific manner (Kumar et al., 1996a,b). Addition of cAMP or forskolin enhanced the expression of the JCV early promoter by two- to threefold only in glial cells (Kumar et al., 1996a,b). However, the sequence of the cAMP response elements (TGA<u>GC</u>TCA) of the JCV promoter deviates by two bases from the consensus CRE sequence (TGA<u>CG</u>TCA), which usually requires a very stringent match (Faisst and Meyer, 1992; Tinti et al., 1997).

GF1 and Sμbp-2. In gel-shift experiments, the central portion of the tandem repeats was found to bind a 45 kDa protein, glial-specific protein (GF1), that activated the late promoter and to a lesser extent the early promoter in glial cells (Ahmed et al., 1990a,b; Kerr and Khalili, 1991; Khalili et al., 1988). Following its cloning and expression, GF1 was found to activate the early promoter weakly in glioma cells. This latter result was interesting because if glial specificity relied solely on the presence or absence of an activating factor such as GF1, ectopic expression of that factor in nonglial cells would be expected to activate the promoter.

Because GF1 appeared to be a partial cDNA for human Sμbp-2, a transfection assay was employed to compare the effects of GF1 and its full-length form Sμbp-2 on JCV regulation (N.N. Chen et al., 1997). Sμbp-2 activated only the JCV late promoter in glial cells and was weaker compared with GF1. In addition, Northern blot analysis (N.N. Chen et al., 1997) showed that GF1 and

Sμbp-2 are expressed in both glial and nonglial cells, suggesting that, if they contributed to cell specificity, they might interact with cell-specific coactivators to induce glial cell-specific expression.

Tst-1/SCIP/Oct-6. Tst-1/SCIP/Oct-6 is a POU domain protein that activates the JCV early promoter in conjunction with MHG-I/Y through sequences adjacent to the TATA box (Leger et al., 1995; Renner et al., 1994; Wegner et al., 1993). Tst-1 also physically interacts with TAg, resulting in a synergistic activation of both early and late viral promoters (Renner et al., 1994). Interestingly, Tst-1 activates the basal promoter region more strongly in glial cells than in nonglial cells. Tst-1 is specifically expressed in myelinating glia, but it is not expressed in many glioma cell lines in which the JCV promoter is strongly and selectively active, nor is it expressed in adult glia. Thus, the role of Tst-1 in glial specificity remains unclear.

TAg. Analysis of the action of TAg on the JCV early promoter has provided some potentially important insights into cell-specific basal promoter regulation. Closely related SV40 TAg represses its own expression by cooperative binding to three sites in the basal region of the SV40 early promoter (Fig. 6.2, LTa I, II, and III) (Rio and Tjian, 1983). TAg repression of the SV40 early promoter is thought to result from blocking the transcription initiation complex through direct DNA binding to the basal promoter (steric hindrance model); however, there is little direct evidence for this mechanism. On the other hand, TAg produces DNA binding-independent activation of other promoters such as the cellular hsp70 promoter, the Rous sarcoma virus (RSV) promoter, and SV40 and JCV late promoters (Rice and Cole, 1993; Taylor et al., 1989). TAg can activate a basal hsp70 promoter consisting of a TATA sequence plus a single upstream element (e.g., Sp1 binding site) in a TATA sequence-dependent manner, and because TAg physically binds to TATA binding protein (TBP), it is likely that it acts during initiation complex formation (Gruda et al., 1993; Taylor et al., 1989).

TAg shares with adenovirus E1A protein the ability to stimulate transcription in the absence of sequence-specific DNA binding. E1A activation also requires a specific TATA sequence (Simon et al., 1988), and there is evidence that E1A activates through derepression (Horikoshi et al., 1995; Kraus et al., 1994). Each of these observations suggests that TAg promoter regulation depends on TATA-specific initiation complexes.

Co-transfection of a JCV TAg expression construct and the MH1luc reporter into U87MG glioma cells revealed that increasing levels of TAg produced fivefold repression of the JCV and SV40 early promoters (Henson et al., 1995; Kim et al., 2000) in a manner similar to that previously shown for the SV40 virus (Hansen et al., 1981). In HeLa cells, by comparison, 100–200-fold transcriptional activation was observed. The effect was specific to the JCV early promoter, as the SV40 promoter was still repressed fivefold in HeLa cells, and TAgs of both JCV and SV40 activated the JCV early promoter to a similar degree. Deletion mutants in a co-transfection assay showed that TAg activation in nonglial cells required only the basal promoter region. Thus, TAg produced glial-specific, divergent regulation of the JCV basal promoter (Henson et al., 1995; Kim et al., 2000).

DNase I footprinting on the MH1 JCV basal promoter region using SV40 or JCV TAg revealed two protected domains (LTa I and LTa II) (Fig. 6.3) (Kim et al., 2000). Site-directed mutagenesis in the area of LTa II indicates that alteration of two specific bases in the second pentanucleotide repeat abolished TAg-induced transactivation while the mutation of the first repeat did not affect either basal or TAg-induced transactivation in HeLa cells (Kim et al., 2000). The change of the TATA to an irrelevant sequence also abolished TAg-induced transactivation. In U87MG cells, the mutations did not alter TAg repression. These results suggest an important role for the second pentanucleotide element and TATA sequence for TAg-induced transactivation in nonglial cells. Functional analysis of three new binding domains on the MH1 JCV promoter, which were identified by footprinting analysis with nuclear extracts, revealed that the Sp1-II and novel sequences are also involved in TAg-induced transactivation (Kim et al., 2000).

The binding of TAg to LTa I and LTa II was not significantly changed by the mutations. These results suggest that TAg regulates the JCV promoter largely by protein–protein interactions surrounding the TATA site rather than by direct DNA binding.

The activation of the JCV early promoter by TAg in nonglial cells could represent either transactivation or derepression. TAg activates a large number of cellular and viral promoters in vitro and in vivo (Rice and Cole, 1993; Taylor et al., 1989). Simple basal promoter regions are sufficient for transactivation, and a wide variety of transcription factor binding sites can cooperate with TAg in activation (Damania and Alwine, 1996; Gruda and Alwine, 1991). TAg lacks a strong activation domain, and DNA binding is not required. TAg can interact with TBP, can discriminate between TATA sequences for transactivation, and can substitute for $TAF_{II}250$ (Gruda and Alwine, 1991), strongly suggesting a role in transcription initiation. Thus, the divergent regulation of the JCV early promoter could reflect cell-specific or TATA-specific TFIID complexes. Indeed, the ability of TATA and pentanucleotide mutations to abolish TAg induction suggests that regions surrounding the TATA box are crucial for this effect. Also, the differences in the footprint analysis over the TATA region between the glial and HeLa nuclear extracts support this hypothesis (Kim et al., 2000).

Recently it was reported that TAg appears to transactivate the Mad-1 JCV late promoter by increasing expression from a basal transcriptional initiation site and through a novel TAg-dependent initiation site (TADI), which is homologous to initiator (Inr) sequences (Raj et al., 1998). The ability of TAg to activate the JCV late promoter might be attributed to the formation of specific protein complexes and increased transcriptional initiation from the TADI site on the late promoter.

Purα, YB-1, and TAg

Purα is a 322 amino acid sequence-specific single-stranded DNA binding protein that has been implicated in the control of both DNA replication and transcription. Purα has been implicated in control of gene transcription involving both viral and cellular promoters, including the JCV early promoter (Chen and Khalili, 1995), the human immunodeficiency virus type 1 promoter (Chepenik et al., 1998), the MBP promoter (Hass et al., 1995), and the neuron-specific FE65 gene promoter (Zambrano et al., 1997).

Purα stimulates transcription of the JCV early promoter up to sixfold. Moreover, TAg attenuates the Purα-induced level of early gene transcription. Although Purα alone has little effect on the late promoter, it is able to decrease TAg-mediated transactivation of the JCV late promoter (Chen and Khalili, 1995). Purα and TAg physically interact and antagonistically modulate each other's function on the JCV promoter (Gallia et al., 1998). According to recent reports, the association of TAg and Purα in vivo appears to downregulate the MBP gene and thus induce hypomyelination in brains of mice transgenically expressing TAg (Tretiakova et al., 1999a,b).

YB-1, a Y-box binding protein, is among the most evolutionarily conserved nucleic acid binding proteins in prokaryotes and eukaryotes (Wolffe, 1994; Wolffe et al., 1992). Y-box binding proteins appear to be involved in a wide variety of biologic functions, including regulation of gene expression at the transcriptional level (Kashanchi et al., 1994; Kerr et al., 1994; Li et al., 1997; Mertens et al., 1998), the translational level (Tafuri and Wolffe, 1993), DNA repair and DNA and RNA condensation (Grant and Deeley, 1993; Wolffe, 1994; Wolffe et al., 1992). The members of the Y-box family proteins are responsive to a wide spectrum of stress-related stimuli, including ultraviolet light radiation (Koike et al., 1997), drug treatment (Bargou et al., 1997), DNA damage-inducing antineoplastic agents (Ise et al., 1999), and interleukin-2 treatment in T cells (Sabath et al., 1990). Because viral infection induces cellular stress, YB-1 may be a candidate for an inducible protein in this setting. Recent data demonstrate that YB-1 is involved in transcriptional regulation of the JCV promoter (Chen and Khalili, 1995; Kerr et al., 1994; Safak et al., 1999a,c). YB-1 activates the JCV late promoter. In Mad-1 JCV, binding of YB-1 to its DNA target within the lytic control element (LCE) is increased by Purα. In contrast, interaction of Purα with the LCE motif is diminished once YB-1 is included in the reaction mixture.

Purα and YB-1 bind to the late (A/G-rich) and early (T/C-rich) strands of the LCE of Mad-1 JCV, respectively, and modulate JCV transcription in glial cells. The LCE is positioned within the enhancer repeat of the JCV promoter in close proximity to the origin of DNA replication and exhibits a remarkable effect on viral gene transcription and DNA replication. This region contains a pentanucleotide repeat sequence, AGGGAAGGGA, in juxtaposition to a poly(dT-dA) tract, which displays a single-stranded configuration. The interplay between Purα, YB-1, and TAg appears to dictate the level of association of these proteins with their target DNA sequences and hence their regulatory action on viral early and late gene transcription (Chen and Khalili, 1995; Gallia et al., 1998; Safak et al., 1999a,c). Based on in vitro binding assays and transfection studies, a model has been proposed for the involvement of Purα, YB-1, and TAg in the transition of early to late gene transcription (Fig. 6.4). According to this model, efficient binding of Purα to the LCE late strand stimulates early gene transcription and facilitates the interaction of YB-1 with its target positioned on the LCE early strand. The association of YB-1 with the DNA, which is concurrent with TAg production and its binding to the origin of DNA replication, may result in dissociation of Purα from the LCE. This

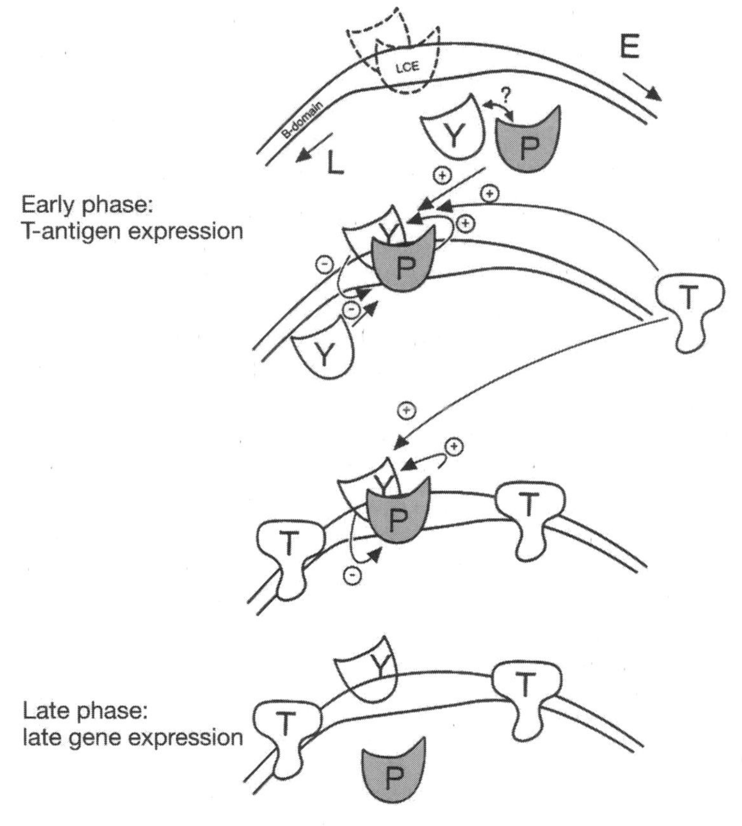

Figure 6.4. Proposed model of the involvement of Purα (P), YB-1 (Y), and TAg (T) in the transition of early to late gene transcription. Efficient binding of Purα to the LCE late strand (L) stimulates early gene transcription and facilitates binding of YB-1 to its target position of the LCE early strand (E). Binding of YB-1 to the DNA, which is concurrent with T antigen production, and its binding to the origin of DNA replication and the B region results in dissociation of Purα from the LCE late strand. This alteration in the pattern of DNA–protein complexes results in a decrease in the level of early gene transcription and an increase in late promoter activity in the late phase of infection.

alteration in the pattern of DNA–protein interaction can lead to a decrease in the level of early gene transcription and an increase in late promoter activity. Interestingly, none of these proteins affects the expression of interacting counterpart proteins (Chen and Khalili, 1995). This result suggests that Purα, YB-1, and TAg modulate JCV transcription through protein–protein interaction. With the concept that YB-1 is ubiquitously expressed in every cell and tissue, one might anticipate the involvement of a negative regulatory mechanism in nonglial cells that interferes with the positive activity of YB-1 on the basal and TAg-induced levels of virus late gene expression.

6. CELL-SPECIFIC REGULATION OF JCV PROMOTER BY SILENCERS

Cell-specific transcriptional regulation can result from selective activation or derepression. A region immediately upstream of the TATA sequence weakly repressed a heterologous promoter (Tada et al., 1991). Within the sequence adjacent to the TATA homology, there is a five base pair repeat (5′ TCCCTTCCCT), and because this region held sequence differences compared with the SV40 basal promoter (see Fig. 6.2), it was considered to be a potential binding site for a transcriptional repressor. Although several proteins have been identified that bind to this sequence, their identity and relevance to glial specificity remains unknown (Sharma and Kumar, 1991; Tada et al., 1991). It was demonstrated that point mutations within the pentanucleotide sequence reduced transcription in glial cells (Kumar et al., 1994). However, the mutations did not increase transcriptional strength in nonglial cells (HeLa cells), as would be expected from a mutation that abolished glial specificity. Recently, one cellular nucleic acid binding protein (Cnbps) was cloned from glial P19 mouse embryonal cells using a pentanucleotide oligonucleotide as a probe (Liu et al., 1998). It negatively regulated only the JCV early promoter and only in glial cells.

Transcriptional activation of the MH1 JCV early promoter by TAg in nonglial cells implies the presence of silencers (Henson et al., 1995; Kim et al., 2000). TAg has been shown to form complexes with pRb or p53 protein (Fanning and Knippers, 1992; Henson et al., 1995; Kim et al., 2000; Tavis et al., 1994). Transient transfection of a p53 expression plasmid with MH1 JCV promoter repressed promoter activity, and the activity was derepressed by TAg in nonglial cells (unpublished data). This result suggests that p53 might be a candidate silencer protein that represses JCV expression in nonglial cells.

7. REGULATION OF JCV TRANSCRIPTION BY CYTOKINES AND HIV

Regulation by Cytokine-Induced Transcription Factors

Immunosuppressive states with accompanying alterations in cytokine profiles have been postulated to play a vital role in the reactivation of viruses from latency. Cytokines regulate gene expression by activating transcription factors via well-characterized signal transduction pathways. One of the factors involved in JCV transcription is GBP-i, a novel GGA/C binding protein (Raj and Khalili, 1994). The expression of GBP-i is induced by phorbol myristate acetate (PMA) and a variety of cytokines and immunomodulators, including interleukin-1β, tumor necrosis factor-α, interferon-γ, and transforming growth factor-β. GBP-i, unlike NFκB, acts as a transcriptional repressor of the JCV late promoter and is ubiquitously expressed in glial and nonglial cells. Interestingly, the GBP-i binding site, GRS, overlaps with the binding site of the JCV TAg and HIV-1 Tat-responsive region. It has been suggested that Tat upregulates the expression of several cytokines and might exert its effect indirectly on the JCV promoter, specifically through the GRS.

In other studies, the importance of the κB motif and the 23–bp sequence that interrupts the LCE motif was suggested in terms of cytokine-mediated regulation of JCV transcription (Safak et al., 1999b). The NFκB motif is located near the early gene translation start site, and the mutation of this motif completely abrogated the basal and PMA-induced levels of JCV late promoter transcription. The 23–bp sequence was critical for the observed inhibitory action. In the early phase of JCV reactivation inducible transcription factors such as NFκB may increase viral early gene transcription indirectly via the 23-bp sequence and to a much lesser degree through the NFκB motif. A 40 kDa protein has been identified that communicates between the κB and LCE regions in controlling viral gene transcription (Mayreddy et al., 1996; Safak et al., 1999b). Results from site-directed mutagenesis indicated that formation of a 40 kDa DNA–protein complex with the 23–bp sequence is critical for the transcriptional activation of the JCV promoter by PMA (Safak et al., 1999b).

Regulation by HIV-Encoded Regulatory Protein Tat and HTLV-1–Encoded Tax

The higher incidence of PML among individuals with AIDS compared with other immunocompromised patients implies that the presence of HIV type 1 (HIV-1) in the brains of infected individuals may directly contribute to the pathogenesis of this disease. In support of this model, earlier in vitro studies have indicated direct intercommunication between HIV-1 and JCV through the HIV-encoded regulatory protein Tat (Chowdhury et al., 1990, 1993). An upstream Tat-responsive DNA element (upTAR) of JCV has been shown to be important for HIV-1 Tat stimulation of the JCV late promoter. Specifically, Tat enhances the ability of Purα to bind the upTAR element and synergistically activates transcription (Krachmarov et al., 1996). Recent data indicate that Tat–Purα are exclusively nuclear and are co-localized in the extranucleolar chromatin structural elements (Wortman et al., 2000). These results also demonstrated that Tat–Purα interaction is direct rather than through an intermediary RNA or DNA molecule, and RNA binding configures Purα for optimal interaction with Tat. It has been postulated that RNA associates with Purα, stabilizing its structure and allowing it to interact with its protein partners (Gallia et al., 1999a,b). Specifically, the two acidic leucine-rich repeats of Purα are involved in the interaction (Krachmarov et al., 1996), and a polypeptide based on one such sequence inhibits binding (Wortman et al., 2000). Because Purα is ubiquitously expressed in human cells and because PUR elements are located near many promoters and origins of replication, the Tat–Purα interaction may be implicated in effects of HIV-1 throughout the full range of HIV-1–infected cells.

Recently it was demonstrated that human T-lymphotropic virus type 1 (HTLV-1)–encoded regulatory protein Tax activates JCV transcription in human neuronal cells but not in non-neuronal cells (Okada et al., 2000). Tax activated the transcription of both early and late promoters of Mad-1 and archetype JCV, and this activation was through the NFκB binding motif. A JCV promoter that lacks the NFκB binding motif could not be activated by Tax, and a Tax mutant lacking the potential for activation via the NFκB pathway did not activate the JCV promoter. From gel-shift assays, it was demonstrated that a Tax-bound protein(s) was present specifically in non-neuronal cells, suggesting the possibility of a repressor or silencer.

8. CHROMATIN AND DNA METHYLATION IN THE REGULATION OF TRANSCRIPTION

In addition to the direct regulation of transcription by nuclear proteins, gene expression is also controlled by molecular and structural modifications of promoter DNA. Chromatin packing and DNA methylation are two such mechanisms. Because these modifications are just beginning to be explored in depth, their significance for neural-specific gene expression remains largely speculative.

Chromatin

Chromatin occurs in two forms: as heterochromatin, which is densely packed and can be stained and visualized by light microscopy in the interphase nucleus; and euchromatin, which cannot be visualized (except in a highly condensed stage during mitosis). A major function of chromatin condensation is the efficient packing of long DNA molecules into the nucleus. However, cells also utilize chromatin as a mechanism to regulate gene expression (Felsenfeld, 1992; Lu et al., 1992; Surridge, 1996; Wolffe and Pruss, 1996). Genes residing within regions of heterochromatin are transcriptionally inactive, whereas actively expressed regions of the genome reside within euchromatin. It is well documented that histones repress gene expression when bound directly over promoters (Croston et al., 1991). There is competition between histones and transcription factors for binding to promoter sequences because preincubation of promoter DNA with either excludes the alternate protein from binding to the DNA. Promoters residing within heterochromatin are presumably inactive because of the stereochemical restraint on transcription factor binding. The state of DNA binding to histones is controlled in part by the degree of histone acetylation (Wolffe and Pruss, 1996). Acetylases and deacetylases modify amino-terminal lysine residues on the outer surface of histone molecules. Acetylation destabilizes heterochromatin, thus providing increased access of transcription factors to promoters and allowing increased gene expression, whereas deacetylation of histones represses gene expression.

Polyomavirus DNA is assembled into a set of approximately 21 nucleosomes, in both the virion and the infected cell, with each nucleosome consisting of an octamer containing two copies of H2A, H2B, H2, and H4. In the infected nucleus, it appears that the histone H1 is associated with at least some of the "minichromosomes" (Bellard et al., 1976). Sequence-specific transcription factors can counteract histone-mediated transcriptional repression by displacing H1 (Croston et al., 1991; Felsenfeld, 1992). Therefore, because JCV basal promoter function is regulated in a cell-specific manner, it is possible that chromatin has a role in promoter specificity.

Methylation

Methylation of cytosine bases at CpG dinucleotides within promoters participates in the regulation of gene expression. Genomic methylation occurs immediately after DNA replication, producing a pattern of methylation that is stably inherited from mother cell to daughter cell (Bestor and Tycko, 1996). Semiconservative replication produces hemimethylated DNA, which is a strong substrate for DNA methyltransferase, thus leading to a fully methylated site.

The majority of genomic DNA contains CpG dinucleotides at a frequency of about 1 pair per 100 bp, which is a fivefold lower frequency than would be expected from the random occurrence of the sequence CpG (Antequera and Bird, 1993; Bird, 1992). These CpG pairs are methylated on both cytosines, and 5-methylcytosine can undergo spontaneous hydrolytic deamination to thymidines, perhaps explaining a loss of genomic CpG dinucleotides over time through mutation, as well as accounting for over one-third of all point mutations in human cancers. Methyl groups can interfere directly with the binding of transcription factors, presumably through steric hindrance or through competition from cellular methyl-CpG binding proteins (e.g., MeCP1 and MeCP2) (Boyes and Bird, 1992; Meehan et al., 1989). Finally, methylated CpG dinucleotides can repress transcription regardless of their location within the promoter, raising the possibility that MeCP1 could act directly as a transcriptional repressor or act indirectly by inhibiting the formation of the stereospecific complex of transcription factors required to activate a tissue-specific promoter.

A much smaller fraction of the genome, estimated at about 2%, contains stretches of approximately 1 kb in which CpG dinucleotides occur at a frequency of 1 pair per 10 bases. These so-called CpG islands are located within the promoter and the first exon at the 5′ end of genes. There are instances in which de novo methylation of CpG islands occurs. Mutation of transcription factor binding sites within one part of a CpG island leads to methylation of the entire island (Macleod et al., 1994).

Three CpG dinucleotides in the JCV promoter occur in the region of the transcription initiation site, suggesting the possibility that JCV promoter cell specificity might be regulated through methylation. In vitro methylation of the MH1 JCV early promoter with SssI methylase leads to very strong repression of transcription in glial cells (unpublished data). Thus, cell-specific methylation is another potential mechanism for cell specificity.

9. JCV AND BKV DNA REPLICATION

It should be readily apparent that a discussion of JCV and BKV DNA replication must start with findings made previously with SV40. At the outset, it should be emphasized that polyomavirus DNA replication requires three distinct components: (1) a single, *trans*-acting viral protein called *large T antigen* (TAg); (2) a *cis*-acting viral DNA element termed the *core origin of replication*, which is necessary and sufficient for initiation of replication in vivo and in vitro; and (3) a collection of factors expressed in a eukaryotic cell known to be permissive (capable of supporting the entire viral life cycle) for the virus in question. Because of the limited coding capacity of the polyomavirus genomes, it is important to note that these viruses must commandeer the replication machinery of the host cell.

The importance of the two viral replication components first became apparent to investigators who infected or transfected cells in culture with mutant virus or viral DNA. Mutations in the TAg gene or the core origin often resulted in complete elimination of viral replication. Later, the development and biochemical characterization of cell-free replication systems allowed researchers to identify proteins in permissive cell extracts that were required for viral DNA replication in vitro. By employing techniques that permitted the detection of protein–protein interactions, it became clear that TAg physically interacted with some of these cellular factors to assemble a replication complex on the viral DNA template.

10. SV40 DNA REPLICATION MODEL

The multifunctional SV40 TAg regulates the initiation and elongation steps of viral DNA replication through a number of its intrinsic biochemical activities (reviewed by Virshup et al., 1992; Stillman, 1994; L. Chen et al., 1997; Smelkova and Borowiec, 1997; Brodsky and Pipas, 1998; Herbig et al., 1999; Kim et al., 1999; Weisshart et al., 1999; Simmons, 2000). To initiate replication, the

708 amino acid (a.a.) protein interacts with three sequence elements contained within the 64 bp viral core origin: (1) a 27–bp central dyad symmetry comprising TAg binding site II (BSII), which includes four copies of the pentameric recognition sequence, GAGGC; (2) an early-side 15–bp imperfect palindrome (IP); and (3) a late-side 17–bp adenine-thymine (AT)-rich sequence (Fig. 6.5) (Deb et al., 1986). Flanking this minimal origin are auxiliary sequences, called *aux*-1 and *aux*-2, that stimulate in vivo DNA replication activity. These elements contain binding sites for viral TAg (BSI) and the cellular transcription factor Sp1. Through its specific DNA binding domain (a.a. 147–246), TAg recognizes BSII and, in the presence of ATP, assembles cooperatively as a double hexamer structure (Fig. 6.6). Each hexamer unit is positioned over one half of the core origin and together they effect the distortion of both the IP and AT-rich regions by melting the former and untwisting the latter sequence. Mutational analyses indicate that these two activities are effected by different parts of TAg and that a.a. 121–135 influence AT untwisting but not IP melting (L. Chen, et al., 1997).

Following local destabilization of the helix, the TAg double hexamer acts as a helicase to promote unwinding of the origin DNA. Each hexamer appears to assemble as a propeller-shaped particle around a channel through which the DNA is reeled (San Martin et al., 1997; Valle et al., 2000). The unwinding reaction requires additional TAg functions, including ATPase activity and the ability to bind/recruit the cellular replication factors. One of these factors, topoisomerase I, may relieve torsional strain during unwinding of the circular genome and increase unwinding specificity (Simmons et al., 1998). A second cellular protein, RPA, binds single-stranded DNA and, together with TAg, recruits the DNA polymerase α–primase complex to the exposed single-stranded DNA of the replication bubble to generate short RNA–DNA primers.

TAg exhibits a 10-fold greater affinity for primate versus bovine DNA polymerase α, suggesting that this interaction contributes to primate-specific replication of SV40 DNA (Dornreiter et al., 1990). Elongation proceeds bidirectionally, with TAg continuing to act as a helicase at the two replication forks. Coupling of these replication forks through the TAg double hexamer structure stimulates the unwinding activity (Smelkova and Borowiec, 1997; Weisshart et al., 1999). Continuous synthesis of the leading strand is accomplished by cellular replication factor C (RFC), proliferating cell nuclear antigen (PCNA), and DNA polymerase δ; discontinuous synthesis of the lagging strand, involving the production of Okazaki fragments, requires the coordinated action of both DNA polymerases α and δ and their associated co-factors (Stillman, 1994). FEN-1, a cellular RNase, removes the short RNA primers, and DNA ligase 1 links the Okazaki fragments together. Termination of replication is not well understood, although it does require topoisomerase activity (Ishimi et al., 1992). This step is rate limiting in the replication process, and it is possible that steric constraints, leading to changes in TAg helicase structure, result in reduced replication fork movement (Smelkova and Borowiec, 1997). Termination oc-

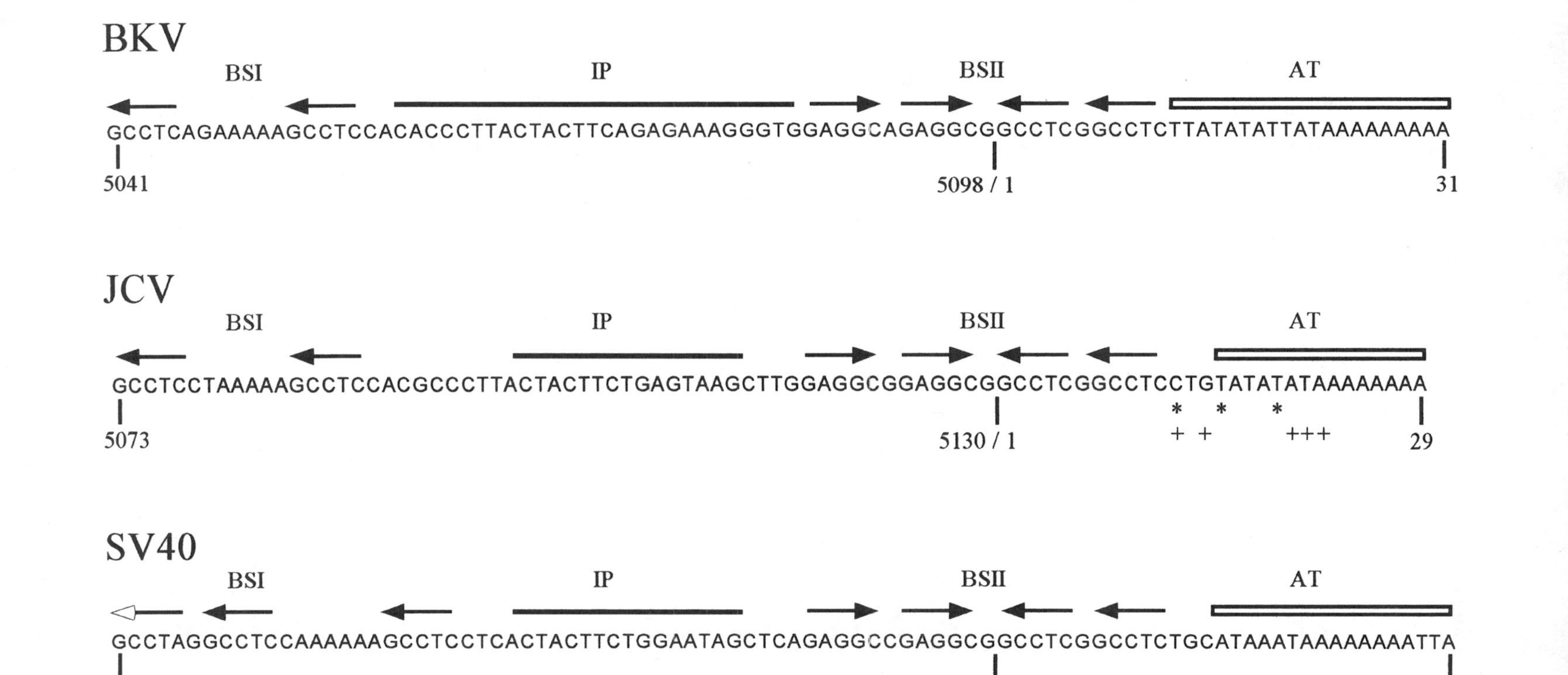

Figure 6.5. Comparisons of the JCV(Mad-1), BKV(AS), and SV40(776) origins of replication. The nucleotide sequences of each viral core origin and a portion of each auxiliary region 1 are shown. BSI and BSII contain multiple copies of the pentanucleotide sequence (denoted by arrows) recognized by TAg; the open arrowhead in SV40 BSI represents a variant consensus sequence. A closed or open bar denotes the imperfect palindrome (IP) or AT-rich region (AT), respectively, of each core origin. Nucleotide positions are indicated beneath each sequence; numbering begins at nucleotide 1 at the center of BSII and continues around the circular genome and ends at nucleotide 5098 (BKV), 5130 (JCV), or 5243 (SV40). Three asterisks (*) and five plus signs (+) mark the nucleotides on the late side of the JCV core origin that differ from the SV40 and BKV sequences, respectively. These differences are hypothesized to affect the ability of JCV TAg to productively interact with each origin.

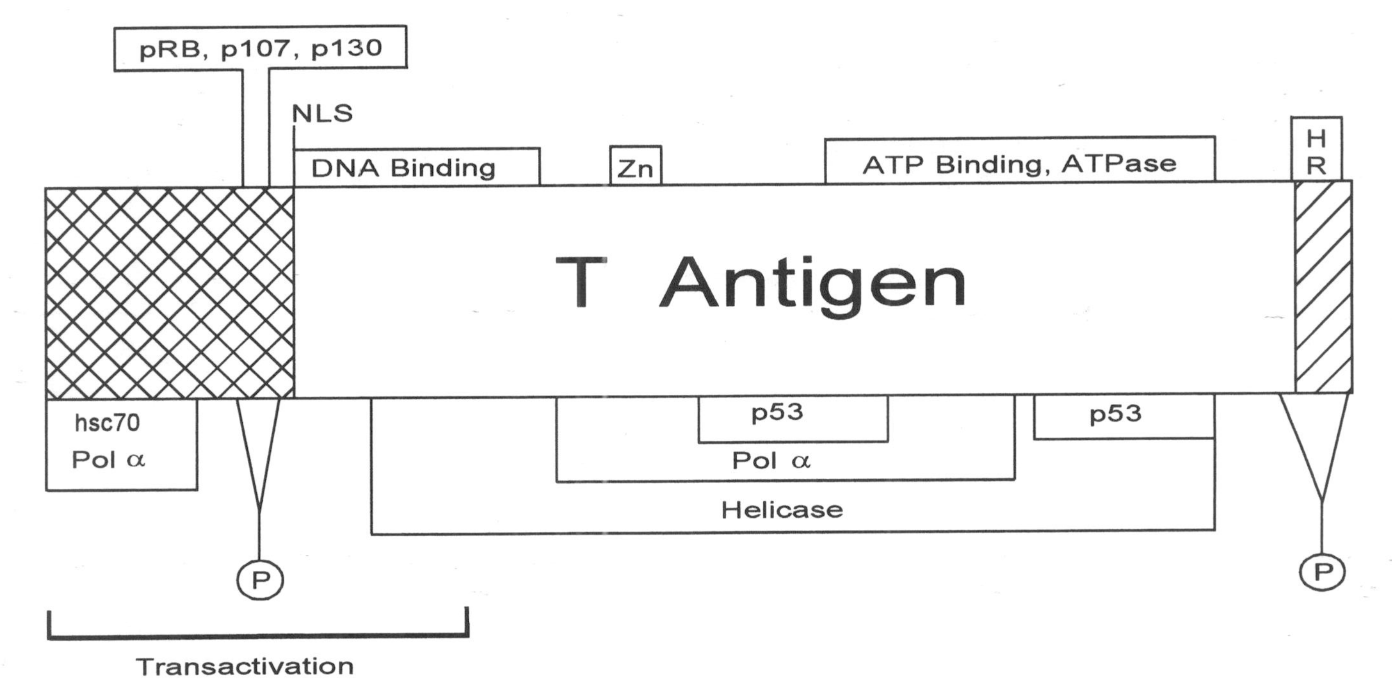

Figure 6.6. Functional domains of the primate polyomavirus TAgs. The JCV(Mad-1), BKV(AS), and SV40(776) TAgs are 688, 691, and 708 amino acids in size, respectively. The amino-terminal sequences of the TAgs (cross-hatched area) are shared with proteins T'_{135}, T'_{136}, T'_{165} (JCV; Trowbridge and Frisque, 1995), T^{*}_{136} (BKV; Prins and Frisque, unpublished data), and 17kT (SV40; Zerrahn et al., 1993) that are encoded by alternatively spliced early viral transcripts. This region of TAg interacts with cellular proteins involved in DNA replication and cellular transformation, including polymerase α (Pol α), hsc70 (via the J domain), and the pRB family of proteins (via the LXCXE domain). Sequences here also localize the proteins to the nucleus (NLS), overlap a transactivation domain, and include serine and threonine residues that are modified by phosphorylation (P). The central region of TAg binds the DNA sequence GAGGC, zinc (Zn), ATP, and the tumor suppressor p53 and exhibits ATPase and helicase activities. The carboxy-terminal 33 amino acids of the JCV TAg (diagonal lines) represent sequences shared with T'_{165}; this region includes a host range domain (HR) as well as a second set of phosphorylation sites.

curs approximately 180° around the circle from the core origin; specific DNA sequences do not appear to signal this event.

The DNA replication functions of TAg depend on the phosphorylation status of serine (Ser) and threonine (Thr) residues in the amino-terminal portion of the protein. Phosphorylation of Thr 124 causes a modest increase in binding of TAg to BSII and is required for origin unwinding and replication of the SV40 genome. In contrast, phosphorylation of TAg residues Ser 120 and Ser 123 interferes with double hexamer formation at the core origin, which correlates with inhibition of DNA replication. Dephosphorylation of these sites by the cellular phosphatase PP2A restores efficient replication function (Scheidtmann et al., 1991c; Virshup et al., 1992).

Several cellular factors involved in transcriptional activation are known to impact SV40 DNA replication. TBP binds the TATA box within the AT-rich region of the core origin; it also binds TAg. Herbig and co-workers (1999) have reported that the interaction of the viral and cellular proteins interferes with viral DNA replication, perhaps because TBP complexes associate with a TAg surface involved in unwinding origin sequences. Similarly, the transcription factor Oct1 has also been shown to bind the SV40 AT-rich region and to negatively regulate origin unwinding by TAg (Kilwinski et al., 1995). On the other hand, a number of transcription factors, including NF1, AP1, and Sp1, bind the SV40 *aux*-2 and enhancer elements and stimulate DNA replication (Cheng and Kelly, 1989; Guo and Depamphilis, 1992). The means by which this stimulation is achieved is not completely understood, although several mechanisms have been proposed (reviewed by Herbig et al., 1999). Recent evidence favors the hypothesis that direct interaction of certain transcription factors (e.g. NF1) with histone H3 activates SV40 DNA replication by altering the chromatin structure at the origin (Muller and Mermod, 2000).

TAg interacts with two additional classes of cellular factors, tumor suppressor proteins and molecular chaperones, which impact the DNA replication process. The former includes the pRB family of proteins that regulates a cell's entry into S phase of the cell cycle (reviewed by Kaelin, 1999). TAg overrides the normal function of these proteins through two domains at its amino terminus. Via its LXCXE domain (Fig. 6.6), TAg binds to pRB and the related proteins p107 and p130, and, in conjunction with its J domain, TAg effects S phase progression by causing the release of a heterodimeric complex composed of members of the E2F and DP families of transcription factors (Nevins, 1992; Zalvide et al., 1998; Kaelin, 1999). The J domain interacts with, and activates the ATPase activity of, the molecular chaperone hsc70. Mutation of a highly conserved HPD sequence in this domain alters TAg's ability to bind hsc70; the mutation also negatively influences SV40 DNA replication (Campbell et al., 1997). It is possible that efficient viral replication depends on the ability of hsc70 to alter the conformation of a replication initiation or elongation complex. In addition, the failure of the mutant TAg to effect release of E2F-DP from the pRB family of proteins might compound the replication defect by interfering with the transition of the cell into S phase (Ohtani, 1999).

11. HUMAN POLYOMAVIRUS DNA REPLICATION

Unique Features of JCV and BKV DNA Replication

It is expected that the mechanisms by which the primate polyomaviruses replicate their genomes will be highly conserved. However, specific differences have been observed between JCV, BKV, and SV40 DNA replication, and attempts have been made to explain the basis for this variability. For example, the host and tissue specificity of JCV replication is highly restricted relative to that of BKV and SV40. In addition, the TAg produced by each virus varies in its ability to support replication from the homologous and heterologous viral origins. Furthermore, considerable variation occurs in the ability of auxiliary sequences flanking the core origins of these three polyomaviruses to influence DNA replication efficiency. These differences suggest that modifications to the SV40 DNA replication model will be necessary to accurately describe the replication process for the human polyomaviruses.

Replication of the JCV genome has been examined more extensively than that of the BKV genome, and the following discussion centers on the former virus. Where appropriate, findings obtained with the BKV system are presented. Furthermore, it should be noted that most JCV studies cited here were conducted with the Mad-1 variant of JCV.

JCV Origin of Replication

The 67–bp core origin of JCV encompasses nucleotides 5094 to 30 (Lynch and Frisque, 1990; Sock et al., 1991, 1993) and includes the IP, BSII, and AT-rich elements described above for SV40 (Fig. 6.5). Although the sequences comprising the core origin of replication are highly conserved among the primate polyomaviruses (Deb et al., 1986; Deyerle et al., 1989; Lynch and Frisque, 1990), JCV's AT-rich element differs from that of SV40 with regard to its degree of DNA bending (Amirhaeri et al., 1988) and its interaction with TAg (Lynch and Frisque, 1990). The *aux*-1 and especially the *aux*-2 elements of the JCV replication origin differ significantly in sequence and in function relative to those of BKV and SV40. Dependence on these auxiliary elements for efficient DNA replication is most apparent for JCV and least apparent for BKV. A pentanucleotide repeat called the lytic control element (LCE; 5′-AGGGA-AGGGA-3′) is located immediately adjacent to the late side of the JCV core origin in *aux*-2 and has been shown to influence viral transcriptional activity (e.g., the transcription factors YB-1 and Purα bind this sequence) and to enhance replication behavior mediated by JCV, but not SV40 or BKV, TAg (Lynch and Frisque, 1990; Tada et al., 1991; Sock et al., 1993; Chang et al., 1994). A related sequence, *rep*, occurs at the same position in the BKV genome, and it too influences BKV replication and transcription (Del Vecchio et al., 1989).

Several approaches have been taken to examine how the JCV origin sequences contribute to replication differences observed between JCV and the other two primate polyomaviruses. These approaches have relied on the characterization of mutant, chimeric, and naturally occurring variant replication origins in in vivo and in vitro assay systems. Early development of an in vitro DNA replication system indicated that a plasmid containing the JCV replication origin could be propagated in a reaction utilizing primate cell extracts and SV40 TAg, thus demonstrating that a productive interaction between the SV40 regulatory protein and the JCV *cis*-acting regulatory sequences occurred (Li and Kelly, 1985). A similar conclusion was reached in a number of studies conducted in vivo. Some of these experiments involved the transfection of a JCV origin-containing vector into monkey or human cells expressing SV40 TAg either constitutively (e.g., in COS and cPOS cells) or transiently (from a cotransfected vector) (Lynch and Frisque, 1990, 1991; Sock et al., 1991, 1993). Other experiments relied on the transfection of primate cells (primary human fetal glial [PHFG] and CV-1) with complete chimeric viruses composed of SV40 coding regions and the JCV regulatory region (Chuke et al., 1986; Lynch and Frisque, 1991; Lynch et al., 1994). These in vivo studies utilized a DpnI replication assay (Peden et al., 1980) that discriminates between input and replicated viral DNA on the basis of susceptibility to the restriction enzyme *Dpn*I. Because the JCV and SV40 TAgs and core origins exhibit a high degree of sequence similarity (Frisque et al., 1984), it was not surprising to find that SV40 TAg promoted replication of the JCV origin, although the ability of SV40 TAg to support this replication more efficiently than did the JCV TAg was unexpected (Chuke et al., 1986; Lynch and Frisque, 1991; Lynch et al., 1994). More surprising was the result obtained when attempts were made to mediate SV40 origin replication with the JCV TAg. In the initial study, a chimera composed of the JCV coding regions and the SV40 regulatory region failed to replicate in PHFG or CV-1 cells (Chuke et al., 1986). Subsequent experiments revealed that this chimera could replicate in cells that constitutively expressed SV40 TAg (COS cells), but not in cells that expressed JCV TAg (POJ cells) (Lynch and Frisque, 1991). Similarly, the BKV TAg was found to activate replication of the JCV origin, but the JCV TAg again failed to drive replication of the heterologous (BKV) origin (Lynch and Frisque, 1990; Sock et al., 1993). These observations have served as the impetus for a number of experiments designed to identify the basis for JCV TAg's ability to discriminate between the polyomavirus origins. This work suggests that sequence differences in the TAgs and replication origins of these three polyomaviruses contribute to this observation.

Replication of JCV Origins from Naturally Occurring Variants. Two types of sequence variation have been detected in the genome of JCV isolates. Single nucleotide polymorphisms in the coding sequences have been identified in JCV isolates obtained from different human populations, and these variations are used to classify JCV into specific genotypes (reviewed by Agostini et al., 1999 and Chapter 18). A second type of variation involves the JCV transcriptional control region (TCR) and reveals two forms of the virus in the human host. Archetype JCV is detected in the kidneys and urine of its host, and it is thought to be the form of the virus that circulates in the population (see Chapter 7). Upon entry into a new susceptible host, it has been hypothesized that the archetype TCR undergoes a rearrangement process involving the deletion and duplication of promoter/enhancer sequences that yields a rearranged variant. The precise location of the deletion and duplication boundaries varies so that a large number of rearranged TCRs may be generated. These variants, compared with archetype, are detected at a wider variety of locations in the body (Newman and Frisque, 1997, 1999; Jensen and Major, 1999). JCV(Mad-1) is an example of a rearranged variant in which a 23–bp and a 66–bp sequence found in the archetype TCR have been deleted and a duplication of 98–bp has occurred. Differences between the archetype and rearranged TCRs primarily affect the *aux*-2 element; the core origin and *aux*-1 sequences remain largely unaffected by the rearrangement process.

One element that influences SV40 DNA replication, the Sp1 recognition sequence or GC box, is present in the *aux*-2 element of archetype, but in only a subset of rearranged variants. Constructs containing replication origins from archetype or rearranged forms of JCV exhibit equivalent replication efficiencies upon transfection into TAg-expressing cells, indicating that differences in the *aux*-2 elements have little influence on this activity (Lynch and Frisque, 1990; Daniel et al., 1996; Sock et al., 1996; Ault, 1997). However, transfection of intact archetype and rearranged JCV genomes into PHFG cells does reveal significant differences in replication behavior (Daniel et al., 1996). These data highlight the difficulty in separating transcription from replication effects when comparing overall accumulation of viral genomes in infected cells. Variations in the *aux*-2 element alter JCV transcriptional efficiency, thereby affecting the levels of early and late protein production. If this efficiency is reduced, lower levels of TAg and the capsid proteins are produced. This in turn leads to reduced replication because of the direct role that TAg plays in this process and because of the indirect role the capsid proteins play in facilitating secondary viral infections to yield biological amplification and spread of viral genomes during an infection.

Replication of JCV–SV40 Chimeric Origins. In an effort to delineate the JCV and SV40 origin sequences recognized differentially by the JCV TAg, JCV–SV40 chimeras were created. The first step in the construction scheme was to alter the only nucleotide that differed in the TAg BSII of JCV and SV40. SV40 nucleotide 5237 (a C) is included within the recognition site for the restriction enzyme *Bgl*I. The corresponding nucleotide in the JCV sequence (5124, a G) was converted to a C to create a *Bgl*I site at the center of the JCV origin (Lynch and Frisque, 1991). This mutation did not alter JCV replication potential (Lynch and Frisque, 1990). Both origin DNAs were cleaved with *Bgl*I, and the resulting restriction fragments were swapped to create two chimeras in which the early half of one origin was joined to the late half of the other origin (Lynch and Frisque, 1990). Analysis of the replication potential of the two chimeras in POJ and COS cells indicated that SV40 sequences in the late half of the origin (including the AT-rich and *aux*-2 elements) were responsible for the inability of the JCV TAg to interact with the SV40 origin. In addition, sequences representing the early half of each origin (including the IP and *aux*-1 elements) only enhanced replication about twofold in the presence of the homologous TAg.

Vacante et al. (1989) also employed a chimera approach to investigate JCV host- and tissue-specific replication behavior. These investigators joined SV40 promoter/enhancer signals (nucleotides 37–270) to the JCV genome at nucleotide 268 within the regulatory region. Propagation of this chimera in PHFG cells resulted in the deletion of JCV nucleotides 90–268 and SV40 nucleotides 37–145 to yield M1-SVE(Δ). Compared with JCV(Mad-1), this stable chimeric virus exhibited accelerated growth kinetics and increased virus titers when propagated in PHFG cells. The host range of M1-SVE(Δ) was also expanded to include monkey cells. Presumably, acquisition of these new properties was a result of enhanced transcriptional activity leading to increased levels of TAg rather than direct alterations to the replication machinery.

Replication of Mutant JCV Origins. Mutational analyses of the JCV regulatory region have identified elements critical to understanding (1) TAg's ability to discriminate between homologous versus heterologous origins, (2) the relative contributions of core and auxiliary origin sequences to DNA replication, and (3) the influence of host/tissue specificity on JCV replication (Table 6.1). The first mutations introduced into the JCV core origin were small deletions that abolished DNA replication (Mandl et al., 1987). These mutants, like similar SV40 mutants used to create COS cells (Gluzman, 1981), were transfected into PHFG cells to generate TAg-expressing POJ cells (Mandl et al., 1987).

To examine the influence of the JCV *aux*-1, *aux*-2, and enhancer elements on the efficiency of DNA replication, two groups created three series of mutants by progressively deleting sequences from the outer boundaries of the JCV replication origin toward the core origin using exonuclease III (Lynch and Frisque, 1990) or Bal31 (Sock et al., 1991, 1993) digestion. Analysis of the *aux*-1 element indicated that TAg BSI stimulated JCV TAg-mediated replication severalfold relative to the minimal core origin, a finding paralleling that made with the SV40 system (Sock et al., 1993). In their study of the *aux*-2 and enhancer elements, Lynch and Frisque (1990) assessed the replication potentials of deletion mutants in POJ, COS, and cPOS cells. The results were compared with those obtained with the JCV(Mad-1) and M1(Δ98) origins. The latter construct is missing one of the two copies of the JCV(Mad-1) enhancer.

The replication patterns of the mutants were similar in the monkey and human cells (COS and cPOS) expressing SV40 TAg; replication activity decreased as the size of the deletion increased. Once the deletion reached the AT-rich region of the core origin, replication was abolished in these cells. In contrast, transfection of the mutant origins into human cells (POJ) expressing the

Table 6.1. JCV Regulatory Region Mutants

Name[a]	Sequence(s) Altered[b]	Type[c]	Location[d]	Replication[e]
1. M1(ΔNco)	RR, T/t/T′	Del	nt 4980–275	—
2. S-3	Core	Del	nt 5072–5130	NT
3. S-8	Core	Del	nt 5072–5	NT
4. S-15	Core	Del	nt 5093–5127	—
5. S-19	Core	Del	nt 5108–5118	—
6. S-27	Core	Del	nt 5107–5116	—
7. M1(*Bgl*I)	Core	Sub	nt 5124, G/C	+
8. M1(Δ98)	*aux*-2	Del	nt 57–154	↓
9. 3A	*aux*-2	Sub	nt30, G/T	↓
			nt 35, G/C	
10. 4A	*aux*-2	Sub	nt 31, G/T	↑
			nt 36, G/C	
11. 5A	*aux*-2	Sub	nt 32, G/T	↓
			nt 37, G/C	
12. pM1(Δ98)o	*aux*-2	Del	nt 57–154	↑
13. d56	*aux*-2	Del	nt 57–275	↑
14. d45	*aux*-2	Del	nt 46–275	↑
15. d38	*aux*-2	Del	nt 39–275	↑
16. d30	*aux*-2	Del	nt 31–275	↓
17. d17	Core, *aux*-2	Del	nt 18–275	—
18. B38C	Core, *aux*-2	Del	nt 39–275	↑
		Sub	nt 5124, G/C	
19. B33C	Core, *aux*-2	Del	nt 34–275	↑
		Sub	nt 5124, G/C	
20. B31C	Core, *aux*-2	Del	nt 32–275	+
		Sub	nt 5124, G/C	
21. B31C(TT)	Core, *aux*-2	Del	nt 32–275	+
		Sub	nt 5124, G/C	
			nt 30, G/T	
			nt 31, G/T	
22. B31C(dl)	Core, *aux*-2	Del	nt 12, 32–275	↓
		Sub	nt 5124, G/C	
23. B31C(SV)	Core, *aux*-2	Del	nt 12, 32–275	↓
		Sub	nt 5124, G/C	
			nt 15, T/C	
			nt 19, T/A	
			nt 30, G/T	
			nt 31, G/T	
24. Δ46	EL	Del	nt 4981–5026	+
25. Δ48	EL	Del	nt 4981–5028	+
26. Δ80	EL	Del	nt 4981–5060	+
27. Δ91	EL	Del	nt 4981–5071	+
28. Δ112	*aux*-1	Del	nt 4981–5092	↓
29. Δ131	*aux*-1, core	Del	nt 4981–5111	—
30. Δ136	*aux*-1, core	Del	nt 4981–5116	—
31. Δ146	*aux*-1, core	Del	nt 4981–5126	—

Table 6.1. (*Continued*)

Name[a]	Sequence(s) Altered[b]	Type[c]	Location[d]	Replication[e]
32. Δ179	*aux*-1, core	Del	nt 4981–30	—
33. Δ48*	EL, *aux*-2	Del	nt 4981–5028 nt 31–279	↓
34. Δ91*	EL, *aux*-2	Del	nt 4981–5071 nt 31–279	↓
35. Δ112*	*aux*-1, 2	Del	nt 4981–5092 nt 31–279	↓
36. Δ131*	Core, *aux*-1, -2	Del	nt 4981–5111 nt 31–279	—
37. pJC264	*aux*-2	Del	nt 111–279	↓
38. pJC264Δpl	*aux*-2	Del Sub	nt 111–279 nt 42, C/A nt 43, T/A nt 44, G/A	↓
39. pJC264Δnf	*aux*-2	Del Sub	nt 111–279 nt 49, C/A nt 50, C/A	↓
40. pJC264Δnp	*aux*-2	Del Sub	nt 111–279 nt 39, T/G nt 40, G/T nt 41, G/T	↓
41–76. 36 mutants[f]	Core, *aux*-1, -2	Del	nt 5090–279 to nt 241–279	—/↓/+

[a]Replication mutants are described by Chuke et al. (1986; mutant 1), Mandl et al. (1987; mutants 2–6), Lynch and Frisque (1991; mutant 7), Daniel et al. (1996; mutant 8), Chang et al. (1994; mutants 9–11), Lynch and Frisque (1990; mutants 12–23), Sock et al. (1993; mutants 24–36), and Sock et al. (1991; mutants 37–76).

[b]Mutations altered one or more of the following JCV coding or regulatory sequences: TAg (T), tAg (t), all three T′ proteins (T′), entire regulatory region (RR), early leader (EL), auxiliary replication sequences 1 and 2 (*aux*-1, -2), and core origin (core).

[c]Types of mutations include deletions (Del) and substitutions (Sub). Substitution mutants 7, 21, 22, and 23 were created to alter the JCV sequence to an SV40 sequence at the specified mutated position(s).

[d]The sites of the mutations are identified by nucleotide (nt) numbers (Frisque et al., 1984). Substitutions are denoted with the wild-type nucleotide shown first, followed by a slash and then the replacement nucleotide. Some mutants contain multiple alterations (e.g., mutant 21 has three substitutions and one deletion).

[e]Replication activity was measured in vivo using the *Dpn*I assay. Mutants either failed to replicate (—) or exhibited reduced (↓), elevated (↑), or similar (+) activity relative to the wild-type JCV genome or to isolated origin sequences. Mutations were introduced into the intact JCV genome to create mutants 1–8; however, an intact or partial JCV origin that had been cloned into a plasmid vector was altered to generate the majority of the mutants (9–76). Replication of the mutant origin–containing plasmids was tested in cells expressing JCV TAg.

[f]Mutants 41–76, which are not listed individually, contain deletions that begin on the late side of the JCV origin (nt 241–279; smallest deletion) and extend through the *aux*-2 sequences first, then the core origin, and finally the *aux*-1 sequences (nt 5090–279; largest deletion). Replication activity of this series of mutants varied from equal to wild type to undetectably low levels.

JCV TAg led to about a twofold increase in replication behavior until the deletion reached the pentanucleotide repeat (LCE; AGGGAAGGGA) immediately adjacent to the core origin, and then replication fell five- to sixfold. As seen in COS and cPOS cells, replication was abolished when the deletion extended into the AT-rich sequence of the core origin. These studies indicate that the LCE sequence stimulates JCV, but not SV40, TAg-mediated replication. Furthermore, effects on replication by other sequences within *aux*-2 and the enhancer also depended on the source of the TAg that was tested. Point mutations introduced into the LCE signal reduce replication potential in the presence of JCV, but not SV40, TAg (Chang et al., 1994). In addition, these mutations altered the structure of the adjacent AT-rich region. It is possible that the AGGGA repeat contributes to the unusual DNA structure detected previously in the JCV AT-rich region and that these alterations are important to JCV TAg's ability to function as a helicase (Amirhaeri et al., 1988; Chang et al., 1994).

Sock et al. (1991) also employed a deletion mutagenesis approach to examine the contributions of the *aux*-2 and enhancer elements to JCV replication. A recognition site for NF-1 partially overlaps the LCE sequence described above, and deletion of this site reduces replication three- to fourfold in COS cells (Lynch and Frisque, 1990; Sock et al., 1991). When these mutants were tested in a human glioblastoma cell line co-transfected with a JCV TAg-expressing plasmid, a reduction in replication was again observed. However, using the complementing POJ cell system, Lynch and Frisque (1990) had not observed this reduction in activity using similar mutants (see above). Because the experimental protocols followed by these two laboratories differed in a number of ways, it is not possible to explain the discrepancies at this time. Sock et al. (1991) did go on to perform site-specific mutagenesis of the NF-1 site that prevented detectable NF-1 binding. These mutants again exhibited reduced SV40 TAg-mediated DNA replication of the JCV sequences (Sock et al., 1991). In contrast to these in vivo results, NF-1 did not have an effect on in vitro DNA replication, leading to the suggestion that this transcription factor exerts its influence via effects on chromatin accessibility (Sock et al., 1991). Finally, several forms of NF-1 have been identified in human tissues. NF-1/AT1, produced in human glial cells, binds JCV DNA and may influence tissue-specific replication behavior (Sumner et al., 1996).

To pursue the observation that the failure of the JCV TAg to support SV40 replication involves sequences within the late half of the origin, cassette mutagenesis was employed. The late half of the JCV and SV40 core origins differ at three positions; nucleotides 12, 15, and 19 (JCV numbering; Fig. 6.5). The JCV origin was converted to an SV40-like origin (Lynch and Frisque, 1990), and this mutant sequence replicated nearly 10-fold less efficiently than the corresponding JCV origin in the presence of JCV TAg; in the presence of SV40 TAg both origins displayed similar levels of amplification. These results identify three specific nucleotide differences in the late halves of the two viral core origins that are in large part responsible for JCV TAg's ability to discriminate between the JCV and SV40 *cis*-acting replication signals. It is interesting to speculate that the basis for this discrimination involves an Oct-1 binding site found in the late half of the SV40 core origin. The Oct-1 recognition sequence is altered in the corresponding region of the JCV origin (nucleotides 10–21), perhaps precluding the transcription factor from recognizing the JCV sequences. A second octamer-binding transcription factor, Tst-1 (also called Oct-6 or SCIP), does bind the JCV origin (nucleotides 10–25) and JCV TAg (Wegner et al., 1993; Renner et al., 1994; Sock et al., 1999). Because Oct-1 interferes with SV40 TAg's ability to unwind the SV40 origin, the JCV TAg may be unable to compete with Oct-1 bound to the SV40 sequences and thus fails to unwind the origin and initiate replication. The JCV protein may not face a similar barrier with its own origin if, in fact, it is true that the Oct-1 recognition site is missing.

JCV Replication Protein(s)

The multifunctional TAg is necessary and sufficient to mediate JCV DNA replication (Tavis and Frisque, 1991; Nesper et al., 1997). However, three recently discovered JCV early proteins, T'_{135}, T'_{136}, and T'_{165}, produced by alternative splicing of the early precursor mRNA (Fig. 6.1), modulate TAg-mediated replication (Trowbridge and Frisque, 1995; Prins and Frisque, 2001). The ability of JCV tAg, a fifth early protein, to influence DNA replication has not been investigated, although studies of the corresponding SV40 protein suggest that it may affect this process indirectly via effects on cellular gene expression and interaction with PP2A (Loeken et al., 1988; Scheidtmann et al., 1991b; Yang et al., 1991). Most of the studies that have examined the role of the JCV early proteins in DNA replication have relied on the extensive SV40 literature to predict functional domains and to target specific sequences for mutagenesis and chimeric approaches.

Replication Induced by TAgs from Naturally Occurring Variants. JCV variants are classified into several genotypes and subtypes based on nucleotide polymorphisms detected in the coding sequences. These point changes are associated with variants isolated from specific human populations and are thought to have arisen as the virus co-evolved with its host. While considerable effort has been invested in determining the influence of TCR variations on JCV replication and transcription behaviors, few studies have investigated whether variation in the coding sequences influence these processes. It is known that archetype and rearranged JCV variants exhibit different replication potentials. Chimeras in which the regulatory and coding regions have been exchanged indicate that changes in the former region are chiefly responsible for these differences, although the data do suggest that alterations in the latter region may have some influence as well. For example, the Mad-1 variant replicates efficiently, whereas the archetype CY strain and the rearranged Mad-8 strain exhibit barely detectable activities. Chimeras containing a CY or Mad-8 regulatory region linked to the Mad-1 coding region exhibit behavior similar to the parental CY and Mad-8 genomes, highlighting the importance of the TCR sequences. On the other hand, replication of chimeric genomes in which the Mad-1 regulatory region is linked to the CY or Mad-8 coding region is readily detectable, although it is not restored to Mad-1 levels. This result suggests that the CY and Mad-8 proteins function less efficiently than those of Mad-1 (Daniel et al., 1996) and that coding region variations are responsible, in part, for altered replication potential of JCV variants. At this time the specific CY and Mad-8 protein(s) and the specific sequence alterations that effect reduced replication activity have not been identified.

A zinc finger motif in the SV40 TAg (Fig. 5.6) contributes to viral replication and transformation activity (Loeber et al., 1989). Agostini and colleagues (1997, 1999) have suggested that sequence variations within the JCV TAg zinc finger domain may influence the replication behavior of the Type 2B form of the virus. Unlike the Asian Type 2A and 2C forms of JCV that have a hydrophobic leucine residue at a.a. position 301 in TAg's zinc finger motif, Type 2B has a hydrophilic glutamine residue. Although this genotypic variant has not yet been shown to exhibit altered replication activity relative to the other Type 2 genotypes, it does appear to be associated with PML more often than would be expected based on its overall prevalence in the population (Agostini et al., 1999).

Major and coworkers (1984, 1987) isolated, cloned, and partially sequenced episomal JCV DNA that was present in cultured cells derived from an owl monkey tumor induced by JCV (Mad-1). The regulatory region of this cloned genome, called JCV-586, contained a 19–bp deletion characteristic of the JCV(Mad-4) variant. In addition, changes in the carboxy terminus of TAg apparently had occurred because the TAg exhibited altered reactivity with several monoclonal antibodies and with cellular p53. JCV-586 is the only JCV isolate reported to replicate in monkey brain cells, and it is possible that changes in the viral TAg contribute to this extended host range property.

Replication Induced by JCV-SV40 Chimeric TAgs. The regulatory regions of primate polyomavirus genomes were exchanged to produce the first JCV–SV40 and JCV–BKV chimeras (Chuke et al., 1986; Bollag et al., 1989; Haggerty et al., 1989). This work suggested that the JCV early coding region, not just the regulatory region, contributed significantly to the unique behavior of JCV; therefore a series of early region chimeras were constructed (Haggerty et al., 1989). To create these JCV–SV40 chimeras, two restriction endonuclease recognition sites (for *Bst*XI and *Nsi*I) found at the corresponding position in each genome were utilized. Each chimeric early region was linked to either a JCV or SV40 regulatory region. As a consequence of the construction scheme, sequences in the VP1 coding region (at an *Eco*RI site) were also swapped, an alteration that precluded virion production but not DNA replication capability. The DNA replication potential of these eight chimeric genomes were compared with the activities of the two parental genomes and the two regulatory region chimeras (Lynch et al., 1994). To ensure that the comparisons were equivalent, sequences were inserted into the VP1 gene of the latter four genomes to prevent virion production. Transfection of the 12 DNAs into PHFG and CV-1 cells yielded the following observations relevant to TAg replication function: (1) the JCV(Mad-1) TCR promoted expression of the SV40 TAg at levels sufficient to support viral DNA replication in the monkey cells; (2) the failure of the JCV TAg to interact with the SV40 origin mapped to its central region (a.a. 82–411), which includes DNA binding, zinc finger, and helicase domains (Fig. 6.6); (3) the ability to replicate in monkey cells was limited to constructs expressing TAgs that contained the carboxy-terminal host range domain of SV40 (Fig. 6.6); and (4) SV40 DNA replication was 10-fold higher than that of JCV in PHFG cells. This replication difference was much less than that observed using intact, parental viral genomes, presumably because replication measurements for the latter DNAs are influenced by the rapid secondary infection of cells by viable SV40.

Replication Induced by Mutant JCV TAgs

Mutation of the Specific DNA Binding Domain of JCV TAg. Mutagenesis of the JCV early coding region has been guided by using the SV40 TAg as a template and by the earlier studies utilizing variant and chimeric viruses (Table 6.2). Initial experiments targeted the specific DNA binding domain of JCV TAg because its function is critical to replication initiation and because its sequence in the JCV and SV40 TAgs differs. Furthermore, JCV TAg had already been shown to bind less efficiently than SV40 TAg to each viral origin (Lynch and Frisque, 1991). Interestingly, JCV TAg interacts with both DNAs with similar efficiencies, ruling out the possibility that differential binding activity is the basis for JCV TAg's inability to stimulate replication of the SV40 origin. Tavis and Frisque (1991) created 10 JCV TAg mutants by introducing single or multiple mutations into the DNA binding domain (a.a. 145, 149, 157, 159, 162, and 168) or the nuclear localization signal (NLS; a.a. 131). Nine of these alterations converted the JCV sequence to an SV40 sequence. The tenth mutation, a lysine (Lys) to arginine (Arg) change at a.a. 168, was created in an initial step of the mutagenesis scheme and did not result in a JCV TAg that was more SV40-like (the corresponding residue in the SV40 TAg is also a Lys). The 10 mutant TAgs bound JCV and SV40 BSI with efficiencies ranging from 44% to 301% of the wild-type JCV protein. Nine of the mutants replicated in PHFG cells (22–220% of wild-type activity); the Arg 168 mutant, which exhibited the most efficient DNA binding activity, was replicative negative. In contrast to previous reports, wild-type JCV, as well as the nine viable TAg mutants, was shown to replicate in CV-1 cells. Although highly inefficient, this activity could be enhanced by propagating the cells in medium containing calf serum rather than fetal bovine serum. Finally, the mutant TAgs were tested to see if any had acquired the ability to drive replication of the SV40 origin. Again a new observation was made—wild type JCV TAg did mediate repli-

Table 6.2. JCV Early Region Mutants

Name[a]	Protein(s) Altered[b]	Type[c]	Location[d]	SV40-like[e]	Replication[f]
1. K-2	T/t/T′	Dup	nt 4915–4918	N	NT
2. S-1	T/t/T′	Del	nt 4871–4937	N	—
3. S-18	T/t/T′	Del	nt 4908–4922	N	NT
4. S-36	T/t/T′	Del	nt 4912–4924	N	NT
5. K-1	t	Dup	nt 4499–4502	N	NT
6. S-6	t	Del	nt 4467–4538	N	NT
7. S-21	t	Del	nt 4486–4511	N	NT
8. S-44	t	Del	nt 4498–4505	N	NT
9. M1(B-B)	T/T′	Del	nt 4244–4309	N	—
10. JCVΔT′	T′	Sub	nt 4274, G/A	N	↓
11. JCV-H42Q	T/t/T′	Sub	a.a. 42, H/Q	N	NT
12. JCTAg-RbS	T/T′	Sub	a.a. 104, D/N	Y	↓
			a.a. 108, H/S		
			a.a. 112, F/P		
			a.a. 113, A/S		
			a.a. 118, N/A		
			a.a. 120, G/D		
		Ins	a.a. 119/120, A		
13. JCTAg-RbN	T/T′	Sub	a.a. 104, D/V	N	↓
			a.a. 105, L/N		
			a.a. 107, C/A		
			a.a. 109, E/Q		
		Ins	a.a. 119/120, A		
14. M1(R131-T)	T/T′	Sub	a.a. 131, K/R	Y	+
15. M1(S145-T)	T	Sub	a.a. 145, A/S	Y	+
16. M1(H149-T)	T	Sub	a.a. 149, Q/H	Y	↓
17. M1(L157-T)	T	Sub	a.a. 157, V/L	Y	↑
18. M1(C159-T)	T	Sub	a.a. 159, S/C	Y	↓
19. M1(I162-T)	T	Sub	a.a. 162, V/I	Y	↓
20. M1(R168-T)	T	Sub	a.a. 168, K/R	Y	—
21. M1(HL-T)	T	Sub	a.a. 149, Q/H	Y	↓
			a.a. 157, V/L		
22. M1(HLC-T)	T	Sub	a.a. 149, Q/H	Y	↑
			a.a. 157, V/L		
			a.a. 159, S/C		
23. M1(SHLCI-T)	T	Sub	a.a. 145, A/S	Y	+
			a.a. 149, Q/H		
			a.a. 157, V/L		
			a.a. 159, S/C		
			a.a. 162, V/I		
24. JCV-N316K	T	Sub	a.a. 316, N/K	Y	+
25. JCV-H317Y	T	Sub	a.a. 317, H/Y	Y	+
26. JCV-NHKY	T	Sub	a.a. 316, N/K	Y	+
			a.a. 317, H/Y		
27. JCV-T125A	T/T′	Sub	a.a. 125, T/A	N	—
28. JCV-T664A	T/T'_{165}	Sub	a.a. 664, T/A	N	+

Table 6.2. (*Continued*)

Name[a]	Protein(s) Altered[b]	Type[c]	Location[d]	SV40-like[e]	Replication[f]
29. JCV-T664S	T/T'_{165}	Sub	a.a. 664, T/S	Y	+
30. JCV-E666A	T/T'_{165}	Sub	a.a. 666, E/A	N	+
31. JCV-E666S	T/T'_{165}	Sub	a.a. 666, E/S	Y	+

[a]Mutants are described by Mandl et al. (1987; mutants 1–8), Lynch and Frisque (1991; mutant 9), Trowbridge and Frisque (1995; mutant 10), Kelley and Georgopoulus (1997; mutant 11), Tavis et al. (1994; mutants 12 and 13), Tavis and Frisque (1991; mutants 14–23), and Swenson et al. (1996; mutants 24–31).

[b]Mutations alter one or more of the following JCV early proteins: TAg (T), tAg (t), all three T′ proteins (T′), or a single T′ protein (T'_{165}). Introducing insertions or deletions at convenient restriction enzyme cleavage sites yielded mutants 1–9. Specific functional domains or sequences were targeted in mutants 10–31 and included the shared donor splice site for T'_{135}, T'_{136}, and T'_{165} (mutant 10), J domain (mutant 11), pRB binding and amino-terminal phosphorylation domains (mutants 12 and 13), NLS (mutant 14), specific DNA binding domain (mutants 15–23), zinc finger domain (mutants 24–26), amino-terminal phosphorylation site (mutant 27), and potential carboxy-terminal phosphorylation sites (mutants 28–31).

[c]Types of mutations include duplications (Dup), deletions (Del), substitutions (Sub), and insertions (Ins).

[d]The sites of the mutations are identified either by nucleotide (nt) or amino acid (a.a.) numbers (Frisque et al., 1984). Nucleotide and amino acid substitutions are denoted using single letter codes, with the wild-type residue shown first, followed by a slash and then the replacement residue. Some mutants contain multiple mutations (e.g., mutant 12 has six substitutions and one insertion).

[e]Mutants created to alter the JCV sequence to an SV40 sequence at a specific position are identified as SV40-like (Y).

[f]Replication activity was measured in vivo using the *Dpn*I assay. Mutants either failed to replicate (—) or exhibited reduced (↓), elevated (↑), or similar (+) activity relative to wild-type JCV. NT = not tested. The mutation in mutant 11 was not introduced into an intact JCV genome, but was created in a clone encoding an *E. coli* DnaJ chaperone protein, thus precluding an analysis of replication potential.

cation of the SV40 origin, but at a very low level that was not altered significantly by changing its specific DNA binding domain to be more similar to that of SV40 TAg. This finding does not alter the earlier conclusion that the interaction between the JCV TAg and the SV40 origin is defective.

Mutation of the pRB Binding Domain of JCV TAg. In a second mutagenesis study, Tavis et al. (1994) investigated the observations that JCV TAg binds the pRB tumor suppressor protein and transforms cells in culture inefficiently. Again the approach was to convert the JCV TAg to an SV40-like protein, this time within the pRB binding domain. JCV and SV40 TAg differ at seven positions between residues 103 and 120, although the critical LXCXE sequence is conserved in both proteins. A JCV TAg mutant altered at each of these seven positions was tested in several assays relevant to either transforming activity or DNA replication behavior. Relative to wild type, the mutant TAg bound JCV and SV40 BSI more efficiently, but surprisingly bound pRB less efficiently. Furthermore, while transforming activity of the mutant was unaffected, DNA replication was decreased 25- to 50-fold and virus production was not detected. In addition to altering the pRB binding domain, the mutations also affected the predicted amino-terminal phosphorylation domain of JCV TAg. Serine residues 106 and 111 in this region of the SV40 TAg are phosphorylated, whereas the corresponding amino acids in the JCV TAg, a histidine (His) and an alanine (Ala), are not expected to be post-translationally modified. Because the pRB mutant JCV TAg was constructed by converting the JCV sequences to SV40 sequences, the mutant TAg may be phosphorylated at these serines. It is known that changes in the phosphorylation pattern of SV40 TAg alter its replication activity and its interaction with BSII; therefore, changes in JCV TAg phosphorylation may have been responsible for enhanced binding of the mutant TAg to BSI. Unfortunately, it was not possible to determine whether the mutant TAg, obtained from cell extracts, was altered in its ability to bind BSII DNA, an interaction that would be more relevant to understanding the basis for reduced DNA replication potential of the mutant. Another possible explanation for the reduced replication potential of the mutant is that precise regulation of pRB activity in the slowly growing PHFG cells may be critical to replication in this system, and the mutant TAg might be defective in this regard.

Mutation of Residues Within the Two Clusters of Phosphorylation Sites in the JCV TAg. Phosphorylation sites have been mapped to Ser and Thr residues within the amino- and carboxy-terminal portions of the JCV TAg (Fig. 6.6; Swenson and Frisque, 1995). Relative to SV40 TAg, JCV TAg is modified at fewer residues, suggesting that the absence of phosphorylation at critical sites might contribute to the less robust activity of the JCV protein. To test this possibility, three residues in the JCV TAg corresponding to three phosphorylation sites in the SV40 TAg were mutated (Swenson and Frisque, 1995; Swenson et al., 1996). Phosphorylation of Ser 677 in the SV40 TAg influences the modification of several amino-terminal regulatory phosphorylation sites, which in turn affects the DNA binding and replication activities of TAg (Schneider and Fanning, 1988; Scheidtmann et al., 1991a). Mutation of its counterpart in the JCV TAg, Thr 664, to a Ser or Ala indicated that this site is not phosphorylated and does not significantly influence DNA replication activity. A second carboxy-terminal phosphorylation site in the SV40 TAg, serine 679, is a glutamic acid (Glu) (a.a. 666) in JCV TAg. Converting the latter residue to a Ser or Ala again did not alter replication behavior, and the Ser 666 mutant did not become phosphorylated. These results suggest that important differences exist in the way the two TAgs are regulated by this post-translational modification. A critical phosphorylation site in the amino-terminal region of SV40 TAg, Thr 124, contributes to stable origin binding and is essential for unwinding and DNA replication (Scheidtmann et al., 1984, 1991a; McVey et al., 1993; Moarefi et al., 1993). Mutation of the conserved Thr 125 residue in the JCV TAg to

Ala abolishes DNA replication and viability, suggesting that phosphorylation of this site plays a similar key role in JCV TAg function.

Mutation of the Zinc Finger Domain of JCV TAg. The zinc finger motif is highly conserved in polyomavirus TAgs (Pipas, 1992), and mutational analysis of this domain in the SV40 TAg (a.a. 302-320) demonstrates its importance in the formation of stable hexamers and modulation of specific DNA binding and origin unwinding (Arthur et al., 1988; Hoss et al., 1990; Loeber et al., 1991). Alterations to any of the predicted zinc coordinating residues or the nearby His residue in the major loop (a.a. 302, 305, 313, 317, and 320) diminish transforming activity and abolish viral infectivity (Loeber et al., 1989). Mutations within the carboxy-terminal region of the major loop (a.a. 312–316) reduce the replication and transforming activities of SV40 TAg. These zinc finger mutants exhibit phenotypes similar to those displayed by wild-type JCV TAg. JCV TAg differs from SV40 TAg at four of the five positions in the major loop (JCV a.a. 313–317). Based on previous SV40 work, three JCV TAg mutants were made by converting residues 316, 317, or 316 + 317 to the corresponding SV40 sequence. Contrary to expectation, these mutations did not lead to enhanced JCV DNA replication activity or to the ability of JCV TAg to interact with the SV40 origin (Swenson et al., 1996). The latter possibility had been examined because the zinc finger domain lies within the central region of the JCV TAg known to negatively regulate its interaction with the SV40 origin.

Biochemical Analyses Relevant to JCV TAg Replication Functions. Biochemical approaches have been employed to examine JCV TAg's stability, phosphorylation status, and ability to bind cellular proteins and viral DNA. This work has utilized TAg obtained from extracts of infected and transformed cells or purified from bacterial, insect, and mammalian sources. Information derived from these experiments is relevant to our understanding of JCV DNA replication.

JCV TAg Stability. Early studies (Cikes et al., 1977; Bollag et al., 1989; Haggerty et al., 1989) suggested that JCV TAg was more labile than its SV40 counterpart. Pulse-chase experiments indicated that the half-life of the SV40 protein was twice as long as that of the JCV protein isolated from transformed human cells (Lynch and Frisque, 1991). Greater instability, coupled with reduced expression from JCV transcription signals, is responsible for the lower levels of JCV TAg relative to SV40 TAg in infected and transformed cells and likely contributes to the less robust activity of the JCV replication protein. In addition, the appearance of a 17 kD T′ protein in JCV-transformed cells (Bollag et al., 1989; Haggerty et al., 1989) was attributed initially to the proteolytic degradation of the large T protein. However, additional pulse-chase experiments comparing the degradation profiles of TAg and the T′ protein suggested that the latter protein was not a breakdown product of the former protein (Trowbridge and Frisque, 1995).

Oligomerization of JCV TAg. Interactions of the SV40 TAg with viral DNA and cellular proteins involve higher order structures of the viral protein, and sequences within the zinc finger and phosphorylation domains influence the formation of these oligomers (Arthur et al., 1988; Schneider and Fanning, 1988; Mastrangelo et al., 1989; Hoss et al., 1990; Loeber et al., 1991; Parsons et al., 1991; Scheidtmann et al., 1991c). Tavis and co-workers (1994) employed sucrose density gradient centrifugation to investigate the quaternary structure of JCV and SV40 TAgs present in extracts of transformed cells. They demonstrated that the unpurified JCV protein oligomerized less efficiently than the SV40 protein under nonreplication conditions; JCV TAg sedimented predominantly as monomers/dimers, whereas SV40 TAg yielded a mixture of monomers/dimers and higher oligomers. Because the TAgs differed significantly in this important property, additional experiments were conducted using purified JCV and SV40 TAgs under replication conditions (Bollag et al., 1996). In the presence of ATP and at 37°C, JCV TAg did form hexamers and double hexamers, albeit less efficiently than did SV40 TAg. In a second experiment, the oligomerization and DNA binding functions of TAg were analyzed in a DNA mobility shift assay using purified proteins and origin-containing DNA fragments. The baculovirus-expressed JCV and SV40 TAgs both behaved in a highly cooperative manner to form double hexamers on the origin DNA.

Binding of JCV TAg to BSI and BSII. The binding of the polyomavirus TAgs to BSII in the core origin is necessary but not sufficient to initiate DNA replication. It should be noted that the arrangement of TAg binding sites differs in the JCV and SV40 origins; JCV TAg BSI contains two, instead of three, pentanucleotide recognition sequences (Fig. 6.5), and a third binding site (BSIII), present in the SV40 origin, appears to be missing altogether. These differences were predicted to affect interactions between JCV TAg and DNA, and this possibility was first investigated using the McKay assay (McKay, 1981) with unpurified protein from transformed cell extracts (Lynch and Frisque, 1991). The JCV TAg exhibited lower binding activity than did the SV40 TAg, especially to BSII, suggesting one possibility for reduced DNA replicating potential of JCV. Both viral proteins bound to BSI with higher affinity than to BSII in this assay, and both bound the SV40 BSI with twofold higher affinity than the JCV site. The latter observation likely reflects differences in the sequences of the two sites. An expectation in these studies was that reduced binding activity was responsible for the nonproductive interaction between JCV Ag and the SV40 origin. However, the JCV protein recognized both origins with similar efficiencies, leading to the hypothesis that JCV TAg fails to promote replication of an SV40 origin because of a step subsequent to specific binding. Similar findings were made in a second study in which the predicted DNA binding domain of JCV TAg was mutated (Tavis and Frisque, 1991). Because binding of purified SV40 TAg to BSII is stimulated by ATP (Dean et al., 1987; Deb and Tegtmeyer, 1987; Borowiec and Hurwitz, 1988;

Mastrangelo et al., 1989), a more thorough investigation of JCV TAg binding to BSII awaited the availability of purified protein.

Windl and Dörries (1995) examined the binding activity of JCV TAg purified from human 293 cells infected with a recombinant adenovirus Ad5-JCVTAg. With a modified McKay assay and JCV TAg immunoprecipitated from the infected cell extracts, the viral protein was shown to interact with DNA fragments containing the JCV BSI and BSII or BSI alone. Reduced binding to the latter fragment was taken as evidence that BSII contributed to the overall binding pattern, but binding to a fragment containing only BSII could not be demonstrated. Thus the results were similar to those obtained previously with unpurified viral protein. In a second approach using purified TAg, Bollag and co-workers (1996) infected Sf9 insect cells with the recombinant baculovirus B-JCT. JCV TAg was purified by immunoaffinity chromatography and tested for specific DNA binding in the presence of ATP and at 37°C (McVey et al., 1993). Under these conditions, JCV TAg was shown capable of interacting with BSII and in fact exhibited a greater affinity for this site than for BSI of either viral origin. In agreement with earlier studies, SV40 TAg preferentially bound BSI (Tjian, 1978; Shalloway et al., 1980; Tegtmeyer et al., 1981), and both TAgs exhibited a slight preference for the SV40 origin. JCV TAg also bound nonorigin DNA (lacking the GAGGC sequence) more efficiently than did the SV40 protein. One could speculate that JCV TAg's altered behavior relative to that of SV40 TAg reflects an adaptation to lower levels of TAg produced during an infection (Bollag et al., 1996). To enhance initiation, JCV TAg may need to display a higher affinity to the core origin sequences, and to ensure efficient elongation the viral protein may require a tighter association with sequences beyond the origin region.

More recently, Nesper et al. (1997) re-examined the interaction between purified JCV TAg and JCV BSI and BSII. Under nonreplication conditions, the TAg was found to bind predominantly to BSI, in agreement with the work of Windl and Dörries (1995). Under replication conditions, however, TAg was found to bind tightly to BSII, confirming the work of Bollag et al. (1996).

Interactions Between JCV TAg and Cellular Proteins. Although not as extensive as the analyses conducted with SV40 TAg, the JCV protein has been shown to bind several cellular factors, including members of the pRB family of tumor suppressor proteins (pRB, p107, p130; Dyson et al., 1989, 1990; Howard et al., 1998); a second tumor suppressor, p53 (Bollag et al., 1989); the molecular chaperone hsc70 (Bollag and Frisque, unpublished data); and Tst-1, a member of the POU family of transcription factors (Renner et al., 1994; Sock et al., 1999). The latter interaction does not appear to alter DNA replication directly; rather, by stimulating early viral transcription, TAg levels increase, thereby elevating replication activity (Wegner et al., 1993). Based on SV40 studies (Campbell et al., 1997), the interaction between the J domain of the JCV TAg and the DnaK protein hsc70 might also be expected to enhance viral DNA replication, possibly by altering the conformation of the replication complex at the core origin. A great deal of effort has been expended examining SV40 TAg's binding to the cell cycle regulatory proteins p53 and the pRB family members. SV40 TAg, like the oncoproteins encoded by a number of other DNA tumor viruses, promotes S phase progression and inhibits apoptosis of cells by binding and inactivating these cellular proteins (reviewed by Brodsky and Pipas, 1998). Although these events have not been demonstrated to enhance SV40 DNA replication directly, most of the experiments have been conducted in exponentially growing cells. Replication of JCV occurs in cells that may be quiescent or slowly growing, and one might speculate that the interaction of JCV TAg with cellular tumor suppressor proteins is critical to ensuring the proper environment to support viral DNA replication.

Contribution of Other JCV Early Proteins to DNA Replication. Five proteins are encoded by JCV mRNAs generated by alternative splicing of the early precursor mRNA (Trowbridge and Frisque, 1995). In addition to large and small T antigens (TAg, tAg), three T′ proteins are expressed in infected human cells (Fig. 6.1). These latter proteins are translated from mRNAs in which two introns have been removed. The intron proximal to the 5′ end of the message is the same one removed from the TAg mRNA. The distal intron excised from each T′ mRNA utilizes a shared donor site but a unique acceptor site. This arrangement results in the production of three T′ proteins that share their amino-terminal 132 a.a. with TAg and have unique carboxy termini (either 3, 4, or 33 a.a. in T'_{135}, T'_{136}, or T'_{165}, respectively). The overlapping sequences include a number of functional domains important to TAg's transforming and replication activities, including sequences predicted or shown to (1) bind Tst-1, DNA polymerase α, pRB, p107, p130, and hsc70; (2) localize the protein to the nucleus (NLS); and (3) transactivate viral and cellular genes (Fig. 6.6). It would be reasonable to expect that the three T′ proteins would exhibit some of these TAg characteristics.

To investigate the possibility that T'_{135}, T'_{136}, and/or T'_{165} influences DNA replication, the shared distal donor splice site of the T′ mRNAs was mutated while preserving the authentic TAg a.a. sequence. This mutant, JCV(ΔT′), failed to express the three T′ proteins and was reduced 10-fold in DNA replicating activity compared with JCV(Mad-1) (Trowbridge and Frisque, 1995). Seven additional T′ mutants were constructed in which the three unique acceptor sites were targeted individually or in combination (Prins and Frisque, 2001). The triple mutant JCV($\Delta T'_{135/136/165}$) had a replication phenotype similar to the donor site mutant, whereas the three single mutants and one double acceptor site mutant ($\Delta T'_{135/165}$) replicated with nearly normal efficiency. The other two double acceptor site mutants ($\Delta T'_{135/136}$ and $\Delta T'_{136/165}$) exhibited partial replication defects, suggesting that T'_{136} might influence replication behavior to a greater extent than does T'_{135} and T'_{165}. These results support the hypothesis that JCV T′ proteins encode TAg replication functions. Similar "truncated" TAgs are also translated from alternatively spliced SV40, BKV, and mouse polyoma virus (PyV) early transcripts (Fig. 6.1; Zerrahn et al., 1993; Riley et al., 1997; Prins and Frisque, unpublished data), but the role of these T′-like proteins in viral DNA replication has not been established. It should be emphasized that there are important differences between TAg and the T′ proteins, including altered structures, phosphorylation status, and temporal and cell-specific expression patterns that are likely to result in functional differences (Swenson and Frisque, 1995; Trowbridge and Frisque, 1995; Prins, Jones and Frisque, unpublished data). This expectation was confirmed recently when these four early proteins where found to differ in the affinity with which they bound members of the pRB family (Bollag et al., 2000).

Cells Permissive for JCV DNA Replication

The initial success in recovering JCV from PML brain tissue was due to the identification of PHFG cells as a culture system that would support the complete lytic cycle of the virus. This heterogeneous population of glial cells remains the most permissive system in which to propagate the virus, although transformed derivatives of these cells (e.g., POJ, SVG, and POS cells) and human tonsillar stromal cells support moderate to efficient production of infectious virions. Additional human cell types have been found that support low levels of virus replication. Jensen and Major (1999) provide an extensive list of permissive and semipermissive cells in their recent review of JCV biology.

A large number of studies have been conducted to uncover the basis for the restricted growth of JCV. Feigenbaum et al. (1987) proposed that the restricted host range behavior of the virus is the result of regulation at the levels of transcription and replication. Their model suggests that because signals within the JCV TCR are glial cell specific (Kenney et al., 1984), early gene expression is limited to these cells, and, because JCV TAg interactions with host cell replication machinery is species (primate) specific, DNA replication is restricted to primate cells. Based on this model, one would predict that JCV replication would be confined to primate glial cells. Given the available data, some modifications to this model are necessary. For example, JCV replicates in cells outside the human CNS in vivo (Jensen and Major, 1999), and in vitro a chimera with a JCV regulatory region linked to the SV40 coding regions replicates efficiently in monkey kidney (CV-1) cells (Chuke et al., 1986; Lynch and Frisque, 1991; Lynch et al., 1994). Both examples indicate that TAg is expressed via the JCV TCR in amounts adequate to support multiplication of the virus in a nonglial cell. Furthermore, in the in vitro experiment, replication of intact JCV was highly inefficient in CV-1 cells, suggesting that JCV TAg, unlike SV40 TAg, interacted only weakly with the replication machinery of the monkey cells. Therefore, while most investigators agree that JCV's restricted host range is a function of regulation at both the transcription and replication levels, the former is not limited to (but is most efficient in) glial cells and the latter is most effective with (although not limited to) the replication machinery of human cells.

In Vitro DNA Replication

The development of cell-free replication systems has permitted investigators to identify cellular factors required for replication of polyomavirus genomes. Because naked viral DNA has been used routinely in this work, the influence of chromatin structure can not be fully assessed in in vitro systems, and therefore some differences in replication requirements in vivo versus in vitro have been recognized (Cheng and Kelly, 1989; Bullock et al., 1997; Halmer and Gruss, 1997; Nesper et al., 1997; Herbig et al., 1999; Muller and Mermod, 2000).

Using a soluble cell-free system derived from primate cells, Li and Kelly (1985) tested the replication of the SV40, JCV, and BKV origins in the presence of SV40 TAg. In extracts of human HeLa cells, constructs containing the core origins and auxiliary sequences of the two human polyomavirus genomes replicated with an efficiency 10–20% that of a plasmid containing the corresponding SV40 sequences. When monkey cell extracts (COS cells) were used, replication of the JCV and BKV origin plasmids was reduced another 5–10-fold relative to that of the SV40 origin-containing vector. In contrast, the SV40 construct replicated with similar efficiency in both the human and monkey cell extracts.

More recently Nesper and colleagues (1997) described two in vitro DNA replication systems containing purified JCV TAg and either purified cellular replication factors or crude extracts of human cells. Replication initiation was examined using the system containing DNA polymerase α-primase, RPA, and topoisomerase I, while replication elongation was measured in the system employing HeLa cell extracts. Sequence requirements for JCV DNA replication were examined with these assays. Deletion of TAg BSI resulted in approximately a twofold reduction in replication, whereas constructs lacking the IP or BSII elements of the core failed to exhibit any significant replication activity. Similar findings were reported when the BSI and BSII mutants were tested in an in vivo replication assay. In addition, a construct that lacked one copy of the JCV(Mad-1) 98 bp enhancer element replicated in vitro slightly better than the intact construct, in agreement with an earlier in vivo study (Lynch and Frisque, 1990). Replacement of the JCV TAg with either the SV40 or PyV TAgs in the in vitro system confirmed that the SV40 protein was a functional substitute, but yielded the new information that the PyV protein did not interact productively with the JCV sequences.

Other Factors Influencing JCV DNA Replication

A limited number of studies have identified factors that inhibit JCV DNA replication. Kerr and colleagues (1993) confirmed their prediction that the antitumor drug camptothecin would block replication of JCV DNA in vivo. The primary target of the drug appears to be topoisomerase I of the cell (Li and Liu, 2001). Pulse treatment of glial cells with nontoxic doses of camptothecin specifically inhibited viral replication; repeated pulse treatment of the cells with the drug was required to maintain the inhibitory effect. The use of camptothecin in the treatment of PML was suggested. In a second study, Staib et al. (1996) determined that exogenous addition of the human and murine tumor suppressor protein p53 strongly inhibited JCV DNA replication in vivo. Their results suggest that the basis for the inhibition was the binding of p53, via its highly conserved central region, to JCV TAg. It was postulated that this interaction might interfere with the ability of TAg to unwind the JCV origin during replication initiation. In work relevant to JCV reactivation in immunocompromised individuals, Chang and co-workers (1996) detected soluble factor(s) secreted by activated T cells that suppress replication of JCV DNA in glial cells. The factor(s) were found to be heat labile and to range in size from 30 to 100 kDa. Finally, N.N. Chen et al. (1997) identified a cellular factor, Sμbp-2, that stimulated JCV late transcription but decreased JCV DNA replication activity in glial cells. A truncated version of Sμbp-2, GF-1, induced JCV promoter function more efficiently than the intact human protein and also had a slight stimulatory effect on JCV DNA replication in vivo.

Evidence has been presented that virus–virus interactions may influence replication of JCV in its human host. The relatively high incidence of PML in HIV-infected individuals was recognized early in the AIDS pandemic, and several studies have indicated that the Tat protein of HIV-1 stimulates JCV transcriptional activity in vitro (Chowdhury et al., 1990, 1993; Tada et al., 1990). Recently, Mock and co-workers (1999) reported that JCV and human herpesvirus 6 (HHV6) co-infect oligodendrocytes within and around PML lesions. These investigators suggest that HHV6 might contribute to central nervous system demyelination, and one could speculate that such an effect might be the result of a direct action of HHV6 on the glial cells or an indirect action via stimulation of JCV replication. A clear demonstration of activation of JCV DNA replication by a second human virus is provided in the study of Heilbronn et al. (1993). These investigators hypothesized that human cytomegalovirus (HCMV) might serve as a helper virus for JCV replication in human fibroblasts, a cell type considered nonpermissive for JCV growth. Indeed, JCV replication was activated in HCMV-infected fibroblasts in culture. Furthermore, ganciclovir-induced inhibition of HCMV replication led to the concomitant inhibition of JCV DNA replication. These results might be relevant to mechanisms of JCV reactivation in immunocompromised, HCMV-infected individuals.

12. CONCLUSIONS

JCV transcription is regulated by complicated mechanisms. Many cell-specific and general transcription factors appear to regulate the JCV promoter via protein–protein interaction or by binding to promoter DNA elements. We have summarized in Figure 6.3 the transcription factor binding sites on the JCV promoter that have been reported. Multiple transcription factors interact with each other and regulate the transcription of JCV. In addition, chromatin structure and DNA methylation might be other factors controlling cell-specific expression. In the future, identification of novel aspects of JCV neurotropism should shed light not only on our understanding of the biochemical mechanism of virus infection, but also on the therapy of glial-specific diseases such as PML, multiple sclerosis, and other demyelinating diseases.

SV40 serves as a model for studying DNA replication in the eukaryotic cell. Its study has also provided a foundation on which to investigate this process in the human polyomaviruses. JCV and BKV are closely related to this monkey virus, and the sequences of the *cis* (origin) and *trans* (TAg) acting replication components of each virus are highly conserved. In addition, all three viruses replicate in primate cells, so it is not surprising that key features of the replication process are shared. However, differences in replication functions and potential do exist, thereby contributing to the unique biology of each virus. Significant variation has been observed in the ability of each TAg to interact with the homologous and heterologous core and auxiliary origin sequences. These variations may reflect differences in the communication between TAg and the viral DNA or TAg and permissive cell factors; they might also signal differences in the ability of core and auxiliary origin sequences to bind one or more cellular factors. Identifying the outcomes of these interactions will be critical to our understanding of how these viruses establish an asymptomatic infection or a pathogenic state. The recent discovery of multiple "truncated" versions of the polyomavirus TAgs that exhibit alterations in sequence and expression levels adds yet another dimension to the regulation of viral DNA replication that investigators have only recently begun to assess.

Acquiring a better understanding of viral DNA replication will be central to our ability to control the pathogenic potential of JCV and BKV in their human host. Preliminary studies based on such information have already been successful in identifying inhibitors that limit DNA replication of these viruses in cell culture systems. Furthermore, as we begin to understand the mechanisms by which other viruses activate JCV and BKV replication, we should be better able to devise the means to prevent the pathogenic consequences of these coinfections.

ACKNOWLEDGMENTS

This work was supported in part by National Institutes of Health grant NS35735 to J.W.H. and grant CA44970 to R.J.F.

REFERENCES

Agostini HT, Jobes DV, Chima SC, Ryschkewitsch CF, Stoner GL (1999): Natural and pathogenic variation in the JC virus genome. Recent Res Dev Virol 1:683–701.

Agostini HT, Ryschkewitsch CF, Mory R, Singer EJ, Stoner GL (1997): JC virus (JCV) genotypes in brain tissue from patients with progressive multifocal leukoencephalopathy (PML) and in urine from controls without PML: Increased frequency of JCV Type 2 in PML. J Infect Dis 176:1–8.

Ahmed S, Chowdhury M, Khalili K (1990a): Regulation of a human neurotropic virus promoter, JCV(E): Identification of a novel activator domain located upstream from the 98 bp enhancer promoter region. Nucleic Acids Res 18:7417–7423.

Ahmed S, Rappaport J, Tada H, Kerr D, Khalili K (1990b): A nuclear protein derived from brain cells stimulates transcription of the human neurotropic virus promoter, JCVE, in vitro. J Biol Chem 265:13899–13905.

Amemiya K, Traub R, Durham L, Major EO (1989): Interaction of a nuclear factor-1–like protein with the regulatory region of the human polyomavirus JC virus. J Biol Chem 264:7025–7032.

Amemiya K, Traub R, Durham L, Major EO (1992): Adjacent nuclear factor-1 and activator protein binding sites in the enhancer of the neurotropic JC virus. J Biol Chem 267:14204–14211.

Amirhaeri S, Wohlrab F, Major EO, Wells R (1988): Unusual DNA structure in the regulatory region of the human papovavirus JC virus. J Virol 62:922–931.

Antequera F, Bird A (1993): Number of CpG islands and genes in human and mouse. Proc Natl Acad Sci USA 90:11995–11999.

Arthur AK, Hoss A, Fanning E (1988): Expression of simian virus 40 T antigen in *Escherichia coli*: Localization of T-antigen origin DNA-binding domain to within 129 amino acids. J Virol 62:1999–2006.

Åström K-E, Mancall EL, Richardson EP (1958): Progressive multifocal leukoencephalopathy. Brain 81:93–111.

Ault GS (1997): Activity of JC virus archetype and PML-type regulatory regions in glial cells. J Gen Virol 78:163–169.

Bargou RC, Jurchott K, Wagener C, Bergmann S, Metzner S, Bommert K, Mapara MY, Winzer KJ, Dietel M, Dorken B, Royer HD (1997): Nuclear localization and increased levels of transcription factor YB-1 in primary human breast cancers are associated with intrinsic MDR1 gene expression. Nat Med 3:447–450.

Beggs AH, Frisque RJ, Scangos GA (1988): Extinction of JC virus tumor-antigen expression in glial cell-fibroblast hybrids. Proc Natl Acad Sci USA 85:7632–7636.

Bellard M, Oudet P, Germond JE, Chambon P (1976): Subunit structure of simian-virus-40 minichromosome. Eur J Biochem 70:543–553.

Bestor TH, Tycko B (1996): Creation of genomic methylation patterns. Nat Genet 12: 363–367.

Bird A (1992): The essentials of DNA methylation. Cell 70:5–8.

Bollag B, Chuke WF, Frisque RJ (1989): Hybrid genomes of the polyomaviruses JC virus, BK virus, and simian virus 40: Identification of sequences important for efficient transformation. J Virol 63:863–872.

Bollag B, MacKeen PC, Frisque RJ (1996): Purified JC virus T antigen derived from insect cells preferentially interacts with binding site II of the viral core origin under replication conditions. Virology 218:81–93.

Bollag B, Prins C, Snyder EL, Frisque RJ (2000): Purified JC virus T and T′ proteins differentially interact with the retinoblastoma family of tumor suppressor proteins. Virology 274:165–178.

Borowiec JA, Hurwitz J (1988): ATP stimulates the binding of simian virus 40 (SV40) large tumor antigen to the SV40 origin of replication. Proc Natl Acad Sci USA 85: 64–68.

Boyes J, Bird A (1992): Repression of genes by DNA methylation depends on CpG density and promoter strength: Evidence for involvement of a methyl-CpG binding protein. EMBO J 11:327–333.

Brodsky JL, Pipas JM (1998): Polyomavirus T antigens: Molecular chaperones for multiprotein complexes. J Virol 72:5329–5334.

Bullock PA, Joo WS, Sreekumar KR, Mello C (1997): Initiation of SV40 DNA replication in vitro: Analysis of the role played by sequences flanking the core origin on initial synthesis events. Virology 227:460–473.

Campbell KS, Mullane KP, Aksoy IA, Stubdal H, Zalvide J, Pipas JM, Silver PA, Roberts TM, Schaffhausen BS, DeCaprio JA (1997): DnaJ/hsp40 chaperone domain of SV40 large T antigen promotes efficient viral DNA replication. Genes Dev 11: 1098–1110.

Chang CF, Tada H, Khalili K (1994): The role of a pentanucleotide repeat sequence, AGGGAAGGGA, in the regulation of JC virus DNA replication. Gene 148:309–314.

Chang CF, Otte J, Kerr DA, Valkkila M, Calkins CE, Khalili K (1996): Evidence that the soluble factors secreted by activated immune cells suppress replication of human neurotropic JC virus DNA in glial cells. Virology 221:226–231.

Chen L, Joo WS, Bullock PA, Simmons DT (1997): The N-terminal side of the origin-binding domain of simian virus 40 large T antigen is involved in A/T untwisting. J Virol 71:8743–8749.

Chen NN, Kerr D, Chang CF, Honjo T, Khalili K (1997): Evidence for regulation of transcription and replication of the human neurotropic virus JCV genome by the human S(mu)bp-2 protein in glial cells. Gene 185:55–62.

Chen NN, Khalili K (1995): Transcriptional regulation of human JC polyomavirus promoters by cellular proteins YB-1 and Purα in glial cells. J Virol 69:5843–5848.

Cheng L, Kelly TJ (1989): Transcriptional activator nuclear factor I stimulates the replication of SV40 minichromosomes in vivo and in vitro. Cell 59:541–551.

Chepenik LG, Tretiakova AP, Krachmarov CP, Johnson EM, Khalili K (1998): The single-stranded DNA binding protein, Purα, binds HIV-1 TAR RNA and activates HIV-1 transcription. Gene 210:37–44.

Chowdhury M, Kundu M, Khalili K (1993): GA/GC-rich sequence confers Tat responsiveness to human neurotropic virus promoter, JCVL, in cells derived from central nervous system. Oncogene 8:887–892.

Chowdhury M, Taylor JP, Tada H, Rappaport J, Wong-Staal F, Amini S, Khalili K (1990): Regulation of the human neurotropic virus promoter by JCV-T antigen and HIV-1 Tat protein. Oncogene 15:1737–1742.

Chuke WF, Walker DL, Peitzman LB, Frisque RJ (1986): Construction and characterization of hybrid polyomavirus genomes. J Virol 60:960–971.

Cikes M, Beth E, Guignard N, Walker DL, Padgett BL, Giraldo G (1977): Purification of simian virus 40 and JC T-antigens from transformed cells. J Natl Cancer Inst 59: 889–894.

Croston GE, Kerrigan LA, Lira LM, Marshak DR, Kadonaga JT (1991): Sequence-specific antirepression of histone H1-mediated inhibition of basal RNA polymerase II transcription. Science 251:643–649.

Damania B, Alwine JC (1996): TAF-like function of SV40 large T antigen. Genes Dev 10:1369–1381.

Daniel AM, Swenson JJ, Mayreddy RPR, Khalili K, Frisque RJ (1996): Sequences within the early and late promoters of archetype JC virus restrict viral DNA replication and infectivity. Virology 216:90–101.

Dean FB, Dodson M, Echols H, Hurwitz J (1987): ATP-dependent formation of a specialized nucleoprotein structure by simian virus 40 (SV40) large tumor antigen at the SV40 replication origin. Proc Natl Acad Sci USA 84:8981–8985.

Deb S, DeLucia AL, Baur CP, Koff A, Tegtmeyer P (1986): Domain structure of the simian virus 40 core origin of replication. Mol Cell Biol 6:1663–1670.

Deb SP, Tegtmeyer P (1987): ATP enhances the binding of simian virus 40 large T antigen to the origin of replication. J Virol 61:3649–3654.

Del Vecchio AM, Steinman RA, Ricciardi RP (1989): An element of the BK virus enhancer required for DNA replication. J Virol 63:1514–1524.

Devireddy LR, Kumar KU, Pater MM, Pater A (2000): BAG-1, a novel Bcl-2–interacting protein, activates expression of human JC virus. J Gen Virol 81:351–357.

Deyerle KL, Sajjadi FG, Subramani S (1989): Analysis of origin of DNA replication of human papovavirus BK. J Virol 63:356–365.

Dornreiter I, Hoss A, Arthur AK, Fanning E. (1990): SV40 T antigen binds directly to the large subunit of purified DNA polymerase alpha. EMBO J 9:3329–3336.

Dyson N, Bernards R, Friend SH, Gooding LR, Hassell JA, Major EO, Pipas JM, Vandyke T, Harlow E (1990): Large T antigens of many polyomaviruses are able to form complexes with the retinoblastoma protein. J Virol 64:1353–1356.

Dyson N, Buchkovich K, Whyte P, Harlow E (1989): The cellular 107K protein that binds to adenovirus E1A also associates with the large T antigens of SV40 and JC virus. Cell 58:249–255.

Faisst S, Meyer S (1992): Compilation of vertebrate-encoded transcription factors. Nucleic Acids Res 20:3–26.

Fanning E, Knippers R (1992): Structure and function of simian virus 40 large tumor antigen. Annu Rev Biochem 61:55–85.

Feigenbaum L, Hinrichs SH, Jay G (1992): JC virus and simian virus 40 enhancers and transforming proteins: Role in determining tissue specificity and pathogenicity in transgenic mice. J Virol 66:1176–1182.

Feigenbaum L, Khalili K, Major E, Khoury G (1987): Regulation of the host range of human papovavirus JCV. Proc Natl Acad Sci USA 84:3695–3698.

Felsenfeld G (1992): Chromatin as an essential part of the transcriptional mechanism. Nature 355:219–224.

Frisque RJ, Bream GL, Cannella MT (1984): Human polyomavirus JC virus genome. J Virol 51:458–469.

Gallia GL, Darbinian N, Johnson EM, Khalili K (1999a): Self-association of Purα is mediated by RNA. J Cell Biochem 74:334–348.

Gallia GL, Darbinian N, Tretiakova A, Ansari SA, Rappaport J, Brady J, Wortman MJ, Johnson EM, Khalili K (1999b): Association of HIV-1 tat with the cellular protein, Purα, is mediated by RNA. Proc Natl Acad Sci USA 96:11572–11577.

Gallia GL, Safak M, Khalili K (1998): Interaction of the single-stranded DNA-binding protein Purα with the human polyomavirus JC virus early protein T-antigen. J Biol Chem 273:32662–32669.

Gluzman Y (1981): SV40 transformed cells support the replication of early SV40 mutants. Cell 23:175–182.

Graff JR, Herman JG, Myohanen S, Baylin SB, Vertino PM (1997): Mapping patterns of CpG island methylation in normal and neoplastic cells implicates both upstream and downstream regions in de novo methylation. J Biol Chem 272:22322–22329.

Grant CE, Deeley RG (1993): Cloning and characterization of chicken YB-1: Regulation of expression in the liver. Mol Cell Biol 13:4186–4196.

Gruda MC, Alwine JC (1991): Simian virus 40 (SV40) T-antigen transcriptional activation mediated through the Oct/SPH region of the SV40 late promoter. J Virol 65: 3553–3558.

Gruda MC, Zabolotny JM, Xiao JH, Davidson I, Alwine JC (1993): Transcriptional activation by simian virus large T antigen: Interactions with multiple components of the transcription complex. Mol Cell Biol 13:961–969.

Guo ZS, Depamphilis ML (1992): Specific transcription factors stimulate simian virus 40 and polyomavirus origins of DNA replication. Mol Cell Biol 12:2514–2524.

Haggerty S, Walker DL, Frisque RJ (1989): JC virus-simian virus 40 genomes containing heterologous regulatory signals and chimeric early regions: Identification of regions restricting transformation by JC virus. J Virol 63:2180–2190.

Halmer L, Gruss C (1997): Accessibility to topoisomerases I and II regulates the replication efficiency of simian virus 40 minichromosomes. Mol Cell Biol 17:2624–2630.

Hansen U, Tenen DG, Livingston DM, Sharp PA (1981): T antigen repression of the SV40 early transcription from two promoters. Cell 27:603–612.

Hass S, Thatikunta P, Steplewski A, Johnson EM, Khalili K, Amini S (1995): A 39-kD DNA-binding protein from mouse brain stimulates transcription of myelin basic protein gene in oligodendrocytes. J Cell Biol 130:1171–1179.

Heilbronn R, Albrecht I, Stephan S, Burkle A, Zurhausen H (1993): Human cytomegalovirus induces JC virus DNA replication in human fibroblasts. Proc Natl Acad Sci USA 90:11406–11410.

Henson JW (1994): Regulation of the glial-specific JC virus early promoter by the transcription factor Sp1. J Biol Chem 269:1046–1050.

Henson JW (1996): Coupling JC virus transcription and replication. J Neurovirol 2:57–59.

Henson J, Saffer J, Furneaux H (1992): The transcriptional factor Sp1 binds to the JC virus promoter and is selectively expressed in glial cells in human brain. Ann Neurol 32:72–77.

Henson JW, Schnitker BL, Lee T-S, McAllister J (1995): Cell-specific activation of the glial-specific JC virus early promoter by large T antigen. J Biol Chem 270:13240–13245.

Herbig U, Weisshart K, Taneja W, Fanning E (1999): Interaction of the transcription factor TFIID with simian virus 40 (SV40) large T antigen interferes with replication of SV40 DNA in vitro. J Virol 73:1099–1107.

Horikoshi N, Usheva A, Chen J, Levine AJ, Weinmann R, Shenk T (1995): Two domains of p53 interact with the TATA-binding protein, and the adenovirus 13S E1A protein disrupts the association, relieving p53-mediated transcriptional repression. Mol Cell Biol 15:227–234.

Hoss A, Moarefi IF, Fanning E, Arthur AK (1990): The finger domain of simian virus 40 large T antigen controls DNA-binding specificity. J Virol 64:6291–6296.

Howard CM, Claudio PP, Gallia GL, Gordon J, Giordano GG, Hauck WW, Khalili K, Giordano, A. (1998). Retinoblastoma-related protein pRb2/p130 and suppression of tumor growth in vivo. J Natl Cancer Inst 90:1451–1460.

Ise T, Nagatani G, Imamura T, Kato K, Takano H, Nomoto M, Izumi H, Ohmori H, Okamota T, Ohga T, Uchimi T, Kuwano M, Kohno K (1999): Transcription factor Y-box binding protein 1 binds preferentially to cisplatin-modified DNA and interacts with proliferating cell nuclear antigen. Can Res 59:342–346.

Ishimi Y, Sugasawa K, Hanaoka F, Eki T, Hurwitz J (1992): Topoisomerase-II plays an essential role as a swivelase in the late stage of SV40 chromosome replication in vitro. J Biol Chem 267:462–466.

Itoyama Y, Webster HdeF, Sternberger NH, Richardson EP, Walker DL, Quarles RH, Padgett BL (1981): Distribution of papovavirus, myelin-associated glycoprotein, and myelin basic protein in progressive multifocal leukoencephalopathy lesions. Ann Neurol 11:396–407.

Jensen PN, Major EO (1999): Viral variant nucleotide sequences help expose leukocytic positioning in the JC virus pathway to the CNS. J Leukocyte Biol 65:428–438.

Kaelin WG (1999): Functions of the retinoblastoma protein. Bioessays 21:950–958.

Kashanchi F, Duvall JF, Dittmer J, Mireskandari A, Reid RL, Gitlin SD, Brady JN (1994): Involvement of transcription factor YB-1 in human T-cell lymphotropic virus type I basal gene expression. J Virol 68:7637–7643.

Kelley WL, Georgopoulos C (1997): The T/t common exon of simian virus 40, JC, and BK polyomavirus T antigens can functionally replace the J-domain of the *Escherichia coli* DnaJ molecular chaperone. Proc Natl Acad Sci. USA 94:3679–3684.

Kenny S, Natarajan V, Strike D, Khoury G, Salzman NP (1984): JC virus enhancer–promoter active in human brain cells. Science 226:1337–1339.

Kerr D, Chang CF, Chen N, Gallia G, Raj G, Schwartz B, Khalili K (1994): Transcription of a human neurotropic virus promoter in glial cells: Effect of YB-1 on expression of the JC virus late gene. J Virol 68:7637–7643.

Kerr DA, Chang CF, Gordon J, Bjornsti MA, Khalili K (1993): Inhibition of human neurotropic virus (JCV) DNA replication in glial cells by camptothecin. Virology 196:612–618.

Kerr D, Khalili K (1991): A recombinant cDNA derived from human brain encodes a DNA binding protein that stimulates transcription of the human neurotropic virus JCV. J Biol Chem 266:15876–15881.

Khalili K, Rappaport J, Khoury G (1988): Nuclear factors in human brain cells bind specifically to the JCV regulatory region. EMBO J 7:1205–1210.

Kilwinski J, Baack M, Heiland S, Knippers R (1995): Transcription factor Oct1 binds to the AT-rich segment of the simian virus 40 replication origin. J Virol 69:575–578.

Kim HS, Goncalves NM, Henson JW (2000): Glial cell-specific regulation of the JC virus early promoter by large T antigen. J Virol 74:755–763.

Kim HY, Barbaro BA, Joo WS, Prack AE, Sreekumar KR, Bullock PA (1999): Sequence requirements for the assembly of simian virus 40 T antigen and the T-antigen origin binding domain on the viral core origin of replication. J Virol 73:7543–7555.

Koike K, Uchiumi T, Ohga T, Toh S, Wada M, Kohno K, Kuwano M (1997): Nuclear translocation of the Y-box binding protein by ultraviolet irradiation. FEBS Letter 417:390–394.

Krachmarov CP, Chepenik LG, Barr-Vagell S, Khalili K, Johnson EM (1996): Activation of the JC virus Tat-responsive transcriptional control element by association of Tat protein of human immunodeficiency virus 1 with cellular protein Purα. Proc Natl Acad Sci USA 93:14112–14117.

Kraus VB, Inostroza JA, Yeung K, Reinberg D, Nevins JR (1994): Interaction of the Dr1 inhibitory factor with the TATA binding protein is disrupted by adenovirus E1A. Proc Natl Acad Sci USA 91:6279–6282.

Krebs CJ, Dey B, Kumar G (1996): The cerebellum-enriched form of nuclear factor I is functionally different from ubiquitous nuclear factor I in glial-specific promoter regulation. J Neurochem 66:1354–1361.

Krebs CJ, McAvoy MT, Kumar G (1995): The JC virus minimal core promoter is glial cell specific in vivo. J Virol 69:2434–2442.

Kumar G, Srivastava DK, Tefera W (1994): A 70- to 80-kDa glial cell protein interacts with the AGGGAAGGGA domain of the JC virus early promoter only in the presence of the neighboring cis elements. Virology 203:116–124.

Kumar KU, Reddy DLN, Pater MM, Pater A (1996a): Human JC virus cAMP response elements functional for enhanced glial cell expression in differentiating embryonal carcinoma cells. Virology 215:178–185.

Kumar KU, Tang SC, Pater MM, Pater A (1996b). Glial and muscle embryonal carcinoma cell-specific independent regulation of expression of human JC virus early promoter by cyclic AMP response elements and adjacent nuclear factor 1 binding sites. J Med Virol 49:199–204.

Leger H, Sock E, Renner K, Grummt F, Wegner M (1995): Functional interaction between the POU domain protein Tst-1/Oct-6 and the high-mobility-group protein HMG-I/Y. Mol Cell Biol 15:3738–3747.

Li JJ, Kelly TJ (1985): Simian virus 40 DNA replication in vitro: Specificity of initiation and evidence for bidirectional replication. Mol Cell Biol 5:1238–1246.

Li WW, Hsiung Y, Wong V, Galvin K, Zhou Y, Shi Y, Lee AS (1997): Suppression of grp78 core promoter element-mediated stress induction by the dbpA and dbpB (YB-1) cold shock domain proteins. Mol Cell Biol 17:61–68.

Li TK, Liu LF (2001): Tumor cell death induced by topoisomerase-targeting drugs. Annu Rev Pharmacol 41:53–77.

Liu M, Kumar KU, Pater MM, Pater A (1998): Identification and characterization of a JC virus pentanucleotide repeat element binding protein: Cellular nucleic acid binding protein. Virus Res 58:73–82.

Loeber G, Dörries K (1988): DNA rearrangement in organ-specific variants of polyomavirus JC strain GS. J Virol 62:1730–1735.

Loeber G, Parsons R, Tegtmeyer P (1989): The zinc finger region of simian virus 40 large T antigen. J Virol 63:94–100.

Loeber G, Stenger JE, Ray S, Parsons RE, Anderson ME, Tegtmeyer P (1991): The zinc finger region of simian virus-40 large T-antigen is needed for hexamer assembly and origin melting. J Virol 65:3167–3174.

Loeken M, Bikel I, Livingston DM, Brady J (1988): Trans-activation of RNA polymerase II and III promoters by SV40 small t antigen. Cell 55:1171–1177.

Lu H, Zawel L, Fisher L, Egly J-M, Reinberg D (1992): Human general transcription factor IIH phosphorylates the C-terminal domain of RNA polymerase II. Nature 358: 641–645.

Lynch KJ, Frisque RJ (1990): Identification of critical elements within the JC virus DNA replication origin. J Virol 64:5812–5822.

Lynch KJ, Frisque RJ (1991): Factors contributing to the restricted DNA replicating activity of JC virus. Virology 180:306–317.

Lynch KJ, Haggerty S, Frisque RJ (1994): DNA replication of chimeric JC virus–simian virus 40 genomes. Virology 204:819–822.

Macleod D, Charlton J, Mullins J, Bird AP (1994): Sp1 sites in the mouse aprt gene promoter are required to prevent methylation of the CpG island. Genes Dev 8:2282–2292.

Major EO, Mourrain P, Cummins C (1984): JC virus–induced owl monkey glioblastoma cells in culture: Biological properties associated with the viral early gene product. Virology 136:359–367.

Major EO, Vacante DA, Traub RG, London WT, Sever JL (1987): Owl monkey astrocytoma cells in culture spontaneously produce infectious JC virus which demonstrates altered biological properties. J Virol 61:1435–1441.

Mandl C, Walker DL, Frisque RJ (1987): Derivation and characterization of POJ cells, transformed human fetal glial cells that retain their permissivity for JC virus. J Virol 61:755–763.

Mastrangelo IA, Hough PVC, Wall JS, Dodson M, Dean FB, Hurwitz J (1989): ATP-dependent assembly of double hexamers of SV40 T antigen at the viral origin of DNA replication. Nature 338:658–662.

Mayreddy RPR, Safak M, Razmara M, Zoltick P, Khalili K (1996): Transcription of the JC virus archetype late genome: Importance of the kB and the 23-base-pair motifs in late promoter activity in glial cells. J Virol 70:2387–2393.

McKay RDG (1981): Binding of simian virus 40 T antigen-related protein to DNA. J Mol Biol 145:471–488.

McVey D, Ray S, Gluzman Y, Berger L, Wildeman AG, Marshak DR, Tegtmeyer P (1993): cdc2 phosphorylation of threonine 124 activates the origin-unwinding functions of simian virus 40 T antigen. J Virol 67:5206–5215.

Meehan RR, Lewis JD, McKay S, Kleiner EL, Bird AP (1989): Identification of a mammalian protein that binds specifically to DNA containing methylated CpGs. Cell 58:499–507.

Mertens PR, Alfonso-Jaume MA, Steinmann K, Lovett DH (1998): A synergistic interaction of transcription factors AP2 and YB-1 regulates gelatinase A enhancer–dependent transcription. J Biol Chem 273:32957–32965.

Moarefi IF, Small D, Gilbert I, Hopfner M, Randall SK, Schneider C, Russo AAR, Ramsperger U, Arthur AK, Stahl H, Kelly TJ, Fanning E (1993): Mutation of the cyclin-dependent kinase phosphorylation site in simian virus 40 (SV40) large T-antigen specifically blocks SV40 origin DNA unwinding. J Virol 67:4992–5002.

Mock DJ, Powers JM, Goodman AD, Blumenthal SR, Ergin N, Baker JV, Mattson DH, Assouline JG, Bergey EJ, Chen BJ, Epstein LG, Blumberg BM (1999): Association of human herpesvirus 6 with the demyelinative lesions of progressive multifocal leukoencephalopathy. J Neurovirol 5:363–373.

Muller K, Mermod N (2000): The histone-interacting domain of nuclear factor I activates simian virus 40 DNA replication in vivo. J Biol Chem 275:1645–1650.

Nesper J, Smith RWP, Kautz AR, Sock E, Wegner M, Grummt F, Nasheuer HP (1997): A cell-free replication system for human polyomavirus JC DNA. J Virol 71:7421–7428.

Nevins JR (1992): E2F—A link between the Rb tumor suppressor protein and viral oncoproteins. Science 258:424–429.

Newman JT, Frisque RJ (1997): Detection of archetype and rearranged variants of JC virus in multiple tissues from a pediatric PML patient. J Med Virol 52:243–252.

Newman JT, Frisque RJ (1999): Identification of JC virus variants in multiple tissues of pediatric and adult PML patients. J Med Virol 58:79–86.

Ohtani K (1999): Implication of transcription factor E2F in regulation of DNA replication. Frontiers Biosci 4:793–804.

Okada Y, Sawa H, Tanaka S, Takada A, Suzuki S, Hasegawa H, Umemura T, Fujisawa J, Tanaka Y, Hall WW, Nagashima K (2000): Transcriptional activation of JC virus by human T-lymphotropic virus type 1 Tax protein in human neuronal cell lines. J Biol Chem 275:17016–17023.

Ondek B, Gloss L, Herr W (1988): The SV40 enhancer contains two distinct levels of organization. Nature 333:40–45.

Padgett BL, Walker DL, ZuRhein GM, Eckroade RJ, Dessel BH (1971): Cultivation of a papova-like virus from human brain with progressive multifocal leukoenchephalopathy. Lancet 1:1257–1260.

Parsons RE, Stenger JE, Ray S, Welker R, Anderson ME, Tegtmeyer P (1991): Cooperative assembly of simian virus-40 T-antigen hexamers on functional halves of the replication origin. J Virol 65:2798–2806.

Peden KWC, Pipas JM, Pearson-White S, Nathans D (1980): Isolation of mutants of an animal virus in bacteria. Science 209:1392–1396.

Pipas JM (1992): Common and unique features of T-antigens encoded by the polyomavirus group. J Virol 66:3979–3985.

Prins C, Frisque RJ (2001): JC virus T′ proteins encoded by alternatively spliced early mRNAs enhance T antigen-mediated viral DNA replication in human cells. J Neurovirol 7:250–264.

Raj GV, Khalili K (1994): Identification and characterization of a novel GGA/C-binding protein, GBP-i, that is rapidly inducible by cytokines. Mol Cell Biol 14:7770–7781.

Raj GV, Gallia GL, Chang CF, Khalili K (1998): T-antigen–dependent transcriptional initiation and its role in the regulation of human neurotropic JC virus late gene expression. J Gen Virol 79:2147–2155.

Renner K, Leger H, Wegner M (1994): The POU domain protein Tst-1 and papovaviral large tumor antigen function synergistically to stimulate glia-specific gene expression of JC virus. Proc Natl Acad Sci USA 91:6433–6437.

Rice PW, Cole CN (1993): Efficient transcriptional activation of many simple modular promoters by simian virus 40 large T antigen. J Virol 76:6689–6697.

Riley MI, Yoo W, Mda NY, Folk WR (1997): Tiny T antigen: An autonomous polyomavirus T antigen amino-terminal domain. J Virol 71:6068–6074.

Rio DC, Tjian R (1983): SV40 T antigen binding site mutations that affect autoregulation. Cell 32:1227–1240.

Sabath DE, Podolin PL, Comber PG, Prystowsky MB (1990): cDNA cloning and characterization of interleukin 2-induced genes in a cloned T helper lymphocyte. J Biol Chem 265:12671–12678.

Safak M, Gallia GL, Ansari SA, Khalili K (1999a): Physical and functional interaction between the Y-box binding protein YB-1 and human polyomavirus JC virus large T antigen. J Virol 73:10146–10157.

Safak M, Gallia GL, Khalili K (1999b): A 23-bp sequence element from human neurotropic JC virus is responsive to NF-kB subunits. Virology 262:178–189.

Safak M, Gallia GL, Khalili K (1999c): Reciprocal interaction between two cellular proteins, Purα and YB-1, modulates transcriptional activity of JCVCY in glial cells. Mol Cell Biol 19:2712–2723.

San Martin MC, Gruss C, Carazo JM (1997): Six molecules of SV40 large T antigen assemble in a propeller-shaped particle around a channel. J Mol Biol 268:15–20.

Scheidtmann KH, Buck M, Schneider J, Kalderon D, Fanning E, Smith AE (1991a): Biochemical characterization of phosphorylation site mutants of simian virus 40 large T-antigen: Evidence for interaction between amino- and carboxy-terminal domains. J Virol 65:1479–1490.

Scheidtmann KH, Hardung M, Echle B, Walter G (1984): DNA binding activity of simian virus 40 large T antigen correlates with a distinct phosphorylation state. J Virol 50:1–12.

Scheidtmann KH, Mumby MC, Rundell K, Walter G (1991b): Dephosphorylation of simian virus-40 large-T antigen and p53 protein by protein phosphatase-2A: Inhibition by small-T antigen. Mol Cell Biol 11:1996–2003.

Scheidtmann KH, Virshup DM, Kelly TJ (1991c): Protein phosphatase-2A dephosphorylates simian virus-40 large T-antigen specifically at residues involved in regulation of DNA-binding activity. J Virol 65:2098–2101.

Schneider J, Fanning E (1988): Mutations in the phosphorylation sites of simian virus 40 (SV40) T antigen alter its origin DNA-binding specificity for site I or II and affect SV40 DNA replication activity. J Virol 62:1598–1605.

Shalloway D, Kleinberger T, Livingston DM (1980): Mapping of SV40 DNA replication origin region binding sites for the SV40 T antigen by protection against exonuclease III digestion. Cell 20:411–422.

Sharma AK, Kumar G (1991): A 53 kDa protein binds to the negative regulatory region of the JC virus early promoter. FEBS 281:272–274.

Simmons DT (2000): SV40 large T antigen functions in DNA replication and transformation. Adv Virus Res 55:75–134.

Simmons DT, Trowbridge PW, Roy R (1998): Topoisomerase I stimulates SV40 T antigen-mediated DNA replication and inhibits T antigen's ability to unwind DNA at nonorigin sites. Virology 242:435–443.

Simon MC, Fisch TM, Benecke BJ, Nevins JR, Heintz N (1988): Definition of multiple, functionally distinct TATA elements, one of which is a target in the hsp70 promoter for E1A regulation. Cell 52:723–729.

Smelkova NV, Borowiec JA (1997): Dimerization of simian virus 40 T-antigen hexamers activates T-antigen DNA helicase activity. J Virol 71:8766–8773.

Sock E, Enderich J, Wegner M (1999): The J domain of papovaviral large tumor antigen is required for synergistic interaction with the POU-domain protein Tst-1/Oct6/SCIP. Mol Cell Biol 19:2455–2464.

Sock E, Renner K, Feist D, Leger H, Wegner M (1996): Functional comparison of PML-type and archetype strains of JC virus. J Virol 70:1512–1520.

Sock E, Wegner M, Fortunato EA, Grummt F (1993): Large T-antigen and sequences within the regulatory region of JC virus both contribute to the features of JC virus DNA replication. Virology 197:537–548.

Sock E, Wegner M, Grummt F (1991): DNA replication of human polyomavirus-JC is stimulated by NF-I in vivo. Virology 182:298–308.

Staib C, Pesch J, Gerwig R, Gerber JK, Brehm U, Stangl A, Grummt F (1996): p53 inhibits JC virus DNA replication in vivo and interacts with JC virus large T-antigen. Virology 219:237–246.

Stillman B (1994): Smart machines at the replication fork. Cell 78:725–728.

Sumner C, Shinohara T, Durham L, Traub R, Major EO, Amemiya K (1996): Expression of multiple classes of the nuclear factor-1 family in the developing brain: Differential expression of two classes of NF-1 genes. J Neurovirol 2:87–100.

Surridge C (1996): Transcription. The core curriculum. Nature 380:287–288.

Swenson JJ, Frisque RJ (1995): Biochemical characterization and localization of JC virus large T antigen phosphorylation domains. Virology 212:295–308.

Swenson JJ, Trowbridge PW, Frisque RJ (1996): Replication activity of JC virus large T antigen phosphorylation and zinc finger domain mutants. J Neurovirol 2:78–86.

Tada H, Lashgari MS, Khalili K (1991): Regulation of JCVL promoter function: Evidence that a pentanucleotide "silencer" repeat sequence AGGGAAGGGA downregulates transcription of the JC virus late promoter. Virology 180:327–338.

Tada H, Lashgari M, Rappaport J, Khalili K (1989): Cell type-specific expression of JC virus early promoter is determined by positive and negative regulation. J Virol 63:463–466.

Tada H, Rappaport J, Lashgari M, Amini S, Wong-Staal F, Khalili K (1990): Transactivation of the JC virus late promoter by the Tat protein of type 1 human immunodeficiency virus in glial cells. Proc Natl Acad Sci USA 87:3479–3483.

Tafuri SR, Wolffe AP (1993): Selective recruitment of masked maternal mRNA from messanger ribonuclearprotein particles containing FRGY2(mRNP4). J Biol Chem 268:24255–24261.

Tavis JE, Frisque RJ (1991): Altered DNA binding and replication activities of JC virus T-antigen mutants. Virology 183:239–250.

Tavis JE, Trowbridge PW, Frisque RJ (1994): Converting the JCV T antigen Rb binding domain to that of SV40 does not alter JCV's limited transforming activity but does eliminate viral viability. Virology 199:384–392.

Taylor IAC, Solomon W, Weiner BM, Paucha P, Bradley M, Kingston RE (1989): Stimulation of the human heat shock 70 promoter in vitro by simian virus 40 large T antigen. J Biol Chem 264:16160–16164.

Tegtmeyer P, Anderson B, Shaw SB, Wilson VG (1981): Alternative interactions of the SV40 A protein with DNA. Virology 115:75–87.

Tinti C, Yang C, Seo H, Conti B, Kim C, Joh TH, Kim KS (1997): Structure/function relationship of the cAMP response element in tyrosine hydroxylase gene transcription. J Biol Chem 272:19158–19164.

Tjian R (1978): Protein–DNA interactions at the origin of simian virus 40 DNA replication. Cold Spring Harbor Symp Quant Biol 43:655–662.

Trapp BD, Small JA, Pulley M, Khoury G, Scangos GA (1988): Dysmyelination in transgenic mice containing JC virus early region. Ann Neurol 23:38–48.

Tretiakova A, Otte J, Croul SE, Kim JH, Johnson EM, Amini S, Khalili K (1999a): Association of JC virus large T antigen with myelin basic protein transcription factor(MEF-1/Purα) in hypomyelinated brains of mice transgenically expressing T antigen. J Virol 73:6076–6084.

Tretiakova A, Steplewski A, Johnson EM, Khalili K, Amini S (1999b): Regulation of myelin basic protein gene transcription by Sp1 and Purα: Evidence for association of Sp1 and Purα in brain. J Cell Physiol 181:160–168.

Trowbridge PW, Frisque RJ (1995): Identification of three new JC virus proteins generated by alternative splicing of the early viral mRNA. J Neurovirol 1:195–206.

Vacante DA, Traub R, Major EO (1989): Extension of JC virus host range to monkey cells by insertion of a simian virus 40 enhancer into the JC virus regulatory region. Virology 170:353–361.

Valle M, Gruss C, Halmer L, Carazo JM, Donate LE (2000): Large T-antigen double hexamers imaged at the simian virus 40 origin of replication. Mol Cell Biol 20:34–41.

Virshup DM, Russo AAR, Kelly TJ (1992): Mechanism of activation of simian virus 40 DNA replication by protein phosphatase-2A. Mol Cell Biol 12:4883–4895.

Wegner M, Drolet DW, Rosenfeld MG (1993): Regulation of JC virus by the POU-domain transcription factor Tst-1: Implications for progressive multifocal leukoencephalopathy. Proc Natl Acad Sci USA 90:4743–4747.

Weisshart K, Taneja P, Jenne A, Herbig U, Simmons DT, Fanning E (1999): Two regions of simian virus 40 T antigen determine cooperativity of double-hexamer assembly on the viral origin of DNA replication and promote hexamer interactions during bidirectional origin DNA unwinding. J Virol 73:2201–2211.

Windl O, Dörries K (1995): Expression of human polyomavirus JC T antigen by an adenovirus hybrid vector and its binding to DNA sequences encompassing the JC virus origin of DNA replication. J Gen Virol 76:83–92.

Wolffe AP (1994): Structural and functional properties of the evolutionarily ancient Y-box family of nucleic acid binding proteins. Bioessays 16:245–251.

Wolffe AP, Pruss D (1996): Targeting chromatin disruption: Transcription regulators that acetylate histones. Cell 84:817–819.

Wolffe AP, Tafuri S, Ranjan M, Familari M (1992): The Y-box factors: A family of nucleic acid binding proteins conserved from *Escherichia coli* to man. New Biol 4: 290–298.

Wortman MJ, Krachmarov CP, Kim JH, Gordon RG, Chepenik LG, Brady JN, Gallia GL, Khalili K, Johnson EM (2000): Interaction of HIV-tat with Purα in nuclei of human glial cells: Characterization of RNA-mediated protein–protein binding. J Cell Biochem 77:65–74.

Yang SI, Lickteig RL, Estes R, Rundell K, Walter G, Mumby MC (1991): Control of protein phosphatase-2A by simian virus-40 small-t antigen. Mol Cell Biol 11:1988–1995.

Yogo Y, Kitamura T, Sugimoto C, Ueki T, Aso Y, Hara K, Taguchi F (1990): Isolation of a possible archetypal JC virus DNA sequence from nonimmunocompromised individuals. J Virol 64:3139–3143.

Zalvide J, Stubdal H, Decaprio JA (1998): The J domain of simian virus 40 large T antigen is required to functionally inactivate RB family proteins. Mol Cell Biol 18: 1408–1415.

Zambrano N, DeRenzis S, Minopoli G, Faraonio R, Donini V, Scaloni A, Cimino F, Russo T (1997): DNA-binding protein Purα and transcription factor YY1 function as transcription activators of neuron-specific FE65 gene promoter. Biochem J 328: 293–300.

Zerrahn J, Knippschild U, Winkler T, Deppert W (1993): Independent expression of the transforming amino-terminal domain of SV40 large T-antigen from an alternatively spliced third SV40 early messenger RNA. EMBO J 12:4739–4746.

Zu Rhein GM, Chou SM (1965): Particles resembling papova viruses in human cerebral demyelinating disease. Science 148:1477–1479.

7

THE ARCHETYPE CONCEPT AND REGULATORY REGION REARRANGEMENT

YOSHIAKI YOGO, PH.D., and CHIE SUGIMOTO, PH.D.

1. HYPERVARIABLE PML-TYPE REGULATORY SEQUENCES

JC virus (JCV) was first isolated in 1971 from the brain of a patient with a fatal demyelinating disease in the brain, progressive multifocal leukoencephalopathy (PML) (Padgett et al., 1971). Since then, JCV has repeatedly been isolated from the brains of PML patients, and it is now established as the etiologic agent of PML (Berger and Major, 1999). Here we designate JCV isolates from the brains of PML patients as PML-type isolates. The complete sequence of the genome of a PML-type isolate (Mad-1) was reported in 1984 by Frisque et al. To date, the complete sequences of six PML-type isolates have been reported, and analyses of these sequences have revealed that the overall genome organization of JCV is consistent among PML-type JCV isolates (Agostini et al., 1998a,d; Frisque et al., 1984; Loeber and Dörries, 1988; Kato et al., 2000).

However, the promoter/enhancer region (designated here as the regulatory region) located between the DNA replication origin and the agnoprotein gene (see Chapter 5) is hypervariable among PML-type isolates. Variation in the regulatory region of PML-type JCVs was first demonstrated in 1981 by Rentier-Delrue et al. To exclude the possibility that the changes were introduced into

Human Polyomaviruses: Molecular and Clinical Perspectives, Edited by Kamel Khalili and Gerald L. Stoner.
ISBN 0-471-39009-7

JCV DNAs during propagation of JCV in culture, they extracted viral DNA from the brain tissues of two PML patients. Each of the viral genomes was cloned intact in *Escherichia coli*, using a plasmid vector. The two JCV genomes were approximately 50 bp larger or approximately 50 bp smaller than Mad-1 DNA. Analysis of the restriction endonuclease cleavage fragments of these two DNAs and Mad-1 DNA revealed that the slight differences were mapped to within the regulatory region. Two years later, Grinnell et al. (1983) directly cloned JCV DNAs from the brains of 10 PML patients and compared the restriction enzyme cleavage profiles among them. They concluded that a genome area corresponding to the regulatory region is hypervariable. Furthermore, Martin et al. (1985) determined the regulatory sequences of several PML-type JCVs and confirmed the hypervariation in the regulatory sequence among PML-type JCVs.

2. ARCHETYPE REGULATORY SEQUENCE

Sero-epidemiologic studies have shown that JCV circulates in humans without any obvious symptoms (Padgett and Walker, 1973, 1976). Indeed, JCV is frequently excreted in the urine of healthy individuals as well as patients without PML (Kitamura et al., 1990, 1994a; Agostini et al., 1996). Yogo et al. (1990) cloned JCV DNA directly from two healthy volunteers and eight nonimmunocompromised patients. Restriction enzyme analysis confirmed that the overall genome structure of the urine-derived isolates was identical with that of PML-type isolates. The regulatory region sequences of the 10 urine-derived isolates were then analyzed. The basic structure of the regulatory region was identical among clones derived from all individuals, with a few nucleotide mismatches (the regulatory sequence of a clone from a healthy individual, CY, is shown in Fig. 7.1). Where multiple clones from the same individual were examined, the structure of the regulatory region was identical among clones examined, with a single exception. Thus, it was concluded that the regulatory regions of JCVs derived from the urine of nonimmunocompromised individuals are highly homogeneous, in marked contrast with the hypervariable regulatory regions of PML-type JCVs.

The regulatory region of the JCV DNA cloned by Yogo et al. (1990) lacked any repetition of a sequence of significant length (Fig. 7.1). It contained 23 and 66 bp sequences, which were inserted into the 98 bp sequence present in a tandem repeat in Mad-1. As a result, the 98 bp sequence was split into three portions of 25, 55, and 18 bp (Fig. 7.1). The 23 bp sequence had been identified in a majority of PML-derived variants, although it was absent in a few, including Mad-1 (Frisque et al., 1984; Martin et al., 1985). As described below, it was later found that a significant number of isolates from the brain and cerebrospinal fluid (CSF) of PML patients retained a region encompassing the 66 bp sequence (Ault and Stoner, 1993; Agostini et al., 1997a; Sugimoto et al., 1998). The regulatory sequence depicted in Figure 7.1 was designated as

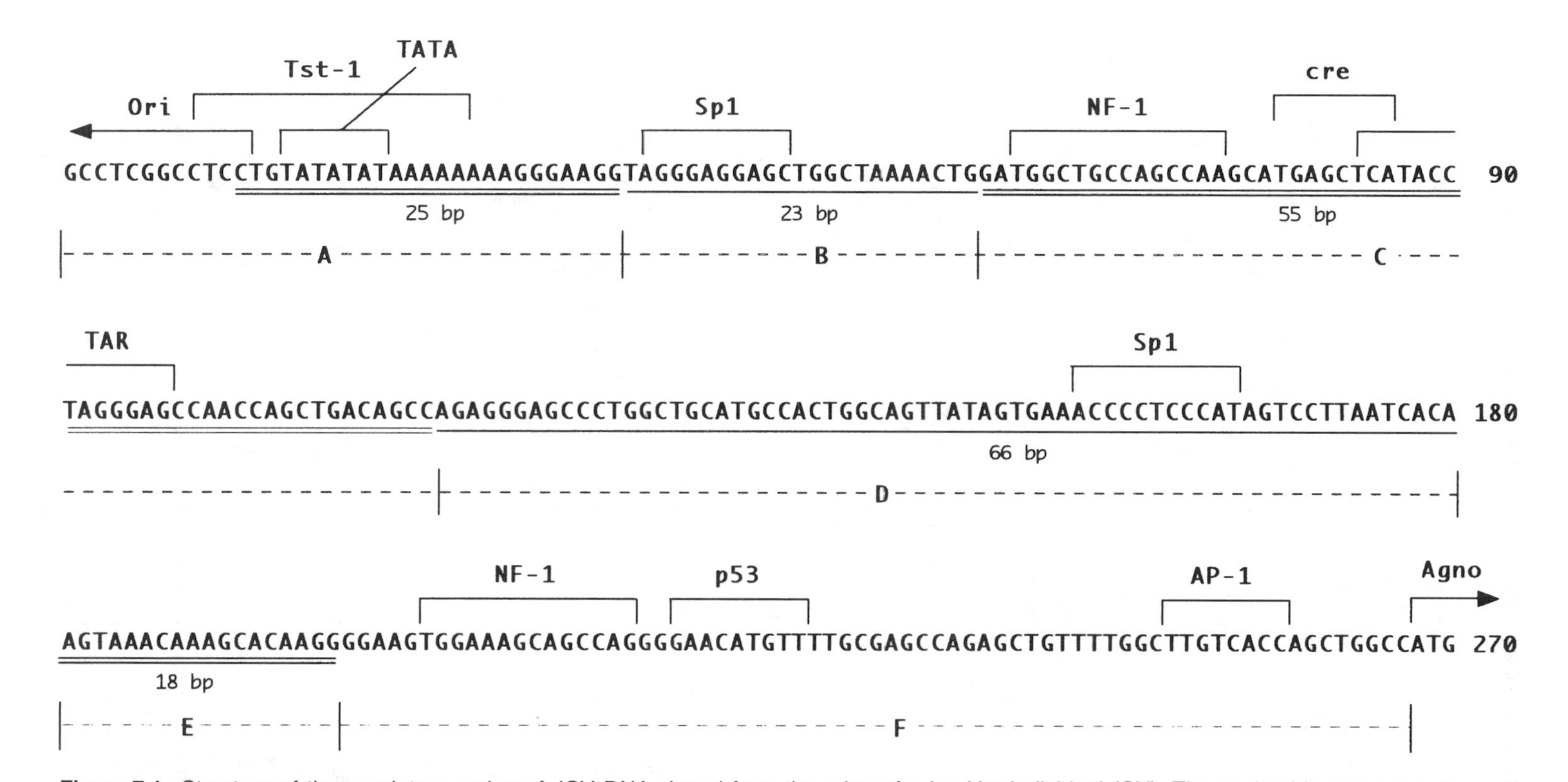

Figure 7.1. Structure of the regulatory region of JCV DNA cloned from the urine of a healthy individual (CY). The nucleotide sequence shown is from the origin of DNA replication (Ori) to the start site of the late leader protein, agnoprotein (Agno). The 25, 55, and 18 bp sequences present in the 98 bp repeated element of Mad-1 (Frisque et al., 1984) are shown by double straight lines, and the 23 and 66 bp sequences not present in Mad-1 are shown by single straight lines. Regions A–F are shown below the sequence (Ault and Stoner, 1993). TATA sequence and several transcriptional factor binding sites or homologies are indicated above the sequence.

the archetype regulatory sequence because, as described below, this sequence contained all sequences required to generate various PML-type sequences (Yogo et al., 1990).

Many complete JCV DNA clones have been established from urine collected in various areas of the Old World, and their regulatory region structures were examined (Yogo et al., 1991a,b; Guo et al., 1986). Furthermore, JCV regulatory regions were amplified by polymerase chain reaction (PCR) from urine collected in the United States (Markowitz et al., 1991; Agostini et al., 1996), Europe (Flægstad et al., 1991), and Taiwan (Chang et al., 1996a,b, 1999; Tsai et al., 1997), and their sequences were examined. These studies indicated that, without exception, JCV DNAs derived from the urine collected throughout the world carry the archetype regulatory sequence or archetype-like regulatory sequences that deviate only slightly from the archetype.

Tominaga et al. (1992) examined whether JCV persisting in normal human kidney tissue contains the archetype regulatory region. Renal medulla, cortex, and tumor tissue from 32 patients bearing renal tumors were screened for JCV DNA by Southern blot hybridization. Viral DNA was detected in the medulla in 13 cases (41%), in the cortex in 2 cases (6%), but in none of the tumor tissue specimens. Representative JCV DNA-positive specimens were used for PCR amplification and sequence analysis of the JCV regulatory regions. Structures of the regulatory regions from all specimens were, with a few nucleotide variations, that of the archetype. These observations indicated that JCVs persisting in the kidney have the archetype regulatory region and that in adults these JCVs actively replicate and excrete progeny in the urine.

In summary, (1) the archetype regulatory sequence is highly conserved, in marked contrast to PML-type regulatory sequences; (2) it contains all sequences present in various PML-type regulatory sequences, with a single exception (see below); (3) JCV DNAs derived from the urine collected throughout the world carry the archetype regulatory sequence or archetype-like regulatory sequences that deviate only slightly from the archetype; and (4) JCVs persisting in the kidney have the archetype regulatory sequence. Thus, it is very likely that JCV with the archetype regulatory sequence (or archetype-like sequences deviating slightly from it) represents the JCV circulating in the human population.

3. ALIGNMENT OF VARIOUS PML-TYPE REGULATORY SEQUENCES WITH THE ARCHETYPE

Several complete DNA clones of PML-type JCVs have been obtained in the United States, Japan, and Germany, and the regulatory sequences of these clones have been clarified (Kato et al., 1994; Loeber and Dörries, 1988; Martin et al., 1985; Matsuda et al., 1987; Takahashi et al., 1992; Yogo et al., 1994). Furthermore, owing to the advent of the PCR technology, many regulatory region structures of PML-type JCVs have also been determined (Agostini et al., 1997a, 1998a; Ault and Stoner, 1993; Chima et al., 1999; Newman and

Frisque, 1997, 1999; Stoner et al. 1998; Wakutani et al., 1998). In addition, some regulatory sequences have been detected from CSF of PML patients for diagnostic purposes (Sugimoto et al., 1998). All of these regulatory sequences (77 in total) were aligned with the archetype sequence by placing duplicated segments on separate lines and leaving gaps where sequences were deleted relative to the archetype (Fig. 7.2).

Each of these rearranged regulatory sequences was unique, that is, no identical regulatory sequences have occurred in the brain or CSF of different PML patients. Nevertheless, they could be categorized into a few groups according to their structural features (Fig. 7.2). To classify the 77 rearranged forms we use here the system introduced by Ault and Stoner (1993). They divided the archetypal regulatory region into five blocks of sequences, A to F (Fig. 7.1), based on sequences in the archetype that are lacking from, or duplicated in, the Mad-1 regulatory region.

The first pattern, designated as "long duplicate," shows duplication of region C and deletion of region D (Fig. 7.2A). The regulatory sequences with this pattern account for about half of the rearranged regulatory sequences shown in Figure 7.2. The second pattern, the "short triplicate" pattern, is characterized by having the first half of region C duplicated or triplicated, the second half of C and all of D deleted, and region E present in three or four copies (Fig. 7.2B). The main feature of the third pattern, "D retaining," was retention of region D with a duplication of sequences before or after it (Fig. 7.2C). We also included regulatory sequences, retaining region D but not having any duplication, in the "D-retaining" group. Regulatory sequences with miscellaneous sequences not included in any of the three groups are shown in Figure 7.2D.

Some regulatory regions (designated as pseudo-archetypes) belonging to the "D-retaining" group had only deletions (see Fig. 7.2C). These deletions destroyed some transcriptional factor binding elements (Table 7.1). In contrast, these elements are conserved in archetype-like regulatory regions (see above) found in the JCV isolates from the urine of non-PML individuals throughout the world (Agostini et al., 1996; Chang et al., 1996a,b; Guo et al., 1996; Yogo et al., 1990, 1991a,b). Thus, pseudo-archetype regulatory regions can be distinguished from the archetype-like regulatory region.

4. ARCHETYPE SEQUENCE CAN GENERATE VARIOUS REARRANGED FORMS

The line drawings in Figure 7.2 indicate that each of the rearranged regulatory sequences detected in the brain or CSF of PML patients can be produced from the archetype by deletions and duplications (or, in one case, by recombination with a coding region of the JCV genome). For example, "long duplicate" structures usually contain one or two identical deletions, with exactly the same 5′ and 3′ end points. This can simply be explained by hypothesizing that, in the archetype sequence, deletion occurred first and the segment carrying the

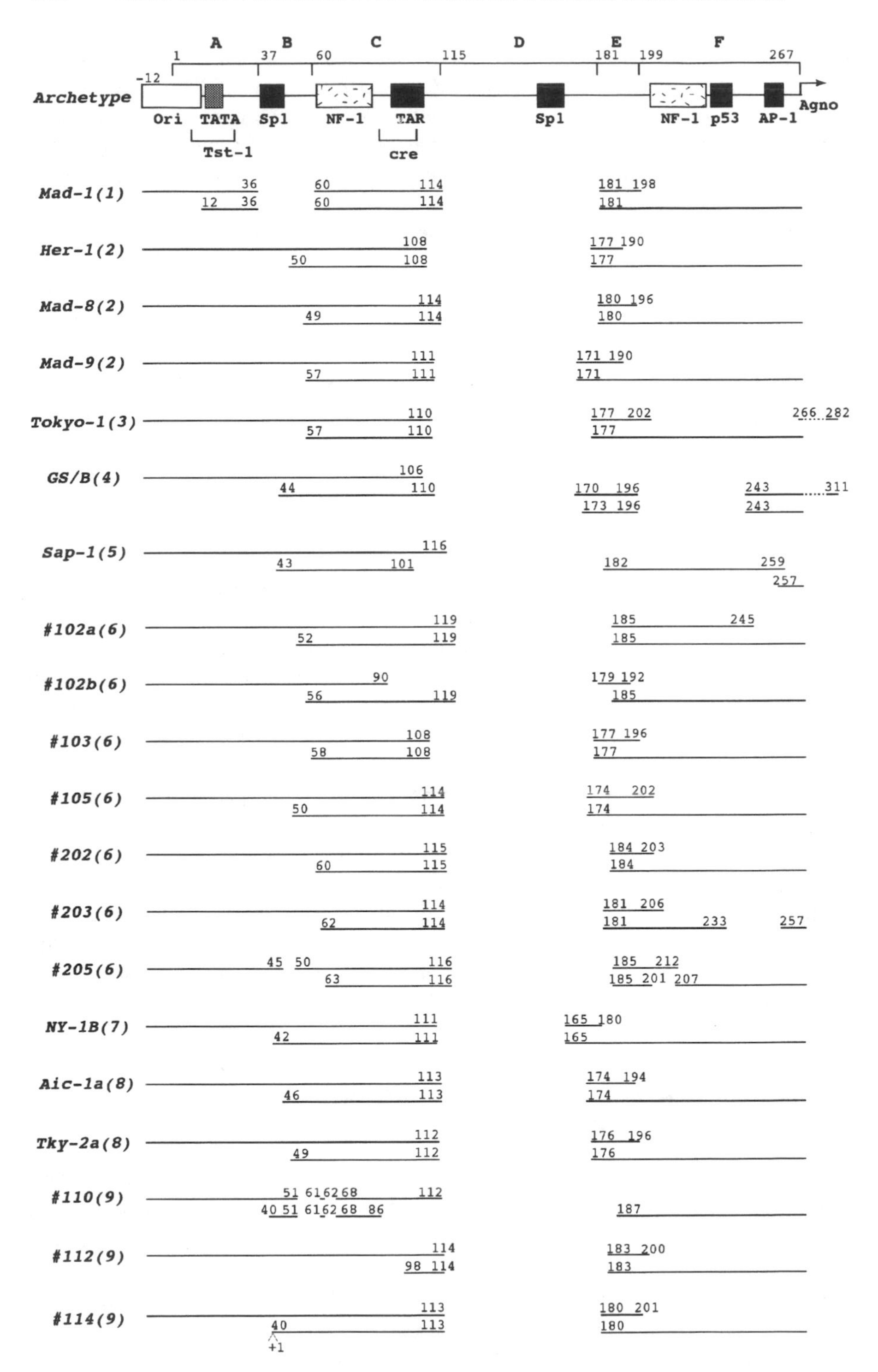
A
B
C
D
E
F
Archetype
Ori
TATA
Sp1
NF-1
TAR
Sp1
NF-1
p53
AP-1
Agno
Tst-1
cre
Mad-1(1)
Her-1(2)
Mad-8(2)
Mad-9(2)
Tokyo-1(3)
GS/B(4)
Sap-1(5)
#102a(6)
#102b(6)
#103(6)
#105(6)
#202(6)
#203(6)
#205(6)
NY-1B(7)
Aic-1a(8)
Tky-2a(8)
#110(9)
#112(9)
#114(9)

Figure 7.2A

deletion was subsequently duplicated (the example of Mad-1 is shown in Fig. 7.3A). Many "long duplicate" structures appear to have been complicated by further deletions or duplications. In general, "short triplicate" structures may have been produced by the duplication of segments containing previously generated deletions and duplications (the example of Matsue-1 is shown in Fig. 7.3B). Likewise, most of the rearranged regulatory sequences of "D-retaining" and miscellaneous groups (Fig. 7.3C,D), except for the NY-1A regulatory sequence, could have been generated from the archetype by deletions and duplications (or only by deletions). As shown in Figure 7.3C, it is very likely that the NY-1A regulatory sequence was generated by recombination between the archetype regulatory region and the T-antigen gene.

Martin et al. (1985) argued that the hypervariation of PML-type regulatory regions arose by complex alterations of the 98 bp repeat of Mad-1 DNA (Fig. 7.2). More recently, Monaco et al. (1998), who reported the detection of JCV DNA in human tonsil tissue, concluded that the Mad-1 regulatory sequence can rearrange to generate multiple genotypes observed in other infected host cells. Nevertheless, there is no simple way in which the various structures shown in Figure 7.2 can be generated from any of the arrangements other than the archetype.

5. SITE SELECTION FOR DNA BREAKAGE AND REJOINING

The presence of the major patterns of deletions and duplications (Fig. 7.2) suggests that there are preferred areas for DNA breakage and rejoining (Ault

Figure 7.2. Alignment of various PML-type regulatory sequences with the archetype. The structure of the archetypal regulatory region is schematically shown at the top. Regions A–F and some transcriptional factor binding sites are indicated. The origin of replication (Ori) and the start site of the late leader protein, agnoprotein (Agno), are indicated. The structures of various rearranged regulatory regions derived from the brain and CSF are shown below, with deletions relative to the archetype shown as gaps. On reading from left to right, when a repeat is encountered, the linear representation is displaced to the line below and to positions corresponding to the sequence of the archetype. Numbers indicate end points of segments present in each regulatory region. Arrowheads represent the insertion of sequences, the sizes of which are indicated below in base pairs. The regulatory regions classified as "long duplicate," "short triplicate," and "D retaining" are shown in **A**, **B**, and **C**, respectively, and those with miscellaneous sequences not included in any of the three groups are shown in **D**. The last three structures in **D** were tentatively designated as AC/Br-1, AC/Br-2, and AC/Br-3, and the last structure in **A** was tentatively designated as AH/Br because they were not named in the original paper (Newman and Frisque, 1999). References for regulatory regions are shown in parentheses by numbers 1, Frisque et al. (1984); 2, Martin et al. (1985); 3, Matsuda et al. (1987); 4, Loeber and Dörries (1988); 5, Takahashi et al. (1992); 6, Ault and Stoner (1993); 7, Yogo et al. (1994); 8, Kato et al. (1994); 9, Agostini et al. (1997a); 10, Newman and Frisque (1997); 11, Agostini et al. (1998a); 12, Wakutani et al. (1998); 13, Stoner et al. (1998); 14, Sugimoto et al. (1998); 15, Chima et al. (1999); 16, Newman and Frisque (1999).

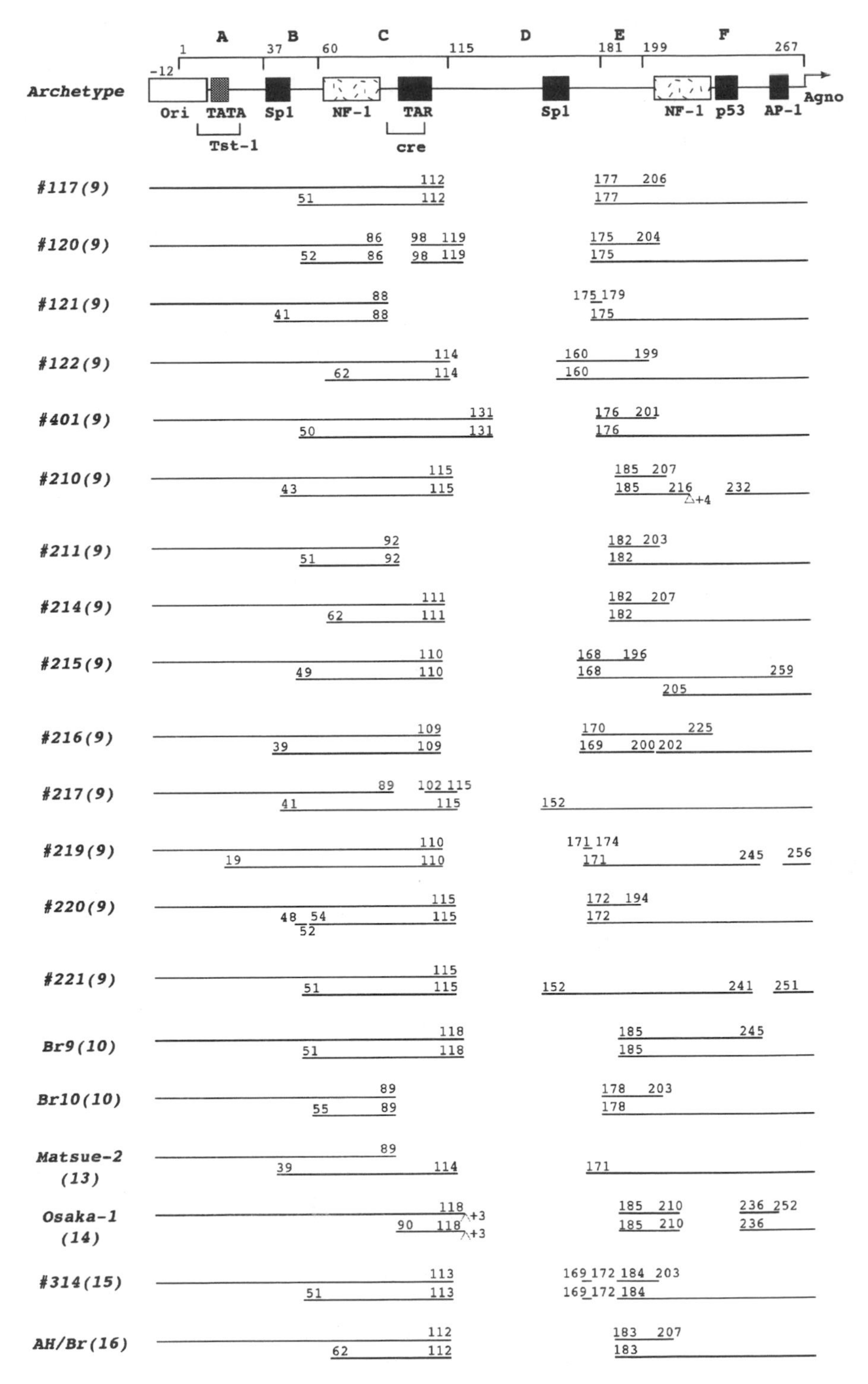

Figure 7.2A. (*Continued*)

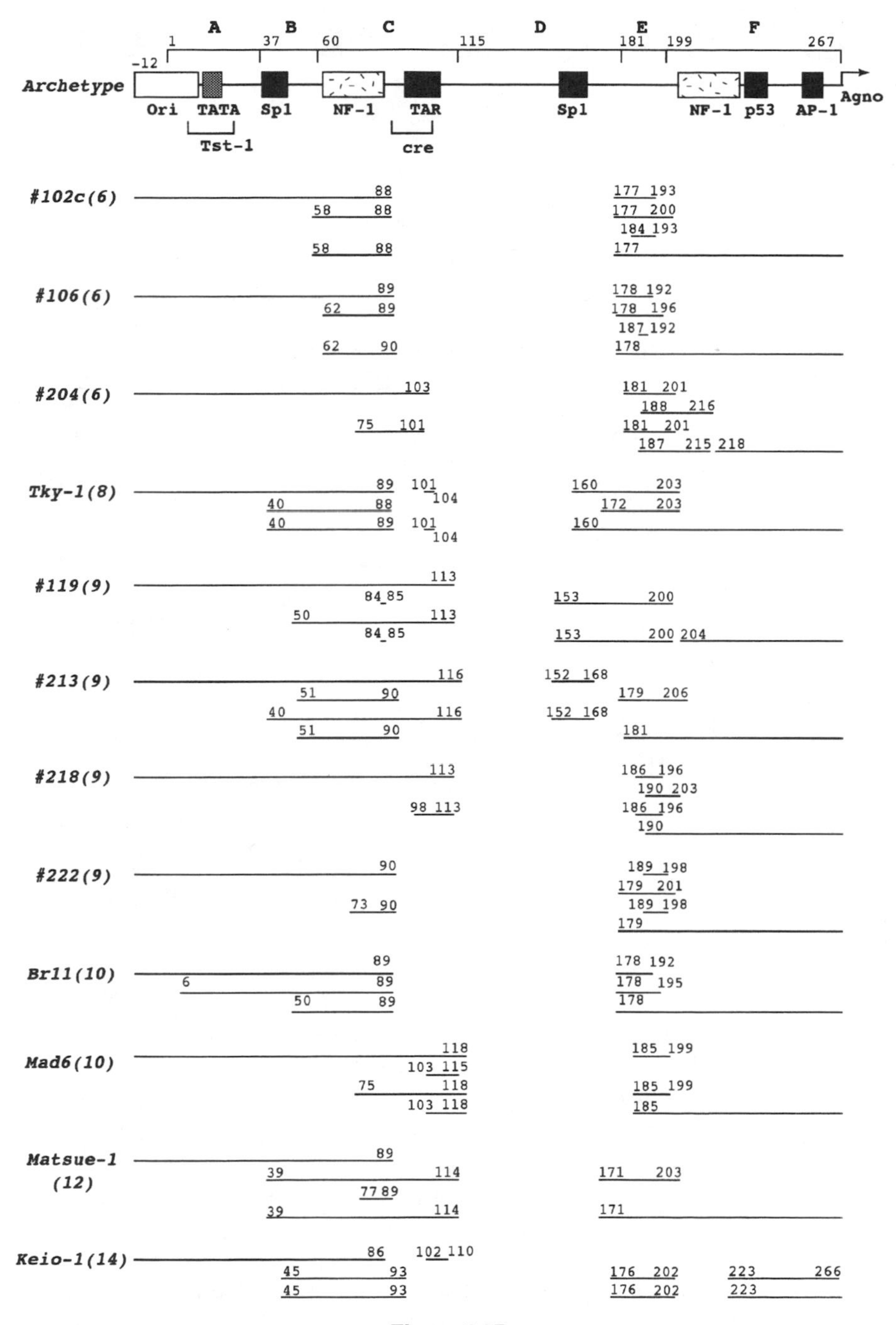
A
B
C
D
E
F
1
37
60
115
181
199
267
-12
Archetype
Ori
TATA
Sp1
NF-1
TAR
Sp1
NF-1
p53
AP-1
Agno
Tst-1
cre
#102c(6)
88
58 88
58 88
177 193
177 200
184 193
177
#106(6)
89
62 89
62 90
178 192
178 196
187 192
178
#204(6)
103
75 101
181 201
188 216
181 201
187 215 218
Tky-1(8)
89 101
104
40 88
40 89 101
104
160 203
172 203
160
#119(9)
113
84 85
50 113
84 85
153 200
153 200 204
#213(9)
116
51 90
40 116
51 90
152 168
179 206
152 168
181
#218(9)
113
98 113
186 196
190 203
186 196
190
#222(9)
90
73 90
189 198
179 201
189 198
179
Br11(10)
89
6 89
50 89
178 192
178 195
178
Mad6(10)
118
103 115
75 118
103 118
185 199
185 199
185
Matsue-1
(12)
89
39 114
77 89
39 114
171 203
171
Keio-1(14)
86 102 110
45 93
45 93
176 202
176 202
223 266
223

Figure 7.2B

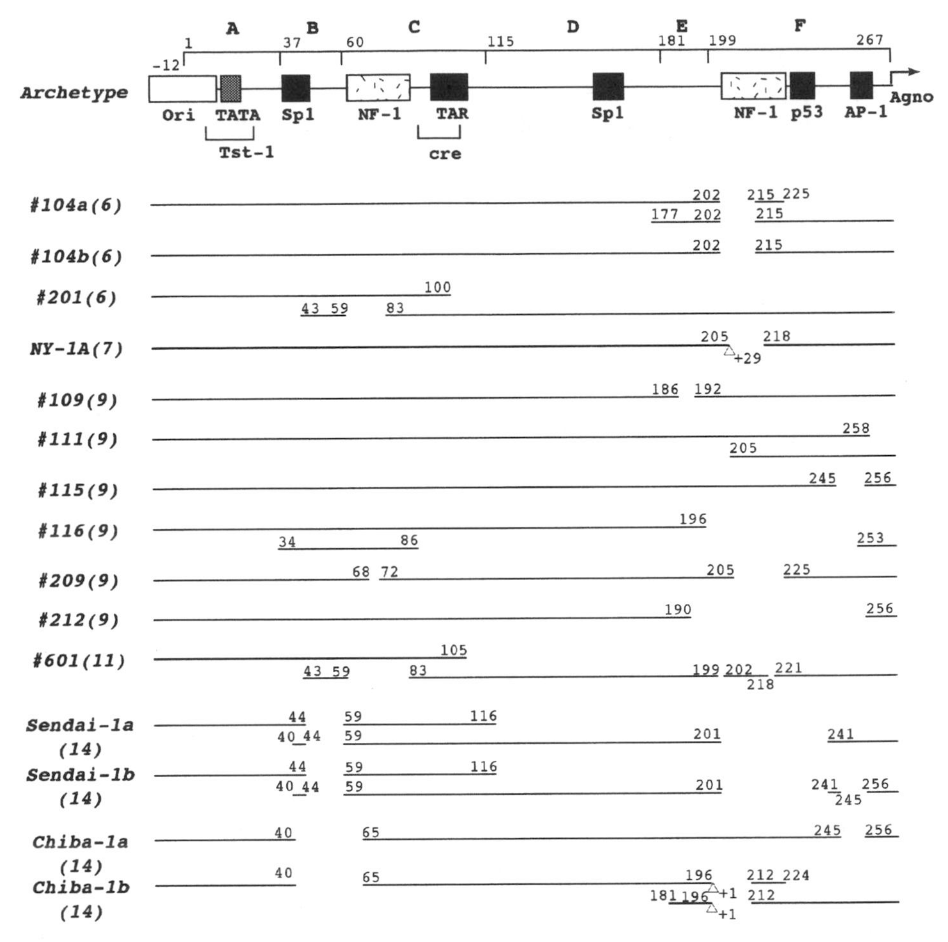

Figure 7.2C

and Stoner, 1993). Thus, most frequently targeted are the junctions of regions B/C, C/D, D/E, E/F, and the middle of region C. One reason for the selection of these particular areas is the pressure to retain sequences necessary for promoter function. The TATA box and the origin-proximal half of region C appear to be indispensable (Fig. 7.2). The latter contains an NF-1 binding site and a CRE-like element (Fig. 7.1). Other transcription factor recognition elements exist in other parts of the archetype regulatory region as well, but are not consistently retained (see Chapter 6).

The existence of preferred breakage and rejoining may be conferred by one of the cellular recombination activities, as discussed by Ault and Stoner (1993). Nevertheless, it is possible that junctions between B/C and C/D are not preferentially targeted, as suggested by Kitamura et al. (1994b). Thus, variants in

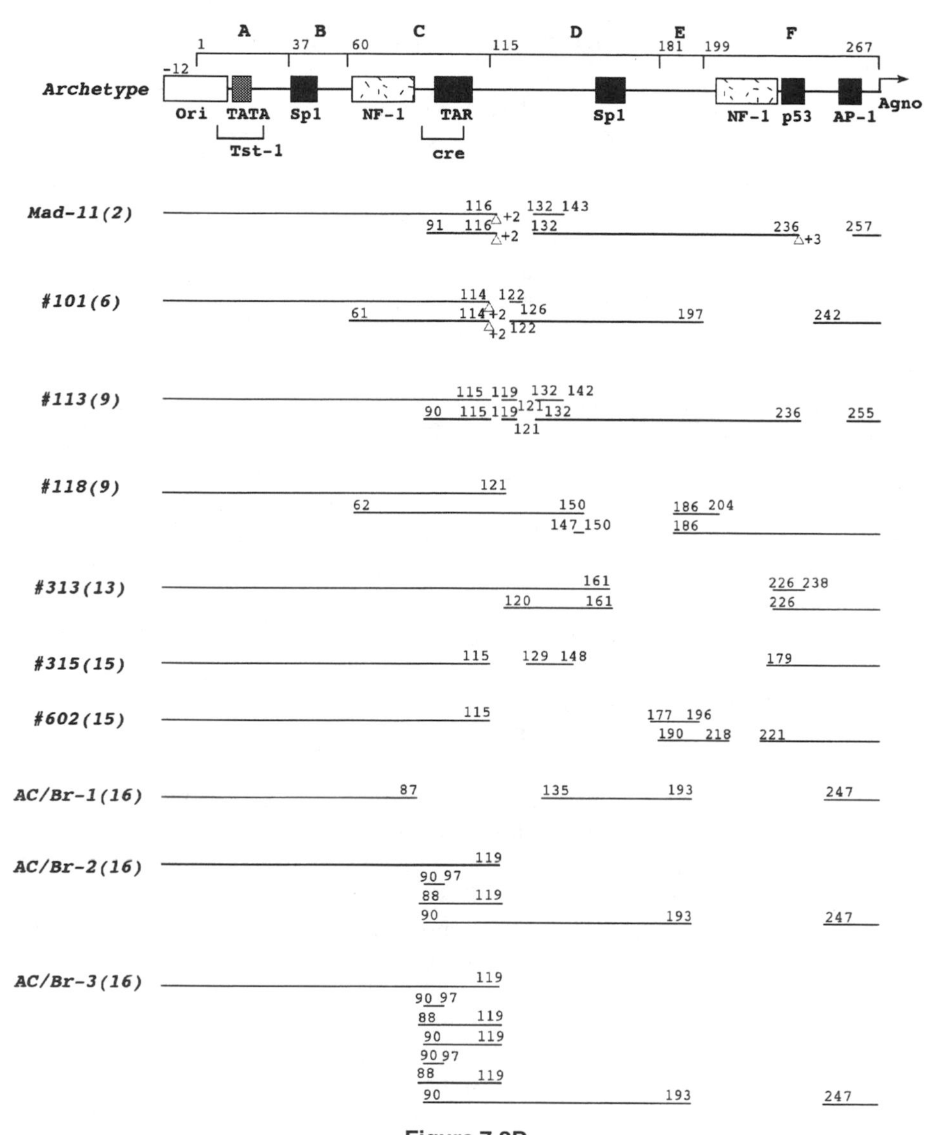

Figure 7.2D

which region C is duplicated and region D is deleted may be generated as a rare event. Once generated, however, these variants may become a major population, if they have selective replication advantage over the others in the target tissue (brain).

Ault and Stoner (1993) suggested that total length constraints may limit the lengths of segments to be duplicated or deleted. This factor may not be very

Table 7.1. Affected Elements in PML-Type Regulatory Regions Carrying Only Deletions

Isolate	Deletions[a]	Affected Elements[b]	Reference for Isolate
104b	nt 203–214	Second NF-1	Ault and Stoner (1993)
109	nt 187–191	Unknown	Agsotini et al. (1997a)
115	nt 246–255	AP-1	Agsotini et al. (1997a)
209	nt 69–71	First NF-1	Agsotini et al. (1997a)
	nt 206–224	Second NF-1, p53	
212	nt 191–255	Second NF-1, p53, AP-1	Agsotini et al. (1997a)
Chiba-1a	nt 41–64	First Sp-1, first NF-1	Sugimoto et al. (1998)
	nt 246–255	AP-1	

[a]Deletion end points are shown. See Figure 7.1 for nucleotide numbers.
[b]The position of each element is shown in Figure 7.1.

rigid, however, because the length of a rearranged regulatory region (213) was increased by 172 bp and that of another (212) was decreased by 65 bp (Fig. 7.3).

6. MOLECULAR MECHANISM OF REGULATORY REGION REARRANGEMENT

It remains unclear what molecular mechanism operates to generate rearranged regulatory regions. As described above, however, it is very likely that the NY-1A regulatory region was produced by nonhomologous recombination between two JCV DNA molecules. Whether JCV usually uses a host recombination system to cause deletions and duplications is not clear.

Yoshiike and Takemoto (1986) proposed an interesting mechanism of BK virus (BKV) regulatory region rearrangement. JCV and BKV are related polyomaviruses infecting only humans, and their overall genome structures are very similar (see Chapters 6 and 14). The archetype regulatory region was found in BKV earlier than in JCV (Rubinstein and Harley 1989; Rubinstein et al., 1987;

Figure 7.3. (**A**) A possible pathway that might have generated Mad-1 regulatory region. The two bracketed regions of the archetype were first deleted to generate the intermediate structure. The bracketed region of the intermediate was then duplicated to generate the Mad-1 regulatory region. (**B**) A possible pathway that might have generated the Matsue-1 regulatory region. The bracketed region of the archetype was first deleted to generate intermediate 1. The bracketed region of intermediate 1 was then duplicated to generate intermediate 2. Finally, the bracketed region of intermediate 1 was duplicated to generate Matsue-1 regulatory region. (**C**) A possible pathway that might have generated NY-1A regulatory region. Nonhomologous recombination occurred between the archetype regulatory region and the large T antigen gene at the indicated sites.

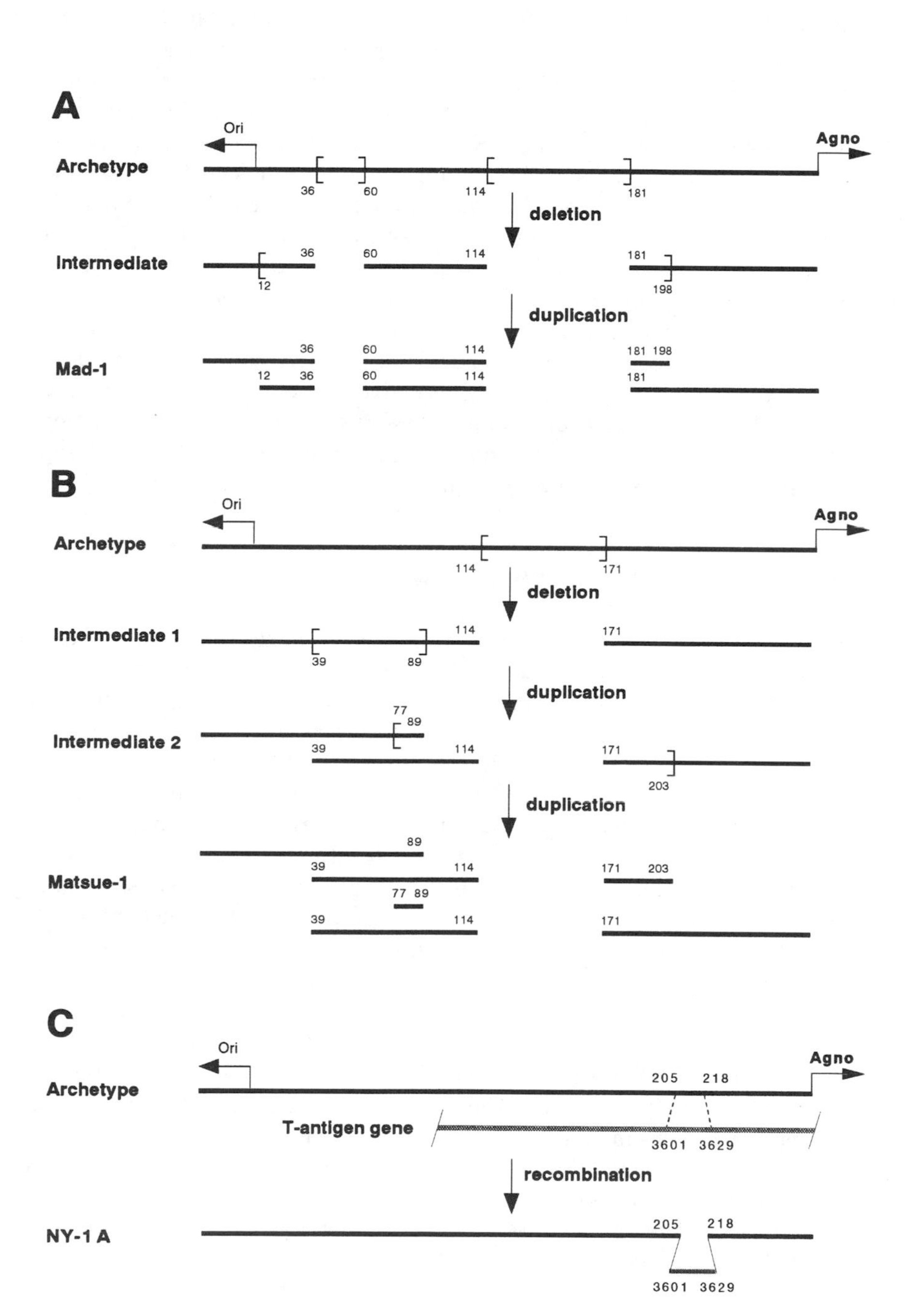
A
Ori
Agno
Archetype
36
60
114
181
deletion
Intermediate
36
60
114
181
12
198
duplication
Mad-1
36
60
114
181 198
12 36
60
114
181
B
Ori
Agno
Archetype
114
171
deletion
Intermediate 1
114
171
39
89
duplication
77
89
Intermediate 2
39
114
171
203
duplication
89
39
114
171 203
Matsue-1
77 89
39
114
171
C
Ori
Agno
Archetype
205 218
T-antigen gene
3601 3629
recombination
NY-1 A
205 218
3601 3629

Sugimoto et al., 1989, 1990; Yoshiike and Takemoto, 1986). The BKV archetype undergoes deletions and duplications, particularly during passage of viruses in cell culture. The mechanism suggested by Yoshiike and Takmoto (1986) is as follows. Recombination may occur between two newly synthesized daughter segments of a replicating DNA molecule at nonhomologous points. The resulting molecule will be a dimer composed of one molecule with deletion and the other with duplication. Dimers may be converted to monomers by homologous recombination that may subsequently occur at any two points within a dimer. It is possible that this mechanism also functions in JCV regulatory region rearrangement.

Another possible mechanism is an error in viral DNA replication. If a replicating point is shifted downstream along the template and replication is restarted, a deletion will be generated. Alternatively, if a replicating point goes back along the template and replication is restarted, a duplication will be generated. However, it is unlikely that this mechanism operates in JCV DNA replication that uses the strictly controlled host DNA replication system.

7. PML-TYPE JCV EVOLVE WITHIN PATIENTS

In the argument given above, we assumed that the genesis of JCV with rearranged regulatory sequences occurred within the patients. This assumption was examined by Iida et al. (1993). They constructed a phylogenetic tree for seven archetype and seven PML-type strains using DNA sequence data on the VP1 (major capsid protein) gene. However, it was recently shown that the whole-genome approach to phylogeny reconstruction offers significant improvement over earlier studies that were limited to partial JCV sequences (Hatwell and Sharp, 2000; Jobes et al., 1998). We therefore performed phylogenetic analysis of 28 JCV isolates, including 6 PML types, complete DNA sequences of which were reported previously (Agostini et al., 1998a,d; Frisque et al. 1994; Kato et al., 2000; Loeber and Dörries, 1988). A phylogenetic tree (Fig. 7.4) constructed by the neighbor-joining method (Saitou and Nei, 1987) revealed that the 28 isolates diverged into eight previously described genotypes (EU, Af1, Af2, SC, CY, MY, B1-b, and B1-c) (Guo et al., 1998; Jobes et al., 1998; Sugimoto et al. 1997). Five (EU, Af1, CY, MY, and B1-c) contained both archetype and PML-type isolates. Thus, we concluded that PML-type isolates are polyphyletic in origin and do not constitute a unique lineage. This conclusion supports the view that PML-type JCVs are generated from archetype JCVs during persistence in the host.

Of the three genotypes (Af2, SC, and B1-b) that contained only archetypes in the phylogenetic tree (Fig. 7.4), a PML-type JCV was detected from the genotype Af2 (Type 3) by analyzing its partial genome sequences (Stoner et al., 1998). However, there have been no reports of the isolation of PML-type JCVs belonging to genotypes Af3, SC, B1-a, B1-b or B1-d. It remains to be clarified whether archetype JCVs belonging to these genotypes are resistant to

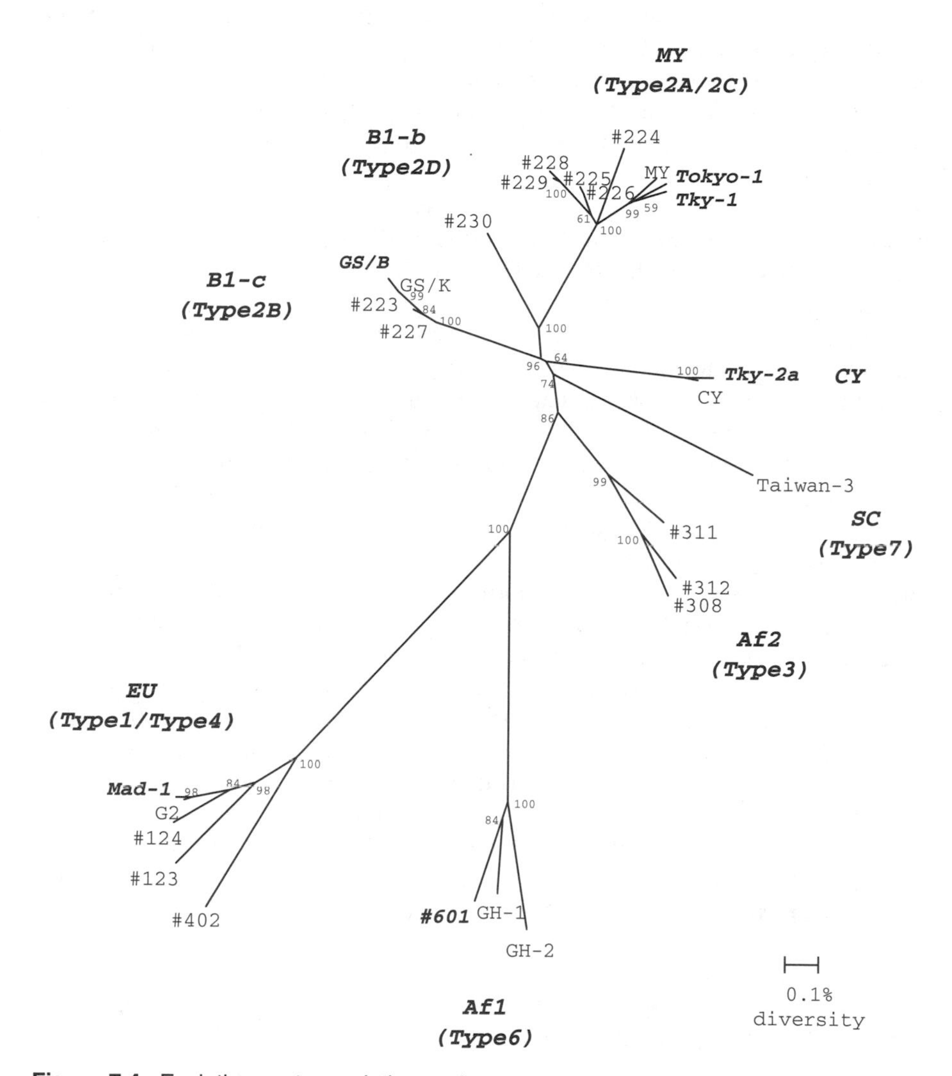

Figure 7.4. Evolutionary tree relating archetype and PML-type JCV isolates. From reported complete JCV DNA sequences, an unrooted evolutionary tree was constructed using the neighbor-joining method (Saitou and Nei, 1987). PML-type isolates are shown in boldface and italics. A genotype is indicated for each cluster according to the classification system of Sugimoto et al. (1997) and Guo et al. (1998) and that of Jobes et al. (1998) (the latter is indicated in parentheses). Numbers at the nodes show bootstrap confidence levels obtained by 100 replicates. The analyzed sequences are from Agostini et al. (1997b, 1998a,b,d), Kato et al. (2000), Loeber and Dörries (1988), and Ou et al. (1997).

conversion to PML-type counterparts. In this connection, it is relevant to mention the study by Agostini et al. (1998c), who reported that the proportion of the JCV genotype B1-c (or Type 2B) in the PML brain (36%) was significantly increased relative to its occurrence in control urine samples (5.9%). This finding suggested that the change of the JCV regulatory region may be influenced by the genotypes of JCV themselves or by the human populations carrying JCV.

8. LACK OF DISEASE-SPECIFIC AMINO ACID CHANGES IN THE VIRAL PROTEINS OF PML-TYPE JCVS

Although the structural differences in the regulatory region between PML-type and archetype isolates are striking, it is possible that PML-type JCVs undergo some functionally significant mutations in their protein-coding regions. Loeber and Dörries (1988) molecularly isolated two JCV variants in the same PML patient, one from the kidney (variant GS/K) and the other from the brain (variant GS/B). GS/K carried an archetype regulatory region, while GS/B carried a unique rearrangement that may have been produced from the GS/K regulatory region. There were two nucleotide differences in their coding regions, one in the VP1 gene and the other in a portion shared by the small t and large T antigen genes. Although the latter nucleotide difference caused no amino acid difference, the former caused a change from Leu to GS/K to Phe in GS/B. This amino acid change does not appear to be associated with PML because another PML-type isolate (Mad-1) did not carry this change. Kato et al. (2000) further examined whether PML-associated amino acid changes occurred in the viral proteins of PML-type JCV. Amino acid sequences of individual viral proteins were deduced from complete DNA sequences and were compared among 16 isolates (6 PML types and 10 archetypes). From the data obtained, it was concluded that PML-associated amino acid changes did not occur in the viral proteins of PML-type JCV.

9. BIOLOGICAL SIGNIFICANCE OF REGULATORY REGION REARRANGEMENT

It has been postulated that the nonpathogenic archetype form is converted to the pathogenic PML-type, which acquires a higher growth capacity in glial cells (i.e., oligodendrocytes) in the brain (Iida et al., 1993; Loeber and Dörries, 1988; Shah, 1996; Yogo et al., 1990, 1991b). A few studies were carried out to examine this hypothesis.

Sock et al. (1996) compared, in glial and nonglial cells, the expression of a reporter gene driven by early- and late-gene promoters present in three typical archetype regulatory regions and two PML-type regulatory regions (Mad-1 and GS/B) with "long duplicate" structures (Fig. 7.2A). Surprisingly, early- and late-gene promoters from not only PML-type but also archetype strains exhibited significantly higher activity in glial than in nonglial cells. Sock et al. (1996) concluded that the archetype is more or less competent to infect the brain without the need for modification of the viral regulatory region. This conclusion argues against any hypothesis postulating that regulatory region rearrangements would be required for infection in oligodendrocytes in the brain. This conclusion, however, does not exclude the possibility that JCV is endowed with more active promoters by regulatory region rearrangement that may occur during viral replication in vivo in glial cells.

Ault (1997) compared the replicating and promoter activities in human glial cells between one archetype and four PML-type regulatory regions (Mad-1, Nos. 104a, 203, and 102). According to the classification system of Ault and Stoner (1993), the regulatory regions of Mad-1/203, 102, and 104a belong to "long duplicate," "short triplicate," and "D retaining" (see Fig. 7.2). All five regulatory regions demonstrated similar levels of DNA replicating activity (this was also reported by Sock et al. [1996] and Daniel et al. [1996]). Surprisingly, the reporter gene expression differed within a fivefold range, and the archetype was intermediate in strength to the PML-type regulatory regions. This finding indicated that some PML types are more active, but some are less active, than the archetype promoter.

10. DUAL REARRANGEMENT MODEL

Major et al. (1992) proposed a hypothesis that some peripheral blood lymphocytes (PBL) harbor JCV in a latent form and that upon activation they carry JCV to the brain. Extending this hypothesis, we propose a dual rearrangement model that can explain both the presence of a variety of PML-type regulatory regions (Fig. 7.2) and the findings reported by Sock et al. (1996) and Ault (1997).

After primary infection, JCV multiplies without severe restriction in infected hosts. This viral growth generates JCV variants with rearranged regulatory regions. Of these variants, those with decreased growth capacity persist in PBL without being detected by the immune surveillance established by stimulation of initial viral growth. Rearranged regulatory regions produced by the initial event are probably those carrying only deletions.

Upon a decrease in host immune activity, the persisting JCV variants begin to replicate actively, resulting in expansion of JCV infection in the peripheral blood. Some of the infected lymphocytes enter the central nervous system (CNS), and transmission of JCV from lymphocytes to more favorable cells (i.e., oligodendrocytes) occurs. In the course of viral replication in the CNS, secondary regulatory region rearrangement frequently, but not always, occurs to generate variants with more complicated regulatory regions that now acquire more active promoters for early and late transcription. This highly pathogenic variant causes rapid expansion of brain lesions. Rearranged regulatory regions produced by the second event in the CNS include those belonging to "long duplicate" and "short triplicate" groups (Fig. 7.3A,B).

11. CONCLUSIONS

The findings described in this chapter can be summarized as the archetype concept as formulated below:

1. JCV with the archetype regulatory sequence is circulating in the human population.
2. The archetype regulatory sequence is highly conserved, in marked contrast to the hypervariable regulatory sequences of PML-derived isolates.
3. Each of the PML-type regulatory sequences is produced from the archetype by deletion and duplication or by deletion alone.
4. The shift of the regulatory region from archetype to PML type occurs during persistence in the host.
5. PML-type JCVs never return to the human population.

With respect to the biologic significance of regulatory region rearrangement, we have presented a dual rearrangement model postulating two independent rearrangements, one occurring in an immunocompetent state and the other in an immunosuppressed state. This model remains to be tested in future studies.

REFERENCES

Agostini HT, Ryschkewitsch CF, Stoner GL (1996): Genotype profile of human polyomavirus JC excreted in urine of immunocompetent individuals. J Clin Microbiol 34:159–164.

Agostini HT, Ryschkewitsch CF, Singer EJ, Stoner GL (1997a): JC virus regulatory region rearrangements and genotypes in progressive multifocal leukoencephalopathy: Two independent aspects of virus variation. J Gen Virol 78:659–664.

Agostini HT, Ryschkewitsch CF, Brubaker GR, Shao J, Stoner GL (1997b): Five complete genomes of JC virus Type 3 from Africans and African Americans. Arch Virol 142:637–655.

Agostini HT, Ryschkewitsch CF, Stoner GL (1998a): Complete genome of a JC virus genotype Type 6 from the brain of an African American with progressive multifocal leukoencephalopathy. J Hum Virol 1:267–272.

Agostini HT, Ryschkewitsch CF, Stoner GL (1998b): JC virus Type 1 has multiple subtypes: Three new complete genomes. J Gen Virol 79:801–805.

Agostini HT, Ryschkewitsch CF, Singer EJ, Baumhefner RW, Stoner GL (1998c): JC virus Type 2B is found more frequently in brain tissue of progressive multifocal leukoencephalopathy patients than in urine from controls. J Hum Virol 1:200–206.

Agostini HT, Shishido-Hara Y, Baumhefner RW, Singer EJ, Ryschkewitsch CF, Stoner GL (1998d): JC virus Type 2: Definition of subtypes based on DNA sequence analysis of ten complete genomes. J Gen Virol 79:1143–1151.

Ault GS (1997): Activity of JC virus archetype and PML-type regulatory regions in glial cells. J Gen Virol 78:163–169.

Ault GS, Stoner GL (1993): Human polyomavirus JC promoter/enhancer rearrangement patterns from progressive multifocal leukoencephalopathy brain are unique derivatives of a single archetypal structure. J Gen Virol 74:1499–1507.

Berger JR, Major EO (1999): Progressive multifocal leukoencephalopathy. Semin Neurol 19:193–200.

Chang D, Sugimoto C, Wang M, Tsai RT, Yogo Y (1999): JC virus genotypes in a Taiwan aboriginal tribe (Bunun): Implications for its population history. Arch Virol 144:1081–1090.

Chang D, Tsai RT, Wang M, Ou WC (1996a): Different genotypes of human polyomaviruses found in patients with autoimmune diseases in Taiwan. J Med Virol 48:204–209.

Chang D, Wang M, Ou WC, Lee MS, Ho HN, Tsai RT (1996b): Genotypes of human polyomaviruses in urine samples of pregnant women in Taiwan. J Med Virol 48: 95–101.

Chima SC, Agostini HT, Ryschkewitsch CF, Lucas SB, Stoner GL (1999): Progressive multifocal leukoencephalopathy and JC virus genotypes in West African patients with acquired immunodeficiency syndrome: A pathologic and DNA sequence analysis of 4 cases. Arch Pathol Lab Med 123:395–403.

Daniel AM, Swenson JJ, Mayreddy RPR, Khalili K, Frisque RJ (1996): Sequences within the early and late promoters of archetype JC virus restrict viral DNA replication and infectivity. Virology 216:90–101.

Flaegstad T, Sundsfjord A, Arthur RR, Pedersen M, Traavik T, Subramani S (1991): Amplification and sequencing of the control regions of BK and JC virus from human urine by polymerase chain reaction. Virology 180:553–560.

Frisque RJ, Bream GL, Cannella MT (1984): Human polyomavirus JC virus genome. J Virol 51:458–469.

Gallia GL, Houff SA, Major EO, Khalili K (1997): Review: JC virus infection of lymphocytes—Revisited. J Infect Dis 176:1603–1609.

Grinnell BW, Padgett BL, Walker DL (1983): Comparison of infectious JC virus DNAs cloned from human brain. J Virol 45:299–308.

Guo J, Sugimoto C, Kitamura T, Ebihara H, Kato A, Guo Z, Liu J, Zheng SP, Wang YL, Na YQ, Suzuki M, Taguchi F, Yogo Y. (1998): Four geographically distinct genotypes of JC virus are prevalent in China and Mongolia: Implications for the racial composition of modern China. J Gen Virol 79:2499–2505.

Guo J, Kitamura T, Ebihara H, Sugimoto C, Kunitake T, Takehisa J, Na YQ, Al-Ahdal MN, Hallin A, Kawabe K, Taguchi F, Yogo Y (1996): Geographical distribution of the human polyomavirus JC virus types A and B and isolation of a new type from Ghana. J Gen Virol 77:919–927.

Hatwell JN, Sharp PM (2000): Evolution of human polyomavirus JC. J Gen Virol 81: 1191–1200.

Iida T, Kitamura T, Guo J, Taguchi F, Aso Y, Nagashima K, Yogo Y (1993): Origin of JC polyomavirus variants associated with progressive multifocal leukoencephalopathy. Proc Natl Acad Sci USA 90:5062–5065.

Jobes DV, Chima SC, Ryschkewitsch CF, Stoner GL (1998): Phylogenetic analysis of 22 complete genomes of the human polyomavirus JC virus. J Gen Virol 79:2491–2498.

Kato K, Guo J, Taguchi F, Daimaru O, Tajima M, Haibara H, Matsuda J, Sumiya M, Yogo Y (1994): Phylogenetic comparison between archetypal and disease-associated JC virus isolates in Japan. Jpn J Med Sci Biol 47:167–178.

Kato A, Sugimoto C, Zheng H-Y, Kitamura T, Yogo Y (2000): Lack of disease-specific amino acid changes in the viral proteins of JC virus isolates from the brain with progressive multifocal leukoencephalopathy. Arch Virol 145:2173–2182.

Kitamura T, Aso Y, Kuniyoshi N, Hara K, Yogo Y (1990): High incidence of urinary JC virus excretion in nonimmunosuppressed older patients. J Infect Dis 161:1128–1133.

Kitamura T, Kunitake T, Guo J, Tominaga T, Kawabe K, Yogo Y (1994a): Transmission of the human polyomavirus JC virus occurs both within the family and outside the family. J Clin Microbiol 32:2359–2363.

Kitamura T, Satoh K, Tominaga T, Taguchi F, Tajima A, Suzuki K, Aso Y, Yogo Y (1994b): Alteration in the JC polyomavirus genome is enhanced in immunosuppressed renal transplant patients. Virology 198:341–345.

Loeber G, Dörries K (1988). DNA rearrangements in organ-specific variants of polyomavirus JC strain GS. J Virol 62:1730–1735.

Major EO, Amemiya K, Tornatore CS, Houff SA, Berger JR (1992): Pathogenesis and molecular biology of progressive multifocal leukoencephalopathy, the JC virus-induced demyelinating disease of the human brain. Clin Microbiol Rev 5:49–73.

Markowitz RB, Eaton BA, Kubik MF, Latorra D, McGregor JA, Dynan WS (1991): BK virus and JC virus shed during pregnancy have predominantly archetypal regulatory regions. J Virol 65:4515–4519.

Martin JD, King DM, Slauch JM, Frisque RJ (1985): Differences in regulatory sequences of naturally occurring JC virus variants. J Virol 53:306–311.

Matsuda M, Jona M, Yasui K, Nagashima K (1987): Genetic characterization of JC virus Tokyo-1 strain, a variant oncogenic in rodents. Virus Res 7:159–168.

Monaco MCG, Jensen PN, Hou J, Durham LC, Major EO (1998): Detection of JC virus DNA in human tonsil tissue: Evidence for site of initial viral infection. J Virol 72:9918–9923.

Newman JT, Frisque RJ (1997): Detection of archetype and rearranged variants of JC virus in multiple tissues from a pediatric PML patient. J Med Virol 52:243–252.

Newman JT, Frisque RJ (1999): Identification of JC virus variants in multiple tissues of pediatric and adult PML patients. J Med Virol 58:79–86.

Ou WC, Tsai RT, Wang M, Fung CY, Hseu TH, Chang D (1997): Genomic cloning and sequence analysis of Taiwan-3 human polyomavirus JC virus. J Formos Med Assoc 96:511–516.

Padgett BL, Walker DL (1973): Prevalence of antibodies in human sera against JC virus, an isolate from a case of progressive multifocal leukoencephalopathy. J Infect Dis 127:467–470.

Padgett BL, Walker DL (1976): New human papovaviruses. Prog Med Virol 22:1–35.

Padgett BL, Walker DL, Zu Rhein GM, Eckroade RJ, Dessel BH (1971): Cultivation of papova-like virus from human brain with progressive multifocal leukoencephalopathy. Lancet 1:1257–1260.

Rentier-Delrue F, Lubiniecki A, Howley PM (1981): Analysis of JC virus DNA purified directly from human progressive multifocal leukoencephalopathy brains. J Virol 38: 761–769.

Rubinstein R, Harley EH (1989): BK virus DNA cloned directly from human urine confirms an archetypal structure for the transcriptional control region. Virus Genes 2:157–165.

Rubinstein R, Pare N, Harley EH (1987). Structure and function of the transcriptional control region of nonpassaged BK virus. J Virol 61:1747–1750.

Saitou N, Nei M (1987): The neighbor-joining method: A new method for reconstructing phylogenetic trees. Mol Biol Evol 4:406–425.

Shah KV (1996): In Fields BN, Knipe DM, Howley PM, Eds. Fields Virology. Lippincott-Raven: New York, pp 2027–2043.

Sock E, Renner K, Feist D, Leger H, Wegner M (1996): Functional comparison of PML-type and archetype strains of JC virus. J Virol 70:1512–1520.

Stoner GL, Agostini HT, Ryschkewitsch CF, Mazlo M, Gullotta F, Wamukota W, Lucas S (1998): Detection of JC virus in two African cases of progressive multifocal leukoencephalopathy including identification of JCV Type 3 in a Gambian AIDS patient. J Med Microbiol 47:733–742.

Sugimoto C, Hara K, Taguchi F, Yogo Y (1989): Growth efficiency of naturally occurring BK virus variants in vivo and in vitro. J Virol 63:3195–3199.

Sugimoto C, Hara K, Taguchi F, Yogo Y (1990): Regulatory DNA sequence conserved in the course of BK virus evolution. J Mol Evol 31:485–492.

Sugimoto C, Ito D, Tanaka K, Matsuda H, Saito H, Sakai H, Sakai H, Fujihara K, Itoyama Y, Yamada T, Kira J, Matsumoto R, Mori M, Nagashima K, Yogo Y (1998): Amplification of JC virus regulatory DNA sequences from cerebrospinal fluid: Diagnostic value for progressive multifocal leukoencephalopathy. Arch Virol 143: 249–262.

Sugimoto C, Kitamura T, Guo J, Al-Ahdal MN, Shchelkunov SN, Otova B, Ondrejka P, Chollet J-Y, El-Safi S, Ettayebi M, Grésenguet G, Kocagöz T, Chaiyarasamee S, Thant KZ, Thein S, Moe K, Kobayashi N, Taguchi F, Yogo Y (1997): Typing of urinary JC virus DNA offers a novel means of tracing human migrations. Proc Natl Acad Sci USA 94:9191–9196.

Takahashi H, Yogo Y, Furuta Y, Takada A, Irie T, Kasai M, Sano K, Fujioka Y, Nagashima K (1992): Molecular characterization of a JC virus (Sap-1) clone derived from a cerebellar form of progressive multifocal leukoencephalopathy. Acta Neuropathol 83:105–112.

Tominaga T, Yogo Y, Kitamura T, Aso Y (1992): Persistence of archetypal JC virus DNA in normal renal tissue derived from tumor-bearing patients. Virology 186:736–741.

Tsai RT, Wang M, Ou WC, Lee YL, Li SY, Fung CY, Huang YL, Tzeng TY, Chen Y, Chang D (1997): Incidence of JC viruria is higher than that of BK viruria in Taiwan. J Med Virol 52:253–257.

Wakutani Y, Shimizu Y, Miura H, Nakashima K, Nakano T, Ohama E, Sugimoto C, Yogo Y, Kobayashi Y, Nagashima K (1998): A case of brain-biopsy proven progressive multifocal leukoencephalopathy: Pathological findings and analysis of JC virus regulatory region. Neuropathology 18:347–351.

Yogo Y, Guo J, Iida T, Satoh K, Taguchi F, Takahashi H, Hall WW, Nagashima K (1994): Occurrence of multiple JC virus variants with distinctive regulatory sequences in the brain of a single patient with progressive multifocal leukoencephalopathy. Virus Genes 8:99–105.

Yogo Y, Iida T, Taguchi F, Kitamura T, Aso Y (1991a): Typing of human polyomavirus JC virus on the basis of restriction fragment length polymorphisms. J Clin Microbiol 29:2130–2138.

Yogo Y, Kitamura T, Sugimoto C, Ueki T, Aso Y, Hara K, Taguchi F (1990): Isolation of a possible archetypal JC virus DNA sequence from nonimmunocompromised individuals. J Virol 64:3139–3143.

Yogo Y, Kitamura T, Sugimoto C, Hara K, Iida T, Taguchi F, Tajima A, Kawabe K, Aso Y (1991b): Sequence rearrangement in JC virus DNAs molecularly cloned from immunosuppressed renal transplant patients. J Virol 65:2422–2428.

Yoshiike K, Takemoto KK (1986): Studies with BK virus and monkey lymphtropic papovavirus. The Papovaviridae, Volume 1, The Polyomaviruses; Salzman NP, Ed.; Plenum Press: New York, pp 295–326.

8

SYNTHESIS AND ASSEMBLY OF POLYOMAVIRUS VIRIONS

YUKIKO SHISHIDO-HARA, M.D., PH.D., and
KAZUO NAGASHIMA, M.D., PH.D.

1. INTRODUCTION

The polyomaviruses are nonenveloped viruses whose capsid is composed of three capsid proteins: a major capsid protein, VP1, and minor capsid proteins, VP2 and VP3. The human polyomaviruses, JC (JCV) and BK (BKV) are closely related to simian virus 40 (SV40) and distantly related to murine polyoma virus. By X-ray crystallography, it has been shown that SV40 and murine polyoma virus have a T = 7d icosahedral structure, which is composed of 72 pentamers of VP1, with VP2 and VP3 inside in a specific stoichiometry (Liddington et al., 1991; Stehle et al., 1994, 1996; Stehle and Harrison, 1997).

JCV infects oligodendrocytes of human brain tissues and causes progressive multifocal leukoencephalopathy (PML), a fatal demyelinating disorder in the central nervous system. By electron microscopy, JCV has been identified as round particles and filamentous forms in the nuclei of infected oligodendrocytes (Fig. 8.1) (Silverman and Rubinstein, 1965; Zu Rhein and Chou, 1965). Distribution and arrangement of the viral particles are variable in cells. The round JCV particles, which are relatively uniform in size, are occasionally arranged in crystalloid array, but in other cases they are distributed randomly. Filamentous forms, which are about one-half to two-thirds the diameter of the round

Human Polyomaviruses: Molecular and Clinical Perspectives, Edited by Kamel Khalili and Gerald L. Stoner.
ISBN 0-471-39009-7

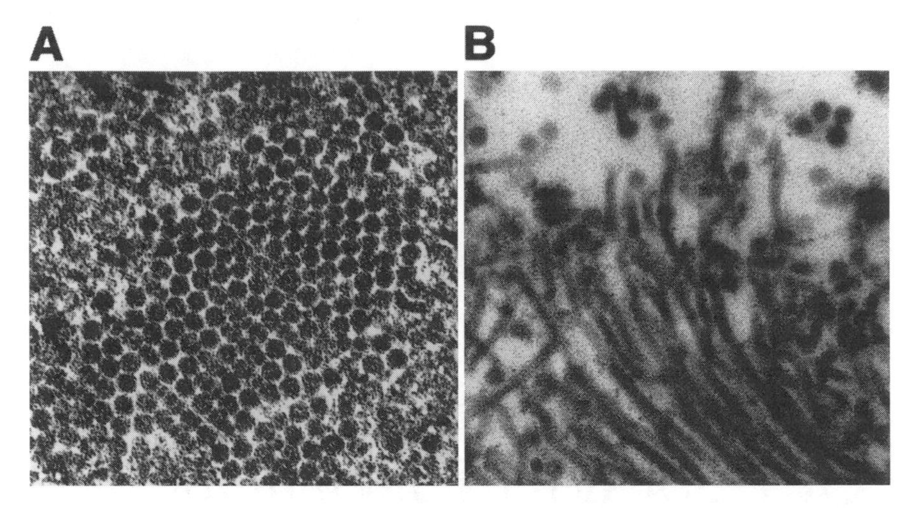

Figure 8.1. Ultrastructural features of JCV virions identified in PML patients' brains. (**A**) JCV round particles arranged in crystalloid array. JCV forms clusters in which the round virions are regularly arranged in a crystalloid array. Within the nucleus, clusters of regularly arranged round virions are isolated or fused with neighboring clusters. The round virions are quite homogeneous in size. (Courtesy of Drs. S. Takeda and H. Takahashi.) (**B**) JCV in filamentous forms. The filamentous forms are more variable in size than the round particles and arranged in strands, in spirals, or scattered irregularly with round particles. ×62,000.

particles, are arranged in strands, in spirals, or scattered irregularly with round particles. In some cells, both round and filamentous forms are distributed near the nuclear membrane. However, structures of the JCV virions and the mechanisms for their assembly are mostly unknown. The precise structures of the round and filamentous profiles have not been analyzed in a higher resolution. During the late stage of the virus replication cycle, the capsid proteins can be synthesized from alternatively spliced RNAs, transported from the cytoplasm to the nucleus, and assembled into progeny virions. Although the round particles are likely composed of the three capsid proteins, VP1, VP2, and VP3, in a specific stoichiometry, it is not known how expression of these component capsid proteins is regulated for efficient assembly into virions.

Recently, we reported unique features of JCV during the late stage of the virus replication cycle (Shishido-Hara et al., 2000). In cell culture systems, JCV replicates more slowly and less efficiently than BKV, SV40, and murine polyoma virus. This slow and inefficient replication of JCV has been a major barrier in studying the late stage of the virus replication cycle. To overcome this difficulty, we have developed a highly efficient eukaryotic expression system for the JCV capsid proteins by using pcDL-SRα296 (Takebe et al., 1988) and COS-7 cells (Gluzman, 1981). With this system, we have elucidated unique features of JCV in production of progeny virions. In this chapter, we review the regulated expression of the JCV capsid proteins, their nuclear transport, and assembly of virions. In comparison with SV40 and murine polyoma virus, the unique features of JCV are described, and possible future studies in elucidating the pathogenesis of PML are discussed.

2. THE JCV LATE PROTEINS: AGNOPROTEIN, VP1, VP2, AND VP3

JCV has a circular DNA genome, which includes two coding regions: the early region and the late region. In the late region, JCV encodes the four late proteins: agnoprotein, VP1, VP2, and VP3 (Fig. 8.2). Agnoprotein is a short peptide, which is encoded upstream of the capsid proteins. Agnoprotein is encoded in JCV, BKV (Rinaldo et al., 1998), and SV40 (Jay et al., 1981). However, agnoprotein has not been reported in murine polyoma virus. The major capsid protein VP1 and the minor capsid proteins VP2 and VP3 are encoded in an overlapping manner. Downstream of the coding sequence for the agnoprotein, the coding sequence for VP2 is encoded, with that for VP3 completely overlapped in the 3′ terminus in the same reading frames. Therefore, the coding sequence for VP3 is identical to two-thirds of the 3′ terminal sequence for VP2. The 3′ termini of the coding sequences for VP2 and VP3 partly overlap the 5′ terminus of that for VP1 in different reading frames.

JCV Tokyo-1 strain has been isolated from the brain tissue of a Japanese patient with PML (Nagashima et al., 1981; Matsuda et al., 1987). JCV Tokyo-1 has a genomic DNA that is 5128 bp in length (Shishido-Hara et al., 2000). The amino acid sequences of the four late proteins have been deduced based on the sequence homology to BKV and SV40: agnoprotein, 71 amino acids, nt 275–490; VP1, 354 amino acids, nt 1467–2531; VP2, 344 amino acids, nt 524–1558; and VP3, 225 amino acids, nt 881–1558. Antibodies against these four late proteins were prepared by immunizing with synthetic peptides whose sequences are identical to a part of the amino acid sequences deduced from the nucleotide sequence data. With these antibodies, expression of the late proteins was analyzed in PML brain tissues by immunohistochemistry. JCV agnoprotein is detected predominantly in the cytoplasm of infected oligodendrocytes (Fig. 8.3A). Subcellular distribution of JCV agnoprotein is similar to

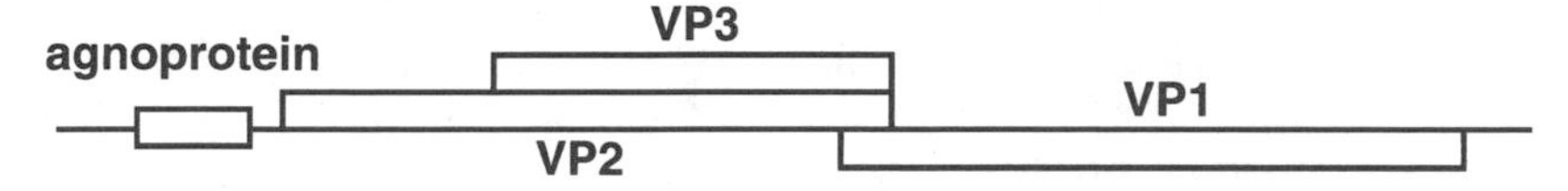

Figure 8.2. JCV genomic organization in the late region encoding the four late proteins: agnoprotein, VP1, VP2, and VP3. The coding sequence for agnoprotein is present in the 5′ end of the late region, followed by those for the three capsid proteins VP1, VP2, and VP3. The coding sequences for the capsid proteins are encoded in an overlapping manner. The coding sequence for VP2 completely includes that for VP3 in the 3′ terminus in the same reading frames. The 3′ termini of the coding sequences for VP2 and VP3 partly overlap the 5′ terminus of VP1 in different reading frames. Agnoprotein is present in JCV, BKV, and SV40, but not in murine polyoma virus.

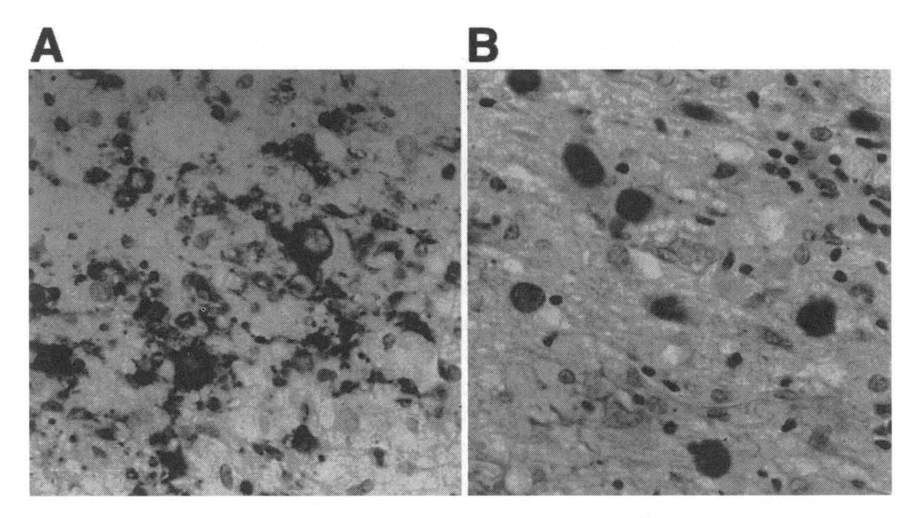

Figure 8.3. Immunostaining of PML brain tissues by using peptide antibodies against agnoprotein and VP1. (**A**) Expression of agnoprotein. Agnoprotein is widely expressed in the glial cytoplasmic processes in the demyelinating PML lesions. (**B**) Expression of VP1. VP1 was detected mostly in the nuclei of the infected glial cells by using the peptide antibodies against the residues in the potential loop structures (see text). Similarly, expression of VP2 or VP3 was also detected mostly in the nuclei of the infected cells.

that of BKV agnoprotein (Rinaldo et al., 1998) and SV40 agnoprotein (Nomura et al., 1983). All three capsid proteins, VP1, VP2, and VP3, are detected predominantly in the nucleus (Fig. 8.3B). Expression of VP2 and that of VP3 are not yet distinguished, because the antibody against VP3 also detects VP2, due to the identical amino acid sequence for VP3 to the carboxy-terminal sequence for VP2. However, these data consistently indicate that JCV encodes the four proteins, agnoprotein, VP1, VP2, and VP3, in the late region and that their amino acid sequences are consistent with those deduced from the nucleotide sequence.

3. AMINO ACID SEQUENCE COMPARISON OF THE LATE PROTEINS FOR JCV TO THOSE FOR THE OTHER POLYOMAVIRUSES: POTENTIAL STRUCTURES AND BIOLOGIC FUNCTIONS OF THE JCV LATE PROTEINS

Although the precise structure is not yet known for JCV virions, it is likely that JCV has some common features to SV40 and murine polyoma virus. The crystal structures of SV40 and murine polyoma virus have been reported (Rayment et al., 1982; Liddington et al., 1991; Stehle et al., 1994, 1996; Yan et al., 1996). In both viruses, the outer shell of the capsid is composed of 360 molecules of VP1 arranged in 12 pentameric and 60 hexameric pentamers on an icosahedral lattice. Inside the capsid, each VP1 pentamer is associated with a single molecule of either VP2 or VP3 (Chen et al., 1998). From the inner surface of the capsid, VP2 or VP3 is extended inwardly and likely anchored to the viral minichromosome (Liddington et al., 1991; Griffith et al., 1992; Stehle et al., 1996). In JCV Tokyo-1 strain, potential functions of the late proteins have been analyzed based on the amino acid sequence comparison with SV40 and murine polyoma virus (Fig. 8.4).

VP1 comprises the capsid surface structures. In SV40, the crystal structure has shown that each VP1 molecule displays elaborate loop structures that emanate outward from the β-sheet framework (Liddington et al., 1991). The external loops were suggested to provide principal antigenic structures, receptor binding sites, or domains responsible for hemagglutination. Based on the amino acid sequence comparison, the VP1 sequence for JCV is divergent from those for other polyomaviruses primarily in the potential loops and is relatively conserved in the potential framework. We selected the peptide sequences from the potential BC, DE, and HI loops of JCV VP1 and prepared antibodies. These antibodies efficiently recognized the enlarged nuclei of the JCV-infected oligodendrocytes of PML brain tissues (Fig. 8.3B). Similarly, Aoki et al. (1996) prepared antibodies against the two peptide sequences from the potential BC loop of JCV VP1, KSISISDTFE (10 residues) and KSISISDTFESDSPNRDM (18 residues; the common residues in JCV and BKV are underlined). The antibody against the 10 residue peptide detected JCV, but not BKV. In contrast, the antibody against the 18 residue peptide cross reacted with both JCV and BKV due to the larger number of common residues. Therefore, in both JCV and BKV, the potential loop structures can provide antigenic domains, and specific sequences in this region can define distinctive antigenic structures.

VP2 and VP3 are internal proteins. These proteins are not yet well analyzed in JCV. In the carboxy-terminal regions for VP2 and VP3, JCV has the corresponding sequences to VP1 interactive domain, a DNA binding domain, and a nuclear localization signal (NLS), which are determined in SV40 or murine polyoma virus (Gharakhanian and Kasamatsu, 1990; Clever and Kasamatsu, 1991; Chang et al., 1992a; Clever et al., 1993; Barouch and Harrison, 1994; Dean et al., 1995). These observations indicate that VP2 and VP3 for JCV has some common functions with those for SV40 and murine polyoma virus in the structure and assembly of virions.

Agnoprotein is not a component of the capsid. Because the SV40 mutant lacking agnoprotein can replicate (Resnick and Shenk, 1986), this protein is not essential for viral replication, but rather regulates efficient production of progeny virions. In JCV, BKV, and SV40, the sequences of agnoprotein are homologous in the first two-thirds of the amino terminus, but are divergent in the carboxy terminal one-third. In SV40, two types of antibodies against agnoprotein were prepared by immunizing the whole agnoprotein and by immunizing a synthetic peptide, VLRRLSRQASVKVR, identical to the 14 residues in the amino-terminal region. In a comparison of these two antibodies, it was noted that SV40 agnoprotein has multiple subpopulations, the majority of

A: VP1

```
JCV  MAPTKRKG.. ......ERKD ...PVQVPKL LIRGGVEVLE VKTGVDSITE VECFLTPEMG DP........ DEHLRGFS.K .SISISDTFE SDSPNKDMLP
BKV  MAPTKRKGEC PGAAPKKPKE ...PVQVPKL LIKGGVEVLE VKTGVDAITE VECFLNPEMG DP........ DENLRGFSLK ..LSAENDFS SDSPERKMLP
SV40 MAPTKRKGSC PGAAPKKPKE ...PVQVPKL VIKGGIEVLG VKTGVDSFTE VECFLNPQMG NP........ DEHQKGLS.K .SLAAEKQFT DDSPDKEQLP
PyV  MAP.KRKSGV SKCETKCTKA CPRPAPVPKL LIKGGMEVLD LVTGPDSVTE IEAFLNPRMG QPPTPESLTE GGQYYGWSRG INLATSDTWI ...PRNNTLP
          disordered       N-arm     β-A   α-A    AB       β-B                           BC loop

JCV  CYSVARIPLP NLNEDLTCGN ILMWEAVTLK TEVLGVTTLM NVHS..NG.Q ATHDNGAGKP VQGTSFHFFS VGGEALELQG VVFNYRTKYP DG......TI
BKV  CYSTARIPLP NLNEDLTCGN LLMWEAVTVQ TEVIGITSML NLHA...GSQ KVHEHGGGKP IQGSNFHFFA VGGEPLEMQG VLMNYRSKYP DG......TI
SV40 CYSVARIPLP NLNEDLTCGN ILMWEAVTVK TEVIGVTAML NLHS...GTQ KTHENGAGKP IQGSNFHFFA VGGEPLELQG VLANYRTKYP AQ......TV
PyV  TWSMAKSSFP CLNEDLTCDT LQMWEAVSVK TEVVGSGSLL DVHGFNKTHR FSKHKGNSTP VEGSQYHVFA GGGEPLDLQG LVTDARTKYK EEGVVTIKTI
        β-C      CD        β-D        α-B            DE loop                β-E                 EF loop

JCV  FPKNATVQSQ VMNTEHKAYL DKNKAYPVEC WVPDPTRNEN TRYFGTLTGG ENVPPVLHIT NTATTVLLDE FGVGPLCKGD NLYLSAVDVC GMFTNRSGS.
BKV  TPKNPTAQSQ VMNTDHKAYL DKNNAYPVEC WVPDPSRNEN ARYFGTFTGG ENVPPVLHVT NTATTVLLDE QGVGPLCKAD SLYVSAADIC GLFTNSSGT.
SV40 TPKNATVDSQ QMNTDHKAVL DKDNAYPVEC WVPDPSKNEN TRYFGTYTGG ENVPPVLHIT NTATTVLLDE QGVGPLCKAD SLYVSAVDIC GLFTNTSGT.
PyV  TKKDMVNKDQ VLNPISKAKL DKDGMYPVEI WHPDPAKNEN TRYFGNYTGG TTAPPVLQFT NTLTTVLLDE NGVGPLCKGE GLYLSCVDIM GWRVTRNYVS
                   EF loop                         β-F       β-G1   β-G2    GH loop      β-H       HI loop

JCV  QQWRGLSRYF KVQLRKRRVK NPYPISFLLT DLINRRTPRV DGQPMYGMDA QVEEVRVFEG TEELPGDPDM MRYVDRYGQL QTKML*
BKV  QQWRGLARYF KIRLRKRSVK NPYPISFLLS DLINRRTQRV DGQPMYGMES QVEEVRVFDG TERLPGDPDM IRYIDKQGQL QTKML*
SV40 QQWKGLPRYF KITLRKRSVK NPYPISFLLS DLINRRTQRV DGQPMIGMSS QVEEVRVYED TEELPGDPDM IRYIDEFGQT TTRMQ*
PyV  SLEKGFPRYF KITLRKRWVK NPYPMASLIS SLFNNMLPQV QGQPMEGENT QVEEVRVYDG TEPVPGDPDM TRYVDRFGKT KTVFPGN*
        β-I          ...        α-C       ...C arm...   β-J    ...C arm...     C-loop
```

B: VP2/VP3

```
     VP2
JCV  MGAALALLGD LVATVSEAAA ATGFSVAE.I AAGEAAATIE VEIASLATVE GITSTSEAIA AIGLTPETYA VITGAPGAV. ..AGFAALVQ TVTGGSAIAQ
BKV  MGAALALLGD LVASVSEAAA ATGFSVAE.I AAGEAAAAIE VQIASLATVE GITSTSEAIA AIGLTPQTYA VIAGAPGAI. ..AGFAALIQ TVSGISSLAQ
SV40 MGAALTLLGD LIATVSEAAA ATGFSVAE.I AAGEAAAAIE VQLASVATVE GLT.TSEAIA AIGLTPQAYA VISGAPAAI. ..AGFAALLQ TVTGVSAVAQ
PyV  MGAALTILVD LIEGLAEVST LTGLS.AEAI LSGEALAALD GEITAL.TLE GVMSSETALA TMGISEEVYG FVSTVPVFVN RTAGAIWLMQ TVQGASTIS.

                            VP3
JCV  LGYRFFADWD HKVSTVGLFQ QPAMALQLFN PEDYYDILFP GVNAFVNNIH YLDPRHWGPS LFSTISQAFW .NLVRDDLPS LTSQ..EI.Q RRTQKLFVET
BKV  VGYKFFDDWD HKVSTVGLYQ QSGMALELFN PDEYYDILFP GVNTFVNNIQ YLDPRHWGPS LFATISQALW .HVIRDDIPS ITSQ..EL.Q RRTERFFRDS
SV40 VGYRFFSDWD HKVSTVGLYQ QPGMAVDLYR PDDYYDILFP GVQTFVHSVQ YLDPRHWGPT LFNAISQAFW .RVIQNDIPR LTSQ..EL.E RRTQRYLRDS
PyV  LGIQRYLHNE E.VPTVNRN. ...MALIPWR DPALLDIYFP GVNQFAH..A LNVVHDWGHG LLHSVGRYVW QMVVQETQHR LEGAVRELTV RQT.HTFLDG

JCV  LARFLEETTW AIVNSP.... .......... VNLYNYISDY YSRLSPVRPS MVRQVAQR.E GTYISFGHSY TQSIDNADSI QEVTQRLDLK N..PNVQ...
BKV  LARFLEETTW TIVNAP.... .......... INFYNYIQQY YSDLSPIRPS MVRQVAER.E GTRVHFGHTY ..SIDDADSI EEVTQRMDLR ....NQQSVH
SV40 LARFLEETTW TVINAP.... .......... VNWYNSLQDY YSTLSPIRPT MVRQVANR.E GLQISFGHTY .DNIDEADSI QQVTERWEAQ SQSPNVQ...
PyV  LARLLENTRW VVSNAPQSAI DAINRGASSV SSGYSSLSDY YRQLG.LNPP QRRALFNRIE GSMGNGGPTP AAHIQDE... .......... ..........

JCV  SGEFIEKSFA PGGANQRSAP QWMLPLLLGL YG..TVTPAL EAYEDGPNKK KRRKE..... ...GPRASSK TSYKRRSRSS RS*
BKV  SGEFIEKTIA PGGANQRTAP QWMLPLLLGL YG..TVTPAL EAYEDGPNQK KRRVSRGSSQ KAKGTRASAK TTNKRRSRSS RS*
SV40 SGEFIEKFEA PGGANQRTAP QWMLPLLLGL YG..SVTSAL KAYEDGPNKK KRKLSRGSSQ KTKGTSASAK ARHKRRNRSS RS*
PyV  SGEVIKFYQA PGGAHQRVTP DWMLPLILGL YGDITPTWAT VIEEDGPQKK KRRL*
```

C: agnoprotein

```
JCV  MVLRQLSRKA SVKVSKTWSG TKKRAQRILI FLLEFLLDFC TGEDSVDGKK RQKHSGLTQQ TYSALPEPKA T*
BKV  MVLRQLSRQA SVKVGKTWTG TKKRAQRIFI FILELLLEFC RGEDSVDGKN K......... STTALPAVKD SVKDS*
SV40 MVLRRLSRQA SVKVRRSWTE SKKTAQRLFV FVLELLLQFC EGEDTVDGKR K.KPERLTEK PES*
```

Figure 8.4. Amino acid sequence comparison of the late proteins of JCV, BKV, SV40, and murine polyoma virus (PyV). (**A**) Comparison of the amino acid sequences for VP1. Potential structure of JCV VP1 is analyzed based on the crystal structure determined in SV40 VP1 (Liddington et al., 1991). JCV VP1 potentially has loop structures, that project from the protein framework. The potential loop structures can provide principal antigenic structures, receptor binding sites, or domains responsible for hemagglutination. (**B**) Comparison of the amino acid sequences for VP2/VP3. The amino acid sequence of VP3 is identical to two thirds of the carboxy terminus of VP2 and therefore is designated as VP2/VP3. The sequences are divergent in the carboxy-terminal regions of VP2/VP3. (**C**) Comparison of the amino acid sequences for agnoprotein. Agnoprotein is encoded in JCV, BKV, and SV40, but not reported in murine polyoma virus. The amino acid sequences of the agnoproteins are highly divergent in the carboxy-terminal one third.

which is not readily detected by the peptide antibody against the amino-terminal 14 residues (Nomura et al., 1983). BKV agnoprotein has been shown to be phosphorylated (Rinaldo et al., 1998). Similarly, JCV agnoprotein may also have multiple isoelectric forms that potentially play different roles in the regulation of the viral replication.

4. STRUCTURES OF THE LATE RNAs ARE DIVERGENT IN JCV, SV40, AND MURINE POLYOMA VIRUS

During the late stage of the virus replication cycle, the capsid proteins are synthesized from the late RNAs generated by alternative splicing. In JCV, SV40, and murine polyoma virus, the structures of the late RNAs are divergent to each other due to different patterns in transcription, extension of the pre-RNAs, and splicing. In particular, the divergence in structures of the 5′ leader sequence affects the presence or absence of the open reading frame (ORF) of agnoprotein. Therefore, the JCV late RNAs and some species of the SV40 late RNAs are in polycistronic structures; in contrast, the murine polyoma virus late RNAs are in monocistronic structures. For convenience in this chapter the 5′ untranslated region (5′-UTR) indicates the sequence upstream of the ORF of agnoprotein, and the term *leader sequence* indicates the sequence upstream of the ORFs of the capsid proteins.

JCV has at least four species of the late RNAs, M1–M4, that encode agnoprotein in the leader sequence (Fig. 8.5) (Shishido-Hara et al., 2000). Transcription of the late RNAs is initiated from the heterogeneous RNA start sites,

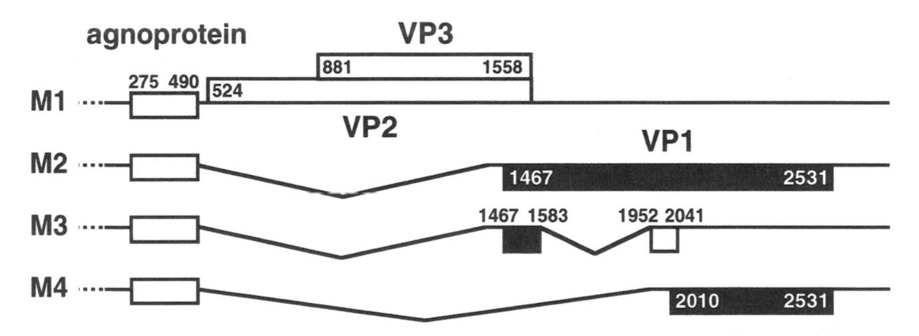

Figure 8.5. Structures of the JCV late RNAs. JCV has at least four species of the late RNAs, M1–M4, in polycistronic structures. JCV has distinctive RNA structures from SV40 and murine polyoma virus. Because most of the late RNAs are not spliced in the leader sequence, the ORF of agnoprotein is encoded. M1 is an unspliced RNA, and M2–M4 are generated by alternative splicing. M1 can produce agnoprotein, VP2, and VP3. M2 can produce agnoprotein and VP1. M3 and M4 can encode potentially the new ORFs in addition to the ORF of agnoprotein. The 5′ terminus of each of the late RNAs has not been analyzed as indicated by dotted lines. The numbering system refers to the JCV Tokyo-1 strain. (Taken with permission from Shishido-Hara et al., 2000.)

as determined in JCV Mad 1 strain by two groups of investigators. Daniel and Frisque (1993) reported the RNA start sites at nt 124–129, 191–192, 200–203, and 5118–5121. Kenney et al. (1986) reported the RNA start sites at nt 5114–5117, 90–98, 198–203, 224, and 236–242. (The sequence numbering follows that of JCV Mad 1 [Frisque et al., 1984].) The JCV late RNAs are generated by alternative splicing, and positions of the splice sites have been mapped in JCV Tokyo-1 strain. M1 is an unspliced RNA. M2 is spliced at nt 490/1425 (934 nt intron); M3, at nt 490/1425 (934 nt intron) and nt 1583/1952 (368 nt introns); and M4 at nt 490/1952 (1461 nt intron). (The sequence numbering system follows JCV Tokyo-1, and paired numbers separated by a slash indicate 5′ and 3′ splice sites.) Most of the JCV late RNAs are not spliced in the leader sequence, and therefore the ORF of agnoprotein can be encoded. M1 can produce agnoprotein, VP2, and VP3. M2 can produce agnoprotein and VP1. M3 and M4 are generated by using splice sites at nt 1583 or nt 1952, and cannot produce intact VP1. The splice sites corresponding to these positions have not been reported in SV40. Thus, M3 and M4 are unique RNA species to JCV and may potentially encode new ORFs in addition to the ORF of agnoprotein. It is not known if each of the M1–M4 RNAs has the same sequence in the 5′-UTR or contains heterogeneous species of different 5′-UTRs. The length of the 5′-UTR can affect translation efficiency of the capsid proteins, as described in detail later.

SV40 has two classes of the late RNAs, 16S and 19S. The majority of the late RNAs encode the ORF of agnoprotein, but some species lack its ORF due to splicing in the leader sequence (Fig. 8.6). Transcription of the SV40 late RNAs is initiated from multiple RNA start sites between nt 28 and 325 (Ghosh et al., 1978; Haegeman and Fiers, 1978; Canaani et al., 1979). However, unlike JCV, transcription of 70–90% of the SV40 late RNAs is initiated from the major RNA start site at nt 325, which is located 25–30 nucleotides downstream of the surrogate TATA signal, TACCTA (Brady et al., 1982; Nandi et al., 1985). The two classes of late RNAs, 16S and 19S, are generated by alternative splicing from common pre-RNAs (Lai et al., 1978; Good et al., 1988b). The 16S RNAs are relatively homogeneous compared with the 19S RNAs. On the 16S RNAs transcription is initiated mostly from the major RNA start site at nt 325. In contrast, on the 19S RNAs it is initiated from multiple RNA start sites between nt 28 and nt 325. The ORF of agnoprotein is encoded on 80% of the 16S RNAs (64% of the total late RNAs) and on 5% of the 19S RNAs (1% of the total late RNAs; Somasekhar and Mertz, 1985b; Good et al., 1988b). Therefore, VP1 is produced from the 16S RNAs, and VP2 and VP3 are produced from the 19S RNAs, in both the presence and absence of the upstream ORF of agnoprotein.

Murine polyoma virus has distinctive RNAs for VP1, VP2, and VP3 (Siddell and Smith, 1978; Hunter and Gibson, 1978). Unlike JCV and SV40, the late RNAs of murine polyoma virus are monocistronic, and the presence of agnoprotein is not reported in the leader sequence. Transcription initiates from the heterogeneous RNA start sites from nt 5075 to 5168 (Flavell et al., 1980; Cowie

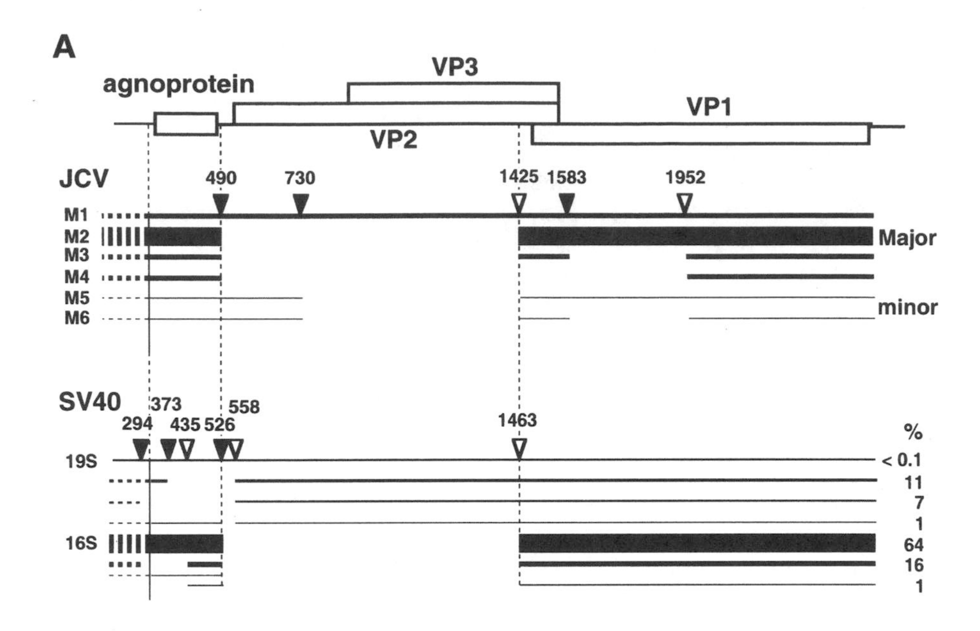

B

5' splice sites

	nt	$^{A}_{C}$AG/GU$^{A}_{G}$AGU
JCV		
▼SV40	294	AAG/GUAccU
JCV	313	AAa GUuAGU
▼SV40	373	AAG/GUucGU
▼ JCV	490	uAG/GUAAGU
▼SV40	526	CuG/GUAAGU
▼ JCV	730	gcu/GUAAuU
SV40	765	gcu GUGAua
▼ JCV	1583	gAG/GUAgaa
SV40	1645	gAG GUGgag

3' splice sites

	nt	UUUUUUUUUUUUUU / CCCCCCCCCCCCC N $^{C}_{U}$AG/G
JCV	375	UUUUaaUUUUUUUgUUAG a
▽SV40	435	UUUUUgUgUUUgUUUUAG/a
JCV	520	UUUUUUUUgUgUUUUCAG G
▽SV40	558	UUUUgUCUUUUaUUUCAG/G
▽ JCV	1425	UgUUgCCUUUaCUUUUAG/G
▽SV40	1463	UgUUgCCUUUaCUUCUAG/G
▽ JCV	1952	ggUggUUUUUaaUUaCAG/a
SV40	2014	UgUgUUagCaaaCUaCAG G

Figure 8.6. Structures and splice sites of the late RNAs of JCV and SV40. (**A**) JCV and SV40 have distinctive structures in the late RNAs despite similar genomic organizations in the late region. The divergence in the RNA structures is due to different patterns in transcription initiation and alternative splicing. In the 5′ terminal region, the heterogeneous sequences resulting from the multiple RNA start sites in SV40 and the sequences that have not been determined for each of the RNA species in JCV are indicated with the dotted lines. The 5′ and 3′ splice sites are indicated by closed and open triangles, respectively. In JCV, M5 and M6 are potential RNAs that were detected in cells transfected with a eukaryotic vector but not in PML brain tissues (Shishido-Hara et al., 2000). The levels of the different species of the SV40 late RNAs are from Good et al. (1988b) and Somasekhar and Mertz (1985b). (**B**) Alignment of nucleotide sequences of the splice sites and the corresponding sequences between JCV and SV40. Nucleotides identical to the consensus sequences defined for splice sites are indicated by bold upper case letters. Exon–intron boundaries are indicated by a slash. The 5′ and 3′ splice sites are marked with closed and open triangles, respectively. (Taken with permission from Shishido-Hara et al., 2000.)

et al., 1981). Pre-RNAs of murine polyoma virus are extended around the circular viral genome multiple times (Acheson, 1978), and are extremely large in size. These giant pre-RNAs are spliced from the leader sequence to the leader sequence. Therefore, the late RNAs processed by leader-to-leader splicing encode multiple copies of the leader sequence and a single ORF of either VP1, VP2, or VP3 (Legon et al., 1979; Kamen et al., 1980). The distinct late RNAs, 16S, 19S, and 18S, respectively encode VP1, VP2, and VP3 (Siddell and Smith, 1978) and represent approximately 80%, 5%, and 15% of the total late RNAs (Batt et al., 1994). The ratios of individual RNAs reflect the ratios of the capsid proteins composed in the virions.

Therefore, in JCV, SV40, and murine polyoma virus, the structures of the late RNAs are highly divergent from each other. The heterogeneity is generated by different patterns in transcription, pre-RNA extension, and splicing. Most importantly, different splicing patterns in the leader sequence determine the presence or absence of the ORF of agnoprotein.

5. THE LEADER SEQUENCE INCLUDING THE ORF OF AGNOPROTEIN *IN CIS* AND AGNOPROTEIN *IN TRANS*, CAN REGULATE VIRUS REPRODUCTION IN MULTIPLE WAYS

JCV has the late RNAs in polycistronic structures, as in some species of the SV40 late RNAs. Polycistronic RNAs are unusual in eukaryotes as most of the eukaryotic RNAs are monocistronic. The polyomaviruses have distinct structures in the late RNAs, especially in the leader sequences. This indicates that expression of the capsid proteins is regulated in a different manner. Although few studies have focused on this aspect of JCV, in SV40 the way in which *cis*-acting DNA or RNA leader sequence or *trans*-acting agnoprotein encoded in this region regulates expression of the capsid proteins has been studied extensively.

In SV40, transcription initiation of the late RNAs was affected by *cis*-acting elements of the DNA sequence. Many investigators reported that the relative frequency in utilization of the heterogeneous RNA start sites was altered by mutating the leader sequence (Piatak et al., 1981, 1983; Ghosh et al., 1982; Haegeman et al., 1979a,b; Somasekhar and Mertz, 1985b). It was also suggested that transcription of the late SV40 RNAs can be attenuated within agnoprotein coding sequence at the U residues downstream of the potential stem-and-loop structure (Hay et al., 1982). The late viral transcription may be also affected by *trans*-acting diffusible factors (Alwine, 1982).

Processing and stability of the SV40 late RNAs was affected by the RNA leader sequence. The relative frequency in utilizing the splice sites was dramatically altered by mutating the leader sequence (Ghosh et al., 1982; Somasekhar and Mertz, 1985a; Good et al., 1988a). It was also suggested that the leader sequence affects the stability of the SV40 late RNAs by splicing (Ryu and Mertz, 1989) and polyadenylation (Chiou et al., 1991).

The translation efficiency from the downstream AUG is generally inefficient in the presence of the upstream AUG on polycistronic RNAs. The translation efficiency of SV40 VP1 was reduced due to the presence of the AUG start codon for agnoprotein on the major 16S RNA (Grass and Manley, 1987; Sedman et al., 1989). Similarly, translation of VP2 and VP3 was reduced in the presence of the AUG for agnoprotein on the 19S RNA (Dabrowski and Alwine, 1988; Good et al., 1988a; Sedman et al., 1989). The minor 16S RNA (16% of the total late RNAs), in which the AUG start codon for agnoprotein is removed by splicing, was more frequently used for VP1 translation than the major 16S RNA encoding the ORF of agnoprotein (64% of the total late RNAs; Barkan and Mertz, 1984). Despite the lower level of RNA expression, the minor 16S RNA produced almost half the amount of VP1 (Sedman et al., 1989).

To analyze functions of SV40 agnoprotein *in trans*, the AUG start codon for agnoprotein was mutated to UUG (Resnick and Shenk, 1986). Detectable differences could not be seen in transcription initiation sites, expression levels of 16S and 19S late RNAs, and expression levels of VP1 protein. Agnoprotein facilitated nuclear transport of VP1 and cell-to-cell spread of virions rather than regulating expression levels of the capsid proteins.

The roles of the SV40 leader sequence *in cis* and agnoprotein encoded in this region *in trans* were controversial in some points. It is likely because expression of the capsid proteins and production of viral progeny are regulated in multiple processes at different levels. Thus, the conclusions of some investigators contradicted each other by using different experimental systems for analysis. However, it has been consistently indicated that the leader sequence greatly affects expression levels of the capsid proteins and it is in this region that agnoprotein, which can regulate virus reproduction *in trans*, is encoded. JCV, SV40, and murine polyoma virus have divergent structures in the late RNAs. The divergence in the leader sequence may be one of the important factors that explain distinct regulatory mechanisms in the production of progeny virions.

6. HOW ARE VP1, VP2, AND VP3 SYNTHESIZED FROM THE POLYCISTRONIC RNAS?

Expression of the three capsid proteins roughly corresponds to the ratio of the capsid proteins incorporated into the virions. Although the mechanism for proportional expression of VP1, VP2, and VP3 is not completely understood, translation can be one of the important processes. Translation on polycistronic RNAs has been well studied with respect to the leaky scanning model (Kozak, 1986, 1996, 1999). In this model, it is proposed that 40S ribosomal subunits bind near the 5′ end of the RNA and scan downstream until they encounter the AUG start codon for translation. The most favorable sequence for translation initiation is GCC $\mathbf{A}^{-3}$CC <u>AUG</u> $\mathbf{G}^{+4}$. Counting A of <u>AUG</u> as +1, a purine, preferably

A at the −3 position (**A^{-3}** or **G^{-3}**) and G at the +4 position (**G^{+4}**) are especially important (as indicated with bold letters). In general, translation is preferentially initiated from the upstream AUG and rarely initiated from the downstream AUG. However, in some cases ribosomes bypass the upstream AUG and efficiently initiate translation from the downstream AUG (bypassing). Ribosomes also can reinitiate translation from the downstream AUG after completing translation of the upstream ORF (reinitiation). However, translation by reinitiation is much less efficient than by bypassing (Fig. 8.7).

On the SV40 bicistronic 16S RNA, it has been shown that ribosomes frequently bypass the AUG start codon for agnoprotein and efficiently synthesize VP1 from the downstream AUG (Grass and Manley, 1987). At least one-third of 40S ribosomal subunits bypass the AUG start codon for agnoprotein despite its strong context for translation, **G^{-3}**CC AUG **G^{+4}** (Sedman et al., 1989). It has been suggested that this is due to the extremely short 5′-UTR of only 10 nucleotides on this 16S RNA. Because the major RNA start site, nt 325, is

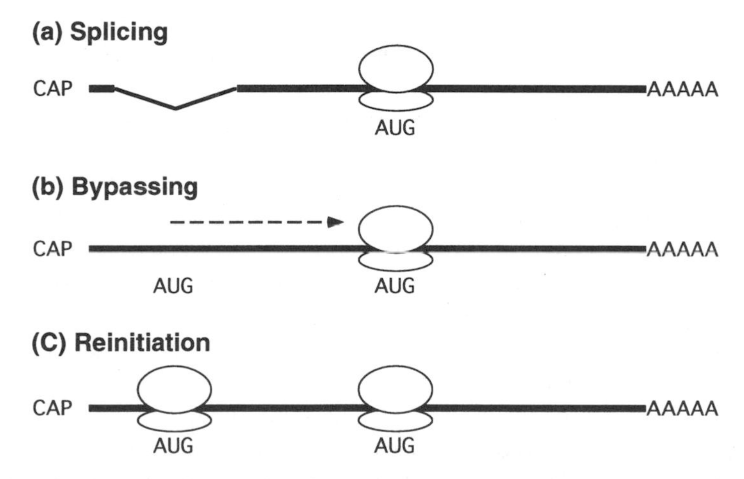

Figure 8.7. Leaky scanning models for translation. According to the model, ribosomes bind to the 5′ terminal region of the RNAs and scan downstream until they encounter the AUG start codon for translation. The presence of the upstream AUG, in general, reduces translation efficiency from the downstream AUG. However, in SV40, the capsid proteins are produced relatively efficiently from the downstream AUG due to the structure of its late RNAs. (**a**) On some species of SV40 late RNAs, the AUG start codon for agnoprotein is removed by splicing in the leader sequence. (**b**) Even if the AUG start codon for SV40 agnoprotein is present in the leader sequence, ribosomes can bypass its AUG, likely due to an extremely short 5′-UTR of only 10 nucleotides. Ribosomes that bypassed the AUG for agnoprotein initiate translation for the capsid proteins from the downstream AUG. (**c**) Ribosomes also re-initiate translation after completing translation of the upstream ORF for agnoprotein. However, it is not as efficient to re-initiate translation from the downstream AUG for the capsid proteins.

predominantly used on the 16S RNA, there are only 10 nucleotides between the 5′ cap site and the AUG start codon for the agnoprotein (Sedman et al., 1990). This extremely short 5′-UTR is quite unique because eukaryotic RNAs generally have 5′-UTR of 20–100 nucleotides in length (Kozak, 1987). When the adenovirus type 2 late promoter was inserted into the SV40 late promoter region, synthesis of SV40 VP1 was dramatically reduced in the presence of the 5′-UTR, which is 33 nucleotides in length (Grass and Manley, 1986). A similar observation has been reported for translation of VP2 and VP3 under the RSV promoter, which generates a 5′-UTR of 40 nucleotides (Dabrowski and Alwine, 1988). Therefore, although SV40 encodes agnoprotein on its genome, the structures of the SV40 late RNAs provide the mechanisms to produce the capsid proteins efficiently. On some species of the late RNAs, the AUG for agnoprotein is removed by splicing. On other species of the late RNAs, even in the presence of the AUG for agnoprotein, ribosomes can easily bypass its AUG under the extremely short 5′-UTR. These mechanisms are more efficient in expression of the capsid proteins than reinitiation of translation after the upstream ORF (Grass and Manley, 1987).

Both SV40 VP2 and VP3 are synthesized from the 19S RNAs by leaky scanning (Sedman and Mertz, 1988). Unlike the 16S RNAs, the 5′-UTRs of the 19S RNAs are heterogeneous in length, which influences utilization of the AUG start codon for agnoprotein (Dabrowski and Alwine, 1988) and those for VP2 and VP3 (Good et al., 1988a). The AUG start codon for VP2 is in a weak context for translation, UCC $\underline{\text{AUG}}$ $\mathbf{G}^{+4}$ compared with that for VP3, $\mathbf{G}^{-3}$ GA $\underline{\text{AUG}}$ $\mathbf{G}^{+4}$. Approximately 70% of the scanning ribosomes bypass the AUG start codon for VP2 and initiate translation from the downstream AUG for VP3. The AUG codon for VP3 was 3.1 times more frequently used for translation than that for VP2 (Dabrowski and Alwine, 1988), corresponding to their ratio in the cytoplasm of infected cells, which ranges from 2.7 to 3.1 (Lin et al., 1984). Therefore, translation is an important process to determine the proportional expression levels of VP2 and VP3.

How is expression of the JCV capsid proteins regulated in the process of translation? It has been suggested that the JCV leader sequence (nt 275–409) or the presence of the ORF for agnoprotein may decrease the expression level of VP1 in the process of translation (Shishido-Hara et al., 2000). Unlike SV40, most of the JCV late RNAs are not spliced in the leader sequence, and the ORF of agnoprotein is encoded upstream of the ORFs of the capsid proteins. In addition, unlike SV40, the 5′-UTRs of the JCV late RNAs are, at minimum, 74 or 35 nucleotides in length, depending on the RNA start sites reported by two groups of investigators (Daniel and Frisque, 1993; Kenney et al., 1986). Therefore, on the JCV polycistronic RNAs, ribosomes can be more frequently intercepted at the AUG for agnoprotein than on the SV40 RNAs before arriving at the downstream AUG for the capsid proteins. Thus, translation efficiency of the capsid proteins may be lower in JCV than in SV40.

7. THE NUCLEAR LOCALIZATION SIGNALS OF VP1, VP2, AND VP3

The polyomaviruses reproduce progeny virions in the nucleus. Thus, after the process of translation, the capsid proteins synthesized in the cytoplasm are transported to the nucleus for subsequent assembly. The proteins targeted to the nucleus, in general, contain a nuclear localization signal (NLS), which is usually a stretch of basic amino acids such as **K** (lysine) or **R** (arginine). These basic amino acids are arranged in either one or two clusters and called a *monopartite* or a *bipartite signal*, respectively. By crystallography, it has been shown that karyopherin α, which recognizes the NLS, has two binding sites for basic amino acids (Conti et al., 1998), suggesting a difference in the binding patterns to the NLS in a monopartite and a bipartite structure. JCV VP1 encodes the basic amino acids in a monopartite structure in the amino-terminal sequence, while SV40 VP1 encodes the basic amino acids in a bipartite structure. A difference in alignment of the basic amino acids caused different efficiencies in nuclear transport between JCV VP1 and SV40 VP1.

JCV VP1 has been shown to be inefficient in nuclear transport (Shishido-Hara et al., 2000). This is due to the unique structure in the amino-terminal sequence. The amino-terminal sequence of JCV VP1, MAPT**KRK**GE**RK**D, encodes the basic amino acids, $\mathbf{K^5R^6K^7}$ and $\mathbf{R^{10}K^{11}}$ interrupted by G^8E^9 (superscripts indicate positions of residues in the amino acid sequence). In comparison to SV40 VP1 and BKV VP1, the eight residues CPGAAPKK are missing between the two clusters of the basic amino acids. JCV VP1 was distributed both in the cytoplasm and in the nucleus when it was expressed in the absence of VP2 and VP3 in COS-7 cells (Fig. 8.8A,D). To analyze the effects of the amino-terminal sequence in nuclear transport, JCV VP1 was mutated by inserting the missing six amino acids CPGAAP between **KRK**GE and **RK**. The VP1 mutant, designated as Mut-1, has the sequence MAPT**KRK**GECP GAAP**RK**D in which **KRK** and **RK** are encoded in a bipartite structure spanning eight residues. When Mut-1 was expressed in COS-7 cells, Mut-1 was transported to the nucleus more efficiently and detected more prominently in the nucleus than wild-type VP1. However, Mut-1 was still distributed in both the cytoplasm and the nucleus (Fig. 8.8A,E). Therefore, the basic amino acids **RK** in Mut-1 were replaced with **KKPK** and designated as Mut-2. Mut-2 has the sequence MAPT**KRK**GECPGAAP**KKPK**D in which **KRK** and **KKPK** are encoded in a bipartite structure spanning eight residues, as in BKV VP1 and SV40 VP1. When Mut-2 was expressed, Mut-2 was localized to the nucleus (Fig. 8.8A,F). Thus, inefficiency in nuclear transport of JCV VP1 is due to the unique amino-terminal sequence **KRK**GE**RK**, a monopartite structure of basic amino acids of **KRK** and **RK**, and low potency of the **RK** residues in nuclear transport compared with **KKPK** (Shishido-Hara et al., 2000).

In contrast, SV40 VP1 is localized to the nucleus in the absence of VP2 and VP3 when it is expressed alone in TC7 cells, a derivative of African green monkey kidney cells like COS-7 cells (Ishii et al., 1996). SV40 VP1 has the

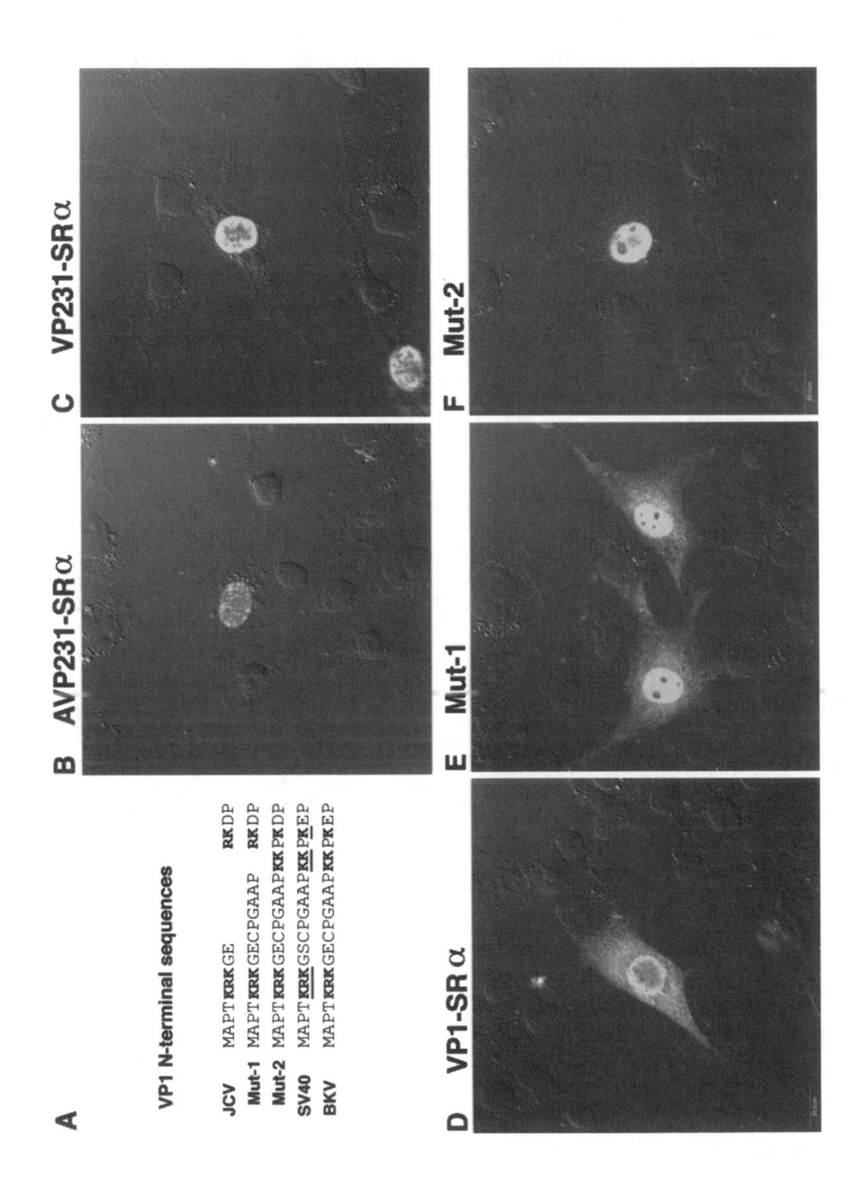
A
VP1 N-terminal sequences
JCV MAPTKRKGE RKDP
Mut-1 MAPTKRKGECPGAAP RKDP
Mut-2 MAPTKRKGECPGAAPKKPKDP
SV40 MAPTKRKGSCPGAAPKKPKEP
BKV MAPTKRKGECPGAAPKKPKEP
B AVP231-SRα
C VP231-SRα
D VP1-SRα
E Mut-1
F Mut-2

sequence MAPT**KRK**GSCPGAAP**KK**P**K**E in which the clusters of the basic amino acids **K**^{5}**R**^{6}**K**^{7} and **K**^{16}**K**^{17}, **K**^{19} are encoded in a bipartite structure spanning the eight residues GSCPGAAP. These two clusters of basic amino acids have been identified as the NLS of SV40 VP1 (Ishii et al., 1996) (Fig. 8.8A). In BKV VP1, a nearly identical sequence MAPT**KRK**GECPGAAP**KK**P**K**E is present in the amino-terminal region, suggesting that BKV VP1 can be transported to the nucleus in the absence of VP2 and VP3 as in SV40 VP1.

For VP2 and VP3, JCV has the putative NLS that corresponds to the NLS determined in SV40 and murine polyoma virus in TC7 cells and COS-7 cells, respectively (Clever and Kasamatsu, 1991; Chang et al., 1992a). In SV40, the NLS of VP2 and VP3 has been determined in the carboxy-terminal region (Gharakhanian et al., 1987; Wychowski et al., 1987), and it has been shown that the sequence of GPNKKKRKL is especially important for nuclear targeting (Clever and Kasamatsu, 1991). In JCV, a nearly identical sequence GPNKKKRRK is present in the corresponding region, suggesting that JCV VP2 and VP3 can be transported to the nucleus autonomously.

Therefore, JCV VP1 is not efficient in nuclear transport, although VP2 and VP3 are likely transported to the nucleus efficiently. However, in oligodendrocytes of PML brain tissues, the JCV capsid proteins, including VP1, are expressed predominantly in the nucleus. Because JCV VP1 was localized predominantly to the nucleus when all the three capsid proteins were expressed together (Fig. 8.8B,C), the presence of VP2 and/or VP3 is likely important for predominant nuclear localization of JCV VP1.

←

Figure 8.8. Nuclear transport of the JCV capsid proteins. (**A**) JCV VP1 has a distinctive sequence from SV40 VP1 and BKV VP1 in the amino-terminal sequence. In JCV VP1, the basic amino acids **KRK** and **RK** are encoded spanning the two residues GE. In contrast, in SV40 VP1, the two clusters of basic amino acids **KRK** and **KK**, **K** are encoded in a bipartite structure spanning eight residues and identified as the NLS (Ishii et al., 1996). A nearly identical sequence is present in BKV VP1. To investigate the effects of this amino-terminal sequence on nuclear transport, two mutants were constructed for JCV VP1. Mut-1 encodes the basic amino acids **KRK** and **RK**, and Mut-2 encodes **KRK** and **KK**P**K** in bipartite structures. VP1 and VP1 mutants were expressed in COS-7 cells, and their distribution was analyzed by immunocytochemistry with a confocal microscope. (**B**) When cells were transfected with an expression vector, AVP231-SRα, which encodes agnoprotein, VP1, VP2, and VP3, VP1 was efficiently transported to the nucleus and identified as speckles in the nucleus. (**C**) When cells were transfected with an expression vector, VP231-SRα, which encodes VP1, VP2, and VP3, VP1 was also efficiently transported to the nucleus and identified as speckles in the absence of agnoprotein. (**D**) When cells were transfected with VP1-SRα, which encodes VP1 alone, VP1 was distributed in both the cytoplasm and the nucleus (**E**) Mut-1 was transported to the nucleus more efficiently and detected more prominently in the nucleus than wild-type VP1. However, Mut-1 was still distributed in both the cytoplasm and the nucleus. (**F**) Mut-2 was efficiently transported to the nucleus and distributed diffusely except for nucleoli. Mut-2 was distributed in the nucleus diffusely. (Taken with permission from Shishido-Hara et al., 2000.)

8. THE VP1 PENTAMERS CAN BE TRANSPORTED TO THE NUCLEUS WITH VP2 OR VP3

How are the capsid proteins transported to the nucleus? It has been thought that VP1 is associated with VP2 or VP3 in the cytoplasm and transported to the nucleus cooperatively. JCV VP1 was distributed in both the cytoplasm and the nucleus in the absence of VP2 and VP3, but predominantly localized to the nucleus in their presence (Fig. 8.8B–D). The NLS-defective SV40 VP1 and murine polyoma virus VP1 were distributed in the cytoplasm, but localized to the nucleus in the presence of wild-type VP2 or VP3 (Ishii et al., 1994; Cai et al., 1994). Similarly, the NLS-defective SV40 VP2 and VP3 was distributed in the cytoplasm, but localized to the nucleus in the presence of wild-type VP1 (Ishii et al., 1994). In insect cells, murine polyoma virus VP2 and VP3 were distributed predominantly in the cytoplasm, but localized to the nucleus in the presence of VP1 (Delos et al., 1993; Forstova et al., 1993). Therefore, the capsid proteins are more efficiently localized to the nucleus when both major and minor capsid proteins are expressed together. Even if the NLS of one of the capsid proteins is defective or functionally weak, the capsid proteins are transported to the nucleus complimentarily. Based on these data, it has been inferred that the major and minor capsid proteins are associated in the cytoplasm and transported to the nucleus as a complex. However, it is not yet demonstrated, in a dynamic state, that the capsid proteins are migrating to the nucleus cooperatively in living cells.

Why are the capsid proteins transported to the nucleus together? In JCV, interaction of VP1 with VP2 or VP3 in the cytoplasm can be a critical step for efficient nuclear transport because VP1 is not efficiently localized to the nucleus in the absence of VP2 and VP3. However, in SV40 and murine polyoma virus, each capsid protein has its own NLS and has a potential to be transported to the nucleus individually (Gharakhanian et al., 1987; Clever and Kasamatsu, 1991; Moreland and Garcea, 1991; Chang et al., 1992a,b; Ishii et al., 1996). In SV40, the ratios of the three capsid proteins were analyzed through the replication cycle. The ratios of the capsid proteins in the nuclear fraction were constantly similar to their ratios incorporated into the virions, suggesting that the amounts of the capsid proteins within the nucleus can be regulated in the process of nuclear transport (Lin et al., 1984). In both SV40 and mouse polyoma virus, VP1 easily interacted with VP2 or VP3 and formed a complex, as demonstrated by co-immunoprecipitation (Gharakhanian et al., 1988, 1990; Barouch and Harrison, 1994; Cai et al., 1994; Forstova et al., 1993; Delos et al., 1995). In murine polyoma virus, interaction of VP1 and VP2 or VP3 within a complex was shown in a specific stoichiometry (Barouch and Harrison, 1994). The crystal structure of this complex indicated that each VP1 pentamer is associated with a single molecule of the carboxy-terminal region of VP2 or VP3 (Chen et al., 1998). Therefore, in the cytoplasm, the capsid proteins are associated and transported to the nucleus likely as a complex in a specific stoichiometry, which might be the VP1 pentamer associated with VP2 or VP3. The complex may be a part of the capsid, which is further assembled into a complete capsid structure, and nuclear transport is likely the initial interaction of the capsid proteins prior to further assembly.

9. PRODUCTION OF PROGENY VIRIONS IN THE NUCLEUS

Where are the viral capsids or virions assembled? According to the crystal structure of the SV40 virions, the NLSs of both VP1 and VP2/VP3 are completely hidden inside the capsid (Liddington et al., 1991). Therefore, it is likely that the VP1 molecules associated with VP2 or VP3 are transported to the nucleus as a complex such as the pentamer that exposes the NLS, but not as the assembled capsid. Assembly may be retarded in the cytoplasm due to a deficiency of calcium ions. When murine polyoma virus VP1 was expressed in insect cells, VP1 was assembled into the virus-like particles in the nucleus, but not in the cytoplasm. However, by adding calcium, the virus-like particles were assembled in both the cytoplasm and the nucleus (Montross et al., 1991). Therefore, the nucleus can provide a more suitable environment for assembly with the proper concentration of calcium.

Agnoprotein is not essential for nuclear transport of the capsid protein, but it likely facilitates their entry. In the absence of agnoprotein, SV40 VP1 was delayed in nuclear transport and localized to the periphery of the nucleus (Resnick and Shenk, 1986; Carswell and Alwine, 1986). SV40 agnoprotein was suggested to interact with VP1 to prevent aberrant cytoplasmic aggregation of the VP1 pentamers and facilitate their nuclear transport (Margolskee and Nathans, 1983; Barkan et al., 1987). The distributions of JCV agnoprotein and SV40 agnoprotein were mostly cytoplasmic, especially in the perinuclear region (Fig. 8.3) (Nomura et al., 1983). Because some intermediate filaments in the perinuclear region are contiguous with the nuclear matrix (Lazarides, 1980; Capco et al., 1982), agnoprotein may facilitate transport of the VP1 pentamers associated with VP2 or VP3 from the perinucleus to the nucleus.

In which regions in the nucleus are the progeny virions generated? In both SV40 and murine polyoma virus, the capsid proteins or virions were associated with the nuclear framework (Ben-Ze'ev et al., 1982; Stamatos et al., 1987). In JCV also, when VP1 was expressed in the presence of VP2 and VP3 in COS-7 cells, VP1 was localized to discrete regions in the nucleus and identified as speckles by confocal microscopy. JCV VP1, identified as speckles, was spread from the nuclear membrane into the center of the nucleus, potentially along subnuclear structures. In contrast, when VP1 was expressed in the absence of VP2 and VP3, VP1 was distributed more diffusely in the nucleus (Fig. 8.8B–D). Therefore, the VP1 pentamers, likely with VP2 and VP3, may accumulate in distinct subnuclear regions and spread along the skeletal frameworks in the nucleus. For encapsidating the viral genome, it was indicated that the VP1 pentamers are added on the viral minichromosome that includes cellular histones possibly guided by VP2 and VP3, and are further assembled into the mature virions (Coca-Prados and Hsu, 1979; Fanning and Baumgartner, 1980; Garber et al., 1980; Garcea and Benjamin, 1983; Blasquez et al., 1983; Ng and Bina, 1984). Because the viral genomic DNA was also associated with the nuclear framework (Buckler-White et al., 1980), the nucleus might be divided into functional domains, some of which may actively support maturation of virions.

Using electron microscopy of PML brain tissues, the JCV round virions were frequently identified as discrete clusters in the nucleus, in which the virions are regularly arranged in crystalloid array (Fig. 8.1A). This unique morphologic feature was observed in the initial phase of virus infection, when the virus was cultured in primary human fetal glial (PHFG) cells (Nagashima et al., 1981). This electron microscopic observation and the speckles identified by confocal microscopy consistently indicate that the three capsid proteins, VP1, VP2, and VP3, accumulate in discrete regions in the nucleus. In these regions, the capsid proteins may be actively assembled into the virions, or the virions assembled elsewhere may gather together and organize clusters.

10. ASSEMBLY OF RECOMBINANT VIRUS-LIKE PARTICLES IN VITRO AND IN VIVO

For the polyomaviruses there are many reports that virus-like particles are generated by molecular manipulation of the viral genome both in vitro and in vivo. Assembly of virus-like particles in defined environmental conditions can give us step-by-step understanding of the highly ordered assembly of virions in the natural viral replication system.

Murine polyoma virus VP1 expressed in *Escherichia coli* was isolated as the VP1 pentamers. In vitro, at high ionic strength, the VP1 pentamers were assembled into virus-like structures, but also into heterogeneous aggregates (Salunke et al., 1986). At low ionic strength and in the presence of calcium, the pentamers were assembled into more homogeneous virus-like structures similar to native virions. At various conditions, the VP1 pentamers were formed into several structures of capsid-like aggregates, such as a 12-pentamer icosahedron and a 24-pentamer octahedron (Salunke et al., 1989). Thus, VP1, even when it is not post-translationally modified in *E. coli*, is assembled into virus-like structures, in the absence of VP2 and VP3 and in appropriate environmental conditions. Conditions of proper ionic strength and the presence of calcium ions are especially important.

In cells VP1 alone expressed in the absence of VP2 and VP3 was also assembled into the virus-like particles. JCV VP1 was isolated as the virus-like particles from *E. coli* (Ou et al., 1999). In the nucleus of insect cells, VP1 expressed by baculovirus vectors was also assembled into virus-like particles in JCV (Chang et al., 1997; Goldmann et al., 1999), in SV40 (Kosukegawa et al., 1996), and in murine polyoma virus (Montross et al., 1991; Gillock et al., 1997). The virus-like particles assembled in cells were much more homogeneous than those assembled in vitro, suggesting additional factors in cells that facilitate assembly (Montross et al., 1991). The virus-like particles isolated from *E. coli* (Ou et al., 1999) and insect cells (Gillock et al., 1997; Goldmann et al., 1999) contained nonviral DNA in the absence of VP2 and VP3. It is not well understood how the viruses encapsidate their own genomic DNA specifically during a natural virus replication.

The three capsid proteins VP1, VP2, and VP3 were expressed together in insect cells by co-infection of the three recombinant baculoviruses. In insect cells, the recombinant capsid proteins were assembled into virus-like particles in SV40 (Colomar et al., 1993; Sandalon and Oppenheim, 1997; Sandalon et al., 1997) and in murine polyoma virus (Delos et al., 1993; Forstova et al., 1993). The proportion of the three capsid proteins in these virus-like particles was comparable to those in natural virions (Forstova et al., 1993). Therefore, even if expression of the capsid proteins is not regulated as in the natural regulatory system, the three capsid proteins are assembled into virus-like particles in an appropriate ratio. In JCV, a viral genomic fragment encoding the overlapping capsid proteins VP1, VP2, and VP3 was located downstream of the SRα promoter, and the plasmid vector was transfected into COS-7 cells (Shishido et al., 1997). The vector-derived RNA was alternatively spliced at the authentic splice sites, and the virus-like structures of both round particles and filamentous forms were assembled in the nucleus (Shishido-Hara et al., 2000). Therefore, the capsid proteins are assembled into virus-like structures, even if their proportional expression is not regulated by the native regulatory system.

Because VP1 alone is spontaneously assembled into the virus-like particles in vitro, the capsid assembly itself is able to occur in a simple environment. However, in a cellular environment, production of progeny virions is regulated in multiple ways, and levels of regulation can be distinctive in each regulatory step. The efficiency of the production of progeny virions influences the viral biologic activities and subsequent occurrence of the virus-induced disease. Therefore, studies on the underlying regulatory mechanism in the production of virions can give us a better understanding of the nature of virus assembly, and clues to the molecular pathogenesis of virus-induced disease.

11. QUESTIONS FOR FUTURE STUDIES

How differently is production of progeny virions regulated in PML patients and in healthy individuals? JCV persists in most healthy individuals in an asymptomatic state and causes PML in some immunodeficient patients. It has been shown that the virus isolated from healthy individuals and the virus from PML patients' brains have divergent sequences in the regulatory region. The virus from healthy individuals has a typical regulatory sequence, which is called the *archetype* (Yogo et al., 1990). In contrast, the viruses from PML patients' brains have divergent regulatory sequences, possibly derived from the archetype by rearrangement involving deletion and duplication. When the two types of viruses were cultured in cells, the virus from a PML patient's brain replicated more efficiently than the urinary virus from a healthy individual (Daniel et al., 1996). The viral transcription activities can be affected by rearrangement in the regulatory region. However, different groups of investigators reported contradictory results when using different experimental systems (Sock et al., 1996; Daniel et al., 1996; Ault, 1997). As we have reviewed in this chapter, production of the progeny virions is regulated in multiple steps during the late stage. It has not yet been investigated how the rearrangement in the regulatory region influences initiation of late RNA transcripts, utilization of splice sites, and translation efficiencies of the capsid proteins in the altered 5′-UTR. The rearrangement might also affect proportional expression levels of the capsid proteins. These subjects remain for future studies.

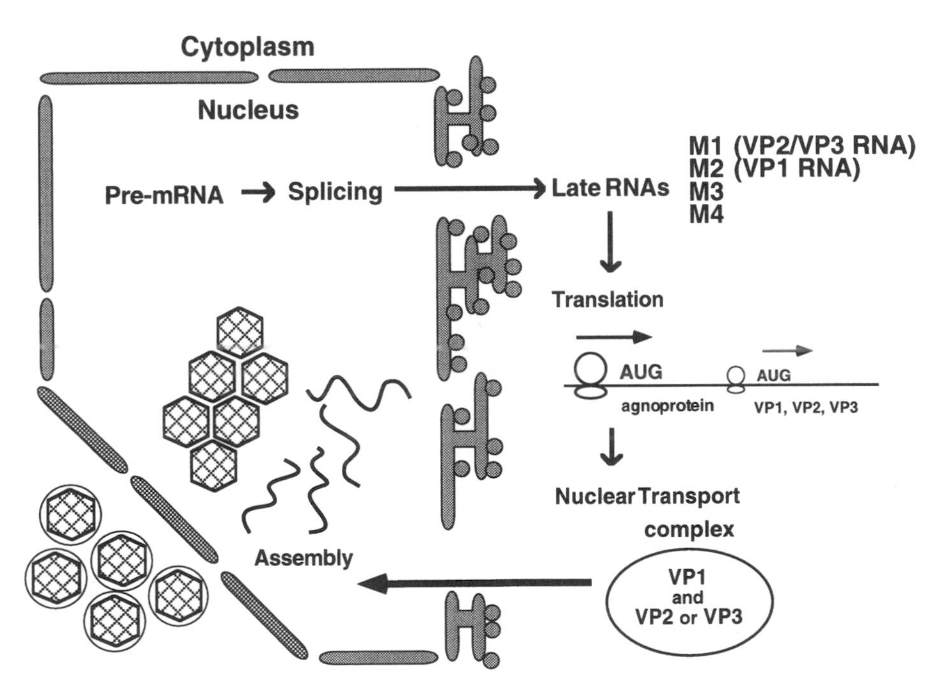

Figure 8.9. Schematic illustration of the late stage in the JCV replication cycle. Transcription occurs in the nucleus, and the pre-mRNAs are generated from heterogeneous RNA start sites. The JCV late RNAs are generated by alternative splicing. M1 can produce agnoprotein, VP2, and VP3. M2 can produce agnoprotein and VP1. Translation of the capsid proteins may be reduced due to the presence of the AUG start codon for agnoprotein in the leader sequence. JCV VP1 is not efficient in nuclear transport by itself, but can be efficiently transported to the nucleus in the presence of VP2 or VP3. In the nucleus, the capsid proteins are likely accumulated to discrete subnuclear regions, which may actively support production of progeny virions. JCV is identified as the round particles and the filamentous forms in the nucleus, and their distribution patterns are variable depending on cells.

In the nucleus of infected oligodendrocytes, JCV is identified as round and filamentous structures by electron microscopy. The proportions of round and filamentous forms are variable depending on the host cells. It has been suggested that filamentous forms result from aberrant assembly (Kiselev and Klug, 1969; Baker et al., 1983). The filamentous structures may be formed due to aberrant interaction of the first two VP1 pentamers (Stehle et al., 1996) or to different phosphorylation patterns of VP1 (Li et al., 1995). It is not known how the viruses in round and filamentous structures are different in nature. However, they likely have different biologic features, and the proportions of the round and filamentous forms may reflect the viral biologic activities as human pathogen. Few studies have focused on JCV virions in different structures.

In conclusion, during the late stage, production of progeny virions is regulated in multiple steps, including transcription, splicing, translation, nuclear transport, and maturation of virions (Fig. 8.9). We have shown in this chapter that JCV has distinctive regulatory mechanism from SV40 and murine polyoma virus. Studies on the underlying molecular mechanisms in the production of progeny virions and their structures can contribute to understanding the nature of JCV and the pathogenesis of PML. Inhibition of critical regulatory steps in the production of progeny virions may provide an efficient therapy for progressive demyelination in the JCV-infected human brain.

ACKNOWLEDGMENTS

We thank Dr. Y. Hara for helpful suggestions, Drs. S. Takeda and H. Takahashi for an electron micrograph, and Dr. S. Ichinose for cooperation in electron microscopy. This work is supported by grants from the Special Coordination Funds (SPSBS), the Science and Technology Agency of the Japanese Government, and the Japanese Ministry of Health, Labour and Welfare.

REFERENCES

Acheson NH (1978): Polyoma virus giant RNAs contain tandem repeats of the nucleotide sequence of the entire viral genome. Proc Natl Acad Sci USA 75:4754–4758.

Alwine JC (1982): Evidence for simian virus 40 late transcriptional control: Mixed infections of wild-type simian virus 40 and a late leader deletion mutant exhibit trans effects on late viral RNA synthesis. J Virol 42:798–803.

Aoki N, Mori M, Kato K, Sakamoto Y, Noda K, Tajima M, Shimada H (1996): Antibody against synthetic multiple antigen peptides (MAP) of JC virus capsid protein (VP1) without cross reaction to BK virus: A diagnostic tool for progressive multifocal leukoencephalopathy. Neurosci Lett 205:111–114.

Ault GS (1997): Activity of JC virus archetype and PML-type regulatory regions in glial cells. J Gen Virol 78:163–169.

Baker TS, Caspar DL, Murakami WT (1983): Polyoma virus "hexamer" tubes consist of paired pentamers. Nature 303:446–448.

Barkan A, Mertz JE (1984): The number of ribosomes on simian virus 40 late 16S mRNA is determined in part by the nucleotide sequence of its leader. Mol Cell Biol 4:813–816.

Barkan A, Welch RC, Mertz JE (1987): Missense mutations in the VP1 gene of simian virus 40 that compensate for defects caused by deletions in the viral agnogene. J Virol 61:3190–3198.

Barouch DH, Harrison SC (1994): Interactions among the major and minor coat proteins of polyomavirus. J Virol 68:3982–3989.

Batt DB, Rapp LM, Carmichael GG (1994): Splice site selection in polyomavirus late pre-mRNA processing. J Virol 68:1797–1804.

Ben-Ze'ev A, Abulafia R, Aloni Y (1982): SV40 virions and viral RNA metabolism are associated with cellular substructures. EMBO J 1:1225–1231.

Blasquez V, Beecher S, Bina M (1983): Simian virus 40 morphogenetic pathway. An analysis of assembly-defective tsB201 DNA protein complexes. J Biol Chem 258: 8477–8484.

Brady J, Radonovich M, Vodkin M, Natarajan V, Thoren M, Das G, Janik J, Salzman NP (1982): Site-specific base substitution and deletion mutations that enhance or suppress transcription of the SV40 major late RNA. Cell 31:625–633.

Buckler-White AJ, Humphrey GW, Pigiet V (1980): Association of polyoma T antigen and DNA with the nuclear matrix from lytically infected 3T6 cells. Cell 22:37–46.

Cai X, Chang D, Rottinghaus S, Consigli RA (1994): Expression and purification of recombinant polyomavirus VP2 protein and its interactions with polyomavirus proteins. J Virol 68:7609–7613.

Canaani D, Kahana C, Mukamel A, Groner Y (1979): Sequence heterogeneity at the 5′ termini of late simian virus 40 19S and 16S mRNAs. Proc Natl Acad Sci USA 76: 3078–3082.

Capco DG, Wan KM, Penman S (1982): The nuclear matrix: Three-dimensional architecture and protein composition. Cell 29:847–858.

Carswell S, Alwine JC (1986): Simian virus 40 agnoprotein facilitates perinuclear–nuclear localization of VP1, the major capsid protein. J Virol 60:1055–1061.

Chang D, Fung CY, Ou WC, Chao PC, Li SY, Wang M, Huang YL, Tzeng TY, Tsai RT (1997): Self-assembly of the JC virus major capsid protein, VP1, expressed in insect cells. J Gen Virol 78:1435–1439.

Chang D, Haynes JID, Brady JN, Consigli RA (1992a): Identification of a nuclear localization sequence in the polyomavirus capsid protein VP2. Virology 191:978–983.

Chang D, Haynes JID, Brady JN, Consigli RA (1992b): The use of additive and subtractive approaches to examine the nuclear localization sequence of the polyomavirus major capsid protein VP1. Virology 189:821–827.

Chen XS, Stehle T, Harrison SC (1998): Interaction of polyomavirus internal protein VP2 with the major capsid protein VP1 and implications for participation of VP2 in viral entry. EMBO J 17:3233–3240.

Chiou HC, Dabrowski C, Alwine JC (1991): Simian virus 40 late mRNA leader sequences involved in augmenting mRNA accumulation via multiple mechanisms, including increased polyadenylation efficiency. J Virol 65:6677–6685.

Clever J, Dean DA, Kasamatsu H (1993): Identification of a DNA binding domain in simian virus 40 capsid proteins Vp2 and Vp3. J Biol Chem 268:20877–20883.

Clever J, Kasamatsu H (1991): Simian virus 40 Vp2/3 small structural proteins harbor their own nuclear transport signal. Virology 181:78–90.

Coca-Prados M, Hsu MT (1979): Intracellular forms of simian virus 40 nucleoprotein complexes. II. Biochemical and electron microscopic analysis of simian virus 40 virion assembly. J Virol 31:199–208.

Colomar MC, Degoumois-Sahli C, Beard P (1993): Opening and refolding of simian virus 40 and in vitro packaging of foreign DNA. J Virol 67:2779–2786.

Conti E, Uy M, Leighton L, Blobel G, Kuriyan J (1998): Crystallographic analysis of the recognition of a nuclear localization signal by the nuclear import factor karyopherin alpha. Cell 94:193–204.

Cowie A, Tyndall C, Kamen R (1981): Sequences at the capped 5′-ends of polyoma virus late region mRNAs: An example of extreme terminal heterogeneity. Nucleic Acids Res 9:6305–6322.

Dabrowski C, Alwine JC (1988): Translational control of synthesis of simian virus 40 late proteins from polycistronic 19S late mRNA. J Virol 62:3182–3192.

Daniel AM, Frisque RJ (1993): Transcription initiation sites of prototype and variant JC virus early and late messenger RNAs. Virology 194:7–109.

Daniel AM, Swenson JJ, Mayreddy RP, Khalili K, Frisque RJ (1996): Sequences within the early and late promoters of archetype JC virus restrict viral DNA replication and infectivity. Virology 216:90–101.

Dean DA, Li PP, Lee LM, Kasamatsu H (1995): Essential role of the Vp2 and Vp3 DNA-binding domain in simian virus 40 morphogenesis. J Virol 69:1115–1121.

Delos SE, Cripe TP, Leavitt AD, Greisman H, Garcea RL (1995): Expression of the polyomavirus minor capsid proteins VP2 and VP3 in *Escherichia coli*: In vitro interactions with recombinant VP1 capsomeres. J Virol 69:7734–7742.

Delos SE, Montross L, Moreland RB, Garcea RL (1993): Expression of the polyomavirus VP2 and VP3 proteins in insect cells: Coexpression with the major capsid protein VP1 alters VP2/VP3 subcellular localization. Virology 194:393–398.

Fanning E, Baumgartner I (1980): Role of fast-sedimenting SV40 nucleoprotein complexes in virus assembly. Virology 102:1–12.

Flavell AJ, Cowie A, Arrand JR, Kamen R (1980): Localization of three major capped 5′ ends of polyoma virus late mRNA's within a single tetranucleotide sequence in the viral genome. J Virol 33:902–908.

Forstova J, Krauzewicz N, Wallace S, Street AJ, Dilworth SM, Beard S, Griffin BE (1993): Cooperation of structural proteins during late events in the life cycle of polyomavirus. J Virol 67:1405–1413.

Frisque RJ, Bream GL, Cannella MT (1984): Human polyomavirus JC virus genome. J Virol 51:458–469.

Garber EA, Seidman MM, Levine AJ (1980): Intracellular SV40 nucleoprotein complexes: Synthesis to encapsidation. Virology 107:389–401.

Garcea RL, Benjamin TL (1983): Isolation and characterization of polyoma nucleoprotein complexes. Virology 130:65–75.

Gharakhanian E, Kasamatsu H (1990): Two independent signals, a nuclear localization signal and a Vp1-interactive signal, reside within the carboxy-35 amino acids of SV40 Vp3. Virology 178:62–71.

Gharakhanian E, Takahashi J, Clever J, Kasamatsu H (1988): In vitro assay for protein–protein interaction: Carboxyl-terminal 40 residues of simian virus 40 structural protein VP3 contain a determinant for interaction with VP1. Proc Natl Acad Sci USA 85:6607–6611.

Gharakhanian E, Takahashi J, Kasamatsu H (1987): The carboxyl 35 amino acids of SV40 Vp3 are essential for its nuclear accumulation. Virology 157:440–448.

Ghosh PK, Piatak M, Mertz JE, Weissman SM, Lebowitz P (1982): Altered utilization of splice sites and 5′ termini in late RNAs produced by leader region mutants of simian virus 40. J Virol 44:610–624.

Ghosh PK, Reddy VB, Swinscoe J, Lebowitz P, Weissman SM (1978): Heterogeneity and 5′-terminal structures of the late RNAs of simian virus 40. J Mol Biol 126:813–846.

Gillock ET, Rottinghaus S, Chang D, Cai X, Smiley SA, An K, Consigli RA (1997): Polyomavirus major capsid protein VP1 is capable of packaging cellular DNA when expressed in the baculovirus system. J Virol 71:2857–2865.

Gluzman Y (1981): SV40-transformed simian cells support the replication of early SV40 mutants. Cell 23:175–182.

Goldmann C, Petry H, Frye S, Ast O, Ebitsch S, Jentsch KD, Kaup FJ, Weber F, Trebst C, Nisslein T, Hunsmann G, Weber T, Luke W (1999): Molecular cloning and expression of major structural protein VP1 of the human polyomavirus JC virus: Formation of virus-like particles useful for immunological and therapeutic studies. J Virol 73:4465–4469.

Good PJ, Welch RC, Barkan A, Somasekhar MB, Mertz JE (1988a): Both VP2 and VP3 are synthesized from each of the alternative spliced late 19S RNA species of simian virus 40. J Virol 62:944–953.

Good PJ, Welch RC, Ryu WS, Mertz JE (1988b): The late spliced 19S and 16S RNAs of simian virus 40 can be synthesized from a common pool of transcripts. J Virol 62:563–571.

Grass DS, Manley JL (1986): Effects of the adenovirus 2 late promoter on simian virus 40 transcription and replication. J Virol 57:129–137.

Grass DS, Manley JL (1987): Selective translation initiation on bicistronic simian virus 40 late mRNA. J Virol 61:2331–2335.

Griffith JP, Griffith DL, Rayment I, Murakami WT, Caspar DL (1992): Inside polyomavirus at 25-Å resolution. Nature 355:652–654.

Haegeman G, Fiers W (1978): Localization of the 5′ terminus of late SV40 mRNA. Nucleic Acids Res 5:2359–2371.

Haegeman G, Iserentant D, Gheysen D, Fiers W (1979a): Characterization of the major altered leader sequence of late mRNA induced by SV40 deletion mutant d1-1811. Nucleic Acids Res 7:1799–1814.

Haegeman G, van Heuverswyn H, Gheysen D, Fiers W (1976b): Heterogeneity of the 5′ terminus of late mRNA induced by a viable simian virus 40 deletion mutant. J Virol 31:484–493.

Hay N, Skolnik-David H, Aloni Y (1982): Attenuation in the control of SV40 gene expression. Cell 29:183–193.

Hunter T, Gibson W (1978): Characterization of the mRNA's for the polyoma virus capsid proteins VP1, VP2, and VP3. J Virol 28:240–253.

Ishii N, Minami N, Chen EY, Medina AL, Chico MM, Kasamatsu H (1996): Analysis of a nuclear localization signal of simian virus 40 major capsid protein Vp1. J Virol 70:1317–1322.

Ishii N, Nakanishi A, Yamada M, Macalalad MH, Kasamatsu H (1994): Functional complementation of nuclear targeting-defective mutants of simian virus 40 structural proteins. J Virol 68:8209–8216.

Jay G, Nomura S, Anderson CW, Khoury G (1981): Identification of the SV40 agnogene product: A DNA binding protein. Nature 291:346–349.

Kamen R, Favaloro J, Parker J (1980): Topography of the three late mRNA's of polyoma virus which encode the virion proteins. J Virol 33:637–651.

Kenney S, Natarajan V, Salzman NP (1986): Mapping 5′ termini of JC virus late RNA. J Virol 58:216–219.

Kiselev NA, Klug A (1969): The structure of viruses of the papilloma-polyoma type. V. Tubular variants built of pentamers. J Mol Biol 40:155–171.

Kosukegawa A, Arisaka F, Takayama M, Yajima H, Kaidow A, Handa H (1996): Purification and characterization of virus-like particles and pentamers produced by the expression of SV40 capsid proteins in insect cells. Biochim Biophys Acta 1290:37–45.

Kozak M (1986): Point mutations define a sequence flanking the AUG initiator codon that modulates translation by eukaryotic ribosomes. Cell 44:283–292.

Kozak M (1987): An analysis of 5′-noncoding sequences from 699 vertebrate messenger RNAs. Nucleic Acids Res 15:8125–8148.

Kozak M (1996): Interpreting cDNA sequences: Some insights from studies on translation. Mamm Genome 7:563–574.

Kozak M (1999): Initiation of translation in prokaryotes and eukaryotes. Gene 234: 187–208.

Lai CJ, Dhar R, Khoury G (1978): Mapping the spliced and unspliced late lytic SV40 RNAs. Cell 14:971–982.

Lazarides E (1980): Intermediate filaments as mechanical integrators of cellular space. Nature 283:249–256.

Legon S, Flavell AJ, Cowie A, Kamen R (1979): Amplification in the leader sequence of late polyoma virus mRNAs. Cell 16:373–388.

Li M, Lyon MK, Garcea RL (1995): In vitro phosphorylation of the polyomavirus major capsid protein VP1 on serine 66 by casein kinase II. J Biol Chem 270:26006–26011.

Liddington RC, Yan Y, Moulai J, Sahli R, Benjamin TL, Harrison SC (1991): Structure of simian virus 40 at 3.8-Å resolution. Nature 354:278–284.

Lin W, Hata T, Kasamatsu H (1984): Subcellular distribution of viral structural proteins during simian virus 40 infection. J Virol 50:363–371.

Margolskee RF, Nathans D (1983): Suppression of a VP1 mutant of simian virus 40 by missense mutations in serine codons of the viral agnogene. J Virol 48:405–409.

Matsuda M, Jona M, Yasui K, Nagashima K (1987): Genetic characterization of JC virus Tokyo-1 strain, a variant oncogenic in rodents. Virus Res 7:159–168.

Montross L, Watkins S, Moreland RB, Mamon H, Caspar DL, Garcea RL (1991): Nuclear assembly of polyomavirus capsids in insect cells expressing the major capsid protein VP1. J Virol 65:4991–4998.

Moreland RB, Garcea RL (1991): Characterization of a nuclear localization sequence in the polyomavirus capsid protein VP1. Virology 185:513–518.

Nagashima K, Yamaguchi K, Yasui K, Ogiwara H (1981): Progressive multifocal leukoencephalopathy. Neuropathology and virus isolation. Acta Pathol Jpn 31:953–961.

Nandi A, Das G, Salzman NP (1985): Characterization of a surrogate TATA box promoter that regulates in vitro transcription of the simian virus 40 major late gene. Mol Cell Biol 5:591–594.

Ng SC, Bina M (1984): Temperature-sensitive BC mutants of simian virus 40: Block in virion assembly and accumulation of capsid-chromatin complexes. J Virol 50: 471–477.

Nomura S, Khoury G, Jay G (1983): Subcellular localization of the simian virus 40 agnoprotein. J Virol 45:428–433.

Ou WC, Wang M, Fung CY, Tsai RT, Chao PC, Hseu TH, Chang D (1999): The major capsid protein, VP1, of human JC virus expressed in Escherichia coli is able to self-assemble into a capsid-like particle and deliver exogenous DNA into human kidney cells. J Gen Virol 80:39–46.

Piatak M, Ghosh PK, Norkin LC, Weissman SM (1983): Sequences locating the 5′ ends of the major simian virus 40 late mRNA forms. J Virol 48:503–520.

Piatak M, Subramanian KN, Roy P, Weissman SM (1981): Late messenger RNA production by viable simian virus 40 mutants with deletions in the leader region. J Mol Biol 153:589–618.

Rayment I, Baker TS, Caspar DL, Murakami WT (1982): Polyoma virus capsid structure at 22.5 Å resolution. Nature 295:110–115.

Resnick J, Shenk T (1986): Simian virus 40 agnoprotein facilitates normal nuclear location of the major capsid polypeptide and cell-to-cell spread of virus. J Virol 60: 1098–1106.

Rinaldo CH, Traavik T, Hey A (1998): The agnogene of the human polyomavirus BK is expressed. J Virol 72:6233–6236.

Ryu WS, Mertz JE (1989): Simian virus 40 late transcripts lacking excisable intervening sequences are defective in both stability in the nucleus and transport to the cytoplasm. J Virol 63:4386–4394.

Salunke DM, Caspar DL, Garcea RL (1986): Self-assembly of purified polyomavirus capsid protein VP1. Cell 46:895–904.

Salunke DM, Caspar DL, Garcea RL (1989): Polymorphism in the assembly of polyomavirus capsid protein VP1. Biophys J 56:887–900.

Sandalon Z, Dalyot-Herman N, Oppenheim AB, Oppenheim A (1997): In vitro assembly of SV40 virions and pseudovirions: Vector development for gene therapy. Hum Gene Ther 8:843–849.

Sandalon Z, Oppenheim A (1997): Self-assembly and protein–protein interactions between the SV40 capsid proteins produced in insect cells. Virology 237:414–421.

Sedman SA, Gelembiuk GW, Mertz JE (1990): Translation initiation at a downstream AUG occurs with increased efficiency when the upstream AUG is located very close to the 5′ cap. J Virol 64:453–457.

Sedman SA, Good PJ, Mertz JE (1989): Leader-encoded open reading frames modulate both the absolute and relative rates of synthesis of the virion proteins of simian virus 40. J Virol 63:3884–3893.

Sedman SA, Mertz JE (1988): Mechanisms of synthesis of virion proteins from the functionally bigenic late mRNAs of simian virus 40. J Virol 62:954–961.

Shishido Y, Nukuzuma S, Mukaigawa J, Morikawa S, Yasui K, Nagashima K (1997): Assembly of JC virus-like particles in COS7 cells. J Med Virol 51:265–272

Shishido-Hara Y, Hara Y, Larson T, Yasui K, Nagashima K, Stoner GL (2000): Analysis of capsid formation of human polyomavirus JC (Tokyo-1 strain) by a eukaryotic expression system: Splicing of late RNAs, translation and nuclear transport of major capsid protein VP1, and capsid assembly. J Virol 74:1840–1853.

Siddell SG, Smith AE (1978): Polyoma virus has three late mRNA's: one for each virion protein. J Virol 27:427–431.

Silverman L, Rubinstein LJ (1965): Electron microscopic observations on a case of progressive multifocal leukoencephalopathy. Acta Neuropathol (Berl) 5:215–224.

Sock E, Renner K, Feist D, Leger H, Wegner M (1996): Functional comparison of PML-type and archetype strains of JC virus. J Virol 70:1512–1520.

Somasekhar MB, Mertz JE (1985a): Exon mutations that affect the choice of splice sites used in processing the SV40 late transcripts. Nucleic Acids Res 13:5591–5609.

Somasekhar MB, Mertz JE (1985b): Sequences involved in determining the locations of the 5′ ends of the late RNAs of simian virus 40. J Virol 56:1002–1013.

Stamatos NM, Chakrabarti S, Moss B, Hare JD (1987): Expression of polyomavirus virion proteins by a vaccinia virus vector: Association of VP1 and VP2 with the nuclear framework. J Virol 61: 516–525.

Stehle T, Gamblin SJ, Yan Y, Harrison SC (1996): The structure of simian virus 40 refined at 3.1 Å resolution. Structure 4:165–182.

Stehle T, Harrison SC (1997): High-resolution structure of a polyomavirus VP1-oligosaccharide complex: Implications for assembly and receptor binding. EMBO J 16: 5139–5148.

Stehle T, Yan Y, Benjamin TL, Harrison SC (1994): Structure of murine polyomavirus complexed with an oligosaccharide receptor fragment. Nature 369:160–163.

Takebe Y, Seiki M, Fujisawa J, Hoy P, Yokota K, Arai K, Yoshida M, Arai N (1988): SR alpha promoter: An efficient and versatile mammalian cDNA expression system composed of the simian virus 40 early promoter and the R-U5 segment of human T-cell leukemia virus type 1 long terminal repeat. Mol Cell Biol 8:466–472.

Wychowski C, Benichou D, Girard M (1987): The intranuclear location of simian virus 40 polypeptides VP2 and VP3 depends on a specific amino acid sequence. J Virol 61:3862–3869.

Yan Y, Stehle T, Liddington RC, Zhao H, Harrison SC (1996): Structure determination of simian virus 40 and murine polyomavirus by a combination of 30-fold and 5-fold electron-density averaging. Structure 4:157–164.

Yogo Y, Kitamura T, Sugimoto C, Ueki T, Aso Y, Hara K, Taguchi F (1990): Isolation of a possible archetypal JC virus DNA sequence from nonimmunocompromised individuals. J Virol 64:3139–3143.

Zu Rhein GM, Chou SM (1965): Particles resembling papova viruses in human cerebral demyelinating disease. Science 148:1477–1479.

9

CELLULAR RECEPTORS FOR THE POLYOMAVIRUSES

WALTER J. ATWOOD, PH.D.

1. INTRODUCTION

Polyomaviruses are widely distributed in nature and are associated with the establishment of lifelong persistent infections in their respective natural hosts. All of the polyomaviruses display a high degree of species specificity. This has generally been linked to factors involved in species-specific DNA replication. Within a given species, the host range, or cell type specificity of the polyomaviruses, can be either very broad or very specific. For example, the mouse polyoma virus (PyV), BK virus (BKV), and simian virus 40 (SV40) all infect a wide variety of cell types both in vivo and in vitro. In contrast, JC virus (JCV) and the lymphotropic papovavirus (LPV) infect a very narrow range of cell types. The presence or absence of cell type-specific transcription factors are important determinants of polyomavirus host range and tissue tropism. It is also becoming clear that the presence or absence of specific polyomavirus receptors and pseudoreceptors contributes to the tropism, spread, and pathogenicity of these viruses. This chapter focuses on past and current efforts to define a role for cellular receptors in mediating infection of cells by polyomaviruses with both wide and narrow cellular tropism. The roles of specific cellular receptors in determining virus tropism are thoroughly examined and discussed. The major focus is on viruses that infect humans; however, this cannot

Human Polyomaviruses: Molecular and Clinical Perspectives, Edited by Kamel Khalili and Gerald L. Stoner.
ISBN 0-471-39009-7

be done without including the excellent and seminal work involving polyomaviruses of other species, most notably the mouse.

2. VIRUS RECEPTORS AND TROPISM

The initial step in the establishment of virus infection is the interaction between the virus and receptors present on the surfaces of cells and tissues. In general, viruses that have a very narrow host range and tissue tropism are often shown to interact with high affinity to a limited number of specific receptors present on susceptible cells (Marsh and Helenius, 1989). In some instances virus tropism is strictly determined by the presence of specific receptors that mediate binding and entry (Dalgleish et al., 1984; Klatzman et al., 1984; Mendelsohn et al., 1989; Racaniello, 1990; Tomassini et al., 1989; Weiss and Tailor, 1995). In other instances, successful entry into a cell is necessary but not sufficient for viral replication (Atwood and Norkin, 1989; Bass and Greenberg, 1992; Mei and Wadell, 1995). In these cases additional permissive factors that interact with viral regulatory elements are required.

The complexity of virus host–cell receptor interactions is apparent when one examines the role of CD4 in mediating HIV-1 infection of susceptible cells. Early observations of the selective loss of CD4-positive T cells and subsequent propagation of HIV-1 in T cells led to the suggestion that CD4 was a specific HIV-1 receptor. It was then shown that human cells lacking CD4 were resistant to infection by HIV-1 and that these cells could be rendered susceptible to infection by the introduction of CD4 (Dalgleish et al., 1984; Klatzman et al., 1984). In contrast, mouse cells expressing human CD4 could not be rendered susceptible to infection with HIV-1. In addition, several primary isolates of HIV-1 failed to infect human CD4-positive T-cell lines. These latter studies concluded that CD4 was necessary, but not sufficient, for infection (Clapham, 1991; James et al., 1996). This discrepancy was resolved when several independent groups discovered that multiple co-factors, in addition to CD4, are required for efficient entry of HIV-1. A seven transmembrane G protein–coupled receptor named CXCR-4 has been identified as a co-factor for entry of several T-cell tropic strains of HIV-1 into diverse cell types (Feng et al., 1996). The β-chemokine receptor CCR-5 functions as the co-factor for entry of many macrophage tropic isolates of HIV-1 (Deng et al., 1996; Dragic et al., 1996). Thus, successful infection of a cell with HIV-1 requires the coordinated participation of several cell surface receptors.

3. THE POLYOMAVIRUS VIRION

Polyomaviruses are small, nonenveloped, double-stranded DNA-containing viruses. The virions range in size from 40 to 50 nm, have sedimentation coefficients of 240S in sucrose gradients, and have a density of 1.34 g/ml in cesium

Table 9.1. Capsid Proteins of the Human Polyomaviruses

Virus	Host	Genome Size	Virus-Encoded Proteins (Amino Acids)		
			VP1	VP2	VP3
JCV	Human	5130	354	344	225
BKV	Human	5133	362 (78)[a]	351 (79)	232 (75)
SV40	Rhesus monkey/ human?	5243	362 (75)	352 (72)	234 (66)

[a]Percent identity with JCV.

chloride. The virions are relatively resistant to heat, surviving for up to 1 hour in water at 55°C. The polyomavirus capsids are assembled from three virus-encoded structural proteins termed VP1, VP2, and VP3. VP1, VP2, and VP3 are relatively conserved among the polyomaviruses JCV, BKV and SV40, sharing approximately 75% amino acid identity (Table 9.1). VP1 is the major capsid protein and represents 80% of the total virion protein in the capsid. VP2 and VP3 are minor components and together represent 20% of the total virion protein in the capsid. The virus particles are icosahedrally symmetric (T = 7d) and consist of 360 copies of VP1 arranged in 72 pentamers with both five- and sixfold axes of rotation (Salunke et al., 1986) (Fig. 9.1). The pentameric structures are held together by the carboxy-terminal segments of each VP1 monomer, which interact with adjoining pentamers in the icosahedral shell. The major receptor binding determinants of the polyomaviruses reside on each of the VP1 monomers, which are exposed at the surface of the virion (Stehle et al., 1994). It has been suggested that VP2 may also play a role in virus entry

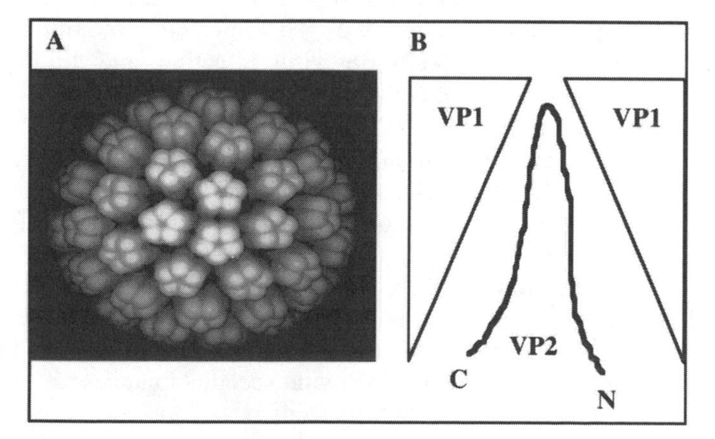

Figure 9.1. (**A**) Computer-generated image of a polyomavirus virion. (**B**) Schematic of the location of VP2 within each VP1 pentamer (only two of the five VP1 molecules are shown).

as the protein extends a loop into the cavity formed within the VP1 pentameric structure (Fig. 9.1) (Chen et al., 1998). VP2 and VP3 may also play a role in anchoring histone-associated viral DNA to VP1 in the capsid; however, virus-like particles (VLPs) consisting of only VP1 can self-assemble and package DNA (Gillock et al., 1997; Salunke et al., 1986). Currently, these VLPs are being developed as gene delivery vehicles. It is unclear whether VLPs share the same cell type specificity as native virions.

4. CELLULAR RECEPTORS FOR THE MOUSE POLYOMA VIRUS

Introduction

The PyV was originally isolated from adenocarcinomas that developed in C3H mice that had been inoculated with extracts prepared from murine leukemia virus–infected AK mice (Gross, 1953). The subsequent demonstration that this virus was capable of inducing a wide range of tumors in experimentally infected newborn mice led to the name *poly-oma* (Stewart et al., 1958). Polyomaviruses have a very broad cellular tropism, infecting more than 30 different cell types (Dawe et al., 1987). This indicates that polyomavirus receptors are either ubiquitously expressed on cells or that polyomaviruses can use more than one receptor. Recent evidence based on the inability to generate monoclonal antibodies that block polyomavirus infection suggest that the latter may be the case (Bauer et al., 1999).

Virus–Receptor Interactions

The interaction between PyV and host cells involves a direct association between the major viral capsid protein VP1 and N-linked glycoproteins bearing specific linkages of sialic acid (Table 9.2). This interaction is critical for viral hemagglutination of red blood cells, for virus infection, and for in vivo tumorigenicity (Cahan et al., 1983; Dubensky et al., 1991; Freund et al., 1991a,b; Fried et al., 1981; Ricci et al., 1992; Sahli et al., 1993).

Early work identified two sialic acid binding variants of mouse polyomavirus, one giving rise to small plaques (RA) and the other to large plaques (PTA) (Diamond and Crawford, 1964). These two strains also differed dramatically in their ability to induce tumors in mice (Dawe et al., 1987). The small plaque strain was found to be very inefficient at inducing tumors in mice, whereas the large plaque strain was highly tumorigenic (Dawe et al., 1987). Early biochemical data demonstrated that a major difference between these two virus strains involved their interaction with specific linkages of sialic acid (Cahan et al., 1983). The small plaque strain was found to recognize both $\alpha(2-3)$-linked and $\alpha(2-6)$-linked sialic acids, whereas the large plaque strain only recognized the $\alpha(2-3)$-linked sugar (Cahan et al., 1983). Subsequent work mapped the ability of these strains to discriminate between $\alpha(2-3)$-linked and

Table 9.2. Polyoma Virus Receptors

Virus[a]	Receptor	Number/ cell	Kd	Tropism	Notes
PyV (LP)	N-linked glycoprotein containing $\alpha(2-3)$-linked sialic acid	25,000	1.8×10^{-11} M	Broad	Mutations in VP1 alter sialic acid recognition and tissue tropism
PyV (SP)	N-linked glycoprotein containing $\alpha(2-3)$-linked sialic acid	25,000	1.8×10^{-11} M	Broad	Also binds to a branched $\alpha(2-6)$-linked sialic acid pseudoreceptor
SV40	MHC class I proteins, also a role for O-linked glycans	90,000	3.76×10^{-12} M	Broad	Only family member that does not interact with sialic acids
LPV	O-linked glycoprotein containing $\alpha(2-6)$-linked sialic acid	1800	2.9×10^{-12} M	Restricted to B cells	Presence of a specific receptor correlates with susceptibility to infection
BKV	Glycolipid containing $\alpha(2-3)$-linked sialic acid	?	?	Broad	
JCV	N-linked glycoprotein containing $\alpha(2-6)$-linked sialic acid	50,000	?	Restricted to glial cells and B cells	Does not share receptor specificity with SV40

[a]LP and SP refer to the large plaque (PTA) and small plaque (RA) isolates of PyV, respectively.

$\alpha(2-6)$-linked sialic acids to a single amino acid polymorphism in the major capsid protein VP1 (Bauer et al., 1995, 1999; Mezes and Amati, 1994; Stehle et al., 1994). The small plaque strain has glycine at position 91 in VP1, which accommodates the presence of a branched $\alpha(2-6)$-linked sialic acid. The large plaque strain has glutamic acid at this position, which prevents binding to the branched $\alpha(2-6)$-linked sugar. It is clear that the $\alpha(2-3)$ linkage is critical for mediating infection of cells by both of these virus strains. The ability of the small plaque strain to recognize $\alpha(2-6)$-linked sialic acids reduces its ability to spread both in vitro and in vivo (Bauer et al., 1999). The $\alpha(2-6)$-linked sugar is therefore considered a pseudoreceptor for this virus.

Interestingly, a third nontumorigenic but highly lethal polyoma virus strain, LID, has glutamic acid at position 91 but has an alanine rather than a valine at position 296 (Bauer et al., 1999). This change leads to less efficient binding to the $\alpha(2-3)$-linked sialic acid, which results in increased viral spread and lethality in the host (Bauer et al., 1999). It is therefore clear that polyoma virus receptor interactions are critical determinants of viral spread and pathogenesis in the host.

Early attempts to identify the proteinaceous component of the PyV receptor used chemical cross-linking and co-immunoprecipitation to identify a 120 kDa protein as a candidate polyoma virus receptor (Griffith and Consigli, 1986; Marriott et al., 1987). Antiserum raised to this protein was subsequently found to block infection of mouse kidney cells by PyV (Griffith and Consigli, 1986; Marriott et al., 1987). The identity of this protein has not been determined. An alternative approach to identifying specific polyomavirus receptors is to generate and screen monoclonal antibodies prepared against permissive cells. This approach was recently attempted, and 2000 hamster monoclonal antibodies directed against mouse 3T3 cells were screened for their ability to block polyoma virus infection (Bauer et al., 1999). Several of the monoclonals were found to bind to the 3T3 cells, but none of them inhibited virus infection. This led to the suggestion that there may be multiple receptor species on 3T3 cells that can be used by polyoma virus to infect the cell (Bauer et al., 1999).

PyV Cell Entry

Early studies examining the entry of PyV into cells found that virions and empty capsids had different receptor specificities (Bolen and Consigli, 1979; Mackay and Consigli, 1976). Virions were found to be selectively targeted to the nucleus, and empty capsids were found to be targeted to phagolysosomes. The binding of virions and empty capsids to mouse cells also differed in their sensitivity to inhibition by neuraminidase. Neuraminidase could, however, inhibit infection of mouse cells with crude virus stocks, and the inhibition could be overcome by incubating the treated cells with β-galactosidase α—2,3-sialyltransferase and CMP-NeuAc (Fried et al., 1981).

Recent data indicate that infectious entry of SV40 and JCV proceeds by different mechanisms (Fig. 9.2). SV40 was found to enter cells by caveolae-

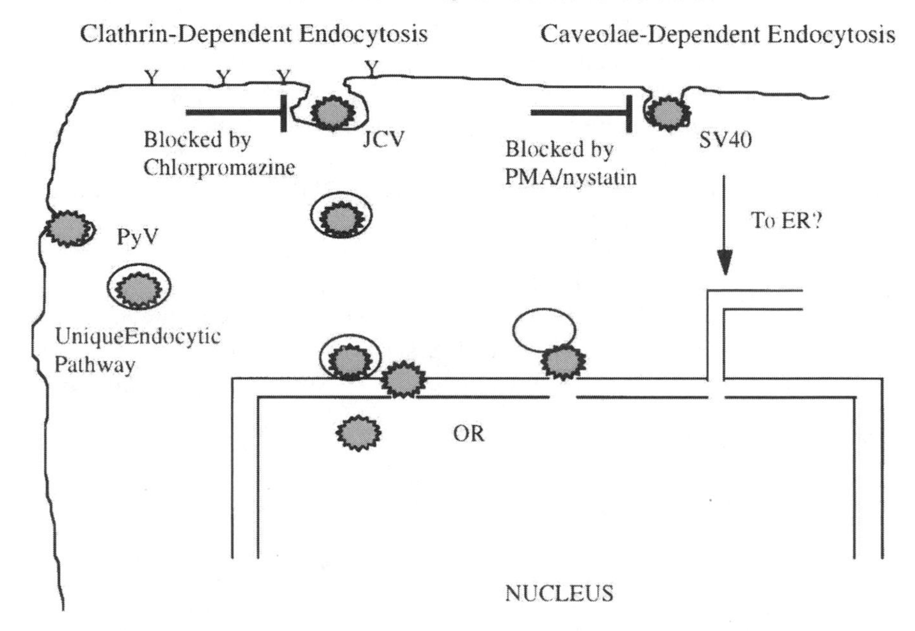

Figure 9.2. Each of the polyomaviruses shown exploits different endocytic pathways to gain access to the cell. JCV uses clathrin-dependent endocytosis, SV40 uses caveolae-dependent endocytosis, and PyV uses a unique pathway involving uncoated vessicles. All three of these viruses must then target their respective genomes to the nucleus. Evidence shows that SV40 virions are targeted at least initially to the endoplasmic reticulum (ER). The virus must then escape this compartment, gain access to the cytoplasm, and deliver the genome to the nucleus via nuclear pores. The mechanisms by which PyV and JCV accomplish this are not known.

mediated endocytosis, and JCV was found to enter cells by classic clathrin-mediated endocytosis (Anderson et al., 1996; Pho et al., 2000; Stang et al., 1997). Similar work with the small plaque isolate of PyV found that this virus uses neither of these pathways to infect baby mouse kidney cells (Gilbert and Benjamin, 2000). It thus appears that each of these related polyomaviruses uses an independent mechanism to enter cells and target their respective genomes to the nucleus (Fig. 9.2).

5. CELLULAR RECEPTORS FOR SV40

Introduction

SV40 was initially isolated from cultures of rhesus monkey kidney cells being used to propagate vaccine strains of poliovirus (Sweet and Hilleman, 1960). At the time of its isolation, several million individuals had received poliovirus vaccines contaminated with significant amounts of this oncogenic virus (Shah and Nathanson, 1976). Fortunately, epidemiologic studies do not show increased rates of cancer among the inoculated individuals (Shah, 2000). However, recent studies suggest that SV40 may in fact circulate in the human population. SV40 has been associated with the development of a number of human cancers, including mesothelioma, osteosarcoma, and choroid plexus papillomas (Barbanti-Brodano et al., 1998; Bergsagel et al., 1992; Carbone et al., 1994, 1996; Lednicky et al., 1995).

SV40 has been the most widely studied polyomavirus, and it has been used as a model system for understanding many aspects of eukaryotic molecular biology, including DNA replication, transcription, and mRNA splicing. SV40 is also the only polyomavirus for which a specific cell surface protein (MHC class I) has been identified as a virus receptor (Atwood and Norkin, 1989; Breau et al., 1992).

Virus–Receptor Interactions

SV40 is unique among polyomaviruses as it does not interact with sialic acids and therefore lacks the ability to hemagglutinate red blood cells. Sialydase treatment of monkey kidney cells also has no effect on either SV40 binding or SV40 infection (Clayson and Compans, 1989). Similarly, treatment of monkey kidney cells with phospholipase C, phospholipase D, trypsin, chymotrypsin, endoglycosidase F, and glycopeptidase F does not inhibit SV40 binding (Clayson and Compans, 1989). The binding of SV40 to monkey kidney cells is saturable and of high avidity. The number of SV40 receptors on cells has been calculated at 9×10^4 with a dissociation constant of 3.76 pM (Table 9.2) (Clayson and Compans, 1989).

The subsequent identification of MHC class I proteins as a principal component of the SV40 receptor came about from experiments designed to determine whether SV40 infection reduced cell surface MHC class I protein expression (Atwood, 1991). It was reasoned that such a mechanism may account for the ability of SV40 to persist in its natural host, the rhesus macaque (Atwood, 1991; Norkin, 1982). During the course of these studies SV40 binding to cells was found to specifically inhibit the subsequent binding of antibodies directed at MHC class I proteins. Conversely, anti-MHC class I antibodies blocked the binding of radiolabeled SV40 to cells and inhibited infection in a dose-dependent manner (Atwood and Norkin, 1989). Subsequent experiments demonstrated that reconstitution of class I protein expression in two separate class I null cell lines could rescue virus binding to these cells but could not rescue infection (Breau et al., 1992). This indicated that MHC class I proteins were necessary but not sufficient for SV40 infection of cells. This was confirmed by a series of studies demonstrating that the distribution and levels of MHC class I protein expression on numerous human cell types did not correlate with their susceptibility to infection (Basak et al., 1992). These data indicate that additional factors contribute to SV40 tropism. These additional factors may include transcription factors and the presence of a co-receptor.

Biochemical data also support a role for MHC class I proteins as cell surface receptors for SV40. For example, the ability of papain to cleave cell surface class I molecules correlates with its ability to inhibit SV40 binding to cells (Clayson and Compans, 1989; Wong et al., 1984). Conversely, MHC class I proteins are resistant to digestion with trypsin and chymotrypsin, and neither of these enzymes inhibits virus binding (Clayson and Compans, 1989; Wong et al., 1984).

The discovery of the SV40 receptor led others to investigate whether related polyomaviruses might also use MHC class I proteins as a receptor. Two polyomaviruses were examined, PyV and JCV, neither of which was found to use MHC class I proteins as a receptor (Liu et al., 1998; Sanjuan et al., 1992).

SV40 Cell Entry

Early work characterizing SV40 entry into cells found that virions internalized into both monopinocytotic vesicles and a larger endosomal compartment (Barbanti-Brodano et al., 1970; Hummeler et al., 1970; Maul et al., 1978). By analogy with similar work in the PyV system, it was assumed that virus in the large endosomes was degraded in phagolysosomes and that virus in monopinocytotic vesicles actually initiated a productive infection (Bolen and Consigli, 1979; Khare and Consigli, 1965; Mackay and Consigli, 1976). Subsequent work studying SV40 entry demonstrated that virus entered via uncoated membrane-bound invaginations that subsequently fused with the endoplasmic reticulum (ER) (Kartenbeck et al., 1989).

These membrane-bound invaginations were subsequently identified as caveolae (Anderson et al., 1996). Caveolae are cholesterol-rich membrane microdomains that have a high concentration of glycosylphosphatidylinositol (GPI)-anchored proteins (Anderson, 1993a,b; Anderson, 1998). They participate in the endocytosis of small-molecular-weight compounds (potocytosis) and in the trafficking of molecules across the cell (transcytosis). Caveolae also participate in the trafficking of plasma membrane components back to the ER (reverse endocytosis). Interestingly, the major cellular receptor for SV40, MHC class I proteins, are enriched in caveolae and participate in localizing SV40 to these membrane domains (Anderson et al., 1998; Stang et al., 1997). The next step in the infectious process is thought to be the release of SV40 from class I heavy chains that is either preceded by or followed by class I heavy chain cleavage by a metalloprotease (Anderson et al., 1998; Chen and Norkin, 1999). The mechanism by which SV40 then targets its genome to the nucleus is not known. The virus presumably escapes into the cytoplasm from the lumen of the endoplasmic reticulum or directly from the endosome. Evidence strongly favors entry of SV40 into the nucleus from the cytoplasm via nuclear pores (Clever et al., 1991; Yamada and Kasamatsu, 1993). It is not clear whether any components of the virion other than nucleic acid enter the nucleus.

6. CELLULAR RECEPTORS FOR THE LYMPHOTROPIC POLYOMAVIRUS

The lymphotropic papovavirus (LPV) was originally isolated from the lymph node of an African green monkey and maintained by continual passage in a B lymphocyte cell line, BJA-B (Brade et al., 1981; zur Hausen and Gissmann, 1979). The host range of LPV is limited to a very narrow subset of human B-cell lines (Takemoto et al., 1982). The restriction to infection of these cells by LPV has been shown to correlate with the presence of specific LPV receptors on permissive cells (Haun et al., 1993; Herrmann et al., 1995, 1997; Keppler et al., 1994, 1995). Binding to, and infection of, human B-cell lines was found to be dependent on the presence of O-linked glycoproteins containing terminal $\alpha(2-6)$-linked sialic acid. The proteinaceous component of LPV receptors has not been determined. The mechanism by which LPV targets its genome to the nucleus has not been studied.

7. CELLULAR RECEPTORS FOR BKV

BKV was originally isolated from the urine of a renal transplant patient and subsequently propagated in Vero cells (Gardner et al., 1971). BKV infects a wide variety of cells, including human and monkey kidney cells, embryonic lung cells, fetal brain cells, fibroblasts, and human foreskin epithelial cells (Yoshiike and Takemoto, 1986). There is little information regarding cellular receptors for BKV or the mechanisms of BKV entry into cells. BKV, like many of the other polyomaviruses, hemagglutinates red blood cells in a sialic acid–dependent manner (Seganti et al., 1981). One report has suggested a role for phospholipids in mediating hemagglutination and infection of Vero cells by BKV (Sinibaldi et al., 1992). The mechanism by which BKV targets its genome to the nucleus has not been studied.

8. CELLULAR RECEPTORS FOR JCV

Introduction

JCV was originally isolated from the brain of a patient with PML and subsequently propagated in cultures of human fetal glial cells (Padgett et al., 1971). As these cells are difficult to obtain, virus isolation and propagation was attempted in a wide variety of other cell types with little or no success (Major et al., 1992). As a result of these early studies, JCV was characterized as having a very narrow host range. In vivo, JCV infection is restricted to oligodendrocytes, astrocytes, and B lymphocytes (Houff et al., 1988; Monaco et al., 1996). This highly restricted cell type specificity is also seen in vitro as virus infection is restricted to primary cultures of human glial cells and to a few established human glial cell lines.

As JCV is the etiologic agent of a fatal central nervous system demeylinating disease, understanding the molecular events governing virus host–cell interactions, including virus–receptor interactions, is critical.

Virus–Receptor Interactions

JCV, like most other polyomaviruses, hemagglutinates red blood cells in a sialic acid–dependent manner. Until recently very little was known about JCV receptors on cells other than erythrocytes. The first suggestion that receptors played a role in tropism came from a comparison of the ability of JCV virions or JCV DNA to initiate early viral gene expression in nonpermissive HeLa cells. In these experiments, JCV virions did not infect HeLa cells, and no early viral gene expression was detected (Schweighardt and Atwood, 2000). In contrast, transfection of JCV DNA into the HeLa cells led to early viral gene expression (Schweighardt and Atwood, 2000). This suggests that one block to infection of nonpermissive cells by JCV is at an early stage in the viral life cycle, perhaps at the level of receptor binding. The nature of the glial cell receptor for JCV was then characterized biochemically. These data demonstrated that infection of glial cells by JCV can be inhibited by treating cells with tunicamycin or by enzymatic removal of $\alpha(2-3)$-linked and $\alpha(2-6)$-linked sialic acids (Liu et al., 1998). A recombinant $\alpha(2-3)$-specific neuraminidase did not inhibit infection of these cells by JCV, suggesting that the $\alpha(2-6)$-linkage is critical. Infection of cells by JCV was also not inhibited by an O-linked glycosylation inhibitor, benzylGalNac, or by trypsin, chymotrypsin, phospholipase A2, or phospholipase C.

These data, when taken together, demonstrate that the JCV receptor is a trypsin resistant N-linked glycoprotein containing $\alpha(2-6)$-linked sialic acid (Liu et al., 1998). These properties distinguish JCV from other polyomaviruses that use sialic acid as cell surface receptors (Table 9.2). It is interesting to note that treatment of glial cells with trypsin reduces virus binding to cells but, paradoxically, leads to increased infectivity (our unpublished observations). This may be due to the elimination of pseudoreceptors for JCV that act to limit accessibility to specific and productive receptor binding sites.

The Relationship Between JCV Binding and Infectivity

Recently, JCV binding to a wide variety of permissive and nonpermissive cells was studied by flow cytometry. JCV bound to all of the cell lines tested regardless of their known susceptibility to infection (Wei et al., 2000). An $\alpha(2-6)$-linked sialic acid–specific lectin, SNA, also bound to all of these cells. It is therefore likely that many cell surface glycolipids and glycoproteins are modified by $\alpha(2-6)$ sialyation but that only a minority of these molecules can serve as a specific receptors for JCV. Interestingly, when JCV binding to primary cells was examined a different story emerged. In these experiments JCV only bound to primary cells that were known to be susceptible to infection.

For example, virus bound to primary human glial cells, to primary tonsillar stromal cells, and to primary human B cells but did not bind to primary human T cells (Wei et al., 2000). This indicates that there may be specificity of JCV binding in vivo that is not apparent when one examines virus binding to tumor cell lines. The identification of a specific glial cell receptor for JCV has remained elusive.

JCV Cell Entry

Recently, the kinetics and mechanisms of JCV entry into human glial cells was compared with that of SV40. These data demonstrated that infectious entry of JCV into human glial cells proceeds more rapidly than infectious entry of SV40 (Pho et al., 2000). The majority of JCV entered into an antibody neutralization resistant compartment within 30 minutes postadsorption, whereas SV40 required 2 hours to enter cells. Specific inhibitors of either caveolae-dependent endocytosis or clathrin-dependent endocytosis were then used to study the mechanism of infectious entry. These experiments confirmed that infectious entry of SV40 is mediated by caveolae- and not by clathrin-dependent endocytosis (Pho et al., 2000). In contrast, infectious entry of JCV was not blocked by inhibitors of caveolae-dependent endocytosis but was blocked by inhibitors of the clathrin-dependent pathway (Pho et al., 2000). These data were confirmed by co-localization of labeled JCV and labeled transferrin in endosomes (Pho et al., 2000). This indicates that infectious entry of JCV proceeds via classic clathrin-dependent endocytosis. These data are also consistent with earlier work demonstrating that JCV and SV40 do not share receptor specificity on human glial cells (Liu et al., 1998).

9. CONCLUSIONS

Sialic acids are major components of cell surface receptors for several human viruses, including orthomyxoviruses, rotaviruses, and all of the polyomaviruses except SV40 (Bass et al., 1991; Chen and Benjamin, 1997; Haun et al., 1993; Ito et al., 1997; Keppler et al., 1995; Mantyjarvi et al., 1972; Willoughby et al., 1990). In each case virus infectivity is mediated by recognition of specific sialic acid linkages to the underlying glycan. In most instances the identities of the proteinaceous components of these receptors is unknown. Interestingly, each of the sialic acid–utilizing polyomaviruses recognizes different linkages of this common carbohydrate moiety.

It is also becoming clear that members of this highly related family of viruses do not share receptor specificity. For example, PyV and SV40 do not compete with each other for binding to cells and do not share MHC class I as a specific cellular receptor. Similarly, JCV and SV40 do not share receptor specificity. It is also becoming apparent that the choice of receptor dictates the endocytic pathway that each virus uses to gain entry into the cells that it infects.

For example, JCV, SV40, and PyV have different receptor specificities, and all exploit different endocytic pathways to target their genomes to the nucleus.

Direct comparisons of JCV with BKV, PyV, or LPV have not been done. This is due to the fact that these viruses all have distinct cellular tropisms that in large part are governed at the transcriptional level. Current efforts to develop polyomavirus pseudovirions that package defined nucleic acids capable of expressing reporter genes in a wide range of cells should allow for these comparisons to be made in the future. Also, efforts are underway to construct chimeric polyomaviruses that only differ with respect to the late-coding region. The identification of receptors for each of these important polyomaviruses should lead to a better understanding of the critical early events in the life cycle of these viruses. This in turn will allow a better understanding of polyomaviral tropism, spread, and pathogenicity in the host. Finally, this may make possible the development of novel therapeutic strategies to prevent infection and spread of these viruses in the immunocompromised host.

Recently, SV40 has been shown to traffic from the plasma membrane to a new subcellular organelle referred to as the caveosome. Transport to the caveosome was shown to be actin dependent, and transport out of the caveosome to smooth ER was shown to be microtubule dependent. (Pelkmans L, Kartenbeck J, Helenius A (2001) Caveolar endocytosis of simian virus 40 reveals a new two-step vesicular-transport pathway to the ER. Nat Cell Biol 3:473–483).

REFERENCES

Anderson HA, Chen Y, Norkin LC (1996): Bound simian virus 40 translocates to caveolin enriched membrane domains, and its entry is inhibited by drugs that selectively disrupt caveolae. Mol Biol Cell 7:1825–1834.

Anderson HA, Chen Y, Norkin LC (1998): MHC class I molecules are enriched in caveolae but do not enter with simian virus 40. J Gen Virol 79:1469–1477.

Anderson RGW (1993a): Caveolae: Where incoming and outgoing messengers meet. Proc Natl Acad Sci USA 90:10909–10913.

Anderson RGW (1993b): Potocytosis of small molecules and ions by caveolae. Trends Cell Biol 3:69–72.

Anderson RGW (1998): The caveolae membrane system. Annu Rev Biochem 67:199–225.

Atwood WJ (1991): Major Histocompatibility Complex Encoded HLA Class I Proteins Are Cell Surface Receptors for the Simian Virus 40. Ph.D. Dissertation, University of Massachusetts.

Atwood WJ, Norkin LC (1989): Class I major histocompatibility proteins as cell surface receptors for simian virus 40. J Virol 63:4474.

Barbanti-Brodano G, Martini F, De Mattei M, Lazzarin L, Corallini A, Tognon M (1998): BK and JC human polyomaviruses and simian virus 40: Natural history of infections in humans, experimental oncogenicity, and association with human tumors. Adv Virus Res 50:69–99.

Barbanti-Brodano G, Swetly P, Koprowski H (1970): Early events in the infection of permissive cells with simian virus 40: Adsorption, penetration, and uncoating. J Virol 6:78–86.

Basak S, Turner H, Compans RW (1992): Expression of SV40 receptors on apical surfaces of polarized epithelial cells. Virology 190:393–402.

Bass DM, Greenberg HB (1992): Strategies for the identification of icosahedral virus receptors. J Clin Invest 89:3–9.

Bass DM, Mackow E, Greenberg HB (1991): Identification and partial characterization of rhesus rotavirus binding glycoprotein on murine enterocytes. Virology 183:602–610.

Bauer PH, Bronson RT, Fung SC, Freund R, Stehle T, Harrison SC, Benjamin TL (1995): Genetic and structural analysis of a virulence determinant in polyomavirus VP1. J Virol 69:7925–7931.

Bauer PH, Cui C, Stehle T, Harrison SC, DeCaprio JA, Benjamin TL (1999): Discrimination between sialic acid containing receptors and pseudoreceptors regulates polyomavirus spread in the mouse. J Virol 73:5826–5832.

Bergsagel DJ, Finegold MJ, Butel JS, Kupsky WJ, Garcea RL (1992): DNA sequences similar to those of simian virus 40 in ependymomas and choroid plexus tumors of childhood. N Engl J Med 326:988–993.

Bolen JB, Consigli RA (1979): Differential adsorption of polyoma virions and capsids to mouse kidney cells and guinea pig erythrocytes. J Virol 32:679–683.

Brade L, Vogl W, Gissman L, zur Hausen H (1981): Propagation of B-lymphotropic papovavirus (LPV) in human B-lymphoma cells and characterization of its DNA. Virology 114:228–235.

Breau WC, Atwood WJ, Norkin LC (1992): Class I major histocompatibility proteins are an essential component of the simian virus 40 receptor. J Virol 66:2037–2045.

Cahan LD, Singh R, Paulson JC (1983): Sialyloligosaccharide receptors of binding variants of polyomavirus. Virology 130:281–289.

Carbone M, Pass HI, Rizzo P, Marinetti MR, Di Muzio M, Mew DJY, Levine AS, Procopio A (1994): Simian virus 40-like sequences in human pleural mesothelioma. Oncogene 9:1781–1790.

Carbone M, Rizzo P, Procopio A, Giuliano M, Pass HI, Gebhardt MC, Mangham C, Hansen M, Malkin DF, Bushart G, Pompetti F, Picci P, Levine AS, Bergsagel JD, Garcea RL (1996): SV40-like sequences in human bone tumors. Oncogene 13:527–535.

Chen MH, Benjamin T (1997): Roles of N-glycans with alpha 2,6 as well as alpha 2,3 linked sialic acid in infection by polyoma virus. Virology 233:440–442.

Chen XS, Stehle T, Harrison SC (1998): Interaction of polyomavirus internal protein VP2 with the major capsid protein VP1 and implications for participation of VP2 in viral entry. EMBO J 17:3233–3240.

Chen Y, Norkin LC (1999): Extracellular simian virus 40 transmits a signal that promotes virus enclosure within caveolae. Exp Cell Res 246(1):83–90.

Clapham P (1991): Human immunodeficiency virus infection of nonhematopoietic cells. The role of CD4 independent entry. Rev Med Virol 1:51–58.

Clayson ET, Compans RW (1989): Characterization of simian virus 40 receptor moieties on the surfaces of Vero C1008 cells. J Virol 63:95–1100.

Clever J, Yamada M, Kasamatsu H (1991): Import of simian virus 40 virions through nuclear pore complexes. Proc Natl Acad Sci USA 88:7333–7337.

Dalgleish A, Beverly PC, Clapham P, Crawford D, Greaves M, Weiss R (1984): The CD4 (T4) antigen is an essential component of the receptor for the AIDS retrovirus. Nature 312:763–767.

Dawe CJ, Freund R, Mandel G, Balmer-Hofer K, Talmage DA, Benjamin TL (1987): Variations in polyomavirus genotype in relation to tumor induction in mice: Characterization of wild type strains with widely differeing tumor profiles. Am J Pathol 127:243–261.

Deng H, Liu R, Ellmeier W, Choe S, Unatmaz D, Burkhart M, DiMarzio P, Marmon S, Sutton RE, Hill CM, Davis CB, Peiper SC, Schall TJ, Littman DR, Landau NR (1996): Identification of a major co-receptor for primary isolates of HIV-1. Nature 381:661–666.

Diamond L, Crawford LV (1964): Some characteristics of large plaque and small plaque lines of polyomavirus. Virology 22:235–244.

Dragic T, Litwin V, Allaway GP, Martin SR, Huang Y, Nagashima KA, Cayanan C, Maddon PJ, Koup RA, Moore JP, Paxton WA (1996): HIV-1 entry into CD4 cells is mediated by the chemokine receptor CC-CKR-5. Nature 381:667–673.

Dubensky TW, Freund R, Dawe CJ, Benjamin TL (1991): Polyomavirus replication in mice: Influences of VP1 type and route of inoculation. J Virol 65:342–349.

Feng Y, Broder CC, Kennedy PE, Berger EA (1996): HIV-1 entry cofactor: Functional cDNA cloning of a seven-transmembrane, G protein–coupled receptor. Science 272: 872–877.

Freund R, Calderone A, Dawe CJ, Benjamin TL (1991a): Polyomavirus tumor induction in mice, effects of polymorphisms of VP1 and large T antigen. J Virol 65:335–341.

Freund R, Garcea RL, Sahli R, Benjamin TL (1991b): A single amino acid substitution in polyomavirus VP1 correlates with plaque size and hemagglutination behavior. J Virol 65:350–355.

Fried H, Cahan LD, Paulson JC (1981): Polyoma virus recognizes specific sialyoligosaccharide receptors. Virology 109:188–192.

Gardner SD, Field AM, Coleman DV, Hulme B (1971): New human papovavirus (B.K.) isolated from urine after renal transplantation. Lancet 1:1253–1257.

Gilbert JM, Benjamin TL (2000): Early steps of polyomavirus entry into cells. J Virol 74:8582–8588.

Gillock ET, Rottinghaus S, Chang D, Cai X, Smiley SA, An K, Consigli RA (1997): Polyomavirus major capsid protein VP1 is capable of packaging cellular DNA when expressed in the baculovirus system. J Virol 71:2857–2865.

Griffith GR, Consigli RA (1986): Cross-linking of a polyoma attachment protein to its mouse kidney cell receptor. J Virol 58:773–781.

Gross L (1953): A filterable agent, recovered from AK leukemic extracts, causing salivary gland carcinomas in C3H mice. Proc Soc Exp Biol Med 83:414.

Haun G, Keppler OT, Bock CT, Herrmann M, Zentgraf H, Pawlita M (1993): The cell surface receptor is a major determinant restricting the host range of the B-lymphotropic papovavirus. J Virol 67:7482–7492.

Herrmann M, Oppenlander M, Pawlita M (1995): Fast and high affinity binding of B-lymphotropic papovavirus to human B-lymphoma cell lines. J Virol 69:6797–6804.

Herrmann M, Wilhelm von der Leith C, Stehling P, Reutter W, Pawlita M (1997): Consequences of subtle sialic acid modification on the murine polyomavirus receptor. J Virol 71(8):5922–5931.

Houff SA, Major EO, Katz DA, Kufta CV, Sever JL, Pittaluga S, Roberts JR, Gitt J, Saini N, Lux W (1988): Involvement of JC virus–infected mononuclear cells from the bone marrow and spleen in the pathogenesis of progressive multifocal leukoencephalopathy. N Engl J Med 318:301–305.

Hummeler K, Tomassini N, Sokol F (1970): Morphological aspects of the uptake of simian virus 40 by permissive cells. J Virol 6:87–93.

Ito T, Suzuki Y, Mitnaul L, Vines A, Kida H, Kawaoka Y (1997): Receptor specificity of influenza A viruses correlates with the agglutination of erythrocytes from different animal species. Virology 227:493–499.

James W, Weiss RA, Simon JH (1996): The receptor for HIV: Dissection of CD4 and studies on putative accessory factors. Curr Top Microbiol Immunol 205:137–158.

Kartenbeck J, Stukenbrok H, Helenius A (1989): Endocytosis of simian virus 40 into the endoplasmic reticulum. J Cell Biol 109:2721–2729.

Keppler OT, Herrmann M, Oppenlander M, Meschede W, Pawlita M (1994): Regulation of susceptibility and cell–surface receptor for the B-lymphotropic papovavirus by N–glycosylation. J Virol 68:6933–6939.

Keppler OT, Stehling P, Herrmann M, Kayser H, Grunow D, Reutter W, Pawlita M (1995): Biosynthetic modulation of sialic acid–dependent virus-receptor interactions of two primate polyoma viruses. J Biol Chem 270:1308–1314.

Khare GP, Consigli RA (1965): Multiplication of polyoma virus: Use of selectively labeled (3H) virus to follow the course of infection. J Bacteriol 90:819–821.

Klatzman D, Champagne E, Chamaret S, Gruest J, Guetard D, Hercend T, Gluckman J, Montagnier L (1984): T lymphocyte T4 molecule behaves as the receptor for human retrovirus LAV. Nature 312:767–768.

Lednicky JA, Garcea RL, Bergsagel DJ, Butel JS (1995): Natural simian virus 40 strains are present in human choroid plexus and ependymoma tumors. Virology 212:710–717.

Liu CK, Hope AP, Atwood WJ (1998a): The human polyomavirus, JCV, does not share receptor specificity with SV40 on human glial cells. J Neurovirol 4:49–58.

Liu CK, Wei G, Atwood WJ (1998b): Infection of glial cells by the human polyomavirus JC is mediated by an N-linked glycoprotein containing terminal alpha (2–6)–linked sialic acids. J Virol 72:4643–4649.

Mackay RL, Consigli RA (1976): Early events in polyoma virus infection: Attachment, penetration, and nuclear entry. J Virol 19:620–636.

Major EO, Amemiya K, Tornatore CS, Houff SA, Berger JR (1992): Pathogenesis and molecular biology of progressive multifocal leukoencephalopathy, the JC virus-induced demyelinating disease of the human brain. Clin Microbiol Rev 5:49–73.

Mantyjarvi RA, Arstila PP, Meurman OH (1972): Hemagglutination by BK virus, a tentative new member of the papovavirus group. Infect Immun 6:824–828.

Marriott SJ, Griffith GR, Consigli RA (1987): Octyl-beta-D-glucopyranoside extracts polyomavirus receptor moieties from the surfaces of mouse kidney cells. J Virol 61: 375–382.

Marsh M, Helenius A (1989): Virus entry into animal cells. Adv Virus Res 36:107–151.

Maul GG, Rovera A, Vorbrodt A, Abramczuk J (1978): Membrane fusion as a mechanism for simian virus 40 entry into different cellular compartments. J Virol 28: 936–944.

Mei YF, Wadell G (1995): Molecular determinants of adenovirus tropism. Curr Top Microbiol Immunol 199:213–228.

Mendelsohn CL, Wimmer E, Racaniello VR (1989): Cellular receptor for poliovirus: Molecular cloning, nucleotide sequence, and expression of a new member of the immunoglobulin superfamily. Cell 56:855–865.

Mezes B, Amati P (1994): Mutations of polyomavirus VP1 allow in vitro growth in undifferentiated cells and modify in vivo tissue replication specificity. J Virol 68: 1196–1199.

Monaco MGC, Atwood WJ, Gravell M, Tornatore CS, Major EO (1996): JCV infection of hematopoietic progenitor cells, primary B lymphocytes, and tonsillar stromal cells: Implication for viral latency. J Virol 70:7004–7012.

Norkin LC (1982): Papovaviral persistent infections. Microbiol Rev 46:384–425.

Padgett BL, Walker DL, Zu Rhein G, Eckroade RJ, Dessel BH (1971): Cultivation of papova-like virus from human brain with progressive multifocal leucoencephalopathy. Lancet 1:1257–1260.

Pho MT, Ashok A, Atwood WJ (2000): JC virus enters human glial cells by clathrin dependent receptor mediated endocytosis. J Virol 74(5):2288–2292.

Racaniello VR (1990): Cell receptors for picornaviruses. Curr Top Microbiol Immunol 161:1–22.

Ricci L, Maione R, Passananti C, Felsani A, Amati P (1992): Mutations in the VP1 coding region of polyomavirus determine differentiating stage specificity. J Virol 66: 7153–7158.

Sahli R, Freund R, Dubensky T, Garcea R, Bronson R, Benjamin T (1993): Defect in entry and altered pathogenicity of a polyoma virus mutant blocked in VP2 myristylation. Virology 192:142–153.

Salunke DM, Caspar DL, Garcea RL (1986): Self–assembly of purified polyomavirus capsid protein VP1. Cell 46:895–904.

Sanjuan N, Zijlstra M, Carroll J, Jaenisch R, Benjamin T (1992): Infection by polyomavirus of murine cells deficient in class I major histocompatibility complex expression. J Virol 66:4587–4590.

Schweighardt B, Atwood WJ (2001): Glial cells as targets of viral infection in the human central nervous system. In Progress in Brain Research, Vol. 132, Castellano Lopez B, Nieto-Sampedro M, Eds; Elsevier Science: Amsterdam, pp. 731–745.

Seganti L, Mastromarino P, Superti F, Sinibaldi L, Orsi N (1981): Receptors for BK virus on human erythrocytes. Acta Virol 25:177–181.

Shah K, Nathanson N (1976): Human exposure to SV40: Review and comment. Am J Epidemiol 103:1–12.

Shah KV (2000): Does SV40 infection contribute to the development of human cancers? Rev Med Virol 10:31–43.

Sinibaldi L, Goldoni P, Pietropaolo V, Cattani L, Peluso C, Di Taranto C (1992): Role of phospholipids in BK virus infection and haemagglutination. Microbiologica 15: 337–344.

Stang E, Kartenbeck J, Parton RG (1997): Major histocompatibility complex class I molecules mediate association of SV40 with caveolae. Mol Biol Cell 8:47–57.

Stehle T, Yan Y, Benjamin TL, Harrison SC (1994): Structure of murine polyomavirus complexed with an oligosaccharide receptor fragment. Nature 369:160–163.

Stewart SE, Eddy BE, Borgese N (1958): Neoplasms in mice inoculated with a tumor agent carried in tissue culture. J Natl Cancer Inst 20:1223.

Sweet B, Hilleman M (1960): The vacuolating virus, SV40. Proc Soc Exp Biol Med 105:420.

Takemoto KK, Furuno A, Kato K, Yoshiike K (1982): Biological and biochemical studies of African green monkey lymphotropic papovavirus. J Virol 42:502–509.

Tomassini JE, Graham D, DeWitt CM, Lineberger DW, Rodkey JA, Colonno RJ (1989): cDNA cloning reveals that the major group rhinovirus receptor on HeLa cells is intercellular adhesion molecule 1. Proc Natl Acad Sci USA 86:4907–4911.

Wei G, Liu CK, Atwood WJ (2000): JC virus binds to primary human glial cells, tonsillar stromal cells, and B-lymphocytes, but not to T-lymphocytes. J Neurovirol 6:127–136.

Weiss RA, Tailor CS (1995): Retrovirus receptors. Cell 82:531–533.

Willoughby RE, Yolken RH, Schnaar RL (1990): Rotaviruses specifically bind to the neutral glycosphingolipid asialo-GM1. J Virol 64:4830–4835.

Wong GH, Bartlett PF, Clark-Lewis I, Battye F, Shrader JW (1984): Inducible expression of H-2 and Ia on brain cells. Nature 310:688–691.

Yamada M, Kasamatsu H (1993): Role of nuclear pore complex in simian virus 40 nuclear targeting. J Virol 67:119–130.

Yoshiike K, Takemoto KK (1986): Studies with BK virus and monkey lymphotropic papovavirus. In The Papovaviridae, Salzman NP, Ed; Plenum Press: New York, Vol 1, pp 295–326.

zur Hausen H, Gissmann L (1979): Lymphotropic papovavirus isolated from African green monkey and human cells. Microbiol Immunol 167:137–153.

10

LATENT AND PERSISTENT POLYOMAVIRUS INFECTION

KRISTINA DÖRRIES, M.D.

1. AFFECTED ORGANS AND CELL TYPES IN POLYOMAVIRUS-ASSOCIATED DISEASE AND PERSISTENT VIRUS INFECTION

The two human polyomaviruses BKV and JCV are associated with persistent infection and diseases of the urogenital tract and the central nervous system (CNS). Induction of disease by the viruses is regularly linked to states of immunoincompetence. The most prominent underlying complications are AIDS and lymphoproliferative disorders. Moreover, iatrogenic immunosuppression in the course of transplantation or therapy of autoimmune disorders can contribute to polyomavirus-induced disease. Clinically overt disease usually correlates with enhanced activity of viral expression in the target organ of viral persistence. Detailed analysis of BKV- and JCV-associated diseases disclosed a variety of organs and cell types to be susceptible to virus infection. These observations fostered multiple studies on the molecular basis of polyomavirus persistence in the infected host. However, despite the availability of very sensitive techniques to discover virions and/or viral products in tissue, body fluids, or cells from infected persons, essential questions with respect to involvement of distinct organs and cell types in polyomavirus persistence are far from being answered unequivocally.

Human Polyomaviruses: Molecular and Clinical Perspectives, Edited by Kamel Khalili and Gerald L. Stoner.
ISBN 0-471-39009-7

Polyomavirus Infection of the Urogenital Tract

Urogenital Diseases Associated with BKV and Asymptomatic Infection. BKV is a urotheliotropic virus, which was originally detected in the urine of a patient with ureteral stenosis after renal transplantation (RT) (Gardner et al., 1971). Nevertheless, studies from recent years suggest that interstitial tubular nephritis is the most frequent BKV-associated disease after RT (Mathur et al., 1997; Pappo et al., 1996; Purighalla et al., 1995). Clinical features may mimic graft rejection or drug toxicity (Binet et al., 1999; Randhawa et al., 1999), but histopathologic examination almost always shows interstitial infiltrates of plasma cells and lymphocytes, interstitial fibrosis, tubular atrophy, and large intranuclear inclusions in tubular epithelial cells. Cells of the transitional bladder epithelium were identified as target cells for BKV infection (Gerber et al., 1980). Virus isolation, DNA detection by polymerase chain reaction (PCR), electronmicroscopy, immunohistologic staining of BKV proteins, and in situ hybridization (ISH) suggests an etiopathologic role of the virus in about 5% of RT patients (Binet et al., 1999; Randhawa et al., 1999). With the introduction of new immunosuppressive strategies, the incidence of BKV-associated complications in RT increased further, thus confirming the close relationship between excessive virus growth and immunologic impairment.

Renal failure due to BKV infection is diagnosed in patients with underlying lymphoma, hereditary immunodeficiencies (de Silva et al., 1995; Rosen et al., 1983), in renal transplant patients (Mathur et al., 1997; Pappo et al., 1996; Purighalla et al., 1995), and with AIDS (Bratt et al., 1999; Nebuloni et al., 1999). Interstitial inflammation and focal necrosis of tubular epithelium with enlargement of tubular epithelial cells and the pelvic urothelium accompany this rare complication. Aggregates of viral particles within nuclei of renal tubular cells (Rosen et al., 1983) confirm susceptibility of the urogenital tract epithelium for BKV infection.

Hemorrhagic cystitis (HC) is a serious BKV-associated complication of bone marrow transplantation (BMT) patients. Prevalence of HC varies from 10% to 68% and leads to severe hemorrhage in about 25% of bone marrow recipients (Arthur et al., 1985; Azzi et al., 1994; Bedi et al., 1995; Chan et al., 1994; Cotterill et al., 1992). Hemorrhage and viruria most likely are due to viral activation in the uroepithelium, as virus particles can be detected in exfoliated urinary cells by means of electron microscopy (Hiraoka et al., 1991). Prolonged hematuria is associated with severe morbidity and increasing viral load in urine (Azzi et al., 1999).

In patients with AIDS, systemic BKV-associated disease involves infection of the CNS, the lung, the eye, and the kidney (Bratt et al., 1999; Hedquist et al., 1999; Nebuloni et al., 1999; Smith et al., 1998; Vallbracht et al., 1993). Concomitantly with multiple lesions in the entire nephron, an interstitial tubulonephritis is seen. Desquamated tubular cells that display focally enlarged, eosinophilic nuclei carry BKV particles and express viral products (Cubukcu-Dimopulo et al., 2000; Nebuloni et al., 1999; Vallbracht et al., 1993). Investigation of tubular cells at the subcellular level revealed the presence of virions in cytoplasmic reticular cisternae (Nebuloni et al., 1999), and virus-specific antigen was located in nuclei and cytoplasm (Nebuloni et al., 1999; Smith et al., 1998). Interstitial clustering of CD68-positive cells with cytoplasmic staining of virus-specific antigens indicated uptake of the virus by phagocytotic cells (Nebuloni et al., 1999). Further analysis at the ultrastructural level detected characteristic viral particles, which occasionally formed dense crystalloid arrays, thus confirming viral presence in these cells (Smith et al., 1998).

In contrast to renal disease under severe immunoincompetence, BKV infection in other patients is most likely in an asymptomatic state. In the course of lymphoproliferative diseases a clinically silent BKV-specific involvement of the urothelial tract has been reported. Virus DNA is distributed in small foci throughout the cortex and medulla of the kidney (Heritage et al., 1981), affecting renal epithelial cells and lining cells of ureter and bladder as demonstrated by BKV-specific immunostaining. In asymptomatic tissue BKV DNA persists in episomal form, and exfoliated cells carry intranuclear inclusions containing BKV antigen (Shinohara et al., 1993; Zu Rhein and Varakis, 1974). The rate of infection varies from 13% (Chesters et al., 1983) to more than 50% (Dörries and Elsner, 1991). This may depend on the study groups, which included patients with underlying diseases ranging from carcinoma and leukemia to inflammatory diseases, coronary heart disease, and multiple injuries from traffic accidents.

Thus far, no data are available on BKV in kidney tissue in the course of pregnancy. However, there is a study addressing the question of transplacental transmission of BKV in humans using tissue specimens from aborted fetuses and placenta. BKV DNA was detected by PCR in 60% of fetal kidney and in 80% of placenta samples. In maternal tissue from a control group with normal pregnancies, BKV was detected in 50% of the samples (Pietropaolo et al., 1998). Besides the finding that BKV might be transmitted vertically, the enhanced detection rates after abortion point to an activated BKV infection and a prevalence of persistent virus infection in about 80% of the population.

Analysis of prostate biopsy specimens for BKV revealed virus DNA in 58% of asymptomatic tissue and in 87% of prostate hyperplasias sampled from prostate carcinoma patients. This was comparable to the rate in bladder tissue. In addition, 70% of cervix and vulvar tissue exhibited viral DNA, and analysis of sperm gave an incidence of 95% for the presence of BKV DNA. In contrast, DNA of the second polyomavirus JC was found in only 5% of cervical and vulvar tissue specimens. Glandular tissue yielded no JCV DNA, whereas 21% of sperm samples were positive for JCV DNA. The high rate of BKV DNA present in asymptomatic tissue and semen suggests that these sites should be considered as locations of polyomavirus persistence (Monini et al., 1995, 1996; Shinohara et al., 1993).

Activation of BKV Infection in the Urogenital Tract. Virus infection of the kidney following viruria probably occurs at the time of primary infection in children (Di Taranto et al., 1997). Beyond childhood, viruria is more likely due to activation of persistent renal infection than to primary infection or re-infection. Virus products in the urine at different stages of immunologic incompetence and a high level of viruria in asymptomatic infection (Arthur et al., 1989; Coleman et al., 1980; Hogan et al., 1980; Kitamura et al., 1990; Reese et al., 1975) are strong indications for a relationship between immunodeficiency and polyomavirus activation.

The role of HIV-related immunodeficiency in BKV-associated viruria was analyzed in patients without BKV-associated diseases. Virus DNA was detected by PCR at a rate of 37–44% in the United Kingdom, Italy, and Tanzania (Agostini et al., 1995; Degener et al., 1997; Jin et al., 1995; Knowles et al., 1999). Lower rates of 20–24% were found in North America and Norway (Markowitz et al., 1993; Shah et al., 1997; Sundsfjord et al., 1994a). However, in all study groups an increase of BKV viruria under AIDS was reported on a basis of less than 8% in normal individuals. Staging of HIV patients according to their immune status revealed an inverse relationship of decreasing T-lymphocyte count and increasing viruria (Jin et al., 1995; Knowles et al., 1999; Markowitz et al., 1993; Shah et al., 1997; Sundsfjord et al., 1994a). From these studies it appears likely that BKV activation in the urinary tract under AIDS is correlated with the state of immunoincompetence. Contrasting reports with 6% BKV viruria in AIDS and AIDS/PML patients as detected by nested PCR technique could probably be explained by technical differences between laboratories (Brouqui et al., 1992; Ferrante et al., 1997).

Viruria in BMT patients was reported as early as 1975 (Reese et al., 1975). Although before transplantation only 1% of patients shed BKV (Arthur et al., 1985), increases have been reported to 22%, 48% (Arthur et al., 1988, 1989; Cotterill et al., 1992; Jin et al., 1993), 67% (Jin et al., 1995), and even 100% of patients in the post-transplant period if classic methods were combined (Gibson et al., 1985). With the advent of the more sensitive PCR, the prevalence of viruria was found to be continuously higher, ranging from 50% to 100%, if multiple samples from one patient were analyzed (Azzi et al., 1994, 1996; Bogdanovic et al., 1994; Flaegstad et al., 1991). BKV viruria often began 2–9 weeks after transplantation. The duration was variable and resolved spontaneously after several weeks of shedding. Virus excretion was not related to graft-versus-host disease, thus revealing that viruria after BMT is a normal asymptomatic activation event of a persistent BKV infection that often is not associated with disease. However, the involvement of BKV in hemorrhagic cystitis in about 20% of allogeneic BMT patients (Azzi et al., 1994) prompted a study of microscopic hematuria (Chan et al., 1994). Viruria in 51% of BMT patients in a 4 month period after transplantation was closely associated with hematuria in about half of the patients (Chan et al., 1994; Jin et al., 1995). Similar to asymptomatic virus shedding, most episodes of hematuria were self-limiting. From these findings it can be suggested that BKV infection after BMT is usually activated to a detectable level of virus load in the urine and might even progress to histologic destruction before virus growth comes under control.

The first carrier of a BKV infection to be described (Gardner et al., 1971) had polyomavirus viruria after RT. Virus particles were characterized by electron microscopy and virus isolation. The virus sometimes affected epithelial cells of both the recipient and donor tissues (Coleman et al., 1973). In contrast to BMT or pregnancy, after renal transplantation primary polyomavirus infection, as defined by antibody titer rises, appears to play a role by introduction of the virus into the recipient by an infected donor (Andrews et al., 1983, 1988; Arthur et al., 1989). The prevalence of polyomavirus viruria ranged from 0% to 47% in early reports, the majority of patients being asymptomatic. Even the use of PCR did not enhance the rate of detection (Sundsfjord et al., 1994a). The discrepancies among the studies might be associated with the lower sensitivities of early methods, but highly variable therapeutic schemes may also contribute to the findings (Lecatsas et al., 1973; Shah et al., 1974). In addition, examination of multiple samples revealed an intermittent and sporadic course of virus excretion as well as a highly variable duration of viruria, ranging from periods of several weeks to years. Observations over more than 3 years revealed a prevalence of 44% viruria in all patients studied. Molecular characterization of the polyomavirus species suggested that BKV viruria was more prominent in RT patients than JCV viruria (Arthur et al., 1989; Boubenider et al., 1999; Gibson and Gardner, 1983). However, viruria often was unrelated to changes in the clinical condition or treatment or even to ureteric obstruction, demonstrating that viruria might not necessarily be associated with RT (Andrews et al., 1988; Arthur et al., 1989).

Under other immunosuppressive conditions viruria often is intermittent, with sparsely distributed infected cells in urine pointing to a rather low rate of activation. Jin et al. (1993) studied a group of patients after cardiac transplantation together with BMT patients. BKV was detected in 50% of the BMT patients and in 25% of the cardiac transplantation group. In all cases excretion was intermittent and sparse and was more often correlated with older age and more aggressive underlying disease. Activating therapeutic influences in these patients could not be stated. Besides the obvious correlation with age, the data further corroborate severe immunoimpairment as a factor for activation and urinary excretion (Hogan et al., 1983; Jin et al., 1993).

In patients with autoimmune diseases in Taiwan, about 40% were found by PCR to be excreters of polyomaviruses. Interestingly, in 15%, double infections were detected, but none was positive for BKV viruria alone (Chang et al., 1996a). Further extension of the study confirmed the lack of BKV viruria in Taiwan (Tsai et al., 1997). In contrast, patients from Scandinavia with systemic lupus erythematosus had higher levels of BK viruria than healthy control subjects, whereas JCV shedding was in the range of the normal sex-matched control group. In a follow-up study, the authors found a high prevalence of intermittent or even continuous shedding of BKV at 1 year. Immunosuppressive drugs such as corticosteroids, azathioprine, cyclophosphamide, and/or methotrexate did not influence the kinetics of virus shedding (Rekvig et al., 1997; Sundsfjord et al., 1999). Clearly, an autoimmune disease is able to activate a persistent BKV infection to an extent that is not observed in normal individuals.

Pregnancy is the most common condition of altered immunocompetence that has been linked to polyomavirus activation (Coleman et al., 1980; Lecatsas et al., 1981). The onset of virus excretion is related to time of gestation, most often occurring late in the second and during the third trimester. Once established, excretion continues intermittently to term and might even extend into the postpartum period (Coleman et al., 1980; Gardner and Knowles, 1994). Serologic studies revealed that excretion in pregnant women is usually the result of virus activation of a persistent infection. Although activation of infection had no clinical significance (Arthur et al., 1989), women excreting polyomaviruses had more illness before and during pregnancy and may have underlying diseases such as diabetes and sarcoidosis (Gardner and Knowles, 1994). When BKV and JCV viruria were differentiated, BKV excretion rates of 15–25% were observed (Jin et al., 1993; Markowitz et al., 1991). The incidence of JCV viruria in the same geographic regions was about 7% (Gardner and Knowles, 1994; Markowitz et al., 1991), clearly demonstrating a higher incidence of BKV than JCV activation. With more advanced PCR techniques, the incidence of BKV viruria was 47% in 40 pregnant women compared to 19 healthy adults with no viruria (Jin et al., 1995). Similar to the findings in patients with systemic lupus erythematosus, in Taiwan an incidence of about 3% of BKV excretion was reported in pregnancy on the basis of 26% JCV positive urine samples and 6.5% of samples with double infections (Chang et al., 1996b; Tsai et al., 1997). This suggests that BK viruria is generally less pronounced in Taiwan than elsewhere.

Asymptomatic immunocompetent patients and healthy individuals were often included as control groups in studies of polyomavirus excretion. In contrast to the above-mentioned basic diseases and the pregnant state, the prevalence of BKV excretion during immunocompetency ranged from 0% (Bogdanovic et al., 1994; Degener et al., 1997; Jin et al., 1993, 1995; Sundsfjord et al., 1994a; Tsai et al., 1997) to about 18% (Arthur et al., 1985, 1989; Azzi et al., 1996; Kitamura et al., 1990; Markowitz et al., 1993; Shah et al., 1997). The exception is a study by Azzi et al. (1999) on BMT and hemorrhagic cystitis patients, who found a prevalence of 40% BKV excretion in 62 immunocompetent patients. The group was not further described with respect to age or possible risk factors, and a study on JCV was not performed. However, in comparison with BMT patients, the virus load in immunocompetent individuals was low (in the range of $<1.2 \times 10^4$ genome copies in 5–10 ml urine). Although the sensitivity of the test system was not higher than that published before, the DNA amount in the urine suggests that asymptomatic BKV activation in immunocompetent adults is limited. This corresponds to a study reporting a low concentration of BKV DNA (<1 pg/ml) in general (Kitamura et al., 1990). Although a correlation between age and BKV excretion in adults was not apparent, in Japan detection rates in those older than 60 years were higher than in younger age groups (Kitamura et al., 1990).

Because of the prevalence of renal infection and urinary excretion it appears likely that common diseases and their associated immune responses play a role in virus activation. Whereas in immunocompetent individuals the rate of infection did not exceed 18%, impairment of the immune response in pregnancy and cardiac transplantation is linked to an activation in about 25% of individuals. In RT and AIDS patients expression is further affected to a rate of 47% at maximum. The most prominent activation processes can be observed in BMT patients. BKV expression after BMT appears to be almost always activated to a high virus load in the urine combined with an asymptomatic state of infection. Due to as yet unknown factors, which may involve host genetics, differences in therapy, or influences by the donor marrow, the infection may further be activated to a stage of cytolytic infection and hematuria without clinical symptoms.

It is conceivable that expression of the virus could be limited by the immune response, or it may, depending on the state of immunoimpairment, proceed unaffected to symptomatic disease. The amount of urinary virus in healthy individuals was regularly lower than that in immunoincompetent patients. This argues for a persistent BKV infection that is progressing stepwise from the latent or attenuated basic state of persistent infection with a rather low virus load that might even be out of the limits of detection. If the virus-specific immune response is impaired, increasing virus load and dissemination could indicate further stages of activation, ultimately leading to fatal disease.

Localization of JCV in the Urinary Tract. In contrast to BKV, JCV was never described as an etiologic agent in a urogenital disorder. Evaluation of the state of renal JCV infection in PML, the only JCV-associated disease, regularly disclosed no histopathologic changes in kidney tissue (Dörries and Elsner, 1991). Nevertheless, a prevalence rate of renal JCV infection in PML patients between 50% and 100% was reported (Ault and Stoner, 1994; Dörries and terMeulen, 1983; Grinnell et al., 1983a; Newman and Frisque, 1997; White et al., 1992). JCV DNA is distributed in small foci throughout renal cortex and medulla (Chesters et al., 1983; Dörries and terMeulen, 1983; Grinnell et al., 1983b; McCance, 1983; Tominaga et al., 1992). The major site of infection is localized to the epithelial cells lining the collective tubules (Dörries and terMeulen, 1983). Isolated cells carry virions in nuclei and cytoplasm. Southern blot analyses in combination with cloning experiments revealed episomal JCV genomes in affected cells without evidence for integrated DNA (Chesters et al., 1983; Dörries and Elsner, 1991).

Compared with PML patients, randomly selected individuals and cancer patients had renal infections less often, and their viral loads in the organ were considerably lower (Chesters et al., 1983; Dörries and terMeulen, 1983; Grinnell et al., 1983a). The presence of JCV DNA in about 50% of kidney samples is contrasted with only 2% of samples being positive for JCV protein. This most likely reflects differences in the sensitivities of the techniques applied rather than a true difference in the activation rates of the virus. Recent PCR analyses suggest that JCV DNA is regularly detected in the kidney of more than 50% of individuals (Aoki et al., 1999). This finding and the fact that primary renal infection with JCV occurs during childhood (Bordin et al., 1997; Grinnell et al., 1983a; Newman and Frisque, 1997, 1999) support the thesis that JCV persistence is most likely established during primary infection followed by an accumulation of virus in the tissue by repeated activation throughout life.

Activation of JCV Renal Infection. Like BKV, JCV DNA is more frequently detected in the kidneys of immunoincompetent individuals than in immunologically healthy persons. Under HIV infection urinary excretion was found to be a frequent event, ranging from 16% to 38% of patients in Europe, North America, and Africa (Agostini et al., 1995, 1997; Degener et al., 1997; Ferrante et al., 1997; Knowles et al., 1999; Markowitz et al., 1993; Shah et al., 1997; Sundsfjord et al., 1994a). In contrast to BKV activation, the incidence of JCV viruria parallels that in the normal population. In most reports, the pattern of JCV shedding was not influenced by the AIDS status or aggressive chemotherapy (Markowitz et al., 1993; Shah et al., 1997; Sundsfjord et al., 1994a). In general, the excretion rate was found comparable to that of normal individuals, being either stable or transient with identical genotypes shed in a period of 1–6 month. The frequency of excretion with increasing age did not differ significantly from that in HIV-negative individuals (Agostini et al., 1997). Obviously, JCV activation in the urinary tract is not influenced by HIV-induced immunoimpairment or therapeutic intervention for AIDS.

Similarly, JCV excretion is an uncommon event after BMT, occurring in about 5% of patients (Arthur et al., 1988; Gardner and Knowles, 1994; Myers et al., 1989; O'Reilly et al., 1981). Even sensitive PCR techniques did not change the basic findings (Azzi et al., 1994, 1999; Bogdanovic et al., 1994). Given the low frequency of viruria, JCV infection is even less active in BMT-associated immunoimpairment than BKV infection under comparable clinical conditions (Arthur et al., 1988; Gardner and Knowles, 1994). Compared with BMT, the frequency of JCV viruria after RT is higher, ranging from 18% to 57% (Gardner and Knowles, 1994; Hogan et al., 1980; Sundsfjord et al., 1994a; Yogo et al., 1991). Although highly variable, is seems likely that the rate of JCV activation after RT is comparable with that of BKV. Whether this is due to alteration of the JCV expression activity by factors related to the disease process or to the number of patients with activated infection is not known.

To analyze the role of CNS diseases other than PML in the activation of renal JCV infection, viruria was studied in a group of patients with multiple sclerosis (MS). PCR analysis revealed an excretion rate of 30–41% in chronic progressive MS (Agostini et al., 2000; Stoner et al., 1998). Because a control group of family members exhibited the same excretion rate, a regular influence of MS on JCV renal infection is rather unlikely. In conclusion, it can be assumed that pathologic changes in the CNS do not necessarily influence JCV activation (Elsner and Dörries, 1992; Stoner et al., 1998).

Analysis of immunocompetent patients demonstrated a higher rate of renal activation with JCV than BKV. Lower polyomavirus excretion rates were in the range of 0–13% (Arthur et al., 1989; Degener et al., 1997; Jin et al., 1993, 1995; Tsai et al., 1997), whereas higher rates, between 20% and 52% (Agostini et al., 1997; Azzi et al., 1996; Bogdanovic et al., 1994; Kato et al., 1997; Kitamura et al., 1990; Markowitz et al., 1993; Shah et al., 1997; Stoner et al., 1998; Sundsfjord et al., 1994a), have also been reported. Although the broad range of excretion rates may reflect technical differences, in 1994 it became clear that urinary excretion of JCV in the normal asymptomatic population also depends on the age of the individuals analyzed (Kitamura et al., 1994). The rate was clearly higher in groups with higher mean ages (Agostini et al., 1996, 1997). Whereas younger adults shed up to $<10^2$ fg JCV DNA/ml, this amount increased with age to more than 5×10^3 fg JCV DNA/ml (Kitamura et al., 1994; Markowitz et al., 1993). The highest level of JCV DNA in urine specimens of the oldest age group reached more than 10^5 fg JCV DNA/ml (Kitamura et al., 1994). Extent and duration of excretion can be either highly variable or stable (Agostini et al., 1997). Most remarkably, the same JCV strains were identified in urine specimens from healthy persons and from patients with malignancies, cerebrovascular disease, or urologic complications over periods up to 6 years (Agostini et al., 1997; Kitamura et al., 1997). From these data it can be unequivocally concluded that JCV urinary excretion is caused by activation of a persistent virus infection.

Renal activation of JCV by diseases that are not related to severe alterations of the immune system clearly demonstrates that JCV viruria is influenced by more than the state of immunocompetence. This is consistent with the "rule" that in PML/non-AIDS patients concomitant JCV viruria is as frequent as in the normal population (Arthur et al., 1989; Ferrante et al., 1997; Koralnik et al., 1999a; Markowitz et al., 1993). However, the level of JCV urinary excretion in immunocompetent individuals is noticeably higher than that of BKV, which rarely exceeds a concentration of 3 fg/ml urine (500 genome copies). Although excretion rates of both BKV and JCV seem tightly linked to increasing age (Kitamura et al., 1990), shedding of BKV and JCV occurs independently (Markowitz et al., 1993). This assumption is strongly supported by the fact that urinary coactivation of both viruses is a rare event (Azzi et al., 1999; Kitamura et al., 1990). Nevertheless, although not concurrently, both viruses were repeatedly detectable in urine from the same individual, suggesting that, once polyomavirus infection is established, activation processes may be induced independently and virus multiplication might then continue intermittently throughout life or at a sustained basal level.

Polyomavirus Infection in the Central Nervous System

CNS Disease Associated with BKV. In early studies, it was reported that fetal brain cells in vitro can be permissive for BKV (Takemoto et al., 1979).

However, even after the molecular detection of BKV DNA in the brain tissue of asymptomatic patients, the neurotropism of the virus was controversial (Elsner and Dörries, 1992). Soon after, the description of BKV-associated CNS diseases, subacute meningoencephalitis, and encephalitis in HIV patients confirmed the CNS as another site of BKV infection (Bratt et al., 1999; Vallbracht et al., 1993; Voltz et al., 1996). BKV could be isolated from the CSF in tissue culture, and the specificity of BKV in autopsy tissue was proven by molecular characterization and DNA sequencing. Histopathologic examination by immunohistochemistry and ISH demonstrated BKV DNA and antigen in affected cells (Vallbracht et al., 1993). BKV infection is associated with blood vessels and fibrocytes in thickened leptomeninges. In cortex and adjoining white matter, reactive astrocytes were affected. The ventricular walls exhibited focal degeneration of the ependymal layer and affected astrocytes in subjacent brain tissue. Target cells for BKV were fibrocytes of the connective tissue, ependymal cells, endothelial and smooth muscle cells of blood vessels, infiltrating macrophages, and astrocytes, the only glial cell type involved in BKV CNS infection (Bratt et al., 1999; Vallbracht et al., 1993). It is of note that the variability of cell types involved in BKV infection is remarkably higher than that of JCV. This suggests a broader cell specificity for BKV and suggests that a large number of different cell types can be involved in BKV persistence.

Asymptomatic BKV Infection and Activation. The assumption that BKV may reach its target organs before disease suggested the presence of BKV DNA and persistent infection in the CNS of healthy and immunocompetent individuals. However, in early analyses brain tissue appeared to be free of virus (Aksamit et al., 1986; Barbanti-Brodano et al., 1987; Chesters et al., 1983; Grinnell et al., 1983a). Likewise, brain tissue from AIDS patients found positive for JCV DNA did not reveal a trace of BKV DNA (Ferrante et al., 1995; Perrons et al., 1996). In contrast, other laboratories repeatedly reported the presence of BK viral DNA in samples from the CNS (De Mattei et al., 1995; Elsner and Dörries, 1992; Vago et al., 1996). These findings were supported by sequencing of new genomic TCR subtypes and cloning of complete virus genomes from a normal brain gene library (Elsner and Dörries, 1992). Nevertheless, compared with the frequency of JCV infection, the presence of BKV DNA in asymptomatic brain infection appears to be considerably lower. Definitive evidence on the localization of BKV in the brain and the cell type involved is not yet available.

Facts about the putative activation of BKV infection in the CNS are limited. BKV is shed in the CSF of patients with BKV-associated CNS disease, and the presence of BKV in the CSF is a diagnostic marker (Bratt et al., 1999; Vallbracht et al., 1993; Voltz et al., 1996). Spinal taps from PML patients and patients at risk for PML were screened for the diagnostic significance of JCV and were found positive for BKV DNA. Co-infection of patients with BKV and JCV is a frequent event and may be detected by PCR with a common primer pair. Because these persons were free of a typical BKV CNS disease, it is conceivable that a persistent BKV infection in the CNS might be subclinically activated under conditions comparable to those of JCV. Nevertheless, the overall frequency of such an event seems to be very low (Gibson et al., 1993; Hammarin et al., 1996; Perrons et al., 1996; Vago et al., 1996).

JCV in Progressive Multifocal Leukoencephalopathy. PML is a demyelinating disease occurring as a late complication of persistent or primary infection with JCV in the course of a basic immunosuppressive disease (Berger and Concha, 1995; Dörries, 1998; Major et al., 1992; Walker and Padgett, 1983; Weber and Major, 1997). Before the AIDS era about half of the PML cases involved lymphoproliferative diseases (Brooks and Walker, 1984). Nonlymphoproliferative malignancies made up the background in most of the residual cases. The course of PML and the cell types affected are essentially the same regardless of the immune impairment or basic disorder (Berger and Concha, 1995; Major et al., 1992; Walker and Padgett, 1983). Several years of treatment regularly precede PML in rheumatoid arthritis, chronic asthma, sarcoidosis, systemic lupus erythematosus (SLE), chronic polymyositis, and renal transplantation. These diseases are also suspected to be involved in the activation of persistent JCV infections.

The most important feature of PML is the striking alteration of cytolytically infected oligodendrocytes at the rim of defined lesions. Activated pleomorphic microglia were sometimes described in early lesions (Zu Rhein, 1969). The central demyelinated area is essentially composed of reactive astrocytes. Involvement of neurons, ependymal cells, or endothelial cells has never been confirmed. Numerous virus particles are localized in the nucleus of oligodendrocytes extending into the cytoplasm in the degenerating cell. Astrocytes may also contain virus particles or virus DNA, but the number is considerably lower than that in oligodendrocytes. From the data at present available, it cannot be determined whether astrocytes are permissive for JCV (Samorei et al., 2000). In one case a few mononuclear cells were found in Virchow-Robin spaces of several brain sections that may have contained JCV DNA and protein (Houff et al., 1988). Whether this represents infection or phagocytosis that might be followed occasionally by infection is not yet clarified.

Asymptomatic JCV Infection of the CNS. Despite the fact that JCV particles and/or viral products are easily detected in PML autopsy material, it is still a matter of controversy whether PML is the result of a cytolytic invasion of the tissue in the course of long-lasting immunosuppression or the consequence of a preceding persistent infection. Consequently, efforts were undertaken to demonstrate JCV in the CNS of patients in the absence of PML. Many attempts to detect JCV DNA in the brains of patients without signs of PML by methods like Southern blot, in situ hybridization (ISH), immunohistochemistry (Aksamit et al., 1986; Chesters et al., 1983; Heinonen et al., 1992; McCance, 1983; Stoner et al., 1988), and PCR (Buckle et al., 1992; Henson et al., 1991; Moret et al., 1993; Perrons et al., 1996; Telenti et al., 1990) failed. From these data, it was assumed that JCV persists exclusively at peripheral sites of the body.

The first instance of JCV DNA detected by ISH in elderly individuals without neurologic diseases in Japan was judged as unspecific binding of the probe to cell structures (Heinonen et al., 1992; Mori et al., 1992). However, detection of JCV in the CNS tissue of patients without evidence of PML was confirmed by PCR or ISH. The prevalence ranged from over 30% (Caldarelli-Stefano et al., 1999; Elsner and Dörries, 1992; Ferrante et al., 1995; Mori et al., 1992; Quinlivan et al., 1992; Vago et al., 1996) to almost 70% (White et al., 1992) of randomly selected patients without PML. In contrast to PML brain samples, JCV DNA sequences were often detected only once in up to eight CNS specimens from one patient without disease (Elsner and Dörries, 1992; White et al., 1992). The identity of the cell type that is affected in persistent infection has not yet been determined. However, in brain sections of HIV patients, JCV-positive isolated nuclei were characterized as oligodendrocytes on the grounds of immunohistochemical staining. The presence of JCV DNA was not associated with myelin loss or cytopathic changes (Vago et al., 1996).

Similar to the disseminated distribution of PML lesions, JCV-specific PCR products in the asymptomatic brain could not be localized to a specific topographic region (Elsner and Dörries, 1992; Vago et al., 1996). Therefore, it may be supposed that JCV has no regional preference for particular CNS structures. Examination of virus DNA in healthy tissue resulted exclusively in free circular JCV genomes, suggesting that persistent virus infection is closely related to an episomal state of the DNA (Elsner and Dörries, 1992). The major difference between asymptomatic and diseased tissue is the amount of virus present in the samples. In brain specimens from patients with PML, thousands of genome equivalents per cell are regularly found (Dörries et al., 1979; Quinlivan et al., 1992; Walker and Padgett, 1983). In contrast, the estimated range in asymptomatic individuals is between 1 and 100 JCV genome equivalents per 20 cells (Elsner and Dörries, 1992; Quinlivan et al., 1992; White et al., 1992). Thus, the presence of JCV DNA in nondiseased brain most likely represents a persistent infection and not early stages of disease.

To date, divergent detection rates of polyomavirus in the nonaffected brain are intensively discussed. At least in part, the observed discrepancies can be explained by technical and methodologic variables. Sensitivity as well as specificity of applied technologies differ between laboratories. In addition, laboratory-specific strategies for sampling, processing, extraction, and storage of specimens might have a serious impact on the success of viral DNA detection (Ferrante et al., 1995; Mori et al., 1992; Quinlivan et al., 1992; Vago et al., 1996). A potential problem that could be associated with the detection of lymphotropic viruses in brain tissue is the contamination of tissue specimens with blood. This is especially true for studies that use the most sensitive PCR technologies. However, PCR examination of highly vascularized liver tissue was consistently negative for JCV DNA (Mori et al., 1992; Quinlivan et al., 1992), making contamination of the JCV-positive brain specimens with blood cell–

associated virus rather unlikely. Demonstration of JCV by less sensitive methods as Southern blot analysis or direct cloning without PCR adds further support to the idea that JCV DNA is present in brain tissue of non-PML persons (Elsner and Dörries, 1992). However, besides detection in bulk DNA, for an ultimate tissue-specific localization of JCV in the course of subclinical persistence, histologic association of cell nuclei with virus DNA or expression products in situ is necessary.

Investigations of the influence of basic disease or of an immunoimpaired state on the frequency of JCV DNA detection in normal brain revealed enhanced rates in patients with HIV infection or with proliferative diseases (Elsner and Dörries, 1992; Quinlivan et al., 1992; Vago et al., 1996). Analysis of multiple areas of the brain disclosed that JCV DNA is similarly distributed as typically described in PML tissue (Walker and Padgett, 1983). However, the amount of virus-specific DNA in the samples is consistently lower than in PML tissue, suggesting that persistent polyomavirus infection is restricted to a few isolated cells. In cases of severe immunosuppression, virus infection appears to be activated (Kleihues et al., 1985), an event that probably contributes to viral spread and an increase in infected cells. This is reflected in the higher prevalence of polyomavirus DNA in multiple CNS specimens of patients with malignant diseases (Hogan et al., 1983; Mori et al., 1992; Rieckmann et al., 1994) and in the higher frequency of activated JCV infection in the elderly (Kitamura et al., 1990).

Activation of Asymptomatic JCV Infection in the CNS. The presence of JCV DNA in CNS tissues of patients whose deaths were attributable to diseases other than PML supports the thesis that asymptomatic persistent JCV infection of the CNS can occur. An increased prevalence of polyomavirus DNA in multiple CNS specimens of patients with impaired immunity and the absence of JCV DNA in fetal brain further suggest that disturbances of immunocompetence are a risk factor not only for the development of CNS disease but also for a higher rate of activation (Elsner and Dörries, 1992; Mori et al., 1992; Quinlivan et al., 1992).

One of the earliest reports on activation of JCV infection in the absence of PML was published by Mori et al. (1992). They described virus-specific protein in a limited number of oligodendrocytes and astrocytes in brain sections. This points to an active expression of viral genes very likely associated with an increase in infected cells, DNA load, and transcription products. Nevertheless, persistent infection remains asymptomatic, probably indicating that the actual number of infected cells and the virus load is low and difficult to detect. In contrast, polyomavirus disease in the CNS correlates strongly with viral presence in CSF (Dörries, 1996; Dörries et al., 1998; Eggers et al., 1999; Vallbracht et al., 1993; Weber and Major, 1997). Consequently, detection of JCV DNA in CSF can serve as an indirect marker for activation of a persistent virus infection.

PCR on CSF of non-PML patients with and without neurologic symptoms was often performed as a control for the establishment of the PML diagnosis (Agostini et al., 2000; Bogdanovic et al., 1998; de Luca et al., 1996; Gibson et al., 1993; Henson et al., 1991; Moret et al., 1993; Perrons et al., 1996; Sugimoto et al., 1998; Vignoli et al., 1993; Weber et al., 1994). Based on the results of all these studies, one could conclude that JCV DNA is not present in the CSF of patients without PML. This is in agreement with the presence of persistent JCV infection in 31% of samples from HIV CNS tissue but the absence of JCV DNA in the corresponding CSF samples (Vago et al., 1996). However, longitudinal determination of the virus load in CSF samples from PML patients over a period of several months demonstrated that, even with PML, virus might be shed to the CSF only intermittently (Drews et al., 2000). Consequently, it appears likely that episodic activation of JCV infection in the absence of PML could be missed if only a single CSF sample is analyzed. Although the presence of detectable JCV DNA in CSF is probably a rare event, it emphasizes the possible activation of virus infection in the CNS of non-PML patients (Dörries, 1996, 1998).

Early evidence for the activation of JCV infection in the brain came from PCR analyses. The prevalence ranged from 0.22% to 8% of patients with neurologic symptoms and/or HIV infection (Cinque et al., 1996; Ferrante et al., 1997; Fong et al., 1995; Koralnik et al., 1999a; McGuire et al., 1995). In other studies, the rate of JCV DNA detection increased from 9% to 100% (Ciappi et al., 1999; Dörries et al., 1998; Ferrante et al., 1998; Matsiota-Bernard et al., 1997). Most of the reports concentrated on HIV infection and other diseases at risk for PML. Association of a specific neurologic disorder or type of immune impairment with an increasing prevalence could not be deduced. However, activation of JCV DNA in the CSF of patients with lymphoproliferative diseases and HIV/AIDS appears to be more frequent than in immunocompetent individuals with other neurologic symptoms. An influence of the immunosuppressive state as defined by the CD4-positive T-lymphocyte count in AIDS patients was not discovered (Dörries et al., 1998).

Due to a possible role of early virus infection for the development of MS, several studies focused on these patients (Agostini et al., 2000; Bogdanovic et al., 1998; Dörries et al., 1998; Ferrante et al., 1998; Koralnik et al., 1999a). Most did not detect virus DNA by PCR, even though a single study of patients from Italy using nested PCR found rates of 7.2% in relapsing remitting courses, 16% in primary chronic progressive MS (CPMS), and 13% in secondary CPMS (Ferrante et al., 1998). Patients with and without other neurologic diseases who were analyzed for control purposes were found negative for JCV DNA in CSF.

Although the variable results on the presence of JCV DNA in CSF cannot be simply explained, several aspects should be envisaged: (1) due to the lack of standardized procedures, the performance of PCR varies considerably from laboratory to laboratory; (2) the heterogeneity of patients is high, and epidemiologic as well as anamnestic data are often not given; (3) time of sampling in the course of basic disease may play a role; and (4) therapeutic regimens may influence the activation of a persistent virus infection. However, cloning and sequencing of new JCV subtypes from the CSF of non-PML individuals (Ciappi et al., 1999) strongly argues for asymptomatic JCV activation in persistently infected brain tissue followed by the occasional presence of JCV DNA in the CSF. The divergence of data confirms that activation of the infection must be a rare event. It is very likely that expression is transient and virus is only infrequently present in the CSF. Whether activation is followed by lifelong accumulation of JCV in the CNS (McGuire et al., 1995) is a question that can probably only be answered in the primate animal model of SV40 and simian PML.

Polyomaviruses in the Lymphatic System

Persistent polyomavirus infection has been suspected to affect a variety of different organs. Dissemination to a wide range of body compartments and the distinct cell specificity of the human polyomaviruses led to the question of routes of viral spread within the host. Multifocal distribution of JCV in PML brain tissue opened the prospect that peripheral blood cells might be a vehicle for hematogeneous spread of polyomaviruses. This assumption was supported by an early report on polyomavirus particles in lymphocytes of immunocompetent children with measles virus infection (Lecatsas et al., 1976) and gave rise to a large number of studies.

Association of BKV with Cells of the Immune System. A study of children with respiratory diseases revealed that the tonsils were affected by BKV infection in more than 40%. Detection of BKV DNA in a throat washing (Jin et al., 1993; Sundsfjord et al., 1994b) pointed to the oropharynx, specifically the lymphoid tissues of Waldayer's ring, as the initial site of BKV infection (Goudsmit et al., 1982; Portolani et al., 1985). BKV DNA in PBLs and tonsils was exclusively in the episomal state (Dörries et al., 1994; Goudsmit et al., 1982). These findings strongly support the idea of lymphoid cell involvement in polyomavirus infection.

In early studies, the stimulatory effects of BKV on human lymphocytes in culture (Lecatsas et al., 1977) were described. BKV-specific receptors on the surface of PBLs argued for their susceptibility to infection (Possati et al., 1983). Indeed, multiplication of BKV was reported in B- and T-lymphocyte cultures from peripheral blood. The rate of infected cells in the presence of virus growth remained stable even for weeks of culture. The amount of virus yield was reported to be 100 times less than in human embryonic fibroblasts, suggesting a restricted growth of BKV in lymphocytes. In monocytes, virus attachment and penetration were observed without subsequent expression, suggesting that monocytes are involved in degradation rather than in multiplication of engulfed virus particles in vivo (Portolani et al., 1985). However, treatment of monocyte cultures with BKV-specific antiserum was followed by virus replication. Therefore, it was assumed that circulating monocytes or tissue resident macrophages in the normal individual might be permissive for polyomavirus infection (Traavik et al., 1988).

In peripheral blood mononuclear cells (PBMC) of individuals after BMT, either no (Schneider and Dörries, 1993) or an increasing prevalence for BKV was reported in more than 60% of patients at later times after transplantation (Azzi et al., 1996; Bogdanovic et al., 1996). As an explanation, a two-step process was assumed. In the course of pretransplantation treatment, loss of lymphocytes and clearance of polyomavirus from the blood are followed by the gradual invasion of the virus after transplantation from persistently infected sites such as the kidney (Azzi et al., 1996). However, in HIV-infected patients the findings were similarly variable. Two laboratories did not find BKV DNA in the PBMC of any patient, including those with PML (Ferrante et al., 1997; Perrons et al., 1996). Other studies reported rates of less than 10% (Degener et al., 1997; Sundsfjord et al., 1994a, 1999). These findings are contrasted by rates of 66% (Azzi et al., 1996) and 100% (Degener et al., 1999). No virus was detected in autoimmune SLE patients (Sundsfjord et al., 1999) or in the cord blood of newborn children (Dörries et al., 1994). In healthy individuals the incidences ranged from none detected (Degener et al., 1997) to 53%, 71%, and 94%, respectively (Azzi et al., 1996; De Mattei et al., 1995; Dörries et al., 1994), whereas in immunocompetent individuals with neurologic diseases an incidence of 82% was reported (Dörries et al., 1994). Two recent reports with refined PCR techniques, including extensive precautions for contamination, demonstrated BKV DNA at a rate of 26% and 55% in blood donors and healthy individuals from Italy and the United States (Chatterjee et al., 2000; Dolei et al., 2000). These conflicting data can mean either that BKV does not persistently infect blood cells or that the amount of virus DNA and/or the number of cells affected must be exceedingly low. Most of these studies were performed by nested PCR techniques, often at the limits of detection, conditions carrying the greatest risk for contamination. The amount of BKV DNA was indeed estimated to be very low compared with that of JCV in peripheral blood cells (Chatterjee et al., 2000; Dörries et al., 1994).

A subtype-specific analysis of peripheral blood cells that may enhance the detection rate has not yet been reported. However, in a study of PBMC in adult blood donors early BKV mRNA expression was detected by RTPCR in 100% of the samples that were positive at a rate of 71% by PCR for BKV DNA (De Mattei et al., 1995). The detection rate by RTPCR is in line with an activated virus infection, thus providing further evidence for a persistent BKV infection in lymphoid cells.

At present, an influence of immunoimpairment on BKV infection in blood cells cannot be assumed on the grounds of published data. However, a study of BKV in the plasma of renal allograft recipients recently revealed the presence of BKV DNA in serum samples from a patient with persistent viral nephropathy (Nickeleit et al., 2000). Provided that lymphoid infection contributes to BKV viremia, it is likely that therapeutic immunoimpairment is a promoting factor for BKV infection in the lymphoid compartment.

The divergence of results from studies of the lymphotropism of polyomaviruses appears to be higher than for other compartments of the host. This might be linked to a variable cell type susceptibility in each blood donor. Alternatively, it may depend on the half-life of the affected cell type or the state of infection at the time of sampling. Moreover, it may also be due to the clinical specimen itself because serum may contain more inhibitors for PCR than other body fluids. Taken together, the high variability of results makes a decision on the role of BKV in blood cell infection rather difficult. However, including the data on BKV-specific ISH in PBMC (Dörries et al., 1994), the detection of mRNA in PBMC (De Mattei et al., 1995), and the indirect evidence for BKV susceptibility of monocytes (Traavik et al., 1988), there is now a wide body of evidence for a regular lymphotropism of BKV in the host.

JCV in Lymphoid Organs and Blood Cells. The pronounced cell specificity of JCV for glial cells in tissue culture was not suggestive of an interaction of the virus with lymphoid cell types. However, general searches for peripheral organs involved in persistent JCV infection rendered spleen and lymph nodes occasionally positive for JCV (Grinnell et al., 1983a). JCV DNA was detected in lymphoid tissue from two children with PML in the course of primary infection (Grinnell et al., 1983a; Newman and Frisque, 1997). Subsequent studies of an adult PML patient demonstrated virus DNA in spleen and lymph node in the persistently infected host (Newman and Frisque, 1999). JCV DNA was also found in the spleens of PML/AIDS patients (Houff et al., 1988). Although JCV DNA could not be detected in the spleens of HIV patients and controls (Caldarelli-Stefano et al., 1999), a 40% incidence in PML patients suggests that virus DNA may accumulate in lymphoid organs.

The thesis of lymphoid cells as a reservoir for JCV (Houff et al., 1988) closed a gap in the understanding of viral pathogenesis. It supported the assumption that JCV-infected lymphoid cells can act as a vector for JCV dissemination. The role of lymphoid infection was further clarified by the detection of tonsillar lymphocytes and tonsillar stromal cells as host cells for JCV (Monaco et al., 1996, 1998). These findings not only suggest the involvement of the tonsils in primary infection but also argue for a persistent polyomavirus infection in tonsillar cell types and peripheral blood cells.

Virus infection in bone marrow was demonstrated for the first time in PML patients. JCV DNA and intranuclear virus capsid protein were located in scattered mononuclear cells as detected by ISH and immunohistochemistry (Houff et al., 1988). In the CNS the characteristic lesions were accompanied by an increased density of infected glial cells in the parenchyma adjacent to blood vessels. Occasionally, infected mononuclear cells were found in the Virchow-Robin spaces, and other infected cells were located just beneath the ependymal layer. Although a possible JCV infection of ependymal cells was never examined extensively, this is reminiscent of BKV infection in HIV-associated CNS disease and suggests a possible invasion of the parenchyma by JCV via the ependymal layer of the ventricles. The mononuclear cells carrying JCV

DNA and products of viral expression were found to be B lymphocytes but not T lymphocytes. The presence of JCV DNA at a concentration of more than 200–1000 virus copies per cell suggested ongoing replication (Houff et al., 1988). A second study reported on oligodendrocytes in PML brain that may occasionally react with B-lymphocyte markers by immunohistochemical staining (Aksamit and Leypold, 1993). Whether this discrepancy is due to technical differences between laboratories, to individual differences of patients, or to course of disease remains open.

To study more precisely the role of lymphoid cells in JCV persistence and pathogenesis, it was asked whether JCV DNA is present in circulating peripheral blood cells during episodes of impaired immunocomptence in the healthy individual. A study of virus infection of bone marrow aspirates from PML patients by ISH revealed the presence of JCV in 31% of the mononuclear cells (Katz et al., 1994; Tornatore et al., 1992). In contrast, amplification by PCR detected JCV DNA in 89.5% of the peripheral lymphocytes. Interestingly, the virus was consistently demonstrable in the course of illness, and even in cells of a prolonged PML survivor JCV DNA was present 4 years after diagnosis of PML. PCR has a considerably higher sensitivity than ISH. Therefore, it cannot be determined whether the higher incidence in PBMC was due to the different methods used or represented a true difference in the amount of infected cells (Tornatore et al., 1992). In subsequent studies, most peripheral blood cell samples of PML patients were positive. Altogether, the prevalence of JCV DNA detected by PCR ranges from 30% to 100%. This indicates that lymphoid cell preparations harbor JCV DNA irrespective of the underlying disease, the affected cell type, the JCV burden, or the number of infected cells (Andreoletti et al., 1999; Dörries et al., 1994; Dubois et al., 1997, 1998; Ferrante et al., 1997; Koralnik et al., 1999a). Quantification of virus load by competitive PCR revealed an average of 35 copies per microgram of cellular DNA in the PBMC of PML patients (Koralnik et al., 1999a). This is higher than in most normal persons (Dörries et al., 1994), but a correlation between HIV immune state and copy number was not detected. In summary, the data are divergent with respect to the level of JCV DNA present. Nevertheless, there is a tendency for a regular involvement of lymphoid cells in the disease process rather than for a nonspecific association.

The question whether lymphoid cells play a role in the dissemination of virus infection in the course of persistence and prior to the induction of disease was addressed by studies of immunoincompetent patients. The detection of JCV DNA in leukocytes from bone marrow and blood of leukemia patients added further support to the idea that polyomaviruses are generally lymphotropic. Detection of JCV DNA in almost all patients before conditioning for BMT revealed that leukemia patients are at a high risk for mononuclear cell infection. The incidence of JCV infection early after BMT was about 10% in an Italian study (Azzi et al., 1996) and 12% in Japanese patients (Shimizu et al., 1999). In contrast, at 60 days post-BMT an increase to 60% was reported (Azzi et al., 1996). This corresponds to a rate of 88% when serial samples were analyzed after BMT (Schneider and Dörries, 1993). In the same study the amount of virus DNA was found to be highly variable in serial PBMC specimens studied up to 1 year after BMT. Although an association of JCV with PBMC in leukemia and BMT patients is evident, highly variable amounts of virus in serial samples point to an intermittent rather than to a continuous interaction (Schneider and Dörries, 1993).

HIV-infected patients are the most prominent group at risk for PML. JCV infection in the peripheral blood cells of such persons was highly divergent, ranging from no trace of virus DNA (Perrons et al., 1996; Quinlivan et al., 1992; Sundsfjord et al., 1994a) to rates between 10% and 25% (Andreoletti et al., 1999; Dubois et al., 1998; Ferrante et al., 1997; Koralnik et al., 1999b; Lafon et al., 1998), to a maximum of 60% (Azzi et al., 1996; Degener et al., 1997; Dubois et al., 1996, 1997; Pietzuch et al., 1996; Tornatore et al., 1992). Interestingly, the presence of JCV DNA in blood cells is not linked to the clinical state or immunocompetence of HIV patients (Andreoletti et al., 1999; Dubois et al., 1996; Lafon et al., 1998). In contrast to HIV patients, patients with other neurologic diseases, including MS, Parkinson's, and SLE, had a prevalence well below 10% (Dörries et al., 1994; Ferrante et al., 1998; Shimizu et al., 1999; Sundsfjord et al., 1999; Tornatore et al., 1992). An increased rate of JCV in PBL can also be seen as an indicator for an enhanced virus load, thus suggesting an influence of HIV-related disease on JCV association with peripheral blood cells.

If the human polyomaviruses are lymphotropic agents and lymphocytes are sites of persistent infection, then JCV DNA should regularly be present in the blood cells of the normal immunocompetent individual. This was supported by the detection of episomal JCV DNA in PBMC from healthy laboratory staff at a rate of 30% by Southern blot analyses. PCR demonstrated JCV DNA at a rate of more than 80% in the same group, although the concentration of JCV DNA varied from case to case. Serologic evaluation of the patients revealed high antibody titers, pointing to a pronounced humoral immune response and a persistent rather than a latent virus infection (unpublished findings; Dörries et al., 1994). Based on all study results, the prevalence of virus-specific DNA ranged from 0% to 59%. This included blood donors with 39% (Azzi et al., 1996) and 4% (Lafon et al., 1998); healthy persons with 8% (Dubois et al., 1997), 10% (Ferrante et al., 1998), and 59% (Pietzuch et al. 1996); and unspecified immunocompetent control persons without JCV-specific amplification (Koralnik et al., 1999b; Schneider and Dörries, 1993).

ISH confirmed virus-specific DNA signals in close association with cell nuclei (Dörries et al., 1994). This made an entirely unspecific association of the virus with PBMC rather unlikely. The concentration of virus-specific DNA was estimated to be less than one genome equivalent in 20 cells. This probably explains the low rates of detection or even failure to amplify JCV DNA by PCR, as extraction of target DNA sequences appears to be a critical step in the detection of persisting virus genomes.

Hematopoietic Cell Subsets as Targets for JCV. The rates of polyomaviruses detected in peripheral blood cells have been highly variable over the years. Therefore, it became essential to further analyze the nature of virus–cell interactions in the lymphoid compartment (Jensen and Major, 1999). One of the most important questions concerns the cellular target of JCV in lymphoid tissue or peripheral blood. Initially, kappa light chain–carrying B lymphocytes were supposed to be the only target cell in bone marrow and spleen (Houff et al., 1988). An attempt to detect JCV by PCR in the peripheral blood of a PML patient demonstrated virus DNA in CD19-positive B lymphocytes (Monaco et al., 1996). However, in B lymphocyte–depleted PBMC from an HIV patient, JCV-specific amplification was similarly reported (Dubois et al., 1997). These findings and our own observation of JCV DNA in B- and T-lymphocyte cultures from bone marrow before transplantation suggested that JCV can be associated with different lymphocyte subsets. This was further confirmed by Koralnik et al. (1999b), who analyzed four different cell types of hematopoietic origin (CD3+, CD19+, CD14+, CD16+) from HIV patients and from a control group. The study confirmed the assumption that the population of JCV DNA–carrying persons among PML patients is higher than among HIV-infected patients alone or normal individuals. Unsorted cells from normal persons were negative for JCV, whereas virus DNA could be detected in sorted populations from the same patient. JCV DNA was detectable in B and T lymphocytes, in granulocytes, and in monocytes. Amplification of DNA was achieved either in one cell type alone or in different combinations from each patient. The finding that JCV DNA could be amplified only in sorted normal blood cells corresponds to reports of a low virus load in blood cells from healthy individuals (Dörries et al., 1994; Shimizu et al., 1999). However, JCV DNA was not quantified; therefore, it remains open whether a specific cell type is dominantly affected.

Studies of JCV infection in tissue culture revealed that cells of the CD19-positive B-lymphocyte subsets, CD34-positive hematopoietic progenitor cells including B-lymphoblastoid cell lines, and B lymphocytes from peripheral blood could serve as viral target cells. Infectious virus was produced, although no more than approximately 2% of B lymphocytes were infected (Monaco et al., 1996). Restricted virus growth in lymphoid cells strengthens the idea that viral replication is generally attenuated in lymphoid subpopulations (Atwood et al., 1992; Monaco et al., 1996). Virus specificity for hematopoietic cells at different stages of ontogeny was supported by the loss of JCV susceptibility after differentiation of the CD34-positive progenitor cell line KG-1 to a macrophage-like cell by the phorbol ester PMA. From these results, it appears likely that JCV DNA molecules are permanently present in the cells of different cell lineages in bone marrow and peripheral blood. However, more experiments are necessary to unequivocally determine the cell types susceptible to infection and those that might be able to mediate virus multiplication in vivo.

Activation of JCV Infection in Lymphoid Cells. Attempts to understand the nature of JCV infection in hematopoietic cells in vivo concentrated on virus-specific transcription in blood cells. Due to the low amount of DNA in these cells, studies were performed by RTPCR on nucleic acids from PBMC. JCV expression as detected by cDNA was reported in lymphocytes of 52% of PML patients and at about the same rate in HIV-infected and healthy individuals. In the same study JCV-specific cDNA was demonstrated not only in B lymphocytes but also in the non-B-lymphocyte fraction after magnetic separation (Pietzuch et al., 1996). This supports our own findings of JCV-affected cells showing that virus-specific DNA load and the targeted cell type both are highly variable in blood donors (personal observations). Additionally, it was confirmed that not only B lymphocytes but also other circulating cell types can be involved in JCV infection. Studies of virus expression by analysis of mRNA in peripheral blood cells or in separated B lymphocytes were performed recently. mRNA was found by RTPCR in five of seven PML patients, whereas no mRNA could be detected in the PBLs or B lymphocytes of a large group of HIV patients and blood donors (Andreoletti et al., 1999; Dubois et al., 1997; Lafon et al., 1998). Irrespective of transient activation events, these data imply a latent state of virus infection in peripheral blood cells rather than a persistent infection with consistent attenuated expression. However, confirming studies from other laboratories are needed to unequivocally determine the state of the virus infection in peripheral blood cells.

Comparable to shedding of virus into the CSF, an activated state of the virus infection in PBLs could be reflected by the virus load in serum. PCR studies on this question were performed after lymphotropism of the polyomaviruses had been detected. As anticipated from earlier PCR studies, the rate of detection varied from no virus present to 4% in SLE patients and HIV-negative control groups, including healthy individuals and blood donors (Koralnik et al., 1999a,b; Sundsfjord et al., 1999). In HIV patients with and without other neurologic symptoms, rates from 4% to 23% were reported (Dubois et al., 1998; Koralnik et al., 1999a,b; Lafon et al., 1998). In contrast, in 32% of all HIV/PML patients studied to date, viremia was detected (Dubois et al., 1997; Koralnik et al., 1999a,b; Tornatore et al., 1992), clearly indicating a higher prevalence under PML. Based on the accumulated data, it can be proposed that the amount of JCV DNA in peripheral blood cells is limited and the virus can be detected only after enrichment of lymphoid subtypes by cell separation techniques. Similarly, expression in lymphoid cells may occur at such a low level that it is not detectable by PCR on bulk DNA.

It seems clear now that JCV recirculates cell-associated as well as cell-free in peripheral blood. The crucial question whether association of the virus with peripheral blood cells is due to productive infection, to phagocytosis, or to unspecific binding to cellular membranes is not yet answered. Although detection of mRNA is a strong argument for an activated virus infection, localization of virus DNA to the nucleus of JCV-infected lymphoid subtypes remains to be demonstrated. Moreover, it cannot be excluded that different types of virus–cell interaction may coexist in the host—each of them regulated by different control mechanisms such as cell specificity, hormonal changes, or immune modulation. Besides the problems associated with PCR detection in blood, such a scenario would explain the extraordinary variance in virus presence detected.

Association of the Human Polyomaviruses With Other Organs

The first detection of a human polyomavirus in the eye was BKV with multifocal slowly progressing retinitis in AIDS patients (Bratt et al., 1999; Hedquist et al., 1999). BKV infection precipitates focal changes in the retinal pigment epithelium. Immunohistochemical examination revealed a diffuse virus protein pattern in necrotic lesions with distinct staining in single cells among photoreceptor remnants. Similar to the focal destruction of subependymal astrocytes and ventricular ependyma in BKV encephalitis (Vallbracht et al., 1993), in retinitis BKV was localized in nuclei adjacent to areas of retinal necrosis. However, the authors assumed that the infected cell type in the retina was more likely a photoreceptor cell than a retinal astrocyte.

In primary BKV infection mild respiratory tract disease suggests the presence of BKV-susceptible cells (Goudsmit et al., 1982; Noordaa and Wertheim-van Dillen, 1977). This corresponds to the description of virus-infected cells in an AIDS patient with interstitial pneumonitis (Vallbracht et al., 1993). Two more reports of pneumonia after umbilical cord transplantation and in the course of AIDS established BKV as a cause of respiratory disease (Cubukcu-Dimopulo et al., 2000; Sandler et al., 1997). Alterations in the lung were characterized by aggregates of desquamated pneumocytes and focal interstitial fibrosis. Pneumocytes and bronchiolar epithelium contained inclusion bodies. Virus products were detected in pneumocytes, in epithelial and smooth muscle cells of the bronchioli, and in fibrocytes. In contrast to the kidney, isolated endothelial cells in the lung occasionally carried virus protein (Vallbracht et al., 1993).

In a study of Kaposi's sarcoma (KS) and a possible interaction of BKV as a co-factor for tumor progression, skin samples from AIDS patients and immunocompetent individuals were analyzed by PCR (Monini et al., 1996). The study revealed an astonishingly high prevalence of BKV in KS skin biopsy specimens (100%) that was also reflected in 75% BKV-positive normal skin biopsies. In cases of JCV, the rate of detection decreased to 20% in classic KS and 16% in normal skin. In the same study JCV DNA was also detected in genital tissues and sperm (Monini et al., 1996). The surprisingly high level of polyomavirus in skin, in genital tissues, and in semen demands further studies to evaluate these body compartments as sites of polyomavirus infection either alone or as a co-factor with other viruses.

Other suspected sites of BKV infection include lung, pancreas, and heart tissue. All such tissues proved to be negative by molecular biologic and immunohistologic methods (Dörries and Elsner, 1991; Vallbracht et al., 1993). Occasionally BKV DNA was cloned from liver tissue (Knepper and diMayorca, 1987); however, in other studies of isolated cases the liver did not show BKV DNA (Dörries and Elsner, 1991; Vallbracht et al., 1993).

In contrast to BKV, JCV is only associated with PML, and the detection of JCV DNA in other organs remains a rare event. Specimens from liver (Bordin et al., 1997; Dörries and Elsner, 1991; Grinnell et al., 1983a; Newman and Frisque, 1997, 1999), lung (Caldarelli-Stefano et al., 1999; Dörries and Elsner, 1991; Grinnell et al., 1983a; Newman and Frisque, 1997, 1999), and cardiac muscle were occasionally found positive for JCV DNA by PCR (Newman and Frisque, 1997, 1999; Quinlivan et al., 1992). Recently, a study of colorectal mucosa revealed JCV DNA in normal epithelium and cancer tissue, differing only in the virus load (Laghi et al., 1999) and in the organization of the Mad-1 like regulatory region (Ricciardiello, et al. 2001). Most of these studies were performed on single clinical specimens, in small groups, or even in single patients who were in highly variable states of health. In addition, amplification of a small DNA fragment alone is not sufficient to determine the localization of a virus infection. Therefore, at present, it is almost impossible to differentiate between infection by JCV virus or accidental presence of virus DNA unrelated to an infectious state in all these organs.

2. POSSIBLE MECHANISMS OF ACTIVATION

Persistent polyomavirus infection is characterized by three major states of infection: the latent or attenuated state, a state of limited activation, and activated virus growth accompanied by tissue destruction. Determination of which factors are responsible for virus activation includes not only the role of the immune system but also the possible influence of other viruses that often co-infect immunoimpaired patients.

Cytomegalovirus infection is common in BMT and RT patients, and HIV infection is one of the major risks for PML. Recently, HHV-6 was found to co-localize with JCV in PML lesions. In addition, co-infection with a second polyomavirus may also have an impact on virus expression. The most important candidates for heterologous transactivation are BKV, herpesviruses, and retroviruses, which are known to have viral transactivating capacities. Heterologous transactivation of viruses can happen at different stages of virus–host interaction and virus growth. In the case of polyomaviruses, the most prominent transactivators are regulatory early proteins acting as either replicating enzymes or as transcription activators. Alternatively, the interaction of cellular proteins on the heterologous promoter can be influenced indirectly by changing the pattern of transcription factors within the cell.

Concomitant infections of BKV and JCV in the urogenital tract affect the same target cells as verified by the detection of viruses in kidney tissue and by viruria. It is a common event, frequently established in individuals after renal and bone marrow transplantation, in HIV-infected patients (Markowitz et al., 1993), in pregnant women (Arthur et al., 1989; Markowitz et al., 1991), and in immunocompetent individuals (Arthur et al., 1989; Chesters et al., 1983; Dörries and Elsner, 1991; Flaegstad et al., 1991; Grinnell et al., 1983b;

McCance, 1983; Sundsfjord et al., 1994b). Nonetheless, the influence of double infections on renal symptomatology or extent of viruria has not been reported (Arthur et al., 1989; Sundsfjord et al., 1994a). Even in BMT patients, JCV in close association with BKV viruria (Azzi et al., 1999; Chan et al., 1994) exhibited no particular differences among single excreters and co-excreting patients. Hence co-expression had no noticeable influence on either JCV or BKV viruria and associated disease (Chan et al., 1994).

The presence of virus DNA in the peripheral blood cells of both healthy and immunocompromised persons was established by PCR analyses and ISH with a radioactive virus-specific probe of genomic length. Reduced sensitivity of the in situ technique detected merely single virus-infected cells; however, the cells carried almost identical amounts of BKV and JCV DNA (Dörries et al., 1994). Because both viruses are able to replicate under restrictions in lymphocytes in vitro (Atwood et al., 1992; Portolani et al., 1985), it is conceivable that they are both periodically activated to virus growth in peripheral blood cells. Whether this has consequences or may influence each other's activity is not yet known.

BKV was not expected to invade the CNS at a high rate, and there have been only rare reports of the presence of both virus genomes in brain tissue in either the asymptomatic brain (Elsner and Dörries, 1992; Vago et al., 1996) or the PML patient (Ferrante et al., 1995; Vago et al., 1996). Cell types promoting polyomavirus persistence in the CNS are not yet defined. Nevertheless, the detection rate of genomic virus DNA in study groups from different laboratories revealed that the amount of BKV DNA in tissue specimens was considerably lower than that of JCV DNA (Elsner and Dörries, 1992; Ferrante et al., 1995; Vago et al., 1996; White et al., 1992). This corresponds to PCR analyses of the CSF demonstrating that BKV is rarely detectable in a large number of samples (Gibson et al., 1993; Hammarin et al., 1996; Perrons et al., 1996; Vago et al., 1996). Although concomitant infection with BKV and JCV can frequently be detected in tissue specimens and in all groups of polyomavirus-infected patients, a transactivating mechanism and the resulting effects remain rather unlikely.

Transactivation may also involve herpesviruses, which can act on polyomavirus DNA replication. Cytomegalovirus (CMV) is highly prevalent in the human population and can infect virtually any organ of its host (Sinzger and Jahn, 1996; Tevethia and Spector, 1989). Co-infection can occur in the kidney, lung, CNS, and lymphoid organs. Specifically, epithelial cells, fibroblasts, and endothelial cells are potential common host cells for BKV and CMV. Stromal cells and CD34-positive bone marrow progenitor cells might be cell types that can be co-targeted by JCV and CMV (Mendelson et al., 1996; Sinclair and Sissons, 1996). CMV infection is often activated in AIDS patients after RT, and there is a high incidence of CMV infection in patients with hemorrhagic cystitis (HC) after BMT (Childs et al., 1998). In AIDS patients no co-detection and no correlation between polyomavirus and CMV viruria was observed (Sundsfjord et al., 1994a). Similarly, the high incidence of CMV after kidney transplantation (Tolkoff-Rubin and Rubin, 1997) is not matched by an enhanced activity of polyomavirus infection.

Molecular interaction of CMV with JCV is believed to affect the level of DNA replication (Heilbronn et al., 1993) or possibly transcriptional activity. In contrast, co-infection of CMV and BKV in tissue culture does not result in activation of BKV infection (Goldstein et al., 1984). Lack of interaction among CMV and the human polyomaviruses was confirmed by treatment with acyclovir being able to reduce CMV activation in HC patients, but having no influence on BKV-associated HC. Although BKV T antigen is able to induce the expression of CMV immediate early and early gene expression (Kristoffersen et al., 1997), co-infection of both viruses in the same cell in vivo has not yet been reported and analyses of polyomavirus load in lymphoid subpopulations do not point to an interaction in vivo (personal observations).

At present, the most interesting virus detected in close association with JCV is human herpesvirus type six (HHV6) in oligodendrocytes within PML lesions (Blumberg et al., 2000). Co-localization was detected by in situ PCR, and correlation of polyomavirus infection with that of HHV6 was astonishingly high. HHV6 is ubiquitous, with a high prevalence in the adult population. It establishes life-long infection in the brain, the urogenital tract, the lung, the liver, and peripheral blood cells. It is conceivable that polyomaviruses and HHV6 have a common host cell not only in the CNS but also in peripheral organs. HHV6 activation occurs frequently after transplantation and is often associated with CMV infection. In the adult, it is usually asymptomatic and not associated with severe illness unless accompanied by CMV (Stoeckle, 2000). Due to the transactivation mechanisms that come into effect by infections with other herpesviruses, a comparable interaction of HHV6 with the human polyomaviruses is conceivable. However, it is a single study exclusively using in situ PCR, which demonstrated co-localization of JCV and HHV6 in PML lesions. Moreover, in situ PCR is one of the most sensitive and most difficult methods at present available and as such is inclined to non-specific signals. Although fascinating, the findings need to be confirmed by other methods and have to be proven by further studies on possible molecular mechanisms involved.

The question of whether the retrovirus HIV-1 may transactivate JCV is important because it became clear that PML is one of the life threatening opportunistic infections in AIDS patients. Molecular studies revealed that transactivation occurs in vitro at the level of transcription by HIV–Tat induction of the JCV promoter, and BKV T antigen is able to transactivate the HIV long terminal repeat (LTR). One site that has recently been described for BKV infection is the skin in AIDS-related Kaposi's sarcoma. The detection rate of BKV in the skin appeared to be higher than at other sites (Monini et al., 1996). However, transactivation events of a persistent virus in combination with progression to tumor development could explain the exceedingly high level of BKV (Cavallaro et al., 1996; Corallini et al., 1996). CNS infection of HIV-1 in oligodendrocytes is regularly low (Bagasra et al., 1996); however, Tat protein appears to be secreted by HIV-infected cells and might then be taken up by the JCV-infected oligodendrocyte and induce transcription on Tat-responsive genes. This thesis was examined by Valle et al. (2000), who demonstrated a localization of accumulated Tat protein to the nuclei of JCV-infected oligodendrocytes.

Recently, PML cases were reported without immunosuppression in the setting of co-infection of another retrovirus, HTLV-1 (Okada et al., 2000; Shimizu et al., 1999). This raised the question of whether HTLV-1 could activate JCV expression by interaction of the transactivating protein, Tax. Tax has been shown to interact with other viruses, such as HIV, CMV, and SV40. In vitro studies revealed a glial cell–specific interaction of Tax with the JCV promoter. Therefore, a comparable interaction to that of HIV-1 Tat protein is conceivable (Okada et al., 2000).

In summary, the detection of amplification products belonging to both polyomavirus species is strong evidence for concomitant infection in all tissues found positive for polyomavirus DNA. The number of individuals with simultaneous JCV and BKV infection is high, probably reflecting the true incidence of polyomavirus infection in the population. However, changes in virus-associated expression or the histologic picture due to co-infection have never been reported; therefore, an influence on expression activity by transactivation events appears to be rather unlikely. Similarly, co-infection with other viruses has been reported in a large number of patients and healthy individuals. Although heterologous transactivation may come into effect in individual cases, a common interaction among heterologous viral transactivators with human polyomaviruses cannot be stated. Consequently, transactivating mechanisms probably do not play a general role in the control of the polyomavirus life cycle and pathogenesis.

CONCLUDING REMARKS

Life-long human polyomavirus infection is established early in life. At present, it cannot be determined whether in the healthy individual the infection is in a latent state or is persistent with a continuous or intermittent expression involving restricted reactivation. Based on in vitro studies on virus expression, it can be assumed that promoter elements are able to mediate basic transcription. This function appears to be unrelated to cell specificity or activation processes and could therefore be responsible for a basal activity in any target cell. Limited activation obviously occurs almost always as a result of functional changes in the immune system. The exact pathways leading to expression of virus in vivo are barely defined. However, it can be assumed that impairment of immunocompetency, and hormonal changes as occur during pregnancy, older age, transient inflammatory states, malignant tumor growth, and AIDS, favor viral activation at sites of latent or persistent infection.

Some individuals may undergo an apparently sporadic activation; however, this could be related to their genotype or to accidental transactivation by heterologous viruses. Additionally, differences in the quality of activation signals apparently result in specific infection patterns that may vary according to the target organ as well as to the activity of virus growth. Host genetics, heterologous viruses present in the target organ, and the host-specific immune pattern all play a decisive role in the induction of different states of infection that precede tissue destruction and disease.

REFERENCES

Agostini HT, Brubaker GR, Shao J, Levin A, Ryschkewitsch CF, Blattner WA, Stoner GL (1995): BK virus and a new type of JC virus excreted by HIV-1 positive patients in rural Tanzania. Arch Virol 140:1919–1934.

Agostini HT, Ryschkewitsch CF, Baumhefner RW, Tourtellotte WW, Singer EJ, Komoly S, Stoner GL (2000): Influence of JC virus coding region genotype on risk of multiple sclerosis and progressive multifocal leukoencephalopathy. J Neurovirol 6: S101–108.

Agostini HT, Ryschkewitsch CF, Mory R, Singer EJ, Stoner, GL (1997): JC virus (JCV) genotypes in brain tissue from patients with progressive multifocal leukoencephalopathy (PML) and in urine from controls without PML: Increased frequency of JCV type 2 in PML. J Infect Dis 176:1–8.

Agostini HT, Ryschkewitsch CF, Stoner GL (1996): Genotype profile of human polyomavirus JC excreted in urine of immunocompetent individuals. J Clin Microbiol 34:159–164.

Aksamit AJ, Leypold B (1993): JC virus is not present in B cells in PML brain. Neurology 43:A177.

Aksamit AJ, Sever JL, Major EO (1986): Progressive multifocal leukoencephalopathy: JC virus detection by in situ hybridization compared with immunohistochemistry. Neurology 36:499–504.

Andreoletti L, Dubois V, Lescieux A, Dewilde A, Bocket L, Fleury HJ, Wattre P (1999): Human polyomavirus JC latency and reactivation status in blood of HIV-1–positive immunocompromised patients with and without progressive multifocal leukoencephalopathy. AIDS 13:1469–1475.

Andrews CA, Daniel RW, Shah KV (1983): Serologic studies of papovavirus infections in pregnant women and renal transplant recipients. Prog Clin Biol Res 105:133–141.

Andrews CA, Shah KV, Daniel RW, Hirsch MS, Rubin RH (1988): A serological investigation of BK virus and JC virus infections in recipients of renal allografts. J Infect Dis 158:176–181.

Aoki N, Kitamura T, Tominaga T, Fukumori N, Sakamoto Y, Kato K, Mori M (1999): Immunohistochemical detection of JC virus in nontumorous renal tissue of a patient with renal cancer but without progressive multifocal leukoencephalopathy. J Clin Microbiol 37:1165–1167.

Arthur RR, Beckmann AM, Li CC, Saral R, Shah KV (1985): Direct detection of the human papovavirus BK in urine of bone marrow transplant recipients: Comparison of DNA hybridization with ELISA. J Med Virol 16:29–36.

Arthur RR, Dagostin S, Shah KV (1989): Detection of BK virus and JC virus in urine and brain tissue by the polymerase chain reaction. J Clin Microbiol 27:1174–1179.

Arthur RR, Shah KV, Charache P, Saral R (1988): BK and JC virus infections in recipients of bone marrow transplants. J Infect Dis 158:563–569.

Atwood WJ, Amemiya K, Traub R, Harms J, Major EO (1992): Interaction of the human polyomavirus, JCV, with human B-lymphocytes. Virology 190:716–723.

Ault GS, Stoner GL (1994): Brain and kidney of progressive multifocal leukoencephalopathy patients contain identical rearrangements of the JC virus promoter/enhancer. J Med Virol 44:298–304.

Azzi A, Cesaro S, Laszlo D, Zakrzewska K, Ciappi S, De Santis R, Fanci R, Pesavento G, Calore E, Bosi A (1999): Human polyomavirus BK (BKV) load and haemorrhagic cystitis in bone marrow transplantation patients. J Clin Virol 14:79–86.

Azzi A, De Santis R, Ciappi S, Leoncini F, Sterrantino G, Marino N, Mazzotta F, Laszlo D, Fanci R, Bosi A (1996): Human polyomaviruses DNA detection in peripheral blood leukocytes from immunocompetent and immunocompromised individuals. J Neurovirol 2:411–416.

Azzi A, Fanci R, Bosi A, Ciappi S, Zakrzewska K, de Santis R, Laszlo D, Guidi S, Saccardi R, Vannucchi AM, et al. (1994): Monitoring of polyomavirus BK viruria in bone marrow transplantation patients by DNA hybridization assay and by polymerase chain reaction: An approach to assess the relationship between BK viruria and hemorrhagic cystitis. Bone Marrow Transplant 14:235–240.

Bagasra O, Lavi E, Bobroski L, Khalili K, Pestaner JP, Tawadros R, Pomerantz RJ (1996): Cellular reservoirs of HIV-1 in the central nervous system of infected individuals: Identification by the combination of in situ polymerase chain reaction and immunohistochemistry. AIDS 10:573–585.

Barbanti-Brodano G, Silini E, Mottes M, Milanesi G, Pagnani M, Reschiglian P, Gerna G, Corallini A (1987): DNA probes to evaluate the possible association of papovaviruses with human tumors. In Monoclonals and DNA Probes in Diagnostic and Preventive Medicine; Gallo RC, Della Porta G, Albertini A, Eds; Raven Press: New York, pp 147–155.

Bedi A, Miller CB, Hanson JL, Goodman S, Ambinder RF, Charache P, Arthur RR, Jones RJ (1995): Association of BK virus with failure of prophylaxis against hemorrhagic cystitis following bone marrow transplantation. J Clin Oncol 13:1103–1109.

Berger JR, Concha M (1995): Progressive multifocal leukoencephalopathy: The evolution of a disease once considered rare. J Neurol Virol 1:5–18.

Binet I, Nickeleit V, Hirsch HH, Prince O, Dalquen P, Gudat F, Mihatsch MJ, Thiel G (1999): Polyomavirus disease under new immunosuppressive drugs: A cause of renal graft dysfunction and graft loss. Transplantation 67:918–922.

Blumberg BM, Mock DJ, Powers JM, Ito M, Assouline JG, Baker JV, Chen B, Goodman AD (2000): The HHV6 paradox: Ubiquitous commensal or insidious pathogen? A two-step in situ PCR approach. J Clin Virol 16:159–178.

Bogdanovic G, Brytting M, Cinque P, Grandien M, Fridell E, Ljungman P, Lönnquvist B, Hammarin AL (1994): Nested PCR for detection of BK virus and JC virus DNA. Clin Diagn Virol 2:211–220.

Bogdanovic G, Ljungman P, Wang F, Dalianis T. (1996): Presence of human polyomavirus DNA in the peripheral circulation of bone marrow transplant patients with and without hemorrhagic cystitis. Bone Marrow Transplant 17:573–576.

Bogdanovic G, Priftakis P, Hammarin AL, Soderstrom M, Samuelson A, Lewensohn Fuchs I, Dalianis T. (1998): Detection of JC virus in cerebrospinal fluid (CSF) samples from patients with progressive multifocal leukoencephalopathy but not in CSF samples from patients with herpes simplex encephalitis, enteroviral meningitis, or multiple sclerosis. J Clin Microbiol 36:1137–1138.

Bordin G, Boldorini R, Caldarelli Stefano R, Omodeo Zorini E (1997): Systemic infection by JC virus in non-HIV induced immunodeficiency without progressive multifocal leukoencephalopathy. Ann Ital Med Int 12:35–38.

Boubenider S, Hiesse C, Marchand S, Hafi A, Kriaa F, Charpentier B (1999): Post-transplantation polyomavirus infections. J Nephrol 12:24–29.

Bratt G, Hammarin AL, Grandien M, Hedquist BG, Nennesmo I, Sundelin B, Seregard S. (1999): BK virus as the cause of meningoencephalitis, retinitis and nephritis in a patient with AIDS. AIDS 13:1071–1075.

Brooks BR, Walker DL (1984): Progressive multifocal leukoencephalopathy. Neurol Clin 2:299–313.

Brouqui P, Bollet C, Delmont J, Bourgeade A (1992): Diagnosis of progressive multifocal leucoencephalopathy by PCR detection of JC virus from CSF. Lancet 339: 1182.

Buckle GJ, Godec MS, Rubi JU, Tornatore C, Major EO, Gibbs CJ, Jr, Gajdusek DC, Asher DM (1992): Lack of JC viral genomic sequences in multiple sclerosis brain tissue by polymerase chain reaction. Ann Neurol 32:829–831.

Caldarelli-Stefano R, Vago L, Omodeo-Zorini E, Mediati M, Losciale L, Nebuloni M, Costanzi G, Ferrante P (1999): Detection and typing of JC virus in autopsy brains and extraneural organs of AIDS patients and non-immunocompromised individuals. J Neurovirol 5:125–133.

Cavallaro U, Gasparini G, Soria MR, Maier JA (1996): Spindle cells isolated from Kaposi's sarcoma-like lesions of BKV/tat-transgenic mice co-express markers of different cell types. AIDS 10:1211–1219.

Chan PK, Ip KW, Shiu SY, Chiu EK, Wong MP, Yuen KY (1994): Association between polyomaviruria and microscopic haematuria in bone marrow transplant recipients. J Infect 29:139–146.

Chang D, Tsai RT, Wang M, Ou WC (1996a): Different genotypes of human polyomaviruses found in patients with autoimmune diseases in Taiwan. J Med Virol 48:204–209.

Chang D, Wang M, Ou WC, Lee MS, Ho HN, Tsai RT (1996b): Genotypes of human polyomaviruses in urine samples of pregnant women in Taiwan. J Med Virol 48: 95–101.

Chatterjee M, Weyandt TB, Frisque RJ (2000): Identification of archetype and rearranged forms of BK virus in leukocytes from healthy individuals. J Med Virol 60: 353–362.

Chesters PM, Heritage J, McCance DJ (1983): Persistence of DNA sequences of BK virus and JC virus in normal human tissues and in diseased tissues. J Infect Dis 147: 676–684.

Childs R, Sanchez C, Engler H, Preuss J, Rosenfeld S, Dunbar C, van Rhee F, Plante M, Phang S, Barrett AJ (1998): High incidence of adeno- and polyomavirus-induced hemorrhagic cystitis in bone marrow allotransplantation for hematological malignancy following T cell depletion and cyclosporine. Bone Marrow Transplant 22: 889–893.

Ciappi S, Azzi A, De Santis R, Leoncini F, Sterrantino G, Mazzotta F, Mecocci L (1999): Archetypal and rearranged sequences of human polyomavirus JC transcription control region in peripheral blood leukocytes and in cerebrospinal fluid. J Gen Virol 80:1017–1023.

Cinque P, Vago L, Dahl H, Brytting M, Terreni MR, Fornara C, Racca S, Castagna A, Monforte AD, Wahren B, Lazzarin A, Linde A (1996): Polymerase chain reaction on cerebrospinal fluid for diagnosis of virus-associated opportunistic diseases of the central nervous system in HIV-infected patients. AIDS 10:951–958.

Coleman DV, Field AM, Gardner SD, Porter KA, Starzl TE (1973): Virus-induced obstruction of the ureteric and cystic duct in allograft recipients. Transplant Proc 5: 95–98.

Coleman DV, Wolfendale MR, Daniel RA, Dhanjal NK, Gardner SD, Gibson PE, Field AM (1980): A prospective study of human polyomavirus infection in pregnancy. J Infect Dis 142:1–8.

Corallini A, Campioni D, Rossi C, Albini A, Possati L, Rusnati M, Gazzanelli G, Benelli R, Masiello L, Sparacciari V, Presta M, Mannello F, Fontanini G, Barbanti Brodano G (1996): Promotion of tumour metastases and induction of angiogenesis by native HIV-1 Tat protein from BK virus/tat transgenic mice. AIDS 10:701–710.

Cotterill HA, Macaulay ME, Wong V (1992): Reactivation of polyomavirus in bone marrow transplant recipients. J Clin Pathol 45:445.

Cubukcu-Dimopulo O, Greco A, Kumar A, Karluk D, Mittal K, Jagirdar J (2000): BK virus infection in AIDS. Am J Surg Pathol 24:145–149.

Degener AM, Pietropaolo V, Di Taranto C, Jin L, Ameglio F, Cordiali-Fei P, Trento E, Sinibaldi L, Orsi N (1999): Identification of a new control region in the genome of the DDP strain of BK virus isolated from PBMC. J Med Virol 58:413–419.

Degener AM, Pietropaolo V, Di Taranto C, Rizzuti V, Ameglio F, Cordiali Fei P, Caprilli F, Capitanio B, Sinibaldi L, Orsi N (1997): Detection of JC and BK viral genome in specimens of HIV-1 infected subjects. New Microbiol 20:115–122.

de Luca A, Cingolani A, Linzalone A, Ammassari A, Murri R, Giancola ML, Maiuro G, Antinori A (1996): Improved detection of JC virus DNA in cerebrospinal fluid for diagnosis of AIDS-related progressive multifocal leukoencephalopathy. J Clin Microbiol 34:1343–1346.

De Mattei M, Martini F, Corallini A, Gerosa M, Scotlandi K, Carinci P, Barbanti-Brodano G, Tognon M (1995): High incidence of BK virus large-T-antigen–coding sequences in normal human tissues and tumors of different histotypes. Int J Cancer 61:756–760.

de Silva LM, Bale P, de Courcy J, Brown D, Knowles W (1995): Renal failure due to BK virus infection in an immunodeficient child. J Med Virol 45:192–196.

Di Taranto C, Pietropaolo V, Orsi GB, Jin L, Sinibaldi L, Degener AM (1997): Detection of BK polyomavirus genotypes in healthy and HIV-positive children. Eur J Epidemiol 13:653–657.

Dolei A, Pietropaolo V, Gomes E, Di Taranto C, Ziccheddu M, Spanu MA, Lavorino C, Manca M, Degener AM (2000): Polyomavirus persistence in lymphocytes: Prevalence in lymphocytes from blood donors and healthy personnel of a blood transfusion centre. J Gen Virol 81:1967–1973.

Dörries K (1996): Virus–host interactions and the diagnosis of human polyomavirus associated disease. Intervirology 39:165–175.

Dörries K (1998): Molecular biology and pathogenesis of human polyomavirus infections. Dev Biol Stand 94:71–79.

Dörries K, Arendt G, Eggers C, Roggendorf W, Dörries R (1998): Nucleic acid detection as a diagnostic tool in polyomavirus JC induced progressive multifocal leukoencephalopathy. J Med Virol 54:196–203.

Dörries K, Elsner C (1991): Persistent polyomavirus infection in kidney tissue of patients with diseases other than progressive multifocal leucoencephalopathy. Virol Adv 10:51–61.

Dörries K, Johnson RT, terMeulen V (1979): Detection of polyoma virus DNA in PML-brain tissue by (in situ) hybridization. J Gen Virol 42:49–57.

Dörries K, terMeulen V (1983): Progressive multifocal leucoencephalopathy: Detection of papovavirus JC in kidney tissue. J Med Virol 11:307–317.

Dörries K, Vogel E, Gunther S, Czub S (1994): Infection of human polyomaviruses JC and BK in peripheral blood leukocytes from immunocompetent individuals. Virology 198:59–70.

Drews K, Bashir T, Dorries K (2000): Quantification of human polyomavirus JC in brain tissue and cerebrospinal fluid of patients with progressive multifocal leukoencephalopathy by competitive PCR. J Virol Methods 84:23–36.

Dubois V, Dutronc H, Lafon ME, Poinsot V, Pellegrin JL, Ragnaud JM, Ferrer AM, Fleury HJ (1997): Latency and reactivation of JC virus in peripheral blood of human immunodeficiency virus type 1–infected patients. J Clin Microbiol 35:2288–2292.

Dubois V, Lafon ME, Ragnaud JM, Pellegrin JL, Damasio F, Baudouin C, Michaud V, Fleury HJ (1996): Detection of JC virus DNA in the peripheral blood leukocytes of HIV-infected patients. AIDS 10:353–358.

Dubois V, Moret H, Lafon ME, Janvresse CB, Dussaix E, Icart J, Karaterki A, Ruffault A, Taoufik Y, Vignoli C, Ingrand D (1998): Prevalence of JC virus viraemia in HIV-infected patients with or without neurological disorders: A prospective study. J Neurovirol 1998:539–544.

Eggers C, Stellbrink HJ, Buhk T, Dörries K (1999): Quantification of JC virus DNA in the cerebrospinal fluid of patients with human immunodeficiency virus–associated progressive multifocal leukoencephalopathy—A longitudinal study. J Infect Dis 180: 1690–1694.

Elsner C, Dörries K (1992): Evidence of human polyomavirus BK and JC infection in normal brain tissue. Virology 198:59–70.

Ferrante P, Caldarelli-Stefano R, Omodeo-Zorini E, Cagni AE, Cocchi L, Suter F, Maserati R (1997): Comprehensive investigation of the presence of JC virus in AIDS patients with and without progressive multifocal leukoencephalopathy. J Med Virol 52:235–242.

Ferrante P, Caldarelli-Stefano R, Omodeo-Zorini E, Vago L, Boldorini R, Costanzi G (1995): PCR detection of JC virus DNA in brain tissue from patients with and without progressive multifocal leukoencephalopathy. J Med Virol 47:219–225.

Ferrante P, Omodeo Zorini E, Caldarelli Stefano R, Mediati M, Fainardi E, Granieri E, Caputo D (1998): Detection of JC virus DNA in cerebrospinal fluid from multiple sclerosis patients. Mult Scler 4:49–54.

Flaegstad T, Sundsfjord A, Arthur RR, Pedersen M, Traavik T, Subramani S. (1991): Amplification and sequencing of the control regions of BK and JC virus from human urine by polymerase chain reaction. Virology 180:553–560.

Fong IW, Britton CB, Luinstra KE, Toma E, Mahony JB (1995): Diagnostic value of detecting JC virus DNA in cerebrospinal fluid of patients with progressive multifocal leukoencephalopathy. J Clin Microbiol 33:484–486.

Gardner SD, Field AM, Coleman DV, Hulme B (1971): New human papovavirus (B.K.) isolated from urine after renal transplantation. Lancet I:1253–1257.

Gardner SD, Knowles WA (1994): Human polyomaviruses. In Principles and Practice of Clinical Virology; Zuckerman AJ, Banatvala JE, Pattison JR, Eds; John Wiley: Chichester, pp 635–651.

Gerber MA, Shah KV, Thung SN, Zu Rhein GM (1980): Immunohistochemical demonstration of common antigen of polyomaviruses in routine histologic tissue sections of animals and man. Am J Clin Pathol 73:795–797.

Gibson PE, Gardner SD (1983): Strain differences and some serological observations on several isolates of human polyomaviruses. Prog Clin Biol Res 105:119–132.

Gibson PE, Gardner SD, Porter AA (1985): Detection of human polyomavirus DNA in urine specimens by hybridot assay. Arch Virol 84:233–240.

Gibson PE, Knowles WA, Hand JF, Brown DW (1993): Detection of JC virus DNA in the cerebrospinal fluid of patients with progressive multifocal leukoencephalopathy. J Med Virol 39:278–281.

Goldstein SC, Tralka TS, Rabson AS (1984): Mixed infection with human cytomegalovirus and human polyomavirus (BKV): J Med Virol 13:33–40.

Goudsmit J, Wertheim van Dillen P, van Strien A, van der Noordaa J (1982): The role of BK virus in acute respiratory tract disease and the presence of BKV DNA in tonsils. J Med Virol 10:91–99.

Grinnell BW, Padgett BL, Walker DL (1983a): Distribution of nonintegrated DNA from JC papovavirus in organs of patients with progressive multifocal leukoencephalopathy. J Infect Dis 147:669–675.

Grinnell BW, Padgett BL, Walker DL (1983b): Comparison of infectious JC virus DNAs cloned from human brain. J Virol 45:299–308.

Hammarin AL, Bogdanovic G, Svedhem V, Pirskanen R, Morfeldt L, Grandien M. (1996): Analysis of PCR as a tool for detection of JC virus DNA in cerebrospinal fluid for diagnosis of progressive multifocal leukoencephalopathy. J Clin Microbiol 34:2929–2932.

Hedquist BG, Bratt G, Hammarin AL, Grandien M, Nennesmo I, Sundelin B, Seregard S. (1999): Identification of BK virus in a patient with acquired immune deficiency syndrome and bilateral atypical retinitis. Ophthalmology 106:129–132.

Heilbronn R, Albrecht I, Stephan S, Burkle A, zur Hausen H (1993): Human cytomegalovirus induces JC virus DNA replication in human fibroblasts. Proc Natl Acad Sci USA 90:11406–11410.

Heinonen O, Syrjanen S, Mantyjarvi R, Syrjanen K, Riekkinen P (1992): JC virus infection and Alzheimer's disease: Reappraisal of an in situ hybridization approach. Ann Neurol 31:439–441.

Henson J, Rosenblum M, Armstrong D, Furneaux H (1991): Amplification of JC virus DNA from brain and cerebrospinal fluid of patients with progressive multifocal leukoencephalopathy. Neurology 41:1967–1971.

Heritage J, Chesters PM, McCance DJ (1981): The persistence of papovavirus BK DNA sequences in normal human renal tissue. J Med Virol 8:143–150.

Hiraoka A, Ishikawa J, Kitayama H, Yamagami T, Teshima H, Nakamura H, Shibata H, Masaoka T, Ishigami S, Taguchi F (1991): Hemorrhagic cystitis after bone marrow transplantation: Importance of a thin sectioning technique on urinary sediments for diagnosis. Bone Marrow Transplant 7:107–111.

Hogan TF, Borden EC, McBain JA, Padgett BL, Walker DL (1980): Human polyomavirus infections with JC virus and BK virus in renal transplant patients. Ann Intern Med 92:373–378.

Hogan TF, Padgett BL, Walker DL, Borden EC, Frias Z (1983): Survey of human polyomavirus (JCV, BKV) infections in 139 patients with lung cancer, breast cancer, melanoma, or lymphoma. Prog Clin Biol Res 105:311–324.

Houff SA, Major EO, Katz DA, Kufta CV, Sever JL, Pittaluga S, Roberts JR, Gitt J, Saini N, Lux W. (1988): Involvement of JC virus–infected mononuclear cells from the bone marrow and spleen in the pathogenesis of progressive multifocal leukoencephalopathy. N Engl J Med 318:301–305.

Jensen PN, Major EO (1999): Viral variant nucleotide sequences help expose leukocytic positioning in the JC virus pathway to the CNS. J Leukocyte Biol 65:428–438.

Jin L, Gibson PE, Booth JC, Clewley JP (1993): Genomic typing of BK virus in clinical specimens by direct sequencing of polymerase chain reaction products. J Med Virol 41:11–17.

Jin L, Pietropaolo V, Booth JC, Ward KH, Brown DW (1995): Prevalence and distribution of BK virus subtypes in healthy people and immunocompromised patients detected by PCR-restriction enzyme analysis. Clin Diagn Virol 3:285–295.

Kato A, Kitamura T, Sugimoto C, Ogawa Y, Nakazato K, Nagashima K, Hall WW, Kawabe K, Yogo Y (1997): Lack of evidence for the transmission of JC polyomavirus between human populations. Arch Virol 142:875–882.

Katz DA, Berger JR, Hamilton B, Major EO, Post MJ (1994): Progressive multifocal leukoencephalopathy complicating Wiskott-Aldrich syndrome. Report of a case and review of the literature of progressive multifocal leukoencephalopathy with other inherited immunodeficiency states. Arch Neurol 51:422–426.

Kitamura T, Aso Y, Kuniyoshi N, Hara K, Yogo Y. (1990): High incidence of urinary JC virus excretion in nonimmunosuppressed older patients. J Infect Dis 161:1128–1133.

Kitamura T, Satoh K, Tominaga T, Taguchi F, Tajima A, Suzuki K, Aso Y, Yogo Y (1994): Alteration in the JC polyomavirus genome is enhanced in immunosuppressed renal transplant patients. Virology 198:341–345.

Kitamura T, Sugimoto C, Kato A, Ebihara H, Suzuki M, Taguchi F, Kawabe K, Yogo Y (1997): Persistent JC virus (JCV) infection is demonstrated by continuous shedding of the same JCV strains. J Clin Microbiol 35:1255–1257.

Kleihues P, Lang W, Burger PC, Budka H, Vogt M, Maurer R, Luthy R, Siegenthaler W (1985): Progressive diffuse leukoencephalopathy in patients with acquired immune deficiency syndrome (AIDS): Acta Neuropathol Berl 68:333–339.

Knepper JE, diMayorca G (1987): Cloning and characterization of BK virus–related DNA sequences from normal and neoplastic human tissues. J Med Virol 21:289–299.

Knowles WA, Pillay D, Johnson MA, Hand JF, Brown DW (1999): Prevalence of long-term BK and JC excretion in HIV-infected adults and lack of correlation with serological markers. J Med Virol 59:474–479.

Koralnik IJ, Boden D, Mai VX, Lord CI, Letvin NL (1999a): JC virus DNA load in patients with and without progressive multifocal leukoencephalopathy. Neurology 52:253–260.

Koralnik IJ, Schmitz JE, Lifton MA, Forman MA, Letvin, NL (1999b): Detection of JC virus DNA in peripheral blood cell subpopulations of HIV-1–infected individuals. J Neurovirol 5:430–435.

Kristoffersen AK, Johnsen JI, Seternes OM, Rollag H, Degre M, Traavik T (1997): The human polyomavirus BK T antigen induces gene expression in human cytomegalovirus. Virus Res 52:61–71.

Lafon ME, Dutronc H, Dubois V, Pellegrin I, Barbeau P, Ragnaud JM, Pellegrin JL, Fleury HJ (1998): JC virus remains latent in peripheral blood B lymphocytes but replicates actively in urine from AIDS patients. J Infect Dis 177:1502–1505.

Laghi L, Randolph AE, Chauhan DP, Marra G, Major EO, Neel JV, Boland CR (1999): JC virus DNA is present in the mucosa of the human colon and in colorectal cancers. Proc Natl Acad Sci USA 96:7484–7489.

Lecatsas G, Blignaut E, Schoub BD (1977): Lymphocyte stimulation by urine-derived human polyoma virus (BK): Arch Virol 55:165–167.

Lecatsas G, Boes EG, Horsthemke E (1981): Intermediate size papovavirus particles in pregnancy urine. J Gen Virol 52:359–362.

Lecatsas G, Prozesky OW, Wyk JV, Els HJ (1973): Papova virus in urine after renal transplantation. Nature 241:343–344.

Lecatsas G, Schoub BD, Rabson AR, Joffe M (1976): Papovavirus in human lymphocyte cultures. Lancet 2:907–908.

Major EO, Amemiya K, Tornatore CS, Houff SA, Berger JR (1992): Pathogenesis and molecular biology of progressive multifocal leukoencephalopathy, the JC virus–induced demyelinating disease of the human brain. Clin Microbiol Rev. 5:49–73.

Markowitz RB, Eaton BA, Kubik MF, Latorra D, McGregor JA, Dynan WS (1991): BK virus and JC virus shed during pregnancy have predominantly archetypal regulatory regions. J Virol 65:4515–4519.

Markowitz RB, Thompson HC, Mueller JF, Cohen JA, Dynan WS (1993): Incidence of BK virus and JC virus viruria in human immunodeficiency virus–infected and –uninfected subjects. J Infect Dis 167:13–20.

Mathur VS, Olson JL, Darragh TM, Yen TS (1997): Polyomavirus-induced interstitial nephritis in two renal transplant recipients: Case reports and review of the literature. Am J Kidney Dis 29:754–758.

Matsiota-Bernard P, De Truchis P, Gray F, Flament-Saillour M, Voyatzakis E, Nauciel C (1997): JC virus detection in the cerebrospinal fluid of AIDS patients with progressive multifocal leucoencephalopathy and monitoring of the antiviral treatment by a PCR method. J Med Microbiol 46:256–259.

McCance DJ (1983): Persistence of animal and human papovaviruses in renal and nervous tissues. Prog Clin Biol Res 105:343–357.

McGuire D, Barhite S, Hollander H, Miles M (1995): JC virus DNA in cerebrospinal fluid of human immunodeficiency virus-infected patients: Predictive value for progressive multifocal leukoencephalopathy. Ann Neurol 37:395–399.

Mendelson M, Monard S, Sissons P, Sinclair J (1996): Detection of endogenous human cytomegalovirus in CD34+ bone marrow progenitors. J Gen Virol 77:3099–3102.

Monaco MC, Atwood WJ, Gravell M, Tornatore CS, Major EO (1996): JC virus infection of hematopoietic progenitor cells, primary B lymphocytes, and tonsillar stromal cells: Implications for viral latency. J Virol 70:7004–7012.

Monaco MCG, Jensen PN, Hou J, Durham LC, Major EO (1998): Detection of JC virus DNA in human tonsil tissue: Evidence for site of initial viral infection. J Virol 72:9918–9923.

Monini P, Rotola A, de Lellis L, Corallini A, Secchiero P, Albini A, Benelli R, Parravicini C, Barbanti-Brodano G, Cassai E (1996): Latent BK virus infection and Kaposi's sarcoma pathogenesis. Int J Cancer 66:717–722.

Monini P, Rotola A, di Luca D, de Lellis L, Chiari E, Corallini A, Cassai E (1995): DNA rearrangements impairing BK virus productive infection in urinary tract tumors. Virology 214:273–279.

Moret H, Guichard M, Matheron S, Katlama C, Sazdovitch V, Huraux JM, Ingrand D (1993): Virological diagnosis of progressive multifocal leukoencephalopathy: Detection of JC virus DNA in cerebrospinal fluid and brain tissue of AIDS patients. J Clin Microbiol 31:3310–3313.

Mori M, Aoki N, Shimada H, Tajima M, Kato K (1992): Detection of JC virus in the brains of aged patients without progressive multifocal leukoencephalopathy by the polymerase chain reaction and Southern hybridization analysis. Neurosci Lett 141: 151–155.

Myers C, Frisque RJ, Arthur RR (1989): Direct isolation and characterization of JC virus from urine samples of renal and bone marrow transplant patients. J Virol 63: 4445–4449.

Nebuloni M, Tosoni A, Boldorini R, Monga G, Carsana L, Bonetto S, Abeli C, Caldarelli R, Vago L, Costanzi G (1999): BK virus renal infection in a patient with the acquired immunodeficiency syndrome. Arch Pathol Lab Med 123:807–811.

Newman JT, Frisque RJ (1997): Detection of archetype and rearranged variants of JC virus in multiple tissues from a pediatric PML patient. J Med Virol 52: 243–252.

Newman JT, Frisque RJ (1999): Identification of JC virus variants in multiple tissues of pediatric and adult PML patients. J Med Virol 58:79–86.

Nickeleit V, Klimkait T, Binet IF, Dalquen P, Del Zenero V, Thiel G, Mihatsch MJ, Hirsch HH (2000): Testing for polyomavirus type BK DNA in plasma to identify renal-allograft recipients with viral nephropathy. N Engl J Med 342:1309–1315.

Noordaa JV, Wertheim-van Dillen P (1977): Rise in antibodies to human papova virus BK and clinical disease. BMJ 1:1471.

Okada Y, Sawa H, Tanaka S, Takada A, Suzuki S, Hasegawa H, Umemura T, Fujisawa J, Tanaka Y, Hall WW, Nagashima K (2000): Transcriptional activation of JC virus by human T-lymphotropic virus type I Tax protein in human neuronal cell lines. J Biol Chem 275:17016–17023.

O'Reilly RJ, Lee FK, Grossbard E, Kapoor N, Kirkpatrick D, Dinsmore R, Stutzer C, Shah KV, Nahmias AJ (1981): Papovavirus excretion following marrow transplantation: Incidence and association with hepatic dysfunction. Transplant Proc 13:262–266.

Pappo O, Demetris AJ, Raikow RB, Randhawa PS (1996): Human polyoma virus infection of renal allografts: Histopathologic diagnosis, clinical significance, and literature review. Mod Pathol 9:105–109.

Perrons CJ, Fox JD, Lucas SB, Brink NS, Tedder RS, Miller RF (1996): Detection of polyomaviral DNA in clinical samples from immunocompromised patients: Correlation with clinical disease. J Infect 32:205–209.

Pietropaolo V, Di Taranto C, Degener AM, Jin L, Sinibaldi L, Baiocchini A, Melis M, Orsi N (1998): Transplacental transmission of human polyomavirus BK. J Med Virol 56:372–376.

Pietzuch A, Bodemer M, Frye S, Cinque P, Trebst C, Lüke W, Weber T (1996): Expression of JCV-specific cDNA in peripheral blood lymphocytes of patients with PML, HIV-infection, multiple sclerosis and in healthy controls. J Neurovirology 2: 46.

Portolani M, Piani M, Gazzanelli G, Borgatti M, Bartoletti A, Grossi MP, Corallini A, Barbanti-Brodano G (1985): Restricted replication of BK virus in human lymphocytes. Microbiologica 8:59–66.

Possati L, Rubini C, Portolani M, Gazzanelli G, Piani M, Borgatti M (1983): Receptors for the human papovavirus BK on human lymphocytes. Arch Virol 75:131–136.

Purighalla R, Shapiro R, McCauley J, Randhawa P (1995): BK virus infection in a kidney allograft diagnosed by needle biopsy. Am J Kidney Dis 26:671–673.

Quinlivan EB, Norris M, Bouldin TW, Suzuki K, Meeker R, Smith MS, Hall C, Kenney S (1992): Subclinical central nervous system infection with JC virus in patients with AIDS. J Infect Dis 166:80–85.

Randhawa PS, Finkelstein S, Scantlebury V, Shapiro R, Vivas C, Jordan M, Picken MM, Demetris AJ (1999): Human polyoma virus–associated interstitial nephritis in the allograft kidney. Transplantation 67:103–109.

Reese JM, Reissing M, Daniel RW, Shah KV (1975): Occurrence of BK virus and BK virus–specific antibodies in the urine of patients receiving chemotherapy for malignancy. Infect Immun 11:1375–1381.

Rekvig OP, Moens U, Fredriksen K, Traavik T (1997): Human polyomavirus BK and immunogenicity of mammalian DNA: A conceptual framework. Methods 11:44–54.

Ricciardiello L, Chang DK, Laghi L, Goel A, Chang CL, Boland CR (2001): Mad-1 is the exclusive JC virus strain present in the human colon, and its transcriptional control region has a deleted 98-base pair sequence in colon cancer tissues. J Virol 75:1996–2001.

Rieckmann P, Michel U, Kehrl JH (1994): Regulation of JC virus expression in B lymphocytes. J Virol 68:217–222.

Rosen S, Harmon W, Krensky AM, Edelson PJ, Padgett BL, Grinnell BW, Rubino MJ, Walker DL (1983): Tubulo-interstitial nephritis associated with polyomavirus (BK type) infection. N Engl J Med 308:1192–1196.

Samorei IW, Schmid M, Pawlita M, Vinters HV, Diebold K, Mundt C, von Einsiedel RW (2000): High sensitivity detection of JC-virus DNA in postmortem brain tissue by in situ PCR. J Neurovirol 6:61–74.

Sandler ES, Aquino VM, Goss-Shohet E, Hinrichs S, Krisher K (1997): BK papova virus pneumonia following hematopoietic stem cell transplantation. Bone Marrow Transplant 20:163–165.

Schneider EM, Dörries K (1993): High frequency of polyomavirus infection in lymphoid cell preparations after allogeneic bone marrow transplantation. Transplant Proc 25:1271–1273.

Shah KV, Daniel RW, Strickler HD, Goedert JJ (1997): Investigation of human urine for genomic sequences of the primate polyomaviruses simian virus 40, BK virus, and JC virus. J Infect Dis 176:1618–1621.

Shah KV, Daniel RW, Zeigel RF, Murphy GP (1974): Search for BK and SV40 virus reactivation in renal transplant recipients. Transplantation 17:131–134.

Shimizu N, Imamura A, Daimaru O, Mihara H, Kato Y, Kato R, Oguri T, Fukada M, Yokochi T, Yoshikawa K, Komatsu H, Ueda R, Nitta M (1999): Distribution of JC virus DNA in peripheral blood lymphocytes of hematological disease cases. Intern Med 38:932–937.

Shinohara T, Matsuda M, Cheng SH, Marshall J, Fujita M, Nagashima K (1993): BK virus infection of the human urinary tract. J Med Virol 41:301–305.

Sinclair J, Sissons P (1996): Latent and persistent infections of monocytes and macrophages. Intervirology 39:293–301.

Sinzger C, Jahn G (1996): Human cytomegalovirus cell tropism and pathogenesis. Intervirology 39:302–319.

Smith RD, Galla JH, Skahan K, Anderson P, Linnemann CC, Jr, Ault GS, Ryschkewitsch CF, Stoner GL (1998): Tubulointerstitial nephritis due to a mutant polyomavirus BK virus strain, BKV(Cin), causing end-stage renal disease. J Clin Microbiol 36:1660–1665.

Stoeckle MY (2000): The spectrum of human herpesvirus 6 infection: From roseola infantum to adult disease. Annu Rev Med 51:423–430.

Stoner GL, Agostini HT, Ryschkewitsch CF, Komoly S (1998): JC virus excreted by multiple sclerosis patients and paired controls from Hungary. Mult Scler 4:45–48.

Stoner GL, Soffer D, Ryschkewitsch CF, Walker DL, Webster HD (1988): A double-label method detects both early (T-antigen) and late (capsid) proteins of JC virus in progressive multifocal leukoencephalopathy brain tissue from AIDS and non-AIDS patients. J Neuroimmunol 19:223–236.

Sugimoto C, Ito D, Tanaka K, Matsuda H, Saito H, Sakai H, Fujihara K, Itoyama Y, Yamada T, Kira J, Matsumoto R, Mori M, Nagashima K, Yogo Y (1998): Amplification of JC virus regulatory DNA sequences from cerebrospinal fluid: Diagnostic value for progressive multifocal leukoencephalopathy. Arch Virol 143:249–262.

Sundsfjord A, Flaegstad T, Flo R, Spein AR, Pedersen M, Permin H, Julsrud J, Traavik T (1994a): BK and JC viruses in human immunodeficiency virus type 1–infected persons: Prevalence, excretion, viremia, and viral regulatory regions. J Infect Dis 169:485–490.

Sundsfjord A, Osei A, Rosenqvist H, Van Ghelue M, Silsand Y, Haga HJ, Rekvig OP, Moens U (1999): BK and JC viruses in patients with systemic lupus erythematosus: Prevalent and persistent BK viruria, sequence stability of the viral regulatory regions, and nondetectable viremia. J Infect Dis 180:1–9.

Sundsfjord A, Spein AR, Lucht E, Flaegstad T, Seternes OM, Traavik T (1994b): Detection of BK virus DNA in nasopharyngeal aspirates from children with respiratory infections but not in saliva from immunodeficient and immunocompetent adult patients. J Clin Microbiol 32:1390–1394.

Takemoto KK, Linke H, Miyamura T, Fareed GC (1979): Persistent BK papovavirus infection of transformed human fetal brain cells. I. Episomal viral DNA in cloned lines deficient in T-antigen expression. J Virol 29:1177–1185.

Telenti A, Aksamit A, Jr, Proper J, Smith TF (1990): Detection of JC virus DNA by polymerase chain reaction in patients with progressive multifocal leukoencephalopathy. J Infect Dis 162:858–861.

Tevethia MJ, Spector DJ (1989): Heterologous transactivation among viruses. Prog Med Virol 36:120–190.

Tolkoff-Rubin NE, Rubin RH (1997): Urinary tract infection in the immunocompromised host. Lessons from kidney transplantation and the AIDS epidemic. Infect Dis Clin North Am 11:707–717.

Tominaga T, Yogo Y, Kitamura T, Aso Y (1992): Persistence of archetypal JC virus DNA in normal renal tissue derived from tumor-bearing patients. Virology 186:736–741.

Tornatore C, Berger JR, Houff SA, Curfman B, Meyers K, Winfield D, Major EO (1992): Detection of JC virus DNA in peripheral lymphocytes from patients with and without progressive multifocal leukoencephalopathy. Ann Neurol 31:454–462.

Traavik T, Uhlin-Hansen L, Flaegstad T, Christie KE (1988): Antibody-mediated enhancement of BK virus infection in human monocytes and a human macrophage-like cell line. J Med Virol 24:283–297.

Tsai RT, Wang M, Ou WC, Lee YL, Li SY, Fung CY, Huang YL, Tzeng TY, Chen Y, Chang D (1997): Incidence of JC viruria is higher than that of BK viruria in Taiwan. J Med Virol 52:253–257.

Vago L, Cinque P, Sala E, Nebuloni M, Caldarelli R, Racca S, Ferrante P, Trabottoni G, Costanzi G (1996): JCV-DNA and BKV-DNA in the CNS tissue and CSF of AIDS patients and normal subjects. Study of 41 cases and review of the literature. J Acquir Immune Defic Syndr Hum Retrovirol 12:139–146.

Vallbracht A, Lohler J, Gossmann J, Gluck T, Petersen D, Gerth HJ, Gencic M, Dörries K (1993): Disseminated BK type polyomavirus infection in an AIDS patient associated with central nervous system disease. Am J Pathol 143:29–39.

Valle LD, Croul S, Morgello S, Amini S, Rappaport J, Khalili K (2000): Detection of HIV-1 Tat and JCV capsid protein, VP1, in AIDS brain with progressive multifocal leukoencephalopathy. J Neurovirol 6:221–228.

Vignoli C, De Lamballerie X, Zandotti C, Tamalet C, De Micco P (1993): Detection of JC virus by polymerase chain reaction and colorimetric DNA hybridization assay. Eur J Clin Microbiol Infect Dis 12:958–961.

Voltz R, Jager G, Seelos K, Fuhry L, Hohlfeld R (1996): BK virus encephalitis in an immunocompetent patient. Arch Neurol 53:101–103.

Walker DL, Padgett BL (1983): Progressive multifocal leukoencephalopathy. In Comprehensive Virology, Vol. 18. Virus-Host Interactions, Fraenkel-Conrat H and Wagner RR, Eds. Plenum Press: New York, pp 161–193.

Weber T, Major EO (1997): Progressive multifocal leukoencephalopathy: Molecular biology, pathogenesis and clinical impact. Intervirology 40:98–111.

Weber T, Turner RW, Frye S, Ruf B, Haas J, Schielke E, Pohle HD, Luke W, Luer W, Felgenhauer K, et al. (1994): Specific diagnosis of progressive multifocal leukoencephalopathy by polymerase chain reaction. J Infect Dis 169:1138–1141.

White FD, Ishaq M, Stoner GL, Frisque RJ (1992): JC virus DNA is present in many human brain samples from patients without progressive multifocal leukoencephalopathy. J Virol 66:5726–5734.

Yogo Y, Kitamura T, Sugimoto C, Hara K, Iida T, Taguchi F, Tajima A, Kawabe K, Aso Y (1991): Sequence rearrangement in JC virus DNAs molecularly cloned from immunosuppressed renal transplant patients. J Virol 65:2422–2428.

Zu Rhein GM (1969): Association of papova-virions with a human demyelinating disease (progressive multifocal leukoencephalopathy): Prog Med Virol 11:185–247.

Zu Rhein GM, Varakis J (1974): Letter: Papovavirions in urothelium of treated lymphoma patient. Lancet 2:783–784.

11

CLINICAL PROGRESSIVE MULTIFOCAL LEUKOENCEPHALOPATHY: DIAGNOSIS AND TREATMENT

JOSEPH R. BERGER, M.D., and AVINDRA NATH, M.D.

1. INTRODUCTION

Although described as early as 1930 by the German neuropathologist Hallervorden (1930) in a monograph on unique and nonclassifiable diseases, progressive multifocal leukoencephalopathy (PML) was first clearly crystallized as a distinct clinical entity in 1958 by Åström et al. This neurologic disorder demonstrated a unique combination of three cardinal histopathologic features, namely, demyelination, enlarged nuclei of oligodendrocytes, and bizarre astrocytes (Åström et al., 1958). Initially, the etiology remained uncertain, although the demonstration of inclusion bodies in the nuclei of damaged oligodendrocytes suggested a possible viral etiology (Cavanaugh et al., 1959). On the basis of electron microscopic criteria, Zu Rhein and colleagues proposed that a polyomavirus was responsible (Zu Rhein 1965, 1969). Six years later, isolation of a new polyomavirus, JC virus, in human fetal brain cultures confirmed this hypothesis (Padgett et al., 1971).

Human Polyomaviruses: Molecular and Clinical Perspectives, Edited by Kamel Khalili and Gerald L. Stoner.
ISBN 0-471-39009-7

2. HOST FACTORS AND UNDERLYING DISEASES

In virtually all patients with PML, an underlying immunosuppressive condition has been recognized. Typically, the abnormality is one of cell-mediated immunity. The first three patients described by Åström et al. (1958) had either chronic lymphocytic leukemia or lymphoma as the underlying illness. In a review by Brooks and Walker published in 1984, lymphoproliferative diseases were the most common underlying disorders, accounting for 62.2% of the cases. Other predisposing illnesses included myeloproliferative diseases in 6.5%; carcinoma in 2.2%; granulomatous and inflammatory diseases, such as tuberculosis and sarcoidosis, in 7.4%; and other immune deficiency states in 16.1%. Although AIDS was included in the latter category, there were but two reported cases of PML complicating AIDS at that time (Brooks and Walker, 1984).

Indeed, until the AIDS epidemic, PML remained a rare disease. To most practicing neurologists it was a medical curiosity about which one learned from the textbooks. However, the AIDS pandemic changed the incidence of this formerly rare illness quite remarkably. In the quarter century between the clear definition of the disorder by Åström et al. (1958) to the extensive review of PML by Brooks and Walker in 1984, only 230 previously published cases were identified. Following the advance of the AIDS pandemic, the spectrum of underlying illnesses changed dramatically. From the University of Miami Medical Center and the Broward County Medical Examiner's Office in south Florida (Berger et al., 1998b), 156 cases of PML were identified over a 14-year interval from 1980 to 1994, with all but two of these cases related to HIV infection. When comparing the 4 year intervals (1980–1984) and (1990–1994) in this series, there was a 20-fold increase in the number of cases of PML (Berger et al., 1998b). Overall, AIDS has been estimated to be the underlying cause of immunosuppression in 55% to more than 85% of all current cases of PML (Major et al., 1992); however, these figures may underestimate the true incidence of HIV/AIDS as the underlying immunosuppressive condition predisposing to PML.

3. CLINICAL EPIDEMIOLOGY IN THE ERA OF AIDS

The first description of PML complicating AIDS followed the description of AIDS in 1981 by 1 year (Miller et al., 1982). By the late 1980s, AIDS was reported to be the most common underlying disorder predisposing to the development of PML at institutions in New York (Krupp et al., 1985) and Miami (Berger et al., 1987). As noted above, the subsequent evolution of the AIDS pandemic has significantly changed the epidemiology of PML. Gillespie and colleagues (1991), studying the prevalence of AIDS-related illnesses in the San Francisco Bay area, estimated a prevalence for PML of 0.3%, but acknowledged that this may have been a significant underestimate. Based on death certificate reporting of AIDS to the Centers for Disease Control and Prevention

(CDC) between 1981 and June 1990, 971 (0.72%) of 135,644 individuals with AIDS were reported to have PML (Holman et al., 1991). Due to the the notorious inaccuracies in death certificate reporting (Messite and Stillman, 1996) and the requirement of pathologic confirmation for inclusion in this study, this is also likely a significant underestimate of the true prevalence. Other studies have suggested that the prevalence of PML in AIDS cases is substantially higher than that reported by the CDC, with most estimates ranging between 1% and 5% in clinical studies and as high as 10% in pathologic series (Berger et al., 1987; Krupp et al., 1985; Kuchelmeister et al., 1993; Kure et al., 1991; Lang et al., 1989; Whiteman et al., 1993). In 1987, a large, retrospective, hospital-based, clinical study (Berger et al., 1987) found PML in approximately 4% of patients hospitalized with AIDS.

Four percent of all patients dying with AIDS had PML in a combined series of seven separate neuropathologic studies comprising a total of 926 patients with AIDS (Kure et al., 1991). Two other large neuropathologic series found PML in 7% (Lang et al., 1989) and 9.8% (Kuchelmeister et al., 1993) of autopsied AIDS patients. The authors of the latter study acknowledged that an unusually high estimate may have resulted from numerous referral cases from outside the study center and the inherent bias imposed (Kuchelmeister et al., 1993). However, a study of 548 consecutive, unselected autopsies between 1983 and 1991 performed on HIV seropositive individuals by the Broward County (Florida) Medical Examiner revealed that 29 (5.3%) had PML confirmed at autopsy (Whiteman et al., 1993). Similarly, the Multicenter AIDS Cohort Study also identified a dramatic rise in the incidence of PML over a similar time period.

Specifically, the Multicenter AIDS Cohort Study identified 22 cases of PML among the cohort of AIDS cases studied from 1985 to 1992: The average annual incidence of PML was 0.15 per 100 person-years, with a yearly rate of increase of 24% between 1985 and 1992 (Bacellar et al., 1994). Although these estimates may be susceptible to selection and other biases, there is an indisputable markedly increased incidence and prevalence of PML since the inception of the AIDS pandemic. Indeed, it appears that the incidence of PML complicating HIV/AIDS is higher than that of any other immunosuppressive disorder relative to their frequency. Possible explanations include differences in the degree and duration of the cellular immunosuppression in HIV infection, facilitation of the entry into the brain of JC virus–infected B lymphocytes (Houff et al., 1988) by alterations in the blood–brain barrier due to HIV (Power et al., 1993) or the upregulation of adhesion molecules on the brain vascular endothelium due to HIV infection (Hofman et al., 1994; Sasseville et al., 1992), and the potential for the HIV Tat protein to transactivate JC virus (Tada et al., 1990).

Concomitant with the increase in PML in association with AIDS has been the not unexpected alteration in the demographics of the affected population. Before the AIDS epidemic, males and females were affected in a ratio of 3:2 (Brooks and Walker, 1984). The incidence of PML increased steadily from middle age (Brooks and Walker 1984). Children, regardless of the cause of underlying immunosuppression, rarely develop PML. As exposure to JC virus occurs sometime during childhood, a minority of young children are at risk for the disease. However, PML has been observed in HIV-infected children (Berger et al., 1992b; Morriss et al., 1997; Singer et al., 1993). As cited, lymphoproliferative diseases were the most common underlying etiology (Brooks and Walker, 1984).

Currently, HIV infection is the most common underlying cause of immunosuppression and the disorder chiefly affects homosexual/bisexual men between the ages of 25 and 50 years, with a correspondingly high male to female ratio of 7.6 to 1.0 (Berger et al., 1998b). One should bear in mind that, as the demography of HIV infection changes, a parallel change in the demography of AIDS-related PML would be anticipated. Curiously, there seems to be a higher degree of prevalence of PML in white males than in African-American males (Holman et al., 1998). Additionally, there may be some geographic differences in the prevalence of PML. For example, PML is considered rare in Africa, and a neuropathologic study in southern India suggests an incidence of 1% (Satishchandra et al., 2000). Some of the population differences observed may be the consequence of the nature of medical care rendered. PML is typically observed in advanced HIV infection, and therefore it is not unlikely that patients succumbing to other AIDS-related disorders early in the course of their infection may not live long enough to develop PML.

In 1996, there was an expansion of available antiretroviral therapies with the development of highly active antiretroviral therapies (HAART). Opportunistic infections, for example, cytomegalovirus (Baril et al., 2000; Verbraak et al., 1999) and toxoplasmosis (Maschke et al., 2000), and primary central nervous system (CNS) lymphomas (Sparano et al., 1999) have been reported to have declined significantly following their introduction. The effect of this therapy on the incidence of PML remains uncertain. D'Arminio Monforte and colleagues (2000) detected a 95% risk reduction in all CNS AIDS-related conditions following the adoption of HAART in their cohort. Twenty cases of PML were identified in this study, but a specific analysis for PML was not undertaken. Others (Maschke et al., 2000) have similarly noted a decline in HIV-related CNS disorders, but have had too few cases to comment specifically about PML. In summary, the epidemiology of PML suggests that a dysfunction of the cellular immune response is the single most important determining factor that predisposes to the development of PML, although gender, genetic factors, and viral strains may also play a role. Therapies, in particular, HAART, with a restoration of immune function, may result in a decline in the frequency of this disorder.

4. CLINICAL FEATURES

PML is the heralding AIDS-defining illness in approximately 1% of all HIV-infected persons (Berger et al., 1998b). Therefore, the occasional patient is seen with neurologic features antedating knowledge of HIV infection. This may lead to significant diagnostic confusion in the otherwise healthy individual with unsuspected HIV infection. The treating physician needs to maintain a high index of suspicion for underlying HIV infection when confronted with unusual neurologic illnesses.

The most common symptoms reported by patients with PML or their caregivers are weakness and motor disturbances (Berger et al., 1998b). Other common symptoms included cognitive abnormalities, headaches, gait disorders, visual impairment, speech disturbance and sensory loss. In a large series of more than 150 AIDS-related PML patients (Berger et al., 1998b), each of these symptoms was seen in more than 15% of patients. In general, these symptoms are similar to those identified in a series of non-HIV–associated PML cases. In comparison to the series of Brooks and Walker (1984), headaches were significantly more common in the HIV-infected population and visual disturbances were more common in the non-HIV infected. Seizures occur in up to 10% of patients and are usually focal in nature, although secondary generalization may occur. Seizures in AIDS-associated PML may reflect involvement of the cortical astrocytes by the JC virus (Sweeney et al., 1994) or may be secondary to some other process or HIV infection of the brain itself (Wong et al., 1990).

Limb weakness is the most common sign observed with HIV-associated PML (Berger et al., 1998b). It was observed in over 50% (Berger et al., 1998b). Cognitive disturbances and gait disorders are seen in approximately one fourth to one third of patients (Berger et al., 1998b). Diplopia, noted by 9% of patients, is usually the consequence of involvement of the third, fourth, or sixth cranial nerve and is typically observed in association with other brain stem findings (Berger et al., 1998b). Visual field loss due to involvement of the retrochiasmal visual pathways is significantly more common than diplopia or other visual disturbances (Berger et al., 1998b; Brooks and Walker, 1984; Omerud et al., 1996). Optic nerve disease does not occur with PML, and, although the lesions of PML have been detected in the spinal cords of HIV-infected patients (Henin et al., 1992), clinical myelopathy secondary to PML must be vanishingly rare. PML does not involve the peripheral nervous system. Although the virus resides in the lymphoid tissue and bone marrow, it remains asymptomatic in these reservoirs. Recent evidence of latent infection of tonsillar tissue (Monaco et al., 1998) supports the hypothesis that JC virus infection is initially acquired as an oropharyngeal or respiratory infection. However, despite evidence of seropositivity for JC virus of up to 80% of the population, no clinical illness has been convincingly established with primary infection to date.

5. NEUROIMAGING

In the appropriate clinical context, radiographic imaging may strongly support the diagnosis of PML. Computed tomography (CT) of the brain of a patient with PML reveals hypodense lesions of the affected white matter (Fig 11.1).

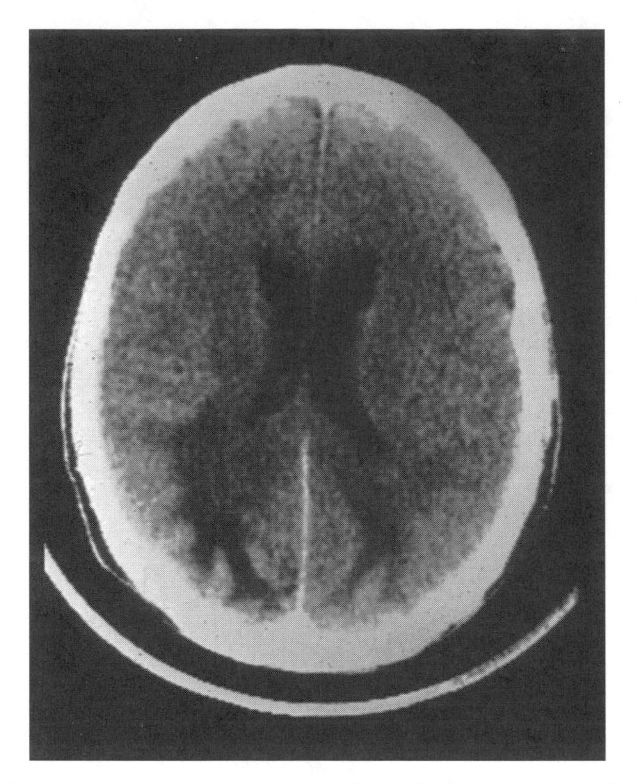

Figure 11.1. Computed tomographic scan showing typical hypodense lesions of PML distributed bilaterally in the parieto-occipital lobes.

On CT scan, the lesions of PML exhibit no mass effect and rarely contrast enhance. A "scalloped" appearance beneath the cortex is noted when there is involvement of the subcortical arcuate fibers (Whiteman et al., 1993). Cranial magnetic resonance imaging (MRI) is far more sensitive to the presence of the white matter lesions of PML than CT scan (Whiteman et al., 1993). MRI shows a hyperintense lesion on T2-weighted images in the affected regions (Fig. 11.2), and, as with CT scan, faint, typically peripheral, contrast enhancement may be observed in 5–10% of cases (Whiteman et al., 1993). The lesions of PML may occur virtually anywhere in the brain. The frontal lobes and parieto-occipital regions are commonly affected, presumably as a consequence of their volume. However, isolated or associated involvement of the basal ganglia, external capsule, and posterior fossa structures (cerebellum and brainstem) may be seen as well (Whiteman et al., 1993).

Other diseases may affect the white matter in a similar manner in association with HIV infection. Particularly notable in this regard are AIDS dementia and cytomegalovirus infection. With respect to AIDS dementia, radiographic distinctions include a greater propensity of PML lesions to involve the subcortical

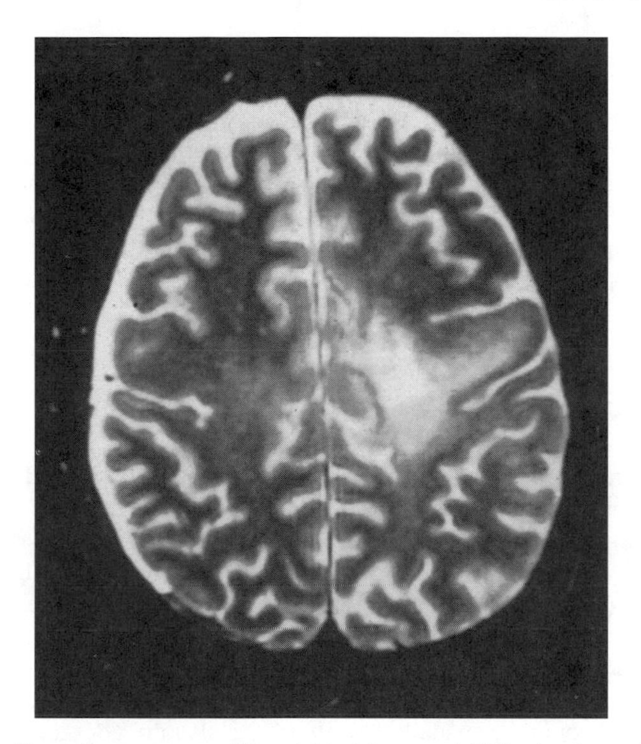

Figure 11.2. Magnetic resonance T2-weighted image of PML showing a hyperintense signal abnormality of the left frontal lobe white matter.

white matter, its hypointensity on T1-weighted images, and its rare enhancement (Whiteman et al., 1993). Cytomegalovirus lesions are typically located in the periventricular white matter and centrum semiovale. Subependymal enhancement is often observed as a consequence of CMV infection. Other potentially HIV-associated disorders that may result in hyperintense signal abnormalities of the white matter resembling PML include varicella-zoster leukoencephalitis (Gray et al., 1994), a multiple sclerosis-like illness (Berger et al., 1992c), acute disseminated encephalomyelitis (Bhigjee et al., 1999; Chetty et al., 1997), CNS vasculitis (Scaravilli et al., 1989), and white matter edema associated with primary or metastatic brain tumors. Almost always, the clinical features, laboratory findings, and associated radiographic features enable the correct diagnosis.

Magnetization transfer MRI studies have been suggested to be an effective means of monitoring the degree of demyelination in PML (Brochet and Dousset, 1999). Magnetic resonance spectroscopy reveals a decrease in *N*-acetylaspartate and creatine and increased choline products, myoinositol, and lactate in the lesions of PML (Chang et al., 1997). These changes were interpreted as reflective of the neuronal loss and cell membrane and myelin breakdown observed in PML (Chang et al., 1997). Cerebral angiography is not routinely performed, but detected arteriovenous shunting and a parenchymal blush in the absence of contrast enhancement on MRI in four of six patients in one study (Nelson et al., 1999).

Pathologic studies suggested that small vessel proliferation and perivascular inflammation were the explanation for these unexpected angiographic features (Nelson et al., 1999). Thallium-201 single-photon emission computed tomography (Tl^{201} SPECT) generally reveals no uptake in the lesions of PML (Iranzo et al., 1999); however, a single case report of a contrast-enhancing lesion with a positive Tl^{201} SPECT has been reported (Port et al., 1999).

6. LABORATORY STUDIES

In the overwhelming majority of HIV-infected patients with PML, severe cellular immunosuppression, as defined by CD4 lymphocyte counts below 200 cells/mm^3, is observed. In three separate series of AIDS-related PML (Berger et al., 1998b; Fong and Toma, 1995; von Einsiedel et al., 1993), the mean CD4 count ranged from 84 to 104 cells/mm^3. However, in the largest series of AIDS-related PML (Berger et al., 1998b), 10% or more of patients had CD4 lymphocyte counts in excess of 200 cells/mm^3.

Cerebrospinal fluid (CSF) examination is very helpful in excluding other diagnoses. Cell counts are usually less than 20 cells/mm^3 (Berger et al., 1998b). In one large study, the median cell count was 2 cells/mm^3, and the mean was 7.7 cells/mm^3 (Berger et al., 1998b). In the same study, 55% had an abnormally elevated CSF protein (Berger et al., 1998b), with the highest recorded value being 208 mg/dl (2.08 gm/L). Hypoglycorrhachia was observed in less than 15%. These abnormalities are not inconsistent with that previously reported to occur with HIV infection alone (Elovaara et al., 1987; Katz et al., 1989; Marshall et al., 1988). Several studies (Fong et al., 1995; McGuire et al., 1995; Weber et al., 1994) demonstrate a high sensitivity and specificity of CSF PCR for JC virus in PML. Many authorities have regarded the demonstration of JC viral DNA coupled with the appropriate clinical and radiologic features sufficiently suggestive of PML to be diagnostic, thus obviating the need for brain biopsy. Quantitative PCR techniques for JC virus in biologic fluids continues to be refined (Drews et al., 2000). Antibody titers in serum or CSF are not useful because most individuals become seropositive for JC virus before adulthood. Additionally, as PML occurs in the context of immunosuppression, the individual may not be able to mount an antibody response.

7. PATHOLOGY

Macroscopically, the cardinal feature of PML is demyelination. Demyelination may, on rare occasions, be monofocal, but it typically occurs as a multifocal process suggesting a hematologic spread of the virus. These lesions may occur in any location in the white matter and range in size from 1 mm to several centimeters (Åström et al., 1958; Richardson, 1970); larger lesions are not infrequently the result of coalescence of multiple smaller lesions. The histopathologic hallmarks of PML are a triad (Åström et al., 1958; Richardson 1970) of multifocal demyelination (Fig. 11.3); hyperchromatic, enlarged oligodendroglial nuclei (Fig. 11.4); and enlarged bizarre astrocytes with lobulated hyperchromatic nuclei (Fig. 11.5). The latter may be seen to undergo mitosis and appear quite malignant. Electron microscopic examination will reveal the JC virus in the oligodendroglial cells. These virions measure 38–45 nm in diameter and appear singly or in dense crystalloid arrays (Richardson, 1970). Less frequently, the virions are detected in reactive astrocytes, and they are uncommonly observed in macrophages that are engaged in removing the affected oligodendrocytes (Mázló and Herndon, 1977; Mázló and Tariska, 1982). The virions are generally not seen in the large, bizarre astrocytes (Mázló and

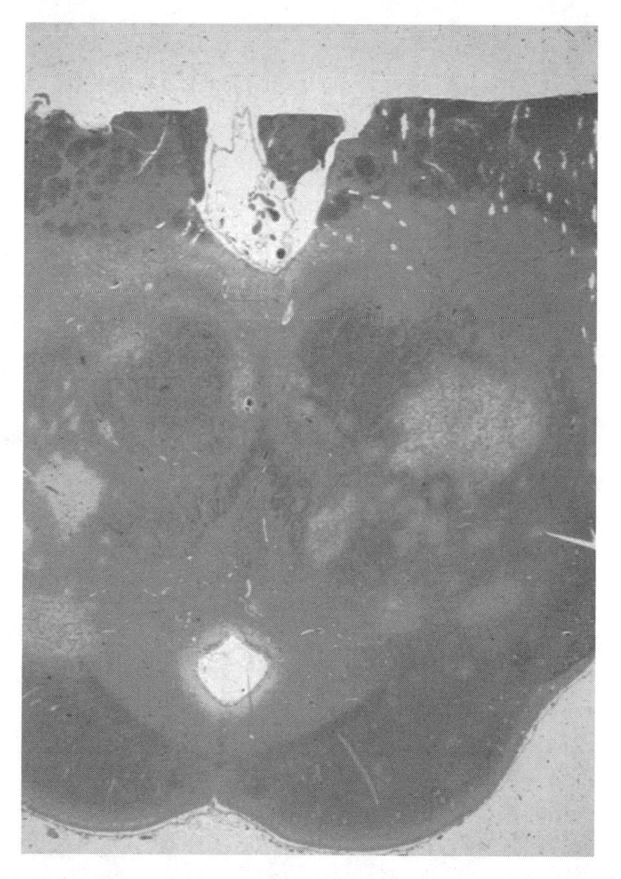

Figure 11.3. Multiple areas of demyelination of the midbrain in PML shown on a hematoxylin and eosin–stained section.

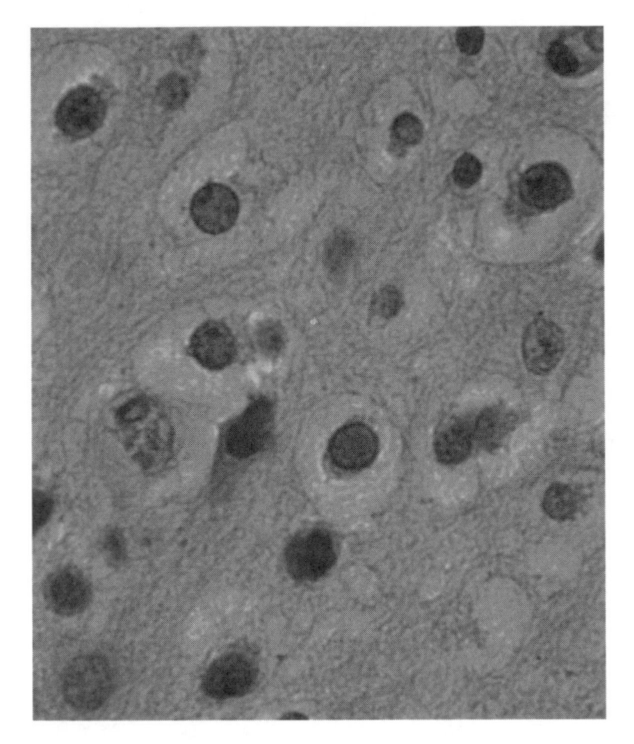

Figure 11.4. Enlarged, hyperchromatic oligodendroglial cell nuclei.

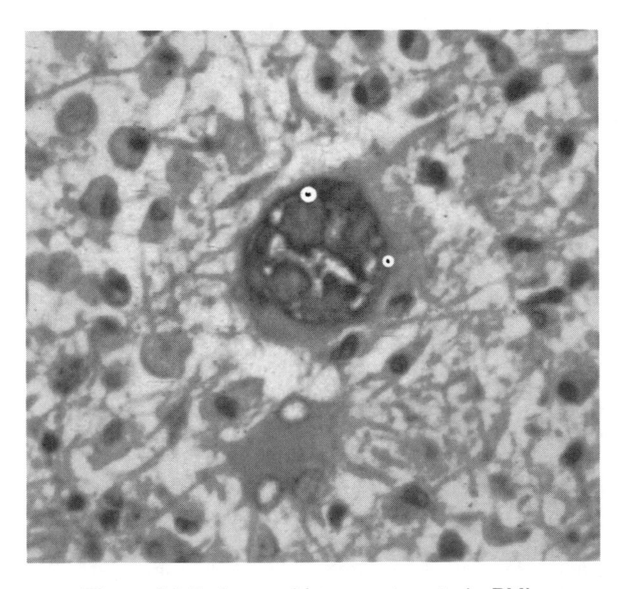

Figure 11.5. Large, bizarre astrocyte in PML.

Tariska, 1982). In situ hybridization and in situ PCR for JCV antigen allows for detection of the virion in the infected cells even in formalin-fixed archival tissue (Samorei et al., 2000).

8. PROGNOSIS

The median survival of PML complicating AIDS is 6 months, and the mode is 1 month (Berger et al., 1998b). The survival of patients with PML with AIDS is not significantly different from that of patients with other immunosuppressive disorders, and survival has not changed measurably in the AIDS era from that in the pre-AIDS era. Recovery of neurologic function, improvement of PML lesions in radiographic imaging, and survival exceeding 12 months have been observed in as many as 10% of patients with AIDS-associated PML (Berger et al., 1998b). Factors that appear to be associated with prolonged survival include PML as the presenting manifestation of AIDS, higher CD4 lymphocyte counts (>300 cells/mm^3), and contrast enhancement on radiographic imaging (Arbusow et al., 2000; Berger et al., 1998a), although another study failed to link any radiologic features with prognosis (Post et al., 1999). Additionally, a correlation between low levels of JC viral DNA in the CSF and prolonged survival has also been demonstrated (De Luca et al., 1999b; Taoufik et al., 1998; Yiannoutsos et al., 1999). The longest survival has been 92 months from onset of illness (Berger et al., 1998b). Our experience and that of others (De Luca et al., 1999b) suggests that the lack of recurrence of PML in some of the patients exhibiting long-term survival and recovery reflects clearance of the JC virus.

9. TREATMENT

Antiretroviral Therapy

Even in the era before the availability of protease inhibitors, antiretroviral therapy had been anecdotally reported to improve PML (Bauer et al., 1973; Conway et al., 1990; Fiala et al., 1988; Martin-Suarez et al., 1994; Singer et al., 1994; von Einsiedel et al., 1993). However, several large series now show that aggressive antiretroviral therapy that includes protease inhibitors is necessary to improve survival in these patients (Albrecht et al., 1998; Berger et al., 1998b; Clifford et al., 1999; Miralles et al., 1998; Tassie et al., 1999). Data compiled by the CDC on 415 patients showed that increased survival time was associated with antiretroviral therapy that contained a protease inhibitor (risk ratio, 0.2; 95% CI, 0.1–0.4) and prescription of other antiretroviral medication (risk ratio, 0.6; 95% CI, 0.5–0.8) compared with no antiretroviral therapy (Dworkin et al., 1999). An independent French study of 246 patients also showed increased survival with antiretroviral therapy containing protease inhibitors (risk ratio, 0.37; 95% CI, 0.22–0.64), while patients without protease inhibitors had a relative risk of 0.62 and a 95% CI of 0.41–0.95. Prolonged survival after HAART for PML is also associated with JC virus clearance from CSF and an increase in antibody response to JC virus proteins (Giudici et al., 2000).

Cytosine Arabinoside

Many other therapies have been suggested for PML (Major et al., 1992). The alleged benefit of cytosine arabinoside (Bauer et al., 1973; Buckman and Wiltshaw, 1976; Conomy et al., 1974; O'Riordan et al., 1990; Peters et al., 1980; Portegies et al., 1991), an agent known to inhibit JC virus replication in vitro (Hou and Major, 1998), was not supported by the results of a large, randomized trial in which cytosine arabinoside was administered either intravenously or intrathecally (Hall et al., 1998). Another European study also failed to confirm the efficacy of cytosine arabinoside given intravenously for PML (Enting and Portegies, 2000). The failure of cytosine arabinoside in this study may simply reflect insufficient levels of the drug at sites of disease. Trials of convection-enhanced delivery of cytosine arabinoside in PML are planned and may prove to be effective in the management of the disorder as experiments in animal models reveal tissue concentration levels adequate to suppress JC viral replication (Groothuis et al., 2000).

Cidofovir

Cidofovir is an acyclic nucleoside phosphonate that has broad-spectrum antiviral effects with activity against all major DNA viruses (Del Clercq 1993). A well-designed, clinical trial of cidofovir in AIDS-related PML sponsored by the NIH has been recently been completed. No improvement in survival was noted (Marra et al., 2001) despite reports of clinical efficacy with cidofovir in single cases and small series (Blick et al., 1998; Brambilla et al., 1999; Cardenas et al., 2001; Chocarro Martinez et al., 2000; De Luca et al., 1999a, 2000; Happe et al., 1999, 2000; Herrero-Romero et al., 2001; Portilla et al., 2000).

Topotecan

Trials of topotecan are currently underway. This drug inhibits DNA topoisomerase and has been shown to inhibit JC viral replication in glial cells in vitro (Kerr et al., 1993).

Alpha-Interferon

Alpha-interferon was proposed as a treatment for AIDS-related PML in 1992 (Berger et al., 1992a). This open-labeled study failed to observe any significant alteration in prognosis compared with historical controls. A subsequent open-labeled study with alpha-interferon before the era of antiretroviral therapy showed increased survival times for patients treated with 3 million units of alpha-interferon daily for a minimum of 3 weeks (Huang et al., 1998).

Splenectomy

A single case report showed remission of PML following splenectomy in a patient with AIDS (Power et al., 1997). It was suggested that removal of the spleen leads to decreased HIV and JC viral load. Several case series in the past have shown that splenectomy causes a decrease in HIV viral load and an increase in CD4 cell count (Bernard et al., 1998) and splenectomy has often been used for treatment of HIV-infected patients with refractory thrombocytopenia with encouraging results (Aboolian et al., 1999).

Novel Therapies

The development of antisense oligonucleotides for PML seems to be logical next step in the treatment of this disorder. However, uncertainty as to the specific mRNA to target with the oligonucleotide and methods of delivery of the molecule to the brain remain formidable problems.

REFERENCES

Aboolian A, Ricci M, Shapiro K, Connors A, LaRaja RD (1999): Surgical treatment of HIV-related immune thrombocytopenia. Int Surg 84:81–85.

Albrecht H, Hoffmann C, Degen O, Stoehr A, Plettenberg A, Mertenskotter T, Eggers C, Stellbrink HJ (1998): Highly active antiretroviral therapy significantly improves the prognosis of patients with HIV-associated progressive multifocal leukoencephalopathy. AIDS 12:1149–1154.

Arbusow V, Strupp M, Pfister HW, Seelos KC, Bruckmann H, Brandt T (2000): Contrast enhancement in progressive multifocal leukoencephalopathy: A predictive factor for long-term survival? J Neurol 247:306–308.

Åström K, Mancall E, Richardson EJ (1958): Progressive multifocal leukoencephalopathy. Brain 81:93–127.

Bacellar H, Muñoz A, Miller EN, Cohen BA, Besley D, Selnes OA, Becker JT, McArthur JC (1994): Temporal trends in the incidence of HIV-1–related neurologic diseases: Multicenter AIDS Cohort Study, 1985–1992. Neurology 44:1892–1900.

Baril L, Jouan M, Agher R, Cambau E, Caumes E, Bricaire F, Katlama C (2000): Impact of highly active antiretroviral therapy on onset of *Mycobacterium avium* complex infection and cytomegalovirus disease in patients with AIDS. AIDS 14:2593–2596.

Bauer WR, Turel AP Jr, Johnson KP (1973): Progressive multifocal leukoencephalopathy and cytarabine. Remission with treatment. JAMA 226:174–176.

Berger JR, Kaszovitz B, Post MJ, Dickinson G (1987): Progressive multifocal leukoencephalopathy associated with human immunodeficiency virus infection. A review of the literature with a report of sixteen cases. Ann Intern Med 107:78–87.

Berger JR, Levy RM, Flomenhoft D, Dobbs M (1998a): Predictive factors for prolonged survival in acquired immunodeficiency syndrome–associated progressive multifocal leukoencephalopathy. Ann Neurol 44:341–349.

Berger JR, Pall L, Lanska D, Whiteman M (1998b): Progressive multifocal leukoencephalopathy in patients with HIV infection. J Neurovirol 4:59–68.

Berger J, Pall L, McArthur J, Hall C, Cimoch P, Evans B, Price R, Feraru E (1992a): A pilot study of recombinant alpha 2a interferon in the treatment of AIDS-related progressive multifocal leukoencephalopathy [abstract]. Neurology 42(Suppl 3):257.

Berger JR, Scott G, Albrecht J, Belman AL, Tornatore C, Major EO (1992b): Progressive multifocal leukoencephalopathy in HIV-1–infected children. AIDS 6:837–841.

Berger JR, Tornatore C, Major EO, Bruce J, Shapshak P, Yoshioka M, Houff S, Sheremata W, Horton GF, Landy H (1992c): Relapsing and remitting human immunodeficiency virus–associated leukoencephalomyelopathy. Ann Neurol 31:34–38.

Bernard NF, Chernoff DN, Tsoukas CM (1998): Effect of splenectomy on T-cell subsets and plasma HIV viral titers in HIV-infected patients. J Hum Virol 1:338–345.

Bhigjee AI, Patel VB, Bhagwan B, Moodley AA, Bill PL (1999): HIV and acute disseminated encephalomyelitis. S Afr Med J 89:283–284.

Blick G, Whiteside M, Griegor P, Hopkins U, Garton T, LaGravinese L (1998): Successful resolution of progressive multifocal leukoencephalopathy after combination therapy with cidofovir and cytosine arabinoside. Clin Infect Dis 26:191–192.

Brambilla AM, Castagna A, Novati R, Cinque P, Terreni MR, Moioli MC, Lazzarin A (1999): Remission of AIDS-associated progressive multifocal leukoencephalopathy after cidofovir therapy. J Neurol 246:723–725.

Brochet B, Dousset V (1999): Pathological correlates of magnetization transfer imaging abnormalities in animal models and humans with multiple sclerosis. Neurology 53: S12–77.

Brooks BR, Walker DL (1984): Progressive multifocal leukoencephalopathy. Neurol Clin 2:299–313.

Buckman R, Wiltshaw E (1976): Letter: Progressive multifocal leucoencephalopathy successfully treated with cytosine arabinoside. Br J Haematol 34:153–158.

Cardenas RL, Cheng KH, Sack K (2001): The effects of cidofovir on progressive multifocal leukoencephalopathy: An MRI case study. Neuroradiology 43:379–382.

Chang L, Ernst T, Tornatore C, Aronow H, Melchor R, Walot I, Singer E, Cornford M (1997): Metabolite abnormalities in progressive multifocal leukoencephalopathy by proton magnetic resonance spectroscopy. Neurology 48:836–845.

Chetty KG, Kim RC, Mahutte CK (1977): Acute hemorrhagic leukoencephalitis during treatment for disseminated tuberculosis in a patient with AIDS. Int J Tuberc Lung Dis 1:579–581.

Chocarro Martinez A, Gonzalez Lopez A, Garcia Garcia I (2000): Successful resolution of progressive multifocal leukoencephalopathy after combination therapy with cidofovir and cytosine arabinoside. Clin Infect Dis 30:234.

Clifford DB, Yiannoutsos C, Glicksman M, Simpson DM, Singer EJ, Piliero PJ, Marra CM, Francis GS, McArthur JC, Tyler KL, Tselis AC, Hyslop NE (1999): HAART improves prognosis in HIV-associated progressive multifocal leukoencephalopathy. Neurology 52:623–625.

Conomy J, Beard N, Matsumoto H, Roessmann U (1974): Cytarabine treatment of progressive multifocal leukoencephalopathy. JAMA 229:1313–1316.

Conway B, Halliday WC, Brunham RC (1990): Human immunodeficiency virus–associated progressive multifocal leukoencephalopathy: Apparent response to 3′-azido-3′-deoxythymidine. Rev Infect Dis 12:479–482.

d'Arminio Monforte A, Duca PG, Vago L, Grassi MP, Moroni M (2000): Decreasing incidence of CNS AIDS-defining events associated with antiretroviral therapy. Neurology 54:1856–1859.

Del Clercq E (1993): Therapeutic potential of HPMPC as an antiviral drug. Rev Med Virol 3:85–96.

De Luca A, Fantoni M, Tartaglione T, Antinori A (1999a): Response to cidofovir after failure of antiretroviral therapy alone in AIDS-associated progressive multifocal leukoencephalopathy. Neurology 52:891–892.

De Luca A, Giancola ML, Ammassari A, Grisetti S, Cingolani A, Paglia MG, Govoni A, Murri R, Testa L, Monforte AD, Antinori A (2000): Cidofovir added to HAART improves virological and clinical outcome in AIDS-associated progressive multifocal leukoencephalopathy. AIDS 14:F117–F121.

De Luca A, Giancola ML, Cingolani A, Ammassari A, Gillini L, Murri R, Antinori A (1999b): Clinical and virological monitoring during treatment with intrathecal cytarabine in patients with AIDS-associated progressive multifocal leukoencephalopathy. Clin Infect Dis 28:624–628.

Drews K, Bashir T, Dörries K (2000): Quantification of human polyomavirus JC in brain tissue and cerebrospinal fluid of patients with progressive multifocal leukoencephalopathy by competitive PCR. J Virol Methods 84:23–36.

Dworkin MS, Wan PC, Hanson DL, Jones JL (1999): Progressive multifocal leukoencephalopathy: Improved survival of human immunodeficiency virus–infected patients in the protease inhibitor era. J Infect Dis 180:621–625.

Elovaara I, Iivanainen M, Valle SL, Suni J, Tervo T, Lahdevirta J (1987): CSF protein and cellular profiles in various stages of HIV infection related to neurological manifestations. J Neurol Sci 78:331–342.

Enting RH, Portegies P (2000): Cytarabine and highly active antiretroviral therapy in HIV-related progressive multifocal leukoencephalopathy. J Neurol 247:134–138.

Fiala M, Cone LA, Cohen N, Patel D, Williams K, Casareale D, Shapshak P, Tourtelotte W (1988): Responses of neurologic complications of AIDS to 3′-azido-3′-deoxythymidine and 9-(1,3-dihydroxy-2-propoxymethyl) guanine. I. Clinical features. Rev Infect Dis 10:250–256.

Fong IW, Britton CB, Luinstra KE, Toma E, Mahony JB (1995): Diagnostic value of detecting JC virus DNA in cerebrospinal fluid of patients with progressive multifocal leukoencephalopathy. J Clin Microbiol 33:484–486.

Fong IW, Toma E (1995): The natural history of progressive multifocal leukoencephalopathy in patients with AIDS. Canadian PML Study Group. Clin Infect Dis 20: 1305–1310.

Gillespie SM, Chang Y, Lemp G, Arthur R, Buchbinder S, Steimle A, Baumgartner J, Rando T, Neal D, Rutherford G, et al (1991): Progressive multifocal leukoencephalopathy in persons infected with human immunodeficiency virus, San Francisco, 1981–1989. Ann Neurol 30:597–604.

Giudici B, Vaz B, Bossolasco S, Casari S, Brambilla AM, Luke W, Lazzarin A, Weber T, Cinque P (2000): Highly active antiretroviral therapy and progressive multifocal leukoencephalopathy: Effects on cerebrospinal fluid markers of JC virus replication and immune response. Clin Infect Dis 30:95–99.

Gray F, Belec L, Lescs MC, Chretien F, Ciardi A, Hassine D, Flament-Saillour M, de Truchis P, Clair B, Scaravilli F (1994): Varicella-zoster virus infection of the central nervous system in the acquired immune deficiency syndrome. Brain 117:987–999.

Groothuis DR, Benalcazar H, Allen CV, Wise RM, Dills C, Dobrescu C, Rothholtz V, Levy RM (2000): Comparison of cytosine arabinoside delivery to rat brain by intravenous, intrathecal, intraventricular and intraparenchymal routes of administration. Brain Res 856:281–290.

Hall CD, Dafni U, Simpson D, Clifford D, Wetherill PE, Cohen B, McArthur J, Hollander H, Yainnoutsos C, Major E, Millar L, Timpone J (1998): Failure of cytarabine in progressive multifocal leukoencephalopathy associated with human immunodeficiency virus infection. AIDS Clinical Trials Group 243 Team. N Engl J Med 338: 1345–1351.

Hallervorden J (1930): Eigennartige und nicht rubriziebare Prozesse. In Bumke O, Ed.; Handbuch der Geiteskranheiten; Springer: Berlin, Vol. XI, pp. 1063–1107.

Happe S, Besselmann M, Matheja P, Rickert CH, Schuierer G, Reichelt D, Husstedt IW (1999): [Cidofovir (vistide): in therapy of progressive multifocal leukoencephalopathy in AIDS. Review of the literature and report of 2 cases]. Nervenarzt 70: 935–943.

Happe S, Lunenborg N, Rickert CH, Heese C, Reichelt D, Schuierer G, Schul C, Husstedt IW (2000): [Progressive multifocal leukoencephalopathy in AIDS. Overview and retrospective analysis of 17 patients]. Nervenarzt 71:96–104.

Henin D, Smith TW, De Girolami U, Sughayer M, Hauw JJ (1992): Neuropathology of the spinal cord in the acquired immunodeficiency syndrome. Hum Pathol 23: 1106–1114.

Herrero-Romero M, Cordero E, Lopez-Cortes LF, de Alarcon A, Pachon J (2001): Cidofovir added to highly active retroviral therapy in AIDS-associated progressive multifocal leukoencephalopathy. AIDS 15:809–809.

Hofman FM, Dohadwala MM, Wright AD, Hinton DR, Walker SM (1994): Exogenous tat protein activates central nervous system–derived endothelial cells. J Neuroimmunol 54:19–28.

Holman RC, Janssen RS, Buehler JW, Zelasky MT, Hooper WC (1991): Epidemiology of progressive multifocal leukoencephalopathy in the United States: analysis of national mortality and AIDS surveillance data. Neurology 41:1733–1736.

Holman RC, Torok TJ, Belay ED, Janssen RS, Schonberger LB (1998): Progressive multifocal leukoencephalopathy in the United States, 1979–1994: Increased mortality associated with HIV infection. Neuroepidemiology 17:303–309.

Hou J, Major EO (1998): The efficacy of nucleoside analogs against JC virus multiplication in a persistently infected human fetal brain cell line. J Neurovirol 4:451–456.

Houff SA, Major EO, Katz DA, Kufta CV, Sever JL, Pittaluga S, Roberts JR, Gitt J, Saini N, Lux W (1988): Involvement of JC virus–infected mononuclear cells from the bone marrow and spleen in the pathogenesis of progressive multifocal leukoencephalopathy. N Engl J Med 318:301–305.

Huang SS, Skolasky RL, Dal Pan GJ, Royal W 3rd, McArthur JC (1998): Survival prolongation in HIV-associated progressive multifocal leukoencephalopathy treated with alpha-interferon: An observational study. J Neurovirol 4:324–332.

Iranzo A, Marti-Fabregas J, Domingo P, Catafau A, Molet J, Moreno A, Pujol J, Matias-Guiu X, Cadafalch J (1999): Absence of thallium-201 brain uptake in progressive multifocal leukoencephalopathy in AIDS patients. Acta Neurol Scand 100:102–105.

Katz RL, Alappattu C, Glass JP, Bruner JM (1989): Cerebrospinal fluid manifestations of the neurologic complications of human immunodeficiency virus infection. Acta Cytol 33:233–244.

Kerr DA, Chang CF, Gordon J, Bjornsti MA, Khalili K (1993): Inhibition of human neurotropic virus (JCV) DNA replication in glial cells by camptothecin. Virology 196:612–618.

Krupp LB, Lipton RB, Swerdlow ML, Leeds NE, Llena J (1985): Progressive multifocal leukoencephalopathy: Clinical and radiographic features. Ann Neurol 17:344–349.

Kuchelmeister K, Gullotta F, Bergmann M, Angeli G, Masini T (1993): Progressive multifocal leukoencephalopathy (PML) in the acquired immunodeficiency syndrome (AIDS). A neuropathological autopsy study of 21 cases. Pathol Res Pract 189:163–173.

Kure K, Llena JF, Lyman WD, Soeiro R, Weidenheim KM, Hirano A, Dickson DW (1991): Human immunodeficiency virus-1 infection of the nervous system: An autopsy study of 268 adult, pediatric, and fetal brains. Hum Pathol 22:700–710.

Lang W, Miklossy J, Deruaz JP, Pizzolato GP, Probst A, Schaffner T, Gessaga E, Kleihues P (1989): Neuropathology of the acquired immune deficiency syndrome (AIDS): A report of 135 consecutive autopsy cases from Switzerland. Acta Neuropathol 77:379–390.

Major EO, Amemiya K, Tornatore CS, Houff SA, Berger JR (1992): Pathogenesis and molecular biology of progressive multifocal leukoencephalopathy, the JC virus–induced demyelinating disease of the human brain. Clin Microbiol Rev 5:49–73.

Marra CM, Rajicic N, Barker DE, Cohen B, Clifford D, and the ACTG 363 Team (2001): Prospective pilot study of cidofovir for HIV-associated progressive multifocal leukoencephalopathy (PML). Proceedings of the 8th Conference on Retroviruses and Opportunistic Infections. Chicago, IL, February 4–8.

Marshall DW, Brey RL, Cahill WT, Houk RW, Zajac RA, Boswell RN (1988): Spectrum of cerebrospinal fluid findings in various stages of human immunodeficiency virus infection. Arch Neurol 45:954–958.

Martin-Suarez I, Aguayo Canela M, Merino Munoz D, Pujol de la Llave E (1994): Prolonged survival in patients with progressive multifocal leukoencephalopathy associated with AIDS treated with high doses of zidovudine. Rev Clin Esp 194:136–138.

Maschke M, Kastrup O, Esser S, Ross B, Hengge U, Hufnagel A (2000): Incidence and prevalence of neurological disorders associated with HIV since the introduction of highly active antiretroviral therapy (HAART). J Neurol Neurosurg Psychiatry 69: 376–380.

Mázló M, Herndon R (1977): Progressive multifocal leukoencephalopathy: Ultrastructural findings in two brain biopsies. Neuropathol Appl Neurobiol 3:323–339.

Mázló M, Tariska I (1982): Are astrocytes infected in progressive multifocal leukoencephalopathy (PML)? Acta Neuropathol 56:45–51.

McGuire D, Barhite S, Hollander H, Miles M (1995): JC virus DNA in cerebrospinal fluid of human immunodeficiency virus–infected patients: Predictive value for progressive multifocal leukoencephalopathy [published erratum appears in Ann Neurol 37:687, 1995]. Ann Neurol 37:395–399.

Messite J, Stillman S (1996): Accuracy of death certificate completion. JAMA 275: 794–796.

Miller JR, Barrett RE, Britton CB, Tapper ML, Bahr GS, Bruno PJ, Marquardt MD, Hays AP, McMurtry JGD, Weissman JB, Bruno MS (1982): Progressive multifocal leukoencephalopathy in a male homosexual with T-cell immune deficiency. N Engl J Med 307:1436–1438.

Miralles P, Berenguer J, García de Viedma D, Padilla B, Cosin J, López-Bernaldo de Quirós JC, Muñoz L, Moreno S, Bouza E (1998): Treatment of AIDS-associated progressive multifocal leukoencephalopathy with highly active antiretroviral therapy. AIDS 12:2467–2472.

Monaco MCG, Jensen PN, Hou J, Durham LC, Major EO (1998): Detection of JC virus DNA in human tonsil tissue: Evidence for site of initial viral infection. J Virol 72:9918–9923.

Morriss MC, Rutstein RM, Rudy B, Desrochers C, Hunter JV, Zimmerman RA (1997): Progressive multifocal leukoencephalopathy in an HIV-infected child. Neuroradiology 39:142–144.

Nelson PK, Masters LT, Zagzag D, Kelly PJ (1999): Angiographic abnormalities in progressive multifocal leukoencephalopathy: An explanation based on neuropathologic findings. AJNR Am J Neuroradiol 20:487–494.

Omerud L, Rhodes R, Gross S, Crane L, Houchkin K (1996): Ophthalmologic manifestations of acquired immune deficiency syndrome–associated progressive multifocal leukoencephalopathy. Ophthalmology 103:899–906.

O'Riordan T, Daly P, Hutchinson M, Shattuck A, Gardner S (1990): Progressive multifocal leukoencephalopathy—remission with cytarabine. J Infect Dis 20:51–54.

Padgett BL, Walker DL, Zu Rhein GM, Eckroade RJ, Dessel BH (1971): Cultivation of papova-like virus from human brain with progressive multifocal leucoencephalopathy. Lancet 1:1257–1260.

Peters AC, Versteeg J, Bots GT, Boogerd W, Vielvoye GJ (1980): Progressive multifocal leukoencephalopathy: Immunofluorescent demonstration of simian virus 40 antigen in CSF cells and response to cytarabine therapy. Arch Neurol 37:497–501.

Port JD, Miseljic S, Lee RR, Ali SZ, Nicol TL, Royal W 3rd, Chin BB (1999): Progressive multifocal leukoencephalopathy demonstrating contrast enhancement on MRI and uptake of thallium-201: A case report. Neuroradiology 41:895–898.

Portegies P, Algra PR, Hollak CE, Prins JM, Reiss P, Valk J, Lange JM (1991): Response to cytarabine in progressive multifocal leucoencephalopathy in AIDS. Lancet 337:680–681.

Portilla J, Boix V, Roman F, Reus S, Merino E (2000): Progressive multifocal leukoencephalopathy treated with cidofovir in HIV-infected patients receiving highly active anti-retroviral therapy. J Infect 41:182–184.

Post MJ, Yiannoutsos C, Simpson D, Booss J, Clifford DB, Cohen B, McArthur JC, Hall CD (1999): Progressive multifocal leukoencephalopathy in AIDS: Are there any MR findings useful to patient management and predictive of patient survival? AIDS Clinical Trials Group, 243 Team. AJNR Am J Neuroradiol 20:1896–1906.

Power C, Kong PA, Crawford TO, Wesselingh S, Glass JD, McArthur JC, Trapp BD (1993). Cerebral white matter changes in acquired immunodeficiency syndrome dementia: Alterations of the blood–brain barrier. Ann Neurol 34:339–350.

Power C, Nath A, Aoki FY, DelBigio M (1997): Remission of progressive multifocal leukoencephalopathy following splenectomy and antiretroviral therapy in a patient with HIV infection. N Engl J Med 336:661–662.

Richardson EJ (1970): Progressive multifocal leukoencephalopathy. In Vinken P, Bruyn G, Eds.; Handbook of Clinical Neurology; Elsevier: New York, Vol. 9, pp 485–499.

Samorei IW, Schmid M, Pawlita M, Vinters HV, Diebold K, Mundt C, von Einsiedel RW (2000): High sensitivity detection of JC-virus DNA in postmortem brain tissue by in situ PCR. J Neurovirol 6:61–74.

Sasseville VG, Newman WA, Lackner AA, Smith MO, Lausen NC, Beall D, Ringler DJ (1992): Elevated vascular cell adhesion molecule-1 in AIDS encephalitis induced by simian immunodeficiency virus. Am J Pathol 141:1021–1030.

Satishchandra P, Nalini A, Gourie-Devi M, Khanna N, Santosh V, Ravi V, Desai A, Chandramuki A, Jayakumar PN, Shankar SK (2000): Profile of neurologic disorders associated with HIV/AIDS from Bangalore, South India (1989–96). Indian J Med Res 111:14–23.

Scaravilli F, Daniel SE, Harcourt-Webster N, Guiloff RJ (1989): Chronic basal meningitis and vasculitis in acquired immunodeficiency syndrome. A possible role for human immunodeficiency virus. Arch Pathol Lab Med 113:192–195.

Singer C, Berger JR, Bowen BC, Bruce JH, Weiner WJ (1993): Akinetic-rigid syndrome in a 13-year-old girl with HIV-related progressive multifocal leukoencephalopathy. Mov Disord 8:113–116.

Singer EJ, Stoner GL, Singer P, Tomiyasu U, Licht E, Fahy-Chandon B, Tourtellotte WW (1994): AIDS presenting as progressive multifocal leukoencephalopathy with clinical response to zidovudine. Acta Neurol Scand 90:443–447.

Sparano JA, Anand K, Desai J, Mitnick RJ, Kalkut GE, Hanau LH (1999): Effect of highly active antiretroviral therapy on the incidence of HIV-associated malignancies at an urban medical center. J Acquir Immune Defic Syndr 21(Suppl 1):S18–22.

Sweeney BJ, Manji H, Miller RF, Harrison MJ, Gray F, Scaravilli F (1994): Cortical and subcortical JC virus infection: Two unusual cases of AIDS associated progressive multifocal leukoencephalopathy. J Neurol Neurosurg Psychiatry 57:994–997.

Tada H, Rappaport J, Lashgari M, Amini S, Wong-Staal F, Khalili K (1990): Transactivation of the JC virus late promoter by the tat protein of type 1 human immunodeficiency virus in glial cells. Proc Natl Acad Sci USA 87:3479–3483.

Taoufik Y, Gasnault J, Karaterki A, Pierre Ferey M, Marchadier E, Goujard C, Lannuzel A, Delfraissy JF, Dussaix E (1998): Prognostic value of JC virus load in cerebrospinal fluid of patients with progressive multifocal leukoencephalopathy. J Infect Dis 178:1816–1820.

Tassie JM, Gasnault J, Bentata M, Deloumeaux J, Boue F, Billaud E, Costagliola D (1999): Survival improvement of AIDS-related progressive multifocal leukoencephalopathy in the era of protease inhibitors. Clinical Epidemiology Group. French Hospital Database on HIV. AIDS 13:1881–1887.

Verbraak FD, Boom R, Wertheim-van Dillen PM, van den Horn GJ, Kijlstra A, de Smet MD (1999): Influence of highly active antiretroviral therapy on the development of

CMV disease in HIV positive patients at high risk for CMV disease. Br J Ophthalmol 83:1186–1189.

von Einsiedel RW, Fife TD, Aksamit AJ, Cornford ME, Secor DL, Tomiyasu U, Itabashi HH, Vinters HV (1993): Progressive multifocal leukoencephalopathy in AIDS: A clinicopathologic study and review of the literature. J Neurol 240:391–406.

Weber T, Turner RW, Frye S, Lüke W, Kretzschmar HA, Lüer W, Hunsmann G (1994): Progressive multifocal leukoencephalopathy diagnosed by amplification of JC virus–specific DNA from cerebrospinal fluid. AIDS 8:49–57.

Whiteman ML, Post MJ, Berger JR, Tate LG, Bell MD, Limonte LP (1993): Progressive multifocal leukoencephalopathy in 47 HIV-seropositive patients: Neuroimaging with clinical and pathologic correlation. Radiology 187:233–240.

Wong MC, Suite ND, Labar DR (1990): Seizures in human immunodeficiency virus infection. Arch Neurol 47:640–642.

Yiannoutsos CT, Major EO, Curfman B, Jensen PN, Gravell M, Hou J, Clifford DB, Hall CD (1999): Relation of JC virus DNA in the cerebrospinal fluid to survival in acquired immunodeficiency syndrome patients with biopsy-proven progressive multifocal leukoencephalopathy. Ann Neurol 45:816–821.

Zu Rhein GM, Chou SM (1965): Particles resembling papova viruses in human cerebral demyelinating disease. Science 148:1477–1479.

Zu Rhein GM (1969): Association of papovavirions with a human demyelinating disease (progressive multifocal leukoencephalopathy). Prog Med Virol 11:185–247.

12

THE NEUROPATHOLOGY AND PATHOGENESIS OF PROGRESSIVE MULTIFOCAL LEUKOENCEPHALOPATHY

MÁRIA MÁZLÓ, M.D., PH.D., HOLLY G. RESSETAR, PH.D.,
and GERALD L. STONER, PH.D.

1. INTRODUCTION

Progressive multifocal leukoencephalopathy (PML) is a primary demyelinating disease in which oligodendrocyte infection results in the death of the myelin-forming cell. As such, it has been of interest to neuropathologists since its recognition as a distinct pathologic entity in 1958 (see Chapter 1). In addition, the pathogenesis of viral demyelinating diseases may shed light on the process of demyelination in related diseases that are still of unknown etiology, such as multiple sclerosis. The other human primary viral demyelinating disease, subacute sclerosing panencephalitis (SSPE), is due to persistence of defective measles virus in the central nervous system (CNS), particularly in oligodendroglia (Allen et al., 1996). Where measles has been eliminated by vaccination, this rare disease is now disappearing completely. Unfortunately, PML is on the increase everywhere the AIDS epidemic is expanding.

While the outlines of the etiology and pathogenesis of PML now seem quite clear, closer examination shows numerous gaps in our knowledge. In this chap-

Human Polyomaviruses: Molecular and Clinical Perspectives, Edited by Kamel Khalili and Gerald L. Stoner.
ISBN 0-471-39009-7

ter we give some historical perspective on the neuropathology and pathogenesis of PML, summarize the understanding of this devastating disease gained in the 40 years since its discovery, and point out additional areas of research that are opening up.

2. LIGHT MICROSCOPY OF PML

Light Microscope Histopathology

The unique characteristic alterations of PML are the hundreds of small (0.1–3 mm diameter), gray lesions anywhere in the white matter of the brain, especially at the cortico–subcortical junction, as well as in the cortical and deep gray matter (Fig. 12.1). These lesions show no tendency towards sparing of the U fibers. The *early*, or *primary*, lesion is characterized by incomplete demyelination (Fig. 12.2), preservation of axis cylinders, a decrease in the number of normal oligodendrocytes, proliferation of the astrocytes and the appearance of activated microglia called *rod cells*. The most pronounced cytologic alteration of the early lesion is the presence of the round or oval, enlarged, amphophylic or deep basophilic nuclei of oligodendrocytes at the margins of the lesion. These "plump" oligodendroglial nuclei are enlarged two to three times and have lost their normal chromatin pattern. The less dense nuclei with margination of the chromatin are easily visible on hematoxylin and eosin (H&E)–stained sections. This alteration corresponds to the viral inclusions seen by electron microscopy (Silverman and Rubinstein, 1965; Zu Rhein and Chou, 1965). Sometimes, irregular, naked nuclei with partial nuclear membrane disintegration can also be found. In some places the middle of the early lesions becomes necrotic, giving a "moth-eaten" appearance of the demyelinated area (Fig. 12.4A). Rarely, inclusions are present in the nuclei of enlarged astrocytes. These usually do not occupy the whole nucleus. In the earliest, very tiny microscopic lesions, where only the fading of the myelin staining is visible, a few peculiar oligodendroglia nuclear inclusions may be located in the middle of the lesion. Proliferation of the astrocytes is already present, but the rod cells have not yet appeared (Fig. 12.2C). Because the hallmark of the early lesion is the presence of the peculiar enlarged oligodendroglia with nuclear inclusions, their appearance is the first sign that can be observed by light microscopy in routine histologic sections stained by luxol fast-blue/hematoxylin–eosin (LFB/H&E) (Fig. 12.5A). For more information about very early alterations found by routine light microscopy and immunocytochemical staining, see the subclinical cases described by (Åström and Stoner, 1994).

It should be emphasized that PML is a disease of both of the white matter and the gray matter. In the early lesions located in the gray matter, the conspicuous oligodendrogial cells are fewer in number, and these oligodendrocytes are mainly in a satellite position (Mázló and Herndon, 1977; Zu Rhein and Chou, 1968). Occasionally, the gray matter is heavily involved (Ledoux et al.,

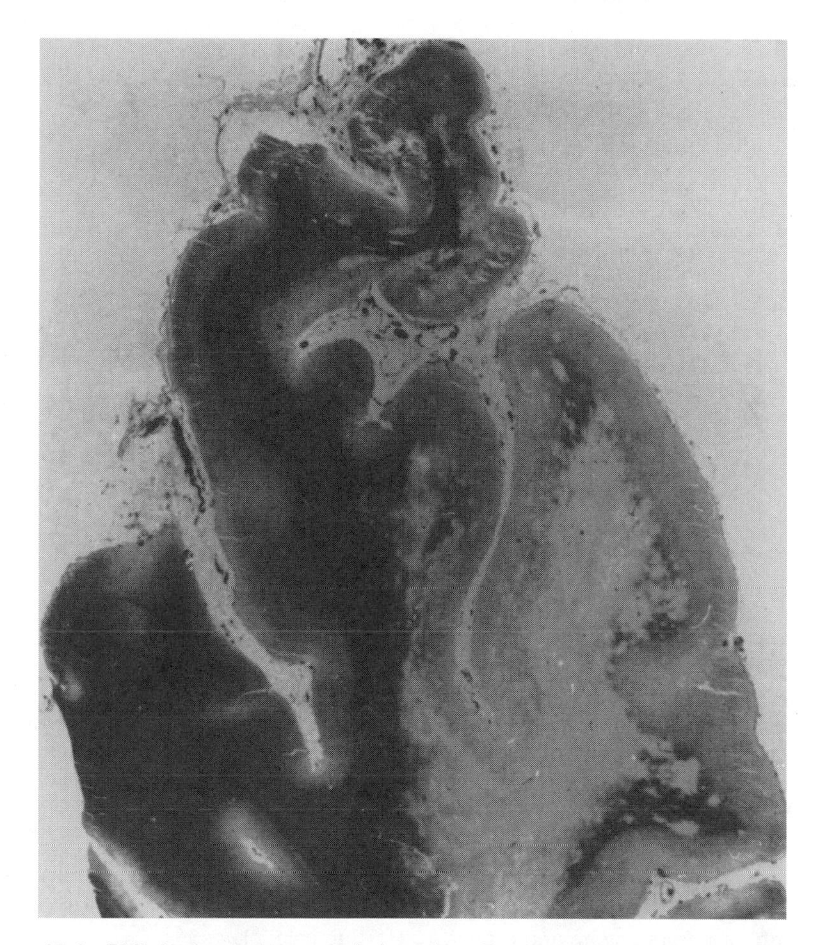

Figure 12.1. PML brain, parieto-occipital section. By the naked eye the staining of the myelin shows many tiny, pinhead-sized as well as large coalescent white matter areas with partial or total loss of myelin. In addition, cortical and subcortical lesions are evident. Woelke's stain.

1989; Sweeney et al., 1994). While glial cells are the primary target, cortical neurons may not always be spared. Recently an unusual case of PML in a 21-year-old man with common variable immunodeficiency syndrome has been described. At autopsy, extensive demyelination and necrosis throughout the white matter were found with apparent alteration of cortical neurons into large polygonal cells resembling "dysplastic or dysmorphic ganglion cells" such as those seen in gangliocytomas or gangliogliomas. These abnormal cells were tentatively termed *dysplastic ganglion-like cells* (DGLC) (Shintaku et al., 2000). JC virus (JCV) infection was shown in DGLC scattered sparsely in the atrophied cortex or forming band-like aggregates with distinct cytoarchitectural derangement of the involved cortex reminiscent of focal cortical dysplasia. In

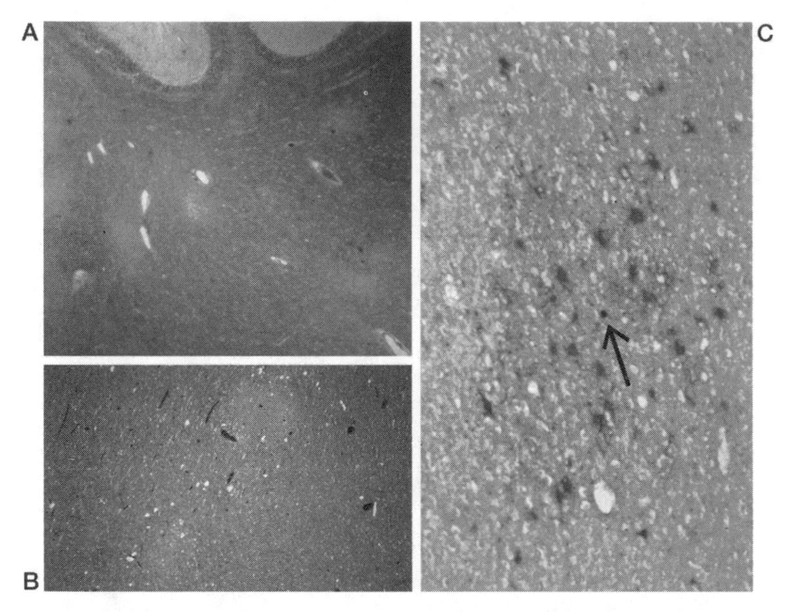

Figure 12.2. Early demyelinated lesions in cerebellar white matter. Luxol fast blue/H&E stain. (**A**) Pallor of the myelin staining and demyelinated foci in the cerebellum. Subacute process. Low magnification. (**B**) Early subcortical demyelinated lesions with reactive astrocytes and few enlarged oligodendrocytes (visible at higher magnification) in a PML case discovered at autopsy. Low magnification. (Åström and Stoner, 1994). (**C**) Immunostained early lesion from same case as in B. Reactive astrocytes throughout the lesion labeled for glial fibrillary acidic protein (GFAP) appear red (monoclonal antibody detected by alkaline phosphatase developed with Vector Red 1), while a single oligodendrocyte in the center (arrow) that is labeled for JCV capsid proteins appears black (rabbit antiserum detected by peroxidase developed with 3,3′-diaminobenzidine and enhanced with $NiCl_2$). Medium magnification. For methodological details, see Åström and Stoner (1994). Taken with permission from Åström and Stoner, 1994. (See color plates.)

DGLC-positive cells immunostaining for T antigen and in situ hybridization for T-antigen mRNA were found, but VP1 as seen in inclusion-bearing oligodendrocytes was not detected.

In a few cases with cerebellar involvement, scattered foci of partial or complete destruction of the granule cell layer have been found (Fig. 12.3). These lesions can be continuous with the underlying demyelinated lesions of the cerebellar white matter. Within the early granular layer lesions, numerous round or slightly oval nuclei with abundant chromatin granules but without stainable cytoplasm are present. These nuclei occur in groups or in clusters resembling the cytoarchitecture of the granule cell layer of the cerebellar cortex. Their appearance and distribution suggest that they are altered granule neurons (Richardson, 1961). The granule neurons are extremely sensitive to selective parenchymal degeneration with no appreciable changes in the Purkinje cells and molecular layer or the underlying white matter and with lack of a glial reaction.

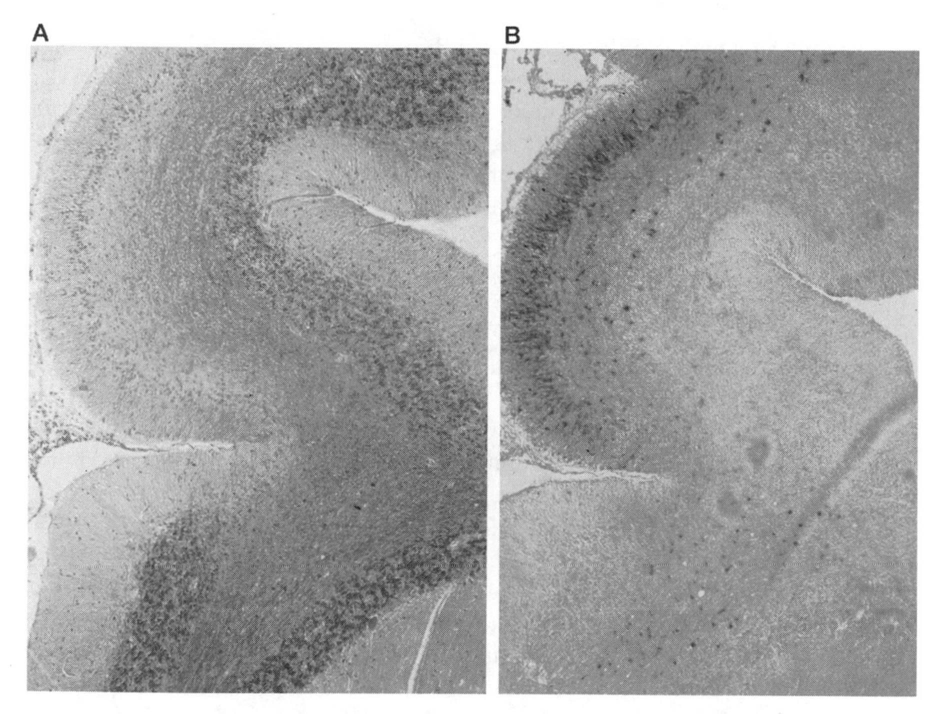

Figure 12.3. Cerebellar folia: Granule cell loss with underlying demyelination. (**A**) Pallor of the folia white matter as well as thinning of granule neurons and in some places complete loss of the granule neurons. LFB/H&E stain. (**B**) Double immunostaining of the parallel section for GFAP and JCV capsid proteins. Here astrocytes in the granular layer lesion are labeled for GFAP and appear black (monoclonal antibody detected by peroxidase developed with 3,3′-diaminobenzidine and enhanced with $NiCl_2$), while JCV-infected oligodendrocytes at the edge of the underlying demyelinated lesion in the white matter show red reaction product in the nucleus (rabbit antiserum detected by alkaline phosphatase developed with Vector Red 1). For details of the method, see Stoner et al. (1998). (See color plates.)

These lesions are widespread and rather symmetric in both hemispheres without any relation to the vascular distribution. Histologically, severe destruction of the granular layer leads to a diffuse, vacuolated, blurred, and washed-out appearance. These alterations may sometimes be the result of fixation artifact because the selective loss of the granule cells usually appears in the central parts of the cerebellum and not in the peripheral folia. There is also a steady increase in the extent of the lesions as the time interval between death and autopsy increases (Ikuta et al., 1963). This specific biologic behavior may represent the peculiar vulnerability these cells to autodigestion. Granule neurons are also sensitive to different toxic diseases and ionizing radiation (Vogel, 1959). Atrophy of the cerebellum is commonly associated with alcoholism. However, shrinkage of the folia occurs particularly in the anterior superior vermis. Quantitatively, there is a significant reduction in the number and size of Purkinje cells with proliferation of the Bergmann glia, which is most marked in the smaller rostral and caudal lobes of the vermis. In addition, there is a reduction in the volume of the molecular and medullary layers in the vermis. The shrinkage of the granular layer is the less severe, 10% compared with 10–40% shrinkage of the molecular layer, which—together with the Purkinje cell loss—appears to be the most vulnerable region in chronic alcoholics (Phillips et al., 1987).

In the cerebellar granular layer alterations illustrated in this chapter, the specimens were taken from the lateral, superficial part of the hemisphere of the cerebellum. Paraffin sections of folia stained with LFB/H&E showed thinning of the granule cells in some places. However, at the deeper part of the layer the remaining granule neurons appear to be better preserved. The folia white matter may be partially demyelinated. Formalin-fixed, paraffin-embedded sections stained by a double-label immunocytochemical method for capsid proteins and GFAP revealed a gliotic region of the granular layer overlying demyelinated folia white matter with infected oligodendrocytes expressing capsid proteins at the margin (Fig. 12.3). Electron microscopy of inclusion-bearing oligodendrocytes and astrocytes showed JCV particles in the nuclei as well as in the cytoplasm of astrocytes, but did not detect JCV particles in the granule neurons. Instead, accumulation of amorphous, fine, dense material was found in the affected granule neurons and in some of the astrocytes (see below). Together, these findings suggest that granule neurons may be capable at most of abortive infection, while oligodendrocytes and astrocytes can express both early and late viral proteins, replicate viral DNA, and assemble virions (Stoner et al., 1988d). The concept of abortive granule neuron infection is supported by the hamster model (see below).

The characteristic advanced PML lesions, formed by coalescence of smaller lesions, seem not to have a constant relation to the blood vessels (Chandor et al., 1965; Richardson, 1961, 1970). Extensive mononuclear cell and/or plasma cell infiltration occurred only in an atypical minority of the pre-AIDS PML cases (Richardson, 1970; Richardson and Johnson, 1975). In another series inflammatory infiltrates were present in 9 of 36 pre-AIDS PML cases (Budka and Shah, 1983). These infiltrates are mainly in the vessel wall, but tissue infiltration can also sometimes occur. In cases where no infiltrating cells or microglial activation can be found despite the tissue degeneration, the myelin breakdown products may be taken up by astrocytes and, on rare occasions, by oligodendrocytes as well (Mázló and Herndon, 1977).

The *late* lesions are characterized by total loss of the myelin in the center of the lesion, the proliferation of plump astrocytes, and, in many cases, macrophages laden with neutral fat. The latter are visible on frozen sections stained by the Oil red O or Sudan methods. The axis cylinders get thinner, and in some places segmental swelling of the axis cylinder can also be observed by Bodian's silver impregnation methods. However, the edge of the late lesions is very similar to the early lesion because at the zone of normal–degenerated myelin the peculiar, large oligodendroglial nuclear inclusions can be found, as well as

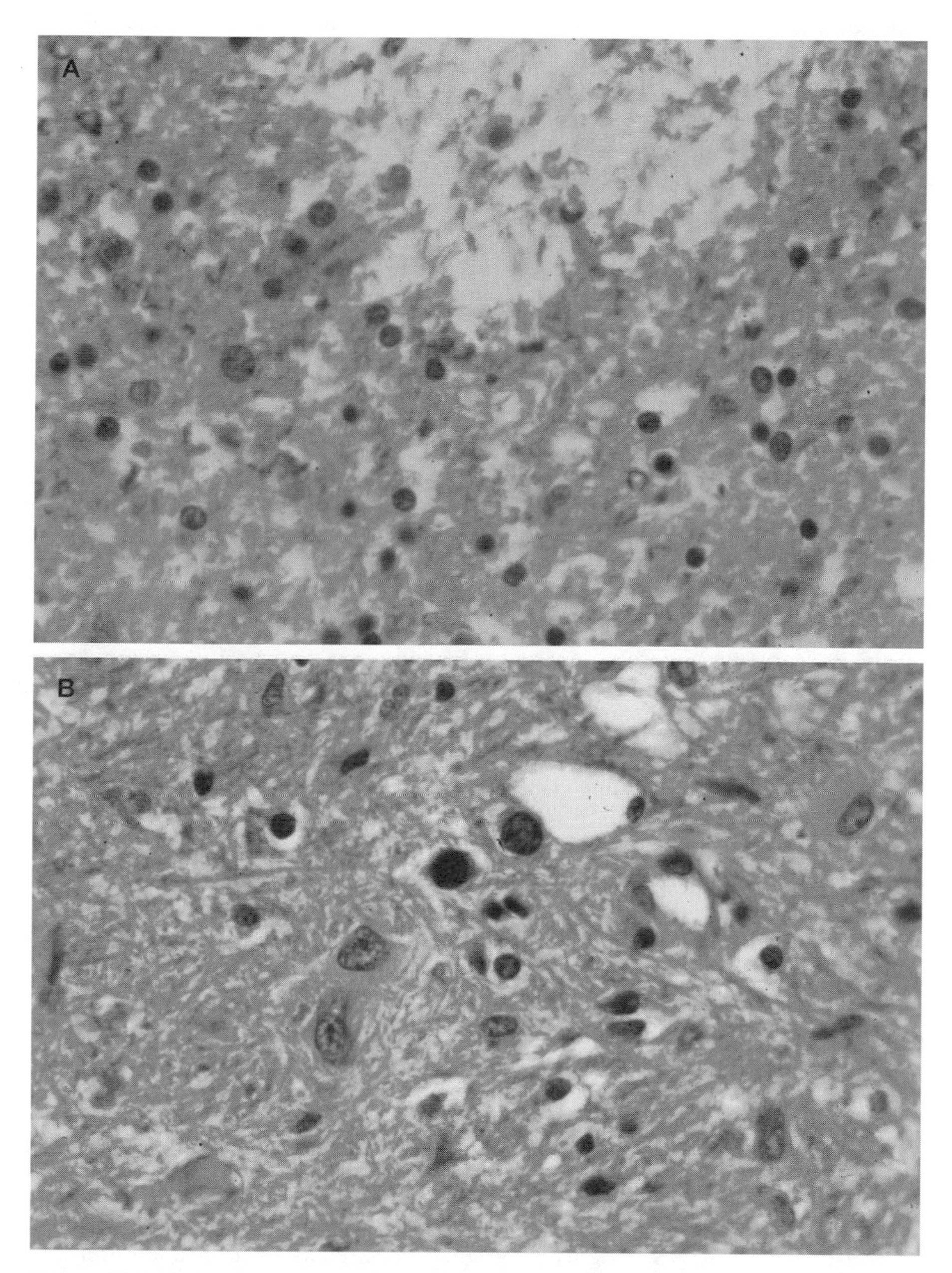

Figure 12.4. Subacute and chronic alterations in a chronic case of PML. LFB/H&E. (**A**) In the radiatio optica, "moth-eaten" myelin loss and oligodendroglial nuclear inclusions are visible around the lesion. High magnification. (**B**) In the cerebellar white matter at the edge of a chronic gliotic focus, enlarged astrocyte nuclei and an oligodendrolgia nuclear inclusion can be found. High magnification.

enlarged astrocytes (Fig. 12.4B). However, the characteristic alteration of the late lesions is the presence of the giant, bizarre astrocytes with enlarged, oval, lobulated, or multiple nuclei and coarse chromatin pattern (Fig. 12.5). These cells are reminiscent of those in malignant glioblastomas, but are, in contrast, scattered and present only singly. Mitosis of large astrocytes is occasionally observed. The giant, bizarre astrocytes are not obligatory residents of the late PML lesions, but are present in about 80% of the cases (Richardson, 1970). Neurons in gray matter contiguous to areas of destruction in the late lesion show shrinkage and pyknosis. These changes may be secondary to the white matter lesions.

Between the early and late lesions there are many gradations. In addition, demyelinating foci tend to enlarge and coalescence, forming larger, confluent lesions (Fig. 12.1). Cyst formation with cavitation can occur. In some chronic cases lesions may appear to be of various ages, with the more recent ones showing the characteristics of the early lesion. On the other hand, chronic cases can lead to "burnt-out" lesions. In a large series of 36 cases, 28 cases had florid lesions and 8 cases had mainly burnt-out lesions with only a few oligodendroglia nuclear inclusions (Budka and Shah, 1983). In the old lesions glial scars resembling those in multiple sclerosis can be found in the sections stained by LFB/H&E or by Holczer's method for fibrous astrocytes. However, in these scars, with careful searching, giant, bizarre astrocytes and more peripherally, in the adjacent, normal looking white matter a few peculiar inclusion-bearing oligodendrocytes can usually be found. Fig. 12.4 and Fig. 12.5C were taken from a chronic PML case with a 10–year duration that was associated with systemic lupus erythematosus and had a multiple sclerosis-like onset (Stoner et al., 1988a).

In summary, at the cellular level PML is a disease of both *oligodendrocytes* and *astrocytes*. The criteria of the histologic diagnosis are (1) the presence of hundreds of foci of demyelination from 0.1 mm to centimeters in diameter scattered throughout the white matter and sometimes the cortical and deep gray matter with relative sparing of the axis cylinders; (2) the absence of the oligodendroglia in the central parts of the lesions and the presence of the large, peculiar oligodendroglia nuclear inclusions at the margins of the early lesions (these two are the explicit criteria of the histological diagnosis); and (3) The presence of the bizarre, giant astrocytes resembling those of glioblastoma cells but scattered singly in the late or chronic lesions in about 80% of the cases. Very late lesions will be accompanied by glial scarring. In addition, in a few cases the granule cells of the cerebellar cortex can be affected. However, in the cerebellum parenchymal degeneration and different toxic etiologies must be considered, as AIDS patients may sometimes resort to drug addiction, including alcohol.

Immunocytochemical Methods and Results

In the 1970s, in addition to the light and electron microscopic methods, immunocytochemical methods were developed and used in the study of PML

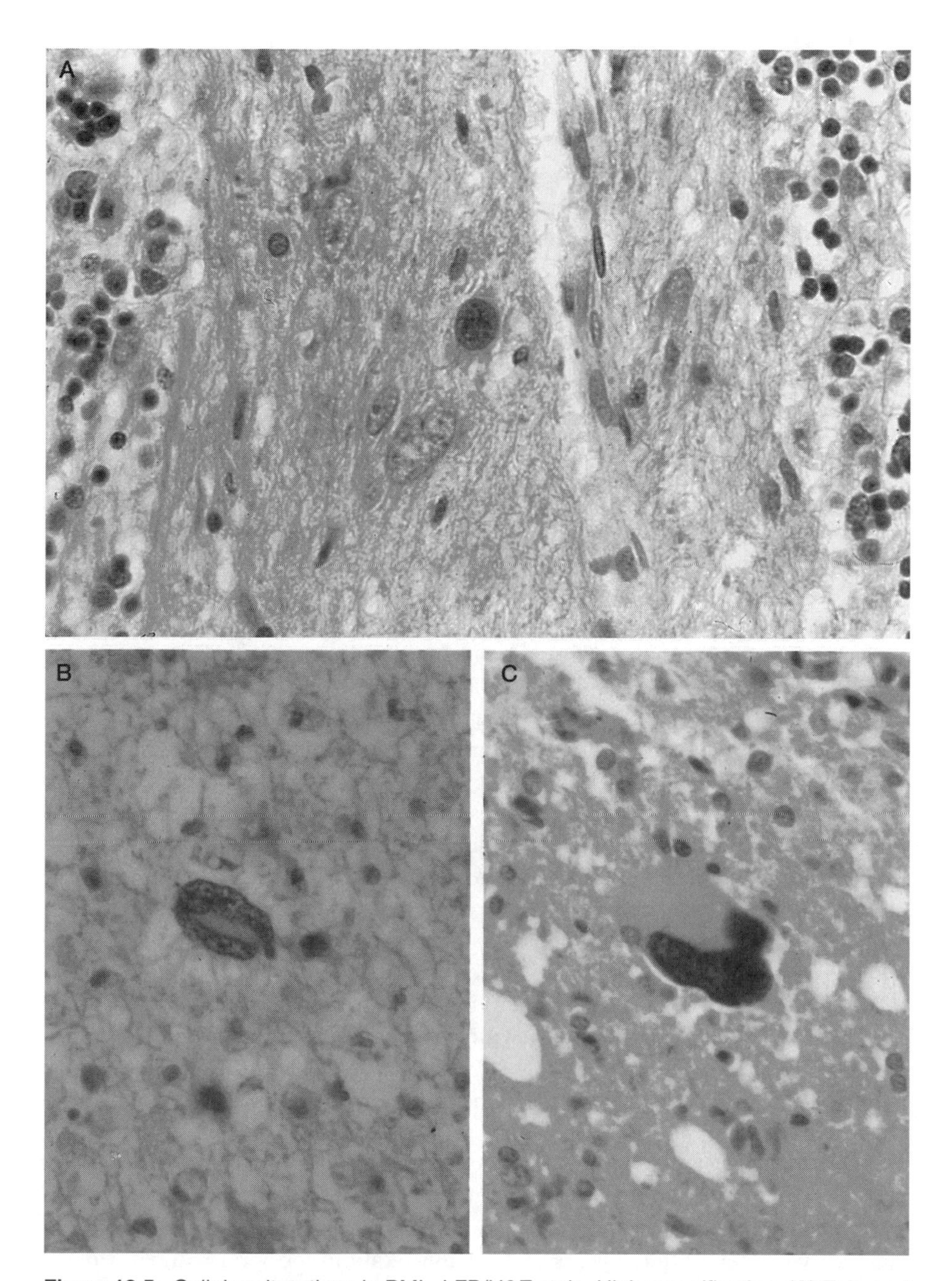

Figure 12.5. Cellular alterations in PML. LFB/H&E stain. High magnification. (**A**) Enlarged oligodendroglial nuclear inclusion and enlarged nucleus of an astrocyte are visible in the demyelinated folia white matter. (**B**) Giant astrocyte with curved nucleus in a necrotic area. (**C**) Highly enlarged giant astrocyte with curved, densely stained basophilic nucleus in completely demyelinated subcortical white matter.

brain infection. For the first time identification of polyomavirus in frozen sections of PML brain biopsy and autopsy material with the fluorescent antibody method using monospecific rabbit antisera to JCV, BK virus (BKV), and simian virus 40 (SV40) was adopted (Padgett and Walker, 1976). However, the usefulness of the immunofluorescent methods is limited because the reaction product is not permanent, and it was not applicable to retrospective studies on the routine, formalin-fixed, and paraffin-embedded material. At the beginning of the 1980s immunoperoxidase methods were developed using the genus-specific antibody of polyomaviruses termed the *common antigen*, which is located on the major capsid protein (VP1). The first genus-specific rabbit antiserum raised to SV40 capsids disrupted by sodium dodecyl sulfate (SDS) was prepared by (Shah et al., 1977) and used to detect polyomavirus capsid antigens by the peroxidase-antiperoxidase (PAP) method in routine histologic sections of formalin-fixed, paraffin-embedded PML brain. This antiserum was capable of reacting with cross-reacting internal capsid antigens of JCV in dilutions of 1:50 to 1:200 (Budka and Shah, 1983; Gerber et al., 1980), which was very useful to investigate large brain sections. The swollen nuclei of oligodendrocytes in a demyelinated area were strongly reactive, showing dark brown reaction product in antigen-containing cells (Gerber et al., 1980).

JCV antiserum was also prepared and used to detect JCV-infected cells at a 1:10 to 1:20 dilution (Itoyama et al., 1982). Most of the reacting cells were oligodendrocytes located in the white matter. In addition, oligodendrocyte nuclear inclusions immunostained in epon-embedded semithin sections contained polyoma virions when an adjacent thin section was examined by electron microscopy. In histologically normal areas, far from demyelinating lesions, staining was limited to nuclei of single, randomly scattered oligodendrocytes being mostly fascicular and rarely in a satellite position to neurons. In transition zones between histologically normal white matter and margins of demyelinated areas, the density of stained oligodendroglia increased, the immunostained nuclei were larger, and staining often extended to the cytoplasm. In central regions of demyelinated zones, relatively few immunostained cells were found; some of them were enlarged astrocytes with densely stained nuclei and punctate–stained cytoplasm (Itoyama et al., 1982). In a larger series of PML brains, the genus-specific, cross-reactive serum indentified viral antigen in oligodendrocytes with swollen nuclei in 35 of 36 cases of definite PML (Budka and Shah, 1983). While the length of the period of storage in the paraffin blocks did not affect the immunoreactivity of the tissue, the time of fixation in formalin was important. When fixation periods were longer than a few weeks, less intense staining was seen. After 3 years of formalin fixation, the viral antigen was not stained (Budka and Shah, 1983). In five brains antigen deposits were prominent in cytoplasm and processes of astrocytes, as was found earlier for virus particles by electron microscopy (Mázló and Tariska, 1982). Giant astrocytes contained viral antigen in six cases. There was no staining of nerve cells or endothelial cells (Budka and Shah, 1983).

Detection of the capsid antigen by immunocytochemical methods was very useful to visualize the distribution of viral antigens in and around early and late PML foci (Budka and Shah, 1983; Itoyama et al., 1982), though cell differentiation and the cellular distribution of viral antigen were better shown by electron microscopy (Mázló and Tariska, 1980; Mázló and Herndon, 1977; Watanabe and Preskorn, 1976). In addition, on the basis of electron microscopic studies the possibility of abortive infection of the astrocytes had been raised (Mázló and Tariska, 1982), but could not be proved. Polyomaviruses may establish a productive or an abortive infection or may transform the cell depending on the host cell type. In each case the DNA binding regulatory protein, large T (tumor) antigen, specified by the early region of the viral genome, is expressed in the cell nucleus (Rigby and Lane, 1983). The T antigen of JCV, BKV, and SV40 share immunologic cross reactivity so that antisera to the T antigen of each virus can be used to label nuclei of cells infected or transformed by these viruses. The monoclonal antibody PAb2000 specifically recognizes the JCV large and small T antigens (Bollag and Frisque, 1992).

The distribution of T antigen in PML brain was not known until 1986, when two laboratories demonstrated the presence of T antigen in the PML brain. Greenlee and Keeney (1986) used rabbit antisera to purified SV40 large T antigen on formalin-fixed, paraffin-embedded sections. Optimal duration for the reaction with primary antibody was found to be 72 hours, in contrast to the usual 30 to 60 minutes or overnight incubation period. Intense staining of oligodendrocyte nuclei at the edges of the demyelinated areas was observed and, less frequently, of oligodendrocyte nuclei within apparently normal brain areas. Nuclear staining of T antigen was also seen in occasional morphologically normal astrocytes surrounding demyelinated areas. Rare atypical astrocytes exhibited intense nuclear staining for T antigen, but the majority of these cells were negative. Nuclear staining specific for T antigen was not detected in neurons, vascular endothelial cells, ependyma, or meninges. Examination of adjacent sections stained with antisera to common structural antigen revealed an identical pattern of immunoenzymatic labeling, indicating that most of the cells expressing T antigen were also expressing structural proteins.

Stoner et al. (1986) obtained a polyclonal antiserum to JCV T antigen from hamsters bearing JCV-induced tumors. Methods for detection of T antigen in JCV-transformed hamster glial cells, as well as SV40-transformed human lung cells, using the highly sensitive PAP technique on acetone-fixed monolayers were developed. The same technique readily detected nuclear T antigen in acetone-fixed frozen sections of five PML brain tissues, including two cases with AIDS. Rabbit hyperimmune antiserum was also used on the parallel sections or cells. Acetone-fixed frozen sections from the tissue blocks of all five cases tested showed staining for both virion antigen and for T antigen. The cells containing virion antigens appeared to be predominantly the large, abnormal oligodendroglia, which showed characteristic nuclear inclusions in sections stained with LFB/H&E. Reaction product could also be observed on the plasma membrane of macrophages invading some of the demyelinated lesions, but the nuclei of macrophages were unstained. Except for this latter cell type, the T antigen staining appeared in the same cells as did staining for virion antigens. In addition, many of the small nuclei of oligodendroglia of normal appearance were heavily stained as were a few cells with large, irregular nuclei. These cells may represent abortively infected astrocytes.

In the acetone-fixed frozen PML sections, the number of cells expressing T antigen in the nucleus clearly exceeded the number of cells expressing virion antigens (Stoner et al., 1986). The findings suggested the presence of T antigen not only in productively infected cells but also in a much wider distribution including many morphologically normal-appearing oligodendrocytes and giant astrocytes. In addition, in the two cases of PML in AIDS patients studied, the distributions of both virion antigens and T antigen throughout the sections were much greater than in the three cases of non-AIDS PML (Fig. 2. in Stoner et al., 1986). The presence of widespread oligodendroglial infection not confined to demyelinated lesions suggested that PML in AIDS may represent a more diffuse infectious process than that usually seen in classic PML.

Monoclonal antibodies provide reagents of high specificity and reproducibility. Detection methods utilizing monoclonal antibodies also serve to confirm the identity of the antigens detected by polyclonal sera. Monoclonal antibody PAb 416 (Harlow et al., 1981) to large T antigen detects the T antigen better than any other antibodies among the 30 tested (Stoner et al., 1988b). Polyclonal mouse serum and polyclonal hamster JCV tumor serum were also used and compared with monoclonal antibodies. The detection system was either the PAP technique or a streptavidin-alkaline phosphatase method. PAb 416 as primary antibody stained all HJC-15 cells and the hamster tumor tissue strongly and stained JCV infected human brain tissue as intensely as did hamster anti-JCV tumor serum. This epitope was present in all 10 infected brain tissues tested (Stoner et al., 1988b).

A new double-label immunocytochemical method was described that detected JCV early (T antigen) and late (capsid) proteins simultaneously in cryostat sections of PML brain tissue (Stoner et al., 1988d) (Fig. 12.6). T antigen was detected by monoclonal antibody (PAb 416) followed by goat antimouse IgG and mouse Clono-PAP, while capsid proteins were detected by a rabbit polyclonal antiserum to capsid proteins followed by biotinylated goat antirabbit IgG and streptavidin-alkaline phosphatase conjugate. The substrates were 3,3′-diaminobenzidine enhanced with $NiCl_2$ and Vector Red I, respectively (Åström and Stoner, 1994). Application of the double-label technique to early (T antigen) and late (capsid) proteins is complicated by the fact that frozen sections are highly advantageous for T antigen detection, but viral capsid proteins are best detected in paraffin-embedded, formalin-fixed tissue. Use of the antigen retrieval technique in formalin-fixed, paraffin-embedded sections using proprietary commercial reagents can markedly enhance the immunostaining for cap-

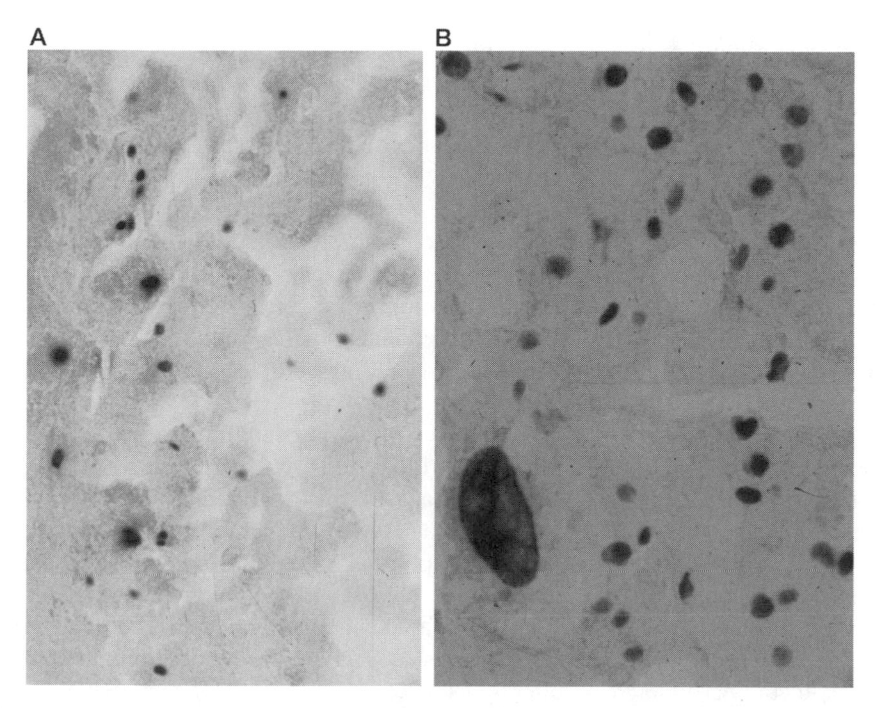

Figure 12.6. JCV protein expression revealed with double-label immunocytochemical staining for early (T antigen) and late (capsid) proteins. Frozen sections with no counterstain. For details of the method, see Stoner et al. (1988d). (**A**) Enlarged oligodendrocytes are double labeled in the nucleus for both T antigen (black) and capsid proteins (red) in a subcortical white matter lesion. Smaller oligodendrocytes and some astrocytes express only T antigen. Medium magnification. (**B**) A giant astrocyte expresses only T antigen in subcortical white matter. High magnification. (See color plates.)

sid proteins, and, with careful choice of conditions, T antigen staining is also possible.

In frozen sections this double-label immunocytochemical method showed that glial cells in and around PML lesions displayed reaction product for either the JCV early protein (T antigen) alone or for both early and late (capsid) proteins, but never for capsid proteins alone (Fig. 12.6). The double-labeled cells expressing both T antigen and capsid proteins were characteristically enlarged oligodendrocytes (Fig. 12.6A). Small oligodendrocytes and giant astrocytes expressed only T antigen (Fig. 12.6B). Compared with the large number of small oligodendrocytes expressing T antigen, the number of double-labeled cells also expressing capsid protein was relatively few, in keeping with earlier findings using single labels on adjacent sections. Glial cells labeled with T antigen alone may be interspersed with capsid–antigen-containing cells. They may be located at the edge of advanced focal lesions, or they may be scattered singly or in clusters in apparently histologically normal tissue. These normal-looking areas containing only cells in which T antigen predominates are good candidates for the early stage of an incipient lesion. As these cells are without the characteristically enlarged, inclusion-bearing nucleus, they would be missed in routine histology. Most of these small, round cells are oligodendrocytes, but the identity of others in the early stage of infection cannot always be determined from their morphology in cryostat section. Some of the cells, which have larger and more irregular nuclei than is characteristic for oligodendrocytes and express T antigen, but not capsid proteins, are very likely to be astrocytes. On the basis of electron microscopic experience, the few cells in which capsid proteins are present in the cytoplasm surrounding a nucleus labeled for T antigen are likely to be astrocytes.

It can be concluded that T antigen expression predominates early, while capsid protein expression predominates following DNA synthesis when infectious virions are being assembled. The fact that many oligodendrocytes can be detected that appear to be in the early stages of a productive infection suggests that the switch from early to late transcription is sometimes delayed or blocked, thus allowing cells in the intermediate stage of infection to accumulate. A temporary block on the part of some oligodendroglia would not be surprising in light of the fact that many infected astrocytes are thought to remain indefinitely in the early stage of infection (abortive infection). The evidence from immunocytochemical staining with the double-label method suggests that expression of T antigen independently of capsid protein synthesis can occur in PML brain. JCV T antigen expression, which is known to be essential for both viral DNA replication and late region expression, may not be sufficient in many oligodendrocytes to turn on virus replication. A second signal is needed whose nature is not yet clear. This fact could help to explain not only the subacute course of PML but also the intermittently progressive and chronic forms, which can run for several years (Hedley-Whyte et al., 1966; Price et al., 1983).

Adaptation of the double-label method to detection of GFAP and JCV capsid proteins (VP1–3) in formalin-fixed, paraffin embedded tissue allowed study of VP1–3 expression in relation to gliotic lesions and individual reactive astrocytes, as well as the subcellular distribution of VP1–3 in astrocytes (Fig. 12.7 and 12.8). The method can be adapted to label GFAP in astrocytes with a red reaction product (using alkaline phosphatase and Vector Red 1) (Åström and Stoner, 1994) or black reaction product (using peroxidase and 3,3′diaminobenzidine with $NiCl_2$ enhancement; Stoner et al., 1998; Fig. 12.3B). JCV capsid proteins are in each case labeled with the contrasting color.

Markers of cellular proliferation and activation are prominently expressed in JCV-infected glial cells. The expression of p53 was readily detected in frozen sections using the monoclonal antibody Pab122. Reports, in addition to p53, have described the presence in JCV-infected cells of proliferating cell nuclear antigen (PCNA) (Ariza et al., 1994; Lammie et al., 1994), as well as the overexpression of Ki-67, a nuclear antigen expressed in cycling cells, and the regulators of cell proliferation, cyclins A and B1 (Ariza et al., 1998).

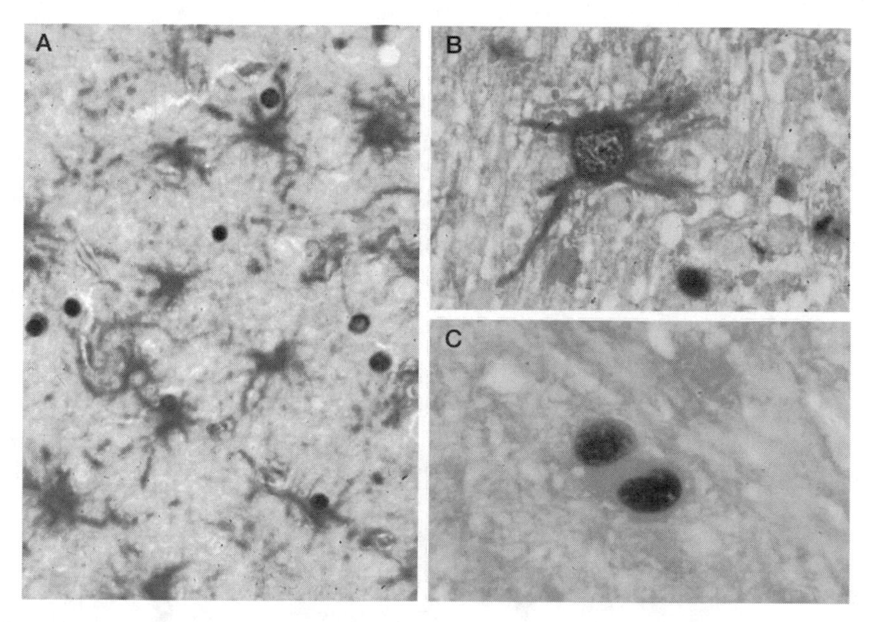

Figure 12.7. JCV protein expression revealed with double-label immunocytochemical staining for late (capsid) proteins and glial fibrillary acidic protein (GFAP). Reactive astrocytes stain red using a monoclonal antibody and alkaline phosphatase detection with Vector Red 1. JCV capsid proteins stain black using a rabbit antibody to VP1–3 with peroxidase detection using 3,3′-diaminobenzidene with $NiCl_2$ enhancement. No counterstain. (**A**) In subcortical white matter of an AIDS-PML patient, oligodendrocytes stain heavily in the nucleus for JCV capsid proteins. Some infected oligodendrocytes are closely embraced by astrocytes not expressing capsid proteins. Medium magnification. (**B**) Rare reactive astrocyte expressing GFAP in the cytoplasm and JCV capsid proteins in the nucleus. Subcortical white matter of AIDS-PML patient. Nuclear staining in the astrocyte is more granular and less dense than in the oligodendrocyte with nuclear staining nearby. High magnification. (**C**) JCV capsid protein expression in the nucleus of a dividing astrocyte in pons basis from the brain of a PML patient without AIDS. Note the absence of GFAP staining in the cytoplasm of this dividing astrocyte. High magnification. (See color plates.)

In Situ Hybridization and Polymerase Chain Reaction

Parallel to the immunocytochemical methods, in situ hybridization methods were developed to label complementary strands of known nucleic acid sequences in tissue sections or tissue culture cells (Woodroofe et al., 1994). Nonradioactive probes may be able to detect on the order of 100–1000 copies per cell and radioactive probes in the range of 10–100 copies per cell. Variability of fixation of tissues, the extent of tissue pretreatment to allow access of the probe, and specific activity differences between different lots of synthesized probe introduce a level of variability in the sensitivity of these techniques. In situ hybridization can be combined with immunocytochemistry on the same tissue section (Ironside et al., 1989), providing information on co-localization

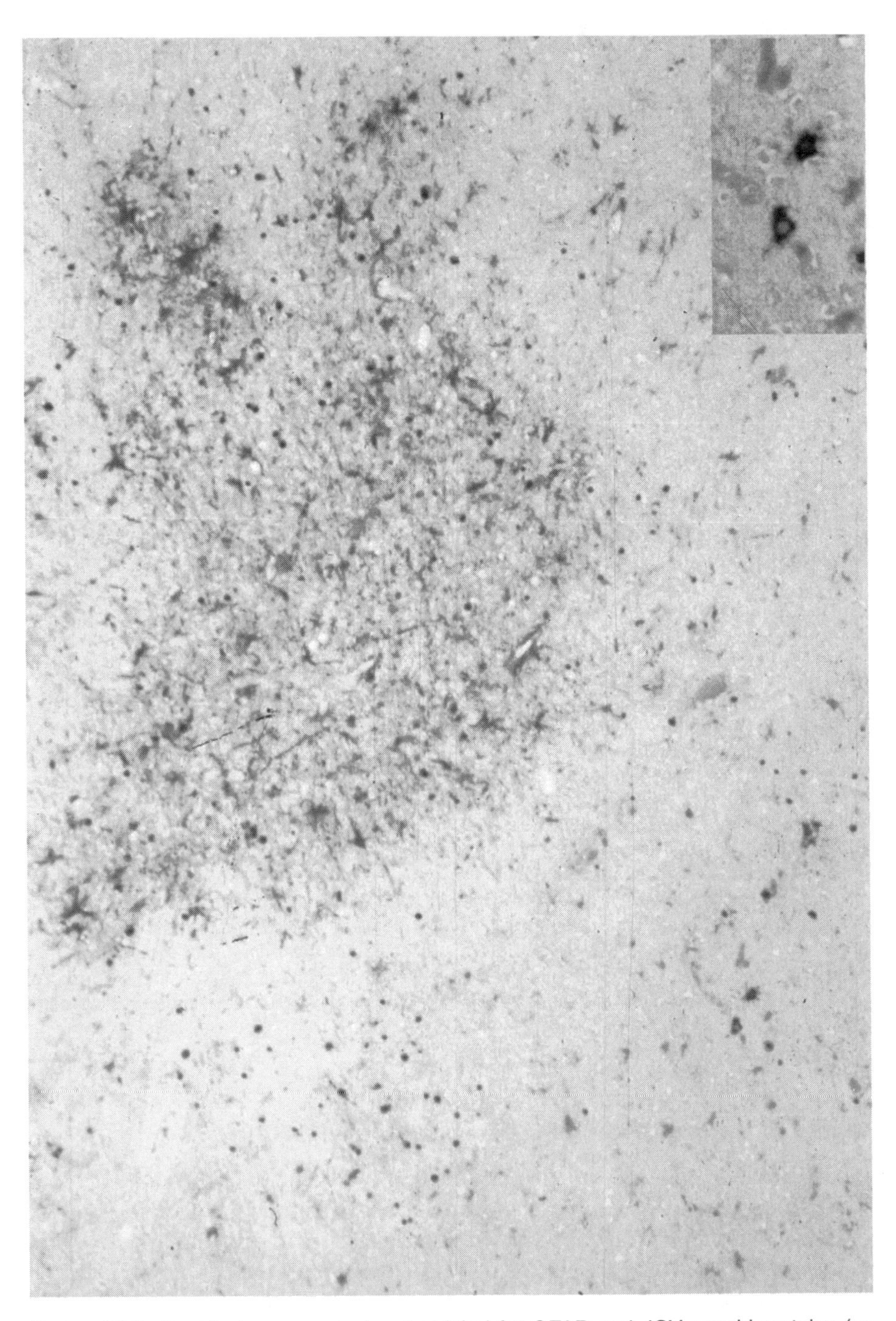

Figure 12.8. Double immunocytochemical label for GFAP and JCV capsid proteins (no counterstain). JCV-containing cells occur in the margins of the mature, gliotic demyelinated lesion and in the incipient lesion below the gliotic lesion with minimal astrocyte reactivity, but with numerous JCV-infected cells, including some with astrocyte morphology. Astrocytes with capsid located cytoplasmically are seen in the lower right corner and are enlarged in the inset. (See color plates.)

of nucleic acid and antigen, but with some sacrifice of sensitivity (Aksamit, 1995). In situ polymerase chain reaction (IS-PCR) methods hold the promise of providing additional answers, as this technique has the potential for identifying only a few copies of viral DNA per cell through in situ amplification and detection techniques. IS-PCR has been used to identify JCV DNA in the characteristic enlarged oligodendrocytes and bizarre astrocytes in PML lesions (Ueki et al., 1994). The number of JCV-positive cells detected by IS-PCR exceeded the in situ hybridization results by two- to threefold (Aksamit, 1995).

Detection of polyomavirus DNA by in situ hybridization with cRNA of JC virus and autoradiography on sections of PML brain was first carried out by Dörries et al. (1979). JCV was extracted from brain with PML and purified, and a highly specific cRNA was generated in vitro. In situ hybridization with the JCV cRNA and autoradiography on sections of the same brain revealed silver grains over many enlarged nuclei of oligodendrocytes, only a minority of astrocytes, and possibly vascular endothelial cells, indicating the presence of JCV DNA in these different cell types. The presence of JCV DNA in the kidney from a patient with PML was also analyzed by in situ methods (Dörries and ter Meulen, 1983). Histologically, the kidney did not show any pathologic lesions, and no virus particles could be detected in the kidney tissue by electron microscopy in contrast to PML brain tissue, where oligodendroglia are filled with virus particles (see below). However, in situ hybridization of frozen kidney sections with a tritiated JCV DNA probe revealed specific label almost exclusively over the nuclei of epithelial cells lining the collecting tubules.

Using the technique of in situ DNA-to-DNA hybridization, a JCV biotinylated DNA probe was developed and applied to formalin-fixed, paraffin-embedded, or unfixed, frozen sections of PML brain tissue (Aksamit et al., 1985). Cell identification based on morphologic appearance was very difficult in the unfixed, frozen sections and was easier in the formalin-fixed, paraffin-embedded sections. Affected oligodendrocytes were found to be the most heavily labeled. Some of the oligodendrocytes were located in the cortex, as was found by electron microscopy (Mázló and Herndon, 1977) and immunocytochemistry (Budka and Shah, 1983; Itoyama et al., 1982). The altered oligodendrocytes were not necessarily perivascular in location. Only some astrocytes were labeled for JCV DNA. The infrequent labeling of reactive astrocytes is compatible with previous immunocytochemical data (Itoyama et al., 1982) and electron microscopic data (Mázló and Tariska, 1982), suggesting limited productive infection in some "nontransformed" astrocytes. No cytoplasmic labeling was detected despite virus particles demonstrated in the cytoplasm of the astrocytes by electron microscopy (Mázló and Ţariska, 1982; Watanabe and Preskorn, 1976) and by immunocytochemistry (Budka and Shah, 1983; Itoyama et al., 1982). The labeling was exclusively intranuclear in the bizarre astrocytes. Macrophages and mononuclear cells, mainly seen in perivascular locations, showed no evidence of labeling. Neurons and ependymal cells did not label. In contrast to the previous study (Dörries et al., 1979), no endothelial cells were labeled (Aksamit et al., 1985, 1986, 1987). This is in agreement with the immunoctyochemical results of Itoyama et al. (1982) and Budka and Shah (1983) and with the electron microscopic results of Mázló and Tariska (1980) and Mázló and Herndon (1977) that endothelial cells did not immunostain or contain virus particles. These findings render the consistent productive infection of endothelial cells in PML unlikely.

The localization of JCV DNA detected by biotin-labeled DNA-to-DNA in situ hybridization were compared with JCV capsid protein detected by immunocytochemistry using a peroxidase labeling method in formalin-fixed, paraffin-embedded brain sections of four cases of PML (Aksamit et al., 1986). Infected oligodendrocytes showed both JCV DNA and JCV late proteins. However, bizarre astrocytes demonstrated JCV capsid protein less often than JCV DNA. This more frequent presence of JCV DNA than viral capsid protein in bizarre astrocytes suggested an abortive infection by JCV. This method was applied to brain biopsy material for confirming the diagnosis of PML in several other cases (Aksamit et al., 1987).

Tritium-labeled cloned JCV DNA probe was used to label JCV-infected cells in formalin-fixed, paraffin-embedded PML brain sections (Shapshak et al., 1986). In early lesions of PML, hybridization was found in the swollen nuclei of oligodendrocytes with minimal labeling in astrocytes. In older lesions the bizarre, giant astrocytes were also variably labeled. The periphery of the advanced lesions exhibited variable levels of hybridization in oligodendrocytes. Unequivocal hybridization of neurons was not seen in frozen or paraffin-embedded material (Shapshak et al., 1986). In addition, hybridization signal was occasionally seen in microglial cells and rarely in endothelial cells or pericytes of a venule or mononuclear cells in the lumen of the small vessels or in the perivascular space. In this study the demyelinated foci were often found to be related to blood vessels. Similar conclusions were reached in the original study of Åström et al. (1958). However, in later investigations Richardson (1970) concluded that the PML foci have no regular relationship to blood vessels. In case 1 of Chandor et al. (1965) in which 40 H&E-stained serial sections were studied in order to learn about the relation of the lesions to vessels no constant relation to vessels was found. The age of the lesions is an important variable in such studies, and the earliest lesions will be the most instructive for questions of pathogenesis.

JCV was found in scattered mononuclear cells of the bone marrow and spleen in the autopsy material of an AIDS patient with PML using in situ DNA hybridization and immunocytochemistry for capsid antigens (Houff et al., 1988). In several brain sections, occasional mononuclear cells located in Virchow-Robin spaces of the temporal lobe and basal ganglia were found to contain JCV DNA and capsid antigen. In another patient who had no known underlying disease, PML was proved by brain biopsy. B lymphocytes infected with JCV were demonstrated in a bone marrow biopsy specimen and aspirates, but not in the brain biopsy specimen (Houff et al., 1988).

In the brain of the AIDS-PML patient, more oligodendrocytes contained JCV DNA than virion antigens. The density of the infected cells, many without inclusions and some that were considered to be astrocytic, appeared to be increased adjacent to the blood vessels. Bizarre astrocytes also contained JCV DNA and, occasionally, virion antigens. This finding was confirmed by the results of Aksamit et al. (1986), who showed that bizarre astrocytes demonstrated capsid protein less often than JCV DNA. As noted above, occasional infected mononuclear cells were found in the Virchow-Robin spaces. The presence of JCV DNA detectable by in situ hybridization and the presence of the capsid antigens in the mononuclear cells suggested ongoing viral replication in this cell type. It was proposed that virus-infected lymphocytes enter the perivascular space of the brain, resulting in infection of the glial cells. This would be one of the possible routes for JCV to reach the CNS. In the kidney JCV DNA, but not capsid protein, was found in the renal medulla.

In the next year several studies using sensitive nonradioisotopic in situ hybridization methods (Boerman et al., 1989; Teo et al., 1989) could not show the presence of JCV DNA in the normal-sized oligodendrocytes or mononuclear cells either intravascularly or in the perivascular space. Like many other investigators, they could not detect JCV DNA in neurons, endothelial cells, or microglia. Negative findings could, however, be the result not only of using different probes but also of using different prehybridization treatments such as pronase or pepsin digestion or different denaturation procedures. Even the time of the fixation of the brain tissue and the components of the fixative could influence the result (Teo et al., 1989) as has been described with immunocytochemical methods (Budka and Shah, 1983; Itoyama et al., 1982; Stoner et al., 1986).

The identification of the cells containing JCV DNA in PML by combined in situ hybridization and immunocytochemical methods was carried out by Ironside et al. (1989). The three autopsy brain samples were fixed in formalin for 3 weeks and the biopsy material for 24 hours in neutral-buffered formalin. All tissues were processed routinely and embedded in paraffin. Sections from each case were immunostained following hybridization. Additional sections from each case were subjected to immunocytochemical methods without either pretreatment and hybridization or with pretreatment but no hybridization. GFAP detected astrocytes, and neurofilament antibody detected neurons. However, immunoreactivitiy for carbonic anhydrase II, a marker for detection of oligodendrocytes, was unsuccessful. Positive nuclear labeling for JCV was seen more frequently in oligodendrocytes both in inclusion-bearing ones and in apparently normal oligodendrocytes in the white matter, which showed that JCV distribution in PML is more extensive than can be seen by routine light microscopy. This result is in agreement with the immunocytochemical results of Stoner et al. (1986) and the immunocytochemical and in situ hybridization results of Aksamit et al. (1986). A positive reaction was also found in the bizarre nuclei of enlarged astrocytes. Most astrocytes, however, have a negative reaction against the JCV probe. Occasionally, mononuclear cells around small blood vessels within areas of demyelination showed nuclear labeling. The identity of these cells and their significance could not be determined. Finally, no endothelial cells or neurons were labeled by these methods.

3. ELECTRON MICROSCOPY OF PML

Fascinated by the nuclear aberrations in glial cells that are unique to this disease, Zu Rhein and Chou (1965) and Silverman and Rubinstein (1965) independently performed electron microscopic studies on PML autopsy brain samples. In the formalin-stored autopsy PML brain material neither plasma membrane nor intracytoplasmic organelles could be well distinguished. However, despite the considerable autolytic damage, the nuclei of the inclusion-bearing cells appeared relatively intact and were packed with many thousands of spherical and elongated virus particles. They usually filled the nucleus in a random distribution, but sometimes the spherical particles were found in a crystalloid arrangement. The center-to-center distance of virions in crystals read consistently as 40 nm (Zu Rhein, 1972). The width of the filamentous forms was one-half to two-thirds the diameter of the spherical ones (Zu Rhein and Chou, 1965). In 1965, Howatson et al. approached the study of PML virions by negative staining methods. The source material used for the virus suspension was formalinized PML brain tissues. By this method it was possible to define the morphologic characteristics of the virus particles more precisely. They were circular or somewhat hexagonal in outline, and their surfaces were studded with small projections resembling viral capsomers (Fig. 12.9). The virus particles were structurally similar to virions of the papovavirus (now called polyomavirus) subgroup. The measured value of the diameter of unfixed polyomavirus particles was 42.5 nm, while that of formalinized particles was 40.5 nm, just slightly less than the unfixed polyomavirus particles. The difference may be explained by slight shrinkage due to prolonged formalin fixation. The average diameter of the polyomavirus particles is appreciably smaller than that of papillomavirus subgroup (52–55 nm) (Howatson et al., 1965).

Correlation of the light and electron microscopic findings of Zu Rhein and Chou (1965) suggested that demyelination resulted from the cytocidal effect of the virus on oligodendroglia. Silverman and Rubinstein (1965) accepted this interpretation and raised the question whether the postulated virus is induced to replicate from a latent resident state in the nervous system or is newly transmitted from person to person. That was more than 35 years ago, and today we still do not have the answer. Ultrastructural studies of many additional cases by other investigators have found the same polyomavirus particles and have agreed with the original interpretation (Dolman et al., 1967; Ikuta, 1969; Muller and Watanabe, 1967; Papadimitriou et al., 1966; Vanderhaeghen and Perier, 1965; Woodhouse et al., 1967; Zu Rhein and Chou, 1965). In autopsy brain materials the autolytic changes of the cells do not allow accurate cellular differentiation. Therefore, most authors have referred to the virus-containing cells as *glial* rather than trying to specify them. However, Zu Rhein (1972) rarely

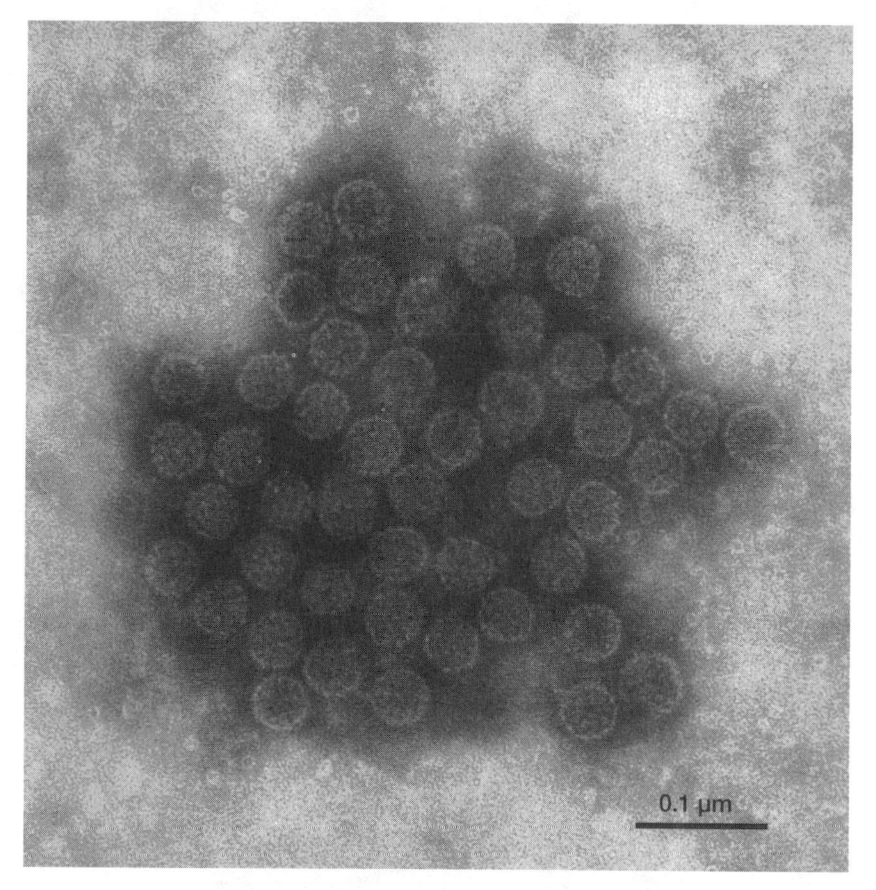

Figure 12.9. JCV particles as seen in the electron microscope by the negative staining method. The capsomers of virus particles are also visible. ×176,800. Bar = 0.1 μm.

found infected astrocytes and was able to describe the disintegration of the virus–containing oligodendroglia nuclear membrane in association with spilling of the virus particles into the cytoplasm and extracellular space (cytolysis). These round virions were never surrounded by individual membranes, and they tended to line up along membranes, including the periphery of the myelin sheath. Virions were not seen between myelin lamellae or inside the sheath. Polyomavirus virions were also demonstrated by Papadimitriou et al. (1966) along the periphery of degenerated myelin sheaths, and their presence was attributed to their pathogenetic role in demyelination. Later, in relatively well-preserved brain material, it was shown that virions can appear in rows between the slightly swollen myelin lamellae of a relatively preserved axon (Fig. 12.10A). However, the separation of the myelin lamellae occurs along the intraperiod line; thus these virions are in fact in an extracellular position and represent only the attraction of the polyomavirus to membranes (Fig. 12.10B).

A

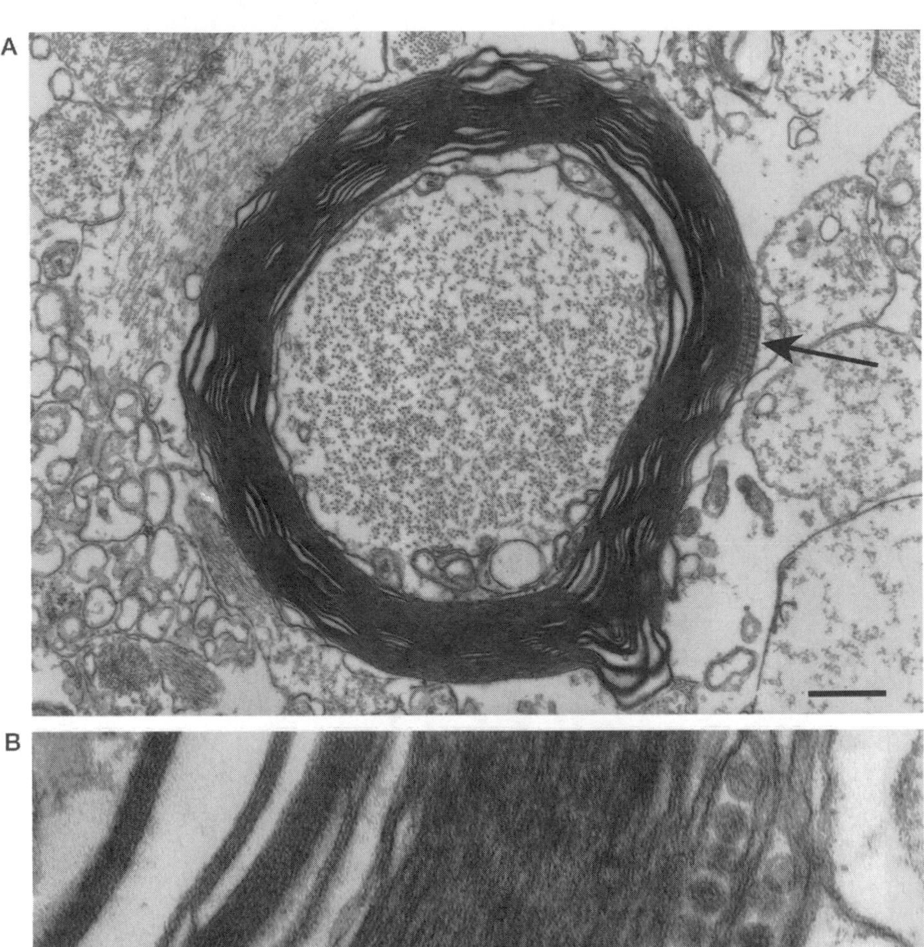

B

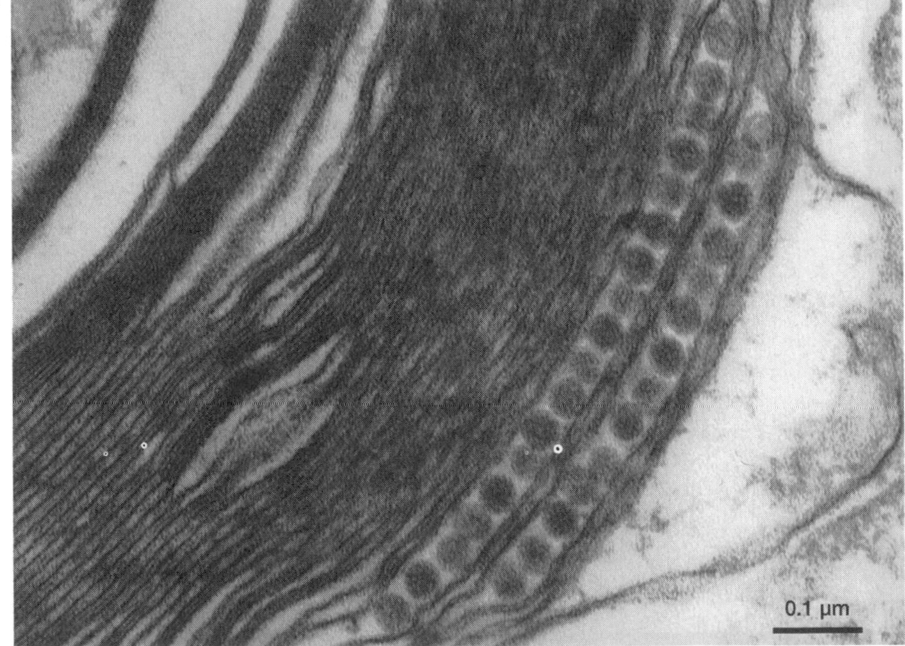

Figure 12.10. (**A**) Rows of JCV particles in slightly loosened myelin lamellae of a relatively well-preserved axon (arrow). ×10,440. Bar = 1 μm. (**B**) At higher magnification the JCV particles are seen between myelin lamellae separated along the intraperiod line. Thus these virions are in an extracellular position. ×118,300. Bar = 0.1 μm. Taken with permission from Mázló and Tariska, 1980.

In subsequent years reports of polyomavirus particles demonstrated by electron microscopy of well-preserved biopsy materials were uncommon (Castaigne et al., 1973; Mathews et al., 1976, case 1). Usually autopsy brain materials were used to verify the postmortem diagnosis of PML.

Good preservation of the ultrastructure of the cells in the brain biopsy specimens and in very fresh autopsy materials has provided the opportunity to study the fine details of the neuropathology of PML. Myelin breakdown in PML lesions can happen in the absence of infiltrating mononuclear cells and is clearly secondary to that of the cytolytic viral infection of oligodendrocytes, as was originally suggested by Zu Rhein and Chou (1965). The oligodendrocytes may remain relatively intact until their nuclear material is completely replaced by virus, when cytolysis occurs and virus is released (Mázló and Herndon, 1977). The myelin sheaths cannot maintain their integrity without being in contact with their supporting oligodendrocyte. A second mechanism by which demyelination might occur would be more subtle damage by metabolic alterations in abortively infected oligodendrocytes expressing only T antigen. There are no data indicating an immunopathologic process of demyelination in "classic" PML.

The availability of well-preserved brain material made it possible to study the relation of the virus to the infected host cell. The cell membrane under the attached JC virions sank, and then the hollow closed above the virion and the vacuole separated off inwardly. In those places where the cell membrane formed deep and narrow invaginations, the virions sank in rows. It is worth mentioning that the virions entering the cells are never seen free in the cytoplasm. They were found either in vesicles and vacuoles or in the cisternae of the endoplasmic reticulum (ER) of the cytoplasm (Fig. 12.11). The entry of JCV and intracellular virus transport could be observed in oligodendrocytes in which the structure of the nucleoplasm showed no alterations attributable to JCV infection, as well as in cells where virus assembly was going on. Rarely, virus particles can be caught in the perinuclear space while being transported to the nucleus (Mázló and Tariska, 1980). The increased volume of the nucleus and cytoplasm of the oligodendrocyte was the first visible change following the disappearance of the first infecting virions into the nucleoplasm. Most of the chromatin seems to disappear from the nucleus and to be replaced by faintly stained, very finely dispersed, dense material in the "immature" inclusion (Fig. 12.12A). Sometimes, more than one nucleolus can be found.

By this time proliferation of the cytoplasmic organelles, especially of the ribosomes, was characteristic. Mild vacuolization appeared in the cytoplasm of the infected oligodendrocytes, which can be ascribed to the production of viral protein of the polyomavirus (Mázló and Tariska, 1980). At the beginning of virus assembly, proliferation of the nuclear membrane could also be observed as indicated by the formation of nuclear membrane invaginations and protrusions (Fig. 12.12). The first virus progeny usually appeared in the vicinity of the nuclear membrane. As long as they were few in number, many elongated virus particles could be seen among the spherical ones in the "immature in-

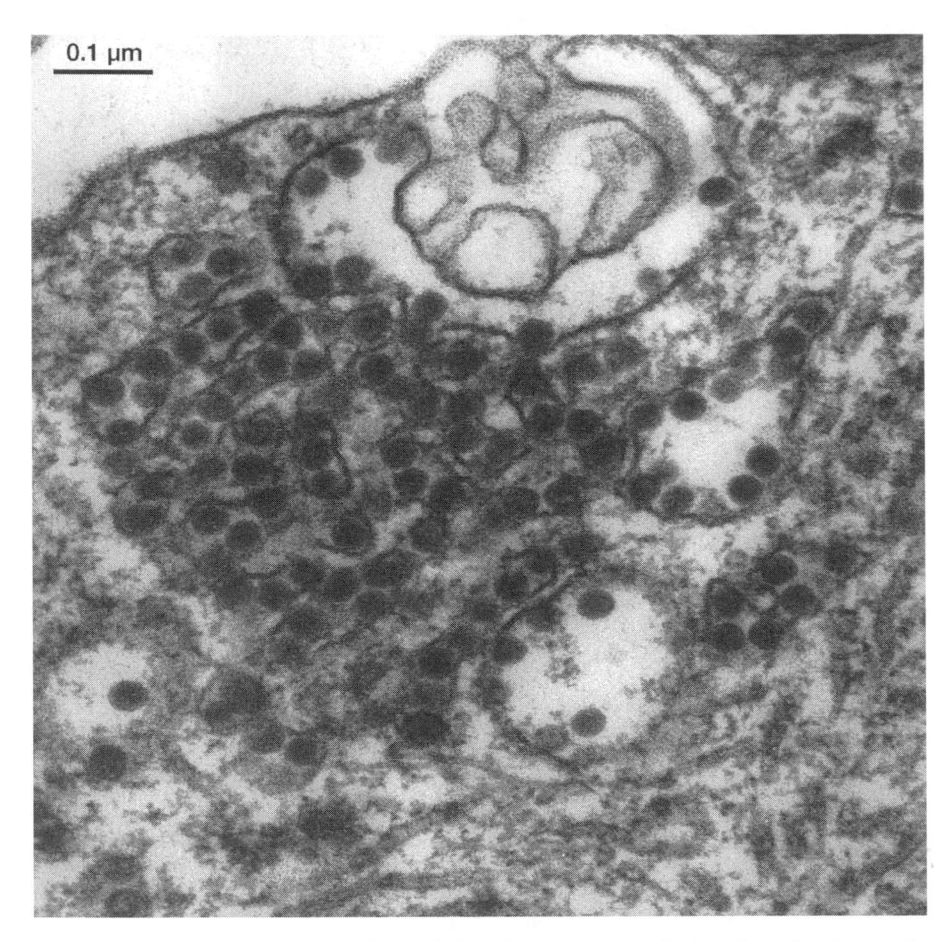

Figure 12.11. Virus uptake and intracellular virus transport. JCV particles in phagocytic vacuoles, vesicles, and cisternae of the endoplasmic reticulum (ER) in the cytoplasm of an oligodendrocyte. ×129,400. Bar = 0.1 μm. Taken with permission from Mázló and Tariska, 1980.

clusion'' (Fig. 12.12). These highly elongated filamentous virus particles often formed small inclusions in the early phase of virus assembly (Mázló and Tariska, 1980). As the number of the spherical particles grew in the nucleus, the number of the elongated ones diminished. Finally, mostly spherical virus particles filled the entire field of the nucleoplasm of the ''mature'' inclusion (Fig. 12.13) containing only a few elongated particles. In the mature inclusion, occasional crystalloid arrangements of the virions can also be observed (Zu Rhein, 1972). The diameter of the virions measured 36–40 nm. At higher magnification their edges looked somewhat paler. The diameter of the elongated particles was 18 to 26 nm, and the capsomers, which were arranged in longitudinal lines on them, exhibited striations. The surface structures of both types of virus particles are basically the same. The difference in shape may indicate an altered mode of self-assembly based on an imbalance between the protein and DNA synthesis (Bancroft et al., 1967). This imbalance seems to exist at the beginning of virus assembly and apparently ceases by the time of its completion.

Parallel to virus assembly, vacuolation of the cytoplasm became more severe and the polyribosomes disintegrated to monosomes. By the time the virions fill the nucleus, the ribosomes disappear and the cristae of the swollen mitochondria and the walls of the smooth and granular endoplasmic reticulum (ER) break up. The cell membrane at this stage was never intact, but was broken up together with the inner membranes (Fig. 12.13A). This was followed by the disintegration of the nuclear membrane, so the progeny virions could reach the severely damaged cytoplasm where they can form crystalloid structures (Zu Rhein, 1972) or can line up along membrane fragments. It would be hard to imagine—at least in the oligodendrocytes—that after the disintegration of the cytoplasmic organelles, including the mitochondria, the cell would have the capacity to envelop the progeny virions, as proposed by some authors. After the nuclear membrane breaks up, some of the virions move directly into the extracellular space. In the necrotic area they can be arranged in crystal-like formation, lining up along membranes or attached to cellular debris (Fig. 12.14). It should be mentioned that in the necrotic area of the biopsy material oligodendrocytes could also be found that died before the whole replication cycle was finished. These oligodendrocytes resembled those found in autopsy material. The edges of these nuclei are sometimes "jagged" (Kepes et al., 1975; Zu Rhein, 1969) in biopsy material (Mázló and Herndon, 1977, Fig. 6), and in the nucleoplasm many elongated virus particles can be found in bundles. In the severely vacuolar and degenerating cytoplasm, membrane-bound JC virions, which could not reach the nucleus, can also be present. However, the effect of these "late arriving" virions in the pyknotic cells and in those cells that are in the late phase of virus assembly is questionable.

Productive infection of *astrocytes* was very rarely found in PML (Boldorini et al., 1993; Mázló and Herndon, 1977; Mázló and Tariska, 1982; Richardson, and Johnson, 1975; Watanabe and Preskorn, 1976). Polyomavirions in the cytoplasm of astrocytes are more frequently seen (Krempien et al., 1972; Mázló and Herndon, 1977; Mázló and Tariska, 1982; Nagashima et al., 1982; Orenstein and Jannotta, 1988; Watanabe and Preskorn, 1976; Zu Rhein, 1972). This observation suggests that in the astrocytes the early phenomenon of virus uptake was not always followed by virus replication. It might be considered that many more astrocytes are infected than can be visualized by electron microscopy, and the infection is only rarely productive. This supposition was confirmed by Stoner et al. (1986), who showed that giant astrocytes express T antigen but not capsid proteins in the nucleus (except possibly during cell division).

The early steps of the human polyomavirus infection in astrocytes were described in 1982 by Mázló and Tariska. The adsorption, penetration, and intracellular virus transport were very similar to those found in the JCV-infected

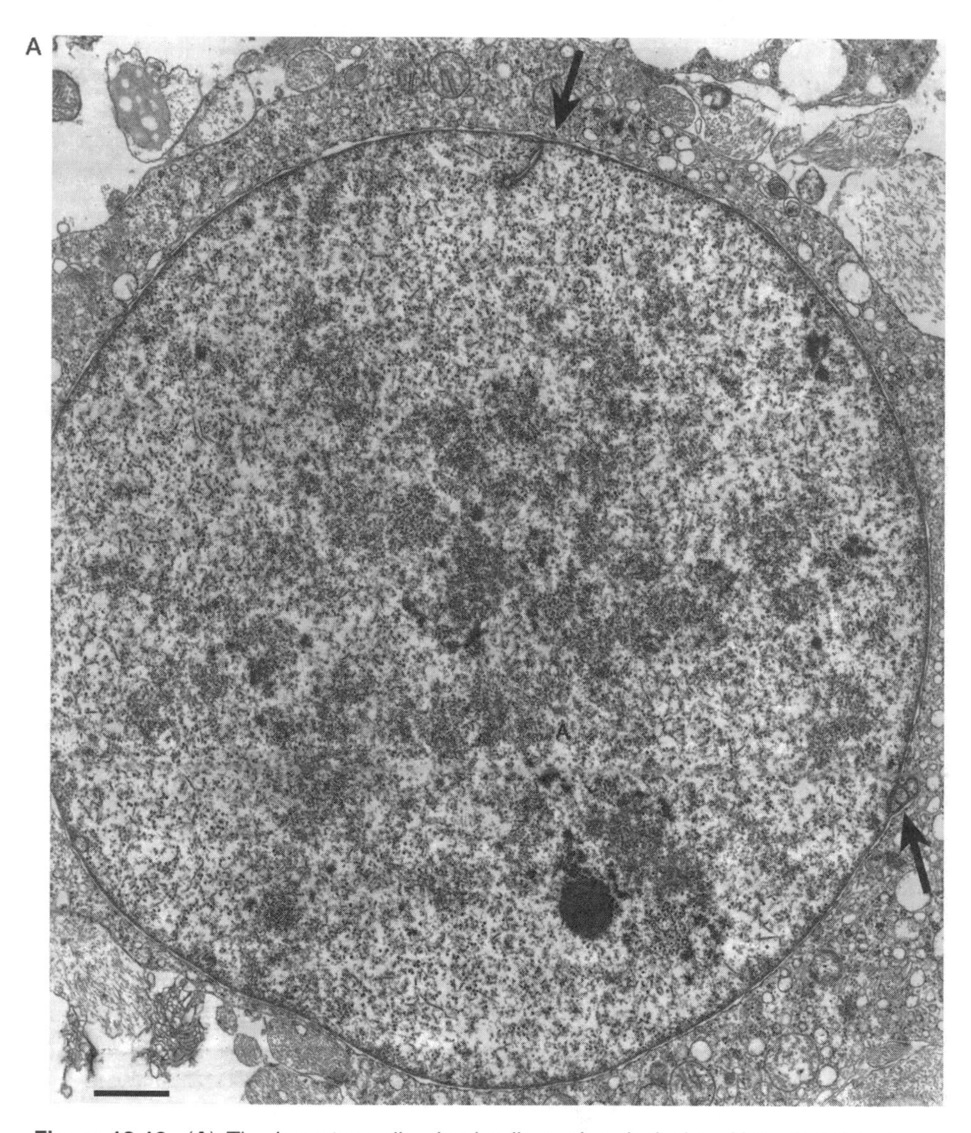

Figure 12.12. (**A**) The immature oligodendroglia nuclear inclusion. Note that the chromatin is replaced by faintly stained dense material, especially in the middle of the nucleus. Nuclear membrane proliferations such as invaginations and protrusions (arrows) can be found. The cytoplasm is slightly vacuolated, and the ribosomes have not yet disappeared. Assembly of JCV particles is beginning, especially in the vicinity of the nuclear membrane. ×10,000. Bar = 1 μm. (**B**) Upper part of the inclusion with nuclear membrane invagination. Note the many elongated JCV particles in the early phase of virus assembly. ×58,000. Bar = 0.2 μm. Taken with permission from Mázló and Tariska, 1980.

B

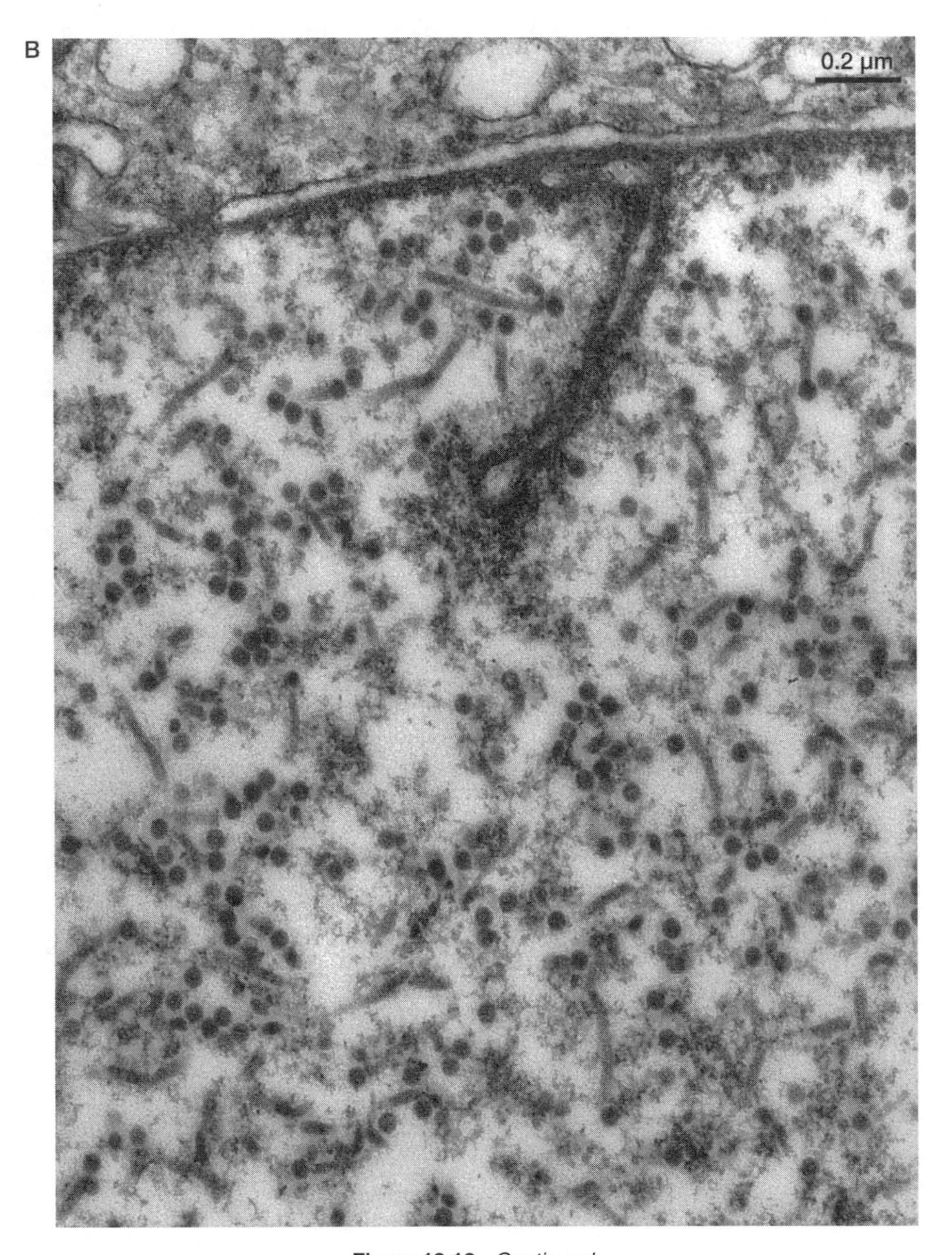

Figure 12.12. *Continued*

A

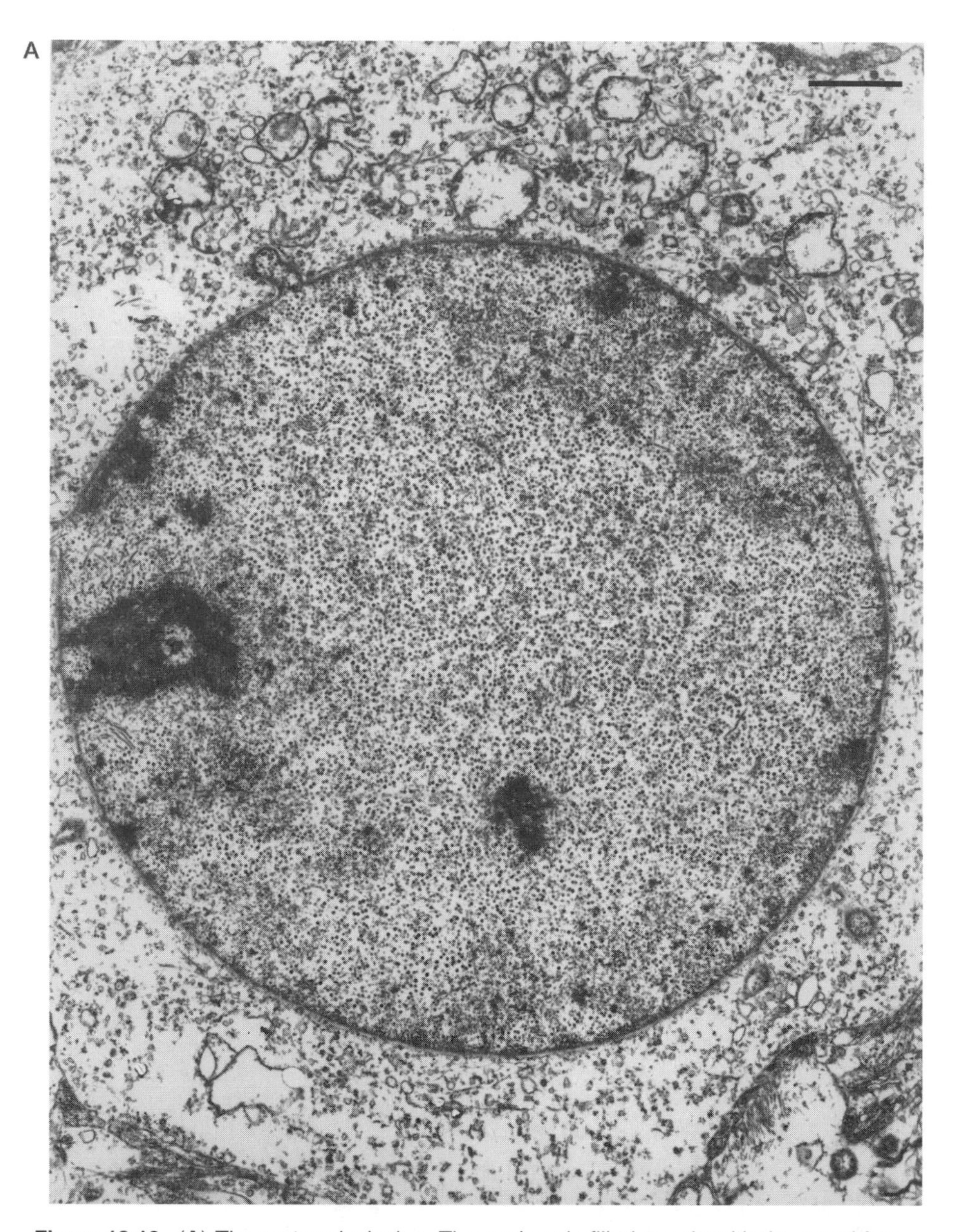

Figure 12.13. (**A**) The mature inclusion. The nucleus is filled mostly with the round form of JCV particles. Only a few elongated particles can be seen. The cytoplasm is severely vacuolated, and the ER and the cell membrane are broken up. The cytolysis is beginning. ×12,000. Bar = 1 μm. (**B**) At higher magnification the round JCV particles are better visible, but the capsomers, seen by negative staining, are not well seen in the fixed tissue section. ×108,000. Bar = 0.1 μm.

B

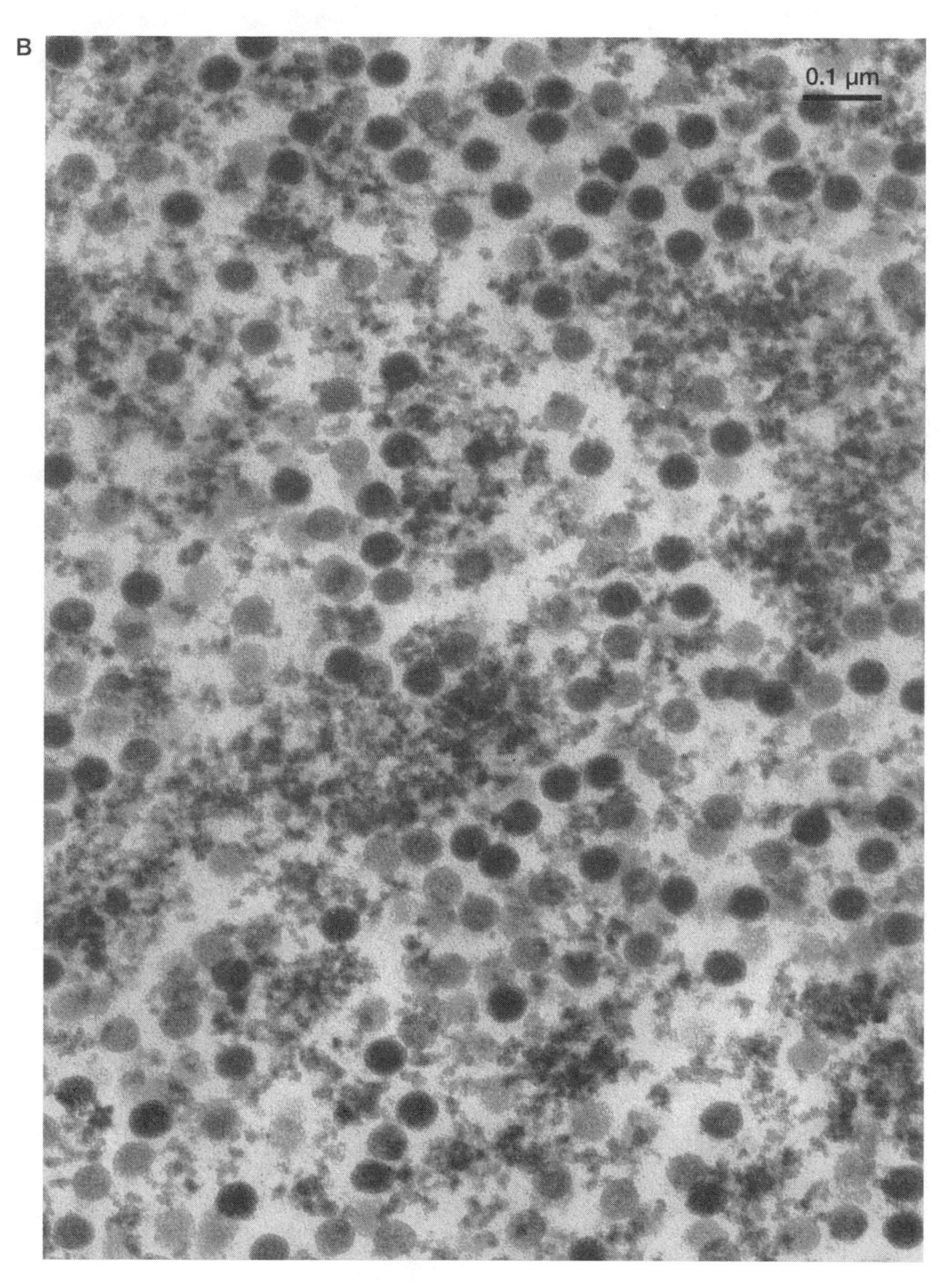

Figure 12.13. *Continued*

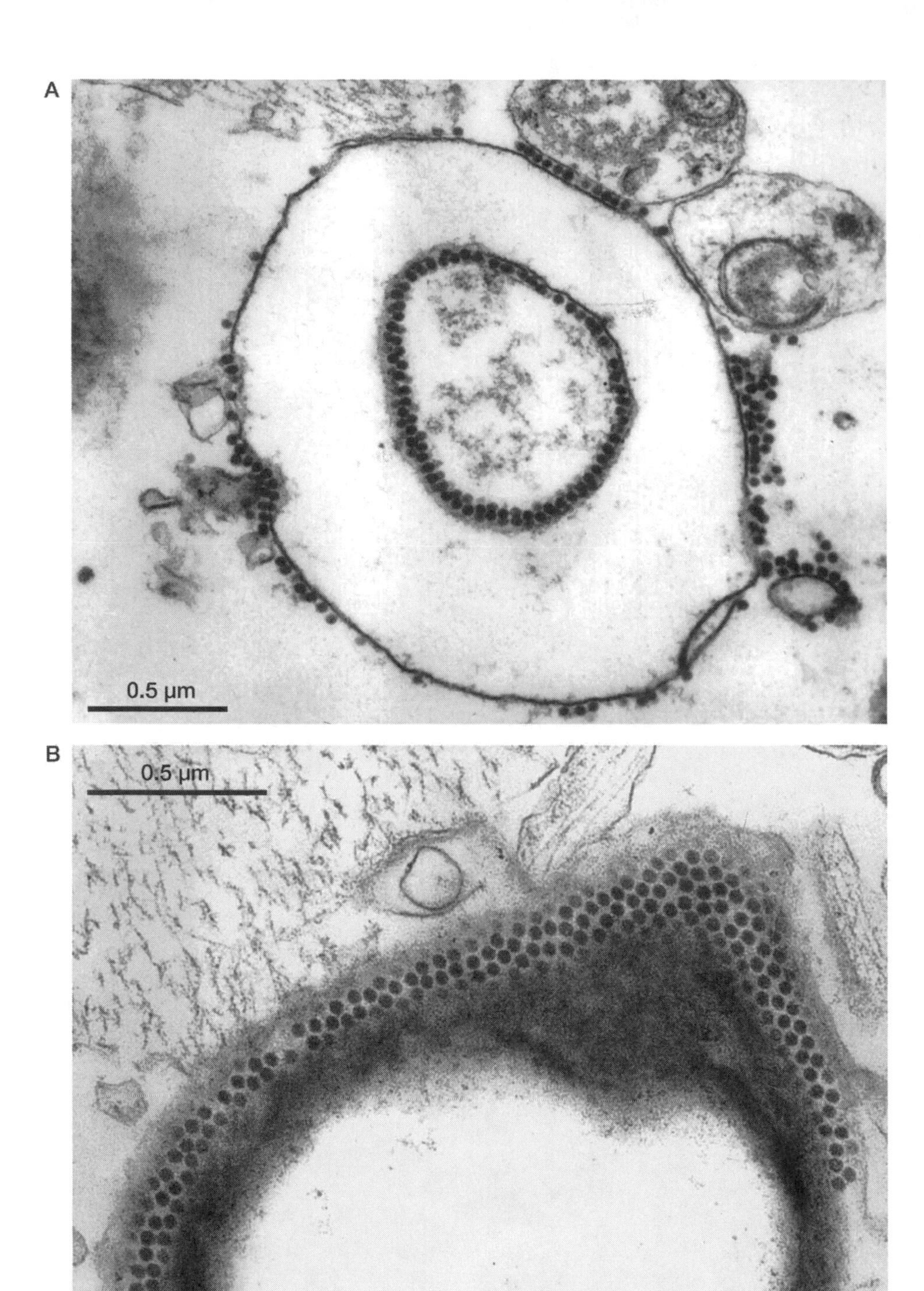

Figure 12.14. JCV particles in extracellular position. (**A**) Attached to disintegrated myelin lamellae. ×44,000. Bar = 0.5 μm. (**B**) In crystal-like position attached to myelin debris. ×57,100. Bar = 0.5 μm. (**C**) Among disintegrated cell membranes and attached to cell processes. ×100,000. Bar = 0.1 μm. Taken with permission from Mázló and Tariska, 1980.

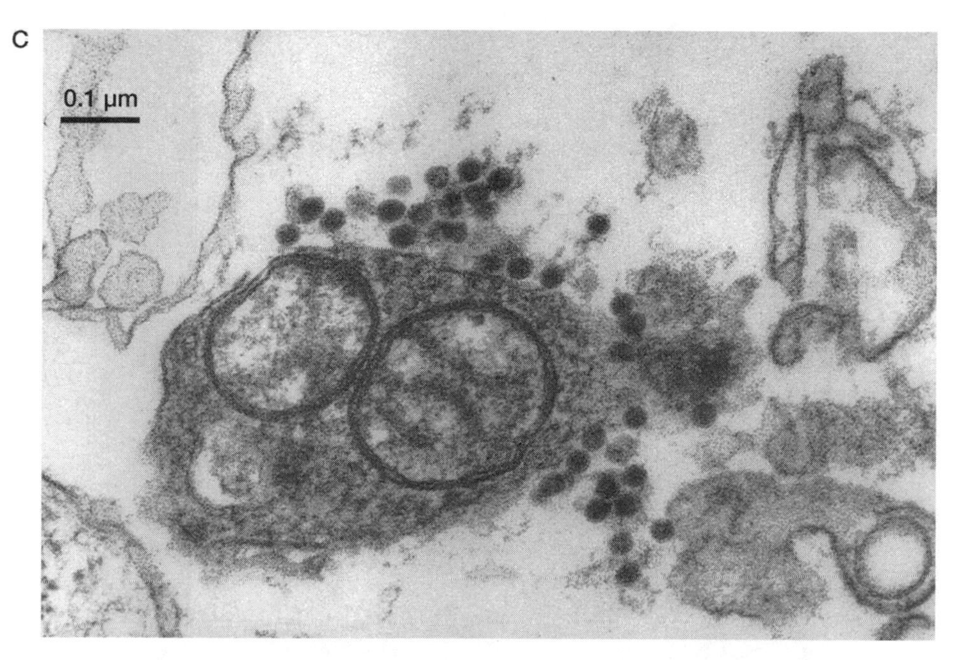

Figure 12.14. *Continued*

oligodendrocytes (Mázló and Tariska, 1980). In light of these studies we know that in the cells with intact nuclei, the virions in the cytoplasm covered by membrane are not progeny virions but infecting ones (Fig. 12.15). These can be found both in the protoplasmic and fibrillary astrocytes. The eclipse phase cannot easily be detected in the astrocytes. The increased volume of the nucleus and cytoplasm of the astrocytes is hard to judge on ultrathin sections. The proliferation of the cytoplasmic organelles, including the ribosomes and glial filaments, is more easily visible in protoplasmic astrocytes (Fig. 12.16). The very finely distributed dense nuclear material is easy to overlook. During virus assembly, the elongated virus particles are used to form small, electron microscopic–sized inclusions in the nucleus (Fig. 12.16). At the end of virus assembly the polyribosomes disintegrate to monosomes and then disappear (Fig. 12.17). When virus assembly is finished, the round virions are randomly distributed in the nucleoplasm (Fig. 12.17). However, their density never reached that seen in the mature oligodendroglial nuclear inclusions (Fig. 12.13A). This is one reason why it is not easy to find intranuclear viral inclusions in astrocytes in biopsy samples. The nuclear membrane proliferation (invagination, protrusion) may draw attention to the nuclear inclusion at low magnification (Fig. 12.17). In the necrotic area in the nucleoplasm of the astrocytes the virions can be found grouped in a small inclusion (Fig. 12.18) where the virions are sometimes also attached to membranes (Mázló and Herndon, 1977). These small intranuclear inclusions are easily visible in the astrocytes by light microscopy.

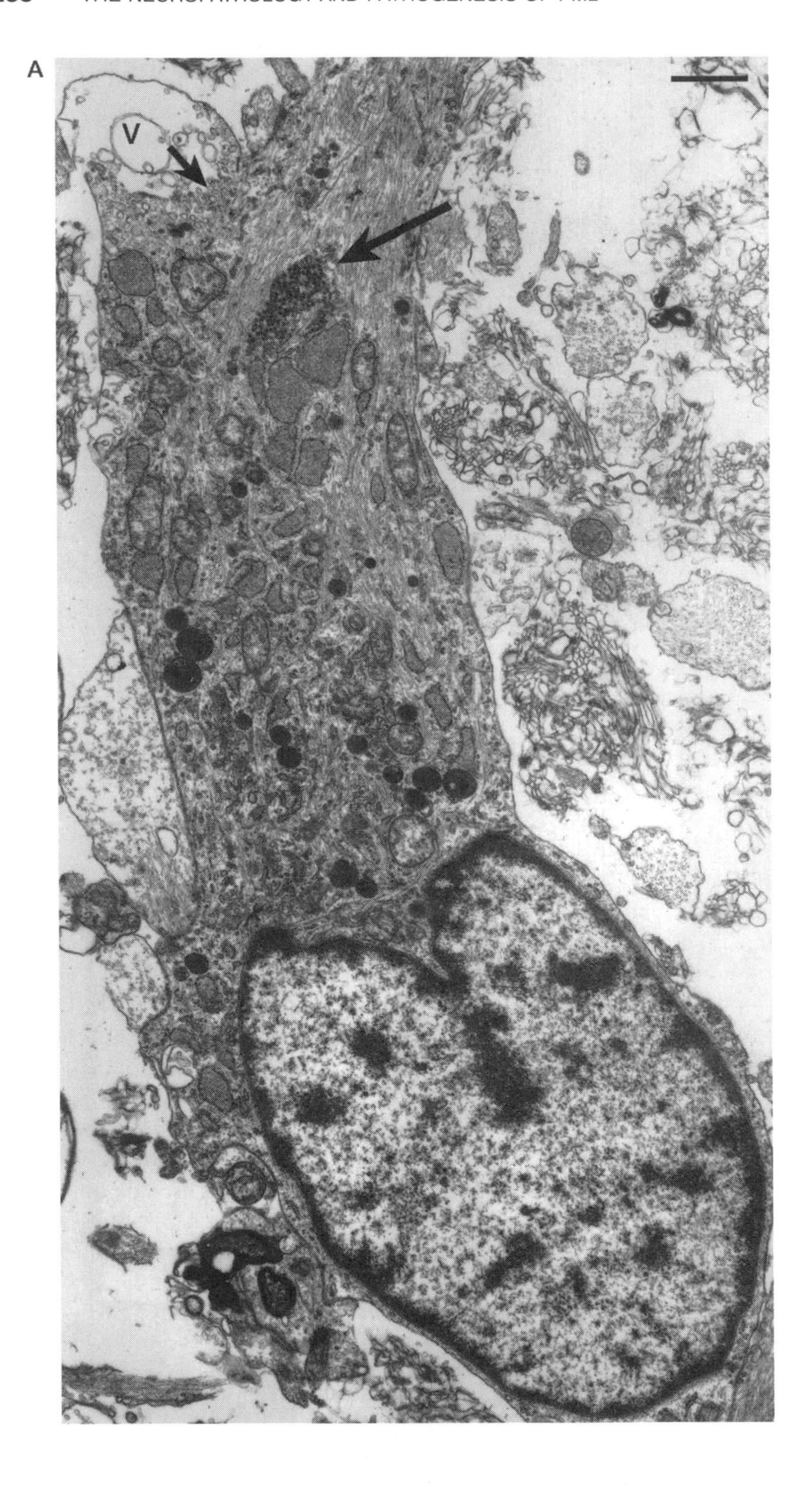
A
V

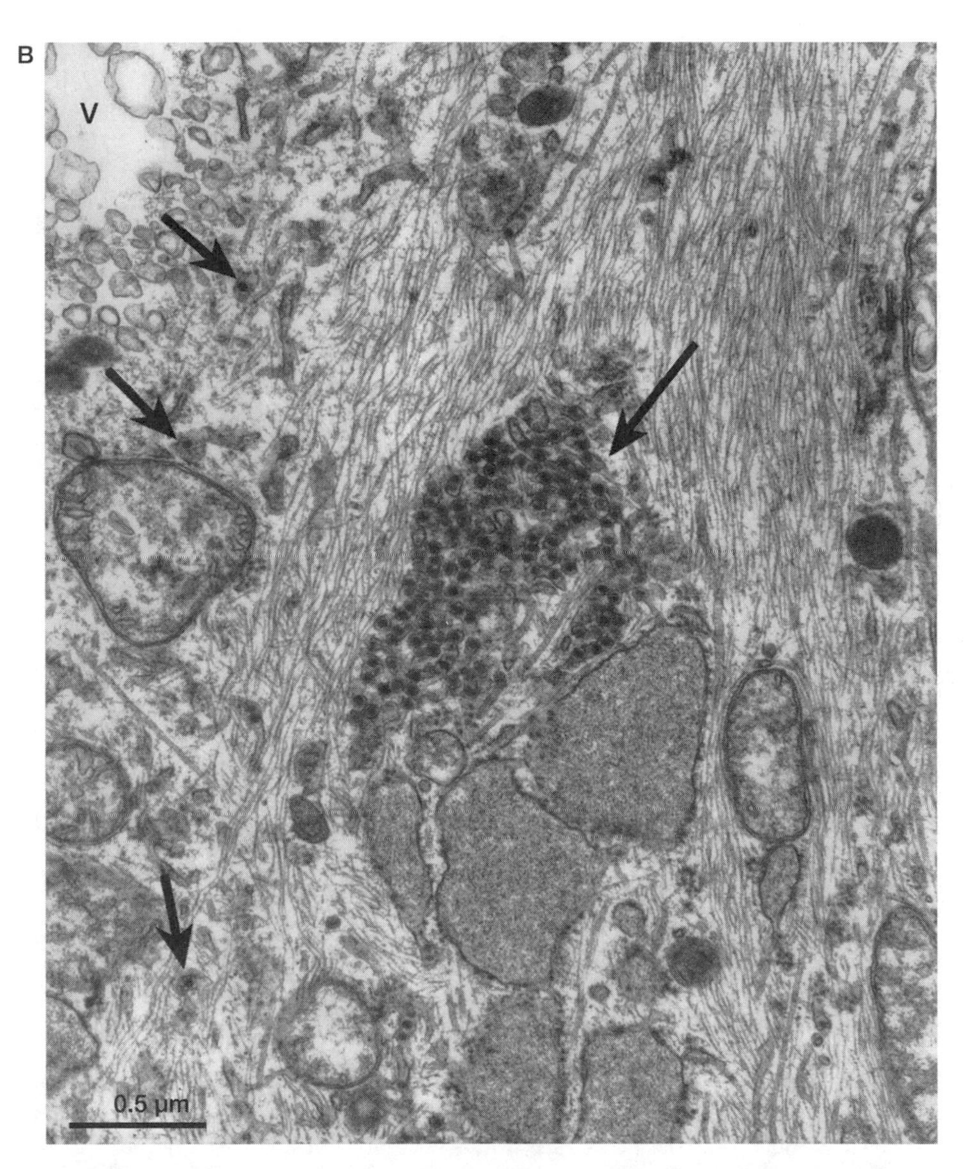

Figure 12.15. (**A**, on opposite page) Fibrillary astrocytes with JCV particles in the cisternae of the endoplasmic reticulum (ER) (large arrow) picked up from the extracellular space. The remnant of a large phagocytic vacuole (v) is also visible at the left upper corner. Small arrow shows the place where at higher magnification single virus particles can be found in vesicles. Note that the nuclei of the astrocytes show no signs of virus replication. ×10,500. Bar = 1 μm. (**B**) At higher magnification virus particles are better seen in the ER cisternae (large arrow). Single JCV particles are also visible in vesicles, close to the cell surface (small arrows). ×35,600. Bar = 0.5 μm. Taken with permisson from Mázló and Tariska, 1980.

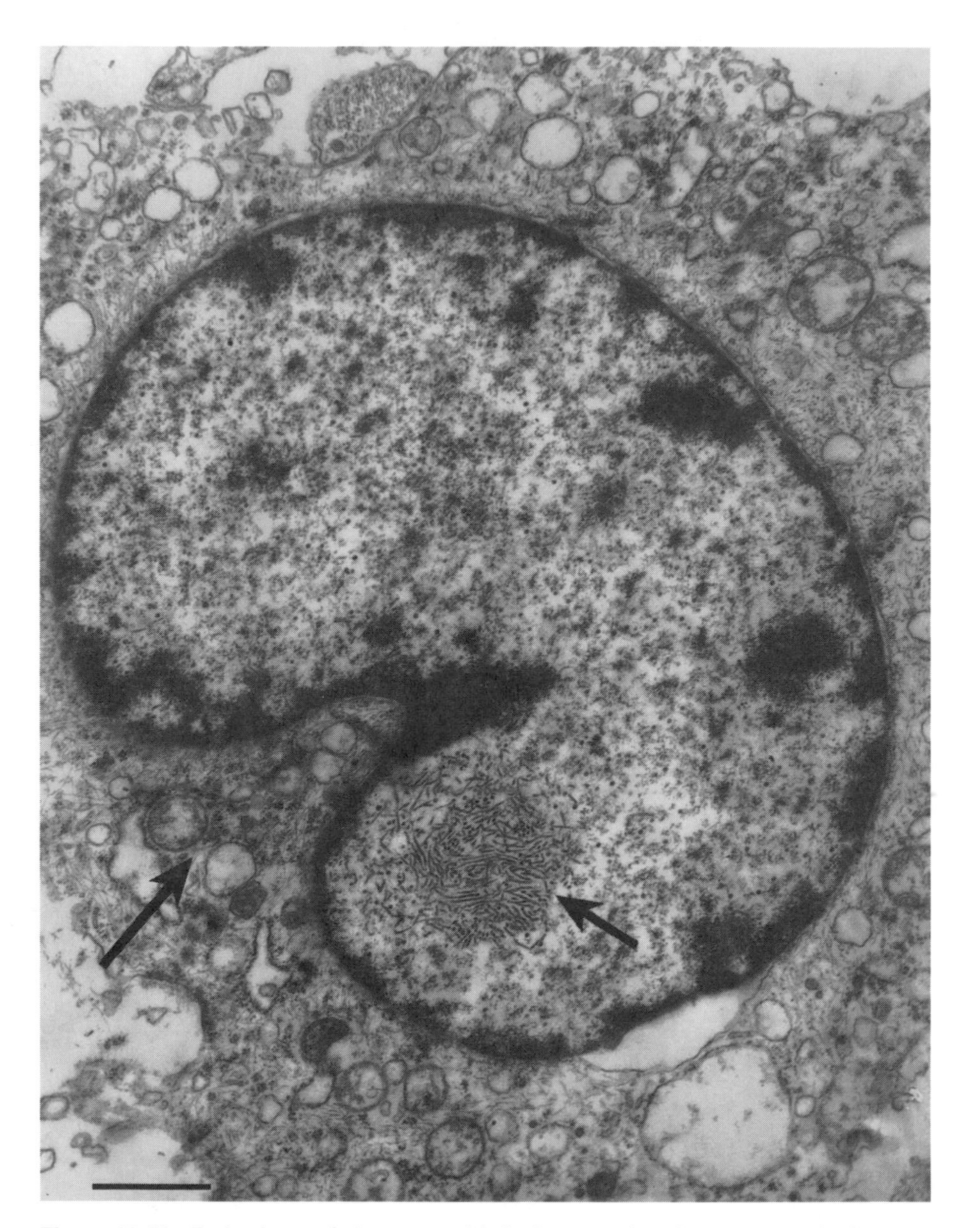

Figure 12.16. Early phase of virus assembly in the protoplasmic astrocytes. In the cytoplasm there are many ribosomes (large arrow) and proliferation of the granular endoplasmic reticulum. In the nucleus a small inclusion (small arrow) of about 1 μm diameter composed of the elongated form of JCV particles is visible. In the nucleoplasm the round JCV particles are loosely arranged. $\times$15,300. Bar = 1 μm.

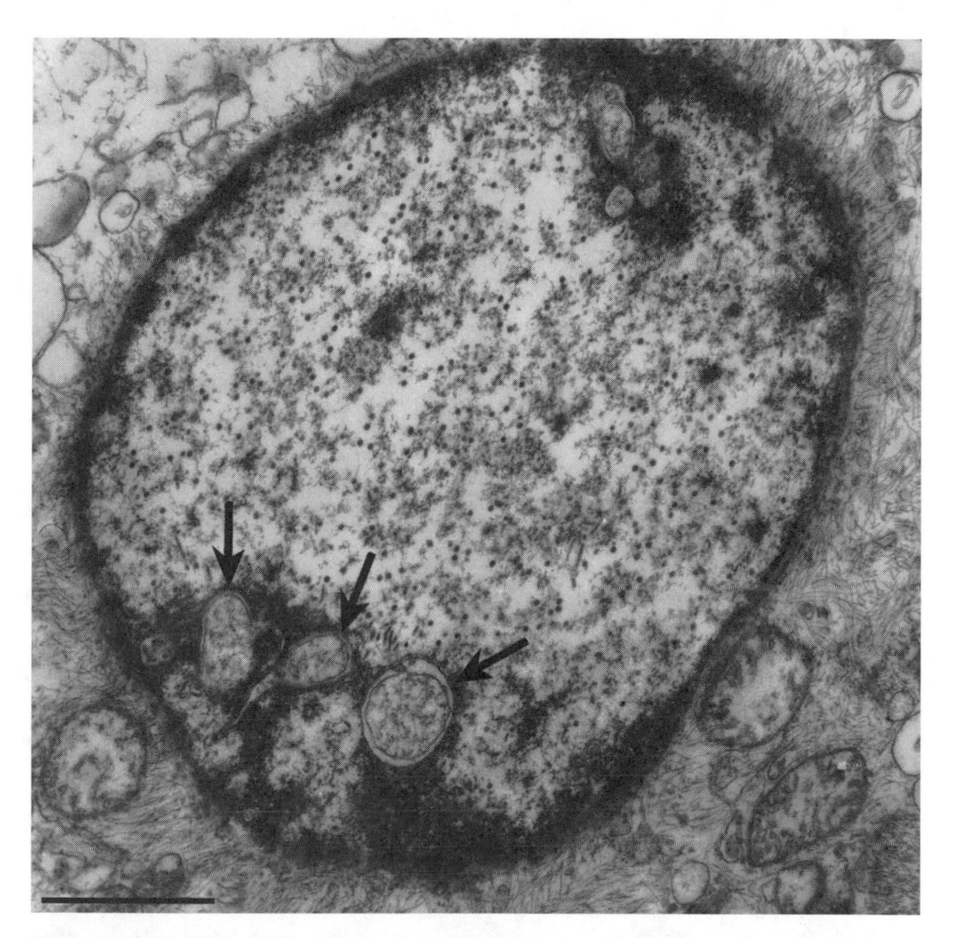

Figure 12.17. The end of virus assembly in astrocytes. Nuclear membrane proliferation (invagination) is visible in the upper and lower parts of the nucleus (arrows). In the nucleoplasm very few elongated virus particles are visible, and the round forms are arranged very loosely (compare the density of virus particles with those seen in the mature oligodendrolgia inclusion in Fig. 12.13A). This is why astrocytic nuclear inclusions are difficult to see by light microscopy. ×23,100. Bar = 1 μm. Taken with permission from Mázló and Tariska, 1982.

Watanabe and Preskorn (1976) have demonstrated the "edematous degeneration" of the productively infected protoplasmic astrocytes, which corresponds to the cytolysis known from JCV infection in oligodendrocytes. It seems that the progeny virions are released from the astrocytes by cytolysis as well. It may be another way to release JC virions from the astrocytes. On the basis of Figure 7 of Watanabe and Preskorn (1976) it was suggested that exocytosis was a possible mechanism by which the virions are released from these protoplasmic astrocytes. However, the possibility of virus uptake (endocytosis) was also taken into consideration. We would favor the latter interpretation, but can-

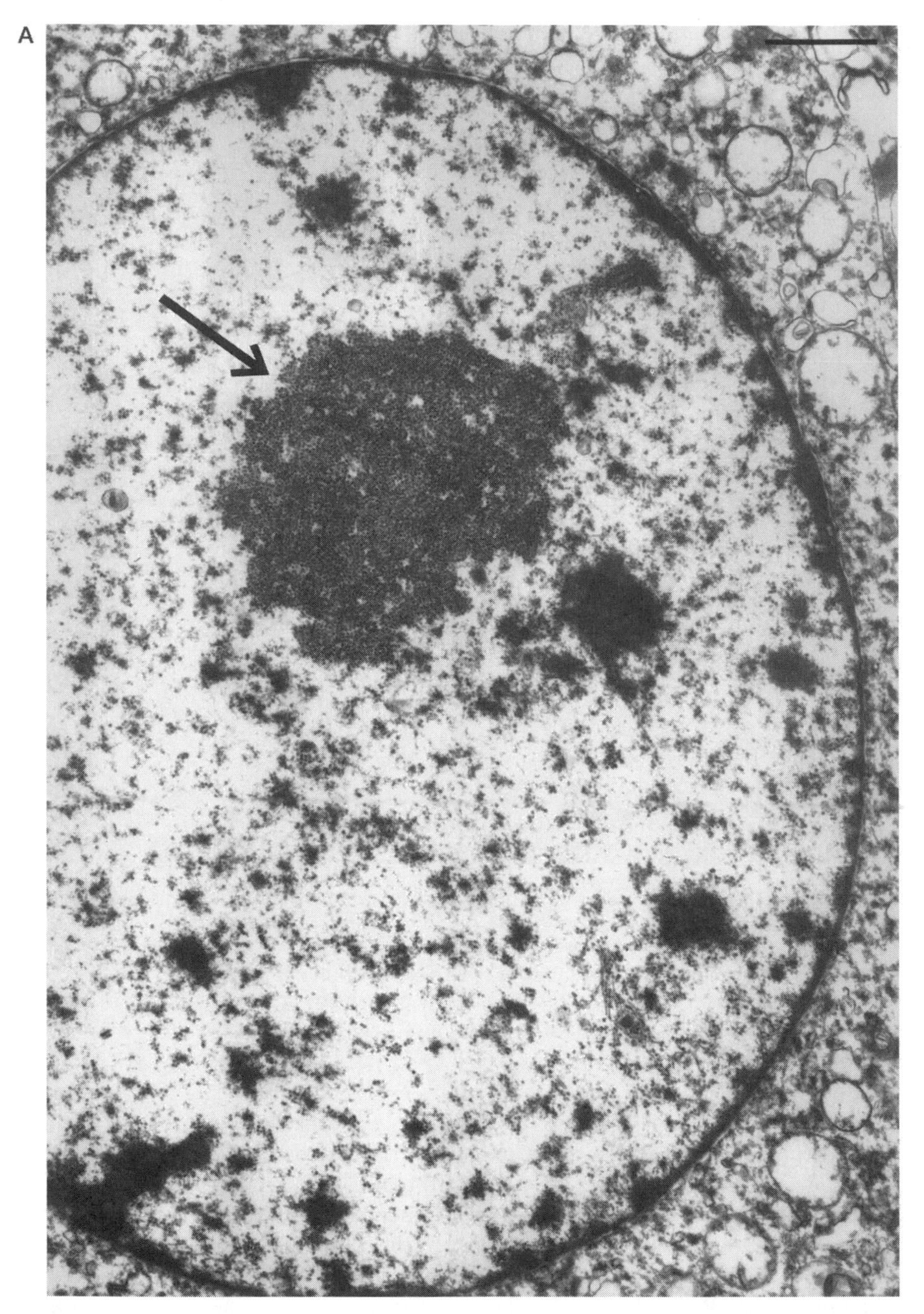

Figure 12.18. (**A**) In the necrotic area, even in biopsy material, the JCV particles form a small, compact inclusion (arrow) in the nucleoplasm of an astrocyte. This is why small nuclear inclusions are visible in the astrocytes of autopsy PML material by light microscopy. ×15,400. Bar = 1 μm. (**B**) At higher magnification the JCV particles are clearly visible. ×29,600. Bar = 1 μm.

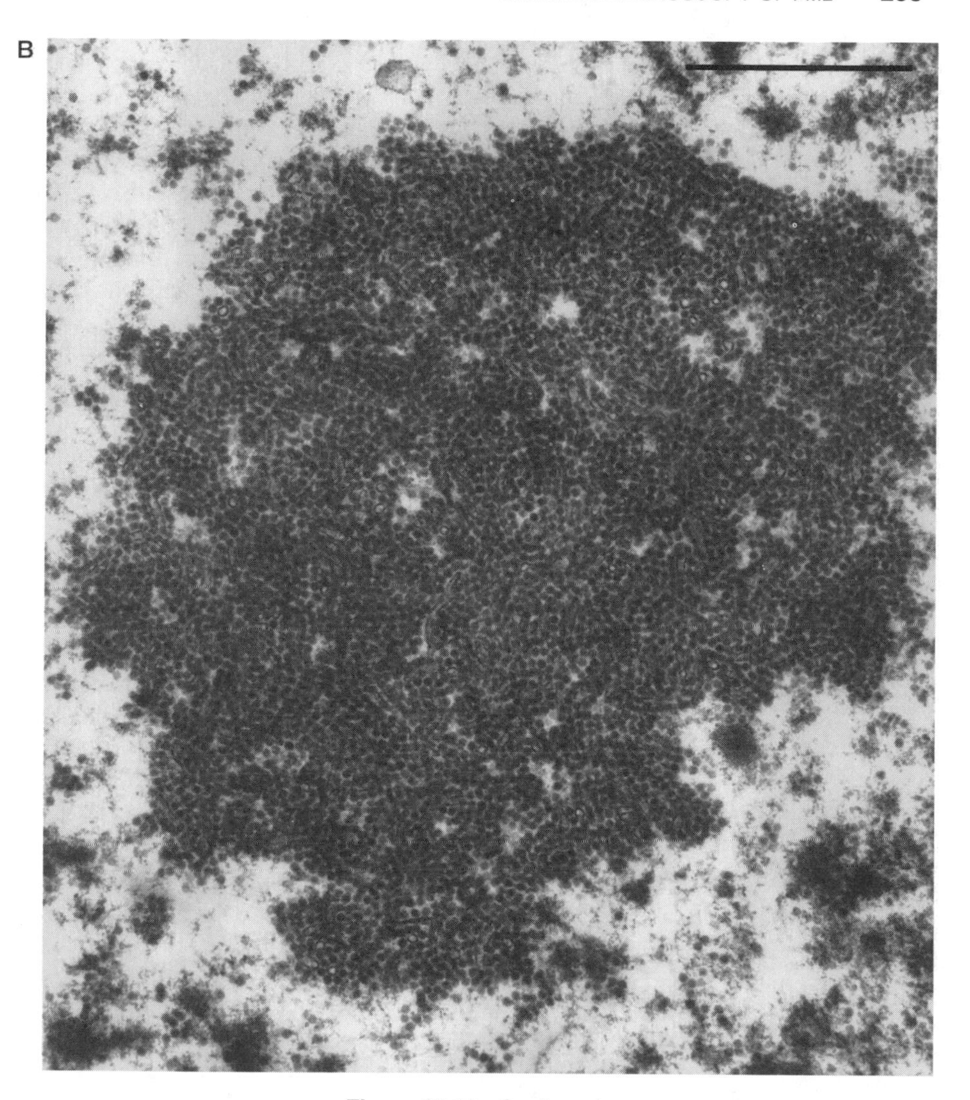

Figure 12.18. *Continued*

not exclude the former. Without seeing the whole cell, especially the nucleoplasm, the question of whether these virus particles are progeny or are infecting virions entering the cell cannot be answered. Thus, the question remains, do the astrocytes have a different mechanism by which to release polyomavirus that does not involve cell lysis? In these astrocytes virus replication might be at such a low level that it can only be detected by extremely sensitive methods such as in situ PCR. These questions need further study.

No evidence of virus replication was found in the nucleus of macrophages by electron microscopy (Silverman and Rubinstein, 1965) or by immunocytochemistry (Mesquita et al., 1992). Phagocytosed oligodendroglial nuclear inclusions in the cytoplasm of a macrophage were first demonstrated by Watanabe and Preskorn (1976) in a brain biopsy sample taken from the middle of an early PML lesion. In the cytoplasm of macrophages, among dense bodies and myelin profiles, vacuoles crowded with polyomavirus particles were also shown (Mesquita et al., 1992). These results are in agreement with the immunocytochemical observations of Stoner et al. (1986) that macrophage nuclei were unstained for JC virion antigens, but reaction product of virion antigens could be observed on the plasma membrane of macrophages. In AIDS-PML cases, the tissue destruction and the macrophage reaction is more pronounced. It is not surprising that numerous JCV particles were found in the necrotic tissue partly attached to myelin debris or within the swollen lamellae of myelin sheaths (Wiley et al., 1988). In the cytoplasm of the macrophages JCV particles together with myelin and cell debris were often found in AIDS-PML cases (Boldorini et al., 1993; Orenstein and Jannotta, 1988; Wiley et al., 1988).

The cell identification in necrotic brain tissue of the AIDS-PML biopsy or autopsy material requires expertise. It also requires knowledge about the biology of cells in the CNS. For example, astrocytes rarely can divide. In the dividing cells in late prophase the chromosomes are close to the equatorial plate, and in this stage of the cell cycle the cell has no nuclear membrane (see Fig. 2 of Scaravilli et al., 1989). Furthermore, the large, "centralized precursor mass" in the nucleus (Fig. 12 of Scaravilli et al., 1989) is entirely like those nuclear bodies demonstrated by Bouteille et al. (1965) in subacute sclerosing leukoencephalitis caused by a defective measles virus. The same type of nuclear body was also found in different tumors as well (Bouteille et al., 1967). Therefore, this "granulo-fibrillary" nuclear body seems to be nonspecific for PML.

Until now, productive infection has never been demonstrated in neurons by electron microscopy (Mázló and Herndon, 1977; Mázló and Tariska, 1982; Silverman and Rubinstein, 1965) or in cultured dorsal root ganglion neurons (Assouline and Major, 1991). The possible infection of cerebellar granule neurons is discussed above. By electron microscopy accumulation of amorphous, dense material was found in the nucleoplasm of the dark granule neurons (Fig. 12.19). In the CNS no endothelial cells contained JCV particles (Mázló and Tariska, 1980; Mázló and Herndon, 1977; Muller and Watanabe, 1967).

Several other viruses are known to infect the CNS in AIDS-PML. First of all, HIV particles were demonstrated by electron microscopy in the cytoplasm of mono- or multinucleated macrophages or budding from them (Budka et al., 1987; Epstein et al., 1985; Koenig et al., 1986; Orenstein and Jannotta, 1988; Rhodes et al., 1988; Wiley et al., 1988). The particles are round to slightly oval double-membraned structures measuring approximately 100 nm in diameter (range 90–130 nm) and contain centrally dense and, depending on the plane of section, round, prismatic, or triangular nucleoids (Budka et al., 1987). Conical and bar-shaped nucleoids measured up to 80 nm in length (Koenig et al., 1986). Particles budding from the surface of the mono- or multinucleated cells

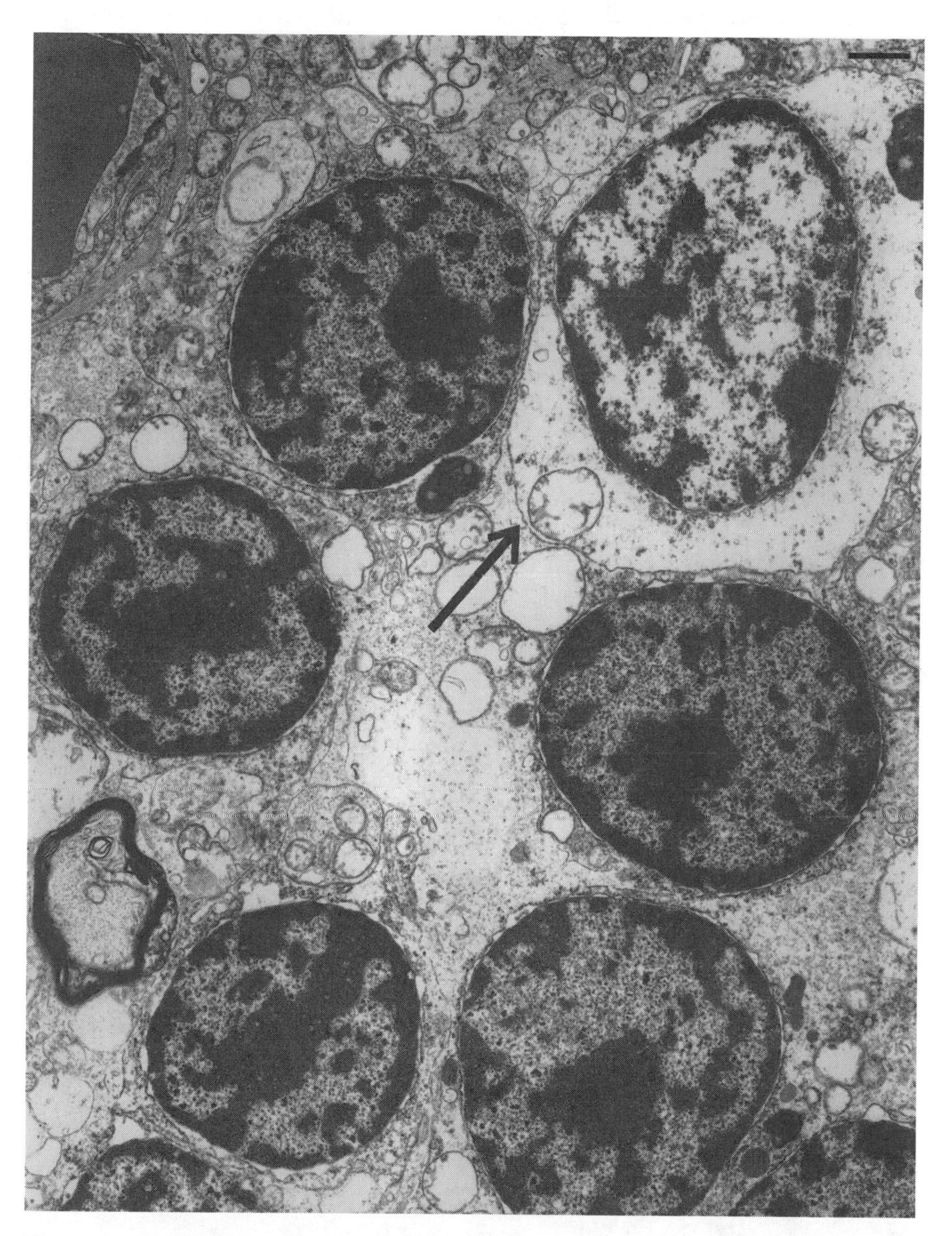

Figure 12.19. Electron microscopic picture of the granule neurons in cerebellar biopsy material. The lightly stained granule neuron (arrow) looks like the granule neurons in autopsy material. In the nucleoplasm of the "dark" granule neurons faintly stained dense material is present among the chromatin clumps. This material has no structure, even at higher magnification. Note that the cytoplasm of the dark granule neurons is better preserved than that of the light one. ×8,100. Bar = 1 μm.

or located extracellularly were also demonstrated (Budka et al., 1987; Koenig et al., 1986).

Reports of HIV particles within the cytoplasm of astrocytes in the CNS should include features to distinguish them from the so-called "Gonatas particles" originally described as virus-like particles in SSPE (Gonatas, 1966; Tellez-Negal and Harter, 1966). These were later retracted as virus particles because they were found in unrelated pathologic conditions (Gonatas et al., 1967; Perier et al., 1967), and their origin remains unidentified. These Gonatas particles were also found in a non-AIDS-PML case in the cytoplasm of reactive astrocytes (Mázló, unpublished data) (Fig. 12.20). The diameter of the Gonatas particles is in about the same range (50–80 nm in diameter, Gonatas, 1966; or 60–110 nm, Tellez-Negal and Harter, 1966) as retrovirus particles. However, the structure of the particles is quite different. Their electron-dense central cores do not show the cone or bar forms. Rather, they are round and fill the "en-

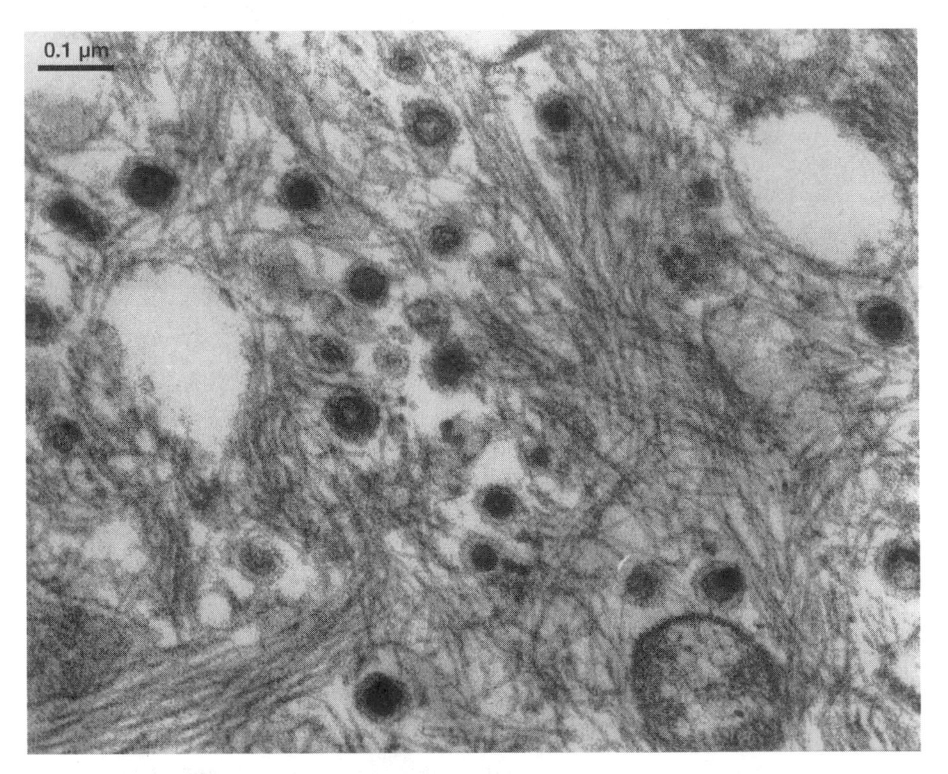

Figure 12.20. "Gonatas particles" among the glial fibrils of a reactive astrocyte. Note the large differences in size and shape among the particles. They measure 50–130 nm in diameter. The electron dense central cores are not uniform, but never show bar-like or cone-like shapes as do the HIV particles. The round cores sometimes have a hollow center. The central core is surrounded by a double unit membrane, but the inner one is sometimes not well defined. ×100,000. Bar = 0.1 μm.

velope." The other difference is that these particles were never found "budding" from the surface of astrocytes or seen in an extracellular location.

CNS infection by herpesviruses, especially cytomegalovirus (CMV), is one of the most common complications of AIDS. Other members of the herpesvirus genus also found are herpes simplex virus-1 (HSV-1), herpes simplex virus-2 (HSV-2), Epstein-Barr virus (EBV), varicella-zoster virus (VZV), and human herpesvirus-6 (HHV-6). It has been suggested, based on immunocytochemical studies, that HHV-6 may be regularly associated with PML, and a role for this virus in the pathogenesis of demyelination has been proposed both in PML and multiple sclerosis based on imunocytochemical and IS-PCR studies (Blumberg et al., 2000; Mock et al., 1999). However, there is no direct electron microscopic evidence on this point. Herpesviruses are rarely demonstrated by electron microscopy in AIDS-PML cases despite the fact that CMV particles were shown by electron microscopy in AIDS-CMV encephalitis (Morgello et al., 1987; Vinters et al., 1989), VZV particles were shown in AIDS-VZV leukoencephalitis (Morgello et al., 1988), and HHV-6 has been demonstrated in a fulminant demyelinating encephalomyelitis (Novoa et al., 1997). It should be noted that the ultrastructure of the members of the herpesvirus genus is similar and application of virus-specific immunocytochemistry, in situ hybridization, or IS-PCR is necessary to differentiate them from each other (see below). Enteroviruses (single-strand positive sense RNA viruses) have never been demonstrated in AIDS or AIDS-PML. Great care and expertise is needed to diagnose a second or third virus, and the possibility of cell components should be excluded. In Figure 6 of Scaravilli et al. (1989), the particles are glycogen rather than enteroviruses.

4. THE PATHOLOGY OF AIDS-ASSOCIATED PML AND OTHER VIRAL CNS DISEASES

Most CNS diseases complicating HIV infection occur in its late phases (Price, 1996). However, PML may occur early or late in the course of the disease and is sometimes the first or only sign of AIDS (Berger and Mucke, 1988; Hair et al., 1992; Jakobsen et al., 1987; Singer et al., 1994; von Einsiedel et al., 1993).

The most frequently occurring cerebral diseases are HIV encephalitis (Budka, 1991), cerebral toxoplasmosis, CMV encephalitis, primary cerebral lymphoma, cryptococcal meningitis, and PML. The incidence of PML in HIV-infected patients varies greatly from study to study, from 8% (Berlin and France) (Table 12.1) to 0.8% (Brazil) (Table 12.2). The incidence of PML has been somewhat higher in Europe (5–8%) than in the United States (2–5%) (Table 12.1). The occurrence of PML in tropical regions falls to 0.8–3% in Africa (Chima et al., 1999; Lucas et al., 1993), Brazil (Chimelli et al., 1992), and Mexico (Gongora-Rivera et al., 2000; Mohar et al., 1992) (Table 12.2). Overall, in the studies cited the incidence in Europe and the United States is 3.5-fold higher than that in the tropics. The reasons for these differences are

Table 12.1. Occurrence of PML in Large AIDS Autopsy Series in Europe and the United States

Region	Study	Total Autopsy Cases	No. PML Cases (%)
New York	Petito et al. (1986)	153	3 (2)
Austria & Italy	Budka et al. (1987)	100	5 (5)
Switzerland	Lang et al. (1989)	135	9 (7)
Bronx, NY	Kure et al. (1991)	221	11 (5)
Berlin	Martinez et al. (1995)	200	16 (8)
Southern France	Hofman et al. (1999)	395	32 (8)
San Diego, CA	Masliah et al. (2000)	390	12 (3)
Vienna	Jellinger et al. (2000)	450	30 (7)
Switzerland	Semela et al. (2000)	314	22 (7)
Total		2358	140 (5.9)

unknown, but may include a lower standard of medical care with shorter survival times during which additional complications such as PML do not have time to develop, as well as differing JCV genotypes in the Hispanic and African populations (see Chapter 18).

Terminology

All authors agree that the multinucleated giant cell is the hallmark of the cerebral HIV infection (Budka, 1986). However, numerous papers and reviews have been published that employ varying terminology (AIDS encephalopathy, subacute encephalitis, subacute encephalitis with multinucleated cells, AIDS dementia complex, AIDS encephalopathy, HIV leukoencephalopathy, progressive diffuse leukoencephalopathy, vacuolar or spongiform leukoencephalopathy, diffuse poliodystrophy) to describe the AIDS-associated structural changes within

Table 12.2. Occurrence of PML in Large AIDS Autopsy Series in Africa and Latin America

Region	Study	Total Autopsy Cases	No. PML Cases (%)
Abidjan, Côte d'Ivoire	Chima et al. (1999), Lucas et al. (1993)	266	4 (1.5)
Brazil	Chimelli et al. (1992)	252	2 (0.8)
Mexico	Mohar et al. (1992)	177	5 (3)
Mexico City	Gongora-Rivera et al. (2000)	149	3 (2)
Combined studies		844	14 (1.7)

the CNS. Identical lesions have often been described and named with different terms, and at other times different lesions have been named with the same term. For example, the term *subacute encephalitis* was used for HIV encephalitis and for CMV encephalitis as well. This lack of standardization has made the comparison of the data from different studies and reviews confusing.

Martinez et al. (1995) in their review of 200 cases used the term *HIV-related encephalopathy* (HIVRE) to designate the entire spectrum of white and gray matter changes consisting of myelin pallor sometimes with spongy, vacuolated appearance and demyelination, whether diffuse, confluent, or patchy, and associated with varying degrees of astrocytosis, microglia proliferation, with or without multinucleated giant cells. HIVRE is thus not a homogeneous disease, but rather appears to show diverse overlapping structural alterations during the clinical course of HIV infection but with a common etiopathogenetic process. For describing PML in AIDS, this term may be too general, because we would like to know which alterations are attributable to PML and which to HIV infection. According to an internationally agreed neuropathology terminology, the HIV-associated CNS diseases are as follows (Budka et al., 1991).

HIV Encephalitis. The multiple disseminated foci are composed of microglia, macrophages, and multinucleated giant cells. If multinucleated giant cells are not found, the presence of HIV antigen or nucleic acids determined by immunocytochemistry or in situ hybridization is required. Sharer et al. (1985) were the first to link the occurrence of multinucleated giant cells to the causal retrovirus of AIDS. The ultrastructural characteristics of the retrovirus particles in the AIDS brain and the resemblance of multinucleated giant cells to macrophages are described by Epstein et al. (1985). Definitive proof of the causal relation between the multinucleated giant cells and HIV encephalitis was provided by Koenig et al. (1986). They found by electron microscopy the budding of retroviral particles from mono- and multinucleated cells whose origin was confirmed by immunocytochemistry and showed production of viral RNA by in situ hybridization. HIV encephalitis is distinct from opportunistic infections and affects the white matter, deep gray matter, and less frequently the cortex (Budka, 1991). Sometimes, central necrosis develops in an encephalitic focus. The multifocal pathology of HIV encephalitis may co-exist or overlap with diffuse white matter damage. The term *HIV encephalitis* includes giant cell encephalitis, multifocal giant cell encephalitis, multinucleated cell encephalitis, multinucleated giant cell encephalopathy, and subacute encephalitis with multinucleated cells.

HIV Leukoencephalopathy. Diffuse damage to white matter includes myelin loss, reactive astrogliosis, the presence of macrophages and multinucleated giant cells, but little or no inflammatory infiltrates. If multinucleated giant cells are not found, demonstration of the presence of HIV antigen or nucleic acids determined by immunocytochemistry or in situ hybridization is required (Budka et al., 1991). HIV leukoencephalopathy includes progressive diffuse leukoencephalopahy (PDL) where myelin staining reveals symmetrical diffuse pallor of varying intensity in the deep cerebral white matter. The U-fibers are usually well preserved. The cerebellar white matter may be affected. Myelin loss and circumscribed areas of demyelination with necrosis also occur.

Vacuolar Myelopathy. The white matter of the spinal cord, predominantly in the dorsolateral spinal tracts, exhibits numerous vacuolar myelin swellings with macrophage infiltration and shows similarity to subacute combined degeneration of the cord (Petito et al., 1985). Secondary axonal degeneration can occur in severely involved cases. The clinical consequence of these alterations is spastic (and ataxic) paralysis. Vacuolar myelopathy is not specific for AIDS and may occur in the absence of HIV infection. Immunocytochemistry and in situ hybridization revealed the presence of HIV in several but not all studies of these lesions of AIDS. In the rare condition called *vacuolar leukoencephalopathy* numerous vacuolar myelin swellings and macrophages are prominent in cerebral white matter, basal ganglia, and brain stem. This alteration may also be present in HIV leukoencephalopathy, but then it does not require an additional diagnosis.

Diffuse poliodystrophy covers reactive astrogliosis and microglial activation involving the cerebral gray matter (Budka, 1991; Budka et al., 1991), especially in the cerebral cortex, basal ganglia, and brain stem nuclei. The pathogenetic relation of lymphocytic meningitis and cerebral vasculitis, including granulomatous angiitis with HIV infection, is not clear (Budka et al., 1991).

Microglial nodular encephalitis is recommended to describe cases with disseminated microglial nodules of unidentified etiology (Budka et al., 1991). This condition, originally named *nonspecific encephalitis with microglial nodules* (Petito et al., 1986), may be distinguished from HIV encephalitis by the different distribution of foci (the cortex being preferentially involved) and by the lack of multinucleated giant cells. Many cases of microglial nodular encephalitis are due to CMV, although cells containing CMV nuclear inclusions may not be present. In other cases with disseminated microglial nodules toxoplasma antigens are detectable (Budka et al., 1987), but there are cases with no detectable opportunistic agent in the nodules.

Differences in AIDS-Associated PML

The first AIDS-associated PML case was described by Miller et al. (1982). AIDS-associated PML is basically the same as described in cases related to other immunosuppressed conditions. However, the pathology is usually more widely distributed, especially in the axis of the brain. Infratentorial lesions could be observed in more than three-fourths of the cases (Kuchelmeister et al., 1993b). In another series cerebellar and brain stem involvement occurred in eight of nine patients (Lang et al., 1989). Parieto-occipital dominance of the lesions may be changed as well. In AIDS-associated PML, fronto-parietal involvement appeared more frequently (von Einsiedel et al., 1993). This fronto-parietal and the upper and lower brain stem involvement are responsible for the change in the order of the first signs and symptoms. The most frequent sign in AIDS-PML cases is motor weakness (Berger et al., 1987). Hemiparesis/plegia is the consequence of white matter lesions not only in the fronto-parietal region but also in the capsula interna, pedunculus cerebri, and sometimes in the medulla oblongata. In two cases PML lesions were also detected in the upper cervical spinal cord (Kuchelmeister ct al., 1993b).

Gray matter involvement in AIDS-PML cases is also increased. Solitary PML lesions in the thalamus diagnosed by MRI and stereotactic brain biopsy and confirmed at autopsy can also occur (Bienfait et al., 1998), as well as thalamic and basal ganglia lesions. The basal ganglia were involved in six of nine patients (Lang et al., 1989). Ledoux et al. (1989) described basal ganglia lesions presenting as movement disorder. As noted above, an unusual clinical presentation with a prominent movement disorder component resulting in a kinetic-rigid syndrome in HIV-related PML was described in a 13-year-old girl (Singer et al., 1993). Exceptionally, widely extended cortical destruction (Sweeney et al., 1994, case 1) as well as thinning of the cerebellar granule layer caused by JCV proved by light microscopy and in situ hybridization (Sweeney et al., 1994, case 2) can also occur, while in the same case immunostaining for other viruses including HIV, bacteria, fungi, or parasites were negative. Thinning of the cerebellar granule cells is not exceptional (Kuchelmeister et al., 1993a; Takahashi et al., 1992; von Einsiedel et al., 1993). Tagliati et al. (1998) have reported cerebellar degeneration associated with HIV infection in ten patients and among them two patients with cerebellar biopsy. The neuropathologic examination of the biopsy sample of patient 1 showed loss of granule neurons and scant perivascular inflammation. There was no evidence of active demyelination or of the presence of CMV, HSV-1 and -2, or toxoplasma infection. In situ hybridization for JCV and Epstein-Barr virus was negative. PCR analysis of the cerebellar tissue samples revealed the presence of JCV DNA. The cerebellar symptoms of patient 1 of Tagliati et al. (1998) may be attributable to JCV infection rather than HIV infection.

At the light microscopic level the mononuclear/macrophage reaction in AIDS-related PML is much more severe than in "classic" PML. Rarely, the other extreme is found: no perivascular infiltration of mononuclear cells occurs (Jakobsen et al., 1987; Ledoux et al., 1989). The origin of the numerous HIV-infected macrophages in the CNS remains unexplained. Resident microglia within the CNS might be infected by HIV that passes through the blood–brain barrier (BBB), but the large number of infected macrophages and their perivascular distribution suggest that these cells arise from systemic monocytes (Wiley and Nelson, 1988).

An unusually prominent inflammatory response was associated with prolonged survival in AIDS-associated PML cases (Berger and Mucke, 1988). Patients with heavy cellular infiltration of PML lesions, whether HIV positive or not, often show stabilization of symptoms and have longer survival times than patients without inflammatory infiltrates (Hair et al., 1992).

The distribution of both T antigen and capsid protein was much greater than in the cases of non-AIDS PML. The presence of widespread oligodendroglial infection not confined to demyelinated lesions suggests that PML in AIDS may represent a more diffuse infectious process than usually seen in "classic" PML. In different studies the JCV immunoreactive cells were seen not only within and at the periphery of demyelinating lesions but also some distance away in the normal-appearing white matter (Lang et al., 1989; Stoner et al., 1986). Burnt-out lesions harbored JCV DNA but not virus capsid antigens (Schmidbauer et al., 1990).

In the two patients reported by Vazeux et al. (1990), both JCV- and HIV-infected areas were closely intermingled. Demyelinated areas induced by JCV were massively invaded by enlarged HIV microglia/macrophage cells. These investigators were unable to demonstrate the replication of both viruses in the same cell by double detection of HIV and JCV antigens or by the expression of JCV RNA in cells replicating HIV. Using immunocytochemistry and in situ hybridization, Vazeux et al. (1990) demonstrated that each virus infects, in a latent or productive fashion, different CNS cell populations. Thus, the coexistence of PML lesions and extensive HIV encephalitis seemed unlikely to be explained by an intracellular transactivation of one virus by the other. However, as the Tat protein is released from HIV-infected cells to act on cells in the vicinity lacking retroviral infection, double infection of the same cell by JCV and HIV is not required in order for the latter to transactivate the former through Tat action.

CNS Pathology of AIDS in Children

The most common histopathologic change in the brain of infants and children with AIDS is *calcification of blood vessels* of all calibers in the basal ganglia and frontal white matter (Dickson et al., 1989; Kure et al., 1991). Progressive calcification in most cases was associated with *brain atrophy*. *HIV encephalitis* with multinucleated giant cells is a frequent neuropathologic finding, mostly in the deep cerebral white matter and in the basal ganglia, thalamus, and brain stem (Dickson et al., 1989). The cerebral white matter in children was often poorly myelinated due to either diffuse lack of myelin or mutifocal areas of myelin loss associated with inflammation. Myelin was also decreased in the corticospinal tract of the spinal cord and the frontobulbar pathways in the cerebral peduncle (Dickson et al., 1989; Kure et al., 1991). Opportunistic infections are rare in infants and children. In a series of 11 cases recognizable opportunistic infection (CMV encephalitis) was limited to one patient (Sharer et al., 1986), and in a series of 26 cases (Dickson et al., 1989) and another of 31 cases (Kure et al., 1991) no toxoplasmosis or PML was found and only two of CMV encephalitis occurred in each series. Despite seroepidemiologic evidence indicating a high prevalence of antibodies to JCV in adolescents (Walker and Padgett, 1983), PML is rare in children with AIDS or any other form of immunodeficiency. In a series of 79 cases of PML with confirmed JCV infection before the AIDS era, PML was observed only in one 5-year-old girl and an 11-year-old boy with immunodeficiency disorders (Brooks and Walker, 1984). The organ distribution and characterization of the virus strain found in the 5-year-old girl (termed Mad-6) has been described (Newman and Frisque, 1997). A 7-year-old boy with congenitally aquired HIV infection had PML confirmed by biopsy using light and electron microscopy and in situ hybridization (Vandersteenhoven et al., 1992).

Berger et al. (1992) reported two HIV-infected children with PML, a 13-year-old girl presumed to be congenitally infected with HIV (confirmed by brain biopsy) and a 10-year-old boy who developed HIV infection from blood transfusion at the age of 3 years. The diagnosis was confirmed at autopsy. Akinetic-rigid syndrome in a 13-year-old girl with HIV-associated PML was reported by Singer et al. (1993). In a large series of 65 pediatric AIDS autopsy cases (Wrzolek et al., 1995), opportunistic infection of the CNS was found in nine cases that included four cases with CMV infection. Only one 12-year-old boy with perinatal HIV infection had PML that coexisted with mycotic encephalitis (*Aspergillus* sp.). PML lesions were found in the cerebellum and in the brain stem (Wrzolek et al., 1995). We conclude that in children with AIDS only isolated cases of PML occur. Interestingly, PML has never been described in infants. Because infants are likely exposed perinatally to JCV excreted frequently by the mother and other family members (see Chapter 18), we presume that the immature brain is not a hospitable environment for viral replication. This is paradoxical in that the virus is best cultured in primary human fetal glial cells.

Other AIDS-Related Infectious CNS Diseases

In many autopsy series CMV encephalomyelitis was the most common AIDS-related opportunistic infection of the CNS, while in other geographical areas it was toxoplasmosis. The pathologic features of CMV encephalitis are characterized by the formation of microglial nodules (Dorfman, 1973) with predominance in the diencephalon and brain stem. The nodules sometimes contain CMV inclusions. Isolated inclusion-bearing cells (neurons, astrocytes, and capillary endothelial cells) unaccompanied by microglial nodules or infiltrating cells were seen in many cases. Focal parenchymal necrosis, ventriculo-encephalitis, and radiculo-myelitis were less frequent. (Morgello et al., 1987). CMV inclusions in ependymal and subependymal cells are also frequent and lead to ulceration of the ependymal lining. Hemorrhagic necrosis can occur in the periventricular area as well as around the aqueductus cerebri and the fourth ventricle (Vinters et al., 1989). In Wiley and Nelson's study (1988), evidence of HIV infection was seen in brains from 37 of the 93 autopsied AIDS patients (40%) detected by immunocytochemistry, while 31 of the 93 brain specimens (33%) contained CMV antigens. In 22 cases the CMV and HIV antigens were found in the same brain. In vitro study showed bidirectional interaction between

HIV-1 and CMV. Co-infection of H9 cells enhances both CMV and HIV-1 productive infection (Skolnik et al., 1988).

Productive varicella-zoster virus (VZV) infection of the CNS was demonstrated in 11 AIDS patients by immunocytochemistry and in situ hybridization (Gray et al., 1994). From this series and 11 other cases appearing in the literature, Gray et al. (1994) identified five pathologic patterns of VZV infection associated with AIDS: (1) Multifocal encephalitis predominantly involving the white matter, (2) ventriculitis with marked vasculitis and necrosis of the ventricular wall, (3) acute hemorrhagic myelo-radiculitis, (4) focal necrotizing myelitis, and (5) vasculopathy involving leptomeningeal arteries and causing cerebral infarcts. This study showed that VZV infection of the CNS should be taken into consideration in the neuropathology of AIDS. Cerebral vasculopathy such as necrotizing arteritis of large vessels and granulomatous angiitis of small arteries with multifocal leukoencephalitis, resembling PML pathologically, had been described earlier (Morgello et al., 1988). In the centrally necrotic demyelinated lesions associated with VZV arteritis, inclusions in oligodendrocytes, astrocytes, and neurons as well as occasional endothelial and vessel wall cells of thickened arteries stained positively for VZV.

It is surprising that other members of the herpesvirus genus are much less frequently associated with HIV infection. Herpes simplex virus (HSV) infection of the CNS is uncommon in AIDS. No evidence of HSV infection was seen in autopsied brains of the 93 AIDS patients investigated by immunocytochemical methods (Wiley and Nelson, 1988). Chrétien et al. (1996) have described an AIDS-associated characteristic HSV-1 encephalitis with limbic distribution in which HSV-1 was demonstrated by electron microscopy, in situ hybridization, and PCR. They reviewed 10 pathologically documented cases appearing in the AIDS literature. In most cases the topography of the lesions was atypical, and the classic limbic encephalitis was observed in only three cases. Nine cases were caused by HSV-1 and two by HSV-2. These two cases showed spinal cord involvement. Most of the cases were associated with CMV encephalitis strictly limited to the periventricular region with numerous cytomegalic cells containing the large "owl's eye" intranuclear inclusion bodies that stained positively for CMV. Productive HIV infection of the CNS was not found in any of these cases. Cinque et al. (1998) tested 918 CSF specimens from HIV-positive patients with neurologic symptoms using a duplex PCR for HSV-1 and HSV-2 DNA. Nineteen (2%) were positive, and all seven of these examined at autopsy had HSV-1 or HSV-2 antigen and DNA in CNS tissue. Again, as in previous studies (Vago et al., 1996), most of the HSV-1 and -2 infections were found in patients with CMV encephalitis.

With the help of different morphologic and molecular biological techniques stereotactic brain biopsy appears to be an effective method in the diagnosis of HIV-associated brain lesions. It must be stressed that biopsy results can have therapeutic consequences, especially for CNS lymphoma patients (Zimmer et al., 1992). Primary central nervous system lymphoma (PCNSL) is a frequent complication in AIDS and can cause problems in the differential diagnosis. PCNSL in AIDS may be generally associated with Epstein-Barr virus (EBV) infection (Clifford, 1999). EBV DNA was detected by PCR in the CSF of 5 of 6 patients with pathologically proven PCNSL and only 1 of 16 controls (Arribas et al., 1995). In another study all 17 patients with primary CNS lymphoma had EBV DNA in the CSF, while EBV DNA was found in the CSF of only 1 of 68 HIV-positive control patients (Cinque et al., 1993). PCR for EBV in the CSF may prove to be useful for diagnosing of AIDS-related PCNSL. CNS infiltration of EBV-infected B lymphocytes in a chronic lymphocytic leukemia patient without HIV but with PML has been reported (Farge et al., 1994). Frontal stereotactic biopsy confirmed the diagnosis of PML by immunocytochemistry, in situ hybridization, and electron microscopy. Leukemic B cells were disseminated in the demyelinated areas. They were labeled for EBV by anti-latent membrane protein and by *Bam*HI W EBV probe after in situ hybridization.

A new member of the herpesvirus genus, human herpesvirus-6 (HHV-6), is the causative agent of the exanthema subitum and febrile illness frequently associated with seizures in children. The HHV-6 B variant has been detected by immunocytochemistry and PCR in the brain of a bone marrow transplant recipient patient with encephalitis (Drobyski et al., 1994) and in an HIV-infected infant (Knox et al., 1995). As noted above, multifocal demyelinating lesions in an adult without overt immunosuppression (Novoa et al., 1997) and demyelinative disease in four of six AIDS patients were closely associated with active HHV-6 infection, with infected cells present only in areas of demyelination. Because other CNS pathogens could be excluded by immunocytochemistry (HIV-1 p24, JCV, CMV, HSV, VZV and measles virus) and EBV by in situ hybridization, the HHV-6–induced white matter disease appeared to be a distinct pathologic syndrome (Knox and Carrigan, 1995).

Using two-step IS-PCR to amplify and detect genomic HHV-6 in formalin-fixed, paraffin-embedded brain tissue, HHV-6–infected cells were frequently detected in PML white matter both within and surrounding demyelinated lesions. The HHV-6 genome was found mainly in oligodendrocytes. Immunocytochemistry for HHV-6 antigens showed actively infected nuclei of swollen cells with oligodendroglial morphology only within the demyelinated lesions of PML. No white matter staining for HHV-6 antigens was seen in either control brains or in brains with HIV-1 encephalopathy but without PML. Double immunocytochemical staining for JCV large T antigen and HHV-6 antigens demonstrated co-labeling of many swollen oligodendrocytes in PML lesions. The evidence suggests that HHV-6 activation in conjunction with JCV infection may be associated with demyelinative lesions of PML (Mock et al., 1999). In a cortical biopsy specimen from an AIDS-related PML patient with meningoencephalitis, HHV-6 genome and antigens were detected in the meningeal and cortical lesions in lymphocytes and microglia as well as in some satellite oligodendrocytes and in rare dead or dying neurons (Ito et al., 2000).

Double or triple infection in AIDS-associated PML is not rare. In many cases HIV and JCV double infection were demonstrated (Kleihues et al., 1985; Wiley et al., 1988, two cases; Rhodes, 1993, six cases). Double infection with JCV and mycotic infection (Wrzolek et al., 1995) and JCV and HHV-6 (Ito et al.,

2000) have also been found. Triple infection with HIV, JCV, and CMV has been reported (Kleihues et al., 1985; Wiley and Nelson, 1988), and triple infection with HIV, JCV, and *Histoplasma capsulatum* meningoencephalitis was shown (Weidenheim et al., 1992). Gray et al. (1987) published a rare AIDS-PML case with HIV, JCV, CMV, and toxoplasma infection in the same brain.

Cerebral toxoplasmosis shows three different patterns. The first is widespread confluent areas of necrosis accompanied by a weak inflammatory response, many times with thrombosis of vessels and abundant tachyzoites at the periphery, and is called *necrotizing toxoplasma encephalitis* (Lang et al., 1989). The second is toxoplasma abscesses that are preferentially located at the border of the gray/white matter and in the basal ganglia. In 27 of 35 toxoplasma-infected patients the basal ganglia were affected (Lang et al., 1989). There is necrosis in the center of the abscess, which is surrounded by a rim of macrophages, lymphocytes, and sometimes polymorphonuclear leukocytes as well. Intra- or extracellular tachyzoites as well as bradyzoites can be observed at the periphery of the abscesses. Microglial nodular encephalitis is the third morphologic manifestation of cerebral toxoplasmosis. It presents as multiple microglial nodules with scattered encysted bradyzoites and a few tachyzoites within or in the vicinity of the microglial nodules.

Toxoplasma abscess is the most common cause of AIDS-related chorea (Navia et al., 1986, case 7; Nath et al., 1987, case 6; Piccolo et al., 1999, case 4). Chorea is a movement disorder characterized by involuntary hyperkinetic movements. However, the causes are variable, and careful evaluation is required to identify them as the etiologies have therapeutic consequences. JCV infection (Piccolo et al., 1999, case 2) and other AIDS-related histopathologic tissue destructions can also be responsible for the basal ganglia symptoms. Other AIDS-related viral encephalitis or bacterial infections such as the rare Whipple's disease may also result in movement disorders (Nath et al., 1987, case 1). Moreover, HIV encephalitis itself can be responsible for basal ganglia symptoms (Gallo et al., 1996).

CNS infection with *Cryptococcus neoformans* usually causes diffuse leptomeningitis, but sometimes meningoencephalitis with granulomas or intracerebral brain abscesses. Aspergillus and candida infections were occasionally found. Bacterial infections can also occur, albeit rarely, causing multiple microabscesses as a result of septicemia, and *Mycobacterium tuberculosis* and *Mycobacterium avium-intracellulare* infections can cause meningitis, tuberculomas, and abscesses.

5. ANIMAL MODELS

Hamster

The discovery and isolation of JCV from the brains of PML patients naturally led to efforts to establish an animal model susceptible to JCV brain infection and exhibiting pathology paralleling the human disease. Instead, the first report of JCV-induced brain pathology in an experimental animal yielded an unexpected discovery by Walker and co-workers (1973): JCV was oncogenic in the hamster. Brain tumors were induced in the golden hamster 3–6 months after intracerebral or subcutaneous inoculation of the Mad-1 strain of JCV into newborns. The findings indicated that JCV induced glioblastomas, medulloblastomas, ependymomas, and other primitive neuroectodermal tumors (PNETs) in hamsters. The induction of tumors rather than demyelination was not entirely unanticipated by these researchers because JCV belonged to the papovavirus subgroup, which included SV40, a virus with known oncogenic properties, and because there were earlier observations that the giant astrocytes in PML brain resembled malignant astrocytes seen in glioblastomas (Åström et al., 1958; Zu Rhein, 1969). While later studies reported the induction of brain tumors in New World primates (see below) and rats (Ohsumi et al., 1986), the hamster became the most reliable and well-described animal model, with JCV consistently inducing a variety of neoplasms. In addition to the first reported tumor types, intracranially injected JCV also induced formation of pineocytomas, meningiomas, neuroblastomas, and other central nervous system tumors (Padgett et al., 1977; Zu Rhein, 1983; Zu Rhein, 1987; Zu Rhein and Varakis, 1979). Specific strains of JCV exhibited different oncogenic capacities, with the prototype Mad-1 strain inducing predominantly medulloblastomas and the Mad-4 variant inducing primarily pineocytomas (Padgett et al., 1977). In addition, when injected intraocularly into newborn hamsters, JCV induced formation of abdominal adrenal neuroblastomas (Ohashi et al., 1978; Varakis et al., 1978).

In the developing hamster brain it was evident that JCV exhibited tropism for undifferentiated and proliferating cell populations, infecting and transforming cells rather than causing a productive lytic infection with subsequent demyelination as seen in PML. However, the characteristic pathology of PML was considered to result from reactivation of latent JCV acquired during initial infection in childhood (Walker et al., 1973). The range of human host cells susceptible to infection by JCV was, and still is, an incomplete list, and the hamster model provided focus on JCV cellular tropism and its multiple pathologic potential. The pathology of JCV-induced hamster neoplasia had been well characterized, but the specific cell populations it targeted for initial infection were extrapolated from tumor morphology. While JCV produced a variety of brain neoplasms in the hamster, the medulloblastoma was the most frequently induced tumor (Zu Rhein and Varakis, 1979). This was significant because medulloblastomas are also one of the most common human childhood tumors (Kadin et al., 1970; Rorke, 1983; Rorke, 1994; Russell and Rubinstein, 1989).

It was postulated that the cells of origin for the hamster medulloblastoma arose in the cerebellar external granular layer (Zu Rhein and Varakis, 1979), and the development of sensitive immunostaining methods to detect JCV proteins permitted confirmation of this theory (Ressetar et al., 1990).

The SV40 and JCV genomes exhibit extensive sequence homology, and some monoclonal antibodies to SV40 have been found to cross react with JCV proteins (Babé et al., 1989; Ball et al., 1984). JCV large T antigen is the early region regulatory protein synthesized prior to, and required for, viral DNA replication (Frisque et al., 1984). In PML, T antigen is the first viral protein expressed in infected cells (Stoner et al., 1986, 1988d), and it is also expressed in cultured hamster glioblastoma cells (Frisque et al., 1980). Using immunostaining methods to detect JCV T antigen, Ressetar et al. (1990) examined the progression of JCV infection in the neonatal hamster brain. These studies revealed that JCV T antigen could be detected in the nuclei of infected brain cells as early as 3 days after intracranial inoculation of newborns. T-antigen-expressing cells were observed in the subventricular zone as well as in areas of postnatal granule neuron production, including the olfactory bulb, hippocampus, and cerebellar external granular layer. In the cerebellum, JCV-infected cells could be sequentially traced through the molecular layer to the internal granular layer. By 30 days after inoculation, JCV-infected cells appeared as more focal clusters at the cerebellar internal graular layer–white matter junction, a prominent site of early medulloblastoma neoplasia. This report, as well as the in situ hybridization studies by Matsuda et al. (1987), confirmed that in the hamster cerebellar medulloblastomas arose from JCV-infected immature granule neurons. JCV infection of granule neurons in PML had been suggested by morphologic granular layer alterations observed in some PML cases (see above), and the hamster studies supported the idea that granule neurons could also serve as host cells to JCV infection in humans.

The hamster model continued to yield information concerning the target cell populations for JCV infection. Improved immunostaining methods permitted detection of JCV T antigen in hamster brain vascular endothelial cells 8–31 days after intracranial inoculation of newborns (Ressetar et al., 1992). Cyclophosphamide-induced immunosuppression of hamsters caused an increase in the number of JCV T-antigen-expressing endothelial cells that was most evident at the cerebeller internal granular layer–white matter junction, a predominant site of focal neoplasia. Blood vessels were intact and T-antigen-expressing endothclial cells appeared histologically normal and expressed the endothelial cell markers for lectin and von Willebrand factor. Immunosuppression did not induce a productive JCV infection in hamsters, and T antigen expression in nonvascular cells in suppressed and nonsuppressed hamsters exhibited similar patterns. Yet the findings of apparent immune-modulated JCV expression in vascular endothelial cells suggested that these cells could play a role in virus distribution and antigen presentation and might ultimately serve as a proliferative host cell population for the persistent virus. A discrepancy between the hamster model and human disease was that JCV infection of endothelial cells has not been widely observed in PML. However, detection of JCV antigens in host cells may be dependent on a variety of factors, including interactions with host cell-cycle regulators (see Chapter 6) that complex with viral proteins and might prevent their reaction with antibodies used in immunostaining procedures. Furthermore, the virus could pass through endothelial cells early in infection without later expression there. In the neonatal hamster, T antigen expression was most pronounced in discrete proliferating cell populations in the external granular layer, the hippocampus, and the olfactory bulb during early infection (Ressetar et al., 1990). Expression of T antigen in brain tumors of 6-month-old hamsters was variable, with expression absent or low in the center of large and well-differentiated tumors and high in the proliferating cells at the margins of tumors that were infiltrating adjacent normal tissue (H.G. Ressetar, unpublished results). Considering these variables, the absence of JCV protein expression in vascular endothelial cells in PML brains does not dismiss the possibility of a nonproductive or semipermissive infection of these cells at some stage of the disease.

The use of polymerase chain reaction (PCR) methods permitted a new level of sensitivity for detection of JCV in infected hamster tissues and addressed the question of viral latency and persistence following initial infection. Using PCR methods, JCV DNA could be detected at high levels in the brains of all 6-month-old hamsters that were intracranially inoculated with JCV as newborns (Ressetar et al., 1997). JCV DNA and T antigen mRNAs were isolated from the brains of JCV-injected hamsters that did not exhibit neoplasia, indicating that latent JCV could persist in brain cells without inducing obvious pathology. Though pathology of nonbrain tissues was not observed in any JCV-infected hamster, JCV DNA was also detected in the urinary bladder, kidney, adrenal gland, and pancreas of most animals. This suggested that intracranially inoculated JCV can enter the systemic circulation and specifically target certain peripheral organs that remain latently infected. While T-antigen-expressing vascular endothelial cells were observed in the brains of hamsters exhibiting neoplasia, they were not observed in any other organs. JCV DNA was either not detected or detected in low incidence in highly vascular organs such as the liver, spleen or reproductive organs. Therefore, vascular endothelial cells or blood do not appear to be the source of JCV DNA in organ samples. Evidence strongly suggests that the kidney serves as a site for JCV latency in humans (Agostini et al., 1996; Kitamura et al., 1990) and the findings of JCV DNA in the kidneys of almost all JCV-infected hamsters indicates that this organ may also harbor JCV in the hamster.

Transgenic Mice

The development of methods to generate transgenic animals permitted more sophisticated manipulation of the polyomavirus genome in the study of demyelination and tumor induction. The early region of the polyomavirus genome consists of the T-antigen-coding region and the regulatory or promoter/enhancer sequences. While the JCV, SV40, and BKV genomes exhibit extensive sequence homology, significant differences in their promoter sequences appeared to affect the host cell specificity of these viruses (Frisque, 1983; Gruss and Khoury, 1981; Rosenthal et al., 1983).

By introducing the entire JCV early region into transgenic mice, two lines of mice were generated that exhibited contrasting pathologies. Founder transgenic mice developed adrenal neuroblastomas that metastasized to other organs (Small et al., 1986a), while the second generation mice and their offspring exhibited central nervous system dysmyelination (Small et al., 1986b; Trapp et al., 1988). These studies demonstrated that the JCV early region sequences, in the absence of the late-coding genes, can direct both neoplastic transformtion and dysmyelination.

Although genetically similar, JCV and SV40 each induces differing pathologies in experimental animals that are specific to the virus (Messing et al., 1988; Palmiter et al., 1985). Later studies in transgenic mice examined the influence of the polyomavirus promoter region versus the T-antigen-coding region on the host cell specificity of these viruses. By exchanging the promoter and T antigen-coding regions between JCV and SV40, chimeric genomes were created and introduced into transgenic mice (Feigenbaum et al., 1992; Ressetar et al., 1993). These transgenic mice displayed a variety of tissue-specific pathologies that could be attributed to either the JCV or the SV40 gene sequences. It was found that the regulatory region sequences of the viruses exerted control over the T-antigen-coding sequences in determining neural and thymic pathology. The SV40 promoter directed JCV T antigen to induce choroid plexus papillomas (Fig. 12.21) and thymomas characteristic of SV40, while the JCV promoter directed the SV40 T antigen to induce adrenal neuroblastomas and contrasting thymic hypoplasia characteristic of JCV. Pathology observed in other tissues appeared to be less influenced by promoter control. SV40 characteristic lymphomas developed in mice even under the control of the JCV promoter. The cells in pathologic tissues also expressed p53, indicating that the interaction of this host cell tumor suppressor with T antigen may influence oncogenicity (Ressetar et al., 1993) by mechanisms well developed in studies of SV40, but for which many questions remain.

These early transgenic studies provided a better understanding of the JCV genomic influence over cellular tropism and induced host cell pathology. For a summary of recent studies showing medulloblastoma induction in transgenic mice bearing the archetypal JCV strain known as CY, see Chapter 15.

Primates and Other Models

While the hamster has proved to be the most productive and well-described animal model with JCV consistently inducing a variety of neoplasms, several other models have been tested. The rat also develops primitive neuroectodermal tumors (PNETs) (Ohsumi et al., 1986).

Two primate models of polyomavirus infection have been developed. One is a model for JCV induction of brain tumors in New World monkeys (owl and squirrel monkeys). The other is a natural model of polyomavirus infection in rhesus monkeys (macaques).

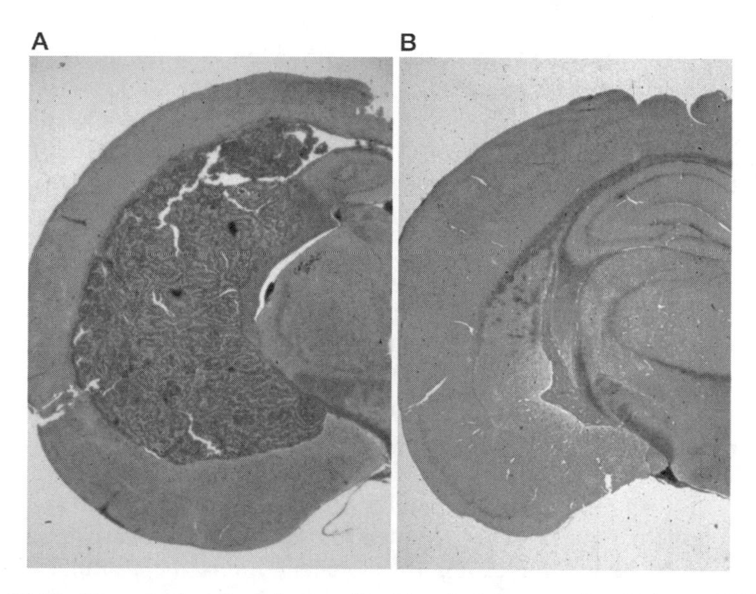

Figure 12.21. Transgenic mouse brain with chimeric transgene in which the JCV T antigen gene is under the control of an SV40 regulatory region. LFB/H&E. Low magnification. (**A**) A well-demarcated choroid plexus papilloma characteristic of SV40 CNS infection filled and distended both lateral ventricles. Many cells in frozen sections could be immunostained for JCV T antigen, but not with antibodies specific for SV40 T antigen (Ressetar et al., 1993). The mouse died at 41 days of age. Frontal section, half brain. (**B**) A control mouse littermate showing morphology of normal choroid plexus and lateral ventricle in a comparable frontal section.

New World Monkeys. Two New World monkeys, the owl monkey and the squirrel monkey, have been utilized to demonstrate the potential of JCV to induce tumors in primates. These models remain the only primate models of brain tumor induction by a neurotropic virus, although SV40 has been associated with a malignant astrocytoma and oligodendroglioma in the SIV-inoculated macaque (see below). The owl monkey (*Aotus trivirgatus*), named for its large eyes and nocturnal habits, which inhabits Amazon rain forests, is one of the few monogamous mammals. The squirrel monkey (*Saimiri sciureus*), another small monkey weighing about 2 pounds (less for females), also inhabits South American forests.

Initial studies reported that intracerebrally injected JCV (Mad-1) induced astrocytomas at 16 months and 25 months in two of four owl monkeys observed for 3 years (London et al., 1978). Uninoculated control monkeys and those similarly injected with BKV or SV40 remained clinically normal during the observation period. JCV DNA was found to be integrated into the cellular genome of all astrocytomas tested by hybridization (Miller et al., 1984). Integration occurred at multiple sites and with multiple tandem copies of the genome in all but one of the tumor tissues. Uninvolved brain tissue was not infected.

Further studies revealed that cerebral astrocytomas were induced by JCV (Mad-1) in four of six inoculated squirrel monkeys at the site of injection with an incubation period of 1–3 years (London et al., 1983). Three of the tumors corresponded to a malignant grade 4 astrocytoma in humans. Multinucleated giant cells were present in all four astrocytomas. Tumors did not occur in control animals or in those inoculated with BKV or SV40 and observed for 3 years.

Until recently, experimental studies of JCV oncogenicity in primates have been limited to the Mad-1 strain, the neurotropic isolate with a rearranged regulatory region obtained from human PML brain. Whether JCV strains with the archetypal regulatory region as excreted in the urine would show similar properties is currently unknown. If, in fact, archetypal JCV is demonstrated to be a co-factor in the induction of brain tumors such as the childhood PNETs (medulloblastomas) in humans (see Chapter 15), the rationale for further application of a primate model to studies of pathogenesis as well as immunotherapy, chemotherapy, or other treatments is strengthened. However, at present the long latent period for tumor development limits the usefulness of this model.

Rhesus Monkeys Naturally Infected With SV40. This is the only animal model that resembles the productive infection seen in human PML, but it is not readily amenable to experimental manipulation. In this model rhesus monkeys naturally or experimentally immunosuppressed by simian immunodeficiency virus (SIV) infection or infection with the chimeric simian human immunodeficiency virus (SHIV) occasionally develop an infection of the brain with SV40 that is a productive glial cell infection. This SV40 infection is a result of the wild-type virus that circulates in primate colonies and that infects macaques naturally. SV40 CNS disease is reactivated in the context of the immunosuppressive retroviral infection, analogous to the occurrence of PML in HIV/AIDS in humans. As in humans, the disease develops as a subacute and fatal glial cell infection. However, unlike the human disease two neuropathologic presentations have been described. One form consisted of a meningoencephalitis (Simon et al., 1999) rather than a leukoencephalopathy (Chrétien et al., 2000; Horvath et al., 1992).

The basis for this unknown, but one can speculate that the former (meningoencephalitis) may follow a primary brain infection occurring during immunosuppression, and the latter might be the result of latency/reactivation analogous to the pathogenesis of human PML. Macaques seropositive for SV40 before SIV infection developed typical leukoencephalopathy, whereas a juvenile monkey that seroconverted to SV40 6 months after SIV inoculation developed a severe bilateral tubulointerstitial nephritis (Horvath et al., 1992). Like tubulointerstitial nephritis, the meningo-encephalitis appears to be a manifestation of a primary SV40 infection. The SV40 DNA sequence in the brains of five SIV-inoculated macaques with SV40-induced lesions could be distinguished from reference strain 776 (Newman et al., 1998).

The possibility exists that SV40 causes not only PML but also brain tumors in its natural host. Inoculation of SIV into the pigtail macaque (*Macaca nemestrina*) resulted in a malignant astrocytoma in the right frontal lobe that contained SV40 DNA (Hurley et al., 1997). The animal was seropositive for SV40 at the time of SIV inoculation, suggesting reactivation of SV40 rather than primary infection. An oligodendroglioma has recently been reported in an SIV-inoculated macaque with PML due to SV40 (Chrétien et al., 2000).

As the macaque model of PML depends on a natural SV40 infection and has a low incidence of CNS disease, its widespread use for experimental studies is precluded. Nevertheless, it is interesting to contemplate that the first report of a PML-like disease due to SV40 in macaques was in 1975 (Gribble et al., 1975; Holmberg et al., 1977), 6 years before the onset of what we now know as the AIDS epidemic. If this outbreak of unusual opportunistic infections in monkeys had been further investigated and explained, the scientific community might have had a head start in the fight to understand and control AIDS in humans. Unfortunately, lacking this foreknowledge, granting agencies were not impressed with the potential of this research and declined at that time to fund it further. This model has now become a useful experimental tool for studying neuro-AIDS, representing as it does the simian homologue of AIDS, including the simian equivalent of AIDS/PML (Petry and Luke, 1997).

Implications of Animal Models

JCV, first known as the agent of a fatal demyelinating disease, has not yet provided a useful infectious experimental model for studying the mechanisms of demyelination. Numerous other viruses provide experimental models delineating the potential mechanisms of demyelination (Stohlman and Hinton, 2001). Several mechanisms of virally induced demyelination can be deduced from these models. The first and most obvious is the death of the acutely infected oligodendrocyte, the myelin-forming cell in the CNS. Second, persistent or abortive infection without infectious viral progeny could fatally alter the metabolism of the cell or render it incapable of maintaining its specialized function of myelin support. In the case of JCV, this condition could occur in the oligodendrocytes expressing T antigen, but not capsid proteins. Third, inflammatory demyelination in which immune mediators are released in response to the virus in white matter could damage the myelin-forming cell. Finally, the inflammatory response, though triggered by a virus, might be directed toward cellular components (autoimmune).

Although JCV provides no experimental model for studies of demyelination (nor do BKV or SV40, except possibly in the natural brain infection by the latter), two clear lessons have been reinforced by several decades of study of JCV in animal models. First is the potential for oncogenesis by this virus, as in the case of the other human polyomaviruses, BKV and SV40. Second, from the transgenic mouse models of second generation (F2) mice, the capability of JCV T antigen expression to interfere with myelination is evident. Molecular correlates of this T antigen effect have been proposed (Devireddy et al., 1996; Tretiakova et al., 1999a,b).

At the same time, JCV T antigen may be able to stimulate the transcription of glial factors capable of promoting its own reactivation (Renner et al., 1996).

6. JCV AND HUMAN TUMORS

The role of polyomaviruses in human tumors, particularly the primitive neuroectodermal tumor (PNET) of childhood, the medulloblastoma, is considered in detail elsewhere (see Chapter 15). Here we merely note that several cases of PML have been associated with glial malignancies (Castaigne et al., 1974; Sima et al., 1983). Furthermore, a case of oligoastrocytoma in an immunocompetent patient has been associated with JCV infection in the absence of PML (Rencic et al., 1996), although other searches for JCV DNA in astrocytomas have been negative (Bogdanovic et al., 1995). Thus, the significance of the coexistence of PML and astrocytomas or gliomas remains uncertain.

7. PATHOGENESIS OF PML

There are five major aspects of JCV infection and PML pathogenesis that remain unsolved. The first concerns the details of virus transmission, including the age, route, and tissue target of the infection. The second concerns the way in which the virus regulatory region rearranges into its neurotropic and neurovirulent forms. The third is when and how the brain becomes infected. The fourth concerns the mechanisms by which the virus reactivates and induces demyelination during active disease. While we assume that a latency/reactivation model remains the most plausible, it has not been proven. There are several reports of JCV detection in normal brain tissue (McGuire et al., 1995; Quinlivan et al., 1992; White et al., 1992), but others have not been able to confirm these observations (Koralnik et al., 1999), and the virus, if present in normal brain, is uncommon (~5%, judging from the prevalence of PML in AIDS patients) and at such a low level as to be near the limits of detection. Importantly, the level is so low that it does not usually interfere with the use of JCV-specific PCR for diagnosis of clinical PML in the CSF (see Chapter 13). The fifth major unsolved problem is the etiologic relation of JCV to human brain tumors, although important progesss in this area is being made (see Chapter 15).

Two cell types are regularly targeted in the brain—astrocytes and oligodendrocytes. The former are semipermissive, and the latter are fully permissive, seeming to prefer, however, not the native, circulating "archetypal" form of the virus with a noniterated regulatory region, but a rearranged form occurring seemingly randomly and without obvious patterns, with each PML brain showing a different predominant rearrangement (Agostini et al., 1997; Ault and

Stoner, 1993; Yogo et al., 1994) (see Chapter 7). While the rearranged configuration is quite variable, there is a tendency in many cases for deletion of a region similar to the 66–bp region (region D) deleted in the Mad-1 strain, while the flanking regions are duplicated. Mad-1 is unique in that an additional 23–bp sequence is deleted and the TATA box is included in the tandem duplication. Archetypal (unrearranged) regulatory regions are also a predominant configuration in some PML brains (Agostini et al., 1997; Ault and Stoner, 1993). No correlations of molecular biology, that is, the nature of the JCV regulatory region rearrangement, have yet been made with the nature or extent of neuropathology or the clinical course of disease.

The infection of the semipermissive astrocyte results in the "transformation" of this cell into giant, bizarre forms in about 80% of PML cases. These cells are invariably expressing T antigen, but not capsid proteins, as noted above. Very rarely, reactive astrocytes expressing GFAP are found to be infected with JCV in the nucleus (Fig. 12.7B). Occasionally dividing astrocytes can be observed to be positive for JCV capsid proteins (Fig. 12.7C).

How can a coherent picture of brain infection by JCV be constructed from the available data? While the entry of JC virus into the brain is still mysterious, we suggest that the primary event may be infection of the semipermissive glial cell, the astrocyte, which is in touch with the basal lamina of small vessels through processes termed *endfeet*. On the luminal surface are the endothelial cells in position to take up JCV from the circulation. Tight junctions between endothelial cells form the *blood–brain barrier* that protects the brain. The hamster model suggests the possibility of CNS endothelial cell infection by JCV. There has been little supporting evidence from studies of the human PML brain (Dörries et al., 1979). However, as noted above, expression of viral proteins in a cell of this nature might be transient. If initial brain infection occurs early in life, the virus might remain latent in these cells until adulthood. A stimulus that provokes viral reactivation would initiate the disease process in the astrocyte. The most frequent co-factor at present is HIV-1 infection. Other immunosuppressive conditions, notably chronic lymphocytic leukemia and lymphoma, less commonly provide the necessary stimulus. This initial event would be evident early in the disease process (Zu Rhein, 1969; Zu Rhein and Chou, 1968), most notably in subclinical PML cases discovered at autopsy (Åström and Stoner, 1994; Shimada et al., 1994).

There are two basic pathologic processes involving astrocytes. These are *astrogliosis* (astrocyte proliferation) and *astrocytosis* (the enlarged "gemistocytic" astrocyte). The initial lesion in PML might be called an *astrogliotic cluster*. In preparation for cell division, astrocytes "revert" back to their immature synthetic state, draw in their processes, cut back on protein production, and concentrate on DNA replication. Any infecting latent DNA virus would likely replicate along with the cell (Fig. 12.7C). GFAP synthesis will be decreased, while vimentin expression will increase (just as in the immature cell). When thinking of PML pathogenesis, one must also distinguish the initiating lesion (the "astrogliotic cluster") from propagation of the lesion, which leads to enlargement and eventual coalescence of contiguous lesions. One involves triggering events in the early astrogliotic cluster, and the other involves susceptibility of the most vulnerable cell, the oligodendrocyte, to either infection through the apposed astrocytes or direct infection from the extracellular milieu due to cytolysis of virus-infected cells nearby.

The occurrence of PML in immunosuppressive disease such as HIV-1 infection suggests a role for retroviral transactivating proteins such as Tat or inflammatory cytokines released from lymphotyes, marcrophages, and microglia. These might include tumor necrosis factor (Selmaj et al., 1991) or interferon (Yong et al., 1991), both of which can stimulate astrocyte proliferation in culture.

Regardless of whether the proliferating astrocyte passes JCV infection to the oligodendrocytes with which it is in contact, or the oligodendrocyte was already latently infected, astrocyte proliferation could have important effects on oligodendrocytes. For example, proliferating astrocytes cut back on their synthesis of trophic factors, which normally maintain the extracellular environment. This might alter the oligodendrocyte, rendering it more susceptible to a reactivating virus infection. In this scenario proliferating astrocytes (the astrogliotic cluster) serve as a pool of semipermissive cells for passage of the virus at low levels. As we noted above, proliferating astrocytes tend to pull back their processes and round up. While their oligodendrocyte contact may be transiently decreased, when reestablished, the virus could make its way into the susceptible oligodendrocyte.

To summarize, astrocyte reaction in PML seems to fall under two differently triggered responses. Early astrogliosis (the astrogliotic cluster) occurs before active JCV replication and cell lysis, and secondary astrogliosis occurs with astrocytosis and is the astrocyte's response to other environmental factors, primarily myelin breakdown products and cytokines. In addition, in the late lesion the infected astrocyte may be "transformed" into the giant, bizarre type expressing T antigen, but not capsid proteins or infectious virus.

When an infected astrocyte begins low-level, nonlytic replication of the virus, infectious virions might spread through the cytoplasm to the surrounding oligodendrocytes with which the astrocyte maintains supportive contact (Stoner, 1993). These permissive cells then replicate the virus to high levels, undergo cytolysis (or apoptosis), with the resulting destruction of the 20–40 myelin internodes that each of them maintained. This level of myelin loss, involving perhaps 10–30 oligodendrocytes and a total of 200–1200 myelin internodes, could provide the characteristic pinhead demyelinated lesions (0.1–0.5 mm diameter) that are the hallmark of the early PML lesion seen on routine H&E-stained sections. We presume that the very early astrogliotic clusters found in a case of asymptomatic PML discovered at autopsy (Åström and Stoner, 1994) are the initial manifestation of this sequence of events (Fig. 12.22A). These small clusters are not associated with evidence of demyelination on adjacent LFB/H&E-stained sections. Evidence for the relation of astrocytes to infected oligodendrocytes in recently established lesions of human PML is illustrated

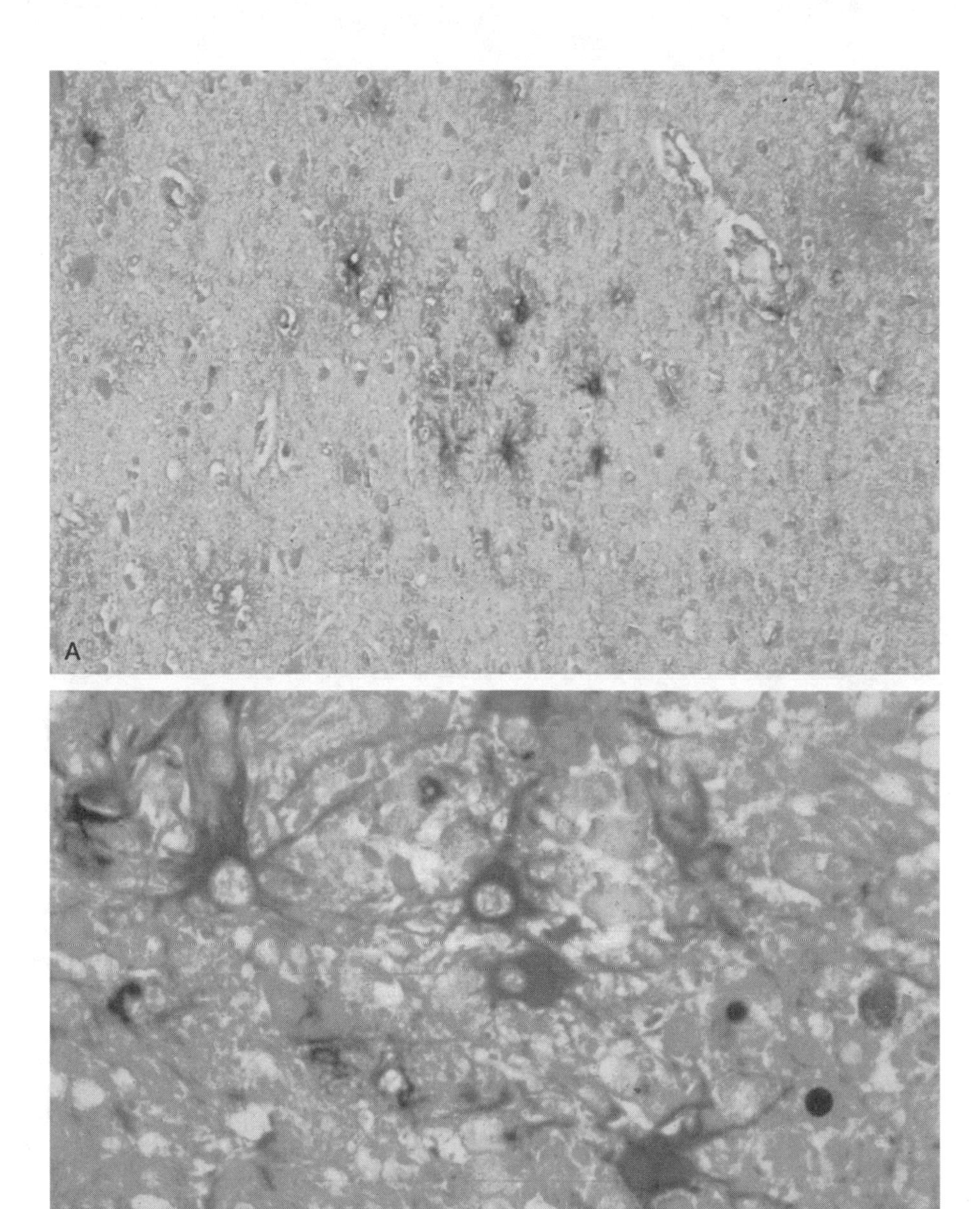

Figure 12.22. Astrocytes and PML pathogenesis. (**A**) Small cluster of enlarged, reactive astrocytes in the subcortical white matter of an early PML case discovered at autopsy (Åström and Stoner, 1994). Cells display GFAP reactivity (red), but no JCV-infected oligodendrocytes with capsid proteins (black) were evident. Method same as in Figure 12.2C. An adjacent section stained with LFB/H&E showed no evidence of myelin destruction (not shown). These astrogliotic clusters were the earliest discernible PML lesions. No counterstain. Medium magnification. Taken with permission from Åström and Stoner, 1994. (**B**) Established lesion in the basal ganglia with reactive astrocytes (red), including one with apparent perinuclear JCV capsid proteins in the cytoplasm (lower right). This cell extends a process to an infected oligodendrocyte with JCV capsid proteins in the nucleus (black). No counterstain. High magnification. (See color plates.)

in Fig. 12.22B. The most dramatic picture of astrocyte-to-oligodendrocyte spread of a polyomavirus comes from SV40 infection in the macaque PML brain (Fig. 12.23). This coordinated infection of many of the oligodendrocytes with which an infected astrocyte is in contact has not been seen with JCV in the human brain. However, there is no reason to believe that simian PML is fundamentally different in its polyomavirus etiology from human PML.

While the scenario we have sketched here is currently a working hypothesis and leaves many questions unanswered, we feel that it begins to accommodate the wealth of pathologic data presently available. The inability of many studies to relate PML lesions to a central blood vessel is consistent with the eccentric relationship that an infected astrocyte would bear to a vessel with which its extended process is in contact. Furthermore, the virus might migrate through repeated cell-to-cell transfers before encountering the conditions that trigger clinically (and pathologically) detectable rounds of viral replication.

Further study of very early PML lesions using markers for astroglial responses, along with the new sensitive methods for JCV detection such as T antigen detection using antigen retrieval methods for T antigen and IS-PCR for viral DNA, will ultimately lead to a consensus of opinion on these crucial questions.

Finally, what kills glial cells in the brain? Is it simply the inexorable buildup of viral particles as first demonstrated by Zu Rhein and Chou in 1964 (Zu Rhein and Chou, 1968) and detailed earlier in this chapter, or does viral initiation of the apoptotic cascade play a discernible role? (Koyama et al., 2000; O'Brien, 1998). Or are still other categories of possibly fatal host cell dysfunction triggered by the virus (Devireddy et al., 1996; Tretiakova et al., 1999a,b)? While the electron micrographs show infected cells in PML brain whose nuclei are loaded with virions, in biopsy material oligodendrocytes were also found that died before the whole replication cycle was finished (see electron microscopic findings, above). Can JCV reprogram the death of the glial cell? The evidence is currently fragmentary, and opinion is divided as to whether the apoptotic cascade contributes significantly to oligodendroglial cell death and demyelination (Sato-Matsumura et al., 1998; Yang and Prayson, 2000). Two things should be considered: the advantage to the virus and the advantage to the infected host cell. With regard to the virus, the dead-end brain infection that kills the host is largely the accidental result of a chance regulatory region rearrangement that has not been selected for in the long history of viral evolution and thus is not directly a part of the adaptation and survival of the virus in humans. Therefore, we can consider that whatever apoptotic mechanisms exist in the brain are unlikely to be part of a specific viral neurologic offensive strategy. On the other hand, the brain is considered to have apoptotic mechanisms as a defense to limit dangerous and potentially fatal viral infections of neurons or glia (Griffin and Hardwick, 1999). Another role of apoptosis may be to promote phagocytosis of dying cells by macrophages, thereby preventing dysregulated inflammatory reactions at the site of virus infection (Koyama et al., 2000). Apoptosis is an intracellular enzymatic cascade mediated by caspases

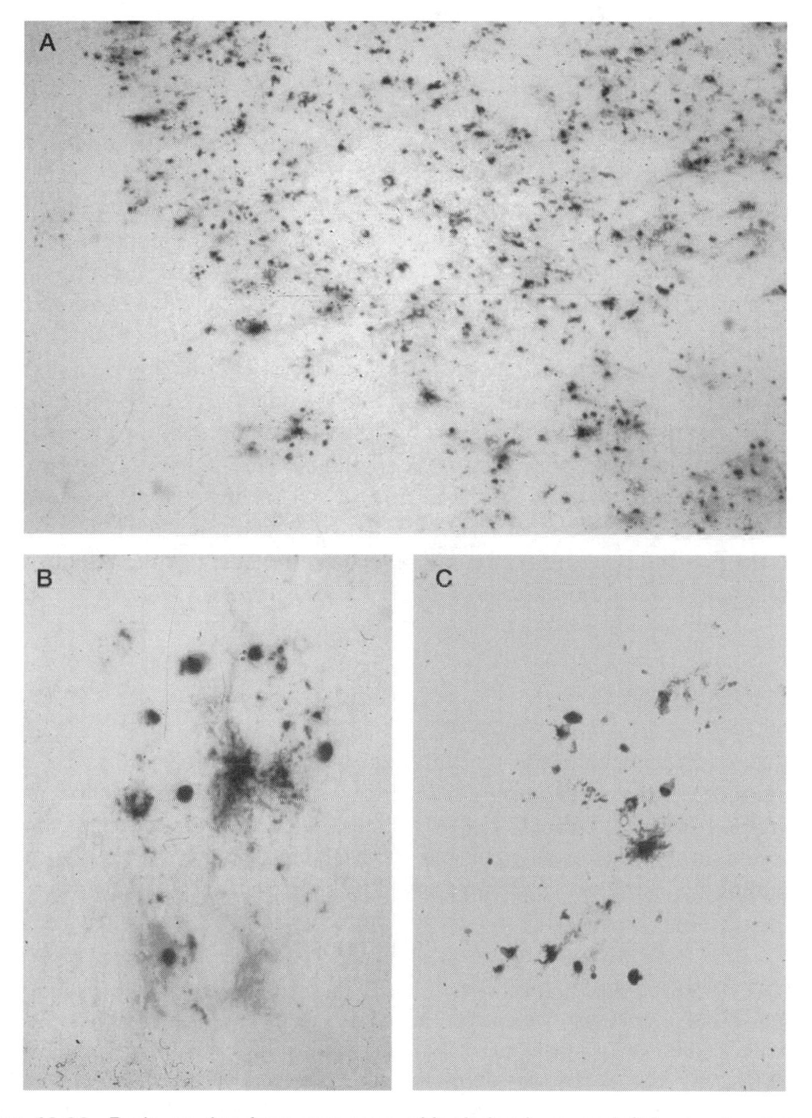

Figure 12.23. Brain section from macaque with simian immunodeficiency virus infection (SAIDS) and PML due to natural SV40 infection. (**A**) Subcortical white matter heavily infected with SV40. Astrocytes with capsid antigen in the cytoplasm are revealed by their stellate morphology. Section is stained with monoclonal antibody BH3 to SV40 major capsid protein VP1 (Babé et al., 1989) and revealed by peroxidase detection with $NiCl_2$ enhancement. No counterstain. Low magnification. (**B**) High power view of cell cluster in lower left corner of A. Infected oligodendrocytes with nuclear capsid protein surround an infected astrocyte with capsid protein VP1 in its cytoplasm and processes. (**C**) The same cell cluster as in B is also seen in the adjacent serial section. This indicates the spherical disposition of the infected oligodendrocytes surrounding this infected astrocyte. Taken with permission from Stoner, 1993.

that is typically initiated by transmembrane signaling at the cell surface. JCV as a latent/reactivating, rather than an acute, infection may be unsuited to trigger this protective mechanism. Otherwise, the cell must recognize the unscheduled or aberrant activation of the cell cycle by viral T antigen (O'Brien, 1998) and eliminate itself by apoptosis. T antigen can be either an inducer or an inhibitor of apoptosis through its interaction with cell cycle regulators, including p53, p300, and the pRb family of proteins (O'Brien, 1998; Simmons, 2000). In transgenic mice SV40 T antigen can block p53-dependent apoptosis, but not p53-independent apoptosis (McCarthy et al., 1994). Clearly, whatever apoptotic mechanisms exist for glial cells in the brain, they fail to contain most of the cell damage, as the disease is usually progressive with a fatal outcome. The published exceptions are few (Berger and Mucke, 1988; Hedley-Whyte et al., 1966; Kepes et al., 1975; Price et al., 1983; Singer et al., 1994). Another possibility would be that CD8-positive cytotoxic T lymphocytes, present in the circulation of long-term surviving PML patients (Koralnik et al., 2001), specifically attack infected glial cells and kill by a mechanism that induces apoptosis in the target. If this is occurring, evidence of apoptosis may be a hopeful sign of the possibility of immunologic rescue with control of the infection and, hopefully, survival of the host.

8. CONCLUSION

The story of PML began with its pathologic description in 1958 (see Chapter 1), continued with visualization of the virus by electron microscopy in glial cells in 1964 (see Chapter 2), and culminated with the cultivation of the etiologic agent in 1971 (see Chapter 3). It has been a story at the interface of neuropathology, virology, and oncology. Dr. Zu Rhein had suggested in 1964 that

> in PML the oligodendroglial cells succumb to the cytocidal effect of a pathogenic virus replicating massively within their nuclei and that demyelination results chiefly from the abolition of a myelin maintaining morphological and functional oligoglial–myelin relationship. . . . The possibility that the discovered particles might represent the virions of a biologically undiscovered primate papova virus should also be considered. This virus might be the analog in the human species to SV40 in monkeys and polyoma in mice. (Zu Rhein and Chou, 1968)

Those predictions have been amply confirmed. Neuropathology has led the way in this disease, but important aspects of PML pathogenesis and JCV biology remain to be elucidated. Perhaps the least understood is the contribution of JCV to neuro-oncology (see Chapter 15). Unfortunately, because of AIDS, PML is on the increase, and our extensive knowledge of this viral demyelinating disease has not yet translated into success in the clinic. The challenge today remains the same as in the 1960s: to find ways to block the neurovirulence of this agent and to augment host defenses in a way that preserves or restores neurologic function. Now we have the tools to succeed. With cooperation between the neurologist, the neuropathologist, and the virologist/immunologist, this 40-year-old challenge can be met.

ACKNOWLEDGMENTS

We thank Dr. D.L. Walker, Dr. P.V. Best, Dr. A.H. Koeppen, Dr. B. Lach, Dr. S.B. Lucas, Dr. L. Matthiessen, Dr. J.R. Miller, and L.J. Rubinstein for providing PML brain tissue for these studies. We also thank Dr. A. Lackner for providing the macaque PML tissue and C.F. Ryschkewitsch for expert technical assistance.

REFERENCES

Agostini HT, Ryschkewitsch CF, Singer EJ, Stoner GL (1997): JC virus regulatory region rearrangements and genotypes in progressive multifocal leukoencephalopathy: Two independent aspects of virus variation. J Gen Virol 78:659–664.

Agostini HT, Ryschkewitsch CF, Stoner GL (1996): Genotype profile of human polyomavirus JC excreted in urine of immunocompetent individuals. J Clin Microbiol 34:159–164.

Aksamit AJ (1995): Progressive multifocal leukoencephalopathy: A review of the pathology and pathogenesis. Microsc Res Tech 32:302–311.

Aksamit AJ, Major EO, Ghatak NR, Sidhu GS, Parisi JE, Guccion JG (1987): Diagnosis of progressive multifocal leukoencephalopathy by brain biopsy with biotin labeled DNA:DNA in situ hybridization. J Neuropathol Exp Neurol 46:556–566.

Aksamit AJ, Mourrain P, Sever JL, Major EO (1985): Progressive multifocal leukoencephalopathy: Investigation of three cases using in situ hybridization with JC virus biotinylated DNA probe. Ann Neurol 18:490–496.

Aksamit AJ, Sever JL, Major EO (1986): Progressive multifocal leukoencephalopathy: JC virus detection by in situ hybridization compared with immunohistochemistry. Neurology 36:499–504.

Allen IV, McQuaid S, McMahon J, Kirk J, McConnell R (1996): The significance of measles virus antigen and genome distribution in the CNS in SSPE for mechanisms of viral spread and demyelination. J Neuropathol Exp Neurol 55:471–480.

Ariza A, Mate JL, Fernandez-Vasalo A, Gomez-Plaza C, Perez-Piteira J, Pujol M, Navas-Palacios JJ (1994): p53 and proliferating cell nuclear antigen expression in JC virus-infected cells of progressive multifocal leukoencephalopathy. Hum Pathol 25:1341–1345.

Ariza A, Mate JL, Isamat M, Calatrava A, Fernández-Vasalo A, Navas-Palacios JJ (1998): Overexpression of Ki-67 and cyclins A and B1 in JC virus–infected cells of progressive multifocal leukoencephalopathy. J Neuropathol Exp Neurol 57:226–230.

Arribas JR, Clifford DB, Fichtenbaum CJ, Roberts RL, Powderly WG, Storch GA (1995): Detection of Epstein-Barr virus DNA in cerebrospinal fluid for diagnosis of AIDS-related central nervous system lymphoma. J Clin Microbiol 33:1580–1583.

Assouline JG, Major EO (1991): Human fetal Schwann cells support JC virus multiplication. J Virol 65:1002–1006.

Åström KE, Mancall EL, Richardson EP Jr (1958): Progressive multifocal leukoencephalopathy: A hitherto unrecognized complication of chronic lymphocytic leukemia and Hodgkin's disease. Brain 81:93–111.

Åström KE, Stoner GL (1994): Early pathological changes in progressive multifocal leukoencephalopathy: A report of two asymptomatic cases occurring prior to the AIDS epidemic. Acta Neuropathol (Berl) 88:93–105.

Ault GS, Stoner GL (1993): Human polyomavirus JC promoter/enhancer rearrangement patterns from progressive multifocal leukoencephalopathy brain are unique derivatives of a single archetypal structure. J Gen Virol 74:1499–1507.

Babé LM, Brew K, Matsuura SE, Scott WA (1989): Epitopes on the major capsid protein of simian virus 40. J Biol Chem 264:2665–2671.

Ball RK, Siegl B, Quelhorst S, Brandner G, Braun DG (1984): Monoclonal antibodies against simian virus 40 nuclear large T tumor antigen: Epitope mapping, papova virus cross-reaction and cell surface staining. EMBO J 3:1485–1491.

Bancroft JB, Hills GJ, Markham R (1967): A study of the self-assembly process in a small spherical virus. Formation of organized structures from protein subunits in vitro. Virology 31:354–379.

Berger JR, Kaszovitz B, Post MJ, Dickinson G (1987): Progressive multifocal leukoencephalopathy associated with human immunodeficiency virus infection. A review of the literature with a report of sixteen cases. Ann Intern Med 107:78–87.

Berger JR, Mucke L (1988): Prolonged survival and partial recovery in AIDS-associated progressive multifocal leukoencephalopathy. Neurology 38:1060–1065.

Berger JR, Scott G, Albrecht J, Belman AL, Tornatore C, Major EO (1992): Progressive multifocal leukoencephalopathy in HIV-1–infected children. AIDS 6:837–841.

Bienfait HP, Louwerse ES, Portegies P, van der Meer JTM (1998): Progressive multifocal leukoencephalopathy presenting as a solitary gray matter lesion. J Neurol 245: 557–558.

Blumberg BM, Mock DJ, Powers JM, Ito M, Assouline JG, Baker JV, Chen B, Goodman AD (2000): The HHV6 paradox: Ubiquitous commensal or insidious pathogen? A two-step in situ PCR approach. J Clin Virol 16:159–178.

Boerman RH, Arnoldus EP, Raap AK, Peters AC, ter Schegget J, van der Ploeg M (1989): Diagnosis of progressive multifocal leucoencephalopathy by hybridisation techniques. J Clin Pathol 42:153–161.

Bogdanovic G, Hammarin AL, Grandien M, Winblad B, Bergenheim AT, Nennesmo I, Dalianis T (1995): No association of JC virus with Alzheimers disease or astrocytomas. Clin Diagn Virol 4:223–230.

Boldorini R, Cristina S, Vago L, Tosoni A, Guzzetti S, Costanzi G (1993): Ultrastructural studies in the lytic phase of progressive multifocal leukoencephalopathy in AIDS patients. Ultrastruct Pathol 17:599–609.

Bollag B, Frisque RJ (1992): PAb 2000 specifically recognizes the large T and small t proteins of JC virus. Virus Res 25:223–239.

Bouteille M, Fontaine C, Vedrenne C, Delarue J (1965): Sur un cas d'encéphalite subaiguë a inclusions. Etude anatomoclinique et ultrastructurale. Rev Neurol (Paris) 113:454–458.

Bouteille M, Kalifat SR, Delarue J (1967): Ultrastructural variations of nuclear bodies in human diseases. J Ultrastruct Res 19:474–486.

Brooks BR, Walker DL (1984): Progressive multifocal leukoencephalopathy. Neurol Clin 2:299–313.

Budka H (1986): Multinucleated giant cells in brain: A hallmark of the acquired immune deficiency syndrome (AIDS). Acta Neuropathol (Berl) 69:253–258.

Budka H (1991): Neuropathology of human immunodeficiency virus infection. Brain Pathol 1:163–176.

Budka H, Costanzi G, Cristina S, Lechi A, Parravicini C, Trabattoni R, Vago L (1987): Brain pathology induced by infection with human immunodeficiency virus (HIV). A histological immunocytochemical, and electron microscopical study of 100 autopsy cases. Acta Neuropathol (Berl) 75:185–198.

Budka H, Shah KV (1983): Papovavirus antigens in paraffin sections of PML brains. In Sever JL, Madden DL, Eds; Polyomaviruses and Human Neurological Disease; Alan R. Liss: New York, pp 299–309.

Budka H, Wiley CA, Kleihues P, Artigas J, Asbury AK, Cho ES, Cornblath DR, Dal Canto MC, DeGirolami U, Dickson D et al. (1991): HIV-associated disease of the nervous system: Review of nomenclature and proposal for neuropathology-based terminology. Brain Pathol 1:143–152.

Castaigne P, Escourolle R, Derouesne C, Cathala F, Hauw JJ, Duclos H, Goust JM (1973): [A case of progressive multifocal leukoencephalopathy. Anatomoclinical and immunological study. Isolation of a new papova strain virus from cerebral biopsy]. Rev Neurol (Paris) 128:85–94.

Castaigne P, Rondot P, Escourolle R, Ribadeau dumas JL, Cathala F, Hauw JJ (1974): Leucoencephalopathie multifocale progressive et "gliomes" multiples. Rev Neurol (Paris) 130:379–392.

Chandor SB, Forno LS, Wivel NA (1965): Progressive multifocal leucoencephalopathy. J Neurol Neurosurg Psychiatry 28:260–271.

Chima SC, Agostini HT, Ryschkewitsch CF, Lucas SB, Stoner GL (1999): Progressive multifocal leukoencephalopathy and JC virus genotypes in West African patients with acquired immunodeficiency syndrome—A pathologic and DNA sequence analysis of 4 cases. Arch Pathol Lab Med 123:395–403.

Chimelli L, Rosemberg S, Hahn MD, Lopes MBS, Barretto Netto M (1992): Pathology of the central nervous system in patients infected with the human immunodeficiency virus (HIV): A report of 252 autopsy cases from Brazil. Neuropathol Appl Neurobiol 18:478–488.

Chrétien F, Bélec L, Hilton DA, Flament-Saillour M, Guillon F, Wingertsmann L, Baudrimont M, De Truchis P, Keohane C, Vital C, Love S, Gray F (1996): Herpes simplex virus type 1 encephalitis in acquired immunodeficiency syndrome. Neuropathol Appl Neurobiol 22:394–404.

Chrétien F, Boche D, De la Grandmaison GL, Ereau T, Mikol J, Hurtrel M, Hurtrel B, Gray F (2000): Progressive multifocal leukoencephalopathy and oligodendroglioma in a monkey co-infected by simian immunodeficiency virus and simian virus 40. Acta Neuropathol (Berl) 100:332–336.

Cinque P, Brytting M, Vago L, Castagna A, Parravivini C, Zanchetta N, Monforte AD, Wahren B, Lazzarin A, Linde A (1993): Epstein-Barr virus DNA in cerebrospinal fluid from patients with AIDS-related primary lymphoma of the central nervous system. Lancet 342:398–401.

Cinque P, Vago L, Marenzi R, Giudici B, Weber T, Corradini R, Ceresa D, Lazzarin A, Linde A (1998): Herpes simplex virus infections of the central nervous system in human immunodeficiency virus–infected patients: Clinical management by polymerase chain reaction assay of cerebrospinal fluid. Clin Infect Dis 27:303–309.

Clifford DB (1999): Opportunistic viral infections in the setting of human immunodeficiency virus. Semin Neurol 19:185–192.

Devireddy LR, Kumar KU, Pater MM, Pater A (1996): Evidence for a mechanism of demyelination by human JC virus: Negative transcriptional regulation of RNA and protein levels from myelin basic protein gene by large tumor antigen in human glioblastoma cells. J Med Virol 49:205–211.

Dickson DW, Belman AL, Park YD, Wiley C, Horoupian DS, Llena J, Kure K, Lyman WD, Morecki R, Mitsudo S, Cho S (1989): Central nervous system pathology in pediatric AIDS: An autopsy study. APMIS Suppl 8:40–57.

Dolman CL, Furesz J, Mackay B (1967): Progressive multifocal leukoencephalopathy: Two cases with electron microscopic and viral studies. Can Med Assoc J 97:8–12.

Dorfman LJ (1973): Cytomegalovirus encephalitis in adults. Neurology 23:136–144.

Dörries K, Johnson RT, ter Meulen V (1979): Detection of polyoma virus DNA in PML-brain tissue by (in situ) hybridization. J Gen Virol 42:49–57.

Dörries K, ter Meulen V (1983): Progressive multifocal leucoencephalopathy: Detection of papovavirus JC in kidney tissue. J Med Virol 11:307–317.

Drobyski WR, Knox KK, Majewski D, Carrigan DR (1994): Brief report: Fatal encephalitis due to variant B human herpesvirus-6 infection in a bone marrow–transplant recipient. N Engl J Med 330:1356–1360.

Epstein LG, Sharer LR, Cho ES, Myenhofer M, Navia BA, Price RW (1985): HTLV-III/LAV-like retrovirus particles in the brains of patients with AIDS encephalopathy. AIDS Res 1:447–454.

Farge D, Herve R, Mikol J, Sauvaget F, Ingrand D, Singer B, Ferchal F, Auperin I, Gray F, Sudaka A, Degos L, Rouffy J (1994): Simultaneous progressive multifocal leukoencephalopathy, Epstein-Barr virus (EBV) latent infection and cerebral parenchymal infiltration during chronic lymphocytic leukemia. Leukemia 8:318–321.

Feigenbaum L, Hinrichs SH, Jay G (1992): JC virus and simian virus 40 enhancers and transforming proteins: Role in determining tissue specificity and pathogenicity in transgenic mice. J Virol 66:1176–1182.

Frisque RJ (1983): Regulatory sequences and virus–cell interactions of JC virus. Prog Clin Biol Res 105:41–59.

Frisque RJ, Bream GL, Cannella MT (1984): Human polyomavirus JC virus genome. J Virol 51:458–469.

Frisque RJ, Rifkin DB, Walker DL (1980): Transformation of primary hamster brain cells with JC virus and its DNA. J Virol 35:265–269.

Gallo BV, Shulman LM, Weiner WJ, Petito CK, Berger JR (1996): HIV encephalitis presenting with severe generalized chorea. Neurology 46:1163–1165.

Gerber MA, Shah KV, Thung SN, Zu Rhein GM (1980): Immunohistochemical demonstration of common antigen of polyomaviruses in routine histologic tissue sections of animals and man. Am J Clin Pathol 73:795–797.

Gonatas NK (1966): Subacute sclerosing leucoencephalitis: Electron microscopic and cytochemical observations on a cerebral biopsy. J Neuropathol Exp Neurol 25:177–201.

Gonatas NK, Martin J, Evangelista I (1967): The osmiophilic particles of astrocytes. Viruses, lipid droplets or products of secretion? J Neuropathol Exp Neurol 26:369–376.

Gongora-Rivera F, Santos-Zambrano J, Moreno-Andrade T, Calzada-Lopez P, Soto-Hernandez JL (2000): The clinical spectrum of neurological manifestations in AIDS patients in Mexico. Arch Med Res 31:393–398.

Gray F, Belec L, Lescs MC, Chretien F, Ciardi A, Hassine D, Flament-Saillour M, De Truchis P, Clair B, Scaravilli F (1994): Varicella-zoster virus infection of the central nervous system in the acquired immune deficiency syndrome. Brain 117:987–999.

Gray F, Gherardi R, Baudrimont M, Gaulard P, Meyrignac C, Vedrenne C, Poirier J (1987): Leucoencephalopathy with multinucleated giant cells containing human immune deficiency virus-like particles and multiple opportunistic cerebral infections in one patient with AIDS. Acta Neuropathol (Berl) 73:99–104.

Greenlee JE, Keeney PM (1986): Immunoenzymatic labelling of JC papovavirus T antigen in brains of patients with progressive multifocal leukoencephalopathy. Acta Neuropathol (Berl) 71:150–153.

Gribble DH, Haden CC, Schwartz LW, Henrickson RV (1975): Spontaneous progressive multifocal leukoencephalopathy (PML) in macaques. Nature 254:602–604.

Griffin DE, Hardwick JM (1999): Perspective: Virus infections and the death of neurons. Trends Microbiol 7:155–160.

Gruss P, Khoury G (1981): The SV-40 tandem repeats as an element of the early promoter. Proc Natl Acad Sci USA 78:943–947.

Hair LS, Nuovo G, Powers JM, Sisti MB, Britton CB, Miller JR (1992): Progressive multifocal leukoencephalopathy in patients with human immunodeficiency virus. Hum Pathol 23:663–667.

Harlow E, Crawford LV, Pim DC, Williamson NM (1981): Monoclonal antibodies specific for simian virus 40 tumor antigens. J Virol 39:861–869.

Hedley-Whyte ET, Smith BP, Tyler HT, Peterson WP (1966): Multifocal leukoencephalopathy with remission and five year survival. J Neuropathol Exp Neurol 25: 107–116.

Hofman P, Saint-Paul MC, Battaglione V, Michiels JF, Loubiere R (1999): Autopsy findings in the acquired immunodeficiency syndrome (AIDS). A report of 395 cases from the south of France. Pathol Res Pract 195:209–217.

Holmberg CA, Gribble DH, Takemoto KK, Howley PM, España C, Osburn BI (1977): Isolation of simian virus 40 from rhesus monkeys (*Macaca mulatta*) with spontaneous progressive multifocal leukoencephalopathy. J Infect Dis 136:593–596.

Horvath CJ, Simon MA, Bergsagel DJ, Pauley DR, King NW, Garcea RL, Ringler DJ (1992): Simian virus 40–induced disease in rhesus monkeys with simian acquired immunodeficiency syndrome. Am J Pathol 140:1431–1440.

Houff SA, Major EO, Katz DA, Kufta CV, Sever JL, Pittaluga S, Roberts JR, Gitt J, Saini N, Lux W (1988): Involvement of JC virus–infected mononuclear cells from the bone marrow and spleen in the pathogenesis of progressive multifocal leukoencephalopathy. N Engl J Med 318:301–305.

Howatson AF, Nagai M, Zu Rhein GM (1965): Polyoma-like virions in human demyelinating disease. Can Med Assoc J 93:379.

Hurley JP, Ilyinskii PO, Horvath CJ, Simon MA (1997): A malignant astrocytoma containing simian virus 40 DNA in a macaque infected with simian immunodeficiency virus. J Med Primatol 26:172–180.

Ikuta F (1969): Virions associated with four Japanese cases of progressive multifocal leukoencephalopathy. Int Arch Allergy Appl Immunol Suppl 36:488–514.

Ikuta F, Hirano A, Zimmerman HM (1963): An experimental study of post-mortem alterations in the granular layer of the cerebellar cortex. J Neuropathol Exp Neurol 22:581–593.

Ironside JW, Lewis FA, Blythe D, Wakefield EA (1989): The identification of cells containing JC papovavirus DNA in progressive multifocal leukoencephalopathy by combined in situ hybridization and immunocytochemistry. J Pathol 157:291–297.

Ito M, Baker JV, Mock DJ, Goodman AD, Blumberg BM, Shrier DA, Powers JM (2000): Human herpesvirus 6–meningoencephalitis in an HIV patient with progressive multifocal leukoencephalopathy. Acta Neuropathol (Berl) 100:337–341.

Itoyama Y, Webster HD, Sternberger NH, Richardson EP Jr, Walker DL, Quarles RH, Padgett BL (1982): Distribution of papovavirus, myelin-associated glycoprotein, and myelin basic protein in progressive multifocal leukoencephalopathy lesions. Ann Neurol 11:396–407.

Jakobsen J, Diemer NH, Gaub J, Brun B, Helweg-Larsen S (1987): Progressive multifocal leukoencephalopathy in a patient without other clinical manifestations of AIDS. Acta Neurol Scand 75:209–213.

Jellinger KA, Setinek U, Drlicek M, Böhm G, Steurer A, Lintner F (2000): Neuropathology and general autopsy findings in AIDS during the last 15 years. Acta Neuropathol (Berl) 100:213–220.

Kadin ME, Rubinstein LJ, Nelson JS (1970): Neonatal cerebellar medulloblastoma originating from the fetal external granular layer. J Neuropathol Exp Neurol 29:583–600.

Kepes JJ, Chou SM, Price LW Jr (1975): Progressive multifocal leukoencephalopathy with 10-year survival in a patient with nontropical sprue. Report of a case with unusual light and electron microscopic features. Neurology 25:1006–1012.

Kitamura T, Aso Y, Kuniyoshi N, Hara K, Yogo Y (1990): High incidence of urinary JC virus excretion in nonimmunosuppressed older patients. J Infect Dis 161:1128–1133.

Kleihues P, Lang W, Burger PC, Budka H, Vogt M, Maurer R, Lüthy R, Siegenthaler W (1985): Progressive diffuse leukoencephalopathy in patients with acquired immune deficiency syndrome (AIDS). Acta Neuropathol (Berl) 68:333–339.

Knox KK, Carrigan DR (1995): Active human herpesvirus (HHV-6) infection of the central nervous system in patients with AIDS. J Acquir Immune Defic Syndr Hum Retrovirol 9:69–73.

Knox KK, Harrington DP, Carrigan DR (1995): Fulminant human herpesvirus six encephalitis in a human immunodeficiency virus–infected infant. J Med Virol 45:288–292.

Koenig S, Gendelman HE, Orenstein JM, Dal Canto MC, Pezeshkpour GH, Yungbluth M, Janotta F, Aksamit A, Martin MA, Fauci AS (1986): Detection of AIDS virus in macrophages in brain tissue from AIDS patients with encephalopathy. Science 233: 1089–1093.

Koralnik IJ, Boden D, Mai VX, Lord CI, Letvin NL (1999): JC virus DNA load in patients with and without progressive multifocal leukoencephalopathy. Neurology 52:253–260.

Koralnik IJ, Du Pasquier RA, Letvin NL (2001): JC virus–specific cytotoxic T lymphocytes in individuals with progressive multifocal leukoencephalopathy. J Virol 75: 3483–3487.

Koyama AH, Fukumori T, Fujita M, Irie H, Adachi A (2000): Physiological significance of apoptosis in animal virus infection. Microbes Infect 2:1111–1117.

Krempien B, Kolkmann F-W, Schiemer HG, Mayer P (1972): Uber die progressive multifokale Leukoencephalopathie (Beitrag zur Differentialdiagnose mit electronenmikroskopischen und cytophotometrischen Untersuchungen). Virchows Arch A 355: 158–178.

Kuchelmeister K, Bergmann M, Gullotta F (1993a): Cellular changes in the cerebellar granular layer in AIDS-associated PML. Neuropathol Appl Neurobiol 19:398–401.

Kuchelmeister K, Gullotta F, Bergman M, Angeli G, Masini T (1993b): Progressive multifocal leukoencephalopathy (PML) in the acquired immunodeficiency syndrome (AIDS). A neuropathological study of 21 cases. Pathol Res Pract 189:163–173.

Kure K, Llena JF, Lyman WD, Soeiro R, Weidenheim KM, Hirano A, Dickson DW (1991): Human immunodeficiency virus-1 infection of the nervous system: An autopsy study of 268 adult, pediatric, and fetal brains. Hum Pathol 22:700–710.

Lammie GA, Beckett A, Courtney R, Scaravilli F (1994): An immunohistochemical study of p53 and proliferating cell nuclear antigen expression in progressive multifocal leukoencephalopathy. Acta Neuropathol (Berl) 88:465–471.

Lang W, Miklossy J, Deruaz JP, Pizzolato GP, Probst A, Schaffner T, Gessaga E, Kleihues P (1989): Neuropathology of the acquired immune deficiency syndrome (AIDS): A report of 135 consecutive autopsy cases from Switzerland. Acta Neuropathol (Berl) 77:379–390.

Ledoux S, Libman I, Robert F, Just N (1989): Progressive multifocal leukoencephalopathy with gray matter involvement. Can J Neurol Sci 16:200–202.

London WT, Houff SA, Madden DL, Fuccillo DA, Gravell M, Wallen WC, Palmer AE, Sever JL, Padgett BL, Walker DL, Zu Rhein GM, Ohashi T (1978): Brain tumors in owl monkeys inoculated with a human polyomavirus (JC virus). Science 201: 1246–1249.

London WT, Houff SA, McKeever PE, Wallen WC, Sever JL, Padgett BL, Walker DL (1983): Viral-induced astrocytomas in squirrel monkeys. Prog Clin Biol Res 105: 227–237.

Lucas SB, Hounnou A, Peacock C, Beaumel A, Djomand G, N'Gbichi J-M, Yeboue K, Honde M, Diomande M, Giordano C, Doorly R, Brattegaard K, Kestens L, Smithwick R, Kadio A, Ezani N, Yapi A, De Cock KM (1993): The mortality and pathology of HIV infection in a West African city. AIDS 7:1569–1579.

Martinez AJ, Sell M, Mitrovics T, Stoltenburg-Didinger G, Iglesias-Rozas JR, Giraldo-Velasquez MA, Gosztonyi G, Schneider V, Cervos-Navarro J (1995): The neuropathology and epidemiology of AIDS. A Berlin experience. A review of 200 cases. Pathol Res Pract 191:427–443.

Masliah E, DeTeresa RM, Mallory ME, Hansen LA (2000): Changes in pathological findings at autopsy in AIDS cases for the last 15 years. AIDS 14:69–74.

Mathews T, Wisotzkey H, Moossy J (1976): Multiple central nervous system infections in progressive multifocal leukoencephalopathy. Neurology 26:9–14.

Matsuda M, Yasui K, Nagashima K, Mori W (1987): Origin of the medulloblastoma experimentally induced by human polyomavirus JC. J Natl Cancer Inst 79:585–591.

Mázló M, Herndon RM (1977): Progressive multifocal leukoencephalopathy: Ultrastructural findings in two brain biopsies. Neuropathol Appl Neurol 3:323–339.

Mázló M, Tariska I (1980): Morphological demonstration of the first phase of polyomavirus replication in oligodendroglia cells of human brain in progressive multifocal leukoencephalopathy (PML). Acta Neuropathol (Berl) 49:133–143.

Mázló M, Tariska I (1982): Are astrocytes infected in progressive multifocal leukoencephalopathy (PML)? Acta Neuropathol (Berl) 56:45–51.

McCarthy SA, Symonds HS, Van Dyke T (1994): Regulation of apoptosis in transgenic mice by simian virus 40 T antigen–mediated inactivation of p53. Proc Natl Acad Sci USA 91:3979–3983.

McGuire D, Barhite S, Hollander H, Miles M (1995): JC virus DNA in cerebrospinal fluid of human immunodeficiency virus–infected patients: Predictive value for progressive multifocal leukoencephalopathy. Ann Neurol 37:395–399.

Mesquita R, Parravicini C, Björkholm M, Ekman M, Biberfeld P (1992): Macrophage association of polyomavirus in progressive multifocal leukoencephalopathy: An immunohistochemical and ultrastructural study. Case report. APMIS 100:993–1000.

Messing A, Pinkert CA, Palmiter RD, Brinster RL (1988): Developmental study of SV40 large T antigen expression in transgenic mice with choroid plexus neoplasia. Oncogene Res 3:87–97.

Miller JR, Barrett RE, Britton CB, Tapper ML, Bahr GS, Bruno PJ, Marquardt MD, Hays AP, McMurtry JG, Weissman JB, Bruno MS (1982): Progressive multifocal leukoencephalopathy in a male homosexual with T-cell immune deficiency. N Engl J Med 307:1436–1438.

Miller NR, McKeever PE, London W, Padgett BL, Walker DL, Wallen WC (1984): Brain tumors of owl monkeys inoculated with JC virus contain the JC virus genome. J Virol 49:848–856.

Mock DJ, Powers JM, Goodman AD, Blumenthal SR, Ergin N, Baker JV, Mattson DH, Assouline JG, Bergey EJ, Chen BJ, Epstein LG, Blumberg BM (1999): Association of human herpesvirus 6 with the demyelinative lesions of progressive multifocal leukoencephalopathy. J Neurovirol 5:363–373.

Mohar A, Romo J, Salido F, Jessurun J, Ponce de León S, Reyes E, Volkow P, Larraza O, Angel Peredo M, Cano C, Gómez G, Sepúlveda J, Mueller N (1992): The spectrum of clinical and pathological manifestations of AIDS in a consecutive series of autopsied patients in Mexico. AIDS 6:467–473.

Morgello S, Block GA, Price RW, Petito CK (1988): Varicella-zoster virus leukoencephalitis and cerebral vasculopathy. Arch Pathol Lab Med 112:173–177.

Morgello S, Cho ES, Nielsen S, Devinsky O, Petito CK (1987): Cytomegalovirus encephalitis in patients with acquired immunodeficiency syndrome: An autopsy study of 30 cases and a review of the literature. Hum Pathol 18:289–297.

Muller J, Watanabe I (1967): Progressive multifocal leukoencephalopathy. A virus disease? Am J Clin Pathol 47:114–123.

Nagashima K, Yamaguchi K, Nakase H, Miyazaki J (1982): Progressive multifocal leukoencephalopathy. A case report and review of the literature. Acta Pathol Jpn 32: 333–343.

Nath A, Jankovic J, Pettigrew LC (1987): Movement disorders and AIDS. Neurology 37:37–41.

Navia BA, Cho ES, Petito CK, Price RW (1986): The AIDS dementia complex: II. Neuropathology. Ann Neurol 19:525–535.

Newman JS, Baskin GB, Frisque RJ (1998): Identification of SV40 in brain, kidney and urine of healthy and SIV-infected rhesus monkeys. J Neurovirol 4:394–406.

Newman JT, Frisque RJ (1997): Detection of archetype and rearranged variants of JC virus in multiple tissues from a pediatric PML patient. J Med Virol 52:243–252.

Novoa LJ, Nagra RM, Nakawatase T, Edwards-Lee T, Tourtellotte WW, Cornford ME (1997): Fulminant demyelinating encephalomyelitis associated with productive HHV-6 infection in an immunocompetent adult. J Med Virol 52:301–308.

O'Brien V (1998): Viruses and apoptosis. J Gen Virol 79:1833–1845.

Ohashi T, ZuRhein GM, Varakis JN, Padgett BL, Walker DL (1978): Experimental (JC virus-induced) intraocular and extraorbital tumors in the Syrian hamster. J Neuropathol Exp Neurol 37:667.

Ohsumi S, Motoi M, Ogawa K (1986): Induction of undifferentiated tumors by JC virus in the cerebrum of rats. Acta Pathol Jpn 36:815–825.

Orenstein JM, Jannotta F (1988): Human immunodeficiency virus and papovavirus infections in acquired immunodeficiency syndrome: An ultrastructural study of three cases. Hum Pathol 19:350–361.

Padgett BL, Walker DL (1976): New human papovaviruses. Prog Med Virol 22:1–35.

Padgett BL, Walker DL, Zu Rhein GM, Varakis JN (1977): Differential neurooncogenicity of strains of JC virus, a human polyoma virus, in newborn Syrian hamsters. Cancer Res 37:718–720.

Palmiter RD, Chen HY, Messing A, Brinster RL (1985): SV40 enhancer and large T-antigen are instrumental in development of choroid plexus tumors in transgenic mice. Nature 316:457–460.

Papadimitriou JM, Kakulas BA, Sadka M (1966): Virus-like particles in proximity to myelin in a case of progressive multifocal leukoencephalopathy. Proc Aust Assoc Neurol 4:133–140.

Perier O, Vanderhaeghen JJ, Pelc S (1967): Subacute sclerosing leuco-encephalitis. Electron microscopic finding in two cases with inclusion bodies. Acta Neuropathol (Berl) 8:362–380.

Petito CK, Cho ES, Lemann W, Navia BA, Price RW (1986): Neuropathology of acquired immunodeficiency syndrome (AIDS): An autopsy review. J Neuropathol Exp Neurol 45:635–646.

Petito CK, Navia BA, Cho ES, Jordan BD, George DC, Price RW (1985): Vacuolar myelopathy pathologically resembling subacute combined degeneration in patients with the acquired immunodeficiency syndrome. N Engl J Med 312:874–879.

Petry H, Luke W (1997): Infection of macaque monkeys with simian immunodeficiency virus: An animal model for neuro-AIDS. Intervirology 40:112–121.

Phillips SC, Harper CG, Kril J (1987): A quantitative histological study of the cerebellar vermis in alcoholic patients. Brain 110:301–314.

Piccolo I, Causarano R, Sterzi R, Sberna M, Oreste PL, Moioli C, Caggese L, Girotti F (1999): Chorea in patients with AIDS. Acta Neurol Scand 100:332–336.

Price RW (1996): Neurological complications of HIV infection. Lancet 348:445–452.

Price RW, Nielsen S, Horten B, Rubino M, Padgett BL, Walker DL (1983): Progressive multifocal leukoencephalopathy: A burnt-out case. Ann Neurol 13:485–490.

Quinlivan EB, Norris M, Bouldin TW, Suzuki K, Meeker R, Smith MS, Hall C, Kenney S (1992): Subclinical central nervous system infection with JC virus in patients with AIDS. J Infect Dis 166:80–85.

Rencic A, Gordon J, Otte J, Curtis M, Kovatich A, Zoltick P, Khalili K, Andrews D (1996): Detection of JC virus DNA sequence and expression of the viral oncoprotein, tumor antigen, in brain of immunocompetent patient with oligoastrocytoma. Proc Natl Acad Sci USA 93:7352–7357.

Renner K, Sock E, Gerber JK, Wegner M (1996): T antigen of human papovavirus JC stimulates transcription of the POU domain factor Tst-1/Oct6/SCIP. DNA Cell Biol 15:1057–1062.

Ressetar HG, Prakash O, Frisque RJ, Webster HD, Re RN, Stoner GL (1993): Expression of viral T-antigen in pathological tissues from transgenic mice carrying JC-SV40 chimeric DNAs. Mol Chem Neuropathol 20:59–79.

Ressetar HG, Walker DL, Webster HD, Braun DG, Stoner GL (1990): Immunolabeling of JC virus large T antigen in neonatal hamster brain before tumor formation. Lab Invest 62:287–296.

Ressetar HG, Webster HD, Stoner GL (1992): Brain vascular endothelial cells express JC virus large tumor antigen in immunocompetent and cyclophosphamide-treated hamsters. Proc Natl Acad Sci USA 89:8170–8174.

Ressetar HG, Webster HD, Stoner GL (1997): Persistence of neurotropic JC virus DNA in hamster tissues six months after intracerebral inoculation. J Neurovirol 3:66–70.

Rhodes RH (1993): Histopathologic features in the central nervous system of 400 acquired immunodefiency syndrome cases: Implications of rates of occurrence. Hum Pathol 24:1189–1198.

Rhodes RH, Ward JM, Walker DL, Ross AA (1988): Progressive multifocal leukoencephalopathy and retroviral encephalitis in acquired immunodeficiency syndrome. Arch Pathol Lab Med 112:1207–1213.

Richardson EP Jr (1961): Progressive multifocal leukoencephalopathy. N Engl J Med 265:815–823.

Richardson EP Jr (1970): Progressive multifocal leukoencephalopathy. In Winken PJ, Bruyn GW, Eds; Handbook of Clinical Neurology. Multiple Sclerosis and Other Demyelinating Diseases; Amsterdam, Vol 9, pp 485–499.

Richardson EP Jr, Johnson PC (1975): Atypical progressive multifocal leukoencephalopathy with plasma-cell infiltrates. Acta Neuropathol Suppl (Berl) Suppl 6:247–250.

Rigby PWJ, Lane DP (1983): Structure and function of simian virus 40 large T-antigen. In Klein G, Ed; Advances in Viral Oncology, Vol. 3, DNA Virus Oncogenes and Their Action; Raven Press: New York, pp 31–57.

Rorke LB (1983): The cerebellar medulloblastoma and its relationship to primitive neuroectodermal tumors. J Neuropathol Exp Neurol 42:1–15.

Rorke LB (1994): Experimental production of primitive neuroectodermal tumors and its relevance to human neuro-oncology. Am J Pathol 144:444–448.

Rosenthal N, Kress M, Gruss P, Khoury G (1983): BK viral enhancer element and a human cellular homology. Science 233:749–755.

Russell DS, Rubinstein LJ (1989): Pathology of tumours of the nervous system. Williams and Wilkins: Baltimore.

Sato-Matsumura K, Hainfellner JA, Budka H (1998): Virus production, cell proliferation and cell death in progressive multifocal leukoencephalopathy. Neuropathology 19: 206–210.

Scaravilli F, Ellis DS, Tovey G, Harcourt-Webster JN, Guiloff RJ, Sinclair E (1989): Unusual development of polyoma virus in the brains of two paitents with the acquired immune deficiency syndrome (AIDS). Neuropathol Appl Neurobiol 15:407–418.

Schmidbauer M, Budka H, Shah KV (1990): Progressive multifocal leukoencephalopathy (PML) in AIDS and in the pre-AIDS era. A neuropathological comparison using immunocytochemistry and in situ DNA hybridization for virus detection. Acta Neuropathol (Berl) 80:375–380.

Selmaj K, Shafit-Zagardo B, Aquino DA, Farooq M, Raine CS, Norton WT, Brosnan CF (1991): Tumor necrosis factor-induced proliferation of astrocytes from mature brain is associated with down-regulation of glial fibrillary acidic protein mRNA. J Neurochem 57:823–830.

Semela D, Glatz M, Hunziker D, Schmid U, Vernazza PL (2000): Causes of death and autopsy findings in patients of the Swiss HIV Cohort Study. Schweiz Med Wochenschr 130:1726–1733.

Shah KV, Daniel RW, Kelly TJ Jr (1977): Immunological relatedness of papovaviruses of the simian virus 40-polyoma subgroup. Infect Immun 18:558–560.

Shapshak P, Tourtellotte WW, Wolman M, Verity N, Verity MA, Schmid P, Syndulko K, Bedows E, Boostanfar R, Darvish M (1986): Search for virus nucleic acid sequences in postmortem human brain tissue using in situ hybridization technology with cloned probes: Some solutions and results on progressive multifocal leukoencephalopathy and subacute sclerosing panencephalitis tissue. J Neurosci Res 16:281–301.

Sharer LR, Cho ES, Epstein LG (1985): Multinucleated giant cells and HTLV-III in AIDS encephalopathy. Hum Pathol 16:760–760.

Sharer LR, Epstein LG, Cho ES, Joshi VV, Meyenhofer MF, Rankin LF, Petito CK (1986): Pathologic features of AIDS encephalopathy in children: Evidence for LAV/HTLV-III infection of brain. Hum Pathol 17:271–284.

Shimada H, Noda K, Mori M, Aoki N, Tajima M, Kato K (1994): Papovavirus detection by electron microscopy in the brain of an elderly patient without overt progressive multifocal leukoencephalopathy. Virchows Arch Int J Pathol 424:569–572.

Shintaku M, Matsumoto R, Sawa H, Nagashima K (2000): Infection with JC virus and possible dysplastic ganglion-like transformation of the cerebral cortical neurons in a case of progressive multifocal leukoencephalopathy. J Neuropathol Exp Neurol 59:921–929.

Silverman L, Rubinstein LJ (1965): Electron microscopic observations on a case of progressive multifocal leukoencephalopathy. Acta Neuropathol (Berl) 5:215–224.

Sima AA, Finkelstein SD, McLachlan DR (1983): Multiple malignant astrocytomas in a patient with spontaneous progressive multifocal leukoencephalopathy. Ann Neurol 14:183–188.

Simmons DT (2000): SV40 large T antigen functions in DNA replication and transformation. Adv Virus Res 55:75–134.

Simon MA, Ilyinskii PO, Baskin GB, Knight HY, Pauley DR, Lackner AA (1999): Association of simian virus 40 with a central nervous system lesion distinct from progressive multifocal leukoencephalopathy in macaques with AIDS. Am J Pathol 154:437–446.

Singer C, Berger JR, Bowen BC, Bruce JH, Weiner WJ (1993): Akinetic-rigid syndrome in a 13-year-old girl with HIV-related progressive multifocal leukoencephalopathy. Mov Disord 8:113–116.

Singer EJ, Stoner GL, Singer P, Tomiyasu U, Licht E, Fahy-Chandon B, Tourtellotte WW (1994): AIDS presenting as progressive multifocal leukoencephalopathy with clinical response to zidovudine. Acta Neurol Scand 90:443–447.

Skolnik PR, Kosloff BR, Hirsch MS (1988): Bidirectional interactions between human immunodeficiency virus type 1 and cytomegalovirus. J Infect Dis 157:508–514.

Small JA, Khoury G, Jay G, Howley PM, Scangos GA (1986a): Early regions of JC virus and BK virus induce distinct and tissue-specific tumors in transgenic mice. Proc Natl Acad Sci USA 83:8288–8292.

Small JA, Scangos GA, Cork L, Jay G, Khoury G (1986b): The early region of human papovavirus JC induces dysmyelination in transgenic mice. Cell 46:13–18.

Stohlman SA, Hinton DR (2001): Viral induced demyelination. Brain Pathol 11:92–106.

Stoner GL (1993): Polyomavirus models of brain infection and the pathogenesis of multiple sclerosis. Brain Pathol 3:213–227.

Stoner GL, Agostini HT, Ryschkewitsch CF, Mázló M, Gullotta F, Wamukota W, Lucas S (1998): Detection of JC virus in two African cases of progressive multifocal leukoencephalopathy including identification of JCV Type 3 in a Gambian AIDS patient. J Med Microbiol 47:733–742.

Stoner GL, Best PV, Mázló M, Ryschkewitsch CF, Walker DL, Webster HD (1988a): Progressive multifocal leukoencephalopathy complicating systemic lupus erythematosus: Distribution of JC virus in chronically demyelinated cerebellar lesions. J Neuropathol Exp Neurol 47:307.

Stoner GL, Ryschkewitsch CF, Walker DL, Soffer D, Webster HD (1988b): A monoclonal antibody to SV40 large T-antigen labels a nuclear antigen in JC virus-transformed cells and in progressive multifocal leukoencephalopathy (PML) brain infected with JC virus. J Neuroimmunol 17:331–345.

Stoner GL, Ryschkewitsch CF, Walker DL, Soffer D, Braun DG, Hochkeppel HK, Webster HD (1988c): Early viral proteins as autoantigens. Evidence from JC virus large T antigen. Ann NY Acad Sci 540:665–668.

Stoner GL, Ryschkewitsch CF, Walker DL, Webster HD (1986): JC papovavirus large tumor (T)-antigen expression in brain tissue of acquired immune deficiency syndrome (AIDS): and non-AIDS patients with progressive multifocal leukoencephalopathy. Proc Natl Acad Sci USA 83:2271–2275.

Stoner GL, Soffer D, Ryschkewitsch CF, Walker DL, Webster HD (1988d): A double-label method detects both early (T-antigen) and late (capsid) proteins of JC virus in progressive multifocal leukoencephalopathy brain tissue from AIDS and non-AIDS patients. J Neuroimmunol 19:223–236.

Sweeney BJ, Manji H, Miller RF, Harrison MJG, Gray F, Scaravilli F (1994): Cortical and subcortical JC virus infection: Two unusual cases of AIDS associated progressive multifocal leukoencephalopathy. J Neurol Neurosurg Psychiatry 57:994–997.

Tagliati M, Simpson D, Morgello S, Clifford D, Schwartz RL, Berger JR (1998): Cerebellar degeneration associated with human immunodeficiency virus infection. Neurology 50:244–251.

Takahashi H, Yogo Y, Furuta Y, Takada A, Irie T, Kasai M, Sano K, Fujioka Y, Nagashima K (1992): Molecular characterization of a JC virus (Sap-1) clone derived from a cerebellar form of progressive multifocal leukoencephalopathy. Acta Neuropathol (Berl) 83:105–112.

Tellez-Negal I, Harter DH (1966): Subacute sclerosing leukoencephalitis: Ultrastructure of intranuclear and intracytoplasmic inclusions. Science 154:899–901.

Teo CG, Wong SY, Best PV (1989): JC virus genomes in progressive multifocal leukoencephalopathy: Detection using a sensitive non-radioisotopic in situ hybridization method. J Pathol 157:135–140.

Trapp BD, Small JA, Pulley M, Khoury G, Scangos GA (1988): Dysmyelination in transgenic mice containing JC virus early region. Ann Neurol 23:38–48.

Tretiakova A, Krynska B, Gordon J, Khalili K (1999a): Human neurotropic JC virus early protein deregulates glial cell cycle pathway and impairs cell differentiation. J Neurosci Res 55:588–599.

Tretiakova A, Otte J, Croul SE, Kim JH, Johnson EM, Amini S, Khalili K (1999b): Association of JC virus large T antigen with myelin basic protein transcription factor (MEF-1/Purα) in hypomyelinated brains of mice transgenically expressing T antigen. J Virol 73:6076–6084.

Ueki K, Richardson EP Jr, Henson JW, Louis DN (1994): In situ polymerase chain reaction demonstration of JC virus in progressive multifocal leukoencephalopathy, including an index case. Ann Neurol 36:670–673.

Vago L, Nebuloni M, Sala E, Cinque P, Bonetto S, Isella A, Ottoni L, Crociati A, Costanzi G (1996): Coinfection of the central nervous system by cytomegalovirus and herpes simplex virus type 1 or 2 in AIDS patients: Autopsy study on 82 cases by immunohistochemistry and polymerase chain reaction. Acta Neuropathol (Berl) 92:404–408.

Vanderhaeghen JJ, Perier O (1965): [Progressive multifocal leukoencephalitis. Demonstration of viral particles by electron microscopy]. Acta Neurol Psychiatr Belg 65: 816–837.

Vandersteenhoven JJ, Dbaibo G, Boyko OB, Hulette CM, Anthony DC, Kenny JF, Wilfert CM (1992): Progressive multifocal leukoencephalopathy in pediatric acquired immunodeficiency syndrome. Pediatr Infect Dis J 11:232–237.

Varakis J, ZuRhein GM, Padgett BL, Walker DL (1978): Induction of peripheral neuroblastomas in Syrian hamsters after injection as neonates with JC virus, a human polyoma virus. Cancer Res 38:1718–1722.

Vazeux R, Cumont M, Girard PM, Nassif X, Trotot P, Marche C, Matthiessen L, Vedrenne C, Mikol J, Henin D, Katlama C, Bolgert F, Montagnier L (1990): Severe encephalitis resulting from coinfections with HIV and JC virus. Neurology 40:944–948.

Vinters HV, Kwok MK, Ho HW, Anders KH, Tomiyasu U, Wolfson WL, Robert F (1989): Cytomegalovirus in the nervous system of patients with the acquired immune deficiency syndrome. Brain 112:245–268.

Vogel FS (1959): Changes in fine structure of cerebellar neurons following ionizing radiation. J Neuropathol Exp Neurol 18:580–589.

Von Einsiedel RW, Fife TD, Aksamit AJ, Cornford ME, Secor DL, Tomiyasu U, Itabashi HH, Vinters HV (1993): Progressive multifocal leukoencephalopathy in AIDS: A clinicopathologic study and review of the literature. J Neurol 240:391–406.

Walker DL, Padgett BL (1983): The epidemiology of human polyomaviruses. In Sever JL, Madden DL, Eds; Polyomaviruses and Human Neurological Diseases; Alan R. Liss: New York, pp 99–106.

Walker DL, Padgett BL, Zu Rhein GM, Albert AE, Marsh RF (1973): Human papovavirus (JC): Induction of brain tumors in hamsters. Science 181:674–676.

Watanabe I, Preskorn SH (1976): Virus-cell interaction in oligodendroglia, astroglia and phagocyte in progressive multifocal leukoencephalopathy. An electron microscopic study. Acta Neuropathol (Berl) 36:101–115.

Weidenheim KM, Nelson SJ, Kure K, Harris C, Biempica L, Dickson DW (1992): Unusual patterns of *Histoplasma capsulatum* meningitis and progressive multifocal leukoencephalopathy in a patient with the acquired immunodeficiency virus. Hum Pathol 23:581–586.

White FA, Ishaq M, Stoner GL, Frisque RJ (1992): JC virus DNA is present in many human brain samples from patients without progressive multifocal leukoencephalopathy. J Virol 66:5726–5734.

Wiley CA, Grafe M, Kennedy C, Nelson JA (1988): Human immunodeficiency virus (HIV) and JC virus in acquired immune deficiency syndrome (AIDS) patients with progressive multifocal leukoencephalopathy. Acta Neuropathol (Berl) 76:338–346.

Wiley CA, Nelson JA (1988): Role of human immunodeficiency virus and cytomegalovirus in AIDS encephalitis. Am J Pathol 133:73–81.

Woodhouse MA, Dayan AD, Burston J, Caldwell I, Adams JH, Melcher D, Urich H (1967): Progressive multifocal leukoencephalopathy: Electron microscope study of four cases. Brain 90:863–870.

Woodroofe MN, Cuzner ML, Ironside JW (1994): In situ hybridization in neuropathology. Neuropathol Appl Neurobiol 20:562–572.

Wrzolek MA, Brudkowska J, Kozlowski PB, Rao C, Anzil AP, Klein EA, Del Rosario C, Abdu A, Kaufman L, Chandler FW (1995): Opportunistic infections of the central nervous system in children with HIV infection: Report of 9 autopsy cases and review of literature. Clin Neuropathol 14:187–196.

Yang B, Prayson RA (2000): Expression of Bax, Bcl-2, and P53 in progressive multifocal leukoencephalopathy. Mod Pathol 13:1115–1120.

Yogo Y, Guo J, Iida T, Satoh KI, Taguchi F, Takahashi H, Hall WW, Nagashima K (1994): Occurrence of multiple JC virus variants with distinctive regulatory sequences in the brain of a single patient with progressive multifocal leukoencephalopathy. Virus Genes 8:99–105.

Yong VW, Moumdjian R, Yong FP, Ruijs TC, Freedman MS, Cashman N, Antel JP (1991): Gamma-interferon promotes proliferation of adult human astrocytes in vitro and reactive gliosis in the adult mouse brain in vivo. Proc Natl Acad Sci USA 88: 7016–7020.

Zimmer C, Marzheuser S, Patt S, Rolfs A, Gottschalk J, Weigel K, Gosztonyi G (1992): Stereotactic brain biopsy in AIDS. J Neurol 239:394–400.

Zu Rhein GM (1969): Association of papova-virions with a human demyelinating disease (progressive multifocal leukoencephalopathy). Prog Med Virol 11:185–247.

Zu Rhein GM (1983): Studies of JC virus–induced nervous system tumors in the Syrian hamster: A review. Prog Clin Biol Res 105:205–221.

Zu Rhein GM (1972): Virions in progressive multifocal leukoencephalopathy. In Minckler J, Ed; Pathology of the Nervous System; McGraw-Hill: New York, Vol 3, pp 2893–2912.

Zu Rhein GM (1987): Human viruses in experimental neurooncogenesis. In Grundmann E, Ed; Cancer Campaign, Vol 10, Experimental Neurooncology, Brain Tumor and Pain Therapy; Gustav Fischer: Stuttgart, pp 19–46.

Zu Rhein GM, Chou SM (1965): Particles resembling papova viruses in human cerebral demyelinating disease. Science 148:1477–1479.

Zu Rhein GM, Chou SM (1968): Papova virus in progressive multifocal leukoencephalopathy. Research Publications. Assoc Res Nervous Mental Dis 44:307–362.

Zu Rhein GM, Varakis JN (1979): Perinatal induction of medulloblastomas in Syrian golden hamsters by a human polyoma virus (JC). Natl Cancer Inst Monogr 51:205–208.

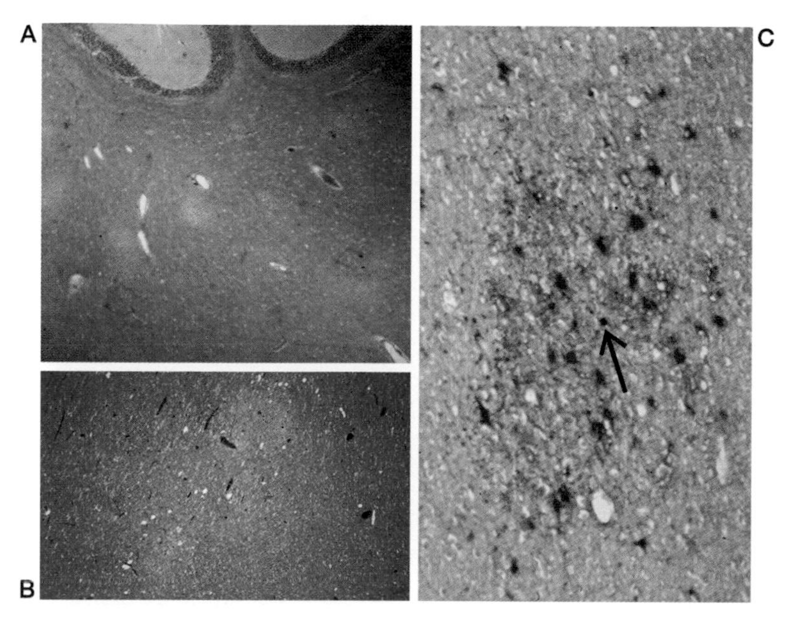

Figure 12.2. Early demyelinated lesions in cerebellar white matter. Luxol fast blue/H&E stain. For full caption, see page 260.

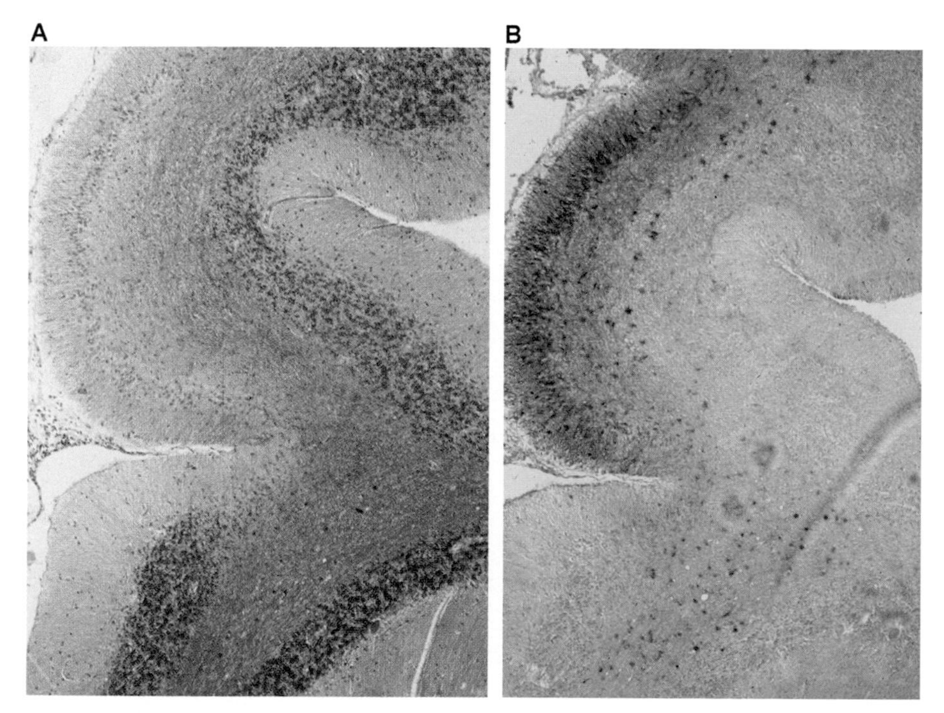

Figure 12.3. Cerebellar folia: Granule cell loss with underlying demyelination. For full caption, see page 261.

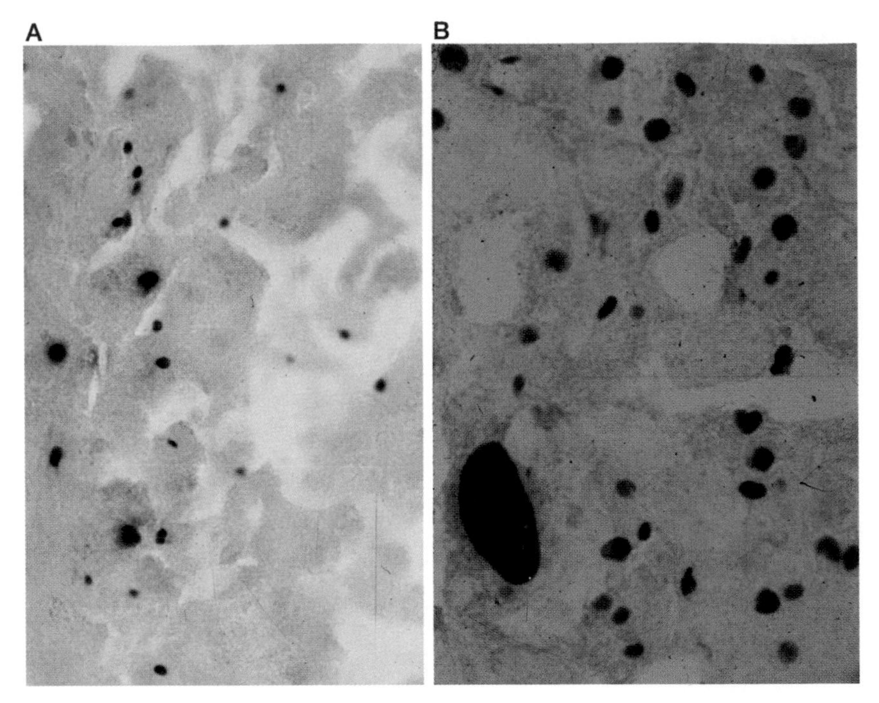

Figure 12.6. JCV protein expression revealed with double-label immunocytochemical staining for early (T antigen) and late (capsid) proteins. For full caption, see page 269.

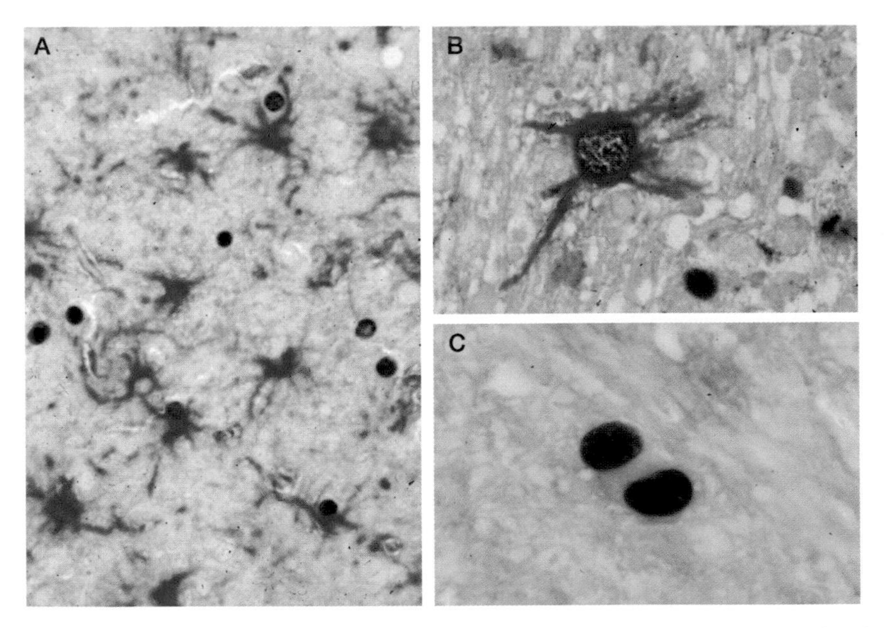

Figure 12.7. JCV protein expression revealed with double-label immunocytochemical staining for late (capsid) proteins and glial fibrillary acidic protein (GFAP). For full caption, see page 271.

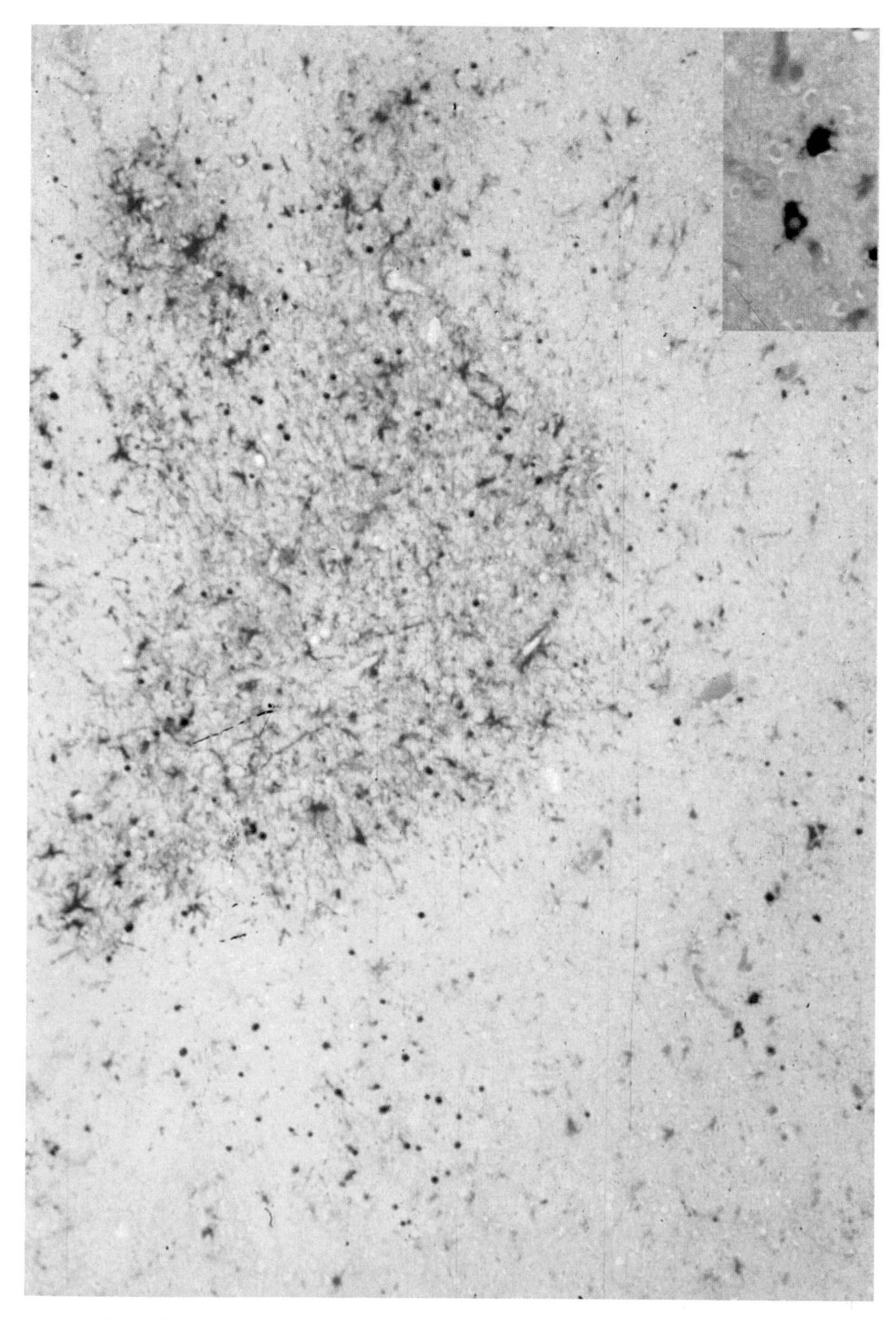

Figure 12.8. Double immunocytochemical label for GFAP and JCV capsid proteins (no counterstain). JCV-containing cells occur in the margins of the mature, gliotic demyelinated lesion and in the incipient lesion below the gliotic lesion with minimal astrocyte reactivity, but with numerous JCV-infected cells, including some with astrocyte morphology. Astrocytes with capsid located cytoplasmically are seen in the lower right corner and are enlarged in the inset.

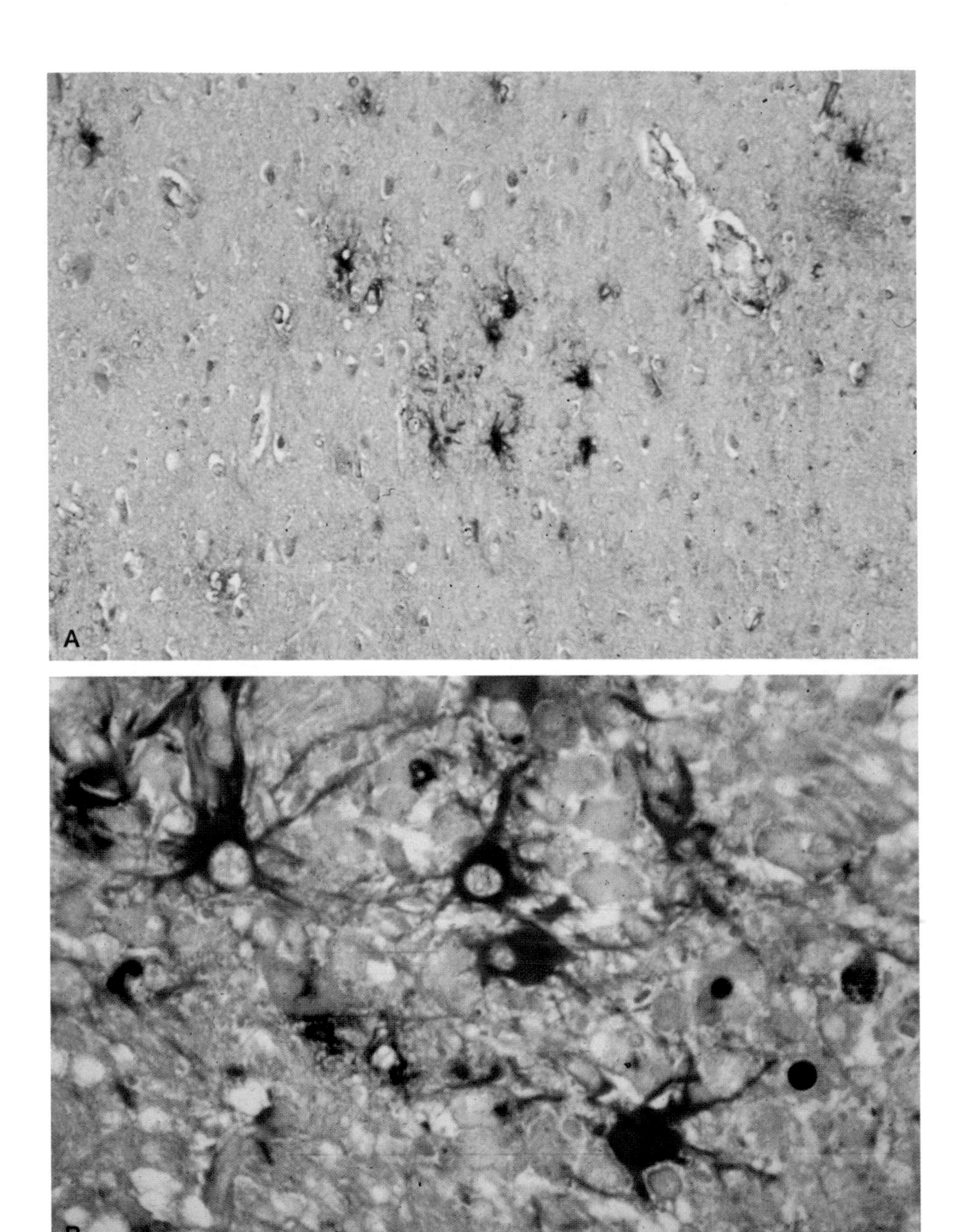

Figure 12.22. Astrocytes and PML pathogenesis. (**A**) Small cluster of enlarged, reactive astrocytes in the subcortical white matter of an early PML case discovered at autopsy (Åström and Stoner, 1994). Cells display GFAP reactivity (red), but no JCV-infected oligodendrocytes with capsid proteins (black) were evident. Method same as in Figure 12.2C. An adjacent section stained with LFB/H&E showed no evidence of myelin destruction (not shown). These astrogliotic clusters were the earliest discernible PML lesions. No counterstain. Medium magnification. Taken with permission from Åström and Stoner, 1994. (**B**) Established lesion in the basal ganglia with reactive astrocytes (red), including one with apparent perinuclear JCV capsid proteins in the cytoplasm (lower right). This cell extends a process to an infected oligodendrocyte with JCV capsid proteins in the nucleus (black). No counterstain. High magnification.

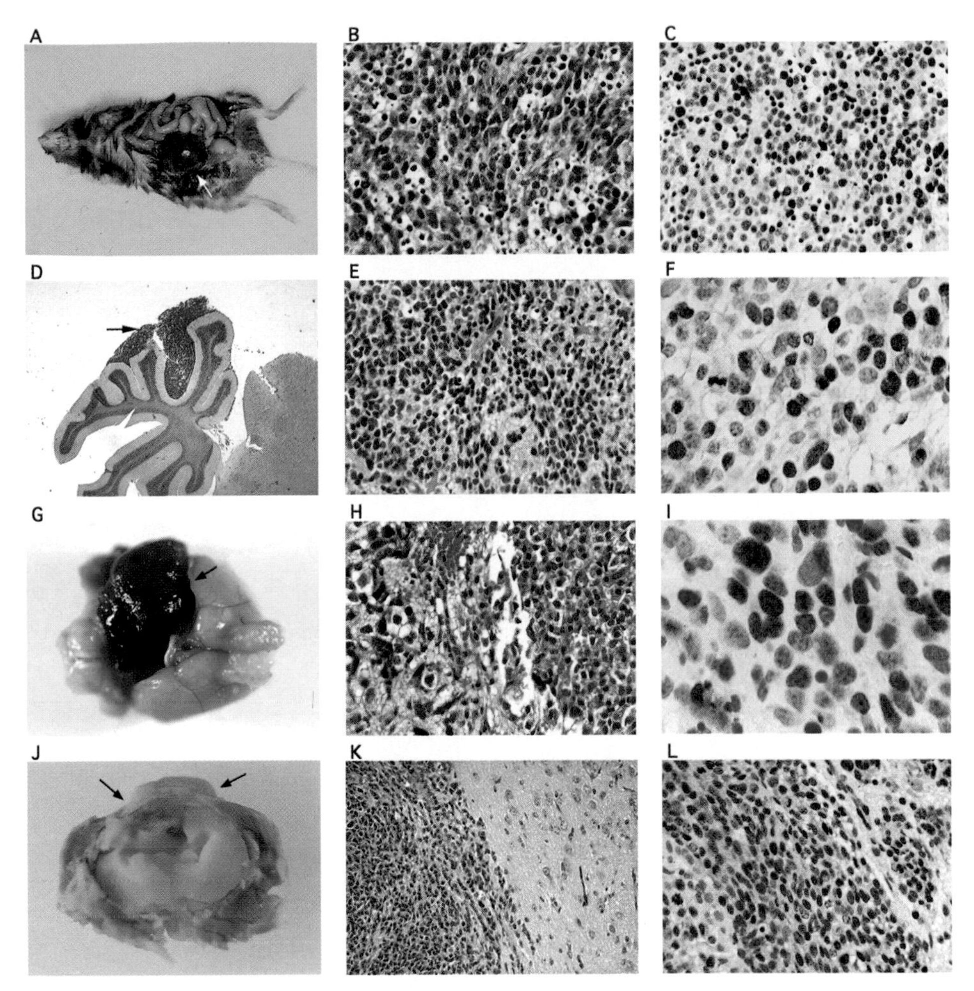

Figure 15.3. Induction of neural origin tumors by JCV T antigen in experimental animals. (**A**) JCV T antigen transgenic mice develop peripheral neuroblastoma, which can present as a solid, well-circumscribed mass, in the abdominal cavity (arrow). (**B**) Neuroblastomas are histologically characterized by densely cellular neoplasms with a high nuclear to cytoplasmic ratio. (**C**) The majority of the nuclei show immunoreactivity to the viral protein T antigen. (**D**) Transgenic mice generated with sequences for T antigen under the control of the JCV archetype promoter develop cerebellar medulloblastomas within the foliæ (arrow). (**E**) Histologically, the tumors appear similar to neuroblastoma. (**F**) JCV T antigen can be detected by immunohistochemistry in tumor cell nuclei. (**G**) T antigen transgenic mice may also develop pituitary tumors, which appear as large masses at the base of the skull (arrow). (**H**) Histology demonstrates a highly pleomorphic tumor (left) adjacent to normal pituitary (right). (**I**) Tumor cells show nuclear staining for JCV T antigen. (**J**) JCV-transformed HJC cells transplanted into the brain of syngeneic Syrian hamsters form an intracranial mass that can protrude through the superior aspect of the skull (arrows). (**K**) A clear line of demarcation is present between the highly cellular pleomorphic cells (left) and the surrounding normal brain parenchyma (right). (**L**) The majority of the tumor cells express nuclear T antigen. A, G, J, original magnification, ×4; D, original magnification, ×10; B, E, H, K, hematoxylin and eosin, original magnification, ×100; C, hematoxylin counterstain, original magnification ×100; F, L, hematoxylin counterstain, original magnification ×200; I, hematoxylin counterstain, original magnification ×400; D. Taken with permission from Krynska et al., 1999.

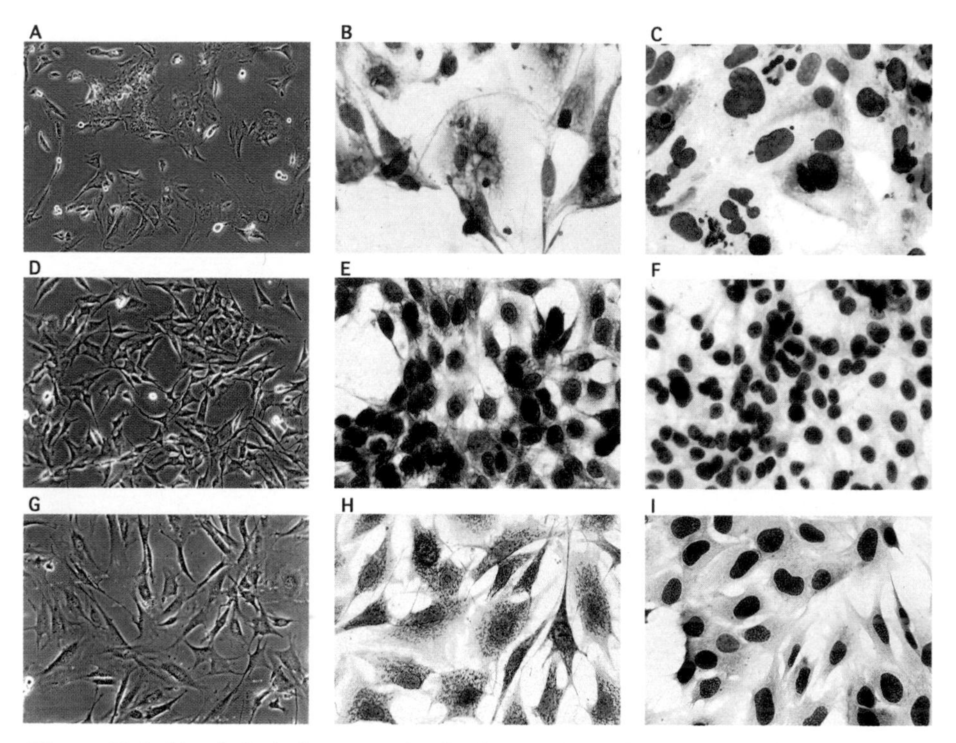

Figure 15.4. Morphologic features of cells derived from JCV-induced tumors in various experimental animals. (**A**) Phase microscopy of the glioblastoma cell line Owl 586 generated upon intracerebral inoculation of an owl monkey with JC virus. (**B**) Immunostaining with the cellular marker GFAP demonstrates that the tumor is of glial origin. (**C**) Immunostaining for JCV T antigen shows that the majority of the cells express T antigen in the nucleus. (**D**) HJC cells cultured from a tumor induced upon intracerebral injection of JCV into newborn Syrian hamsters shown by phase microscopy. (**E**) HJC glial cells are positive for GFAP. (**F**) Nearly all of the cells express JCV T antigen. (**G**) Phase contrast of BS-1 B8 cells derived from JCV T antigen–induced mouse medulloblastoma. (**H**) Synaptophysin staining of BS-1 B8 cells demonstrate the neuronal origin of the tumor cells. (**I**) The majority of the tumor cells show nuclear immunoreactivity for JCV T antigen.

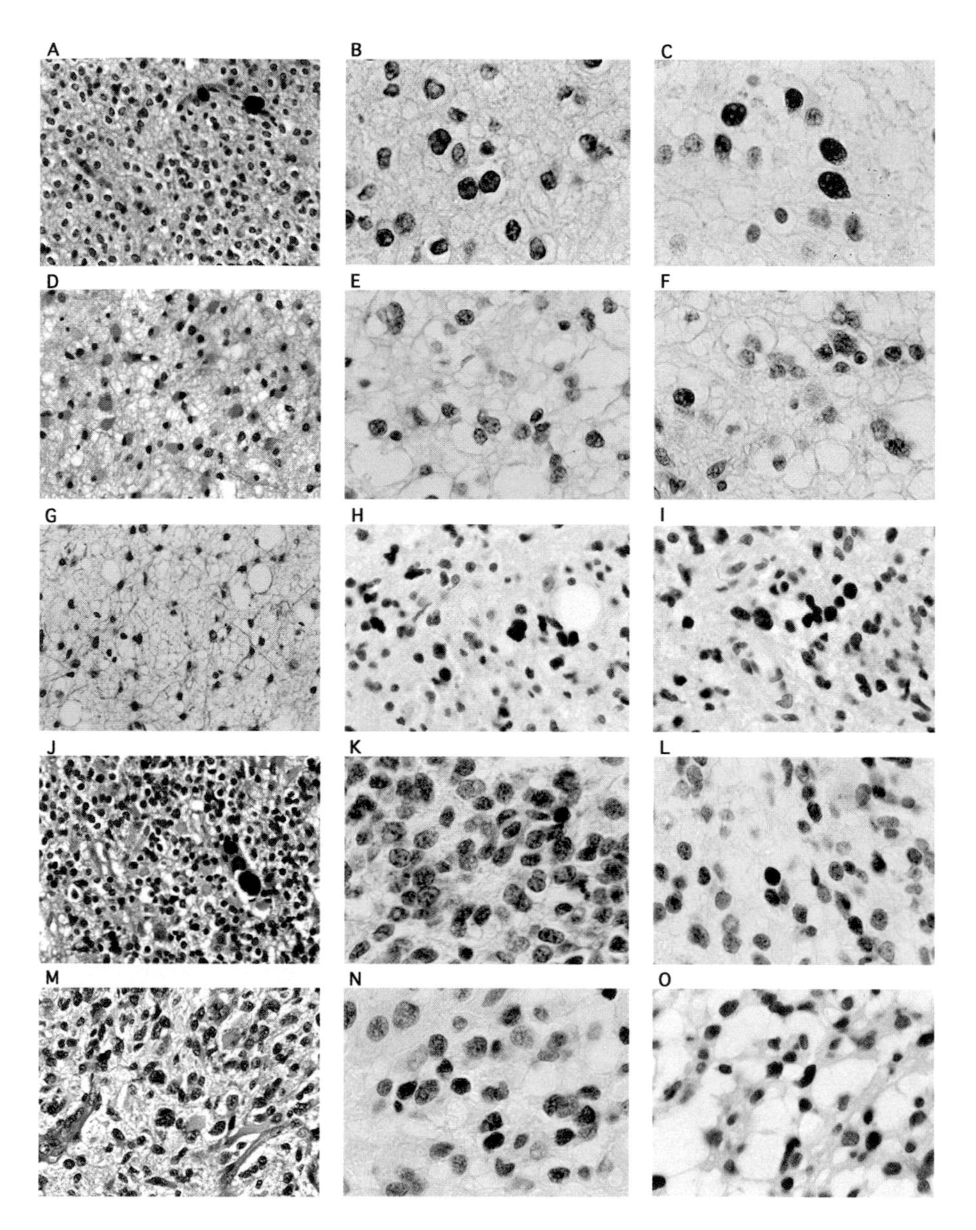

Figure 15.5. Detection of JCV T antigen and the cellular protein p53 in human glial origin tumors. Histologic evaluation and immunohistochemistry for JCV T antigen and the cellular tumor suppressor protein p53 in a number of human glial tumors is shown: oligodendroglioma (**A**, hematoxylin and eosin staining; **B**, immunohistochemical staining for T antigen; **C**, immunostaining for p53); gemistocytic astrocytoma (**D**–**F**); fibrillary astrocytoma (**G**–**I**); anaplastic oligodendroglioma (**J**–**L**); anaplastic astrocytoma (**M**–**O**). A, D, G, J, and M, hematoxylin and eosin, original magnification, ×400; B, C, E, F, H, I, K, L, N, O, hematoxylin counterstain, original magnification, ×1000.

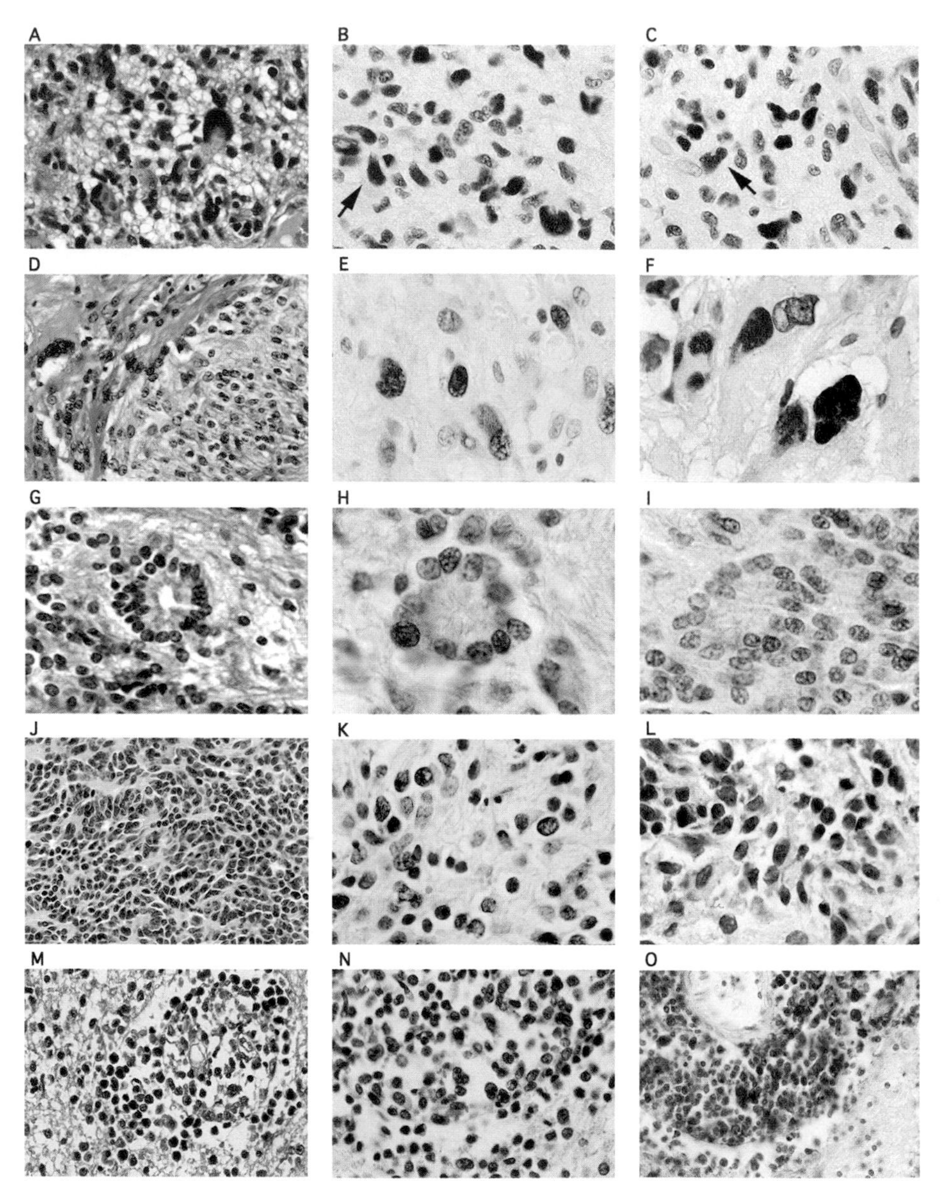

Figure 15.6. Immunohistochemical staining of various human tumors for T antigen expression. JCV T antigen has been detected in a variety of human tumors. Histology and immunohistochemistry for T antigen and cellular proteins are shown for the following tumors: glioblastoma multiforme (**A**, hematoxylin and eosin staining; **B**, immunohistochemical staining for T antigen; **C**, immunostaining for p53); gliosarcoma (**D**–**F**), ependymoma (**G**–**I**), medulloblastoma (**J**–**L**), B-cell lymphoma (**M**, hematoxylin and eosin staining; **N**, immunohistochemical staining for T antigen; **O**, immunostaining for the EBV protein latent membrane protein [LMP]). A, D, G, J, and M, hematoxylin and eosin staining, original magnification, ×400; B, C, E, F, K, L, N, and O, hematoxylin counterstain, original magnification, ×400; H and I, hematoxylin counterstain, original magnification ×1000.

13

LABORATORY FINDINGS IN PROGRESSIVE MULTIFOCAL LEUKOENCEPHALOPATHY

THOMAS WEBER, Dr. MED.

1. INTRODUCTION

Until the advent of the polymerase chain reaction (PCR), diagnosis of progressive multifocal leukoencephalopathy (PML) intra vitam was based on brain biopsy (Major et al., 1992; Walker, 1985) or relayed for confirmation at autopsy (Åström et al., 1958; Richardson, 1961; Budka and Shah, 1983). A sporadic report suggested diagnosis of PML by the detection of viral antigen in cerebrospinal fluid (CSF) (Peters et al., 1980), but these findings could not be reproduced. Antibody detection by hemagglutination inhibition test (HAI) failed to detect JC virus (JCV) capsid-specific antibody in the CSF of nine autopsy-proven cases of PML (Padgett and Walker, 1983).

With an increasing incidence of PML due to the AIDS pandemic and the introduction of PCR, several studies indicated the feasibility of the diagnosis of PML by CSF analysis (Vignoli et al., 1993; Brouqui et al., 1992; Henson et al., 1991; Moret et al., 1993; Gibson et al., 1993; Weber et al., 1994a; McGuire et al., 1995; Cinque et al., 1996; Vago et al., 1996; Perrons et al., 1996; de Luca et al., 1996; d'Arminio Monforte et al., 1997; Antinori et al., 1997; Ferrante et al., 1997, 1998; Matsiota Bernard et al., 1997; Sugimoto et al., 1998). Comparative sequence analyses of clones obtained from the CSF

Human Polyomaviruses: Molecular and Clinical Perspectives, Edited by Kamel Khalili and Gerald L. Stoner.
ISBN 0-471-39009-7

and brain of the same patients showed the preponderance of a single genotype that was identical in both compartments, thus supporting the hypothesis that CSF reflects intraparenchymal JCV replication (Sugimoto et al., 1998). The sensitivity for the diagnosis of PML by PCR of a single CSF sample is about 81%, while the specificity is about 99% (Weber et al., 1996).

With the introduction of highly active antiretroviral therapy (HAART), PML cases with prolonged survival are reported with increased frequency (Miralles et al., 1998; Albrecht et al., 1998; Yiannoutsos et al., 1999; Power et al., 1997; Elliot et al., 1997; Domingo et al., 1997; Baldeweg and Catalan, 1997; Teofilo et al., 1998; Baqi et al., 1997; Dworkin et al., 1999; Gasnault et al., 1999; Tantisiriwat et al., 1999; Inui et al., 1999; Giudici et al., 2000). Current evidence suggests that JC viral load in CSF as measured by semiquantitative or quantitative PCR is a predictor of survival time (Yiannoutsos et al., 1999; De Luca et al., 1999; Taoufik et al., 2000). Survival time has also been linked to the CD4 count at the time of presentation with symptoms suggestive of PML (Dworkin et al., 1999). Patients with a CD4 count below 50 cells/mm^3 have a median survival time ranging from 1 month to at most 5 months (Dworkin et al., 1999).

2. CELLULAR IMMUNE RESPONSE

E.P. Richardson was the first to suggest an altered immune response as the underlying cause leading to the development of PML (Richardson, 1961). In 1969, Ellison reported the first study on the immunologic status of a patient with PML. This patient showed no evidence at all of a delayed-type hypersensitivity (DHT) reaction to tetanus toxoid, diphtheria toxoid, mumps vaccine, or 2,4-dinitro-fluorobenzene (DNFB). Furthermore, immunoglobulin A (IgA) and M (IgM) levels were below the lower limit of normal values. Immunoglobulin G (IgG) was within the normal range.

In 1972, Knight and colleagues reported on a 22-year-old patient who developed PML in the setting of a malabsorption syndrome with long-standing hypogammaglobulinemia of IgG and IgA. This patient also had no DHT reaction and showed a decreased proliferation in one-way mixed leukocyte cultures. In addition, incorporation of ^{3}H-thymidine after stimulation with phytohemagglutinin (PHA) was significantly reduced.

Rockwell and colleagues described in 1976 a 45-year-old female patient with a biopsy-proven PML without evidence for deficits in either cellular or humoral immunity. Skin tests for tuberculin and *Candida* antigens as well as dinitrochlorobenzene were normal. The patient's lymphocytes transformed normally in response to pokeweed and PHA mitogens as well as purified protein derivative of tuberculin antigen. The number and function of T lymphocytes as determined by the sheep erythrocyte rosette test were also normal. In contrast to these findings are those of Willoughby and colleagues (1980), who tested seven patients with PML. Lymphocyte proliferation in response to the mitogens

PHA, concanavalin A (ConA), and pokeweed mitogen (PWM) was significantly reduced for the seven PML patients compared with seven healthy controls (Willoughby et al., 1980). In addition, production of the pleiotropic cytokine leukocyte migration inhibitory factor (LIF) in response to stimulation with JCV antigen was absent.

Katz and colleagues (1994) described a 15-year-old boy with Wiskott-Aldrich syndrome with a CD4 count of 0.12 X 10^9/l and depressed T-cell function as measured by mitogen-induced proliferation to ConA and PHA. In 1995, Owen and colleagues described a 43-year-old man with chronic myeloid leukemia who received allogeneic bone marrow transplantation after immunosuppressin. He underwent cytomegalovirus seroconversion and developed PML. By reactive expansion of CD3-positive large granular lymphocytes humoral and cellular immunities were reduced, and the CD4 count ranged from 0.3 to 0.4×10^9/l (Owen et al., 1995).

Frye et al. (1997) studied the cellular immune response of a PML patient and compared it to the proliferative response of healthy donors. With purified JCV particles as the antigen, stimulation indexes (SI) of 8 and 9 were found for the healthy donors, and an SI of 4 for the PML patient. After stimulation with PHA, healthy donors showed SI values of 15 and 25, whereas no response was seen with the PML patients' peripheral blood mononuclear cells (PBMC). When recombinant VP1 was used as the target antigen, a reduced proliferation of PBMC was seen in PML patients compared with healthy donors (Weber et al., 1998). Interestingly, the PBMC of one HIV-negative patient with biopsy-proven PML who recovered from PML showed an increased production of interferon-γ in response to stimulation with VP1 (Weber et al., 1998).

A cytotoxic T-cell assay showed specific killing of autologous targets expressing recombinant JCV VP1 in the PBMC of four of four HIV-positive PML patients with long-term survival, in one of seven HIV-positive patients without PML, in one of three HIV-negative PML patients, and in none of four healthy donors, indicating a role of CD8-positive effector cells in the control of the progression of PML (Koralnik et al., 2001).

3. HUMORAL IMMUNITY

JCV agglutinates human type O erythrocytes (Padgett and Walker, 1973). Antibodies to JCV can be determined by an HAI test using human type O erythrocytes from donors lacking antibodies against JCV. Padgett and her colleagues used JCV grown in cultures of primary human fetal glial (PHFG) cells and assayed the sera of 406 persons. In their series, seropositivity, defined as an HAI titer of 32 or greater, rose from 10% for those aged 0–4 years to 60% for those aged 20–29 years. The highest rate of seroconversion occurred during the first 14 years. In addition, sera with HAI titers of 32 or greater contained detectable levels of neutralizing antibodies. No sex difference in the incidence of antibodies against JCV was seen (Padgett and Walker, 1973).

Serologic assays by HAI, complement fixation, neutralizing antibodies, or various immunoassays are not diagnostic of PML (Weber et al., 1997b; Knowles et al., 1995; Padgett and Walker, 1976). Almost all patients except those few cases with hypo- or agammaglobulinemia with PML have detectable IgG antibodies in their serum, thus further substantiating JCV as the causative agent of PML.

In 1995, Knowles and her colleagues used an HAI test and a JCV IgM-antibody capture radioimmunoassay (MACRIA) to analyze the humoral immune response to JCV in 28 PML patients and 71 controls. JCV HAI antibodies were detected in the sera of all but one patient with PML (96%) compared with 68% of control patients.

As immunoassays based on recombinant VP1 have several advantages over tissue culture–derived virus, Weber and colleagues (1997b) developed a recombinant antigen-based ELISA. First, sufficient amounts of JCV are difficult to grow. Second, the amount of viral protein may be insufficient for an optimal antigen concentration in an immunoassay. Third, purified JCV is not stable. Fourth, HAI titer decreases with time (Major et al., 1992; Weber et al., 1997b). A quantitative ELISA using recombinant VP1 revealed an overall seroprevalence of VP1-specific IgG antibodies of 84.5% in 155 patients (Weber et al., 1997b) (Fig. 13.1). Of these, 86% of healthy controls (43 of 50), 88% of patients with an impaired blood–brain barrier (29 of 33), 76% of patients with multiple sclerosis (28 of 37), and 89% of HIV-seropositive patients (31 of 35) had detectable antibodies in their sera.

4. CEREBROSPINAL FLUID

The CSF in PML has been reported as normal with a normal protein content (≤50 mg/dl), a normal cell count (≤4 cells/μl), and without intrathecal antibody synthesis as demonstrated by the lack of oligoclonal bands (Walker, 1985). More recently, however, Berger and colleagues (1987) reported a slight increase in CSF protein (range 52–72 mg/dl) in about one fourth, an elevated IgG or IgG–albumin index in about one fifth, and a slight pleocytosis in about one eighth of the patients. Others have seen a pleocytosis in 14%, an increased CSF protein in 54% (median, 54; mean, 52; range, 33–130 mg/dl), and an intrathecal IgG synthesis in 29% of cases (Weber et al., 1996). A more specific indicator of blood–brain barrier function is the albumin quotient Q(Alb) (Reiber and Felgenhauer, 1987; Reiber and Lange, 1991), which is calculated according to the following formula:

$$Q(Alb) = \frac{[Alb_{CSF}] * 10^3}{[Alb_{Serum}]}$$

The normal value is age dependent. Values exceeding 6.5 indicate an impaired blood–brain barrier in patients aged 15–40 years; in those aged 40–60

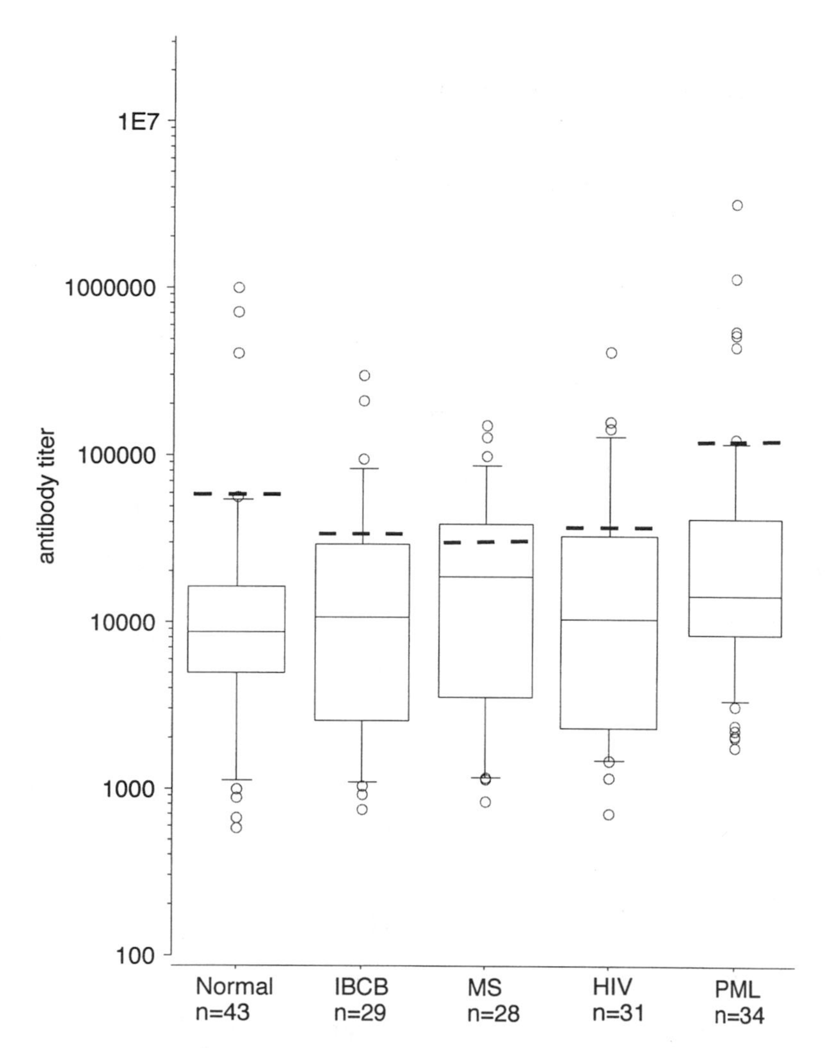

Figure 13.1. The plot shows 10th, 25th, 50th, 75th, and 90th centiles and outliers. The dashed line indicates the mean titer for each group; the solid line inside the box plot indicates the median. The mean titer for normal controls is 59,913; for patients with an impaired blood brain barrier (IBBB) it is 33,791; for multiple sclerosis patients it is 30,658; for HIV-positive patients it is 38,595; and for PML patients it is 118,248. There is no significant difference between the five groups in the mean serum titers (Kruskal-Wallis test, $p = 0.1083$).

years, the upper limit of normal is 8.0 (Reiber and Felgenhauer, 1987). In a series of 62 PML patients, Q(Alb) was normal for 74% (6.5 $\pm$ 4; range, 1.9–22.03) (Weber et al., 1997b). An even higher percentage of normal findings in 87% of PML patients has been reported in another study (Monno et al., 1999). Thus, in 13–25% of patients with PML a moderate blood–brain barrier impairment may be found. In HIV-positive patients in particular, these changes may indicate a concomitant opportunistic infection such as cerebral toxoplasmosis, cryptoccocal meningitis, or cytomegalovirus ventriculitis/encephalitis.

Oligoclonal bands were seen in 42% of PML patients compared with 41% in asymptomatic HIV-positive patients (Weber et al., 1997b). These data suggest that these oligoclonal bands are nonspecific CSF findings due to the underlying HIV infection.

Until 1995 it was widely believed that CSF does not contain antibodies to JCV (Padgett et al., 1976). CSF analysis for JCV-specific antibodies was thus considered nondiagnostic (Major et al., 1992; Major and Ault, 1995). To determine an intrathecal humoral immune response, Knowles and colleagues (1995) calculated an antibody index (AI) according to the following formula:

$$AI = \frac{\text{CSF JCV antibody titer}}{\text{Serum JCV antibody titer}} \div \frac{\text{CSF albumin}}{\text{Serum albumin}}$$

With this approach, a JCV-specific intrathecal immune response was defined by an AI of 2 or greater. In 12 of 18 PML patients (67%) an intrathecal immune response by JCV HI antibodies was found, but none in the controls. As further proof of an intrathecal humoral immune response, JCV-specific IgG bands were detected in seven PML patients. In a comparable approach, the systemic and intrathecal humoral immune responses were analyzed in 62 PML patients and 155 controls (Weber et al., 1997b).

Recombinant JCV VP1 cloned in the baculovirus system served as antigen instead of tissue culture–derived JCV (Goldmann et al., 1999). The recombinant VP1 was used in an ELISA assay for the detection of VP1-specific IgG. Antibody titers in CSF and serum were measured in arbitrary units (E). The VP1-specific CSF/serum ratio was calculated according to the following formula:

$$Q(IgG)spec = \frac{E_{CSF} * 10^3}{E_{Serum}}$$

For the detection of locally synthesized VP1 specific antibodies, the Q(IgG)spec was related to the ratio of CSF to serum IgG (Q(IgG)tot:

$$Q(IgG)tot = \frac{[IgG_{CSF}] * 10^3}{[IgG_{Serum}]}$$

The ratio of Q(IgG)spec to Q(IgG)tot is defined as the antibody specificity index (ASI) (Weber et al., 1991, 1997b; Reiber and Lange, 1991). To compensate for a potential blood–brain barrier leak, Q(IgG)spec has to be related to the individual albumin ratio (Reiber and Felgenhauer, 1987). This is achieved by using the upper limit of the normal range of IgG in relation to the blood–brain barrier function (Q(IgG)lim, which is calculated by

$$Q(IgG)\lim = 0.8 * \sqrt{(QAlb)^2 + 1.5 * 10^{-6}} - 1.8 * 10^{-3}$$

In cases with a normal IgG ratio, that is, without intrathecal synthesis of polyspecific IgG antibodies (Q[IgG]tot < Q[IgG]lim) (Reiber and Lange, 1991), the ASI is calculated as follows:

$$ASI = \frac{Q(IgG)spec}{Q(IgG)tot}$$

In those cases with a polyspecific intrathecal IgG synthesis, that is, Q(IgG)tot > Q(IgG)lim, the ASI is calculated by

$$ASI = \frac{Q(IgG)spec}{Q(IgG)\lim}$$

With this approach, an intrathecal humoral immune response to JCV-specific VP1 ($ASI_{VP1} \geq 1.5$) was found in 78% of PML patients (47 of 62), but in only 3.2% of controls (5 of 155) (Fig. 13.2) (Weber et al., 1997b). By Western blot analysis of paired serum/CSF samples using identical IgG concentrations for each pair, these findings could be supported by the demonstration of more intense bands in those CSF samples from patients with PML than in the respective serum samples (Weber et al., 1997b). Knowles and colleagues (1995) could detect JCV-specific oligoclonal bands in seven PML patients tested.

By antigen-driven immunoblotting (ADI) (Sindic et al., 1994), JCV–VP1-specific oligoclonal bands were detected in 55% of PML cases (10 of 18) and in 6% in controls (2 of 31) (Sindic et al., 1997). Of these patients, one with multiple sclerosis also had an ASI of 2.5, while the second patient with neuroborreliosis had an ASI_{VP1} of 0.64. No oligoclonal bands were detected in four PML patients with an $ASI_{VP1} \leq 1.5$ and in four cases with a moderately elevated ASI_{VP1} ranging from 1.78 to 3.04 (Sindic et al., 1997). In a case of PML without any underlying immunodeficiency, comparative analysis of the intrathecal humoral immune response ADI and VP1–ELISA of serial CSF/serum samples revealed clearly detectable VP1-specific oligoclonal bands with an $ASI_{VP1} \geq 1.5$ (Guillaume et al., 2000).

Taken together, these data suggest that analysis of the intrathecal humoral immune response by ELISA is more sensitive than by ADI. It appears that those cases with an $ASI_{VP1} \geq 3$ are also positive by ADI (Sindic et al., 1997; Guillaume et al., 2000). In AIDS patients with PML receiving HAART, an intrathecal immune response to recombinant VP1 evolved during therapy in three of four patients (Giudici et al., 2000). Neuropathologic analysis in 11 PML patients showed an excellent correlation (r = 0.985) between the plasma cell count in brain tissue and the humoral immune response to VP1 (Weber et al., 1997b). The comparable rates of 67% and 76%, respectively, found for an intrathecal humoral immune response are in close agreement with neuropath-

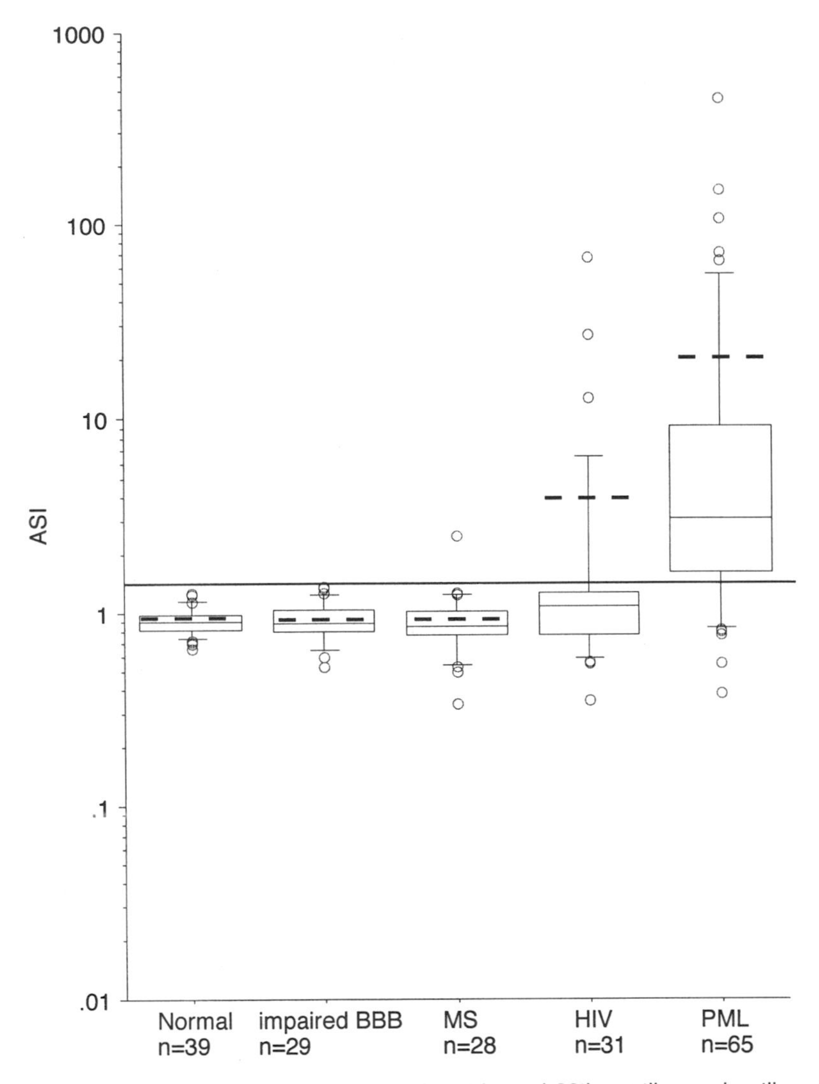

Figure 13.2. The plot shows 10th, 25th, 50th, 75th, and 90th centiles and outliers. The solid line indicates the cut-off at 1.5, above which an intra blood–brain barrier (BBB) synthesis of IgG antibodies to VP1 is present. The mean values are shown as solid lines for each group. In the normal group the mean ASI_{VP1} was 0.92 (±0.16; range, 0.65–1.26); in patients with impaired BBB (IBBB) it was 0.92 (±0.22; range, 0.52–1.36); in multiple sclerosis (MS) patients it was 0.91 (±0.38; range, 0.34–2.49); in HIV-positive patients (HIV-pos) it was 4.35 (±12.71; range, 0.35–67.2); and in the PML group it was 21.84 (±61.91; range, 0.38–451). By the Kruskal-Wallis test there is a significant difference between the five groups ($p < 0.001$). By the Mann-Whitney U-test there is a highly significant difference between PML and normal, IBBB, MS, and HIV-pos ($p < 0.001$), while the difference between MS versus HIV-pos was marginally significant at $p = 0.03$. The differences between normal versus MS ($p = 0.30$), normal versus IBBB ($p = 0.96$), MS versus IBBB ($p = 0.44$), IBBB versus HIV-pos ($p = 0.13$), and normal versus HIV-pos ($p = 0.06$) were not significant.

ologic findings of perivascular infiltrates in about 70% of cases of PML in AIDS patients (Kuchelmeister et al., 1993). These findings clearly indicate the plasma cells as a source of the intrathecally produced IgG and suggest a role for B cells in the development of PML. They further suggest that PML should be considered as an inflammatory disease of the CNS and thus may be better designated as *progressive multifocal leukoencephalitis.*

5. POLYMERASE CHAIN REACTION

With the introduction of PCR, urine, PBMC, and CSF have been evaluated for the presence of JCV DNA and cDNA. These investigations addressed not only the diagnostic sensitivity and specificity but also looked at JCV latency, persistence, and reactivation in these body compartments.

Urine

JCV is frequently shed in the urine of immunocompetent and immunodeficient patients and increases with age (Kitamura et al., 1994). It is found in 9–17% of individuals under the age of 20 years, in 46% of those aged 20–29 years, and increases gradually to 66% for those aged 80 to 89 years (Kitamura et al., 1994). Approximately 33% of HIV-positive patients with PML as well as HIV-positive patients without PML and HIV-negative controls were found to have JCV viruria, suggesting that the detection of JCV in urine has no clinical value (Koralnik et al., 1999a).

In about 38% of HIV-positive patients, JCV replication as measured by a nested cDNA PCR of the VP1 gene can be detected in urine (Lafon et al., 1998). This is in agreement with other studies showing urinary excretion of JCV in 20–45% of immunocompetent and immunodeficient patients (Markowitz et al., 1993; Sundsfjord et al., 1994; Chang et al., 1996; Degener et al., 1997; Tsai et al., 1997; Stoner et al., 1998).

PBMC

In 1988, Houff and colleagues described two patients with PML in whom they identified JCV DNA by in situ hybridization in mononuclear cells. In one case, JCV DNA was identified in spleen B lymphocytes by combined in situ hybridization and immunohistochemistry. Following this observation an initial report suggested a higher frequency of JCV DNA in PBMC in patients with PML than in HIV-1–seropositive patients without PML and patients with Parkinson's disease (Tornatore et al., 1992). Later cross-sectional studies showed a wide variation in the percentage of JCV DNA–positive PBMC in healthy donors, HIV-positive patients without PML, and patients with PML (Table 13.1). The frequency of JCV DNA–positive lymphocytes, however, appears to be much higher in HIV-positive individuals than healthy controls and even

Table 13.1. Percentage of JCV DNA–Positive PBMCs

Healthy Donor	HIV-Positive	HIV-Positive with PML	Study
0/18 (0%)	13/103 (12.6)	5/8 (62.5%)	Koralnik et al. (1999b)
4/50 (8%)	17/72 (23.6%)	5/7 (71.4%)	Dubois et al. (1997)
2/50 (4%)	5/50 (10 %)	2/10 (20%)	Lafon et al. (1998)
—	20/186 (10.8%)	9/19 (47.4%)	Dubois et al. (1998)
14/36 (39%)	15/32 (46.9%)	—	Azzi et al. (1996)
15/18 (83.3%)	—	3/3	Dörries et al. (1994)
—	5/50 (10%)	9/12 (75%)	Ferrante et al. (1997)
—	31/151 (20.5%)	6/10 (60%)	Andreoletti et al. (1999)
0/30	10/26 (38%)	17/19 (83)	Tornatore et al. (1992)
Totals			
5.1%	25.3%	77.3	

higher in patients with PML (Table 13.1). As long as no standardized and quality-controlled prospective studies have been performed, analysis of PBMC or isolated lymphocyte subpopulations cannot be used for the diagnosis of PML.

It appears that separation of lymphocyte subpopulations either by FACS or monoclonal antibody-coated magnetic beads before PCR analysis for JCV DNA or RT-PCR for the detection of JCV-specific cDNA increases the rate of positive results (Lafon et al., 1998; Koralnik et al., 1999b; Pietzuch et al., 1996). Using either magnetic beads to separate B from T cells or FACS, JCV-specific DNA or cDNA was found in both B cells and non-B cells or T cells with about equal frequency (Lafon et al., 1998; Koralnik et al., 1999b; Pietzuch et al., 1996).

In a prospective study, 18.5% of patients (5 of 27) who developed PML within 1 year had JCV DNA detected in their PBMC at entry, while only 1.3% (2 of 159) who had no detectable JCV DNA in their PBMC developed PML (Dubois et al., 1998). Although these data yield a high negative predictive value of 0.99, the low positive predictive value of 0.19 indicates that even serial PCR analysis of PBMC for JCV is not diagnostic of PML. Most investigations, however, failed to demonstrate active replication of JCV in either PBMC or lymphocytes (Lafon et al., 1998; Dubois et al., 1997). Sequencing of the regulatory region of JCV DNA from PBMC revealed the presence of a rearranged transcriptional control region (TCR) in AIDS patients with PML, while AIDS patients without PML and HIV-negative patients without PML had an archetypal TCR (Azzi et al., 1996; Ciappi et al., 1999).

Plasma/Serum

JCV DNA has been reported to occur with high frequency in the plasma of PML patients (Koralnik et al., 1999a). In this study (Koralnik et al., 1999a),

Table 13.2. Percentage of JCV DNA–Positive Plasma Samples

Healthy Donor	HIV-Positive	HIV-Positive with PML	Study
0/13 (0%)	7/32 (21.8%)	3/7	Koralnik et al. (1999a)
2/50 (4%)	12/72 (16.7%)	4/7	Dubois et al. (1997)
0/88 (0%)	20/88 (22.7%)	3/10	Lafon et al. (1998)
—	10/186 (5.4%)	3/19	Dubois et al. (1998)
Totals			
2/151	49/378	13/43	
1.3%	12.9%	30.2%	

JCV DNA was found in the plasma of 3 of 7 HIV-positive patients with PML, in 7 of 32 HIV-positive patients without PML (22%), and in none of 13 controls. These findings are supported by Lafon and colleagues (1998) who found JCV DNA in the plasma of 22.7% of HIV-positive patients but not in 88 controls.

In another study, however, plasma was positive in 5.5% of patients with neurologic diseases without PML (n = 90), in 5.2% of patients without neurologic diseases (n = 96), and in 15.8% (3 of 19) of patients with PML (Dubois et al., 1997). By nested PCR, JCV VP1 cDNA indicative of active viral replication was found only in one HIV-positive patient without PML (Lafon et al., 1998). The combined evidence indicates that JCV DNA is found about two to three times less frequently in plasma than in PBMC (Table 13.2). These data are at best compatible with an intermittent active replication of JCV in PBMC and point to an extravascular reservoir of JCV replication (kidney, urinary tract, bone marrow?).

Cerebrospinal Fluid

There are still some concerns, however, as to whether the presence of JCV DNA in the CSF can always be equated with the diagnosis of PML (Ferrante et al., 1997; Dörries et al., 1998; White et al., 1992). The majority of clinical and pathologic studies on the validity of JCV-specific DNA amplification from brain tissue and CSF suggest that the presence of JCV DNA in CSF correlates with PML (Moret et al., 1993; Gibson et al., 1993; Weber et al., 1994a,b; McGuire et al., 1995; Cinque et al., 1996; Perrons et al., 1996; de Luca et al., 1996; d'Arminio Monforte et al., 1997; Matsiota Bernard et al., 1997; Sugimoto et al., 1998; Koralnik et al., 1999a; Bogdanovic et al., 1995; Chang et al., 1997; Buckle et al., 1992; Fedele et al., 1999; Garcia de Viedma et al., 1999; Hammarin et al., 1996; Mehal et al., 1993; Tachikawa et al., 1999; Telenti et al., 1990). The combined evidence from these studies (Table 13.3) shows a sensitivity of 81% and a specificity of 99% for the diagnosis of PML by CSF PCR.

Table 13.3. Sensitivity and Specificity of JCV PCR in CSF

Sensitivity (%)	Specificity (%)	PML (n)	Controls (n)	Study
61	100	51	249	Aksamit and Kost (1994)
75	—	8	—	Asensi et al. (1999)
90	99	20	67	Antinori et al. (1997)
86	100	14	20	Chang et al. (1997)
72	100	39	500	Cinque et al. (1996)
90	99	19	83	de Luca et al. (1996)
100	93	10	40	Fedele et al. (1999)
92	100	12	52	Ferrante et al. (1997)
90	100	17	20	Garcia de Viedma et al. (1999)
77	100	13	41	Gibson et al. (1993)
100	100	20	191	Hammarin et al. (1996)
93	99	14	92	Koralnik et al. (1999a)
100	75	11	8	Matsiota Bernard et al. (1997)
92	92	26	130	McGuire et al. (1995)
75	100	12	5	Moret et al. (1993)
83	100	23	67	Perrons et al. (1996)
100	100	4	80	Sugimoto et al. (1998)
62	100	13	16	Vago et al. (1996)
80	100	28	82	Weber et al. (1994a,b)
Totals				
81	99	1724	1743	

The specificity of 99% implies a false-positive result in 1 of 100 CSF samples analyzed. This partially explains several reports showing JCV DNA in the CSF of patients with diagnoses other than PML (Moret et al., 1993; McGuire et al., 1995; de Luca et al., 1996; Antinori et al., 1997; Matsiota Bernard et al., 1997; Koralnik et al., 1999a; Vaz et al., 2000).

Detailed analyses of pathologic specimens using primer pairs amplifying segments of the large T and VP1 antigen-encoding sequences of JCV failed to detect these in 19 patients with multiple sclerosis, 1 Alzheimer's disease (AD) patient, and 8 control brain specimens (Buckle et al., 1992). In another study, JCV DNA was only found in 1 of 17 autopsy specimens from AD patients and in none of 26 brain tissues from patients without neurologic diseases (Bogdanovic et al., 1995). The relationship between the influence of the prevalence of a disease, that is, PML (prior probability), on the positive and negative predictive values for a PCR assay with a sensitivity of 80%, 95%, and 99% assuming a specificity of 99% for each is given in Table 13.4. Assuming a sensitivity of 80% and a specificity of 99% for a single CSF PCR, analysis of unselected controls will yield 10 false-positive results if 1000 samples are analyzed (Weber et al., 1996). Based on the analysis of 1000 samples in an unselected AIDS population with a test that is 80% sensitive and 99% specific,

Table 13.4. Predictive Values for the Diagnosis of PML by PCR of CSF

	Predictive Value						
	Positive			Negative			
Prior Probability	S = 0.8 Sp = 0.99	S = 0.95 Sp = 0.99	S = 0.99 Sp = 0.99	S = 0.8 Sp = 0.99	S = 0.95 Sp = 0.99	S = 0.99 Sp = 0.99	Comments
0.000001	0.00008	0.00009	0.0001	1.0	1.0	1.0	Prevalence of PML in the general population
0.05	0.80808	0.83333	0.83898	0.98948	0.99735	0.99947	Prevalence of PML in AIDS
0.5	0.98765	0.98958	0.99	0.83193	0.95192	0.99	MRI consistent with PML
0.9	0.99861	0.99883	0.9988	0.35484	0.68750	0.91667	Brain biopsy specimen positive for PML

S = sensitivity; Sp = specificity.

40 samples are true positive, 9 are false positive, 10 are false negative, and 940 are true negative. Even with an assay that is 99% sensitive and 99% specific, 50 true-positive tests would occur in this population, and there would still be 9 false-positive tests but no false-negative tests.

Although an inhibitory effect of CSF on the PCR assay has been reported (Dennett et al., 1991), a quality control study for the detection of JCV DNA in CSF by PCR has shown no such effect (Weber et al., 1997a). This study showed comparable levels of sensitivity between six laboratories. The sensitivity was between 1 and 10 copies of JCV DNA per 10 μl of CSF. Further analysis showed that the time taken for specimen transport and storage did not reduce sensitivity of the assays used. The limit of detection was also independent of the extraction procedure used (no extraction, boiling, freeze-thawing, commercial extraction kit), or the primer pairs used (VP1, large T, small t).

Although some reports suggest that a rearranged TCR in CSF is diagnostic of PML (Sugimoto et al., 1998; Azzi et al., 1996), a prospective study found a rearranged TCR in all PML cases but also in one case of clinically definite multiple sclerosis, indicating that even sequencing cannot exclude false-positive results (Vaz et al., 2000). Interestingly, 70.5% of PML patients had a type 2 TCR and 29.5% had a type 1 TCR, that is, a sequence identical to or closely related to the prototypical isolate Mad-1 (Vaz et al., 2000; Frisque, 1983).

With the recent report of prolonged survival in AIDS PML patients receiving HAART, attempts have been made to quantify the JCV load in CSF and correlate it with survival (Yiannoutsos et al., 1999; De Luca et al., 1999; Taoufik et al., 2000; Koralnik et al., 1999a; Garcia de Viedma et al., 1999). A semiquantitative assay on the CSF of 15 patients with biopsy-proven PML showed that a low JCV load was more predictive of a longer survival than a high JCV load (Yiannoutsos et al., 1999). In this study semiquantitation was achieved by $\times$10 serial dilutions of JCV DNA template and densitometric tracing of Southern blots. Of the 15 patients analyzed, 11 had detectable DNA levels at baseline. Levels above 50–100 copies of JCV DNA per μl of CSF were strongly associated with increased mortality. The number of copies detected ranged from 1 to 800 per μl of CSF. Those patients classified as having undetectable to low levels of JCV DNA in their CSF had a median survival of 24 weeks, while those above 50–100 copies per μl of CSF had a median survival of 7.6 weeks.

An internal competitor quantitative PCR ELISA analysis of eight patients with PML showed a prolonged survival of 7 months in two with a lower copy number, that is, 46–67 copies per μl of CSF (Koralnik et al., 1999a). Based on a competitive PCR and densitometry of ethidum bromide–stained viral and competitor product, a JCV load ranging from 10 fg to 100 ng (1.2×10^3 to 1.2×10^7 copies per μl of CSF sample) has been reported (Eggers et al., 1999; Drews et al., 2000). Discrepancies in the amount of JCV DNA product reported range from 1 to 800 copies per μl of CSF in two studies (Yiannoutsos et al., 1999; Koralnik et al., 1999a) and from 1200 to 1,200,000 copies per μl of CSF in other reports (Eggers et al., 1999; Drews et al., 2000). These discrepancies may be due to the different methodologies used or to different patient populations studied (e.g., patients receiving HAART versus those not receiving HAART).

ACKNOWLEDGMENTS

The work in the author's laboratory was supported by the Deutsche Forschungsgemeinschaft (DFG We1297/3–1), by the Deutsche Akademische Austauschdienst, and by the Werner Otto Stiftung, Hamburg.

REFERENCES

Aksamit AJ, Kost S (1994): PCR detection of JC virus in PML and control CSF. Neuroscience of HIV Infection, Basic and Clinical Frontiers, 1994, August 2–5, Vancouver, p. 98.

Albrecht H, Hoffmann C, Degen O, Stoehr A, Plettenberg A, Mertenskötter T, Eggers C, Stellbrink HJ (1998): Highly active antiretroviral therapy significantly improves the prognosis of patients with HIV-associated progressive multifocal leukoencephalopathy. AIDS 12:1149–1154.

Andreoletti L, Dubois V, Lescieux A, Dewilde A, Bocket L, Fleury HJ, Wattre P (1999): Human polyomavirus JC latency and reactivation status in blood of HIV-1–positive immunocompromised patients with and without progressive multifocal leukoencephalopathy. AIDS 13:1469–1475.

Antinori A, Ammassari A, De Luca A, Cingolani A, Murri R, Scoppettuolo G, Fortini M, Tartaglione T, Larocca LM, Zannoni G, Cattani P, Grillo R, Roselli R, Iacoangeli M, Scerrati M, Ortona L (1997): Diagnosis of AIDS-related focal brain lesions: A decision-making analysis based on clinical and neuroradiologic characteristics combined with polymerase chain reaction assays in CSF. Neurology 48:687–694.

Asensi V, Carton JA, Maradona JA, Ona M, Melon S, Martin-Roces ER, Asensi JM, Arribas JM (1999): [Progressive multifocal leukoencephalopathy associated with human immunodeficiency virus infection: The clinical, neuroimaging, virological and evolutive characteristics in 35 patients]. Med Clin (Barc) 113:210–214.

Åström K-E, Mancall EL, Richardson EP (1958). Progressive multifocal leukoencephalopathy. A hitherto unrecognized complication of chronic lymphatic leukaemia and Hodgkin's disease. Brain 81:93–111.

Azzi A, De Santis R, Ciappi S, Leoncini F, Sterrantino G, Marino N, Mazzotta F, Laszlo D, Fanci R, Bosi A (1996): Human polyomaviruses DNA detection in peripheral blood leukocytes from immunocompetent and immunocompromised individuals. J Neurovirol 2:411–416.

Baldeweg T, Catalan J (1997): Remission of progressive multifocal leukoencephalopathy after antiretroviral therapy. Lancet 349:1555–1555.

Baqi M, Kucharczyk W, Walmsley SL (1997): Regression of progressive multifocal encephalopathy with highly active antiretroviral therapy. AIDS 11:1526–1527.

Berger JR, Kaszovitz B, Post MJ, Dickinson G (1987): Progressive multifocal leukoencephalopathy associated with human immunodeficiency virus infection. A review of the literature with a report of sixteen cases. Ann Intern Med 107:78–87.

Bogdanovic G, Hammarin A-L, Grandien M, Winblad B, Bergenheim AT, Nennesmo I, Dalianis T (1995): No association of JC virus with Alzheimer's disease or astrocytomas. Clin Diagn Virol 4:223–230.

Brouqui P, Bollet C, Delmont J, Bourgeade A (1992): Diagnosis of progressive multifocal leukoencephalopathy by PCR detection of JC virus from CSF. Lancet 339: 1182.

Buckle GJ, Godec MS, Rubi JU, Tornatore C, Major EO, Gibbs CJJ, Gajdusek DC, Asher DM (1992): Lack of JC viral genomic sequences in multiple sclerosis brain tissue by polymerase chain reaction. Ann Neurol 32:829–831.

Budka H, Shah KV (1983): Papovavirus antigens in paraffin sections of PML brains. Prog Clin Biol Res 105:299–309.

Chang L, Ernst T, Tornatore C, Aronow H, Melchor R, Walot I, Singer E, Cornford M (1997): Metabolite abnormalities in progressive multifocal leukoencephalopathy by proton magnetic resonance spectroscopy. Neurology 48:836–845.

Chang D, Wang M, Ou WC, Lee MS, Ho HN, Tsai RT (1996): Genotypes of human polyomaviruses in urine samples of pregnant women in Taiwan. J Med Virol 48: 95–101.

Ciappi S, Azzi A, De Santis R, Leoncini F, Sterrantino G, Mazzotta F, Mecocci L (1999): Archetypal and rearranged sequences of human polyomavirus JC transcription control region in peripheral blood leukocytes and in cerebrospinal fluid. J Gen Virol 80:1017–1023.

Cinque P, Vago L, Dahl H, Brytting M, Terreni MR, Fornara C, Racca S, Castagna A, Monforte AD, Wahren B, Lazzarin A, Linde A (1996): Polymerase chain reaction on cerebrospinal fluid for diagnosis of virus-associated opportunistic diseases of the central nervous system in HIV-infected patients. AIDS 10:951–958.

d'Arminio Monforte A, Cinque P, Vago L, Rocca A, Castagna A, Gervasoni C, Terreni MR, Novati R, Gori A, Lazzarin A, Moroni M (1997): A comparison of brain biopsy and CSF-PCR in the diagnosis of CNS lesions in AIDS patients. J Neurol 244:35–39.

Degener AM, Pietropaolo V, Di Taranto C, Rizzuti V, Ameglio F, Cordiali Fei P, Caprilli F, Capitanio B, Sinibaldi L, Orsi N (1997): Detection of JC and BK viral genome in specimens of HIV-1 infected subjects. New Microbiol 20:115–122.

de Luca A, Cingolani A, Linzalone A, Ammassari A, Murri R, Giancola ML, Maiuro G, Antinori A (1996): Improved detection of JC virus DNA in cerebrospinal fluid for diagnosis of AIDS-related progressive multifocal leukoencephalopathy. J Clin Microbial 34:1343–1346.

De Luca A, Giancola ML, Cingolani A, Ammassari A, Gillini L, Murri R, Antinori A (1999): Clinical and virological monitoring during treatment with intrathecal cytarabine in patients with AIDS-associated progressive multifocal leukoencephalopathy. Clin Infect Dis 28:624–628.

Dennett C, Klapper PE, Cleator GM, Lewis AG (1991): CSF pretreatment and the diagnosis of herpes encephalitis using the polymerase chain reaction. J Virol Methods 34:101–104.

Domingo P, Guardiola JM, Iranzo A, Margall N (1997): Remission of progressive multifocal leukoencephalopathy after antiretroviral therapy. Lancet 349:1554–1555.

Dörries K, Arendt G, Eggers C, Roggendorf W, Dörries R (1998): Nucleic acid detection as a diagnostic tool in polyomavirus JC induced progressive multifocal leukoencephalopathy. J Med Virol 54:196–203.

Drews K, Bashir T, Dörries K (2000): Quantification of human polyomavirus JC in brain tissue and cerebrospinal fluid of patients with progressive multifocal leukoencephalopathy by competitive PCR. J Virol Methods 84:23–36.

Dörries K, Vogel E, Günther S, Czub S (1994): Infection of human polyomaviruses JC and BK in peripheral blood leukocytes from immunocompetent individuals. Virology 198:59–70.

Dubois V, Dutronc H, Lafon ME, Poinsot V, Pellegrin JL, Ragnaud JM, Ferrer AM, Fleury HJ (1997): Latency and reactivation of JC virus in peripheral blood of human immunodeficiency virus type 1–infected patients. J Clin Microbiol 35:2288–2292.

Dubois V, Moret H, Lafon ME, Janvresse CB, Dussaix E, Icart J, Karaterki A, Ruffault A, Taoufik Y, Vignoli C, Ingrand D (1998): Prevalence of JC virus viraemia in HIV-infected patients with or without neurological disorders: A prospective study. J Neurovirol 4:539–544.

Dworkin MS, Wan PC, Hanson DL, Jones JL (1999): Progressive multifocal leukoencephalopathy: Improved survival of human immunodeficiency virus–infected patients in the protease inhibitor era. J Infect Dis 180:621–625.

Eggers C, Stellbrink HJ, Buhk T, Dörries K (1999): Quantification of JC virus DNA in the cerebrospinal fluid of patients with human immunodeficiency virus–associated progressive multifocal leukoencephalopathy—A longitudinal study. J Infect Dis 180: 1690–1694.

Elliot B, Aromin I, Gold R, Flanigan T, Mileno M (1997): 2.5 year remission of AIDS-associated progressive multifocal leukoencephalopathy with combined antiretroviral therapy. Lancet 349:850.

Ellison GW (1969): Progressive multifocal leukoencephalopathy (PML). I. Investigation of the immunologic status of a patient with lymphosarcoma and PML. J Neuropathol Exp Neurol 28:501–506.

Fedele CG, Ciardi M, Delia S, Echevarria JM, Tenorio A (1999): Multiplex polymerase chain reaction for the simultaneous detection and typing of polyomavirus JC, BK and SV40 DNA in clinical samples. J Virol Methods 82:137–144.

Ferrante P, Caldarelli-Stefano R, Omodeo-Zorini E, Cagni AE, Cocchi L, Suter F, Maserati R (1997): Comprehensive investigation of the presence of JC virus in AIDS patients with and without progressive multifocal leukoencephalopathy. J Med Virol 52:235–242.

Ferrante P, Omodeo-Zorini E, Caldarelli-Stefano R, Mediati M, Fainardi E, Granieri E, Caputo D (1998): Detection of JC virus DNA in cerebrospinal fluid from multiple sclerosis patients. Mult Scler 4:49–54.

Frisque RJ (1983): Nucleotide sequence of the region encompassing the JC virus origin of DNA replication. J Virol 46:170–176.

Frye S, Trebst C, Dittmer U, Petry H, Bodemer M, Hunsmann G, Weber T, Lüke W (1997): Efficient production of JC virus in SVG cells and the use of purified viral antigens for analysis of specific humoral and cellular immune response. J Virol Methods 63:81–92.

Garcia de Viedma D, Alonso R, Miralles P, Berenguer J, Rodriguez-Creixems M, Bouza E (1999): Dual qualitative-quantitative nested PCR for detection of JC virus in cerebrospinal fluid: high potential for evaluation and monitoring of progressive multifocal leukoencephalopathy in AIDS patients receiving highly active antiretroviral therapy. J Clin Microbiol 37:724–728.

Gasnault J, Taoufik Y, Goujard C, Kousignian P, Abbed K, Boue F, Dussaix E, Delfraissy JF (1999): Prolonged survival without neurological improvement in patients with AIDS-related progressive multifocal leukoencephalopathy on potent combined antiretroviral therapy. J Neurovirol 5:421–429.

Gibson PE, Knowles WA, Hand JF, Brown DW (1993): Detection of JC virus DNA in the cerebrospinal fluid of patients with progressive multifocal leukoencephalopathy. J Med Virol 39:278–281.

Giudici B, Vaz B, Bossolasco S, Casari S, Brambilla AM, Lüke W, Lazzarin A, Weber T, Cinque P (2000): Highly active antiretroviral therapy and progressive multifocal leukoencephalopathy: Effects on cerebrospinal fluid markers of JC virus replication and immune response. Clin Infect Dis 30:95–99.

Goldman C, Petry H, Frye S, Ast O, Ebitsch S, Jentsch KD, Kaup FJ, Weber F, Trebst C, Nisslein T, Hunsmann G, Weber T, Lüke W (1999): Molecular cloning and expression of major structural protein VP1 of the human polyomavirus JC virus: Formation of virus-like particles useful for immunological and therapeutic studies. J Virol 73:4465–4469.

Guillaume B, Sindic CJ, Weber T (2000): Progressive multifocal leukoencephalopathy: Simultaneous detection of JCV DNA and anti-JCV antibodies in the cerebrospinal fluid. Eur J Neurol 7:101–106.

Hammarin AL, Bogdanovic G, Svedhem V, Pirskanen R, Morfeldt L, Grandien M (1996): Analysis of PCR as a tool for detection of JC virus DNA in cerebrospinal fluid for diagnosis of progressive multifocal leukoencephalopathy. J Clin Microbiol 34:2929–2932.

Henson J, Rosenblum M, Armstrong D, Furneaux H (1991): Amplification of JC virus DNA from brain and cerebrospinal fluid of patients with progressive multifocal leukoencephalopathy. Neurology 41:1967–1971.

Houff SA, Major EO, Katz DA, Kufta CV, Sever JL, Pittaluga S, Roberts JR, Gitt J, Saini N, Lux W (1988): Involvement of JC virus–infected mononuclear cells from the bone marrow and spleen in the pathogenesis of progressive multifocal leukoencephalopathy. N Engl J Med 318:301–305.

Inui K, Miyagawa H, Sashihara J, Miyoshi H, Tanaka-Taya K, Nishigaki T, Teraoka S, Mano T, Ono J, Okada S (1999): Remission of progressive multifocal leukoencephalopathy following highly active antiretroviral therapy in a patient with HIV infection. Brain Dev 21:416–419.

Katz DA, Berger JR, Hamilton B, Major EO, Post MJ (1994): Progressive multifocal leukoencephalopathy complicating Wiskott-Aldrich syndrome. Report of a case and review of the literature of progressive multifocal leukoencephalopathy with other inherited immunodeficiency states. Arch Neurol 51:422–426.

Kitamura T, Kunitake T, Guo J, Tominaga T, Kawabe K, Yogo Y (1994): Transmission of the human polyomavirus JC virus occurs both within the family and outside the family. J Clin Microbiol 32:2359–2363.

Knight A, O'Brien P, Osoba D (1972): "Spontaneous" progressive multifocal leukoencephalopathy. Immunologic aspects. Ann Intern Med 77:229–233.

Knowles WA, Luxton RW, Hand JF, Gardner SD, Brown DWG (1995): The JC virus antibody response in serum and cerebrospinal fluid in progressive multifocal leukoencephalopathy. Clin Diagn Virol 4:183–194.

Koralnik IJ, Boden D, Mai VX, Lord CI, Letvin NL (1999a): JC virus DNA load in patients with and without progressive multifocal leukoencephalopathy. Neurology 52:253–260.

Koralnik IJ, Du Pasquier RA, Letvin NL (2001): JC virus-specific cytotoxic T lymphocytes in individuals with progressive multifocal leukoencephalopathy. J Virol 75: 3483–3487.

Koralnik IJ, Schmitz JE, Lifton MA, Forman MA, Letvin NL (1999b). Detection of JC virus DNA in peripheral blood cell subpopulations of HIV-1–infected individuals. J Neurovirol 5:430–435.

Kuchelmeister K, Gullotta F, Bergmann M, Angeli G, Masini T (1993): Progressive multifocal leukoencephalopathy (PML) in the acquired immunodeficiency syndrome (AIDS). A neuropathological autopsy study of 21 cases. Pathol Res Pract 189:163–173.

Lafon ME, Dutronc H, Dubois V, Pellegrin I, Barbeau P, Ragnaud JM, Pellegrin JL, Fleury HJ (1998): JC virus remains latent in peripheral blood B lymphocytes but replicates actively in urine from AIDS patients. J Infect Dis 177:1502–1505.

Major EO, Amemiya K, Tornatore CS, Houff SA, Berger JR (1992): Pathogenesis and molecular biology of progressive multifocal leukoencephalopathy, the JC virus-induced demyelinating disease of the human brain. Clin Microbiol Rev 5:49–73.

Major EO, Ault GS (1995): Progressive multifocal leukoencephalopathy: Clinical and laboratory observations on a viral induced demyelinating disease in the immunodeficient patient. Curr Opin Neurol 8:184–190.

Markowitz RB, Thompson HC, Mueller JF, Cohen JA, Dynan WS (1993): Incidence of BK virus and JC virus viruria in human immunodeficiency virus–infected and virus–uninfected subjects. J Infect Dis 167:13–20.

Matsiota Bernard P, De Truchis P, Gray F, Flament Saillour M, Voyatzakis E, Nauciel C (1997): JC virus detection in the cerebrospinal fluid of AIDS patients with progressive multifocal leukoencephalopathy and monitoring of the antiviral treatment by a PCR method. J Med Microbiol 46:256–259.

McGuire D, Barhite S, Hollander H, Miles M (1995). JC virus DNA in cerebrospinal fluid of human immunodeficiency virus-infected patients: Predictive value for progressive multifocal leukoencephalopathy. Ann Neurol 37:395–399.

Mehal WZ, Esiri MM, Lo YM, Chapman RW, Fleming KA (1993): Detection of reactivation and size variation in the regulatory region of JC virus in brain tissue. J Clin Pathol 46:646–649.

Miralles P, Berenguer J, Garcia de Viedma D, Padilla B, Cosin J, Lopez-Bernaldo de Quiros JC, Munoz L, Moreno S, Bouza E (1998): Treatment of AIDS-associated progressive multifocal leukoencephalopathy with highly active antiretroviral therapy. AIDS 12:2467–2472.

Monno L, Zimatore GB, Di Stefano M, Appice A, Livrea P, Angarano G (1999): Reduced concentrations of HIV-RNA and TNF-alpha coexist in CSF of AIDS patients with progressive multifocal leukoencephalopathy. J Neurol Neurosurg Psychiatry 67: 369–373.

Moret H, Guichard M, Matherson S, Katlama C, Sazdovitch V, Huraux J-M, Ingrand D (1993): Virological diagnosis of progressive multifocal leukoencephalopathy: Detection of JC virus DNA in cerebrospinal fluid and brain tissue of AIDS patients. J Clin Microbiol 31:3310–3313.

Owen RG, Patmore RD, Smith GM, Barnard DL (1995): Cytomegalovirus-induced T-cell proliferation and the development of progressive multifocal leukoencephalopathy following bone marrow transplantation. Br J Haematol 89:196–198.

Padgett BL, Walker DL (1973): Prevalence of antibodies in human sera against JC virus, an isolate from a case of progressive multifocal leukoencephalopathy. J Infect Dis 127:467–470.

Padgett BL, Walker DL (1983): Virologic and serologic studies of progressive multifocal leukoencephalopathy. Prog Clin Biol Res 105:107–117.

Padgett BL, Walker DL (1976): New human papovaviruses. Prog Med Virol 22:1–35.

Padgett BL, Walker DL, Zu Rhein GM, Hodach AE, Chou SM (1976): JC papovavirus in progressive multifocal leukoencephalopathy. J Infect Dis 133:686–690.

Perrons CJ, Fox JD, Lucas SB, Brink NS, Tedder RS, Miller RF (1996): Detection of polyomaviral DNA in clinical samples from immunocompromised patients: Correlation with clinical diseases. J Infect 32:205–209.

Peters AC, Versteeg J, Bots GT, Boogerd W, Vielvoye GJ (1980): Progressive multifocal leukoencephalopathy: Immunofluorescent demonstration of simian virus 40 antigen in CSF cells and response to cytarabine therapy. Arch Neurol 37:497–501.

Pietzuch A, Bodemer M, Frye S, Cinque P, Trebst C, Lüke W, Weber T (1996): Expression of JCV-specific cDNA in peripheral blood lymphocytes of patients with PML, HIV infection, multiple sclerosis and in healthy controls. J Neurovirol 2:46.

Power C, Nath A, Aoki FY, Bigio MD (1997): Remission of progressive multifocal leukoencephalopathy following splenectomy and antiretroviral therapy in a patient with HIV infection. N Engl J Med 336:661–662.

Reiber H, Felgenhauer K (1987): Protein transfer at the blood cerebrospinal fluid barrier and the quantitation of the humoral immune response within the central nervous system. Clin Chim Acta 163:319–328.

Reiber H, Lange P (1991): Quantification of virus-specific antibodies in cerebrospinal fluid and serum: Sensitive and specific detection of antibody synthesis in brain. Clin Chem 37:1153–1160.

Richardson EP Jr (1961): Progressive multifocal leukoencephalopathy. N Engl J Med 265:815–823.

Rockwell D, Ruben FL, Winkelstein A, Mendelow H (1976): Absence of immune deficiencies in a case of progressive multifocal leukoencephalopathy. Am J Med 61: 433–436.

Sindic CJ, Monteyne P, Laterre EC (1994): The intrathecal synthesis of virus-specific oligoclonal IgG in multiple sclerosis. J Neuroimmunol 54:75–80.

Sindic CJ, Trebst C, Van Antwerpen MP, Frye S, Enzensberger W, Hunsmann G, Lüke W, Weber T (1997): Detection of CSF-specific oligoclonal antibodies to recombinant JC virus VP1 in patients with progressive multifocal leukoencephalopathy. J Neuroimmunol 76:100–104.

Stoner GL, Agostini HT, Ryschkewitsch CF, Komoly S (1998): JC virus excreted by multiple sclerosis patients and paired controls from Hungary. Mult Scler 4:45–48.

Sugimoto C, Ito D, Tanaka K, Matsuda H, Saito H, Sakai H, Fujihara K, Itoyama Y, Yamada T, Kira J, Matsumoto R, Mori M, Nagashima K, Yogo Y (1998): Amplification of JC virus regulatory DNA sequences from cerebrospinal fluid: Diagnostic value for progressive multifocal leukoencephalopathy. Arch Virol 143:249–262.

Sundsfjord A, Flaegstad T, Flo R, Spein AR, Pedersen M, Permin H, Julsrud J, Traavik T (1994): BK and JC viruses in human immunodeficiency virus type 1–infected persons: Prevalence, excretion, viremia, and viral regulatory regions. J Infect Dis 169:485–490.

Tachikawa N, Goto M, Hoshino Y, Gatanaga H, Yasuoka A, Wakabayashi T, Katano H, Kimura S, Oka S, Iwamoto A (1999): Detection of *Toxoplasma gondii*, Epstein-Barr virus, and JC virus DNAs in the cerebrospinal fluid in acquired immunodeficiency syndrome patients with focal central nervous system complications. Intern Med 38:556–562.

Taoufik Y, Delfraissy JF, Gasnault J (2000): Highly active antiretroviral therapy does not improve survival of patients with high JC virus load in the cerebrospinal fluid at progressive multifocal leukoencephalopathy diagnosis. AIDS 14:758–759.

Tantisiriwat W, Tebas P, Clifford DB, Powderly WG, Fichtenbaum CJ (1999): Progressive multifocal leukoencephalopathy in patients with AIDS receiving highly active antiretroviral therapy. Clin Infect Dis 28:1152–1154.

Telenti A, Aksamit AJJ, Proper J, Smith TF (1990): Detection of JC virus DNA by polymerase chain reaction in patients with progressive multifocal leukoencephalopathy. J Infect Dis 162:858–861.

Teofilo E, Gouveia J, Brotas V, da Costa P (1998): Progressive multifocal leukoencephalopathy regression with highly active antiretroviral therapy. AIDS 12:449.

Tornatore C, Berger JR, Houff SA, Curfman B, Meyers K, Winfield D, Major EO (1992): Detection of JC virus DNA in peripheral lymphocytes from patients with and without progressive multifocal leukoencephalopathy. Ann Neurol 31:454–462.

Tsai RT, Wang M, Ou WC, Lee YL, Li SY, Fung CY, Huang YL, Tzeng TY, Chen Y, Chang D (1997): Incidence of JC viruria is higher than that of BK viruria in Taiwan. J Med Virol 52:253–257.

Vago L, Cinque P, Sala E, Nebuloni M, Caldarelli R, Racca S, Ferrante P, Trabottoni G, Costanzi G (1996): JCV-DNA and BKV-DNA in the CNS tissue and CSF of AIDS patients and normal subjects. Study of 41 cases and review of the literature. J Acquir Immune Defic Syndr Hum Retrovirol 12:139–146.

Vaz B, Cinque P, Pickhardt M, Weber T (2000): Analysis of the transcriptional control region in progressive multifocal leukoencephalopathy. J Neurovirol 6:398–409.

Vignoli C, De-Lamballerie X, Zandotti C, Tamalet C, De-Micco P (1993): Detection of JC virus by polymerase chain reaction and colorimetric DNA hybridization assay. Eur J Clin Microbiol Infect Dis 12:958–961.

Walker DL (1985): Progressive multifocal leukoencephalopathy. In Vinken PJ, Bruyn GW, Klawans HL, Eds; Handbook of Clinical Neurology, Demyelinating Diseases; Elsevier Science Publishers: Amsterdam, pp 503–524.

Weber T, Freter A, Luer W, Haas J, Stark E, Poser S, Felgenhauer K (1991): The use of recombinant antigens in ELISA procedures for the quantification of intrathecally produced HIV-1-specific antibodies. J Immunol Methods 136:133–137.

Weber T, Frye S, Bodemer M, Otto M, Lüke W (1996): Clinical implications of nucleic acid amplification methods for the diagnosis of viral infections of the nervous system. J Neurovirol 2:175–190.

Weber T, Klapper PE, Cleator GM, Bodemer M, Lüke W, Knowles W, Cinque P, Van Loon AM, Grandien M, Hammarin AL, Ciardi M, Bogdanovic G (1997a): Polymerase chain reaction for detection of JC virus DNA in cerebrospinal fluid: A quality control study. European Union Concerted Action on Viral Meningitis and Encephalitis. J Virol Methods 69:231–237.

Weber F, Krämer M, Hunsmann G, Lüke W, Young P, Weber T (1998): Immune response in progressive multifocal leukoencephalopathy. Ann Neurol 44:488–489.

Weber T, Trebst C, Frye S, Cinque P, Vago L, Sindic CJ, Schultz WJ, Kretzschmar HA, Enzensberger W, Hunsmann G, Luke W (1997b): Analysis of the systemic and intrathecal humoral immune response in progressive multifocal leukoencephalopathy. J Infect Dis 176:250–254.

Weber T, Turner RW, Frye S, Lüke W, Kretzschmar HA, Lüer W, Hunsmann G (1994a): Progressive multifocal leukoencephalopathy diagnosed by amplification of JC virus-specific DNA from cerebrospinal fluid. AIDS 8:49–57.

Weber T, Turner RW, Frye S, Ruf B, Haas J, Schielke E, Pohle H-D, Lüke W, Lüer W, Felgenhauer K, Hunsmann G (1994b): Specific diagnosis of progressive multifocal leukoencephalopathy by polymerase chain reaction. J Infect Dis 169:1138–1141.

White FA, Ishaq M, Stoner GL, Frisque RJ (1992). JC virus DNA is present in many human brain samples from patients without progressive multifocal leukoencephalopathy. J Virol 66:5726–5734.

Willoughby E, Price RW, Padgett BL, Walker DL, Dupont B (1980): Progressive multifocal leukoencephalopathy (PML): In vitro cell-mediated immune responses to mitogens and JC virus. Neurology 30:256–262.

Yiannoutsos CT, Major EO, Curfman B, Jensen PN, Gravell M, Hou J, Clifford DB, Hall CD (1999): Relation of JC virus DNA in the cerebrospinal fluid to survival in acquired immunodeficiency syndrome patients with biopsy-proven progressive multifocal leukoencephalopathy. Ann Neurol 45:816–821.

14

MOLECULAR BIOLOGY OF BK VIRUS AND CLINICAL AND BASIC ASPECTS OF BK VIRUS RENAL INFECTION

UGO MOENS, PH.D. and OLE PETTER REKVIG, M.D., PH.D.

1. INTRODUCTION

The first human polyomavirus was detected three decades ago in the urine of a renal allograft recipient who developed ureteric stenosis and was given the initials of the patient BK (BKV) (Gardner et al., 1971). Serologic surveys conducted in several countries have shown that this virus is distributed worldwide and has a high incidence among humans. In children up to 10 years of age, antibodies against the capsid proteins of BKV were detected in 63–100% of those examined (Brown et al., 1975; Dei et al., 1982; Dougherty and DiStefano, 1974; Gardner, 1973; Mäntyjärvi et al., 1973; Portolani et al., 1974; Rziha et al., 1978b; Shah et al., 1973b; Taguchi et al., 1982). In some remote South American Indian tribes and remote populations of Malaysia, no or a very low prevalence of anti-BKV antibodies was measured in the adult population (Brown, et al., 1975; Major and Neel, 1998). Primary infection occurs predominantly during childhood and seems, with few exceptions, to be subclinical. Primary infection is normally followed by a life-long dormant and asymptomatic persistence in immunocompetent individuals (reviewed by Dörries, 1997;

Human Polyomaviruses: Molecular and Clinical Perspectives, Edited by Kamel Khalili and Gerald L. Stoner.
ISBN 0-471-39009-7

Shah, 1996). There are, however, situations in which BKV may be actively expressed. A recent report suggested a role for BKV in the development of encephalitis in an immunocompetent patient (Voltz et al., 1996). Intermittent or chronic BKV reactivation can occur in individuals with perturbed immune conditions such as pregnant women (Bendiksen et al. 2000a; Chang et al., 1996b; Coleman et al., 1980; Markowitz et al., 1991), transplant recipients (reviewed by Dörries, 1997), AIDS patients (Markowitz et al., 1993), and patients suffering from autoimmune diseases (Bendiksen et al., 2000a; Chang et al., 1996a; Rekvig et al., 1997; Sugimoto et al., 1989; Sundsfjord et al., 1999).

BKV and JC virus (JCV) were the first human viruses isolated that strongly resembled the morphology and genomic organization of the previously described oncogenic polyomaviruses simian virus 40 (SV40) and mouse polyoma virus. The relationship with these DNA tumor viruses and the presence of SV40 large tumor-antigen reactive antibodies in human sera (Corallini et al., 1976; Shah et al., 1973a) intensified the research for these viruses. Nevertheless, for many years BKV has lived a scientific existence in the shadow of SV40. However, during recent years, BKV has again drawn the attention of the scientific and medical communities because of its potential to cause human diseases.

This chapter deals with the general molecular biology of BKV with an emphasis on the naturally occurring diversity of the noncoding control region (NCCR) and the role of this region in the regulation of viral gene expression and DNA replication. Furthermore, clinical and basic aspects of renal infection by BKV are reviewed and a possible involvement of BKV in nephropathy and systemic lupus erythematosus (SLE) is discussed in detail. Finally, suggestions on the clinical management of BKV in transplant recipients are proposed.

2. MOLECULAR BIOLOGY OF THE BK VIRUS

Polyomaviruses, and especially SV40, have been intensively studied to examine fundamental processes in eukaryotic molecular biology. Important contributions to the understanding of chromosomal DNA replication, gene regulation, and the action of oncogenes and tumor suppressor genes have been obtained with these viruses as a model system. Because of BKV's close resemblance with SV40, much of the current knowledge of the molecular biology of BKV derives from extrapolations from studies with SV40. The structure and genetic organization of BKV have been excellently described elsewhere (Howley, 1980; Shah, 1996; Yoshiike and Takemoto, 1986) and are only briefly summarized here. Recent contributions to the understanding of the molecular biology of BKV are reviewed in more detail.

Structure of the BK Virion

Biophysical Characteristics. BKV is a nonenveloped virus with an icosahedral capsid of about 40.5–44 nm in diameter. Virions consist of 88% proteins and 12% DNA. The capsid is composed of 72 capsomers, each consisting of three different structural proteins, VP1, VP2, and VP3. The particle has a sedimentation coefficient $S_{20,w}$ of 240S and a molecular mass of about 2.7×10^7 dalton. Infectious particles have a buoyant density in CsCl of 1.34 g/cm^3, while empty, noninfectious particles band at a density of 1.29 g/cm^3 (Howley, 1980; Shah, 1996).

The Viral Genome. The viral genome consists of a single copy of a circular double-stranded DNA molecule of approximately 5 kilobase pairs. The genomes of the BKV strains MM (accession number V01109), Dunlop (accession number J02038), and AS (accession number M23122) have been completely sequenced. The BKV genome shares 75% overall homology with JCV (Cole, 1996) and a 70% overall homology with the SV40 genome (Yoshiike and Takemoto, 1986). In mature virions, viral DNA is associated with the cellular histones H2A, H2B, H3, and H4 with a structure analogous to chromatin or a minichromosome (Meneguzzi et al., 1978). The minichromosome found in SV40-infected cells is also packed with histone H1 (Varshavsky et al., 1976). On the average, each genome contains 21 nucleosomes (Meneguzzi et al., 1978).

The genome can be divided into three functional subregions (Fig. 14.1): (1) The early region encodes the regulatory proteins large tumor (T) antigen (TAg) and small tumor (t) antigen (tAg); (2) the late region encompasses the genetic information for the capsid proteins and the agnoprotein; and (3) the NCCR spans the origin of replication and the sequences involved in the transcriptional regulation of both the early and the late genes. The early region is expressed during the early stages of the viral life cycle and can be expressed later during infection, after the onset of viral DNA replication. The late region of the genome is by definition transcribed only after viral DNA replication begins, although low levels of transcription may occur early after infection as well (reviewed by Cole, 1996).

The different BKV strains display more sequence variability in their NCCR than in other regions of the genome. This variation is due to deletions, duplications, and rearrangement of the enhancer region (Shah, 1996). Whether these alterations reflect the NCCR architecture in vivo is more uncertain because in most cases the NCCR sequences of new isolates were determined after the virus had been passaged in cell culture, a process known to predispose for NCCR rearrangements. However, the development of the polymerase chain reaction (PCR) has made it possible to directly obtain the sequence of an isolate without propagating the virus in vitro. These naturally occurring BKV NCCR are briefly discussed here. An extensive description has recently been published (Moens et al., 1995a). The NCCR of the proposed archetypal BK strain WW has been arbitrarily divided into three transcription factor binding sequence blocks, called P (68 bp), Q (39 bp), and R (63 bp), as an aid to visualize the rearrangements found in the different NCCR variants (Markowitz and Dynan, 1988). We expanded this nomenclature with the O (142 bp) and S (63 bp) blocks (Moens et al., 1995a). The O block spans the sequences between the

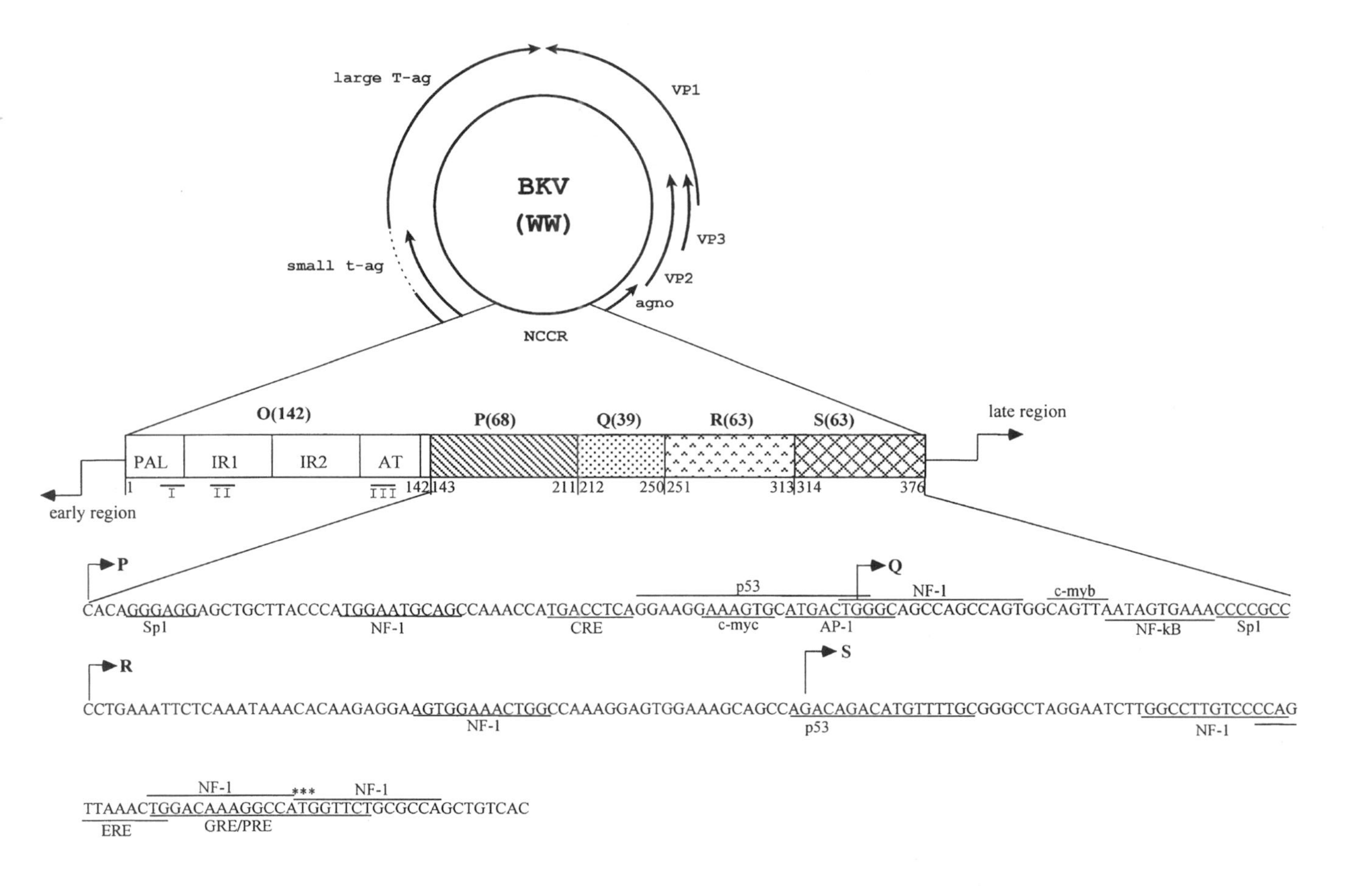
large T-ag
VP1
BKV
(WW)
VP3
small t-ag
VP2
agno
NCCR
O(142)
P(68)
Q(39)
R(63)
S(63)
late region
PAL
IR1
IR2
AT
1
I
II
III
142
143
211
212
250
251
313
314
376
early region
P
Q
p53
NF-1
c-myb
CACAGGGAGGAGCTGCTTACCCATGGAATGCAGCCAAACCATGACCTCAGGAAGGAAAGTGCATGACTGGGCAGCCAGCCAGTGGCAGTTAATAGTGAAACCCCGCC
Sp1
NF-1
CRE
c-myc
AP-1
NF-kB
Sp1
R
S
CCTGAAATTCTCAAATAAACACAAGAGGAAGTGGAAACTGGCCAAAGGAGTGGAAAGCAGCCAGACAGACATGTTTTGCGGGCCTAGGAATCTTGGCCTTGTCCCCAG
NF-1
p53
NF-1
NF-1
***
NF-1
TTAAACTGGACAAAGGCCATGGTTCTGCGCCAGCTGTCAC
ERE
GRE/PRE

start codon for the early genes and the P block and encompasses the origin of replication and the TATA box. The S block represents the sequences between the R block and the translational start for agnoprotein. Application of this nomenclature revealed that archetypal BKV(WW) has the linear O–P–Q–R–S configuration. The naturally occurring BKV strains, based on their NCCR, are summarized in Table 14.1. Almost all NCCR sequences obtained from urinary BKV DNA possessed the BKV(WW) strain configuration, while sequences derived from BKV DNA from other cellular sources displayed distinct NCCRs. This may suggest that cell-specific rearrangements in the NCCR were required to adapt the virus to diverse cell types in vivo.

The correlations of rearrangements in the BKV NCCR with transcriptional activity and transformation efficiency have been well studied. Rearrangements in the NCCR can remove, create, or increase the number of transcription factor binding sites (reviewed by Moens et al., 1995a). Moreover such promoters may be less repressed by large TAg and direct different rates of BKV replication (Deyerle et al., 1989; Yoshiike and Takemoto, 1986). The diversity of the NCCR probably confers a selective advantage for the virus in its host, allowing more efficient replication and transcription of the virus genome.

Viral Proteins

Early Proteins

LARGE T ANTIGEN. SV40 TAg is a multifunctional protein that is involved in DNA replication and transcription of the viral genome. Moreover, the transformation potential of the protein has been well studied (reviewed by Cole, 1996). SV40 TAg can also influence the expression of cellular genes (reviewed by Moens et al., 1997) and can induce the production of anti-DNA antibodies (Moens et al., 1995b; Rekvig et al., 1997). The different functions of TAg are exerted by distinct regions of the protein. These regions of SV40 and BKV TAgs share high sequence homology and display similar functions. The functional domains of BKV TAg and regions with sequence homology to SV40 TAg are discussed below.

Figure 14.1. Schematic of the archetypal BKV(WW) genome. The gene products encoded by the early region (TAg and tAg) and the late region (the agnoprotein, agno, and the capsid proteins VP1, VP2, and VP3) are indicated. A diagram of the noncoding control region (NCCR), using the O–P–Q–R–S nomenclature (Markowitz and Dynan, 1988; Moens et al., 1995a) is shown. This region spans the origin of replication and the early and late promoter/enhancer sequences. The O block consists of a palindromic sequence (PAL), two inverted repeats (IR1 and IR2), and an AT-rich motif (AT). The numbers in parentheses represent the numbers of base pairs in each block. The early and late regions are indicated by a broken arrow. The TAg binding sites I, II, and III are illustrated. Proven and putative binding sites for transcription factors are depicted. The nucleotide numbers forming the borders between the different blocks are shown below the blocks. Numbering starts from the first nucleotide in the O block.

Table 14.1. Sources and Anatomy of Naturally Occurring BKV Noncoding Control Region (NCCR) Variants That Have Been Sequenced Directly Without Passage in Cell Culture

Strain	NCCR Anatomy[a]	Source	References
AO	$\mathbf{P}_{1-68}-\mathbf{Q}_{1-39}-\mathbf{R}_{4-9}-\mathbf{P}_{22-68}-\mathbf{Q}_{1-39}-\mathbf{R}_{1-63}-\mathbf{S}_{1-63}$	Urine of an SLE patient	Rekvig et al. (1997)
Cin	48 bp deletion and 48 bp duplication[b]	Renal biopsy specimen from an HIV-positive patient with tubulointerstitial nephritis	Smith et al. (1998)
DDP, 9a, 30b, 33, 39[c]	$\mathbf{P}_{1-68}-\mathbf{Q}_{1-39}-\mathbf{S}_{1-63}$	Brain and kidney from aborted fetuses and maternal placenta tissue	Pietropaola et al. (1998)
			Degener et al. (1999)
		Peripheral blood mononuclear cells of HIV-positive and -negative patients	Chatterjee et al. (2000), Dolei et al. (2000)
		Peripheral blood mononuclear cells of healthy blood donors	
Dunlop (=BK-17)	$\mathbf{P}_{1-68}-\mathbf{P}_{1-7;26-68}-\mathbf{P}_{1-64}-\mathbf{S}_{1-63}$	Peripheral blood leukocytes of immunocompetent individuals	Dörries et al. (1994)
			Frisque, R[d]
		Kidney of an accident victim	
IR-1	$\mathbf{P}_{1-68}-\mathbf{Q}_{1-31}-\mathbf{P}_{20-50}-\mathbf{P}_{39-68}-\mathbf{Q}_{1-31}-\mathbf{P}_{20-68}-\mathbf{R}_{15-63}-\mathbf{S}_{1-63}$	Human tumor of pancreatic cells	Negrini et al. (1990)
MT-1[e]	$\mathbf{P}_{1-68}-\mathbf{Q}_{1-39}-\mathbf{R}_{1-63}-\mathbf{S}_{1-63}$	Urine of a 17-year-old male SLE patient	Sugimoto et al. (1989)
NP132	$\mathbf{P}_{1-68}-\mathbf{Q}_{1-26}-\mathbf{P}_{20-68}-\mathbf{Q}_{1-39}-\mathbf{S}_{1-63}$	Nasopharyngeal aspirate of a child	Sundsfjord et al. (1994b)
NP164[f]	$\mathbf{P}_{1-68}-\mathbf{P}_{1-7,26-68}-\mathbf{P}_{1-68}-\mathbf{Q}_{1-28}-\mathbf{S}_{7-63}$	Nasopharyngeal aspirate of a child	Sundsfjord et al. (1994b)
		Urine and peripheral blood of HIV-infected persons	Sundsfjord et al. (1994a)
Proto-2	$\mathbf{P}_{1-68}-\mathbf{P}_{1-7,26-68}-\mathbf{P}_{1-68}-\mathbf{Q}_{1-28}-\mathbf{S}_{7-63}$	Urine of HIV-infected patients	Sundsfjord et al. (1994a)
TU	$\mathbf{P}_{1-68}-\mathbf{Q}_{1-39}-\mathbf{R}_{1-12}-\mathbf{P}_{16-68}-\mathbf{Q}_{1-35}-\mathbf{R}_{52-63}-\mathbf{S}_{1-63}$	Urine of children	Sundsfjord et al. (1990)
		Patients with ventricular carcinoma, acute lymphatic leukemia, epilepsia	Flægstad et al. (1991)
URO1	$\mathbf{P}_{1-68}-\mathbf{Q}_{1-26}-\mathbf{P}_{32-68}-\mathbf{Q}_{1-26}-\mathbf{R}_{56-63}-\mathbf{S}_{1-63}$	Prostate hyperplasia, kidney adenocarcinomas, ureter and bladder urothelial carcinoma	Monini et al. (1995)
WW	$\mathbf{P}_{1-68}-\mathbf{Q}_{1-39}-\mathbf{R}_{1-63}-\mathbf{S}_{1-63}$	Urine of a renal transplant recipient	Rubinstein et al. (1987)
		Urine of bone marrow transplant recipient	Flægstad et al. (1991), Negrini et al. (1991), Schätzl et al. (1994)

		Urine from HIV-positive and HIV-negative patients	Degener et al. (1999), Flægstad et al. (1991)
		Urine from an acute myeloid leukemia patient	Schätzl et al. (1994)
		Urine from autoimmune patients	Chang et al. (1996a), Rekvig et al. (1997)
		Urine from pregnant women	Chang et al. (1996b), Markowitz et al. (1991)
		Cerobrospinal fluid of 34-year-old man with encephalitis	Voltz et al. (1996)
		Peripheral blood leukocytes of healthy donors	Chatterjee et al. (2000)
WL-1[g]	$\mathbf{P}_{1-40,61-68}$–$\mathbf{O}_{132-142}$–$\mathbf{P}_{1-40,61-68}$–$\mathbf{Q}_{1-39}$–$\mathbf{R}_{1-49}$–$\mathbf{S}_{3-63}$	Peripheral blood leukocytes of immunocompetent individuals	Dörries et al. (1994)
#14	$\mathbf{P}_{1-31}$–$\mathbf{P}_{9-68}$–$\mathbf{Q}_{1-39}$–$\mathbf{R}_{-1-63}$–$\mathbf{S}_{1-63}$	Urine of bone marrow transplant recipient	Negrini et al. (1991)
#6	$\mathbf{P}_{1-68}$–$\mathbf{Q}_{1-31}$–$\mathbf{P}_{24-68}$–$\mathbf{Q}_{1-30}$–$\mathbf{P}_{12-68}$–$\mathbf{Q}_{1-4}$–$\mathbf{S}_{1-63}$	Peripheral blood leukocytes of healthy donors	Chatterjee et al. (2000)
#7	$\mathbf{P}_{1-68}$–$\mathbf{Q}_{1-39}$–$\mathbf{P}_{17-68}$–$\mathbf{Q}_{1-18,32-39}$–$\mathbf{S}_{1-63}$	Peripheral blood leukocytes of healthy donors	Chatterjee et al. (2000)
#9c	$\mathbf{P}_{1-68}$–$\mathbf{Q}_{1-39}$–$\mathbf{R}_{1-13}$–$\mathbf{P}_{16-68}$–$\mathbf{Q}_{1-34}$–$\mathbf{S}_{1-63}$	Peripheral blood leukocytes of healthy donors	Chatterjee et al. (2000)
#13[h]	$\mathbf{P}_{1-68}$–$\mathbf{Q}_{1-39}$–$\mathbf{R}_{1-13}$–$\mathbf{S}_{1-2}$–$\mathbf{P}_{16-68}$–$\mathbf{Q}_{1-3,9-39}$–$\mathbf{R}_{52-63}$–$\mathbf{S}_{1-63}$	Peripheral blood leukocytes of healthy donors	Chatterjee et al. (2000)
#16a,20	$\mathbf{P}_{1-68}$–$\mathbf{Q}_{1-26}$–$\mathbf{R}_{1-8}$–$\mathbf{S}_{1-63}$	Peripheral blood leukocytes of healthy donors	Chatterjee et al. (2000)
#16b	$\mathbf{P}_{1-68}$–$\mathbf{Q}_{1-39}$–$\mathbf{P}_{18-68}$–$\mathbf{Q}_{1-39}$–$\mathbf{S}_{1-63}$	Peripheral blood leukocytes of healthy donors	Chatterjee et al. (2000)
#17	$\mathbf{P}_{1-68}$–$\mathbf{Q}_{1-39}$–$\mathbf{R}_{1-12}$–$\mathbf{P}_{18-68}$–$\mathbf{Q}_{1-24}$–$\mathbf{S}_{1-63}$	Peripheral blood leukocytes of healthy donors	Chatterjee et al. (2000)
#30a	$\mathbf{P}_{1-68}$–$\mathbf{Q}_{1-26}$–$\mathbf{R}_{1-9}$–$\mathbf{P}_{37-68}$–$\mathbf{Q}_{1-25}$–$\mathbf{R}_{1-9}$–$\mathbf{S}_{1-63}$	Peripheral blood leukocytes of healthy donors	Chatterjee et al. (2000)

[a]Only the anatomies of the P, Q, R, and S blocks are given. The numbers refer to the nucleotides and their positions in the respective blocks, e.g., P_{1-68} indicates that the P block contains nucleotides 1–68, while $P_{1-7,26-68}$ refers to nucleotides 1–7 and 26–68, i.e., a deletion of bp 8–25.

[b]No exact sequence of the NCCR was presented in the article.

[c]Identical to the PQ strain obtained after selection of BKV (Gardner) in Vero cells (Ferguson and Subramani, 1994).

[d]R. Frisque, Personal communication in discussion of Markowitz and Dynan (1988).

[e]Identical with BKV(WW).

[f]Almost identical with proto-2.

[g]This strain has a short duplication of O block sequences.

[h]Almost identical to BKV(TU).

The BKV and SV40 early proteins TAg and tAg mRNAs are derived from a common pre-mRNA by alternative splicing. For SV40, it was shown that the amount of tAg versus TAg mRNA in infected cells seems to be cell specific, and the production of the two splice variants was mediated by a specific protein called ASF (Ge and Manley, 1990). Five to 10 times more TAg mRNA than tAg transcripts were detected in SV40-infected human 293 cells, while a 100-fold excess of TAg mRNA was measured in cell extracts of HeLa cells. The cellular protein ASF promoted the generation of the tAg mRNA from the common precursor for TAg and tAg mRNAs (Ge and Manley, 1990). The mechanism for alternative splice utilization of the BKV early pre-mRNA transcript has not been studied thus far.

BKV TAg consists of 695 amino acids and has an estimated molecular mass ranging from 86 to 97 kDa. The TAgs of SV40 and BKV show extensive homology, and 529 of 695 amino acids are identical. The SV40 TAg protein carries several post-translational modifications such as phosphorylation, O-glycosylation, acylation, palmitylation, adenylation, poly(ADP)-ribosylation, and N-terminal acetylation (reviewed by Cole, 1996). Little is known about the functions of any of these modifications except for phosphorylation, which plays a major role in controlling the activity and function of the protein (Fanning, 1992). BKV TAg is, in accordance with SV40, phosphorylated at multiple residues (Farrell et al., 1978). Acetylation of proteins was first described for histones and was correlated to the transcriptional activity of the chromatin (reviewed by Kornberg and Lorch, 1999). Recently, the number of reports describing acetylation of nonhistone proteins, including transcription factors, has rapidly increased. Acetylation can result in stimulation or disruption of DNA binding of the modified proteins, in regulation of protein–protein interactions, and in protein stability (reviewed by Kouzarides, 2000). Acetylation and its effects on the function of BKV TAg have not been investigated.

Mutational analyses have contributed enormously to unraveling the biologic functions of SV40 TAg. A database summarizing these mutations and their biologic effects is available (http://bigdaddy.bio.pitt.edu/SV40/; Kierstead and Pipas, 1996). The results from studies of these mutations may probably be extended to BKV TAg because of the high homology between the two proteins. BKV TAg possesses several motifs that are identical, or very similar, to SV40 and probably exert the same function (Fig. 14.2) (reviewed by Pipas, 1992). The amino-terminal regions of all polyomavirus large TAgs (and small tAgs) share high homology and possess the absolutely conserved hexapeptide HPDKGG (residues 42 to 47 in BKV). This motif is part of the J domain that shows extensive homology to the DnaJ family of molecular chaperone proteins. The common regions of the TAgs and tAgs of SV40, BKV, and JCV can functionally substitute for the J domain of *Escherichia coli* DnaJ protein (Kelly and Georgopoulos, 1997; Srinivasan et al., 1997). The SV40 J domain is required for efficient viral replication, transformation, specific interaction with hsp70 family member hsc, and turnover of p107 and p130 (Campbell et al., 1997; Sawai and Butel, 1989; Srinivasan et al., 1997; Stubdal et al., 1996).

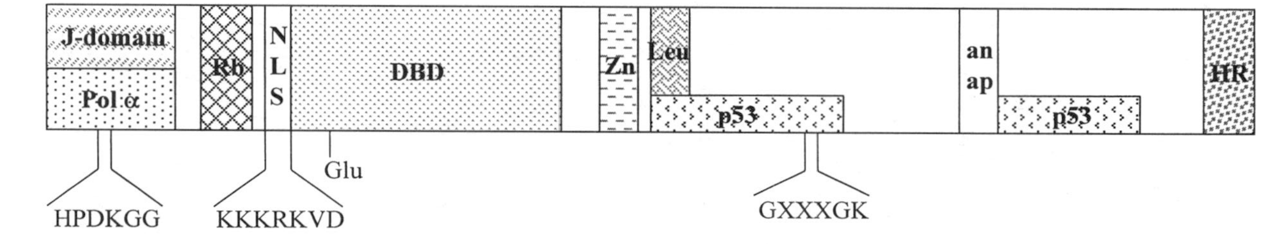

Figure 14.2. Proven and putative functional domains of BKV TAg based on experimental studies and extrapolations of identified functions in the SV40 TAg. Primary sequences of characterized or unknown conserved domains are shown. The one-letter amino acid code is used. Polα, DNA polymerase α binding site; J domain, domain with homology to the Dna J domain; Rb, binding site for the retinoblastoma family members; NLS, nuclear localization signal; DBD, DNA binding domain; Zn, zinc finger; Leu, leucine-rich stretch; p53, p53 binding domain; an.ap., anti-apoptotic domain; HR, host range and adenovirus helper function domain. The conserved Glu residue at position 166 is indicated.

Less is known about the functions of the BKV J domain. Recent studies have shown that BKV TAg can increase the levels of free, transcriptionally active E2F. This induction requires the pRb binding domain as well as an intact J domain (Harris et al., 1998). The J domain seems also to be involved in mediating serum-independent growth of stable BKV TAg-transfected African green monkey BSC-1 cells (Harris et al., 1996). Finally, the amino-terminal domain of SV40 large TAg binds the CBP/p300 proteins, general co-activators of transcription with an intrinsic histone acetyltransferase activity (Avantaggiati et al., 1996; Eckner et al., 1996). The interaction depends on the integrity of a region that includes the pRb binding motif. Amino acids 105–116, which span the pRb site in the SV40 TAg, show high homology with the BKV TAg. It was therefore no surprise to find that high levels of BKV TAg were able to form complexes with pRb105 (Dyson et al., 1990; Harris et al., 1996; Simmons, 1995).

BKV TAg is also able to bind the p107 and p130 members of the retinoblastoma family of proteins both in vivo and in vitro. However, at physiologic levels of TAg at which complexes with p53 could be observed, no association with the pRbs was detected. At these low concentrations, however, a reduction in the amount of pRb, p107, and p130, as well as an increase in hypophosphorylation of these proteins, was observed (Harris et al., 1996). SV40 TAg contains the nuclear localization signal KKKRKVD (residues 127–133), which is completely identical with BKV TAg. A putative DNA binding domain (amino acids 137–251) shows 80% identity with the minimal DNA binding domain of SV40 TAg. Glu-166 of this domain is absolutely conserved among SV40, BKV, and JCV. An SV40 TAg protein with relaxed origin-binding specificity was obtained when this residue was mutated to a Gln or a Lys. The zinc-finger motif CLKCIKKEQPSHYKYHEKH in SV40 TAg is present as CKKCQKKDOPYHFKYHEKH in BKV (residues 304–322). Mutations in this motif in SV40 did not prevent TAg from binding to the origin of replication, but the mutants failed to form hexamers (Loeber et al., 1991). The Leu-rich $LX_7LX_7LX_7L$ stretch in the central part of SV40 TAg is also present in BKV TAg (residues 347–371). The functional significance of this pattern remains unclear.

The most highly conserved domain among all polyomavirus large TAg is the ATPase activity domain (residues 370–661), which is important for ATP binding and for complex formation with p53. BKV TAg binds p53 in vivo (Harris et al., 1996; Nakshatri et al., 1988; Shivakumar and Das, 1996; Simmons, 1995). This region also encompasses the domain essential for transformation of different rodent cells. This domain, spanning nucleotides 356–384, shares only 54% homology with SV40 (Nakshatri et al., 1988). The GXXXXGK motif (amino acids 427–433 in BKV TAg), found in many nucleotide binding proteins, is present in TAg of all polyomaviruses, but its functional role has not been examined. Amino acids 525–541 of SV40 TAg contain antiapoptotic activity independent of inactivation of p53 in REF-52 cells (Conzen et al., 1997). The corresponding BKV region shares almost complete identity with only two amino acid substitutions (Ile to Leu and Ser to Pro). This may suggest an analogous function. The ATPase domain of SV40 large TAg is comprised of the residues 593–661. A substantial portion of this region (residues 640–661) is not present in BKV TAg, and the remaining region shows only 45% homology (Pipas, 1992). It is therefore possible that BKV large TAg lacks ATPase activity or that this activity is located in another functional domain of the protein.

Finally, a functional domain located at the carboxy terminus of the SV40 TAg has been assigned a host range function that seems to be required in viral assembly, and it also possesses the adenovirus helper function allowing productive infection of adenovirus. This region is involved in stimulating expression of the late genes (reviewed by Cole, 1996). Of the 38 SV40 amino acid residues required for host range function, 22 are conserved in BKV. BKV has been shown to provide a helper function for adenovirus, but the genomic region responsible for this function has not been determined (Howley, 1980).

SMALL T ANTIGEN. SV40 tAg possesses several important biologic properties. This protein augments viral and cellular DNA replication, promotes cell cycle progression in different cells, and is also involved in transformation (Cicala et al., 1994; Howe et al., 1998; Manfredi and Prives, 1994; Shenk et al., 1976; Sontag et al., 1993; Whalen et al., 1999). SV40 tAg can *trans*-activate viral and cellular promoters (Moens et al., 1997) and contains an antiapoptotic activity (Kolzau et al., 1999). Few studies have been aimed at characterizing the biologic properties of BKV tAg.

The Cys-rich tAg is located in both the nucleus and the cytoplasm of infected cells (Shah, 1996). tAg is composed of 172 residues and is approximately 17 kDa. Large TAg and small tAg proteins share the amino-terminal 82 amino acids and contain different carboxy-terminal regions. tAgs of BKV and SV40 show 73% overall homology and 59% homology in their unique carboxy-terminal half. The functional domains of BKV tAg have been less well studied. Early studies showed that two cellular proteins (56 and 32 kDa), probably the A and C subunits of PP2A, co-immunoprecipitated with BKV tAg (Rundell et al., 1981). Indeed, both SV40 and BKV tAgs bind the serine–threonine protein phosphatase 2A (PP2A), a heterotrimeric enzyme that consists of a catalytic C subunit, a structural A subunit, and one of a diverse array of regulatory B subunits (reviewed by Mumby, 1995). Two regions of SV40 small tAg are involved in binding to PP2A. The amino acid residues 110–130 are important for binding the A subunit, while the 95–105 region is crucial for interaction with the C subunit (Ruediger et al., 1999). These regions share 65% and 64% (7 residues of 11) identity with the BKV small tAg.

Removal of the first 51 amino acids of SV40 tAg causes a 140-fold reduction in affinity for the AC complex (Mateer et al., 1998). This region exhibits sequence similarity with the J domains of DnaJ proteins (see "Large T Antigen") and is strongly conserved between SV40 and BKV (44 out of 51 residues are identical) (Kelley and Georgopoulus, 1997; Srinivasan et al., 1997). A stretch of Cys residues with conserved spacing ($CX_5CX_7CXCX_2CX_{21}CXCX_2C$), followed by the conserved sequence WFG, is present in the carboxy-terminal region of the tAg of most polyomaviruses, including BKV. The function of this motif remains elusive.

The exact functions of tAg in the life cycle of BKV have not been revealed. tAg seems to be dispensable for viral replication because BKV(MM), which has a deletion in the coding region for tAg, can be successfully propagated in cell culture. The BKV(MM) tAg coding sequence is 216 nucleotides shorter than other sequenced BKV strains and encodes only a 10 kDa tAg (Seif et al., 1979). The amino-terminal 98 amino acids of BKV(MM) tAg are identical with wild-type tAg except for residue 35 (a tyrosine), which is deleted (Seif et al., 1979; Yang and Wu, 1979).

The role of tAg in in vivo transformation remains controversial. BKV(MM) and BKV(IR), which also has a deletion in the tAg sequences, were originally isolated from the reticulum cell sarcoma of the brain and the urine of a patient with Wiskott-Aldrich syndrome and from a tumor of pancreatic islets, respectively (Pagnani et al., 1986; Takemoto et al., 1974). The fact that both viruses were rescued from human tumors may suggest that BKV tAg is not absolutely required for neoplasia. Intracerebral injection in hamsters with BKV(MM) resulted in development of ependymomas (Costa et al., 1976), but BKV(MM)-transformed primary rabbit kidney cells failed to induce tumors when injected in nude mice (Mason and Takemoto, 1977). On the other hand, in vitro transformation studies of baby hamster or baby rat kidney cells indicated an additive role for tAg in the transformation process. Transformation with plasmids expressing intact large TAg, but truncated tAg was reduced 10–35% compared with transformation efficiencies obtained with plasmids expressing both wild-type large T- and small tAgs (Nakshatri et al., 1988). In conclusion, tAg seems to fulfill an auxiliary role for TAg.

Late Proteins. The transcripts for the structural proteins VP1, VP2, and VP3 are generated from a common precursor mRNA by alternative splicing. VP2 (351 residues) and VP3 (232 residues) are translated from the same messenger by using alternative start codons. VP2 contains the entire VP3 sequence at its carboxy terminus and an additional 119 amino acids at its amino terminus. The coding region for the amino terminus of VP1 overlaps that for the carboxy termini of VP2 and VP3, but VP1 is translated from a different reading frame to produce a 362 amino acid long protein. VP1 constitutes between 69% and 84% of the protein mass of the virion and is the major capsid protein (Howley, 1980). The capsid proteins possess three functional domains that facilitate the packing of the viral genome and maturation of the virus particle:

1. Nuclear localization of the capsid proteins is mediated through a nuclear localization signal (NLS). The first 11 NH_2-terminal amino acids from SV40 VP1, when fused to poliovirus VP1, were sufficient to target the fusion protein into the nucleus. These residues contain the conserved MAPTKRKG motif, which is also found in BKV VP1 (Wychowski et

al., 1986). Studies with SV40 mutants have restricted the NLS to the first 19 amino-terminal residues, and the motifs $K_5R_6K_7$ and $K_{16}K_{17}X_{18}K_{19}$ (the numbers refer to the positions in the protein) were independently important for nuclear localization (Ishii et al., 1996). Identical motifs with conserved spacing are present in BKV VP1. Also SV40 VP2 and VP3 contain an NLS. The KKKRK NLS signal in SV40 VP2 and VP3 is conserved as a QKKRR motif in BKV.

2. SV40 VP1, VP2, and VP3 have been shown to bind viral and nonviral DNA in a sequence-independent manner (Clever et al., 1993; Soussi, 1986). The DNA binding domain in SV40 VP1 has not been mapped (Dean et al., 1995). The carboxy-terminal 40 residues of SV40 VP2 and VP3 were sufficient for binding duplex DNA as well as single-stranded DNA and RNA. Thirty-one of these 40 amino acids are conserved in BKV VP2 and VP3, suggesting that BKV VP2 and VP3 also can associate with DNA (Chang et al., 1993; Clever et al., 1993; Soussi, 1986). The carboxy-terminal 13 residues of VP3 are critical for DNA binding. They are 69% identical with BKV VP3 and 64% identical with an 11 residue stretch region near the amino terminus of histone H2A. This region is thought to make contact with the linker DNA connecting adjacent nucleosomes in the viral minichromosome, thereby aiding in their condensation (Arents et al., 1991; Struck et al., 1992).
3. Capsid proteins contain protein–protein interaction domains. Residues 222–234 of SV40 VP3 specify a VP1-interaction determinant (Gharakhanian et al., 1988). This region is 80% identical with the corresponding region in BKV VP3. Moreover, SV40 VP3 is able to bind Sp1 and form ternary complexes with Sp1 and DNA in vitro. This complex can repress transcription from the early promoter. The turn-off of viral expression by VP3 was tightly coupled to viral assembly. These observations suggest a role for VP3 and Sp1 in virus maturation (Gordon-Shaag et al., 1998). The carboxy-terminal 13 amino acids that participated in protein–protein interaction with Sp1 are strongly conserved in BKV VP3.

SV40 and mouse polyoma virus VP2 are myristylated at their amino terminus. This post-translational modification plays a role in the early events of infection (Streuli and Griffin, 1987). Myristylation of BKV VP2 has not been reported thus far. Deletion mutants of SV40 that produce no VP2 are weakly viable, and plaques formed by these mutants appear very slowly and enlarge slowly (Cole et al., 1977). Similar studies in BKV are lacking.

Agnoprotein. The late region of BKV encodes an agnoprotein of 66 amino acids that is found in the perinuclear area of BKV-infected cells. The agnoprotein is a phosphoprotein and co-immunoprecipitates with cellular proteins of approximately 50, 75, and 100 kDa. The identity of these host cell proteins remains unknown (Rinaldo et al., 1998). The BKV agnoprotein seems to be dispensable for the viral life cycle. The BKV strain AS contains a 32 basepair deletion in the late mRNA leader that alters the initiation codon for the agnogene. If this initiation codon is utilized, an additional eight amino acids would be added to the amino-terminal end of the BKV(AS) agnoprotein (Tavis et al., 1989). Despite this mutation in the agnogene, BKV(AS) has been successfully propagated in human fetal kidney, lung, and glial cells (Tavis et al., 1989). Deletions in the agnogene of SV40 have been reported to reduce virus yield and plaque size compared with wild-type SV40 (Barkan and Mertz, 1981; Shenk et al., 1976). It has been suggested that the SV40 agnoprotein facilitates efficient packing of the capsid proteins and enhances virus yield (Carswell and Alwine, 1986). The growth properties of BKV with a deleted agnogene have not been compared with a wild-type BKV strain.

Other Open Reading Frames. Alternative small open reading frames that may encode peptides are found in the BKV genome. BELP, a putative 39 amino acid hydrophobic peptide encoded by the early region corresponds to the previously described *S*V40 *e*arly *l*eader *p*eptide, SELP. SV40 mutants carrying a 23 amino-acid deletion in SELP behaved like wild-type virus (Khalili et al., 1987). Another open reading frame, located at the 3′ end of the early region, may theoretically encode a 75 amino acid peptide (Hey et al., 1994). The existence and role(s) of these two putative proteins in the BKV life cycle have not been determined.

Viral DNA Replication

Replication of BKV DNA occurs in the nucleus of infected cells, proceeds bidirectionally, and terminates at a point 180° from the origin. Several areas in the NCCR are important for DNA replication. This is evident from the studies summarized below. Viral DNA synthesis begins at a unique origin of DNA replication located in the NCCR (Fig. 14.1). The center of the BKV DNA replication origin is determined by the sequence GAGGCC GAGGCC GCCTCT GCCTC. The origin is further characterized by a true palindrome of 17 residues (PAL), followed by two sets of purine-rich inverted repeats (IR1 and IR2) and a stretch of 20 AT residues. This structure is almost identical with the structure of the SV40 origin of DNA replication (Howley, 1980; Yoshiike and Takemoto, 1986). Earlier studies had shown that a plasmid containing the *Hin*dIII fragment encompassing the NCCR of BKV could replicate in BKV-transformed human embryonic fibroblast L603 cells that express TAg. This plasmid did not replicate in human osteosarcoma 143 B cells that lack the gene for TAg expression (Milanesi et al., 1984). These results indicate that the TAg is absolutely required for replication.

Studies in COS-1 cells, an SV40-transformed CV-1 cell line that constitutively expresses the early proteins, using plasmids containing BKV(Dun) sequences showed that the 21 bp at the 5′ end of the P block were indispensable for replication. This element was referred to as the *rep element* (Del Vecchio et al., 1989). Plasmids containing the core element of the BKV origin of replication, that is, the AT-rich stretch, the palindromic sequences, and the TAg binding motifs, but lacking the rep element, did not replicate under these conditions. The remaining 47 bp of the P block in conjunction with this 21 bp rep element increased the level of replication, but these 47 bp alone were not sufficient to support replication. Juxtaposing these 21 bp to the AT-rich motif was required for efficient replication because changing the position between the AT block and the 21 bp abrogated replication. Addition of P or Q blocks did not further enhance replication. Furthermore, mutations in neither the Sp1 site of the rep element nor in the NF-1 site in the central region of the P block had any effect on replication.

The exact function of the rep element in DNA replication remains unsolved, but it could either bind a cellular factor or assume a certain DNA conformation required for replication (Del Vecchio et al., 1989). Deyerle and colleages (1989) used a similar assay to define the exact boundaries of the BKV origin of replication in different cell lines with different sources of TAg. Replication of plasmids with different BKV *ori* sequences was tested in HeLa or CV-1 cells provided with an expression plasmid for the early BKV proteins in *trans* or in l603 cells (BKV-transformed human fibroblast cell line that constitutively expresses the early proteins), or in COS cells. Deletions extending from the early region showed that mutants lacking the high-affinity TAg binding site I replicated in all four cell types. Caputo and co-workers (1986) also found that TAg binding site I was dispensable for supporting replication of plasmids with the NCCR sequences of the Gardner strain in COS cells. Additional deletion of the first 7 bp of IR1 abolished replication in COS cells. Extending the deletion another 8 bp resulted in absence of replication in the human cell lines HeLa and L603. Deletion of 2 bp more totally prevented replication in all cell lines. Deletion studies extending from the late region proved that the complete enhancer was dispensable for replication in HeLa, CV-1, and L603, but not COS, cells. Replication in COS cells demanded the presence of 16 bp at the 5′ end of the P block. These results confirmed the findings of Del Vecchio and coworkers (1989), who identified the importance of the rep element for replication.

Deletions removing the entire AT block completely inhibited replication in all cell types tested. Internal deletion of the first 15 bp of IR1 did not affect replication in HeLa and CV-1 cells, but prevented replication in COS cells. However, complete deletion of IR1 impaired replication in HeLa and CV-1 cells. A linker scan mutation of the TAg binding site II in IR2 had no effect on replication in HeLa cells, but abolished replication in COS cells. Removal of the 3′ end of IR2 plus the AT block completely prevented replication in all cell types. These results indicate that the exact boundaries of the BKV *ori* depend on the cell type and on the viral origin of TAg. The variation in sequence requirement probably reflects differences between SV40 and BKV TAg or/and the availability of specific cellular replication factors.

Replication of SV40 DNA requires, in addition to TAg, numerous cellular proteins, including DNA polymerases α and δ, replication protein A, replication factor C, proliferating cell nuclear antigen, Rnase H, MFI exonuclease, DNA ligase I, and topoisomerases (reviewed by Cole, 1996). The same cellular proteins are most probably also involved in BKV DNA replication, although experimental proof is lacking to support this.

Viral Transcription

Early Transcription. Transcription of the BKV genome extends bidirectionally, with early and late mRNAs being transcribed from opposite strands (Shah, 1996). Conscientious studies have mapped the BKV regulatory sequences required for early and late transcription and the major transcription sites. Most studies have used the BKV(Gardner) strain ($\mathbf{P}_{1-68}$–$\mathbf{P}_{1-7,26-68}$–$\mathbf{P}_{1-68}$–$\mathbf{Q}_{1-39}$–$\mathbf{S}_{1-63}$ configuration) to dissect the sequences involved in early and late expression in different cell types, mostly of animal origin. Unfortunately, similar studies with BKV(WW) in relevant human cell lines are missing. The major early transcription start sites were mapped in the AT block, with minor start sites located towards the early coding region and towards the border of the O and P blocks (Deyerle et al., 1987; Ferguson and Subramani, 1994).

In semipermissive monkey (CV-1) or nonpermissive rodent (BHK, NRK) cells, it was shown that the palindromic motif in NCCR (Fig. 14.1) was dispensable for early promoter activity. Additional deletion of IR1 reduced the promoter strength by 50% and 80% compared with wild type in NRK and CV-1 cells, respectively. Further deletion of IR2 resulted in an early promoter with about 10% strength of the wild-type promoter in the three cell types. Deletion of the complete O block resulted in promoters with very low activity (1–4%) in all cells tested. Internal deletions of the AT block, IR1, or IR1 + IR2 generated a promoter that retained essentially wild-type activity in CV-1 cells. An internal deletion that removed the IR1 + IR2 + AT–rich sequence created a promoter with only 2% activity of the wild-type promoter. Deletion of the S block did not affect promoter activity in CV-1 cells, but increased it by 50% in rodent cells (NRK and BHK). Deletion of the S and Q blocks created promoters with 75% activity of the wild-type promoter in the rodent cells, while in CV-1 cells promoter activity was reduced to 25%. Further removal of the distal P block had almost no effect. However, deletion of the middle truncated P block reduced promoter activity another three- to sixfold. Deleting the proximal P block (i.e., only the O block is retained), on the other hand, gave promoters with <1% of the wild-type activity. Deletions removing the middle or distal P block, or both, reduced promoter activity two- to fourfold. These results show that the BKV early promoter and enhancer are overlapping and require elements both upstream and downstream of the early mRNA start sites (Deyerle et al., 1987; Deyerle and Subramani, 1988). These results are different from those obtained with the SV40 promoter, where deletion of sequences downstream from the early mRNA start sites had no effect on the rate of transcription (Cole, 1996).

Grinnell and co-workers (1988) found that the BKV(P2) enhancer ($\mathbf{P}_{1-68}$–$\mathbf{Q}_{1-26}$–$\mathbf{R}_{57-63}$–$\mathbf{S}_{1-11}$–$\mathbf{P}_{14-68}$–$\mathbf{Q}_{1-26}$–$\mathbf{R}_{57-63}$–$\mathbf{S}_{1-63}$) is under negative regulation in HeLa cells and that the regions at the P–Q borders bind HeLa cell–specific proteins that are involved in this repression. The junctions contain binding sites for AP-1 and NF-1 (Moens et al., 1995a).

Several groups have demonstrated the involvement of AP1, NF-1, CRE, and Sp1 in transcriptional regulation of the BKV promoter. Moreover, exact spacing between the elements is crucial for optimal promoter activity (Chakraborty and Das, 1989, 1991; Deyerle and Subramani, 1988; Ferguson and Subramani, 1994; Grinnell et al., 1988; Markowitz and Dynan, 1988; Moens et al., 1990, 1995a; Nakshatri et al., 1991; Nowock et al., 1985). Deletion of the NF-1 site in the 5′ end of the proximal P block reduced early promoter activity by 50% (Deyerle and Subramani, 1988). However, a mutation in the NF-1 motif in the P block in the context of the BKV(PQ) promoter resulted in a twofold increase in promoter activity (Ferguson and Subramani, 1994). Mutation of the NF-1 site in the Q block of BKV(Gardner) diminished promoter strength by 75%, while it enhanced the BKV(PQ) promoter approximately threefold (Deyerle and Subramani, 1988; Ferguson and Subramani, 1994). Deletion of the CRE motif in the context of a PQ promoter/enhancer reduced early promoter activity by 30% (Ferguson and Subramani, 1994). The corresponding mutant in a BKV(Gardner) promoter was not tested.

These examples suggest that the contribution of a binding site to promoter activity depends on its context. Indeed, it was illustrated that spacing and interaction between multiple binding motifs determined their influence on the transcriptional activity of the promoter. Synergistic and antagonistic interactions between different motifs were observed (Ferguson and Subramani, 1994). Two binding sites (SI and SII) for the tumor suppressor protein p53 were identified in the BKV(Dun) NCCR (Fig. 14.1). Co-expression of murine p53 repressed the BKV(Dun) early promoter in HeLa, H1299 (human non-small cell lung carcinoma), and SaOS2 cells, but activated the promoter in NIH 3T3 and CV-1 cells. No effect was measured in rat HepG2 cells (Das et al., 1995). In H1299 cells, dimer SI sites could induce p53 responsiveness to a basal promoter, while SII could not. A promoter with two copies of both the SI and SII motifs was strongly activated by p53. This p53-dependent *trans*-activation was abrogated by BKV TAg. A p53 mutant unable to bind DNA could also repress the BKV early promoter in H1299 cells, which indicates that the ability of p53 to regulate the BKV promoter is not determined solely by its binding, but is modulated by other incompletely characterized cellular factors. The p53 binding sites are in close proximity to NF-1 and Sp1 binding sites, and in vitro synthesized Sp1, p53, and NF-1 proteins could be co-immunoprecipitated with a monoclonal antibody against p53 when mixed together. Regulation of the BKV promoter by p53 may therefore be caused by complex protein–protein and protein–DNA interactions (Shivakumar and Das, 1996). Insertion of oligonucleotides with AP1 sites strengthened the early promoter, while further addition of NF-1 sites did not increase promoter activity (Markowitz et al., 1990).

Our studies have identified a functional hormone response unit in the S block (Moens et al., 1994, 1999). This hormone response unit consists of a consensus estrogen responsive element (ERE) and a nonconsensus glucocorticoid/progesterone responsive element (GRE/PRE). These elements could bind their cognate nuclear receptor in vitro and could mediate estrogen- and glucocorticoid/progesterone-induced transcription, respectively, when linked to a heterologous promoter. Moreover, these hormones stimulated early and late transcription and enhanced virus yield in cell culture (Moens et al., 1994). Our group later showed that the AP-1 sites at the junctions of the P block also had the potential to mediate estrogen-induced transcription and that concerted expression of TAg and tAg strongly stimulated estrogen receptor–mediated transcription (Moens et al., 1999). The prevalence of BKV reactivation in pregnant women varies between 3.2% and 26%. Reactivation was most frequent in the third trimester of pregnancy, when the concentrations of estrogens and progesterone are at peak levels (Bendiksen et al., 2000a; Chang et al., 1996b; Coleman et al., 1980; Markowitz et al., 1991; Tsai et al., 1997). These observations may suggest a functional role of the hormone response unit for BKV in vivo. However, BKV(AS), keeping in mind that it was originally isolated from the urine of a pregnant women, has a 32 bp deletion in the 3′ end of the S block that removes the ERE and GRE/PRE motifs. This argues against the importance of the hormone response unit in BKV reactivation during pregnancy. Alternatively, deletion of the ERE and GRE/PRE may have occurred during cell culture propagation because BKV(AS) sequences were first obtained after the virus had been passaged in cell cultures (Tavis et al., 1989).

Early promoter activity is autoregulated in the presence of TAg. The BKV(Gardner) promoter/enhancer was activated 4-, 9-, 31-, and 28-fold in COS, PAF, SV-1, and W603 cells, respectively, by TAg. The first three cell lines are SV40-transformed cells, while the latter is a BKV-transformed cell line (Caputo et al., 1986). However, a repression of early promoter activity was measured when BKV TAg was provided in *trans*. This negative autoregulation was strongest in CV-1 cells and weak or absent in BHK and NRK cells, respectively. The reason for this cell-specific effect remains unclear. The degree of repression by TAg increased when P blocks were deleted (Deyerle and Subramani, 1988). Repression was partially relieved with a mutant promoter lacking TAg binding site I (Deyerle et al., 1989). These conflicting results can be explained by a concentration dependence of the TAg effect. Binding of TAg to viral DNA blocks the assembly of a functional transcription complex, thereby repressing early transcription. For SV40 it was shown that low levels of TAg stimulated transcription from its own promoter by a mechanism that does not require direct binding of TAg to viral DNA. Relatively high levels of TAg resulted in repression of viral transcription (Wildeman, 1989).

Late Transcription. The BKV late promoter lacks a canonical TATA box motif, but a TATA-like sequence (TTAAA) is present in the S block that overlaps with the ERE motif. The significance of this motif remains to be determined, but BKV(AS) has a deletion in this region, and this seems not to affect viral replication (Tavis et al., 1989). For SV40 it has been shown that a cellular repressor protein (IBP), present in limited concentrations in HeLa cells, blocks late transcription. As the number of DNA templates increases after replication, this repressor is diluted out and late transcription can proceed. This cellular repressor is an orphan member of the steroid-thyroid hormone nuclear receptor superfamily and binds to the $AGG^{T/T/}_{C\,C}A$ consensus motif (Wiley et al., 1993). The BKV(WW) NCCR contains two putative binding sites for IBP; one overlaps with the CRE motif in the P block, and the other is part of the GRE/PRE motif in the S block.

In vitro transcription assays with HeLa cell extracts mapped minor late transcription initiation sites in BKV(Gardner) in the P block proximal to the late region, in the Q block, and in the S block. In addition, one start site overlapped with the start codon of the agnoprotein, while the major late start site was located within the coding region of the agnoprotein (Chakraborty and Das, 1991). These results differ from in vivo mapping where start sites were only detected in the third P block, the Q block, and the S block. Transcription start sites in the Q block accounted for >85% of the BKV late mRNA, while the third P block, and the S block accounted for <5% and about 10%, respectively (Cassill and Subramani, 1988).

It was also determined which enhancer elements of BKV(Gardner) were necessary for late transcription in CV-1 cells under both replicating and nonreplicating conditions. The late promoter is encompassed within the early enhancer region under both conditions. Deletion of the P block proximal to the early region reduced late promoter activity by 50%. Further deletion of the middle P block or internal deletion removing this P block resulted in a promoter with 35% activity of the wild-type promoter. Deletion of all three P blocks gave a very weak promoter (5% activity). Deletion of just the S and Q blocks strongly reduced promoter activity. Extending this deletion with the P block adjacent to the Q block had no further effect on promoter strength. Similar results were obtained in HeLa cells except for a mutant promoter deprived of the S, Q, and P block closest to the Q block. This mutant promoter had a 2.5-fold increased activity compared with the nonmutated promoter.

Expanding the late promoter with the complete coding region of the agnoprotein did not increase promoter activity. A late promoter consisting of the middle P block and the Q block had a threefold lower activity than a similar promoter but with a complete 68 bp P block. The 18-bp deletion eliminates an Sp1 and NF-1 binding site, indirectly suggesting the importance for late transcription of these binding sites (Cassill and Subramani, 1989). The contribution of single transcription factor binding motifs in late transcription was examined by linker scan mutation analysis. These studies proved the involvement of Sp1, NF1, and CRE motifs in the P block and of AP1, NF-1, and Sp1 motifs in the Q block. Most linker scan mutations did not reduce transactivation by TAg except for mutations in the NF-1, CREB, and p53 sites in the P block. The latter mutation had, however, no effect on promoter activity in the absence of large TAg (Cassill et al., 1989). The 18 bp that are deleted in the middle P block in BKV(Gardner) contain part of the potential binding sites for Sp1 and NF-1. Replacing this truncated P block with a full-length P block decreased late promoter activity in CV-1 cells (Cassill and Subramani, 1989).

BKV TAg stimulated the late promoter about sixfold even in the absence of identifiable TAg binding sites (Cassill and Subramani, 1988). SV40 TAg can substitute functionally for BKV TAg, but may not be competent for switching to late gene functions (Major and Matsumura, 1984).

SV40 tAg has been shown to regulate the transcriptional activity of CREB, AP1, p53, Sp1 and NF-κB (Frost et al., 1994; Garcia et al., 2000; Scheidtmann et al., 1991; Sontag et al., 1997; Westphal et al., 1998; Wheat et al., 1994). Putative or proven binding sites for these transcription factors are present in the BKV NCCR (Fig. 14.1), suggesting that tAg also can contribute to the regulation of the BKV promoter/enhancer.

All these studies on BKV supplemented with observations for SV40 emphasize the crucial role for TAg in the outcome of the life cycle of these polyomaviruses. When present in low concentrations, TAg will stimulate its own expression by activating the early promoter/enhancer. The anatomy of the NCCR determines the strength of this promoter and therefore the amount of TAg transcripts. As the concentration of TAg increases during infection, a switch in TAg's function from transcriptional activator to repressor occurs. TAg will sequentially bind other sites of the NCCR to stimulate replication. Cellular repressors, which are present in low concentration and prevent late transcription, will be diluted out due to increased numbers of viral genomes, and late transcription will be initiated. The capsid proteins can then assemble with the newly replicated viral genomes into mature virions. Thus, the outcome of a BKV infection in vivo (latent or productive infection) depends on the amount of TAg produced as well as on specific host cell transcription factors required for expression of the BKV genes.

3. BIOLOGY OF BKV

Research from recent years has contributed extensively to our knowledge of the molecular biology of BKV. Our understanding of important aspects of BKV infections in humans remains, however, limited. Very little is known about transmission, route of infection, host cells, and mechanisms for spread of this virus. Some of these aspects are briefly reviewed here but are also discussed in Chapters 10, 16, and 19.

Primary infection usually occurs at an early age with the highest incidence between the ages of 1–6 years and appears to be asymptomatic in most individuals. An association of primary infection with mild respiratory tract disease, mild pyrexia, and transient cystitis has been reported (reviewed by Dörries, 1997; Shah, 1996). The route of infection has not been firmly defined. BKV seroconversion has been associated with both tonsillitis and upper respiratory infections (Goudsmit et al., 1981, 1982; Mäntyjärvi et al., 1973). However, PCR-based investigation of nasopharyngeal aspirates from 201 children revealed the presence of BKV DNA in only two samples, and subsequent virus isolation after inoculation of cell cultures was unsuccessful. No BKV DNA could be detected in 70 saliva specimens. These results suggested that the oropharyngeal tract is not the site of entrance (Sundsfjord et al., 1994b).

The demonstration of BKV-specific IgM antibodies in sera derived from umbilical cord blood or the sera of children less than 2 weeks of age support a prenatal infection with BKV (Rziha et al., 1978a; Taguchi et al., 1975). PCR analysis of maternal and fetal materials revealed a high prevalence (50% and 75%, respectively) of BKV DNA (Pietropaola et al., 1998). This suggests that transplacental transmission may be an important route for primary infection and is consistent with the frequent reactivation of BKV in the third trimester of pregnancy. Not all studies support this assumption (Borgatti et al., 1979; Coleman et al., 1980; Shah et al., 1980; Taguchi et al., 1985). BKV DNA has been reported in male and female genital tissues and in human sperm, suggesting possible sexual transmission as an alternative route for primary infection (Monini et al., 1996).

After primary infection, BKV seems to be spread throughout the body by the blood. Indeed, several groups have found BKV DNA sequences in peripheral blood (Chatterjee et al., 2000; De Mattei et al., 1995; Dolei et al., 2000; Dörries et al., 1994). Lymphocytes have been shown to possess BKV-specific receptors on their cell surface (Possati et al., 1983). BKV usually establishes a harmless latent infection. All the target organs for viral persistence have not yet been identified, but viral proteins or nucleic acid sequences have been demonstrated in the kidney (Chester et al., 1983), bladder (Monini et al., 1995), tonsils (Goudsmit et al., 1982), brain (Elsner and Dörries, 1992; De Mattei et al., 1995), liver (Knepper and diMayorca, 1987), lymphocytes (Dörries et al., 1994), and bone tissue (De Mattei et al., 1994) of normal individuals.

The factors that control the balance between the state of latency and reactivation in the human host remain incompletely understood. Reactivation of BKV in the urinary tract has been observed under a wide variety of conditions, including pregnancy (Bendiksen et al., 2000a; Chang et al., 1996b; Coleman et al., 1980; Markowitz et al., 1991; Tsai et al., 1997), kidney and bone marrow transplantations (reviewed by Shah, 1996), heart transplantation (Masuda et al., 1998), immunodeficiency diseases (Markowitz et al., 1993), systemic lupus erythematosus (Rekvig et al., 1997), and immunosuppressive therapy (reviewed by Shah, 1996). Nonhemorrhagic (Padgett et al., 1983) or hemorrhagic cystitis, ureteric stenosis, interstitial nephritis (see section on BKV and renal infection), encephalitis (Bratt et al., 1999; Voltz et al., 1996), transient hepatic dysfunction (O'Reilly et al., 1981), nephrotic syndrome (Nagao et al., 1982), and retinitis (Bratt et al., 1999; Hedquist et al., 1999) have all been observed in association with BKV reactivation. BKV has also been suggested as the cause of interstitial pneumonia in an 8-month-old girl who underwent hematopoietic stem cell transplantation with an unrelated cord blood unit (Sandler et al., 1997). Despite many examples of concurrent BKV reactivation under clinical conditions, only recent conclusive evidence for BKV-related kidney pathology has been recognized. This condition, called *BKV nephropathology* (BKN), is discussed in the next section.

4. BKV AND INFECTION OF THE URINARY TRACT

Studies performed over many years with different methods have indicated a high prevalence of BK viruria. This urinary virus shedding may represent (1) a transient event during asymptomatic primary infection, (2) virus reactivation in healthy or immunocompromised individuals, (3) a pathologic process in the urinary tract in which the virus may play a role, or (4) an indirect indication of virus activity at other sites in the body (Arthur and Shah, 1989). The detection of BKV in urine and the fact that the virus can be successfully propagated in human embryonic and neonatal kidney cells (Sandler et al., 1997; Seehafer et al. 1975) suggested a role for the kidneys in the biology of the virus. The presence of low levels of episomal BKV in 33% of normal renal parenchyma examined from 30 different human sources further strengthened this assumption (Chester et al., 1983). BKV DNA sequences have been detected in renal epithelium (McCance, 1983), and immunofluorescence staining revealed the presence of viral proteins in exfoliated urothelial cells in urine (Hogan et al., 1980). More recent studies have shown BKV particles, viral proteins (VP1, TAg, and agnoprotein) or DNA sequences in the transitional epithelial cells of the renal calyces, in parietal epithelium of Bowman's capsule, in superficial transitional cells of the renal pelvis, as well as in lining epithelial cells and the transitional epithelium of the urinary tract. BKV inclusions and proteins have also been reported in the urinary bladder surface layer and umbrella cells of the transitional epithelium, and a few desquamated cells were also positive for TAg (Bratt et al., 1999; de Silva et al., 1995; Itoh et al., 1998; Nebuloni et al., 1999; Nickeleit et al., 1999; Smith et al., 1998; Shinohara et al., 1993; Vallbracht et al., 1993).

Hemorrhagic Cystitis and Ureteric Stenosis

An association of BKV with renal dysfunction was first suspected in the case of hemorrhagic cystitis (Hashida et al., 1976) and ureteric stenosis (Gardner et al., 1971). A possible etiologic role for BKV and hemorrhagic cystitis in bone marrow recipients and ureteric stenosis in renal transplant patients remains controversial. Conflicting results on the frequency of BK viruria among renal transplant recipients (5–60%) and bone marrow transplants (0–52%) and numerous single case reports have made it difficult to definitively establish a causal role for BKV infection in these pathologic conditions (Andrews et al., 1988; Apperley et al., 1987; Arthur et al., 1986; Azzi et al., 1994; Bogdanovic et al., 1996, 1998; Chan et al., 1994; Coleman et al., 1978; Cottler-Fox et al.,

1989; Gardner et al., 1984; Goddard and Saha, 1997; Hogan et al., 1980; Koss, 1987; Mathur et al., 1997; Nickeleit et al., 1999; O'Reilly et al., 1981; Randhawa and Demetris, 2000; Rice et al., 1985; Sandler et al., 1997; Traystman et al., 1980; Verdonck et al., 1987; Vögeli et al., 1999).

Some observations, however, favor a role for BKV in hemorrhagic cystitis in bone marrow transplant patients. Long-lasting hemorrhagic cystitis occurred four times more frequently in BKV excretors than in nonexcretors. BK viruria often preceded or coincided with the onset of the disease. Moreover, BKV was recovered far more frequently in urine samples collected during the episodes of hemorrhagic cystitis (55%) than in urine specimens collected in hemorrhagic cystitis-free periods (approximately 10%) (Apperley et al., 1987; Arthur et al., 1986; Chan et al., 1994; Rice et al., 1985; Vögeli et al., 1999). In a recent study by Bedi and co-workers (1995), 95 bone marrow transplant recipients were examined for hemorrhagic cystitis and for consecutive BK viruria by polymerase chain reaction. BKV DNA sequences could not be detected in 45 of the patients, while 12 patients had one episode of BKV viruria. The remaining patients (38) had two or more urine samples that were positive for BKV DNA. Thirty-seven patients had hemorrhagic cystitis, and 19 of them had persistent BK viruria. None of the 45 patients free of BK viruria developed hemorrhagic cystosis. A strong temporal correlation was found between the shedding of BKV and the onset of hemorrhagic cystosis ($r = 0.95$) (Bedi et al., 1995).

BKV-associated hemorrhagic cystitis has also been described in a patient infected with human immune deficiency virus type-1 (Gluck et al., 1994). Ureteric stenosis has been reported in renal transplant recipients with active BKV excretion in their urine samples (Coleman et al., 1978; Gardner et al. 1984; Hogan et al., 1980; Traystman et al., 1980). It has been suggested that damage of the ureter by ischemia or by inflammation may reactivate latent polyomavirus infection in the ureteric epithelium and cause ureteric stenosis (Coleman et al., 1978). In situ hybridization studies of the urinary bladders of seven interstitial cystitis patients failed to detect BKV nucleic acid sequences (Hukkanen et al., 1996). Despite this etiologic association between BK viruria and hemorrhagic cystitis and ureteric stenosis in bone marrow and renal transplant recipients, the clinical consequences of BKV infection in these complications remains tentative.

BKV Nephropathy

Most reports indicate a possible causal role for BKV in renal diseases, but the ultimate proof of BKV-induced kidney pathology is lacking. Recently, several groups described a graft dysfunction in renal transplant recipients as a result of tubular necrosis in the context of extensive BKV replication in the tubular epithelium (Binet et al., 1999; Boubenider et al., 1999; Drachenberg et al., 1999; Howell et al., 1999; Mathur et al., 1997; Nickeleit et al., 1999, 2000a,b; Pappo et al., 1996; Randhawa et al., 1999). This complication, BKV nephropathy (BKN), is characterized by deterioration of the graft and may contribute to graft loss. Upregulation of HLA-DR on tubular cells is a typical finding in graft biopsy specimens with cellular reaction. This may be of significance because HLA-DR can stimulate an allogeneic lymphocytic response. However, BKV did not stimulate HLA-DR expression in renal recipient patients with BKN (Nickeleit et al., 2000a). Inflammation is minimal or absent in these patients. It is therefore assumed that the virus does not provoke a pronounced inflammatory response.

Serum creatinine levels were above normal (150% increase) (Howell et al., 1999; Nickeleit et al., 2000b; Randhawa et al., 1999). The morphologic hallmarks of BKN are intranuclear viral inclusions in epithelial cells and focal necrosis of tubular cells. A typical finding in BKN are infected cells that are rounded up and extruded from the epithelial cell layer into the tubular lumen. Cytopathogenic signs are seen along the entire nephron, but they are most abundant in distal tubular segments and collecting ducts. Sporadically infected cells are noted in the epithelium lining Bowman's capsule. Neither urethral obstructions nor hemorrhagic cystitis are features of BKN. BKN is limited to the kidney transplant and the attached ureters.

In most cases, no other viruses were detected (Binet et al., 1999; Nickeleit et al., 2000b). However, co-infection with cytomegalovirus has been reported by one group (Howell et al., 1999). Low levels of JCV DNA sequences were sometimes observed co-existing with BKV sequences (Binet et al., 1999; Nickeleit et al., 2000b; Randhawa et al., 1999). Thus far, no NCCR sequences are available from BKV found in the kidneys of patients suffering BKN, but it would be interesting to see whether specific strains are associated with this disease.

There is no evidence that the native, autologous kidney or other organs of the recipient were affected. An earlier study had shown that primary BKV infection (or reactivation) in renal transplant recipients (n = 496) increased significantly from 7.3% to 33.7% (or 9.5% to 23%) when the kidney donor was BKV seropositive instead of seronegative (Andrews et al., 1988). This finding, in light of BKN, may suggest a reactivation of latent BKV in the transplanted kidney. Alternatively, a primary infection of the transplant kidney originates from BKV harbored in the recipient, or a de novo primary infection of the recipient has occurred. Because most individuals are BKV seropositive, a reactivation of persisting virus in the renal graft is most likely. Polyomavirus interstitial nephritis has also been reported in renal transplant recipient cynomolgus monkeys undergoing immunosuppressive therapy. A new virus, cynomolgus polyomavirus (CPV), which was antigenically and genomically related to SV40, was isolated. The clinical and morphologic symptoms in affected animals had a striking similarity to BKN and to previously published cases of SV40-induced interstitial nephritis in rhesus monkeys (van Gorder et al., 1999). These animal models support the assumption of polyomavirus-induced nephropathy in humans.

Fatal renal tubular injury attributed to BKV has also been reported in children with immunodeficiency (de Silva et al., 1995; Rosen et al., 1983) and in AIDS patients (Bratt et al., 1999; Nebuloni et al., 1999; Vallbracht et al., 1993). A recent report describes the isolation of a BKV mutant from the kidney of a 36-year-old AIDS patient with interstitial nephritis. The NCCR of this strain, BKV(Cin), revealed duplications and deletions (Table 14.1). Moreover, a substitution of Gln169 to Leu in the DNA binding domain of TAg was found. The authors suggested that this mutation promoted the correct folding and/or stabilized the active protein conformation. The mutation, together with rearrangements in the NCCR, may alter viral DNA replication and enhance pathogenicity (Smith et al., 1998). Finally, urine samples of single case reports of patients with a variety of urinary tract diseases (hematuria, urethral stricture, prostatic hypertrophy, renal calculi, bladder cancer, cystocoele, suprapubic pain, and epididymitis) contained BKV DNA but no inclusion-bearing cells (Cobb et al., 1987).

BKV and Systemic Lupus Erythematosus

Systemic lupus erythematosus (SLE), the prototype of a systemic autoimmune syndrome, is characterized by the production of antinuclear antibodies and particularly anti-dsDNA antibodies (Lahita, 1999). The latter antibodies have the potential to induce nephritis, a pathologic process that may be fatal for SLE patients (D'Andrea et al., 1996; Raz et al., 1989; Termaat et al., 1992; Van Bruggen et al., 1997a,b). The etiology of SLE is poorly understood, but genetic and environmental factors as well as stochastic events linked to somatic generation of the repertoire of the immune system and to affinity maturation of the immune responses may contribute to the development of pathogenic anti-dsDNA antibodies and this disease. Viruses such as Herpesviridae members, endogenous retroviruses, and hepatits C virus have been suggested to be involved in SLE, but the molecular and cellular bases for virus-induced autoimmune syndromes remain poorly understood. Molecular mimicry has been proposed as a possible mechanism to explain the association between antiviral antibodies and specific autoantibodies (reviewed by Cooper et al., 1998).

Our group observed antibodies to DNA and histones in the sera of immunologically normal rabbits after inoculation with BKV, but not in their preimmune sera (Flægstad et al., 1988; Fredriksen et al., 1993, 1994). From these initial observations, the following model was constructed: If DNA is regarded immunochemically as a hapten, it needs, in order to act as a functional immunogen, to be complexed with a protein to which T cells are not tolerant. Such complexes may be sufficient to evoke a cognate B-cell– and T-helper-cell–dependent immune response to DNA. The same would be true if DNA itself did not have the property to be presented by HLA class II molecules. Also in this situation, a carrier protein would be crucial. In this hapten-carrier model, a viral DNA binding protein would fulfill several immunologic requirements that would be sufficient to play a role as a carrier protein. Human polyomaviruses encode several DNA binding proteins that could accomplish this function. Both TAg and the capsid proteins VP1, VP2, and VP3 have the potential to bind nucleosomes and DNA, thus rendering DNA immunogenic.

Because of the crucial role of TAg in the biology of BKV, we tested whether TAg expression in vivo was sufficient to elicit production of anti-dsDNA antibodies. Immunization of immunologically normal mice with a plasmid encoding wild-type DNA binding TAg resulted in production of antibodies to this protein. These antibodies were kinetically linked to the production of antibodies to dsDNA, histones, and certain transcription factors like CREB and TBP, but not to other autoantigens (Moens et al., 1995b; Rekvig et al., 1997). Injections with a plasmid expressing the non-DNA binding protein luciferase, or with a plasmid containing large TAg-coding sequences but lacking a eukaryotic promoter did not result in such antibodies. These latter observations also strongly suggested that plasmid DNA itself was not immunogenic. This experiment demonstrated that TAg could (1) bind nucleosomes and (2) following this complex formation, could initiate the production of anti-dsDNA antibodies. Subsequently, we demonstrated that this model had its natural correlate in processes operational in SLE patients as demonstrated by the fact that 16 of 20 (80%) SLE patients had single, intermittent, or continuous episodes of BK or JC viruria over a 1 year observation period (Rekvig et al., 1997).

This vivid tendency for virus reactivation correlated strongly ($p < 0.0001$) with the presence of anti-TAg and anti-DNA antibodies. This correlation indicates that the results of the TAg-expressing plasmid immunization experiment described above reflect processes that are operational in vivo in SLE. The high prevalence of human polyomavirus reactivation could theoretically be associated with a special strain circulating in these patients. Sequencing the NCCR revealed that 15 patients excreted the archetypal BKV(WW) or JCV(CY) strains, while one patient had urinary shedding of the rearranged BKV(AO) strain (Table 14.1) (Rekvig et al., 1997). The NCCR sequences of the virus DNA detected in the urine samples of these patients collected during an extended 3 year observation period remained stable as no or very few point mutations were detected (Sundsfjord et al., 1999). Genotyping of a region of the VP1 coding sequence, which has been used as a marker for strain determination, failed to detect new virus strains in the urine of SLE patients (Bendiksen et al., 2000a). These results indicate that the human polyomavirus strains found in SLE patients are not different from those circulating in the human population. All SLE patients with BK or JC viruria possessed anti-DNA and anti-TAg antibodies, and the production of anti-TAg antibodies coincided with anti-DNA antibodies.

Interestingly, antibodies to dsDNA correlated more strongly to persistent viruria than to infrequent or intermittent episodes of urinary virus shedding (Rekvig et al., 1997). A highly significant correlation ($p < 0.0001$) between anti-TAg and anti-DNA antibodies was also found in patients with other autoimmune diseases, although the frequencies of these linked antibodies were lower than in SLE (Bredholt et al., 1999). In vitro studies again confirmed that complexes of TAg and nucleosomes can create a molecular basis for initiation of anti-DNA and antihistone antibodies. Stimulation of co-cultured autologous B cells and TAg-specific T cells isolated from SLE patients or normal individuals with nucleosome–TAg complexes resulted in the production of antibodies with specificity against TAg and DNA (Andreassen et al., 1999a). In this scenario, generation of functional autoimmune histone-specific T cells could also be detected using cells from normal individuals not responding directly to histones in primary culture (see below) (Andreassen et al., 1999b).

A model explaining how B cell and T cell tolerance to nucleosomes may be terminated during in vivo replication of human polyomaviruses is illustrated in Figure 14.3. During viral replication, TAg forms complexes with nucleosomes of the infected host cells. Such complexes can be presented to the immune system when liberated from infected cells. DNA-specific B cells binding and processing such complexes may simultaneously present peptides derived from both large TAg and histones. T cells not tolerant for large TAg will respond, produce IL-2, and proliferate. In this situation, the B cells may transform into anti-DNA antibody producing plasma cells, and T-cell anergy (or nonresponsiveness) for histones may be terminated. The latter point is in agreement with observations by others who have unequivocally demonstrated that IL-2 has the potential to drive tolerant (presumably anergic) T cells into proliferation with subsequent loss of the nonresponsive state (DeSilva et al., 1991; Jenkins, 1992). Thus, during this process, nonresponsive histone-specific T cells receive two different stimuli: the first as a nonspecific bystander stimulus by IL-2 produced by TAg-specific responder T cells and the other by a conventional antigen-specific stimulus by histone peptides presented by the same B cells as those presenting TAg peptides. Thus, if DNA-specific B cells present both TAg and histones to responder T cells, initiation of autoimmune histone-specific T cells and anti-DNA antibodies will be the final result. If TAg expression is sustained, for example, during a persistent active viral infection, sustained stimulation of anti-DNA–specific B cells may result in affinity maturation that may transform them into high-affinity, potentially nephritogenic, anti-dsDNA antibodies by progressive somatic mutations in their variable regions (Fig. 14.4) (Desai et al., 1993; Krishnan and Marion, 1993; Marion et al., 1997; Shlomchik et al., 1990).

A consequence of this affinity maturation is the production of anti-dsDNA antibodies with the potential to induce nephritis. The capsid proteins VP1, VP2, and VP3 may also function as carrier proteins because of their potential to bind both viral and nonviral DNA. A molecular mimicry mechanism as an additional process for BKV-induced autoantibodies cannot be excluded because the capsid proteins share stretches of amino-acid identity with histones (see "Biophysical Characteristics," above). Although a strong causal correlation between polyomavirus reactivation and SLE occurs, other, yet unknown mechanisms may be responsible for the induction of anti-DNA antibodies in human SLE. One patient in our study produced high levels of anti-dsDNA antibodies but never

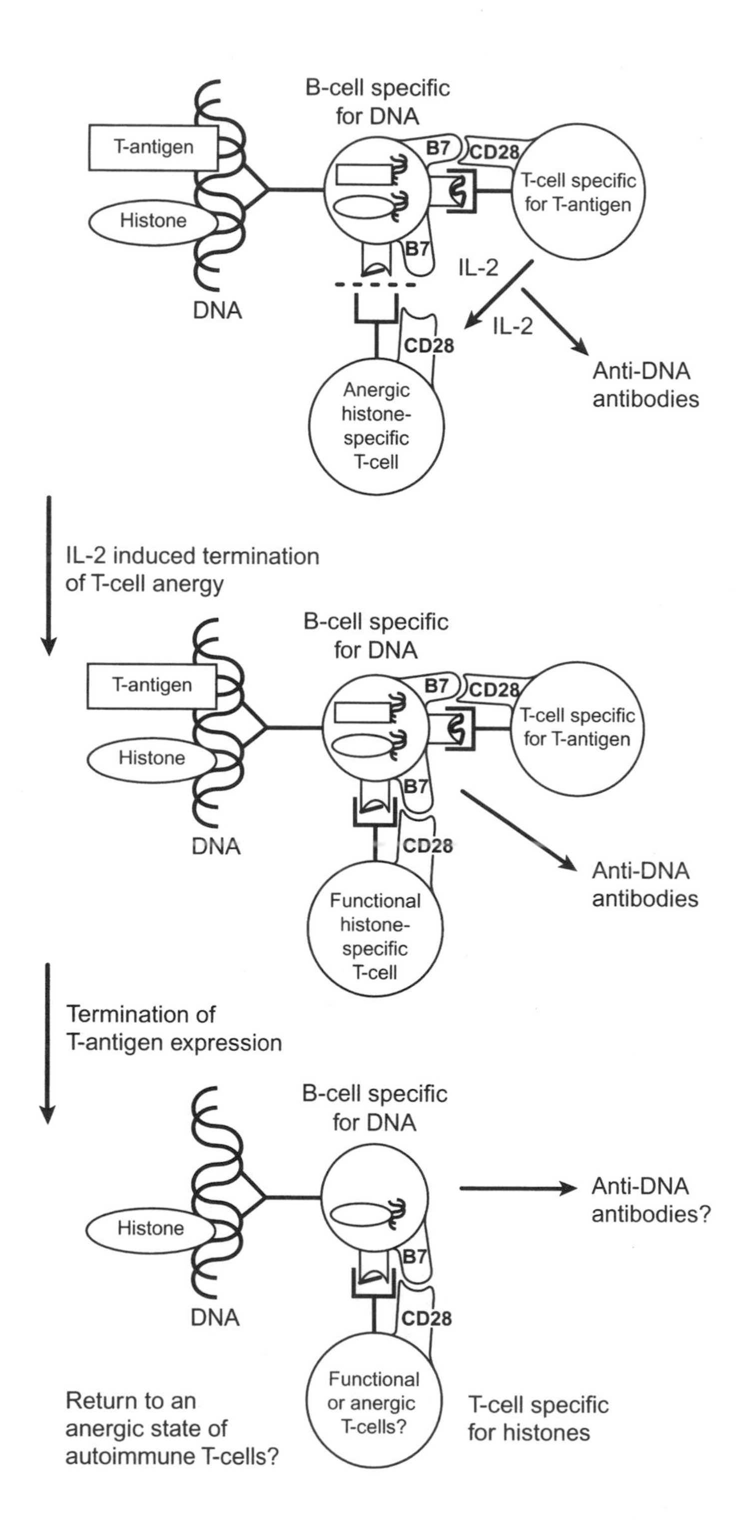
B-cell specific for DNA
T-antigen
Histone
DNA
B7
CD28
T-cell specific for T-antigen
IL-2
IL-2
Anti-DNA antibodies
Anergic histone-specific T-cell
IL-2 induced termination of T-cell anergy
B-cell specific for DNA
T-antigen
Histone
DNA
B7
CD28
T-cell specific for T-antigen
Anti-DNA antibodies
Functional histone-specific T-cell
Termination of T-antigen expression
B-cell specific for DNA
Histone
DNA
Anti-DNA antibodies?
B7
CD28
Functional or anergic T-cells?
Return to an anergic state of autoimmune T-cells?
T-cell specific for histones

showed signs of BK or JC viruria or of anti-TAg antibodies (Rekvig et al., 1997).

Another problem relates to the processes that trigger reactivation of BKV and JCV in SLE patients. If our model for polyomavirus-induced anti-DNA antibodies in SLE is correct, an experimental and descriptive approach to explain the molecular and biologic bases for florid polyomavirus reactivation in SLE is required. To clarify the processes that are responsible for losing control of virus replication may prove to be a characteristic of SLE that may help to elucidate basic dysregulations predisposing to this disease. In another study, serologic analyses for antiviral antibodies and PCR-based detection of viral DNA in urine samples failed to perceive an active human cytomegalovirus infection in the SLE patient group described above during the observation period (Bendiksen et al., 2000b). We cannot, however, exclude a role for other viruses in the etiology of SLE.

BKV and Cancers of the Urinary Tract

Several observations suggest a role for BKV in the induction of renal tumors. BKV can transform human embryonic kidney cells in vitro (Purchio and Fareed, 1979). BKV TAg transgenic mice develop renal tumors (Small et al., 1986; Dalrymple and Beemon, 1990), and viral DNA sequences have also been demonstrated in human neoplastic kidney tissue, urinary tract, and prostate specimens (Monini et al., 1995; Pater et al., 1980). A causal role remains elusive because the prevalence of BKV DNA sequences in normal kidney tissue was comparable with neoplastic tissue (Monini et al., 1995) (see Chapter 16).

5. CLINICAL MANAGEMENT OF BKV INFECTIONS IN TRANSPLANTS

The high prevalence of latent BKV in the human population and the pathologic potential of this virus require a reliable diagnosis and efficient therapeutic drugs or treatments. No drug therapies with a beneficial effect against human polyomaviruses are currently available, and diagnostic tools with high predictive value distinquishing BKV-diseased humans from healthy individuals expressing the virus are lacking. BK viruria, often based on detection of viral DNA by PCR, is not a good indication for the causal association of BKV infection with ureteric stenosis and hemorrhagic cystitis in renal or bone marrow transplants

Figure 14.3. Molecular and cellular bases for TAg-induced production of autoantibodies. The non-self protein TAg, when physically linked to histones through its binding to nucleosomes, may initiate processes that result in induction of antibodies to DNA and also in termination of histone-specific T-cell anergy by a linked presentation of the two ligands by B cells. See text for details.

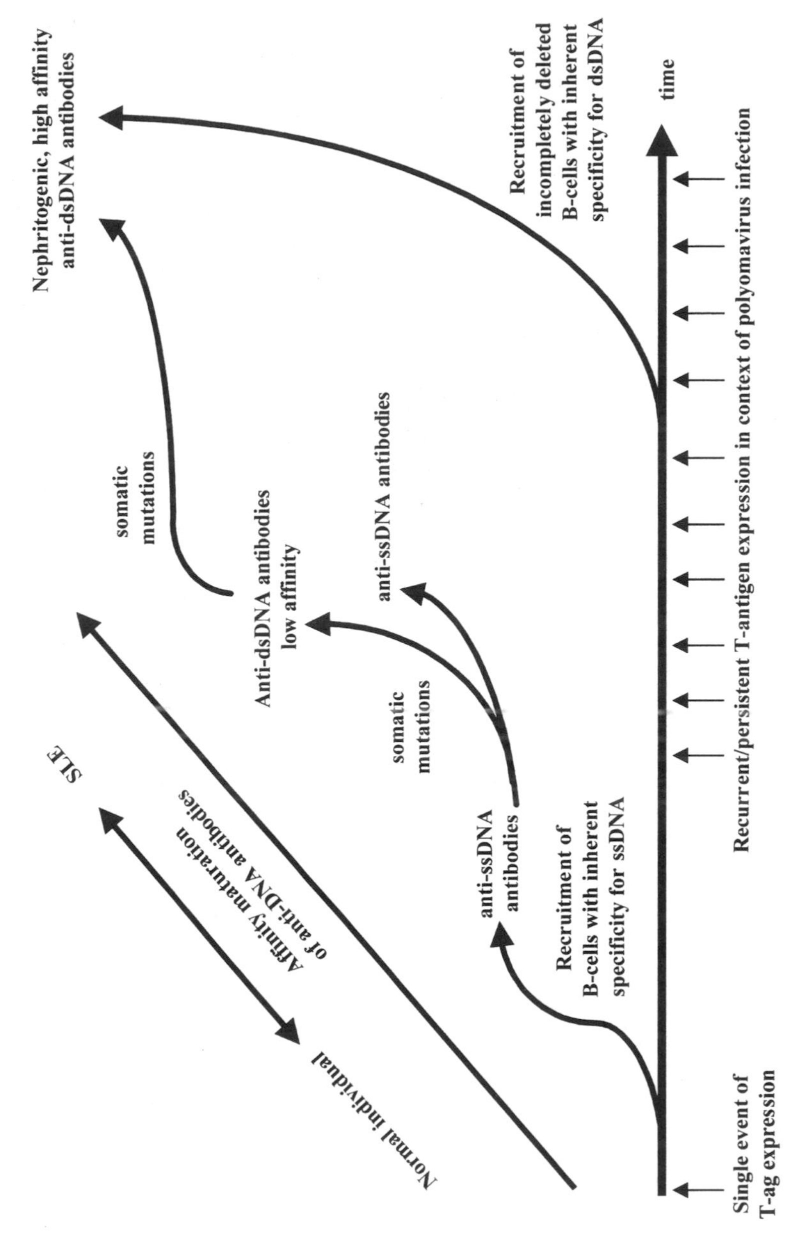
Complex formation of polyomavirus T-antigen with nucleosomes:
A molecular basis for initiation of anti-ssDNA and anti-dsDNA antibodies.
Nephritogenic, high affinity
anti-dsDNA antibodies
somatic
mutations
Anti-dsDNA antibodies
low affinity
anti-ssDNA antibodies
somatic
mutations
anti-ssDNA
antibodies
SLE
Affinity maturation
of anti-DNA antibodies
Normal individual
Recruitment of
B-cells with inherent
specificity for ssDNA
Recruitment of
incompletely deleted
B-cells with inherent
specificity for dsDNA
time
Recurrent/persistent T-antigen expression in context of polyomavirus infection
Single event of
T-ag expression

(Andrews et al., 1988; Apperley et al., 1987; Arthur et al., 1986; Azzi et al., 1994; Bogdanovic et al., 1996, 1998; Chan et al., 1994; Coleman et al., 1978; Cottler-Fox et al., 1989; Gardner et al., 1984; Goddard and Saha, 1997; Hogan et al., 1980; Mathur et al., 1997; Nickeleit et al., 1999; O'Reilly et al., 1981; Randhawa and Demetris, 2000; Rice et al., 1985; Sandler et al., 1997; Traystman et al., 1980; Vögeli et al., 1999).

Inclusion-bearing cells in the urine, also called *decoy cells*, are a characteristic finding in BKN. However, the presence of decoy cells is not unique for BKN and may merely indicate a transient asymptomatic activation not associated with renal dysfunction. Among renal transplant recipients undergoing immunosuppressive therapy, up to 60% have reactivation of polyomavirus accompanied by shedding of decoy cells, while BKV has been implicated as a cause of BKN in about 5% (Randhawa and Demetris, 2000). In fact, the positive predictive value for BKN diagnosis by detecting decoy cells was only 18%, whereas the negative predictive values for the absence of BKN was 100% (Binet et al., 1999). The same researchers suggested that PCR-based detection of BKV DNA in plasma may possess a favorable diagnostic value (Nickeleit et al., 2000b). In the plasma of all patients with BKN tested (n = 9), viral DNA was detected by PCR. BK viremia preceded (n = 3) or coincided with the histologic diagnosis of BKN. BK viremia disappeared in five patients who overcame the disease. Of control transplant recipients without BKN, only 5% (2 of 41) showed BK viremia, while 17 of 17 HIV type-1–infected patients without transplantations were negative for BKV DNA in their plasma samples. This indicates a close association between BK viremia and nephropathy. Nickeleit et al. (2000b) found a specificity of 95% when renal-allograft recipients with BKN (n = 9) were compared with recipients without evidence of BKN (n = 41). However, long-term prospective studies on a larger patient group are required to evaluate the usefulness of BKV viremia to predict the development of BKN.

A disadvantage of the PCR technique is its sensitivity. Qualitative PCR is insufficient because BKV DNA can also be detected in the blood of healthy individuals (Chatterjee et al., 2000; Dolei et al., 2000; Dörries et al., 1994). Therefore, quantitative PCR should be used to monitor the BKV load. Quan-

←

Figure 14.4. Polyomavirus TAg may interfere with the fine balance between generation of antibodies to ssDNA and dsDNA. Antibodies against dsDNA, as opposed to ssDNA, are regarded as clinically important and can develop from two sources. They may originate from a progressive development of antibodies to ssDNA by somatic mutations of anti-ssDNA antibody variable regions (affinity maturation) secondary to continuous stimulation of the immune system. This could be achieved by, for example, a persistent productive human polyomavirus infection. The other source would be the result of stimulation of B-cell clones with an inherent specificity for dsDNA by DNA–TAg complexes. Such clones are normally deleted, but may escape deletion and proliferate if immunogenic DNA competes with the physiologic deletion processes for autoimmune, high avidity B-cell clones. Taken with permission from Andreassen et al., 1999b.

tifying the viral load in the plasma of renal transplant patients may allow us to lower the level of immunosuppression to a point that would minimize viral multiplication and yet prevent graft loss due to rejection.

Raised levels of serum creatinine have also been suggested as a hallmark for renal dysfunction in renal transplant recipients and may thus offer a means of diagnosis. However, no significant differences were found between kidney transplant patients with (110 sera tested) or without (386 sera tested) active BKV infections (Andrews et al., 1988). Elevated serum creatinine concentrations have been measured in all renal transplant recipients with BKN, but only a limited number of sera have been examined (Drachenberg et al., 1999; Howell et al., 1999; Mathur et al., 1997; Nickeleit et al., 1999; Randhawa et al., 1999).

Several studies have shown that careful reduction in the doses of immunosuppressive therapy helped to clear the viral infection in renal transplant recipients with BKN, and in some of these patients the renal function stabilized or improved (Binet et al., 1999; Howell et al., 1999; Mathur et al., 1997; Nickeleit et al., 2000b; Pappo et al., 1996; Randhawa et al., 1999). The reports also suggested that new immunosuppressive drugs like tacrolimus (FK506) and mycophenolate mofetil played an active role in promoting BKN, possibly by interfering with the biology of the virus. This assumption is based on two observations. First, renal transplant recipients with BKN were not encountered until the new immunosuppressive therapies were implemented in therapy. Second, BKN remitted, and no BK viremia could be detected in some patients after high doses of tacrolimus were decreased or replaced by cyclosporine A.

BKN was more frequent in patients receiving quadruple therapy (i.e., antilymphocyte preparations, cyclosporine A, azathioprine, and steroids) than in patients treated with double therapy (cyclosporine A and prednisone) (Binet et al., 1999; Howell et al., 1999; Mathur et al., 1997; Nickeleit et al., 2000a; Randhawa et al., 1999). It is unlikely that new immunosuppressive drugs alone account for the prevalence of BKN. Tissue injury due to acute cellular rejection or drug toxicity may create a microenvironment favorable for viral reactivation (Howell et al., 1999). Renal parenchymal damage promoted both primary infection and viral replication of mouse polyomavirus in an experimental murine model (Atencio et al., 1993). A reduction in immunosuppressive therapy was also successful in a kidney transplant patient with active BKV infection who suffered from ureteric stenosis (Gardner et al., 1984). Reducing immunosuppressive therapy is not always possible because of the state of the underlying disease. Because there are only a few studies on this aspect of treatment, long-term prospective studies are required to prove the usefulness of this method.

Aspects of Specific Drug Therapies

Effective antipolyomavirus drugs are lacking. In vitro studies have shown that 9-β-D-arabinofuranosyl adenine (Taguchi et al., 1980), retinoic acid (Talmage and Listerud, 1994), prokaryotic DNA gyrase (Ferrazzi et al., 1988; Portolani et al., 1988), 5-bromo-2′deoxyuridine (Tarabek et al., 1991), monensin (Pietropaola et al., 1995), and cidofovir (HPMPC) (Andrei et al., 1997) inhibited replication of polyomaviruses. Cidofovir has been successfully used in a bone marrow transplant patient with BK viruria and concomitant hemorrhagic cystitis (Held et al., 2000). Human leukocyte interferon has been unsuccessfully administered as an anti-BKV drug in a clinical trial (Cheeseman et al., 1980). The effect of vidarabine remains uncertain, although some efficacy in the treatment of BKV-associated hemorrhagic cystitis in bone marrow transplant recipients has been reported (Chapman et al., 1991; Kawakami et al., 1997; Vianelli et al., 2000).

Steroid hormones and antilymphocyte preparations have been administered for empirical reasons to kidney recipients to reduce polyomavirus reactivation, but solid proof for their efficacy is lacking. Most patients had little or only a transient response to steroid hormone treatment. In fact, high dose steroids may provoke active BKV infection (Mackenzie et al., 1978). A definite increase in viral load in kidney biopsy specimens was observed in one BKN patient after a course of steroid therapy (Randhawa et al., 1999). Lowering high doses of corticosteroid treatment in renal transplant patients who developed BKN resulted in a drop of serum creatinine levels and a decrease in the number of BKV-infected cells (Drachenberg et al., 1999). Pulse steroid therapy has led to lowered levels of creatinine and transient clinical improvement in a patient with a kidney transplant (Purighalla et al., 1995). This is in accordance with the in vitro observation that steroid hormones enhanced viral transcription and BKV yield in cell cultures (Moens et al., 1994, 1999). A dramatic difference (44% versus 0%) in viruria in renal allograft recipients after receiving antilymphocyte serum also been reported (Lecatsas et al., 1973). Drachenberg and coworkers (1999), however, measured in one kidney transplant patient with BKN a decline in serum creatinine concentrations, and the number of BKV-positive cells in kidney biopsy material after antilymphocyte therapy was reduced.

Prostaglandin E_2 (PGE_2) treatment of bone marrow transplant patients who developed hemorrhagic cystitis has been reported to give variable results. No effect of PGE_2 was observed in a 7-year-old boy who underwent allogeneic bone marrow transplantation (Goddard and Saha, 1997). Several investigators found that some bone marrow transplant recipients who received PGE_2 promptly recovered from hemorrhagic cystitis, but the BKV viruria pattern was not described in their studies (Azzi et al., 1994; Laszlo et al., 1995).

The donor–recipient BKV seroreactivity may form a risk factor for BKV-induced renal dysfunction (Shah, 2000). A study including 496 renal transplant recipients demonstrated that primary BKV infection as well as reactivation in the recipients increased significantly from 7.3% to 33.7%, or 9.5% to 23.3%, respectively, when BKV-seropositive kidney donors were used instead of seronegative donors (Andrews et al., 1988). Serotyping of the donor may therefore be recommended before transplantation. However, the high prevalence (>80%) of BKV antibodies in the human population probably would make it even more difficult to obtain already scarce compatible kidneys from seronegative donors for renal transplantation.

Future Therapeutic Strategies

Vaccination, antisense RNA, or ribozymes may be novel strategies to prevent primary viral infection. To our knowledge, no human trials with human polyomavirus vaccination have been reported, although SV40 was accidentally used in contaminated poliovirus vaccines between 1955 and 1963 (Butel, 2000). Vaccination with recombinant TAg or peptide derivatives, as well as plasmids encoding TAg, have been shown to give protective immunity in experimental animals (Bright et al., 1998; Xie et al., 1999). These results indicate potential clinical utility for a TAg-based vaccine. However, we have shown that inoculation with TAg-encoding plasmid could induce autoimmunity to dsDNA and nucleosomes (Moens et al., 1995b). These antibodies did not, however, have pathophysiologic potential as proteinuria was absent at the end of the observation period (our unpublished results). Inactivated avian polyomavirus vaccines have been successfully used to control polyomavirus outbreaks in birds (Ritchie et al., 1996, 1998). In vitro studies demonstrated that antisense oligodeoxyribonucleotides and ribozymes strongly inhibited expression of TAg (Graessmann et al., 1991; Jennings and Molloy, 1987; Liu et al., 1994; Ottavio et al., 1992; Westermann et al., 1989). Some of these strategies may become valid for selected patient groups at risk for polyomavirus-related diseases.

6. CONCLUSIONS

Since the original isolation of BK virus in 1971, much has been learned about the molecular biology and epidemiology of the virus. BKV is latently present in almost the entire human population. Although its relation with the host usually seems harmless, BKV is not merely an innocent inhabitant. Reactivation of the virus is associated with several clinical manifestations, and a causal role in oncogenesis is becoming apparent. Recently, BKV has been implicated in nephropathy in kidney transplant patients. The high prevalence of latent virus in humans and the pathologic potential of BKV make urgent the development of effective therapeutic drugs or treatments for this virus. Another challenge for future research is to identify the mechanisms responsible for reactivation of latent virus infections as in SLE and immunocompromised individuals. Knowing the bases for viral reactivation, and being able to make an accurate diagnosis, along with having effective treatment should improve the condition of patients suffering from an active BKV infection.

REFERENCES

Andreassen K, Bredholt G, Moens U, Bendiksen S, Kauric G, Rekvig OP (1999a): T cell lines specific for polyomavirus T-antigen recognize T-antigen complexed with nucleosomes: A molecular basis for anti-DNA antibody production. Eur J Immunol 29:2715–2728.

Andreassen K, Moens U, Nossent H, Marion TN, Rekvig OP (1999b): Termination of human T cell tolerance to histones by presentation of histones and polyomavirus T antigen provided T antigen is complexed with nucleosomes. Arthritis Rheum 42: 2449–2460.

Andrei G, Snoeck R, Vandeputte M, De Clerq E (1997): Activities of various compounds against murine and primate polyomaviruses. Antimicrobiol Agents Chemother 41:587–593.

Andrews CA, Shah KV, Daniel RW, Hirsch MS, Rubin RH (1988): A serological investigation of BK virus and JC virus infections in recipients of renal allografts. J Infect Dis 158:176–181.

Apperley JF, Rice SJ, Bishop JA, Chia YC, Krausz T, Gardner SD, Goldman JM (1987): Late-onset hemorrhagic cystitis associated with urinary excretion of polyomaviruses after bone marrow transplantation. Transplant 43:108–112.

Arents G, Burlingame RW, Wang BC, Love WE, Moudrianakis EN (1991): The nucleosomal core histone octamer at 3.1 Å resolution: A tripartite protein assembly and a left-handed superhelix. Proc Natl Acad Sci USA 88:10148–10152.

Arthur RR, Shah KV (1989): Occurrence and significance of papovaviruses BK and JC in the urine. Prog Med Virol 36:42–61.

Arthur RR, Shah KV, Baust SJ, Santos GW, Saral R (1986): Association of BK viruria with hemorrhagic cystitis in recipients of bone marrow transplants. N Engl J Med 315:230–234.

Atencio IA, Shadan FF, Zhou XJ, Vaziri ND, Villarreal LP (1993): Adult mouse kidneys become permissive to acute polyomavirus infection and reactivate persistent infection in response to cellular damage and regeneration. J Virol 67:1424–1432.

Avantaggiati ML, Carbone M, Graesmann A, Nakatani Y, Howard B, Levine AS (1996): The SV40 large T antigen and adenovirus E1a oncoproteins interact with distinct isoforms of the transcriptional co-activator, p300. EMBO J 15:2236–2248.

Azzi A, Fanci R, Bosi A, Ciappi S, Zakrewska K, de Santis R, Laszlo D, Guidi S, Saccardi R, Vannucchi AM, Longo G, Rossi-Ferrini P (1994): Monitoring of polyomavirus BK viruria in bone marrow transplantation patients by DNA hybridization assay and by polymerase chain reaction: An approach to assess the relationship between BK viruria and hemorrhagic cystitis. Bone Marrow Transplant 14:235–240.

Barkan A, Mertz JE (1981): DNA sequence analysis of simian virus 40 mutants with deletions mapping in the leader region of the late mRNAs: Mutants with deletions similar in size and position exhibit varied phenotypes. J Virol 61:3190–3198.

Bedi A, Miller CB, Hanson JL, Goodman S, Ambinder RF, Charache P, Arthur RR, Jones RJ (1995): Association of BK virus with failure of prophylaxis against hemorrhagic following bone marrow transplantation. J Clin Oncol 13:1103–1109.

Bendiksen S, Rekvig OP, Van Ghelue M, Moens U (2000a): VP1 DNA sequences of JC and BK viruses detected in urine of systemic lupus erythematosus patients reveals no differences from strains expressed in normal individuals. J Gen Virol 81:2625–2633.

Bendiksen S, Van Ghelue M, Rekvig OP, Gutteberg T, Haga HJ, Moens U (2000b): A longitudinal study of human cytomegalovirus serology and viruria fails to detect active viral infection in 20 systemic lupus erythematosus patients. Lupus 9:120–126.

Binet I, Nickeleit V, Hirsch HH, Prince O, Dalquen P, Gudat F, Mihatsch MJ, Thiel G (1999): Polyomavirus disease under new immunosuppressive drugs. Transplantation 67:918–922.

Bogdanovic G, Ljungman P, Wang F, Dalianis T (1996): Presence of human polyomavirus DNA in the peripheral circulation of bone marrow transplant patients with and without hemorrhagic cystitis. Bone Marrow Transplant 17:573–576.

Bogdanovic G, Priftakis P, Taemmeraes B, Gustafsson A, Flægstad T, Winiarski J, Dalianis T (1998): Primary BK virus (BKV) infection due to possible BKV transmission during bone marrow transplantation is not the major cause of haemorrhagic cystitis in transplanted children. Pediatr Transplant 2:288–293.

Borgatti M, Corranzo F, Portolani M, Vullo C, Osti L, Masi M, Barbanti-Bordano G (1979): Evidence for reactivation of persistent infection during pregnancy and lack of congenital transmission of BK virus, a human papovavirus. Microbiology 2:173–179.

Boubenider S, Hiesse C, Marchand S, Hafi A, Kriaa F, Charpentier B (1999): Post-transplantation polyomavirus infections. J Nephrol 12:24–29.

Bratt G, Hammarin AL, Gradien M, Hedquist BG, Nennesmo I, Sundelin B, Seregard S (1999): BK virus as the cause of meningoencephalitis, retinitis and nephritis in a patient with AIDS. AIDS 13:1071–1075.

Bredholt G, Olaussen E, Moens U, Rekvig OP (1999): Linked production of antibodies to mammalian DNA and to human polyomavirus large T antigen. Footprints of a common molecular and cellular process? Arthritis Rheum 42:2583–2592.

Bright RK, Shearer MH, Pass HI, Kennedy RC (1998): Immunotherapy of SV40 induced tumours in mice: A model for vaccine development. Dev Biol Stand 94:341–353.

Brown P, Tsai T, Gajdusek DC (1975): Seroepidemiology of human papovaviruses. Discovery of virgin populations and some unusual patterns of antibody prevalence among remote peoples of the world. Am J Epidemiol 102:331–340.

Butel JS (2000): Simian virus 40, poliovirus vaccines, and human cancer: Research progress versus media and public interests. Bull WHO 78:195–198.

Campbell KS, Mullane KP, Aksoy IA, Stubdal H, Zalvide J, Pipas JM, Silver PA, Roberts TM, Schaffhausen BS, DeCaprio JA (1997): DnaJ/hsp40 chaperone domain of SV40 large T antigen promotes efficient viral DNA replication. Genes Dev 11: 1098–1110.

Caputo A, Barbanti-Brodano G, Wang E, Ricciardi RP (1986): Transactivation of BKV and SV40 early promoters by BKV and SV40 T-antigens. Virol 152:459–465.

Carswell S, Alwine JC (1986): Simian virus agnoprotein facilitates perinuclear–nuclear localization of VP1, the major capsid protein. J Virol 60:1055–1061.

Cassill JA, Subramani S (1988): The late promoter of the human papovavirus BK is contained within the early promoter enhancer region. Virology 166:175–185.

Cassill JA, Deyerle KL, Subramani S (1989): Unidirectional deletion and linker scan analysis of the late promoter of the human papovavirus BK. Virology 169:172–181.

Cassill JA, Subramani S (1989): A naturally occurring deletion in the enhancer repeats of the human papovavirus BK optimizes early enhancer function at the expense of late promoter activity. Virology 170:296–298.

Chakraborty T, Das GC (1989): Identification of HeLa cell nuclear factors that bind to and activate the early promoter of human polyomavirus BK in vitro. Mol Cell Biol 9:3821–3828.

Chakraborty T, Das GC (1991): Proteins of the nuclear factor-1 family act as an activator of the late promoter in human polyomavirus BK in vitro. J Gen Virol 72: 1935–1942.

Chan PK, Ip KW, Shiu SY, Chiu EK, Wong MP, Yen KY (1994): Association between polyomaviruria and microscopic haematuria in bone marrow transplant recipients. J Infect 29:139–146.

Chang D, Cai X, Consigili RA (1993): Characterization of the DNA binding properties of polyomavirus capsid proteins. J Virol 67:6327–6331.

Chang D, Tsai RT, Wang ML, Ou WC (1996a): Different genotypes of human polyomaviruses found in patients with autoimmune diseases in Taiwan. J Med Virol 48: 204–209.

Chang D, Wang M, Ou WC, Lee MS, Ho HN, Tsai RT (1996b): Genotypes of human polyomaviruses in urine samples of pregnant women in Taiwan. J Med Virol 48: 95–101.

Chapman C, Flower AJE, Durrant STS (1991): The use of vidarabine in the treatment of human polyomavirus associated acute haemorrhagic cystitis. Bone Marrow Transplant 7:481–483.

Chatterjee M, Weyandt TB, Frisque RJ (2000): Identification of archetype and rearranged forms of BK virus in leukocytes from healthy individuals. J Med Virol 60: 353–362.

Cheeseman SH, Black PH, Rubin RH, Cantell K, Hirsch MS (1980): Interferon and BK papovavirus—Clinical and laboratory studies. J Infect Dis 141:157–161.

Chester PM, Heritage J, McCance DJ (1983): Persistence of DNA sequences of BK virus and JC virus in normal human tissues and in diseased tissues. J Infect Dis 147: 676–684.

Cicala C, Avantaggiati ML, Graesmann A, Rundell K, Levine AS, Carbone M (1994): Simian virus 40 small-t antigen stimulates viral DNA replication in permissive monkey cells. J Virol 68:3138–3144.

Clever J, Dean DA, Kasamatsu H (1993): Identification of a DNA binding domain in simian virus 40 capsid proteins VP2 and VP3. J Biol Chem 268:20877–20883.

Cobb JJ, Wickenden C, Snell ME, Hulme B, Malcolm ADB, Coleman DV (1987): Use of hybridot assay to screen for BK and JC polyomaviruses in non-immunosuppressed patients. J Clin Pathol 40:777–781.

Cole CN (1996): Polyomaviruses: The viruses and their replication. In Fields BN, Knipe DM, Howley PM, Eds; Fields Virology; Lippincott-Raven Publishers: Philadelphia, pp 1997–2025.

Cole CN, Landers T, Goff SP, Manteuil-Brutlag S, Berg P (1977): Physical and genetic characterization of deletion mutants of simian virus 40 constructed in vitro. J Virol 24:277–294.

Coleman DV, Mackenzie EFD, Gardner SD, Poulding JM, Amer B, Russell W (1978): Human polyomavirus (BK) infection and ureteric stenosis in renal allograft recipients. J Clin Pathol 31:338–347.

Coleman DV, Wolfendale MR, Daniel RA, Dhanjal NK, Gardner SD, Gibson PE, Field AM (1980): A prospective study of human polyomavirus infection in pregnancy. J Infect Dis 142:1–8.

Conzen SD, Snay CA, Cole CN (1997): Identification of a novel antiapoptotic functional domain in simian virus 40 large T antigen. J Virol 71:4536–4543.

Cooper GS, Dooley MA, Treadwell EL, St Clair EW, Parks CG, Gilkeson GS (1998): Hormonal, environmental, and infectious risk factors for developing systemic lupus erythematosus. Arthritis Rheum 41:1714–1724.

Corallini A, Barbanti-Brodano G, Portolani M (1976): Antibodies to BK virus structural and tumor antigens in human sera from normal persons and from patients with various diseases, including neoplasia. Infect Immun 13:1684–1691.

Costa J, Lee C, Tralka TS, Rabson AS (1976): Hamster ependymomas produced by intracerebral inoculation of a human papovavirus (MMV). J Natl Cancer Inst 56: 863–864.

Cottler-Fox M, Lynch M, Deeg JH, Koss LG (1989): Human polyomavirus: Lack of relationship of viruria to prolonged or severe hemorrhagic cystitis after bone marrow transplant. Bone Marrow Transplant 4:279–282.

Dalrymple SA, Beemon KL (1990): BK virus T antigens induce kidney carcinomas and thymoproliferative disorders in transgenic mice. J Virol 64:1182–1191.

D'Andrea DM, Coupaye Gerard B, Kleyman TR, Foster MH, Madaio MP (1996): Lupus autoantibodies interact directly with distinct glomerular and vascular cell surface antigens. Kidney Int 49:1214–1221.

Das GC, Shivakumar CV, Todd SD (1995): Cell-specific modulation of the papovavirus promoters by tumor-suppressor protein p53 in the absence of large T-antigen. Oncogene 10:449–455.

Dean DA, Li PP, Lee LM, Kasamatsu H (1995): Essential role of the Vp2 and Vp3 DNA-binding domain in simian virus 40 morphogenesis. J Virol 69:1115–1121.

Degener AM, Pietropaola V, Di Taranto C, Jin L, Ameglio F, Cordiali-Fei P, Trento E, Sinibaldi L, Orsi N (1999): Identification of a new control region in the genome of the DDP strain of BK virus isolated from PBMC. J Med Virol 58:413–419.

Dei R, Marmo F, Corte D, Sampietro MG, Franceschini E, Urbano P (1982): Age-related changes in the prevalence of precipitating antibodies to BK virus in infants and children. J Med Microbiol 15:285–291.

Del Vecchio AM, Steinman RA, Ricciardi RP (1989): An element of the BK virus enhancer required for DNA replication. J Virol 63:1514–1524.

De Mattei M, Martini F, Corallini A, Gerosa M, Scotlandi K, Carinci P, Barbanti-Brodano G, Tognon M (1995): High incidence of BK virus large-T-antigen–coding sequences in normal human tissues and tumors of different histotypes. Int J Cancer 61:756–760.

De Mattei M, Martini F, Tognon M, Serra M, Baldini N, Brabanti-Brodano G (1994): Polyomavirus latency and human tumors. J Infect Dis 169:1175–1176.

Desai DD, Krishnan MR, Swindle JT, Marion TN (1993): Antigen-specific induction of antibodies against native mammalian DNA in nonautoimmune mice. J Immunol 151:1614–1626.

DeSilva DR, Urdahl KB, Jenkins MK (1991): Clonal anergy is induced in vitro by T cell receptor occupancy in the absence of proliferation. J Immunol 147:3261–3267.

de Silva LM, Bale P, de Courcy J, Brown D, Knowles W (1995): Renal failure due to BK virus infection in an immunodeficient child. J Med Virol 45:192–196.

Deyerle KL, Cassill JA, Subramani S (1987): Analysis of the early regulatory region of the human papovavirus BK. Virology 158:181–193.

Deyerle KL, Sajjadi FG, Subramani S (1989): Analysis of origin of DNA replication of human papovavirus BK. J Virol 63:356–365.

Deyerle KL, Subramani S (1988): A linker scan analysis of the early regulatory region of the human papovavirus BK. J Virol 62:3378–3387.

Dolei A, Pietropaola V, Gomes E, Di Taranto C, Ziccheddu M, Spanu MA, Lavorino C, Manca M, Degener AM (2000): Polyomavirus persistence in lymphocytes: Prevalence in lymphocytes from blood donors and healthy personnel of a blood transfusion centre. J Gen Virol 81:1967–1973.

Dörries K (1997): New aspects in the pathogenesis of polyomavirus-induced disease. Adv Virus Res 48:205–261.

Dörries K, Vogel E, Gunther S, Czub S (1994): Infection of human polyomaviruses JC and BK in peripheral blood leukocytes from immunocompetent individuals. Virology 198:59–70.

Dougherty RM, DiStefano HS (1974): Isolation and characterization of a papovavirus from human urine. Proc Soc Exp Biol Med 146:481–487.

Drachenberg CB, Beskow CO, Cangro CB, Bourquin PM, Simsir A, Fink J, Weir MR, Klassen DK, Bartlett ST, Papadimitriou JC (1999): Human polyomavirus in renal allograft biopsies: Morphological findings and correlation with urine cytology. Hum Pathol 30:970–977.

Dyson N, Bernards R, Friend SH, Gooding LR, Hassell JA, Major EO, Pipas JM, Vandyke T, Harlow E (1990): Large T antigens of many polyomaviruses are able to form complexes with the retinoblastoma protein. J Virol 64:1353–1356.

Eckner R, Ludlow JW, Lill NL, Oldread E, Arany Z, Modjtahedi N, DeCaprio JA, Livingston DM, Morgan JA (1996): Association of p300 and CBP with simian virus 40 large T antigen. Mol Cell Biol 16:3454–3464.

Elsner C, Dörries K (1992): Evidence of human polyomavirus BK and JC infection in normal human brain tissue. Virology 191:72–80.

Fanning E (1992): Simian virus 40 large T antigen: The puzzle, the pieces, and the emerging picture. J Virol 66:1289–1293.

Farrell MP, Mäntijärvi RA, Pagano JS (1978): T antigen of BK papovavirus in infected and transformed cells. J Virol 25:871–877.

Ferguson AT, Subramani S (1994): Complex functional interactions at the early enhancer of the PQ strain of BK virus. J Virol 68:4274–4286.

Ferrazzi E, Peracchi M, Biasolo MA, Faggionato O, Stefanelli S, Palu G (1988): Antiviral activity of gyrase inhibitors norfloxacin, coumermycin A_1 and nalidixic. Biochem Pharmacol 37:1885–1886.

Flægstad T, Fredriksen K, Dahl B, Traavik T, Rekvig OP (1988): Inoculation with BK virus may break immunological tolerance to histone and DNA antigens. Proc Natl Acad Sci USA 85:8171–8175.

Flægstad T, Sundsfjord A, Arthur RR, Pedersen M, Traavik T, Subramani S (1991): Amplification and sequencing of the control regions of BK and JC virus from human urine by polymerase chain reaction. Virology 180:553–560.

Fredriksen K, Osei A, Sundsfjord A, Traavik T, Rekvig OP (1994): On the biological origin of anti-double-stranded (ds) DNA antibodies: Systemic lupus erythematosus–related anti-dsDNA antibodies are induced by polyomavirus BK in lupus-prone (NZB × NZW) F1 hybrids, but not in normal mice. Eur J Immunol 24:66–70.

Fredriksen K, Skogsholm A, Flægstad T, Traavik T, Rekvig OP (1993): Antibodies to dsDNA are produced during primary BK virus infection in man, indicating that anti-dsDNA antibodies may be related to virus replication in vivo. Scand J Immunol 38: 401–406.

Frost JA, Alberts AS, Sontag E, Guan K, Mumby MC, Feramisco JR (1994): Simian virus 40 small t antigen cooperates with mitogen activated kinases to stimulate AP-1 activity. Mol Cell Biol 14:6244–6252.

Garcia A, Cereghini S, Sontag E (2000): Protein phosphatase 2A and phosphatidylinositol 3-kinase regulate the activity of Sp1-responsive promoters. J Biol Chem 275: 9385–9389.

Gardner SD (1973): Prevalence in England of antibody to human polyomavirus (B.K.). BMJ 1:77–78.

Gardner SD, Field AM, Coleman DV, Humle B (1971): New human papovavirus (B.K.) isolated from urine after renal transplantation. Lancet 1:1253–1257.

Gardner SD, Mackenzie EF, Smith C, Porter AA (1984): Prospective study of the human polyomaviruses BK and JC and cytomegalovirus in renal transplant recipients. J Clin Pathol 37:578–586.

Ge H, Manley JL (1990): A protein factor, ASF, controls cell-specific alternative splicing of SV40 early pre-mRNA in vitro. Cell 62:25–34.

Gharakhanian E, Takahashi J, Clever J, Kasamatsu H (1988): In vitro assay for protein–protein interaction: Carboxyl-terminal 40 residues of simian virus 40 structural protein VP3 contain a determinant for interaction with VP1. Proc Natl Acad Sci USA 85:6607–6611.

Gluck TA, Knowles WA, Johnson MA, Brook MG, Pillay D (1994): BK virus–associated haemorrhagic cystits in an HIV-infected man. AIDS 8:391–392.

Goddard AG, Saha V (1997): Late-onset hemorrhagic cystitis following bone marrow transplantation: A case report. Pediatr Hematol Oncol 14:273–275.

Gordon-Shaag A, Ben-Nun-Shaul O, Kasamatsu H, Oppenheim AB, Oppenheim A (1998): The SV40 capsid protein VP3 cooperates with the cellular transcription factor Spl in DNA-binding and in regulating viral promoter activity. J Mol Biol 275:187–195.

Goudsmit J, Baak ML, Slaterus KW, van der Noordaa J (1981): Human papovavirus isolated from urine of a child with acute tonsillitis. BMJ 283:1363–1364.

Goudsmit J, Wertheim-van Dillen P, van Strien A, van der Noordaa J (1982): The role of BK virus in acute respiratory tract disease and presence of BKV DNA in tonsils. J Med Virol 10:91–99.

Graessmann M, Michaels G, Berg B, Graessmann A (1991): Inhibition of SV40 gene expression by microinjected small antisense RNA and DNA molecules. Nucleic Acids Res 19:53–58.

Grinnell BW, Berg DT, Walls JD (1988): Negative regulation of the human polyomavirus BK enhancer involves cell-specific interaction with a nuclear repressor. Mol Cell Biol 8:3448–3457.

Harris KF, Christensen JB, Imperiale MJ (1996): BK virus large T antigen: Interactions with the retinoblastoma family of tumor suppressor proteins and effects on cellular growth control. J Virol 70:2378–2386.

Harris KF, Christensen JB, Radany EH, Imperiale MJ (1998): Novel mechanisms of E2F induction by BK virus large-T antigen: Requirement of both the pRb-binding and the J domains. Mol Cell Biol 18:1746–1756.

Hashida Y, Gaffney PC, Yunis EJ (1976): Acute haemorrhagic cystitis of childhood and papovavirus-like particles. J Pediatr 89:85–87.

Hedquist BG, Bratt G, Hammarin AL, Grandien M, Nennesmo I, Sundelin B, Seregard S (1999): Identification of BK virus in a patient with aquired immune deficiency syndrome and bilateral atypical retinitis. Ophthalmology 106:129–132.

Held TK, Biel SS, Nitsche A, Kurth A, Chen S, Gelderblom HR, Siegert W (2000): Treatment of BK virus–associated hemorrhagic cystits and simultaneous CMV reactivation with cidofovir. Bone Marrow Transplant 26:347–350.

Hey AW, Johnsen JI, Johansen B, Traavik T (1994): A two fusion partner system for raising antibodies against small immunogens expressed in bacteria. J Immunol Methods 173:149–156.

Hogan TF, Borden EC, McBain JA, Padgett BL, Walker DL (1980): Human polyomavirus infections with JC virus and BK virus in renal transplant patients. Ann Intern Med 92:373–378.

Howe AK, Gaillard S, Bennett JS, Rundell K (1998): Cell cycle progression in monkey cells expressing simian virus 40 small t antigen from adenovirus vectors. J Virol 72: 9637–9644.

Howell DN, Smith SR, Butterly DW, Klassen PS, Krigman HR, Burchette JL, Miller SE (1999): Diagnosis and management of BK polyomavirus interstitial nephritis in renal transplant recipients. Transplantation 68:1279–1288.

Howley PM (1980): Molecular biology of SV40 and the human polyomaviruses BK and JC. In Klein G, Ed; Viral Oncology; Raven Press: New York, pp 489–550.

Hukkanen V, Haarala M, Nurmi M, Klemi P (1996): Viruses and interstitial cystitis: Adenovirus genomes cannot be demonstrated in urinary bladder biopsies. Urol Res 24:235–238.

Ishii N, Minami N, Chen EY, Medina AL, Chico MM, Kasamatsu H (1996): Analysis of a nuclear localization signal of simian virus 40 major capsid protein Vp1. J Virol 70:1317–1322.

Itoh S, Irie K, Nakamura Y, Ohta Y, Haratake A, Morimatsu M (1998): Cytologic and genetic study of polyomavirus-infected or polyomavirus-activated cells in human urine. Arch Pathol Lab Med 122:333–337.

Jenkins MK (1992): The role of cell division in the induction of clonal anergy. Immunol Today 13:69–73.

Jennings PA, Molloy PL (1987): Inhibition of SV40 replicon function by engineered antisense RNA transcribed by RNA polymerase III. EMBO J 6:3043–3047.

Kawakami M, Ueda S, Maeda T, Karasuno T, Teshima H, Hiraoka A, Nakamura H, Tanaka K, Masaoka T (1997): Vidarabine therapy for virus-associated cystitis after allogeneic bone marrow transplantation. Bone Marrow Transplant 20:485–490.

Kelly WL, Georgopoulos C (1997): The T/t common exon of simian virus 40, JC and BK polyomavirus T antigens can functionally replace the J-domain of the *Escherichia coli* DnaJ molecular chaperone. Proc Natl Acad Sci USA 94:3679–3684.

Khalili K, Brady J, Khoury G (1987): Translational regulation of SV40 early mRNA defines a new viral protein. Cell 48:639–645.

Kierstad TD, Pipas JM (1996): Database of mutations that alter the large tumor antigen of simian virus 40. Nucleic Acids Res 24:125–126.

Knepper JE, diMayorca G (1987): Cloning and characterization of BK virus–related DNA sequences from normal and neoplastic tissues. J Med Virol 21:289–299.

Kolzau T, Hansen RS, Zahra D, Reddel RR, Braithwaite AW (1999): Inhibition of SV40 large T antigen induced apoptosis by small t antigen. Oncogene 18:5598–5603.

Kornberg RD, Lorch Y (1999): Twenty-five years of the nucleosome, fundamental particle of the eukaryote chromosome. Cell 98:285–294.

Koss LG (1987): BK viruria and hemorrhagic cystitis. N Engl J Med 316:108–109.

Kouzarides T (2000): Acetylation: A regulatory modification to rival phosphorylation? EMBO J 19:1176–1179.

Krishnan MR, Marion TN (1993): Structural similarity of antibody variable regions from immune and autoimmune anti-DNA antibodies. J Immunol 150:4948–4957.

Lahita RG (1999): Systemic Lupus Erythematosus, 3rd ed; Academic Press: New York.

Laszlo D, Bosi A, Guidi S, Saccardi R, Vannucchi AM, Lombardini L, Longo G, Fanci R, Azzi A, De Santis R (1995): Prostaglandin E2 bladder instillation for the treatment of hemorrhagic cystitis after allogeneic bone marrow transplantation. Haematology 80:421–425.

Lecatsas G, Prozesky OW, van Wyk J, Els HJ (1973): Papovavirus in urine after renal transplantation. Nature 241:343–344.

Liu Z, Batt DB, Carmichael GG (1994): Targeted nuclear antisense RNA mimics natural antisense-induced degradation of polyoma virus early RNA. Proc Natl Acad Sci USA 91:4258–4262.

Loeber G, Stenger JE, Ray S, Parsons RE, Anderson ME, Tegtmeyer P (1991): The zinc finger region of simian virus 40 large T antigen is needed for hexamer assembly and origin melting. J Virol 65:3167–3174.

Mackenzie EF, Poulding JM, Harrison PR, Amer B (1978): Human polyomavirus (HPV)—A significant pathogen in renal transplantation. Proc Eur Dial Transplant Assoc 15:352–360.

Major EO, Matsumura P (1984): Human embryonic kidney cells: Stable transformation with an origin-defective simian virus 40 DNA and use as hosts for human papovavirus replication. Mol Cell Biol 4:379–382.

Major EO, Neel JV (1998): The JC and BK human polyomaviruses appear to be recent introductions to some South American Indian tribes: There is no serological evidence of cross-reactivity with the simian polyomavirus SV40. Proc Natl Acad Sci USA 95:15525–15530.

Manfredi JJ, Prives C (1994): The transforming activity of simian virus 40 large tumor antigen. Biochim Biophys Acta 1198:65–83.

Mäntyjärvi RA, Meurman OH, Vihma L, Berglund B (1973): A human papovavirus (B.K.), biological properties and seroepidemiology. Ann Clin Res 5:283–287.

Marion TN, Krishnan MR, Desai DD, Jou NT, Tillman DM (1997): Monoclonal anti-DNA antibodies: Structure, specificity, and biology. Methods 11:3–11.

Markowitz RB, Eaton BA, Kubik MF, Latorra D, McGregor JA, Dynan WS (1991): BK virus and JC virus shed during pregnancy have predominantly archetypal regulatory regions. J Virol 65:4515–4519.

Markowitz RB, Dynan WS (1988): Binding of cellular proteins to regulatory region of BK virus. J Virol 62:3388–3398.

Markowitz RB, Thompson HC, Mueller JF, Cohen JA, Dynan WS (1993): Incidence of BK virus and JC virus viruria in human immunodeficiency virus–infected and –uninfected subjects. J Infect Dis 167:13–20.

Markowitz RB, Tolberts S, Dynan WS (1990): Promoter evolution in BK virus: Functional elements are created at sequence junctions. J Virol 64:2411–2415.

Mason DH, Takemoto KK (1977): Transformation of rabbit kidney cells by BKV(MM) human papovavirus. Int J Cancer 19:391–395.

Masuda K, Akutagawa K, Yutani C, Kishita H, Ishibashi-Ueda H, Imakita M (1998): Persistent infection with human polyomavirus revealed by urinary cytology in a patient with heart transplantation. Acta Cytol 42:803–806.

Mateer SC, Federov SA, Mumby MC (1998): Identification of structural elements involved in the interaction of simian virus 40 small tumor antigen with protein phosphatase 2A. J Biol Chem 273:35339–35346.

Mathur VS, Olson JL, Darragh TM, Yen BTS (1997): Polyomavirus-induced interstitial nephritis in two renal transplant recipients: Case report and review of the literature. Am J Kidney Dis 29:754–758.

McCance DJ (1983): Persistence of animal and human papovavirus in renal and nervous tissue. Prog Clin Biol Res 105:343–357.

Meneguzzi G, Pignatti PF, Barbanti-Brodano G, Milanesi G (1978): Minichromosome from BK virus as a template for transcription in vitro. Proc Natl Acad Sci USA 75: 1126–1130.

Milanesi G, Barbanti-Brodano G, Negrini M, Lee D, Corallini A, Caputo A, Grossi MP, Ricciardi RP (1984): BK virus–plasmid expression vector that persists episomally in human cells and shuttles into *Escherichia coli*. Mol Cell Biol 4:1551–1560.

Moens U, Johansen T, Johnsen JI, Seternes, OM, Traavik T (1995a): Noncoding control region of naturally occurring BK virus variants: Sequence comparison and functional analysis. Virus Gen 10:261–275.

Moens U, Seternes OM, Hey AW, Silsand Y, Traavik T, Johansen B, Rekvig OP (1995b): In vivo expression of a single viral DNA-binding protein generates systemic lupus erythematosus–related autoimmunity to double-stranded DNA and histones. Proc Natl Acad Sci USA 92:12393–12397.

Moens U, Seternes OM, Johansen B, Rekvig OP (1997): Mechanisms of transcriptional regulation of cellular genes by SV40 large T- and small t-antigens. Virus Gen 15: 135–154.

Moens U, Subramaniam N, Johansen B, Johansen T, Traavik T (1994): A steroid hormone response unit in the late leader of the noncoding control region of the human polyomavirus BK confers enhanced host cell permissivity. J Virol 68:2398–2408.

Moens U, Sundsfjord A, Flægstad T, Traavik T (1990): BK virus early RNA transcripts in stably transformed cells: Enhanced levels induced by dibutyryl cyclic AMP, forskolin and 12-*O*-tetradecanoylphorbol-13-acetate. J Gen Virol 71:1461–1471.

Moens U, Van Ghelue M, Johansen B, Seternes OM (1999): Concerted expression of BK virus large T- and small t-antigens strongly enhances oestrogen receptor–mediated transcription. J Gen Virol 80:585–594.

Monini P, Rotola A, de Lellis L, Corallini A, Secchiero P, Albini A, Benelli R, Parravicini C, Barbanti-Brodano G, Cassai E (1996): Latent BK virus infection and Kaposi's sarcoma pathogenesis. Int J Cancer 66:717–722.

Monini P, Rotola A, Di Luca D, De Lellis L, Chiari E, Corallini A, Cassai E (1995): DNA rearrangements impairing BK virus productive infection in urinary tract tumors. Virology 214:273–279.

Mumby M (1995): Regulation by tumour antigens defines a role for PP2A in signal transduction. Semin Cancer Biol 6:229–237.

Nagao S, Iijima S, Suzuki H, Yokota T, Shigeta S (1982): BK virus-like particles in the urine of a patient with nephrotic syndrome: An electron microscopic observation. Fukushima J Med Sci 29:45–49.

Nakshatri H, Pater MM, Pater A (1988): Functional role of BK virus tumor antigens in transformation. J Virol 62:4613–4621.

Nakshatri H, Pater MM, Pater A (1991): Activity and enhancer binding factors for BK virus regulatory elements in differentiating embryonal carcinoma cells. Virology 183: 374–380.

Nebuloni M, Osoni A, Boldorini R, Monga G, Carsana L, Bonetto S, Abeli C, Cladarelli R, Vago L, Costanzi G (1999): BK virus renal infection in a patient with the acquired immunodeficiency syndrome. Arch Pathol Lab Med 123:807–811.

Negrini M, Rimessi P, Mantovani C, Sabbioni S, Corallini A, Gerosa MA, Barbanti-Brodano G (1990): Characterization of BK virus variants rescued from human tumours and tumour cell lines. J Gen Virol 71:2731–2736.

Negrini M, Sabbioni S, Arthur RR, Castagnoli A, Barbanti-Brodano G (1991): Prevalence of the archetypal regulatory region and sequence polymorphisms in nonpassaged BK virus variants. J Virol 65:5092–5095.

Nickeleit V, Hirsch HH, Binet IF, Gudat F, Prince O, Dalquen P, Thiel G, Mihatsch MJ (1999): Polyomavirus infection of renal allograft recipients: From latent infection to manifest disease. J Am Soc Nephrol 10:1080–1089.

Nickeleit V, Hirsch HH, Zeiler M, Gudat F, Prince O, Thiel G, Mihatsch MJ (2000a): BK-virus nephropathy in renal transplants—tubular necrosis, MHC-class II expression and rejection in a puzzling game. Nephrol Dial Transplant 15:324–332.

Nickeleit V, Klimkait T, Binet IF, Dalquen P, Del Zenero V, Thiel G, Mihatsch MJ, Hirsch HH (2000b): Testing for polyomavirus type BK DNA in plasma to identify renal-allograft recipients with viral nephropathy. N Engl J Med 342:1309–1315.

Nowock J, Borgmeyer U, Puschel AW, Rupp RAW, Sippel AE (1985): The TGGCA protein binds to the MMTV LTR, the adenovirus origin of replication, and the BK virus enhancer. Nucleic Acids Res 13:2045–2061.

O'Reilly RJ, Lee FK, Grossbard E, Kapoor N, Kirkpatrick D, Dinsmore R, Stutzer C, Shah KV, Nahmias AJ (1981): Papovavirus excretion following marrow transplantation: Incidence and association with hepatic dysfunction. Transplant Proc 13:262–266.

Ottavio L, Sthandier O, Ricci L, Passananti C, Amati P (1992): Constitutive synthesis of polyoma antisense RNA renders cell immune to virus infection. Virology 189: 812–816.

Padgett BL, Walker DL, Desquitado MM, Kim DU (1983): BK virus and non-haemorrhagic cystitis in a child. Lancet 1:770.

Pagnani M, Negrini M, Reschiglian P, Corallini A, Balboni PG, Scherneck S, Macino G, Milanesi G, Barbanti-Brodano G (1986): Molecular and biological properties of BK virus–IR, a BK virus variant isolated from a human tumor. J Virol 59:500–505.

Pappo O, Demetris AJ, Raikow RB, Randhawa PS (1996): Human polyomavirus infection of renal allografts: Histopathologic diagnosis, clinical significance, and literature review. Mod Pathol 9:105–109.

Pater MM, Pater A, Fiori M, Slota J, di Mayorca G (1980): BK virus DNA sequences in human tumors and normal tissues and cell lines. In Essex M, Todaro G, zur Hausen H, Eds; Viruses in Naturally Occurring Cancers; Cold Spring Harbor Laboratory: New York, pp 329–341.

Pietropaola V, Degener AM, Di Taranto C, Sinibaldi L, Orsi N (1995): The in vitro effect of monensin on BK polyomavirus replication. New Microbiol 18:341–349.

Pietropaola V, Di Taranto C, Degener AM, Jin L, Sinibaldi L, Baiochini A, Melis M, Orsi N (1998): Transplacental transmission of human polyomavirus BK. J Med Virol 56:372–376.

Pipas JM (1992): Common and unique features of T antigens encoded by the polyomavirus group. J Virol 66:3979–3985.

Portolani M, Marzocchi A, Brabanti-Brodano G, LaPlaca M (1974): Prevalence in Italy of antibodies to a new human papovavirus (BK virus). J Med Micro 7:543–546.

Portolani M, Pietrosemoli P, Cermelli C, Mannini-Palenzona A, Grossi MP, Paolini L, Barbanti-Bordano G (1988): Suppression of BK virus replication and cytopathic effect by inhibitors of prokaryotic DNA gyrase. Antiviral Res 9:205–218.

Possati L, Rubini C, Portolani M, Gazzanelli G, Piani M, Borgatti M (1983): Receptors for the human papovavirus BK on human lymphocytes. Arch Virol 75:131–136.

Purchio AF, Fareed GC (1979): Transformation of human embryonic kidney cells by human papovavirus BK. J Virol 29:763–769.

Purighalla R, Shapiro R, McCauley J, Randhawa P (1995): BK virus infection in a kidney allograft diagnosed by needle biopsy. Am J Kidney Dis 26:671–673.

Randhawa PS, Demetris AJ (2000): Nephropathy due to the polyomavirus type BK. N Engl J Med 342:1361–1362.

Randhawa PS, Finkelstein S, Scantlebury V, Shapiro R, Vivas C, Jordan M, Picken MM, Demetris AJ (1999): Human polyomavirus-associated interstitial nephritis in the allograft kidney. Transplantation 67:103–109.

Raz E, Brezis M, Rosemann E, Eilat D (1989): Anti-DNA antibodies bind directly to renal antigens and induce kidney dysfunction in the isolated perfused rat kidney. J Immunol 142:3076–3082.

Rekvig OP, Moens U, Sundsfjord A, Bredholt G, Osei A, Haaheim H, Traavik T, Arnesen E, Haga HJ (1997): Experimental expression in mice and spontaneous expression in human SLE of polyomavirus T-antigen: A molecular basis for induction of antibodies to DNA and eukaryotic transcription factors. J Clin Invest 99:2045–2054.

Rice SJ, Bishop JA, Apperley J, Gardner SD (1985): BK virus as a cause of haemorrhagic cystitis after bone marrow transplantation. Lancet ii:844–845.

Rinaldo HC, Traavik T, Hey A (1998): The agnoprotein of the human polyomavirus BK is expressed. J Virol 72:6233–6236.

Ritchie BW, Niagro FD, Latimer KS, Pritchard N, Campagnoli RP, Lukert PD (1996): An inactivated avian polyomavirus vaccine is safe and immunogenic in various Psittaciformes. Vaccine 14:1103–1107.

Ritchie BW, Vaughn SB, Leger JS, Rich GA, Rupiper DJ, Forgey G, Greenacre CB, Latimer KS, Pesti D, Campagnoli R, Lukert PD (1998): Use of an inactivated virus vaccine to control polyomavirus outbreaks in nine flocks of psittacine birds. J Am Vet Med Assoc 212:685–690.

Rosen S, Harmon W, Krensky AM, Edelson PJ, Padgett BL, Grinell BW, Rubino MJ, Walker DL (1983): Tubulo-interstitial nephritis associated with polyomavirus (BK type) infection. N Engl J Med 308:1192–1196.

Rubinstein R, Pare N, Harley EH (1987): Structure and function of the transcriptional control region of nonpassaged BK virus. J Virol 61:1747–1750.

Ruediger R, Fields K, Walter G (1999): Binding specificity of protein phosphatase 2A core enzyme for regulatory B subunits and T antigens. J Virol 73:839–842.

Rundell K, Major EO, Lampert M (1981): Association of cellular 56,000- and 32,000-molecular weight proteins with BK virus and polyoma virus t-antigens. J Virol 37: 1090–1093.

Rziha HJ, Belohradsky BH, Schneider U, Schwenk HU, Bornkamm GW, zur Hausen H (1978a): BK virus. II. Serologic studies in children with congenital disease and patients with malignant tumors and immunodeficiencies. Med Microbiol Immunol 165:83–92.

Rziha HJ, Bornkamm GW, zur Hausen H (1978b): BK virus: I. Seroepidemiologic studies and serologic response to viral infection. Med Microbiol Immunol 165:73–81.

Sandler ES, Aquino VM, Goss-Shohet E, Hinrichs S, Krisher K (1997): BK papova virus pneumonia following hematopoietic stem cell transplantation. Bone Marrow Transplant 20:163–165.

Sawai ET, Butel JS (1989): Association of a cellular heat shock protein with simian virus 40 large T antigen in transformed cells. J Virol 63:3961–3973.

Schätzl HM, Sieger E, Jäger G, Nitschko H, Bader L, Ruckdeschel G, Jäger G (1994): Detection by PCR of human polyomaviruses BK and JC in immunocompromised individuals and partial sequencing of control regions. J Med Virol 42:138–145.

Scheidtmann KH, Mumby MC, Rundell K, Walter G (1991): Dephosphorylation of simian virus 40 large-T antigen and p53 protein by protein phosphatase 2A: Inhibition by small-t antigen. Mol Cell Biol 11:1996–2003.

Seehafer J, Salmi A, Scraba DG, Colter JS (1975): A comparative study of BK and polyoma viruses. Virology 66:192–205.

Seif I, Khoury G, Dhar R (1979): The genome of the human papovavirus BKV. Cell 18:963–977.

Shah KV (1996): Polyomaviruses. In Fields BN, Knipe DM, Howley PM, Eds; Fields Virology; Lippincott-Raven: Philadelphia, pp 2027–2043.

Shah KV (2000): Human polyomavirus BKV and renal disease. Nephrol Dial Transplant 15:754–755.

Shah KV, Daniel R, Madden D, Stagno S (1980): Serological investigations of BK papovavirus infection in pregnant women and their offspring. Infect Immun 30:29–35.

Shah KV, Daniel RW, Murphy G (1973a): Antibodies reacting to simian virus 40 T antigen in human sera. J Natl Cancer Inst 51:687–690.

Shah KV, Daniel RW, Warszawski RM (1973b): High prevalence of antibodies to BK virus, an SV40-related papovavirus, in residents of Maryland. J Infect Dis 128:784–787.

Shenk TE, Carbon J, Berg P (1976): Construction and analysis of viable deletion mutants of simian virus 40. J Virol 18:664–672.

Shinohara T, Matsuda M, Cheng SH, Marshall J, Fujita M, Nagashima K (1993): BK virus infection of the human urinary tract. J Med Virol 41:301–305.

Shivakumar CV, Das GC (1996): Interaction of human polyomavirus BK with the tumor-suppressor protein p53. Oncogene 13:323–332.

Shlomchik M, Mascelli M, Shan H, Radic MZ, Pisetsky D, Marshak Rothstein A, Weigert M (1990): Anti-DNA antibodies from autoimmune mice arise by clonal expansion and somatic mutation. J Exp Med 171:265–292.

Simmons DT (1995): Transformation by polyomaviruses. Role of tumor suppressor proteins. In Barbanti-Brodano G, Bendinelli N, Friedman H, Eds; DNA Tumor Viruses: Oncogenic Mechanisms; Plenum Press: New York, pp 27–50.

Small JA, Khoury G, Jay G, Howley PM, Scangos GA (1986): Early regions of JC virus and BK virus induce distinct and tissue-specific tumours in transgenic mice. Proc Natl Acad Sci USA 83:8288–8292.

Smith RD, Galla JH, Skahan K, Anderson P, Linnemann CC, Ault GS, Ryschkewitsch CF, Stoner GL (1998): Tubulointerstitial nephretis due to a mutant polyomavirus BK virus strain, BKV(Cin), causing end-stage renal disease. J Clin Microbiol 36:1660–1665.

Sontag E, Federov S, Kamibayashi C, Robbins D, Cobb M, Mumby M (1993): The interaction of SV40 small tumor antigen with protein phosphatase 2A stimulates the MAP kinase pathway and induces cell proliferation. Cell 75:887–897.

Sontag E, Sontag, JM, Garcia A (1997): Protein phosphatase 2A is a critical regulator of protein kinase Cζ signaling targeted by SV40 small t to promote cell growth and NF-κB activation. EMBO J 18:5662–5671.

Soussi T (1986): DNA-binding properties of the major structural protein of simian virus 40. J Virol 59:740–742.

Srinivasan A, McClellan AJ, Vartikar J, Marks I, Cantalupo P, Li Y, Whyte P, Rundell K, Brodsky JL, Pipas JM (1997): The amino-terminal transforming region of simian virus 40 large T and small t antigens functions as a J domain. Mol Cell Biol 17: 4761–4773.

Streuli CH, Griffin BE (1987): Myristic acid is coupled to a structural protein of polyomavirus and SV40. Nature 326:619–622.

Struck MM, Klug A, Richmond TJ (1992): Comparison of X-ray structures of the nucleosome core particle in two different hydration states. J Mol Biol 224:253–264.

Stubdal H, Zalvide J, DeCaprio JA (1996): Simian virus 40 large T antigen alters the phosphorylation state of the RB-related proteins p130 and p107. J Virol 70:2781–2788.

Sugimoto C, Hara K, Taguchi F, Yogo Y (1989): Growth efficiency of naturally occurring BK variants in vivo and in vitro. J Virol 63:3195–3199.

Sundsfjord A, Flægstad T, Flø R, Spein AR, Pedersen M, Permin H, Julsrud J, Traavik T (1994a): BK and JC viruses in human immunodeficiency virus type 1–infected persons: prevalence, excretion, viremia, and viral regulatory regions. J Infect Dis 169:485–490.

Sundsfjord A, Johansen T, Flægstad T, Moens U, Villand P, Subramani S, Traavik T (1990): At least two types of control regions can be found among naturally occurring BK virus strains. J Virol 64:3864–3871.

Sundsfjord A, Osei A, Rosenqvist H, Van Ghelue M, Silsand Y, Haga HJ, Rekvig OP, Moens U (1999): BK and JC viruses in patients with systemic lupus erythematosus: Prevalent and persistent BK viururia, sequence stability of the viral regulatory regions, and nondetectable viremia. J Infect Dis 180:1–9.

Sundsfjord A, Spein AR, Lucht E, Flægstad T, Seternes OM, Traavik T (1994b): Detection of BK virus DNA in nasopharyngeal aspirates from children with respiratory infections but not in saliva from immunodeficient and immunocompetent adult patients. J Clin Microbiol 32:1390–1394.

Taguchi F, Imatani Y, Nagaki D, Nakagawa A, Omura S (1980): Selective antiviral activity of the antibiotic 2′-amino-2′-deoxyribofuranosyl adenine. J Antibiot 34:313–316.

Taguchi F, Kajioka J, Miyamura T (1982): Prevalence rate and age of acquisition of antibodies against JC virus and BK virus in human sera. Microbiol Immunol 26: 1057–1064.

Taguchi F, Kajioka J, Shimada N (1985): Presence of interferon and antibodies to BK virus in amniotic fluid of normal pregnant women. Acta Virol 29:299–304.

Taguchi F, Nagaky D, Saito M, Haruyama C, Iwasaki K, Suzuky T (1975): Transplacental transmission of BK virus in human. Jpn J Microbiol 19:395–398.

Takemoto KK, Rabson AS, Mullarkey MF, Blaese RM, Garon CF, Nelson D (1974): Isolation of papovavirus from brain tumor and urine of a patient with Wiskott-Aldrich syndrome. J Natl Cancer Inst 53:1205–1207.

Talmage DA, Listerud M (1994): Retinoic acid suppresses polyomavirus transformation by inhibiting transcription of the c-*fos* proto-oncogene. Oncogene 9:3557–3563.

Tarabek J, Zemla J, Bacik I (1991): Northern blot hybridization analysis of polyomavirus-specific RNA synthesized under the block of virus replication by 5-bromo-2′deoxyuridine. Acta Virol 35:305–312.

Tavis JE, Walker DL, Gardner SD, Frisque RJ (1989): Nucleotide sequence of the human polyomavirus AS virus, an antigenic variant of BK virus. J Virol 63:901–911.

Termaat RM, Assmann KJ, Dijkman HB, van Gompel F, Smeenk RJ, Berden JH (1992): Anti-DNA antibodies can bind to the glomerulus via two distinct mechanisms. Kidney Int 42:1363–1371.

Traystman MD, Gupta PK, Shah KV, Reissig M, Cowles LT, Hillis WD, Frost JK (1980): Identification of viruses in the urine of renal transplant recipients by cytomorphology. Acta Cytol 24:501–510.

Tsai RT, Wang M, OU WC, Lee YL, Li SY, Fung CY, Huang YL, Tzeng TY, Chen Y, Chang D (1997): Incidence of JC viruria is higher than that of BK viruria in Taiwan. J Med Virol 52:253–257.

Vallbracht A, Lohler J, Gossmann J, Gluck T, Petersen D, Gerth HJ, Gencic M, Dörries K (1993): Disseminated BK type polyomavirus infection in an AIDS patient associated with central nervous system disease. Am J Pathol 143:29–39.

Van Bruggen MC, Kramers C, Walgreen B, Elema JD, Kallenberg CG, van den Born J, Smeenk RJ, Assmann KJ, Muller S, Monestier M, Berden JH (1997a): Nucleosomes and histones are present in glomerular deposits in human lupus nephritis. Nephrol Dial Transplant 12:57–66.

Van Bruggen MC, Walgreen B, Rijke TP, Tamboer W, Kramers K, Smeek RJ, Monestier M, Fournie GJ, Berden JH (1997b): Antigen specificity of anti-nuclear antibodies complexed to nucleosomes determines glomerular basement membrane binding in vivo. Eur J Immunol 27:1564–1569.

Van Gorder MA, Pelle PD, Henson JW, Sachs DH, Cosimi AB, Colvin RB (1999): Cynomolgus polyoma virus infection. A new member of the polyomavirus family causes interstitial nephritis, ureteritis, and enteritis in immunosuppresses cynomolgus monkeys. Am J Pathol 154:1273–1284.

Varshavsky AJ, Bakayev VV, Chumackov PM, Georgiev GP (1976): Minichromosome of simian virus 40: Presence of H1. Nucleic Acids Res 3:2101–2113.

Verdonck LF, Dekker AW, Rozenberg-Arska M, Kaspenberg JG, van de Avoort HGAM, de Gast GC (1987): BK viruria and hemorrhagic cystitis. N Engl J Med 316:109.

Vianelli N, Renga, M, Azzi A, De Santis R, Bandini G, Tosi P, Tura S (2000): Sequential vidarabine infusion in the treatment of polyomavirus-associated acute haemorrhagic cystitis late after allogeneic bone marrow transplantation. Bone Marrow Transplant 25:319–320.

Vögeli TA, Peinemann F, Burdach S, Ackermann R (1999): Urological treatment and clinical course of BK polyomavirus-associated hemorrhagic cystitis in children after bone marrow transplantation. Eur Urol 36:252–257.

Voltz R, Jäger G, Seelos, K, Fuhry L, Hohfeld R (1996): BK virus encephalitis in an immunocompetent patient. Arch Neurol 53:101–103.

Westermann P, Gross B, Hoinkis G (1989): Inhibition of expression of SV40 virus large T-antigen by antisense oligodeoxyribonucleotides. Biomed Biochim Acta 48:85–93.

Westphal RS, Anderson KA, Means AR, Wadzinski BE (1998): A signaling complex of Ca^{2+}-calmodulin–dependent protein kinase IV and protein phosphatase 2A. Science 280:1258–1261.

Whalen B, Laffin J, Friedrich TD, Lehman JM (1999): SV40 small t antigen enhances progression to >G2 during lytic infection. Exp Cell Res 251:121–127.

Wheat WH, Roesler WJ, Klemm DJ (1994): Simian virus 40 small tumor antigen inhibits dephosphorylation of protein kinase A–phosphorylated CREB and regulates CREB transcriptional stimulation. Mol Cell Biol 14:5881–5890.

Wildeman AG (1989): Transactivation of both early and late simian virus 40 promoters by large tumor antigen does not require nuclear localization of the protein. Proc Natl Acad Sci USA 86:2123–2127.

Wiley SR, Kraus RJ, Zuo F, Murray EE, Loritz K, Mertz JE (1993): SV40 early-to-late switch involves titration of cellular transcriptional repressors. Genes Dev 7:2206–2219.

Wychowski C, Benichou D, Girard M (1986): A domain of SV40 capsid polypeptide VP1 that specifies migration into the cell nucleus. EMBO J 5:2569–2576.

Xie YC, Hwang C, Overwijk W, Zeng Z, Eng MH, Mulé JJ, Imperiale MJ, Restifo NP, Sanda MG (1999): Induction of tumor antigen-specific immunity in vivo by a novel vaccinia vector encoding safety-modified simian virus 40 T antigen. J Natl Cancer Inst 91:169–175.

Yang CA, Wu R (1979): Comparative study of papovavirus DNA: BKV(MM), BKV(WT) and SV40. Nucleic Acids Res 3:651–668.

Yoshiike K, Takemoto KK (1986): Studies with BK virus and monkey lymphotropic papovavirus. In Salzman NP, Ed; The Papovaviridae; Plenum: New York, pp 295–326.

15

JC VIRUS IN EXPERIMENTAL AND CLINICAL BRAIN TUMORIGENESIS

LUIS DEL VALLE, M.D., JENNIFER GORDON, PH.D.,
PASQUALE FERRANTE, M.D., and KAMEL KHALILI, PH.D.

1. INTRODUCTION

JC virus (JCV) is a member of the polyomavirus family of DNA tumor viruses, which includes BK virus and the well-known simian virus 40 (SV40). JCV exists within the human population, as about 80% of adults worldwide exhibit JCV-specific antibodies (Major et al., 1992). Infection with the virus is thought to be subclinical and occur in early childhood. The virus is presumed to remain latent in the kidney until it reactivates under immunosuppressive conditions to result in progressive multifocal leukoencephalopathy (PML).

Before the AIDS epidemic, PML was considered a rare disorder associated with immunocompromising diseases such as lymphomas or was seen in renal transplant and chemotherapy patients as a complication of immunosuppressive therapies. However, recent reports indicate that 70% of all HIV-1–infected patients will exhibit neurologic disorders, and as many as 10% of all HIV-1–infected patients will develop PML (Berger and Concha, 1995).

In addition to having a role in the development of PML, recent evidence points to the association of JCV with human cancer, most notably, brain tumors. Several clinical studies report cases of concomitant PML and central nervous

Human Polyomaviruses: Molecular and Clinical Perspectives, Edited by Kamel Khalili and Gerald L. Stoner.
ISBN 0-471-39009-7

system (CNS) neoplasms in patients with long-standing immunosuppressive disorders such as leukemia, lymphoma, and AIDS (for review, see Gallia et al., 1998a). The oncogenicity of JCV has been repeatedly demonstrated by several laboratories in multiple animal models. For example, JCV induces glioblastoma multiforme, a malignant astrocytoma in new world primates: owl and squirrel monkeys. In rodents, intracerebral injection of JCV induces a variety of tumors, including astrocytoma, medulloblastoma, and ependymoma. Several in vitro cell culture and in vivo transgenic animal models have demonstrated that the oncogenicity of JCV is attributed to its multipotent early gene product, the large T antigen, which is capable of interaction with tumor suppressor proteins, such as p53 and the pRb family. T antigen is thought to functionally inactivate these tumor suppressor genes in neural cells, resulting in the formation of neural origin tumors in these animals. In this chapter, we review the transforming ability of JCV in vitro, discuss the tumorigenicity of JCV in several experimental animal models, and review the current information on the association of JCV with human cancer.

2. JCV EARLY PROTEIN, T ANTIGEN, AND T ANTIGEN TRANSFORMING ACTIVITY

Several studies have demonstrated that JCV can transform cells in culture, although the efficiency of transformation by JCV is not as efficient as those by other polyomaviruses, for example, SV40. JCV-transformed cells expressing T antigen have common features, including distinct morphologic changes, rapid replication, prolonged life spans, and the ability to form dense foci in culture. Human primary fetal glial cells and human vascular endothelial cells were among the first human cells used in JCV transformation assays (Fareed et al., 1978; Walker and Padgett 1978). Transformation of primary hamster brain cells by various strains of JCV, including Mad-1 and Mad-4, results in rapidly growing cells with several characteristics of neoplastic cells, including growth in low serum, enhanced production of plasminogen activator, and anchorage-independent growth. In addition, JCV can transform baby hamster kidney cells and rat fibroblasts with low frequency (Bollag et al., 1989; Haggerty et al., 1989). Using an origin-defective mutant of JCV, primary human fetal glial cells were transformed, leading to the development of POJ cells, which constitutively produce JCV T antigen (Mandl et al., 1987).

While the precise mechanism responsible for the cellular transformation by JCV has not been fully explored, it is believed that T antigen may mediate the transforming potential of JCV through its interaction with host cellular factors.

T antigen is the major early gene product of the primate polyomavirus, SV40, and the human JC and BK viruses. While our knowledge on JCV T antigen is not as extensive as that on SV40, the homology between JCV and SV40 T antigen can assist us in gaining information on the function of JCV T antigen. It has been well established that SV40 T antigen exhibits great oncogenic potential due to its ability to alter several aspects of normal cellular function. JCV T antigen shares more than 70% sequence homology with SV40 T antigen, the area of greatest difference being the carboxy-terminal region, which is responsible for host range effect. It is believed that, similar to SV40 T antigen, the JCV early protein has multifunctional activities, including ATP-ase, helicase, DNA binding, and α-polymerase. Furthermore, results from co-immunoprecipitation assays have revealed that, similar to SV40 T antigen, JCV T antigen has the capacity to interact with p53 and members of the pRb family (Bollag et al., 1989; Dyson et al., 1989, 1990; Fanning and Knippers, 1992; Haggerty et al., 1989; Krynska et al., 1999b; Ludlow, 1993 Ludlow and Skuse, 1995). JCV has also been associated with several other cellular proteins such as YB-1 and Purα (Gallia et al., 1998b; Safak et al., 1999) (Fig. 15.1). T antigen is known to undergo many post-translational modifications, and the strong effect of SV40 T antigen on viral and host gene expression has been demonstrated in its ability to induce cellular transformation in many cell culture systems and several animal models. In particular, SV40 T antigen is capable of immortalizing a broad range of cells in vitro as well as inducing tumorigenesis in many experimental animal models.

The transforming ability of JCV T antigen appears limited to neural origin tissue, as observed both in vitro and in many experimental animal models (see below). This cell-type specificity is due in part to the structure of the viral regulatory region, which differs significantly from that of SV40 (Frisque and White, 1992).

The ability of T antigen to physically interact with p53 and pRb family members provides evidence of mechanisms by which it may induce cellular transformation. p53 and pRb are two essential cellular proteins that exert their effects on cellular proliferation in many ways, most notably via their effect on cell growth and differentiation demonstrated by their regulation of the cell cycle. Both of these proteins exert their control on the cell cycle during the

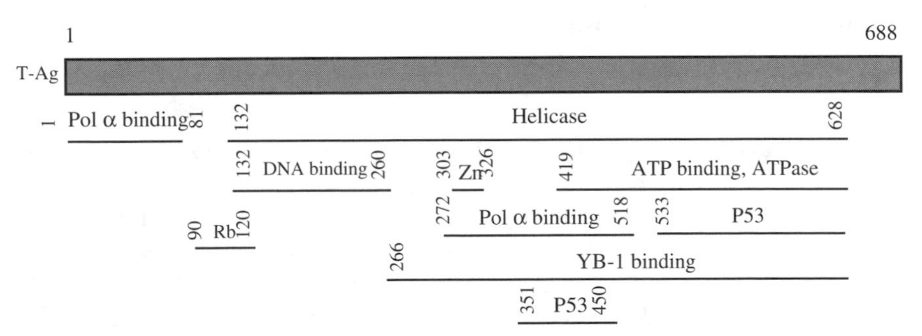

Figure 15.1. Structural organization of polyomaviruses, including JCV early protein T antigen. Linear structure of T antigen depicting the various functional domains as well as the sites for interaction of this protein with other cellular proteins.

G1 to S-phase transition. The deregulation of p53 and pRb by T antigen alters the normal function of these proteins, leading to their inactivation, resulting in unchecked cell division that results in uncontrolled cellular proliferation or tumorigenesis.

Loss of expression of pRb through deletions or rearrangements has been observed in certain human cancers, including retinoblastoma, a neural origin tumor that has been shown to be induced by JCV in several animal models. The pRb protein, like p53, is a nuclear phosphoprotein, and pRb is tightly regulated by phosphorylation during the cell cycle. More specifically, hypophosphorylated pRb is found to have a negative growth effect by providing a block to a cell's exit from the G1 phase of the cell cycle, thereby preventing cells from progressing to S phase. This effect results in part by hypophosphorylated Rb's ability to bind the cellular transcriptional activator, E2F-1. However, the phosphorylation of pRb by cyclins and cyclin-dependent kinases (cdks) results in the liberation of E2F-1 from the complex, allowing E2F-1 to exert its downstream effects, which promote the transition from the G1 to S phase of the cell cycle. The region of pRb that is bound by T antigen is apparently a critical one, as mutations in the pRb protein that are found in human cancers abrogate the mutant protein's ability to bind to T antigen (Manfredi and Prives, 1994).

p53, often referred to as "the gate-keeper of the genome" and the gene most often found altered in human cancers, was initially characterized by its ability to bind to SV40 T antigen (Levine, 1997). p53, a nuclear phosphoprotein that is capable of binding DNA in a sequence-specific manner to activate transcription, has been shown to be stabilized in cells transformed by SV40 T antigen. The increased stability of p53 may result from its association with T antigen. Furthermore, increased levels of p53 may result from a cellular process in response to T-antigen expression (Ludlow, 1993). When in complex with T antigen, p53 is thought to remain sequestered and unable to carry out its normal function of maintaining cells in the G1 phase of the cell cycle. p53 utilizes downstream factors such as p21WAF1 in order to maintain cells in check in that high levels of p21WAF1 can inhibit cyclin:cdk activity, which is required to induce progression of the cell cycle (Fig. 15.2).

3. MECHANISMS OF CELL CYCLE DEREGULATION BY T-ANTIGEN

One can envision a model whereby T-antigen, through its interaction with p53 and pRb, may dysregulate cell cycle pathways at several levels. The association of JCV T antigen with p53 may block the ability of p53 to induce p21WAF1, a protein that inhibits cyclin:cdk activity. Deregulation of the participant cyclins and their associated kinases may in turn lead to the phosphorylation of pRb and the liberation of E2F-1. The release of E2F-1 from the pRb:E2F-1 complex

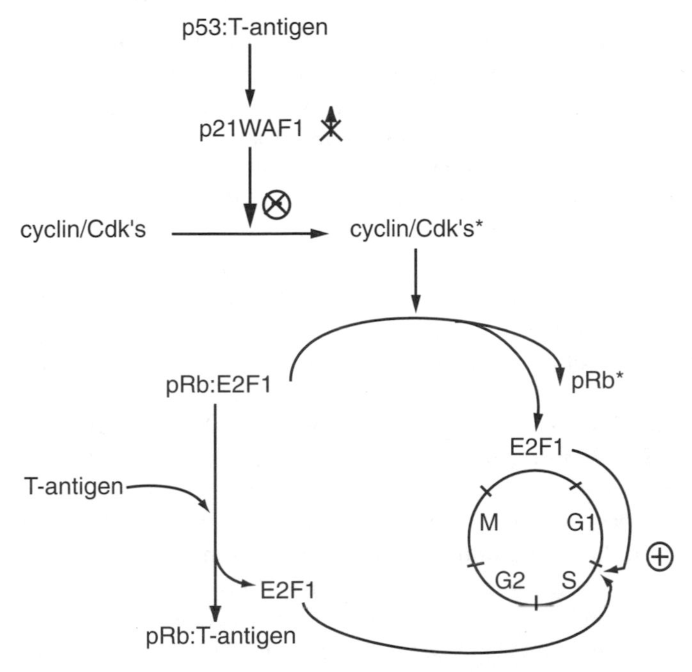

Figure 15.2. Schematic model for the de-regulation of the cell cycle pathway upon interaction of JCV T antigen with p53 and/or pRb. The association of T antigen with p53 inactivates p53, thus interfering with the ability of p53 to induce p21WAF-1, a protein that blocks kinase activity of cyclin/cdks and retains E2F-1 in a nonfunctional state through its association with unphosphorylated pRb. Phosphorylation of pRb and/or association of pRb with JCV T antigen can lead to liberation of E2F-1, allowing free E2F-1 to stimulate cell cycle progression at the G1/S stage of the cell cycle.

may also be accomplished by the association of JCV T antigen with pRb. As levels of E2F-1 increase in cells, E2F-1 may induce its own gene expression, as well as expression of other S-phase-specific promoters such as PCNA, thereby promoting rapid entry of cells into S phase.

Importantly, the delicate balance of these proteins is key to understanding T antigen's role in tumorigenesis. T antigen may interact with p53 in an experimental tumor, for example, but the relative levels of the two proteins must be considered as these proteins are able to interact with numerous cellular proteins and can affect multiple cellular pathways simultaneously. T antigen may induce cellular transformation, but if p53 protein levels or activity are greater than the amount of T antigen, p53 may continue to function at the G1 checkpoint of the cell cycle. Likewise, the inactivation of pRb may require high levels of T antigen to exert an effect.

While a great deal of attention has been focused on the complexation of T antigen with p53 and pRb, additional levels of regulation of these proteins must be considered. Several recent publications have demonstrated a role for T antigen in the regulation of cellular proteins that is distinct from its binding ability. For example, SV40 T antigen has been shown to regulate p53-mediated transcription without interacting with p53 (Rushton et al., 1997). In addition, SV40 T antigen itself can alter the phosphorylation levels of the pRb family members (Stubal et al., 1996). Furthermore, recent data indicate that BKV T antigen may induce the transcriptional transactivator E2F-1 without evidence of T antigen binding to pRb (Harris et al., 1998). Likewise, it is well known that p21WAF1 is regulated by both p53-dependent as well as p53-independent mechanisms. Finally, recent evidence has shown that SV40 T antigen interacts with hyper- and not hypophosphorylated p107 and pRb-2 (Knudsen and Wang, 1998).

As p53 function is important for genomic stability, the association of T antigen with p53 may result in genomic instability and mutation in various important genes, including p53 itself. In support of this idea, our recent studies have indicated that a mutation in exon 4 of the p53 gene in JCV-induced tumor cells results in the expression of a truncated version of p53 with no amino acid residues of exon 4 (Khalili et al., unpublished observations). The ability of T antigen to dysregulate tumor suppressor proteins supports its potential role in cellular transformation in humans. Clearly, a cell contains redundancies and overlapping pathways in order to control such an important event as cell growth. Recent data have pointed to several novel ways in which T antigen may interfere with these pathways on many levels. While a great deal of attention has been focused on the interaction of T antigen with cellular proteins, additional studies will continue to reveal mechanisms by which such a viral protein may upset the delicate balance of factors that determine a cell's decision to proliferate.

4. OTHER POTENTIAL MECHANISMS OF T-ANTIGEN TRANSFORMATION

In addition to its ability to derange cell cycle regulatory proteins, JCV has also been demonstrated to possess mutagenic activity (Theile and Grabowski, 1990). Additional studies (Neel et al., 1996; Lazutka et al., 1996) indicate that JCV is associated with lymphocytes that contain multiple chromosome-type aberrations. Perhaps more relevant in light of earlier studies indicating that JCV is able to transform primary human fetal glial cells (Mandl et al., 1987) is recent evidence that JCV can induce chromosomal damage in these cells (Neel et al., 1996). In further support of this concept, cytometric analysis of cell lines derived from JCV T antigen–induced medulloblastomas that express T antigen show evidence of aneuploidy, whereas cells derived in parallel that do not express T antigen demonstrate a diploid pattern (Krynska et al., unpublished observations).

5. ASSOCIATION OF JCV WITH TUMORS

Laboratory Animal Studies

The highly oncogenic potential of polyomaviruses, including JCV, has been well established in several animal models (Table 15.1). In an extensive series of studies, newborn Golden Syrian hamsters were shown to develop a broad range of tumors approximately 6 months after brain inoculation with JCV, among which were medulloblastoma (Zu Rhein and Varakis 1979), astrocytoma, glioblastoma multiforme, primitive neuroectodermal tumors, and peripheral neuroblastomas (Walker et al., 1973; Zu Rhein 1983). Tumors of the CNS have been detected in more than 85% of newborn hamsters inoculated intracerebrally with JCV, clearly demonstrating its oncogenic potential in neural origin tissue. Further studies have demonstrated that JCV T antigen was detected in tumor tissue, but no evidence of viral replication or persistent infection could be found. Similarly, injection of JCV into the brains of newborn rats has induced undifferentiated neuroectodermal origin tumors in the cerebrum of 75% of the animals (Ohsumi et al., 1985, 1986).

JCV represents the only polyomavirus that induces tumors in nonhuman primates. Intracerebral inoculation of JCV into owl and squirrel monkeys results in the development of astrocytomas, glioblastomas, and neuroblastomas 16–24 months after intracerebral, subcutaneous, or intravenous inoculation (London et al., 1978, 1983). Again, as in the hamster models, studies of the monkey tumor tissue revealed the expression of the JCV early protein T antigen, but virion antigens could not be detected in the animals. With one exception, tissue from JCV-infected monkeys was unable to release virus when grown in cell culture (Major et al., 1992), indicating that primate-origin cells may not be permissive for JCV infection.

Perhaps some of the most interesting observations on the oncogenicity of JCV relates to studies on several lines of transgenic mice that have been generated containing the entire gene for JCV T antigen under the control of its own promoter. Because these mice do not contain any of the viral late genes, the phenotypes observed are solely dependent on the expression of JCV T antigen. Earlier studies by Small et al. (1986a,b) resulted in T antigen transgenic mice that developed one of two phenotypes, the appearance of adrenal neuroblastomas, or less frequently, abnormal formation of myelin sheaths in the CNS (dysmyelination). We have generated additional mice utilizing the same JCV T-antigen transgene and have established a line of mice that exhibits tumors of primitive neuroectodermal origin (Franks et al., 1996). Two interesting conclusions can be drawn from these studies: first, that T antigen, in the absence of JCV and therefore viral replication, can alter myelin formation to induce dysmyelination and, second, that JCV T antigen is able to transform cells and induce tumors of neural origin.

Our group has described the development of a transgenic animal model using the early region of the genome of the human ubiquitous JCV, which

Table 15.1. JCV Induces Neural Origin Tumors in Several Animal Models[a]

Species	Tumor Type	References
Owl and squirrel monkeys	Astrocytoma	London et al. (1978, 1983)
	Glioblastoma multiforme	
Syrian hamster	Medulloblastoma	Walker et al. (1973), Zu Rhein (1983)
	Astrocytoma	
	Glioblastoma multiforme	
	Primitive neuroectodermal	
	Ependymoma	
	Pineocytoma	
Rat	Primitive neuroectodermal	Ohsumi et al. (1985)
JCV TAg transgenic mouse		
Mad-4 promoter	CNS dysmyelination	Small et al. (1986b); Franks et al. (1996)
	Adrenal neuroblastoma	Small et al. (1986a); Franks et al. (1996)
	Pituitary adenoma	Gordon et al. (2000)
	Malignant peripheral nerve sheath	Gordon et al. (unpublished observation)
Archetype promoter	Medulloblastoma	Krynska et al. (1999b)

[a]All animal models except transgenic mice were induced by intracerebral inoculation of JCV. Transgenic mice express the entire gene for the JC viral early protein T antigen under the control of the viral promoter derived from either the Mad-4 or archetype strain of JCV.

contains the viral early promoter expressing the early gene encoding the oncoprotein JCV T antigen (Krynska et al., 1999b). Histologic examination of the CNS revealed no features of hypomyelination, an established pathology that is seen upon expression of JCV T antigen in the brain (Small et al., 1986b). Unexpectedly, cerebellar tumors, which closely resemble human medulloblastomas in location, histologic appearance, and expression of differentiation markers, were found in the affected animals. In another line of transgenic mice, also expressing JCV T antigen under the control of its own promoter, nearly 50% of the animals developed large, solid masses within the base of the skull by 1 year of age. Evaluation of the location as well as histologic and immunohistochemical studies demonstrate that the tumors arise from the pituitary gland (Gordon et al., 2000) (Fig. 15.3). Several cell lines have been developed from various JCV-induced tumors in different experimental animal models. These cells, which exhibit common features of transformed cells, including rapid cell growth, express T antigen in the nuclei (Fig. 15.4).

As mentioned above, T antigen is known to interact with several cell cycle regulators; therefore, the neoplasms were analyzed for the presence of the tumor suppressor protein p53. Immunoprecipitation/Western blot analysis demonstrated overexpression of wild-type, but not mutant, p53 within tumor tissue of both medulloblastoma and pituitary animal models. In addition, co-immunoprecipitation established an interaction between p53 and T antigen and overexpression of the p53 downstream target protein p21/WAF1 in pituitary tumors. While T antigen was found to be expressed in almost every tissue of the animals by RT-PCR, detectable levels of the protein were observed only in the tumor tissue. Co-immunoprecipitation experiments revealed that T antigen is complexed with p53 in protein extracts prepared from tumor tissue, and, as expected, low levels of p21/WAF were detected in the tumor. In medulloblastoma-bearing animals, the levels and kinase activities of the protein cyclin E and its cellular partner Cdk2 were examined and revealed to be highly active in the tumor compared with other tissues. Finally, another important regulator, Rb, was evaluated and found to be complexed with T antigen in a phosphorylated state, which liberates its cellular partner E2F to induce proliferation. These observations demonstrate a possible regulatory pathway whereby JCV T antigen can induce tumors of neural origin in vivo.

Association of JCV with Human Tumors

Brain Tumors. Since the first description of PML, several reports have pointed to an association between the human polyomavirus JCV and human brain tumors. The first of these cases was noted by Richardson (1961), who first described PML. In this case, the post-mortem examination of a 58-year-old man with chronic lymphocytic leukemia and PML incidentally revealed an oligodendroglioma (Table 15.2). There have also been cases described in which PML was associated with multiple astrocytomas (Sima et al., 1983). Castaigne et al. (1974) described a 68-year-old man who suffered a long-lasting immu-

Table 15.2. Association of JCV with Human CNS Neoplasms

Brain Tumor	PML	Detection Method[a]	Positive Total	References
Astrocytoma	Yes	EM	1/1	Sima et al. (1983)
	No	PCR/Seq	4/10	Caldarelli-Stefano et al. (2000)
CNS lymphoma	Yes	EM	1/4	Davies et al. (1973)
	Yes	EM/IF	1/1	GiaRusso and Koeppen (1978)
	Yes	EM	1/1	Egan et al. (1980)
	Yes	EM/IHC	1/1	Ho et al. (1980)
	Yes	ND	1/1	Liberski et al. (1982)
	Yes	ND	1/1	Williamson et al. (1989)
Dysplastic ganglion-like cells	Yes	IHC/ISH	1/1	Shintaku et al. (2000)
Ependymoma	No	PCR/Seq	1/5	Caldarelli-Stefano et al. (2000)
Glioma	Yes	HA/EM	1/1	Castaigne et al. (1974)
	Yes	ISH	1/1	Gullotta et al. (1992)
	No	PCR/IHC	1/1	Del Valle et al. (2000)
Medulloblastoma	No	PCR	8/11	Khalili et al. (1999)
Oligoastrocytoma	No	IHC/WB/PCR	1/1	Rencic et al. (1996)
Oligodendroglioma	Yes	ND	1/10	Richardson (1961)
	No	PCR/Seq	1/5	Caldarelli-Stefano et al. (2000)
Xanthoastrocytoma	No		1/1	Boldorini et al. (1998)

[a]HA = hemagglutination assay; EM = electron microscopy; IF = immunofluorescence; IHC = immunohistochemistry; ISH = in situ hybridization; PCR = polymerase chain reaction; Seq = sequencing; ND = not determined.

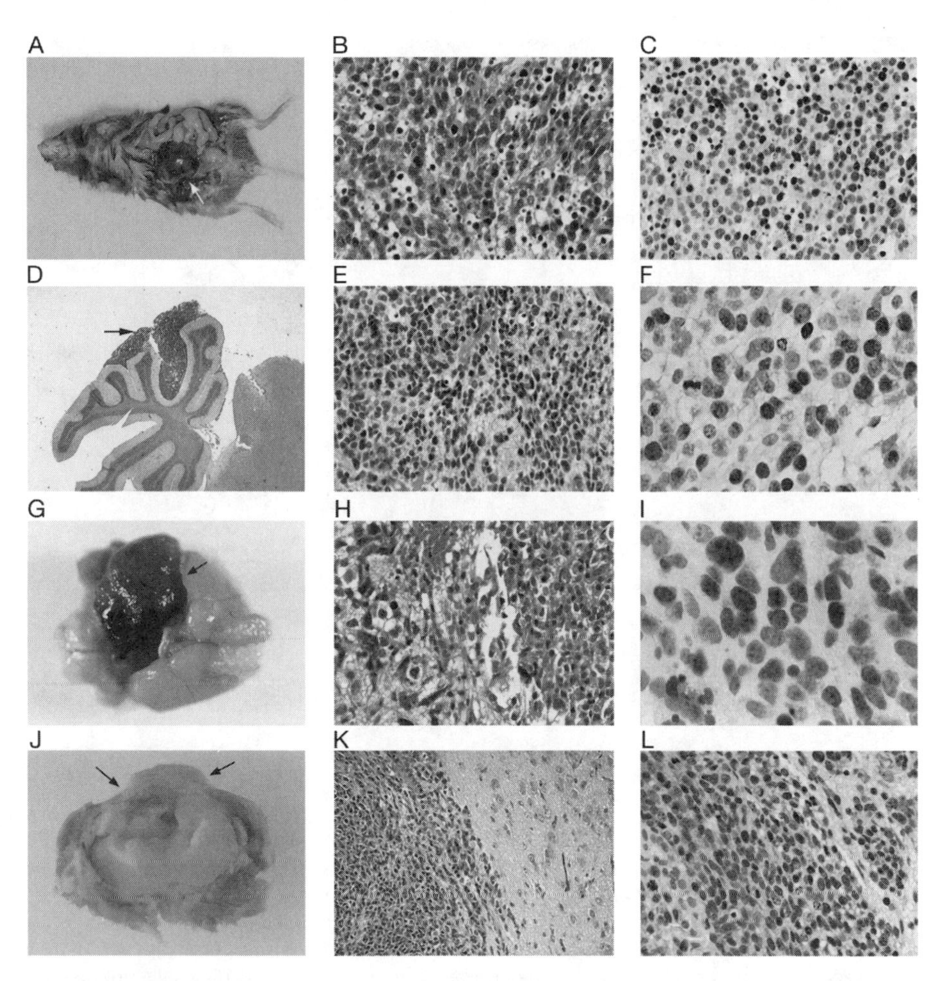

Figure 15.3. Induction of neural origin tumors by JCV T antigen in experimental animals. (**A**) JCV T antigen transgenic mice develop peripheral neuroblastoma, which can present as a solid, well-circumscribed mass, in the abdominal cavity (arrow). (**B**) Neuroblastomas are histologically characterized by densely cellular neoplasms with a high nuclear to cytoplasmic ratio. (**C**) The majority of the nuclei show immunoreactivity to the viral protein T antigen. (**D**) Transgenic mice generated with sequences for T antigen under the control of the JCV archetype promoter develop cerebellar medulloblastomas within the foliæ (arrow). (**E**) Histologically, the tumors appear similar to neuroblastoma. (**F**) JCV T antigen can be detected by immunohistochemistry in tumor cell nuclei. (**G**) T antigen transgenic mice may also develop pituitary tumors, which appear as large masses at the base of the skull (arrow). (**H**) Histology demonstrates a highly pleomorphic tumor (left) adjacent to normal pituitary (right). (**I**) Tumor cells show nuclear staining for JCV T antigen. (**J**) JCV-transformed HJC cells transplanted into the brain of syngeneic Syrian hamsters form an intracranial mass that can protrude through the superior aspect of the skull (arrows). (**K**) A clear line of demarcation is present between the highly cellular pleomorphic cells (left) and the surrounding normal brain parenchyma (right). (**L**) The majority of the tumor cells express nuclear T antigen. A, G, J, original magnification, ×4; D, original magnification, ×10; B, E, H, K, hematoxylin and eosin, original magnification, ×100; C, hematoxylin counterstain, original magnification ×100; F, L, hematoxylin counterstain, original magnification ×200; I, hematoxylin counterstain, original magnification ×400; D. Taken with permission from Krynska et al., 1999. (See color plates.)

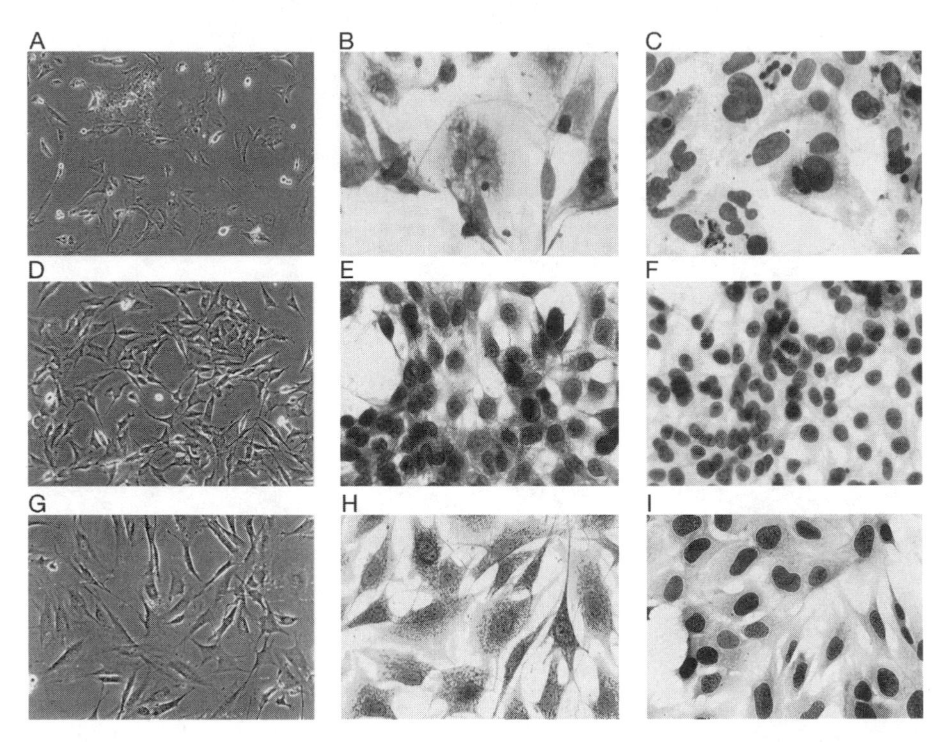

Figure 15.4. Morphologic features of cells derived from JCV-induced tumors in various experimental animals. (**A**) Phase microscopy of the glioblastoma cell line Owl 586 generated upon intracerebral inoculation of an owl monkey with JC virus. (**B**) Immunostaining with the cellular marker GFAP demonstrates that the tumor is of glial origin. (**C**) Immunostaining for JCV T antigen shows that the majority of the cells express T antigen in the nucleus. (**D**) HJC cells cultured from a tumor induced upon intracerebral injection of JCV into newborn Syrian hamsters shown by phase microscopy. (**E**) HJC glial cells are positive for GFAP. (**F**) Nearly all of the cells express JCV T antigen. (**G**) Phase contrast of BS-1 B8 cells derived from JCV T antigen–induced mouse medulloblastoma. (**H**) Synaptophysin staining of BS-1 B8 cells demonstrate the neuronal origin of the tumor cells. (**I**) The majority of the tumor cells show nuclear immunoreactivity for JCV T antigen. (See color plates.)

nodeficiency syndrome. Postmortem examination revealed, in addition to PML demyelinating lesions, numerous foci of anaplastic astrocytes. Ultrastructural analysis of these lesions demonstrated viral particles in both oligodendrocytes and astrocytes within the PML foci, but not in the neoplastic astrocytes. Interestingly, no type of immunodeficiency was revealed in this 68-year-old man either clinically or at autopsy. Gullota et al. (1992) also reported the concomitant occurrence of PML and CNS gliomas. In this case, a 30-year-old HIV-1-negative PML patient, autopsy revealed the presence of gliomatous foci. The authors noted a pleomorphic astrocytoma in the brain and diffuse neoplastic infiltration in the brain stem. In neither of these lesions was JCV detected by in situ hybridization. Recently, Shintaku and colleagues (2000) reported a case of dysplastic ganglion like-cells in a patient who succumbed to PML. The brain contained a large number of dysplastic or dysmorphic ganglion-like cells in the cerebral cortex that showed properties of neurons. Detailed immunohistochemical studies revealed that the neurons were infected with JCV and expressed T antigen but not the capsid protein VP1.

In addition to the cases of concomitant PML and cerebral neoplasm, JCV has been associated with human brain tumors in the absence of any PML lesions. Rencic et al. (1996) detected DNA from the JCV noncoding region and T antigen DNA, RNA, and protein in tumor tissue from an immunocompetent HIV-1–negative patient with an oligoastrocytoma. Moreover, this patient did not clinically or microscopically exhibit any features of PML. Interestingly, Boldorini et al. (1998) reported the presence of JCV DNA in the brain of an immunocompetent 9 year old with a pleomorphic xantoastrocytoma. These two cases have demonstrated the presence of JCV in human brain tumors of immunocompetent non-PML patients. Although in all these cases a causal relationship between JCV and the development of brain tumors may be considered speculative, these intriguing findings raise the question of whether or not JCV is involved in tumor pathogenesis in the CNS and prompted several attempts to establish the association of JCV with different types of brain tumors in humans.

In one series of studies, to test the association of JCV with human brain tumors of glial origin, DNA amplification techniques using a pair of primers that amplify the amino-terminus of JCV T antigen between nucleotides 4255 and 4427 were utilized on 85 samples of tumor tissue (Del Valle et al., 2001). Results from multiple experiments revealed that 57.1% of oligodendrogliomas, 76.9% of astrocytomas, 80% of pilocytic astrocytomas, 62.5% of oligoastrocytomas, 66% of anaplastic oligodendrogliomas, 83.3% of anaplastic astrocytic tumors, 57.1% of glioblastoma multiforme, and 83.3% of ependymomas contained the JCV early gene sequence. Also the only cases of gliomatosis cerebri, gliosarcoma, and subependymoma that were available for this study were positive for JCV DNA sequences and the expression of T antigen.

The detection of the JCV early gene sequence in various human brain tumors provided a rationale to examine production of the early gene product T antigen by immunohistochemical analysis. T-antigen–positive nuclei were found by immunohistochemical staining of oligodendroglioma, anaplastic oligodendroglioma, pilocytic astrocytoma, astrocytoma, glioblastoma multiforme, and ependymoma (Del Valle et al., 2001). Immunohistochemical studies to detect VP1 demonstrated no evidence for expression of the viral late proteins in any of the samples. A discrepancy was noted in the number of JCV-positive samples as determined by PCR amplification compared with those obtained by immunohistochemistry. For instance, the results from PCR showed that while four of five pilocytic astrocytomas contained the JCV DNA sequence, only one of six samples was positive for T antigen expression, suggesting that perhaps mutations in the viral genome abrogated expression of T antigen in tumor cells.

As mentioned earlier, JCV T antigen has the ability to interact with and functionally inactivate the p53 tumor suppressor protein. This association increases the steady-state level of p53 and allows its detection in cancer cells. Examination of the level of p53 in various human tumor cells exhibiting positive staining for T antigen using an anti-p53 antibody that recognizes both wild-type and mutant human p53 showed that, with the exception of the oligoastrocytomas, glioblastoma multiforme, and ependymoma, the number of tumor samples that were positive for T antigen closely paralleled those that showed p53 immunoreactivity in the nuclei (Del Valle et al., 2001). It is likely that in oligoastrocytoma, anaplastic astrocytoma, and glioblastoma multiforme mutations in the p53 gene can stabilize this protein and permit its detection in T-antigen–negative cells. Double labeling immunostaining for T antigen and p53 was performed in selected cases and demonstrated the presence of both proteins in the nuclei of the same neoplastic cells. In a different series of studies the presence of JCV was examined in samples from 25 glial-derived tumors consisting of 10 astrocytomas, 5 ependymomas, 5 oligodendrogliomas, and 5 glioblastomas by molecular and histologic approaches. Results from gene amplification showed the presence of JCV in six cases (four astrocytomas, one oligodendroglioma, and one ependymoma) (Caldaraelli-Stefano et al., 2000).

The development of medulloblastoma in transgenic mice prompted studies to examine the presence of JCV in human primitive neuroectodermal tumors (PNETs), including medulloblastomas. With PCR techniques, it was demonstrated that 87% of human medulloblastomas contain DNA sequences corresponding to the JCV T antigen and VP1 genes in tumor tissue. Expression of JCV T antigen was detected in the nuclei of some, but not all, tumor samples that were positive for JCV sequences (Khalili et al., 1999; Krynska et al., 1999a). Altogether, results from gene amplification and immunohistochemistry of the various brain tumor samples point to the possible association of JCV with a wide variety of CNS neoplasias (Figs. 15.5 and 15.6).

In addition to these group studies, there have been several interesting case reports that further link JCV to human brain tumors. In one case, a 24-year-old immunocompetent man with no medical history experienced headaches behind the left eye that began to increase in intensity. MRI of the brain revealed a large, partially cystic mass that was classified as glioblastoma multiforme (GBM). Evaluation of the primary GBM, and its recurrence, for association with JCV showed the presence of the JCV early genome in both primary and recurrent samples, yet expression of T antigen was observed only in the recurrent tissue. Interestingly, the levels of several inducible transcription factors that have the ability to stimulate transcription of JCV promoter activity was found in the primary tumor, suggesting that postoperative induction of these regulatory proteins may lead to the reactivation of JCV gene expression that resulted in the development of the tumor in the brain (Del Valle et al., 2000).

In another case, clinical investigation of an immunocompetent patient with a neurologic disorder revealed the concurrence of multiple sclerosis plaques in the white matter and a glioblastoma multiforme in the thalamus. Histologic

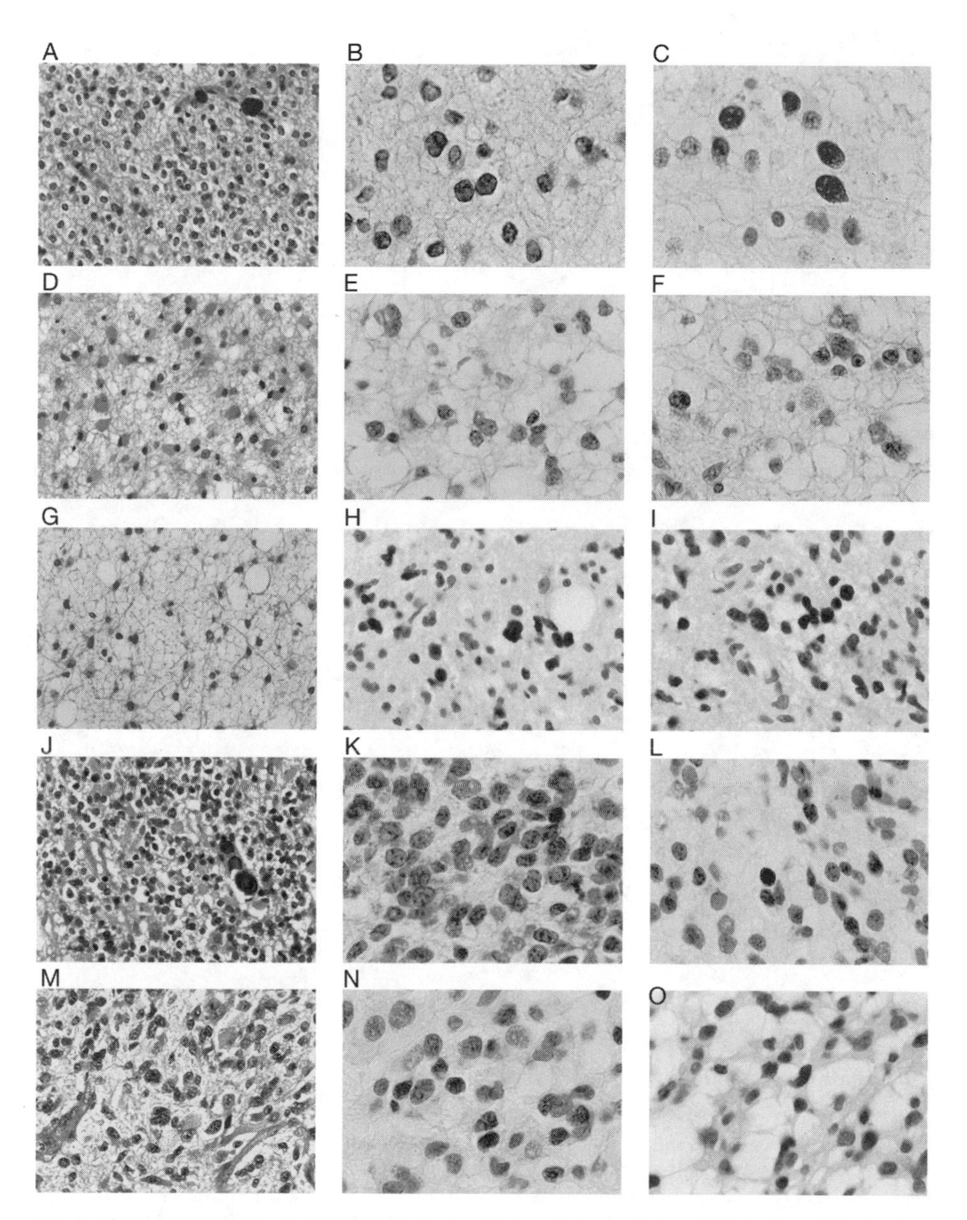

Figure 15.5. Detection of JCV T antigen and the cellular protein p53 in human glial origin tumors. Histologic evaluation and immunohistochemistry for JCV T antigen and the cellular tumor suppressor protein p53 in a number of human glial tumors is shown: oligodendroglioma (**A**, hematoxylin and eosin staining; **B**, immunohistochemical staining for T antigen; **C**, immunostaining for p53); gemistocytic astrocytoma (**D**–**F**); fibrillary astrocytoma (**G**–**I**); anaplastic oligodendroglioma (**J**–**L**); anaplastic astrocytoma (**M**–**O**). A, D, G, J, and M, hematoxylin and eosin, original magnification, ×400; B, C, E, F, H, I, K, L, N, O, hematoxylin counterstain, original magnification, ×1000. (See color plates.)

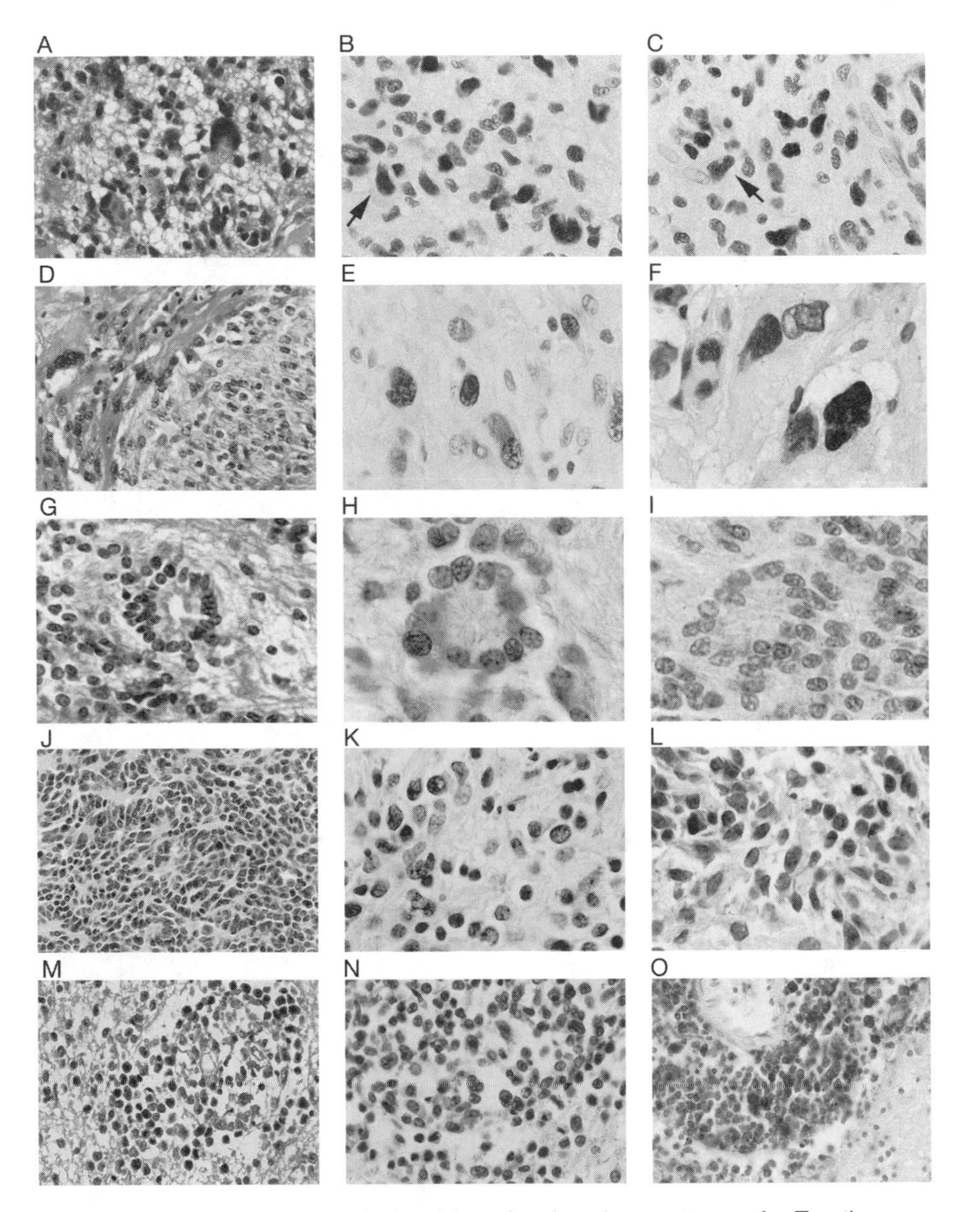

Figure 15.6. Immunohistochemical staining of various human tumors for T antigen expression. JCV T antigen has been detected in a variety of human tumors. Histology and immunohistochemistry for T antigen and cellular proteins are shown for the following tumors: glioblastoma multiforme (**A**, hematoxylin and eosin staining; **B**, immunohistochemical staining for T antigen; **C**, immunostaining for p53); gliosarcoma (**D**–**F**), ependymoma (**G**–**I**), medulloblastoma (**J**–**L**), B-cell lymphoma (**M**, hematoxylin and eosin staining; **N**, immunohistochemical staining for T antigen; **O**, immunostaining for the EBV protein latent membrane protein [LMP]). A, D, G, J, and M, hematoxylin and eosin staining, original magnification, ×400; B, C, E, F, K, L, N, and O, hematoxylin counterstain, original magnification, ×400; H and I, hematoxylin counterstain, original magnification ×1000. (See color plates.)

examination revealed the presence of nude axons and severe myelin loss in the plaque areas. The tumor was composed of a highly cellular neoplasm with atypical pleomorphic cells and areas of necrosis. PCR analysis of DNA from demyelinated and tumor areas using primers derived from specific regions of the JC virus genome revealed the presence of viral DNAs, corresponding to the viral early and late genes. Results from immunohistochemistry showed the detection of the viral early protein T antigen and the cellular tumor suppressor protein p53 in the nuclei of neoplastic cells. Interestingly, expression of T antigen, but not p53, was observed in the white matter areas juxtaposed to the multiple sclerosis lesions. These observations provide the first molecular and clinical evidence of an association of JCV with a brain tumor in a patient with multiple sclerosis and expression of JCV in white matter neurons.

Non-brain Tumors. In addition to glial cells (i.e., oligodendrocytes and astrocytes) and other neural cells including the external granular layer of the cerebellum, JCV has also been detected in B lymphocytes and mononuclear cells located within brain vascular spaces of PML patients, suggesting that JCV can infect B lymphocytes.

Given the presence of JCV in B lymphocytes and the oncogenic potential of this virus, a collection of well-characterized CNS lymphoma samples were tested for the presence of the JCV genome and expression of the viral T antigen by gene amplification and immunohistochemical techniques, respectively. Eighteen paraffin-embedded tumor samples were used for genomic DNA isolation and DNA amplification as well as immunohistochemical staining using specific antibodies. Results from gene amplification using a pair of primers that recognize the viral T antigen gene sequence and hybridization with a JCV-specific DNA probe showed that greater than 60% of samples contain the T antigen DNA sequence. Expression of T antigen was detected in 27% of CNS lymphomas. No evidence for the production of viral capsid proteins was obtained, ruling out productive replication of JCV in these cells. Imunohistologic examination of the tumors for the presence of Epstein-Barr virus (EBV), which is commonly associated with B-cell malignancies, revealed the presence of the viral latent membrane protein (LMP) in 16 samples. Four of the LMP-positive samples showed positive nuclear staining for T antigen. Altogether, this study shows that significant numbers of the CNS lymphomas contain JCV genomic DNA corresponding to the portion of the early coding sequence and that nearly 20% of these tumors express the viral oncoprotein T antigen in the nuclei of the cells. This observation is consistent with earlier ideas of the association of JCV in "rogue" cells, lymphocytes that exhibit chromosomal damage, presumably due to the inactivation of p53. The simultaneous presence of EBV and JCV in the tumor cells is an interesting observation, suggesting that cross-talk between JCV and EBV may occur during tumorigenesis of human CNS lymphoma.

JCV has also been detected in the gastrointestinal tract and solid non-neural tumors, including colorectal cancers (Laghi et al., 1999; Ricciardiello et al.,

2000). Also, examination of tissue, urine, and blood samples from patients with malignant and nonmalignant pleural disease revealed that 2 of 23 malignant mesotheliomas are positive for the JCV sequence, and JCV was detected in 7 out of 15 urinary sediments (Strizzi et al., 2000).

6. CONCLUDING REMARKS

The results of the many in vitro cell culture studies, in vivo animal models, and clinical sample studies described above strongly point to the involvement of JCV in human tumors of neural origin and open a new avenue of thought toward the association of JCV with non-neural tumors, in particular B-cell lymphoma, and perhaps colorectal tumors. JCV is the established etiologic agent of the human demyelinating disease PML. This virus induces demyelination of the CNS upon productive replication in and cytolytic destruction of oligodendrocytes, the myelin-producing cells of the CNS. Also, JCV can abortively replicate in astrocytes and results in the formation of bizarre astrocytic cells in PML brains. Productive replication of JCV in oligodendrocytes begins with expression of the viral early protein T antigen, which can stimulate viral DNA replication and late gene expression, leading to the formation of virions. Several rounds of productive infection in cells eventually results in a growing PML lesion. Should expression of JCV T antigen occur in the absence of viral DNA replication and/or late gene expression, the result would be the accumulation of T antigen in the affected cells. Under these circumstances T antigen may interact with host factors that control cell proliferation, such as p53 and pRb, and result in inactivation of their function and uncontrolled proliferation.

The expression of JCV T antigen in the absence of productive viral replication may stem from various biologic events, including mutation in the key regions of JCV that are important for initiation of viral DNA replication or the synthesis of viral late gene products, the capsid proteins, and the lack of many other regulatory factors that are crucial for viral replication in these cells. The availability of animal models, in particular transgenic mice, should help to decipher the molecular mechanism that dictates the involvement of JCV, upon its reactivation, in inducing PML and allow further study of the association of JCV with human brain tumors.

The important question that is periodically raised by many virologists and cancer researchers is whether JCV actually induces brain tumors in humans or if JCV gains an opportunity to express its genome in tumor cells and tissues. While the answer to this question may not be easily addressed in a clinical setting, one may refer to well-controlled animal models in which expression of the JCV genome induces tumors in vivo. In particular, earlier studies in owl and squirrel monkeys are important in showing that the exposure of these animals to JCV results in the development of glioblastoma.

It is highly likely that the lack of sensitive techniques during the last two decades has hampered many efforts to study the association of JCV with human cancer. However, in light of new data, it is possible that by using modern techniques many laboratories may put forth major efforts toward studying the role of human polyomaviruses, in particular JCV, in various tumors.

ACKNOWLEDGMENTS

The authors thank past and present members of the Center for Neurovirology and Cancer Biology for their insightful discussion and sharing of ideas and reagents. We also thank C. Schriver for editorial assistance and manuscript preparation.

REFERENCES

Berger JR, Concha M (1995): Progressive multifocal leukoencephalopathy: The evolution of a disease once considered rare. J Neurovirol 1:5–18.

Boldorini R, Caldarelli-Stefano R, Monga G, Zocchi M, Mediati M, Tosoni A, Ferrante P (1998): PCR detection of JC virus DNA in the brain tissue of a 9 year-old child with pleomorphic xantoastrocytoma. J Neurovirol 2:242–245.

Bollag B, Chuke WF, Frisque RJ (1989): Hybrid genomes of the polyomaviruses JC virus, BK virus, and simian virus 40: Identification of sequences important for efficient transformation. J Virol 63:863–872.

Caldarelli-Stefano R, Boldorini R, Monga G, Meraviglia E, Zorini EO, Ferrante P (2000): JC virus in human glial-derived tumors. Hum Pathol 31:394–395.

Castaigne P, Rondot P, Escourolle R, Ribadeau Dumas JL, Cathala F, Hauw JJ (1974): Progressive multifocal leukoencephalopathy and multiple gliomas. Rev Neurol Paris 9–10:379–392.

Davies JA, Hughes JT, Oppenheimer DR (1973): Richardson's disease (progressive multifocal leukoencephalopathy). Q J Med 167:481–501.

Del Valle L, Azizi SA, Krynska B, Enam S, Croul SE, Khalili K (2000): Reactivation of human neurotropic JC virus expressing oncogenic protein in a recurrent glioblastoma multiforme. Ann Neurol 48:932–936.

Del Valle L, Gordon J, Assimakopolou M, Enam S, Geddes JF, Varakis J, Katsetos C, Croul SE, Khalili K (2001): Detection of JC virus DNA sequences and expression of the viral regulatory protein, T-antigen, in tumors of the central nervous system. Cancer Res 61:4287–4293.

Dyson N, Bernards R, Friend SH, Gooding LR, Hassell JA, Major EO, Pipas JM, van Dyke T, Harlow E (1990): Large T antigens of many polyoma viruses are able to form complexes with the retinoblastoma protein. J Virol 64:1353–1356.

Dyson N, Buchkovich K, Whyte P, Harlow E (1989): The cellular 107K protein that binds to adenovirus E1A also associates with the large T antigens of SV40 and JC virus. Cell 58:249–255.

Egan JD, Ring BL, Reding MJ, Wells IC, Shuman RM (1980): Reticulum cell sarcoma and progressive multifocal leukoencephalopathy following renal transplantation. Transplantation 29:84–86.

Fanning E, Knippers R (1992): Structure and function of simian virus 40 large tumor antigen. Annu Rev Biochem 61:55–85.

Fareed GC, Takemoto KK, Gimbrone MA Jr (1978): Interaction of simian virus 40 and human papovaviruses, BK and JC, with human vascular endothelial cells. In Schlessinger D, Ed; Microbiology 1978; American Society for Microbiology: Washington, DC, pp 427–431.

Franks RR, Rencic A, Gordon J, Zoltick PW, Curtis M, Knobler RL, Khalili K (1996): Formation of undifferentiated mesenteric tumors in transgenic mice expressing human neurotropic polyomavirus early protein T-antigen. Oncogene 12:2573–2578.

Frisque RJ, White FA III (1992): The molecular biology of JCV, causative agent of progressive multifocal leukoencephalopathy. In RP Roos, Ed; Molecular Neurovirology; Humana Press: Totowa, NJ, pp 25–158.

Gallia GL, Gordon, J, Khalili K (1998a): Tumor pathogenesis of human neurotropic JC virus in the CNS. J Neurovirol 4:175–181.

Gallia GL, Safak M, Khalili K (1998b): Interaction of single-stranded DNA binding protein, Purα, with human polyomavirus, JCV, early protein, T-antigen. J Biol Chem 273:32662–32669.

GiaRusso MH, Koeppen AH (1978): Atypical progressive multifocal leukoencephalopathy and primary cerebral malignant lymphoma. J Neurol Sci 35:391–398.

Gordon J, Del Valle L, Otte J, Khalili K (2000): Pituitary neoplasia induced by expression of human neurotropic virus JCV, early genome in transgenic mice. Oncogene 19:4840–4846.

Gullota F, Masini T, Scarlato G, Kuchelmeister K (1992): Progressive multifocal leukoencephalopathy and gliomas in a HIV negative patient. Patho Res Pract 188:964–972.

Haggerty S, Walker DL, Frisque RJ (1989): JC virus–siman virus 40 genomes containing heterologous regulatory signals and chimeric early regions: Identification of regions restricting transformation by JC virus. J Virol 63:2180–2190.

Harris KF, Christensen JB, Radany EH, Imperiale MJ (1998): Novel mechanisms of E2F induction by BK virus large T antigen: Requirement of both the pRb-binding and the J domains. Mol Cell Biol 18:1746–1756.

Ho K-C, Garancis JC, Paegle RD, Gerber MA, Borkowski WJ (1980): Progressive multifocal leukoencephalopathy and malignant lymphoma of the brain in a patient with immunosuppressive therapy. Acta Neuropathol 52:81–83.

Khalili K, Krynska B, Del Valle L, Katsetos C, Croul SE (1999): Medulloblastomas and the human neurotropic polyomavirus, JCV. Lancet 353:1152–1153.

Knudsen ES, Wang JYJ (1998): Hyperphosphorylated p107 and p130 bind to T-antigen: Identification of critical regulatory sequence present in Rb but not in p107/p130. Oncogene 16:1655–1663.

Krynska B, Del Valle L, Croul S, Gordon J, Katsetos K, Carbone M, Giordano A, Khalili K (1999a): Detection of human neurotropic JC virus DNA sequence and expression of the viral oncogenic protein in pediatric medulloblastomas. Proc Natl Acad Sci USA 96:11519–11524.

Krynska B, Otte J, Franks R, Khalili K, Croul C (1999b): Human ubiquitous JCV-CY T-antigen gene induces brain tumors in experimental animals. Oncogene 18:39–46.

Laghi L, Randolph AE, Chauman DP, Marra G, Major EO, Neel JV, Boland CR (1999): JC virus DNA is present in the mucosa of the human colon and in colorectal cancer. Proc Natl Acad Sci USA 96:7484–7489.

Lazutka JR, Neel JV, Major EO, Dedonyte V, Mierauskine J, Slapsyte G, Kesminiene A (1996): High titles of antibodies to two human polyomaviruses, JCV and BKV, correlate with increased frequency of chromosomal damage in human lymphocytes. Cancer Lett 109:177–183.

Levine AJ (1997): p53, the cellular gatekeeper for growth and division. Cell 88:323–331.

Liberski PP, Alwasiak J, Wegrzyn Z (1982): Atypical progressive multifocal leukoencephalopathy and primary cerebral lymphoma. Neuropat Pol 20(3–4):413–419.

London WT, Houff SA, Madden DL, Fuccillo DA, Gravell M, Wallen WC, Palmer AE, Sever JL, Padgett BL, Walker DL, Zu Rhein GM, Ohashi T (1978): Brain tumors in owl monkeys inoculated with a human polyomavirus (JCV). Science 201:1246–1249.

London WT, Houff SA, McKeever PE, Wallen WC, Sever JL, Padgett BL, Walker DL (1983): Viral-induced astrocytomas in squirrel monkeys. In Polyomaviruses and Human Neurological Diseases; Alan R. Liss, Inc: New York, pp 227–237.

Ludlow JW (1993): Interactions between SV40 large-tumor antigen and the growth suppressor proteins pRB and p53. FASEB J 7:866–871.

Ludlow JW, Skuse GR (1995): Viral oncoprotein binding to pRb, p107, p130, and p300. Virus Res 35:113–121.

Mandl C, Walker DL, Frisque RJ (1987): Derivation and characterization of POJ cells, transformed human fetal glial cells that retain their permissivity for JC virus. J Virol 61:755–763.

Manfredi JJ, Prives C (1994): The transforming activity of simian virus 40 large tumor antigen. Biochim Biophys Acta 1198:65–83.

Major EO, Amemiya K, Tornatore CS, Houff SA, Berger JR (1992): Pathogenesis and molecular biology of progressive multifocal leukoencephalopathy, the JC virus induced demyelinating disease of the human brain. Clin Micro Rev 5(1):49–73.

Neel JV, Major EO, Awa AA, Glover T, Burgess A, Traub R, Curfman B, Satoh C (1996): Hypothesis: "Rogue" cell-type chromosomal damage in lymphocytes is associated with infection with the JC human polyoma virus and its implications for oncogenesis. Proc Natl Acad Sci USA 93:2690–2695.

Ohsumi S, Ikehara I, Motoi M, Ogawa K, Nagashima K, Yasui K (1985): Induction of undifferentiated brain tumors in rats by a human polyomavirus (JCV). Jpn J Cancer Res 76:429–431.

Ohsumi S, Motoi M, Ogawa K (1986): Induction of undifferentiated tumors by JC virus in the cerebrum of rats. Acta Pathol Jpn 36:815–825.

Rencic A, Gordon J, Otte J, Curtis M, Kovatich A, Zoltick P, Khalili K, Andrews D (1996): Detection of JCV DNA sequence and expression of the viral oncoprotein, tumor antigen, in brain of immunocompetent patient with oligoastrocytoma. Proc Natl Acad Sci USA 93:7352–7357.

Ricciardiello L, Laghi L, Ramamirthan P, Chang CL, Chang DK, Randolph AE, Boland CR (2000): JC virus DNA sequences are frequently present in the human upper and lower gastrointestinal tract. Gastroenterology 119:1228–1235.

Richardson EP Jr (1961): Progressive multifocal leukoencephalopathy. N Engl J Med 265:815–823.

Rushton JJ, Srinivasan A, Pipas JM, Robbins PD (1997): Simian virus 40 T antigen can regulate p53-mediated transcription independent of binding p53. J Virol 71: 5620–5623.

Safak M, Gallia GL, Ansari SA, Khalili K (1999): Physical and functional interaction between the Y-box binding protein YB-1 and human polyomavirus JV virus large T-antigen. J Virol 73:10146–10157.

Shintaku M, Matsumoto R, Sawa J, Nagashima K (2000): Infection with JC virus and possible dysplastic ganglion-like transformation of the cerebral cortical neurons in a case of progressive multifocal leukoencephalopathy. J Neuropathol Exp Neurol 59:921–929.

Sima AAF, Finkelstein SD, McLachlan DR (1983): Multiple malignant astrocytomas in a patient with spontaneous progressive multifocal leukoencephalopathy. Ann Neurol 14:183–188.

Small JA, Khoury G, Jay G, Howley PM, Scangos GA (1986a): Early regions of JC virus and BK virus induce distinct and tissue-specific tumors in transgenic mice. Proc Natl Acad Sc USA 83:8288–8292.

Small JA, Scangos GA, Cork L, Jay G, Khoury G (1986b): The early region of human papovavirus JC induces dysmyelination in transgenic mice. Cell 46:13–18.

Strizzi L, Vianale G, Giuliiano M, Sacco R, Tassi F, Chiodera P, Casalini P, Procopio A (2000): SV40, JC and BK expression in tissue, urine, and blood samples from patients with malignant and non-malignant pleural disease. Anticancer Res 20:885–889.

Stubal H, Zalvide J, De Caprio JA (1996): Simian virus 40 large T antigen alters the phosphorylation state of the Rb-related proteins p130 and p107. J Virol 70:2781–2788.

Theile M, Grabowski G (1990): Mutagenic activity of BKV and JCV in human and other mammalian cells. Arch Virol 113:221–233.

Walker DL, Padgett BL (1978): Biology of JC virus, a human papovairus. In Schlessinger D, Ed; Microbiology 1978; American Society for Microbiology: Washington, DC, pp 432–434.

Walker DL, Padgett BL, Zu Rhein GM, Albert AE, Marsh RF (1973): Human papova virus (JC): Induction of brain tumors in hamsters. Science 181:674–676.

Williamson PJ, Allan NC, McIntyre MA (1989): Cerebral Hodgkin's disease and progressive multifocal leukoencephalopathy. Clin Lab Haematol 11:281–285.

Zu Rhein GM (1983): Studies of JCV-induced nervous system tumors in the Syrian hamster: A review. In Polyomaviruses and Human Neurological Diseases; Alan R. Liss, Inc: New York, pp 205–221.

Zu Rhein GM, Varakis JN (1979): Perinatal induction of medulloblastomas in Syrian golden hamsters by a human polyomavirus (JC). Natl Cancer Inst Monogr 51:205–208.

16

EVIDENCE FOR BK VIRUS AS A HUMAN TUMOR VIRUS

ALFREDO CORALLINI, PH.D., MAURO TOGNON, PH.D.,
MASSIMO NEGRINI, PH.D., and
GIUSEPPE BARBANTI-BRODANO, PH.D.

1. INTRODUCTION

The human polyomavirus BK (BKV) is recognized as a ubiquitous human commensal that is found in many normal human tissues, mostly from immune-compromised individuals (Coleman et al., 1980). BKV is oncogenic in rodents and is able to transform in vitro cells from different species to the neoplastic phenotype. Due to these characteristics, BKV is considered as a potential viral agent in the etiology of human tumors, although formal proof for its role in human oncogenesis remains to be obtained. In this chapter we review BKV, including its general properties, natural history of infection, experimental transformation, and oncogenicity, as well as the evidence for an association between BKV and human tumors.

2. GENERAL CHARACTERISTICS OF BKV

Several reviews of BKV have been published (Padgett and Walker, 1976; Walker and Frisque, 1986; Yoshiike and Takemoto, 1986; Monini et al., 1995b; Barbanti-Brodano et al., 1998; Lednicky and Butel, 1999; Imperiale, 2000;

Human Polyomaviruses: Molecular and Clinical Perspectives, Edited by Kamel Khalili and Gerald L. Stoner.
ISBN 0-471-39009-7

Shah, 2000). BKV belongs to the genus *Polyomavirus* of the family Papovaviridae, an acronym proposed by Melnick and obtained by fusing the names of the three representative viruses *Pa*pilloma, *Po*lyoma, and *Va*cuolating agent. The virion, composed of 72 skewed capsomeres, is a 45 nm icosahedral particle with a density of 1.34–1.35 g/cm^3, and the genome is a circular, double-stranded DNA molecule. Like JC virus (JCV) and SV40, two other polyomaviruses, the BKV genome codes for six viral proteins, two from the early region and four from the late region. The two early proteins are nonstructural polypeptides, the large tumor antigen (TAg) and the small tumor antigen (tAg). The four late proteins are the capsid proteins VP1, VP2, and VP3 and the agnoprotein, probably involved in processing of late mRNA and in assembly of viral particles (Alwine, 1982; Hay et al., 1982; Ng et al., 1985). The early and late genes are transcribed from different DNA strands of the viral genome.

The transcription is divergent from the origin of replication, located in the regulatory region, and terminates within DNA sequences containing the polyadenylation signals. The TAg of BKV strongly cross reacts with sera to JCV and SV40 TAgs (Takemoto and Mullarkey, 1973; Walker et al., 1973). Although only a little cross reactivity is observed in most structural antigenic determinants, a genus-specific capsid antigen, located on viral peptide VP1, has been identified (Shah et al., 1977). The DNA sequences of BKV share 75% homology with JCV (Frisque et al., 1984), while the homology with SV40 is 70% (Yang and Wu, 1979). The greatest homology is found in the early region, whereas a low homology is detected in the regulatory region, which represents the most variable sequence in the viral genome. This probably reflects adaptation to in vitro cell culture (Martin et al., 1983; Shinohara et al., 1989; Markowitz et al., 1991; Rubinstein et al., 1991; Yogo et al., 1993), and most laboratory strains may have evolved from a common, natural archetype (Rubinstein et al., 1987; Yogo et al., 1990; Flaegstad et al., 1991; Negrini et al., 1991; Tominaga et al., 1992). However, analyses of independent isolates by either direct cloning or sequencing of products obtained by polymerase chain reaction (PCR) amplification show that different arrangements of the regulatory region are often detected in vivo (Loeber and Dörries, 1988; Sundsfjord et al., 1990; Yogo et al., 1991; Ault and Stoner, 1992, 1993). Selection of variants with a particular cell specificity or transformation potential was proposed as a possible outcome of such variability (Loeber and Dörries, 1988; Negrini et al., 1990; Yogo et al., 1991).

3. BKV INFECTION IN HUMANS

Epidemiology

BKV was first isolated in 1971 from the urine of a renal allograft recipient who developed ureteric stenosis (Gardner et al., 1971). Subsequently, it was found that BKV is ubiquitous and has a worldwide distribution in the human population (Padgett and Walker, 1976). Primary infection takes place during childhood. At 3 years of age BKV antibodies are detected in 50% of children, and, with the exception of some segregated populations living in isolated regions of Brazil, Paraguay, and Malaysia (Brown et al., 1975), almost all individuals are infected by the age of 10 years (Gardner, 1973; Mantyjarvi et al., 1973; Shah et al., 1973; Portolani et al., 1974).

Primary Infection, Latency, Reactivation, Relationship to Inflammatory Pathology, and Route of Transmission

Primary infection with BKV is usually inapparent and only occasionally may be accompanied by mild respiratory illness or urinary tract disease (Hashida et al., 1976; Goudsmit et al., 1981; Mininberg et al., 1982; Padgett et al., 1983). During primary infection, viremia occurs and the virus spreads to several organs of the infected individual, where it remains in a latent state. Clinical studies, carried out in immunosuppressed and immunocompetent persons, indicate that the reactivation of BKV from latency is mainly associated with immunologic impairment (Padgett and Walker, 1976).

Virus isolation and Southern blot hybridization analyses established that the kidney is the main site of BKV latency in healthy individuals (Heritage et al., 1981; Chesters et al., 1983; Grinnel et al., 1983; McCance, 1983). By these technical approaches, BKV sequences were also detected in other organs, such as liver, stomach, lungs, parathyroid glands, and lymph nodes (Israel et al., 1978; Pater et al., 1980). Polyomavirus virions were detected in peripheral blood cells (PBC) (Lecatsas et al., 1976a; Schneider and Dörries, 1993), whereas Goudsmit et al. (1982) have demonstrated a role of BKV in acute respiratory tract disease. The detection of BKV DNA in tonsils suggests that the oropharynx may be the initial site of BKV infection. In addition, BKV replication was shown in human lymphocytes (Portolani et al., 1985); therefore, the lymphoid tissue of Waldayer's ring may be involved in primary BKV infection. Infected lymphocytes could then carry BKV to the bloodstream, allowing virus transport to other organs.

PCR studies of BKV disclosed the presence of its DNA- and RNA-specific sequences in a variety of normal human tissues. BKV was detected by PCR in brain and peripheral blood mononuclear cells (PBMC) (Elsner and Dörries, 1992; Dörries et al., 1994; De Mattei et al., 1994, 1995). The prevalence of positive samples ranged from 30% in the brains of patients with neurologic diseases (Elsner and Dörries, 1992) to 100% in normal brains (De Mattei et al., 1994, 1995). In the study by Elsner and Dörries (1992), the great majority of the samples positive for JCV were found to be co-infected by BKV, suggesting a specific competence of certain brain cells for infection by human polyomaviruses or a need for cooperation or interference between the two viruses to establish latent infection in the brain. The amount of latent BKV DNA in the brain, corresponding to 1–10 genome equivalents in 100 cells, was less than the amount of latent JCV DNA (1–500 genome equivalents in 100 cells), suggesting a reduced viral activity of BKV compared with JCV in the central nervous system (CNS). Cloning of BKV DNA sequences from latently infected brain tissues led to isolation of full-length viral genomes, indicating that polyomavirus DNA is usually in an episomal state in human CNS. BKV DNA sequences were detected by PCR in kidney, bladder, prostate, uterine cervix, vulva, lips, and tongue (Monini et al., 1996b). The frequency of positive samples ranged from 40% to 83% in different tissues. Moreover, the percentage of samples positive for viral sequences in lymphocytes was 94.2% (Dörries et al., 1994). The results, obtained by PCR, indicate that BKV can establish latent infection in many more organs and cells than previously thought. This evidence may have important consequences on the routes and mechanisms of virus transmission, as well as on the epidemiology and the reactivation of BKV latent infection.

Inflammatory syndromes affecting several organs were detected after BKV reactivation in both immunosuppressed and immunocompetent individuals. Reactivation of BKV was detected in the urine of renal and bone marrow transplant recipients undergoing immunosuppressive therapy (Coleman et al., 1973; Lecatsas et al., 1973; Hogan et al., 1980; Traystman et al., 1980; O'Reilly et al., 1981; Gardner et al., 1984; Rice et al., 1985; Arthur et al., 1986, 1988; Apperley et al., 1987; Chan et al., 1994), in the urine of pregnant women (Coleman et al., 1977, 1980, 1983; Shah et al., 1980), and in patients with both hereditary and acquired immunodeficiency syndromes (Takemoto et al., 1974; Rhiza et al., 1978a; Snider et al., 1983; Flaegstad et al., 1988; Wiley et al., 1988; Vazeux et al., 1990; Gillespie et al., 1991; Quinlivan et al., 1992; Markowitz et al., 1993; de Silva et al., 1995). BKV was also detected by PCR in 3.8% of urine samples of 211 community children, in 6% of 33 HIV-1–infected children (Di Taranto et al., 1997), and in 62% of adult patients with AIDS (Jin et al., 1995). Cubukcu-Dimopulo et al. (2000) reported a 14-year-old boy with AIDS who developed a BKV infection of the kidney and lung that progressed to diffuse alveolar damage and death. In a hemophilia patient with AIDS, a severe systemic disease due to BKV reactivation was described, consisting of tubulointerstitial nephropathy, interstitial desquamative pneumonitis, and subacute meningoencephalitis (Vallbracht et al., 1993). In another patient with AIDS who was affected by encephalitis, nephritis, and retinitis, BKV DNA was found in the brain, eye tissue, cerebrospinal fluid, urine, and PBMC (Hedquist et al., 1999; Bratt et al., 1999).

Interstitial nephritis due to BKV occurred in 2.5% and ureteral obstruction in 71.5% of patients receiving renal transplants; in some cases viral infection induced transplant dysfunction and graft rejection (Howell et al., 1999). An association between hemorrhagic cystitis and BKV was shown in bone marrow transplant recipients (Apperley et al., 1987; Arthur et al., 1988; Azzi et al., 1994). The incidence of the disease was from 20% to 44%. BKV reactivation occcurred in 55% of BKV-seropositive patients, and BKV viruria began 2–8 weeks after transplantation (Arthur et al., 1988). Hemorrhagic cystitis and encephalitis are also caused by BKV in immunocompetent people (Saitoh et al.,

1993; Vago et al., 1996; Voltz et al., 1997; Andrey et al., 1997). Furthermore, a reversible BKV encephalitis has been reported in an apparently healthy patient, possibly due to a late primary infection (Voltz et al., 1997). BKV reactivation was also demonstrated in patients affected by a number of other diseases, only some of which are related to immunosuppression: neoplastic disease (lymphoma and carcinoma) (Hogan et al., 1983), systemic lupus erythematosus (Taguchi and Nagaki, 1978; Sundsfjord et al., 1999), various forms of anemia (Lecatsas et al., 1976b, 1997; Lecatsas and Bernard, 1982), nephrotic syndrome (Nagao et al., 1982), and Guillain-Barrè syndrome (van der Noordaa and Wartheim-van Dillen, 1977).

Little is known about the modality of virus transmission and the ways of access to susceptible tissues, though induction of upper respiratory disease by BKV and detection of latent BKV DNA in tonsils (Goudsmit et al., 1982) indicate a possible oral or respiratory route of transmission. The identification of polyomaviruses in the urine of pregnant women suggested reactivation from latency during pregnancy with the possibility of congenital transmission. However, early reports showing the presence of virus-specific IgM in umbilical cord sera (Taguchi et al., 1975; Rhiza et al., 1978b) were not confirmed by other studies (Borgatti et al., 1979; Coleman et al., 1980; Shah et al., 1980; Daniel et al., 1981). In a recent investigation, the presence of BKV DNA in both maternal and fetal tissues was demonstrated by PCR, suggesting transplacental BKV infection (Pietropaolo et al., 1998).

4. IN VITRO TRANSFORMATION BY BKV

Transformation of Rodent Cells

BKV, complete BKV DNA, and subgenomic BKV DNA fragments containing the early region are able to transform embryonic fibroblasts and cells cultured from kidney and brain of hamster, mouse, rat, rabbit, and monkey (Major and Di Mayorca, 1973; Portolani et al., 1975; van der Noordaa, 1976, 1979; Tanaka et al., 1976; Takemoto and Martin, 1976; Costa et al., 1977; Mason and Takemoto, 1977; Seehafer et al., 1977; Bradley and Dougherty, 1978; Portolani et al., 1978; Seehafer et al., 1979; Grossi et al., 1982b; Watanabe and Yoshiike, 1982a). The efficiency of in vitro transformation depends on the genetic features of the viral strain and does not necessarily correspond to the oncogenic potential of the viral isolate (Watanabe and Yoshiike, 1982a).

A recombinant DNA containing the BKV TAg gene and c-Ha-*ras* oncogene (pBK/c-*ras*A) induced neoplastic transformation of early-passage hamster embryo cells with greater efficiency than each of the two genes transfected independently. In addition, cells transformed by pBK/c-*ras*A have a more definite transformed phenotype and more advanced growth characteristics than cells transformed separately by each of the two genes (Pagnani et al., 1988). Transformation of rat pancreatic islet cells, a natural target of BKV tumorigenesis in rodents, has been described (Haukland et al., 1992). However, human pancreatic islet cells persistently infected with BKV did not display a transformed phenotype (van der Noordaa et al., 1986), although BKV is frequently present in human islet cell tumors (Corallini et al., 1987b).

Transformation of Human Cells

Transformation of human cells by BKV is inefficient and often abortive (Shah et al., 1976; Portolani et al., 1978). BKV-infected or -transfected human cells generally do not display a completely transformed phenotype, characterized by immortalization, anchorage independence, and tumorigenicity in nude mice, although they show morphologic alterations and an increased life span (Purchio and Fareed, 1979; Grossi et al., 1982a). A fully transformed phenotype was observed in human embryo kidney (HEK) cells transfected with a recombinant plasmid containing BKV early region and the adenovirus 12 E1A gene (Vasavada et al., 1986). These cells grow as a continuous cell line, suggesting that, at least in human cells, BKV TAg is competent to contribute only a partially transformed phenotype and must interact with other oncogene functions to induce a complete transformation. The recombinants pBK/c-*ras*A and pBK/c-*myc* induced transformation of human embryo fibroblasts and HEK cells, but transformed cells were not immortalized (Pater and Pater, 1986; Corallini et al., 1991).

Tumorigenic cell lines were established only from transformed human glial brain cells persistently infected by BKV or after transfection of pBK/c-*ras*A or pBK/c-*myc* in HEK-T cells (human embryo kidney cells from a fetus with Turner's syndrome) (Takemoto et al., 1979; Corallini et al., 1991). Fetal brain cells had all the characteristics of transformed cells and retained viral DNA in an episomal state but, unlike HEK-T cells, they were negative for TAg expression (Takemoto et al., 1979). The reason for the higher susceptibility of HEK-T cells to transformation is unclear, but it may depend on genes suppressing the expression or the activity of BKV Tag, c-*ras* or c-*myc* or on the X chromosome missing in HEK-T cells. Indeed, cytogenetic analysis indicates that loss of X and Y chromosomes is a frequent and nonrandom alteration in several human solid tumors and leukemias, suggesting that both sex chromosomes probably harbor tumor suppressor genes (Sandberg, 1983, 1985; Wolman et al., 1988). Moreover, experimental evidence exists that X and Y chromosome may regulate cellular growth rate because their loss in cultured human fibroblasts confers on cells a selective advantage leading to a greater cloning efficiency compared with normal diploid cells (Barlow, 1972; Sandberg, 1985). In addition, clonal cytogenetic alterations of the long arm of chromosome 6 were detected in HEK-T cells transformed by pBK/c-*ras*A recombinants (Corallini et al., 1991). Further evidence that BKV-dependent transformation can be abolished by human chromosomes has recently been presented. BKV-transformed mouse and hamster cells were reduced or suppressed in both anchorage independence and tumorigenicity after transferring of human chromosomes 6 or 11

(Negrini et al., 1992; Gualandi et al., 1994; Sabbioni et al., 1994). Interestingly, one clone that lost the tumorigenic phenotype, but continued to grow in soft agar, had deleted the short arm of chromosome 11 while maintaining the long arm intact, suggesting that different human genes may control separate functions in BKV transformation (Negrini et al., 1992).

Presence and State of BKV DNA in Transformed Cells

The presence and physical state of BKV DNA in transformed cells has been studied by several authors (Howley and Martin, 1977; Chenciner et al., 1980; Beth et al., 1981; Meneguzzi et al., 1981; Grossi et al., 1982c). BKV DNA is generally present in an integrated state in rodent cells, often in a tandem array, together with variable amounts of free episomal viral genomes. In one hamster osteosarcoma induced by BKV, viral DNA was exclusively found as monomeric and polymeric extrachromosomal defective genomes (Yogo et al., 1980). Unlike in rodent cells, in BKV-transformed human cells viral DNA sequences were found mostly free in an episomal state (Purchio and Fareed, 1979; Takemoto et al., 1979; Grossi et al., 1982a).

5. BK VIRUS EXPERIMENTAL ONCOGENICITY

The oncogenic potential of BKV has been well documented by several reports. Young or newborn mice, rats, and hamsters, inoculated with BKV through different routes, developed tumors. BKV-induced tumors contained integrated and free BKV DNA sequences and expressed TAg (Corallini et al., 1977, 1978, 1982; Uchida et al., 1979; Chenciner et al., 1980). Fusion of tumor cells with permissive monkey or human cells yielded infectious virus (Corallini et al., 1977, 1978). The frequency of tumor induction in hamsters is strictly dependent on the route of injection. In fact, BKV was weakly oncogenic when inoculated subcutaneously (Nase et al., 1975; Shah et al., 1975; van der Noordaa, 1976; Corallini et al., 1977), but induced tumors in the range of 73–88% of hamsters inoculated intracerebrally or intravenously (Costa et al., 1976; Uchida et al., 1976, 1979; Corallini et al., 1977, 1978, 1982). Tumors induced in BKV-injected hamsters belong to a variety of histotypes, such as ependymoma, neuroblastoma, pineal gland tumors, tumors of pancreatic islets, fibrosarcoma, osteosarcoma (Shah et al., 1975; Nase et al., 1975; van der Noordaa, 1976; Costa et al., 1976; Uchida et al., 1976, 1979; Dougherty, 1976; Greenlee et al., 1977; Corallini et al., 1977, 1978, 1982; Watanabe et al., 1979, 1982b; Noss et al., 1981; Noss and Stauch, 1984). The same tumors were induced by BKV inoculated intravenously in hamsters immunosuppressed with antilymphocyte serum or methylprednisolone acetate alone or in association with γ-radiation (Corallini et al., 1982).

In both immunocompetent and immunosuppressed hamsters, however, ependymomas, tumors of pancreatic islets, and osteosarcomas were the most frequent histotypes, suggesting that BKV has a tropism for specific organs. Tumors induced by BKV in mice and rats were fibrosarcoma, liposarcoma, ostesarcoma, nephroblastoma, glioma, and choroid plexus papilloma, the latter arising only in mice (Corallini et al., 1977; Noss et al., 1981; Noss and Stauch, 1984). Moreover, transgenic mice expressing BKV TAg developed hepatocellular carcinoma, renal tumors, and lymphoproliferative disease (Small et al., 1986; Dalrymple and Beemon, 1990). Gardner's BKV strain seems to be more oncogenic than other isolates, such as MM, BKV-IR, or RF (Costa et al., 1976; Dougherty, 1976; Caputo et al., 1983). It has been shown that the induction of different tumors may reflect the presence of several viral variants in the same inoculum (Uchida et al., 1979). In particular, an insulinoma-inducing variant has been associated with a viable deletion mutant originated in a Gardner BKV stock after several passages in culture (Watanabe et al., 1979, 1982b). A key role for viral genetic heterogeneity is further suggested by the oncogenic properties of BKV-IR, a variant rescued from a human insulinoma (Caputo et al., 1983) and associated with other human tumors (Negrini et al., 1990). Tumors induced by BKV-IR in hamsters develop at a lower frequency but display a more malignant phenotype than tumors induced by wild-type BKV (Negrini et al., 1990).

Purified BKV DNA is not oncogenic after intravenous or subcutaneous inoculation and induces tumors at a very low frequency when inoculated intracerebrally in rodents (Corallini et al., 1982). It displays, however, a strong synergism with activated oncogenes. The recombinant pBK/c-*ras*A, made up of the BKV TAg and the activated c-Ha-*ras* oncogene, produced highly undifferentiated, rapidly growing, malignant sarcomas by subcutaneous inoculation in newborn hamsters, whereas the two genes independently or BKV DNA linked to normal c-Ha-*ras* protoncogene were not tumorigenic (Corallini et al., 1987a). The same recombinant, pBK/c-*ras*A induced brain tumors upon intracerebral inoculation in newborn hamsters (Corallini et al., 1988).

BKV DNA sequences were found in all tumors examined, either intregrated into the cell genome or in an episomal state. The quantity of viral DNA present varied from tumor to tumor, and no generalized pattern was described (Yogo et al., 1980). The amount ranged from 1 to about 80 viral genome equivalents per diploid cell genome (Chenciner et al., 1980; Grossi et al., 1981; Yogo et al., 1980, 1981). Viral genomes were randomly integrated, and several different integration sites were utilized in the cellular and viral DNA. Tandemly integrated, full-length genomes were observed either with or without free episomal DNA sequences in all tumors of different histotypes (Chenciner et al., 1980; Corallini et al., 1982). Analysis of the mode of integration of viral DNA in virus-induced tumors offered a valuable tool to investigate the problem of the clonal origin of tumor cells. Chenciner et al. (1980) observed that hybridization patterns obtained in tumor tissues were more complex than hybridization patterns of cultured tumor cells and that single-cell clones isolated from such cultured tumor cells had even simpler patterns. These results are compatible with a polyclonal origin of BKV-induced tumors and with selection of certain types of tumor cells during cell culture.

6. ROLE OF BKV LARGE T AND SMALL T ANTIGENS IN TRANSFORMATION AND ONCOGENICITY

It has been demonstrated that BKV TAg, like JCV and SV40 TAg, interacts with p53 and p105RB1 tumor suppressor proteins (Dyson et al., 1990; Kang and Folk, 1992; Harris et al., 1996; Shivakumar and Das, 1996). Furthermore, an interaction with the RB-related p107 protein has been described for TAg of several transforming polyomaviruses, including BKV (Dyson et al., 1989; Harris et al., 1996). Consistent with a continuous need for such interactions, persistent expression of a functional TAg has been shown to be required for maintenance of BKV transformation (Nakashatri et al., 1988). Expression of antisense TAg RNA in BKV-transformed rodent cells resulted in the abrogation of anchorage-independent growth (Nakashatri et al., 1988).

As to the ability of BKV TAg to bind the p105RB1, p107, and p130RB2, the amount of BKV TAg normally produced in BKV-transfected simian BSC-1 cells seems to be too low to bind a significant number of proteins of the pRB family. BKV TAg binds the p53 protein available in the cell and induces serum independence, but is unable to allow anchorage-independent growth in semisolid medium (Harris et al., 1996). These data support the idea that BKV TAg can affect cellular growth control mechanisms, but other events are required for full transformation of simian cells by BKV. It was shown recently that the complex of SV40 TAg with mouse p53 completely blocks the transactivating effect of the p53 protein, whereas the same complex in human cells allows human p53 to exert its transcriptional activity, thereby activating expression of genes inducing apoptosis and growth arrest (Sheppard et al., 1999). Due to the high homology between the TAg of SV40 and of human polyomaviruses BK and JC, it is possible that the difficulty in inducing transformation of human cells by SV40 and human polyomaviruses is due to the inability of their TAgs to completely block the effect of human tumor suppressor proteins pRB and p53.

BKV can occasionally transform cells independently from the expression of TAg. Indeed, a BKV-induced choroid plexus papilloma contained only one copy per cell of the BKV genome integrated within the early region, implying that expression of a functional TAg was no longer possible (Yogo et al., 1981). Moreover, BKV may transform cells via a "hit and run" mechanism, as proposed for herpesviruses (Galloway and McDougall, 1983). In fact, hamster cells were transfected with purified DNA obtained from a human tumor containing BKV DNA sequences. Although transfection resulted in the appearance of transformed cells, BKV DNA was absent in most cell clones (Brunner et al., 1989). Therefore, either BKV was irrelevant to the pathogenesis of this human tumor or genetic changes fixed in human cells after initiation of the oncogenic process by BKV were sufficient to maintain expression of the transformed phenotype. These changes may be due to a mutagenic activity of the virus because BKV induces mutations in rodent and human cells (Theile and Grabowski, 1990).

Like SV40 TAg, which is clastogenic in human fibroblasts (Ray et al., 1990; Stewart and Bacchetti, 1991), BKV TAg induces numerical and structural chromosomal alterations characterized by gaps, breaks, dicentric and ring chromosomes, deletions, duplications, and translocations (Fig. 16.1). Chromosome damage in cells transfected with BKV early region was evident before the appearance of immortalization and of the morphologically transformed phenotype (Trabanelli et al., 1998), suggesting that it is a cause rather as a consequence of transformation. Similar alterations were observed in cell lines from human glioblastoma multiforme, harboring the TAg-coding sequences of both BKV and SV40 (Tognon et al., 1996). The molecular mechanism of the clastogenic effect of BKV TAg is not clear at present. It may reside in its ability to bind topoisomerase I (Simmons et al., 1996) or in its helicase activity (Dean et al., 1987), which might induce chromosome damage when unwinding the two strands of cellular DNA. Moreover, because BKV TAg binds the p53 protein (Shivakumar and Das, 1996), inactivating its functions, the direct clastogenic effect of the viral oncoprotein may be enhanced because it inhibits p53-induced apoptosis and allows DNA-damaged cells to survive, increasing their probability to transform and to acquire immortality. The clastogenic and mutagenic activities of BKV may hit genes that are crucial in oncogenesis, such as oncogenes, tumor suppressor genes, and DNA repair genes. Because BKV is latent in many human organs, the expression of BKV TAg in host cells during the lifelong period of latency may induce chromosomal damage responsible for the initiation or progression of an oncogenic process.

Knowledge of BKV tAg functions is rather limited. tAg cooperates with TAg in transformation by reducing serum dependence of transformed cells (Bouck et al., 1978; Sleigh et al., 1978; Martin et al., 1979; Frisque et al., 1980). Indeed, a low efficiency of stable transformation and a reduced incidence of tumors with a longer latency period was observed with BKV-IR (Pagnani et al., 1986), a BKV strain bearing a deletion of 253 bp that removes sequences coding for the carboxy terminus of tAg (Caputo et al., 1983). Moreover, BKV tAg binds the catalytic (36 kDa) and regulatory (63 kDa) subunits of protein phophatase 2A (PP2A) (Rundell et al., 1981), inactivating their function. PP2A is a serine–threonine phosphatase that regulates the phosphorylation signal activated by protein kinases (Baysal et al., 1998) and has recently been shown to be a tumor suppressor gene involved in lung, colon, breast carcinoma, and melanoma (Wang et al., 1998; Calin et al., 2000). PP2A binding may be an important function related to the mechanisms of viral transformation because it is conserved in small DNA tumor viruses. Indeed, polyoma virus tAg and middle T antigens as well as SV40 tAg (which shows 80% homology to BKV tAg) and adenovirus protein E4 binds PP2A (Pallas et al., 1990; Walter et al., 1990; Beck et al., 1998). In the case of polyoma virus and SV40, it was shown

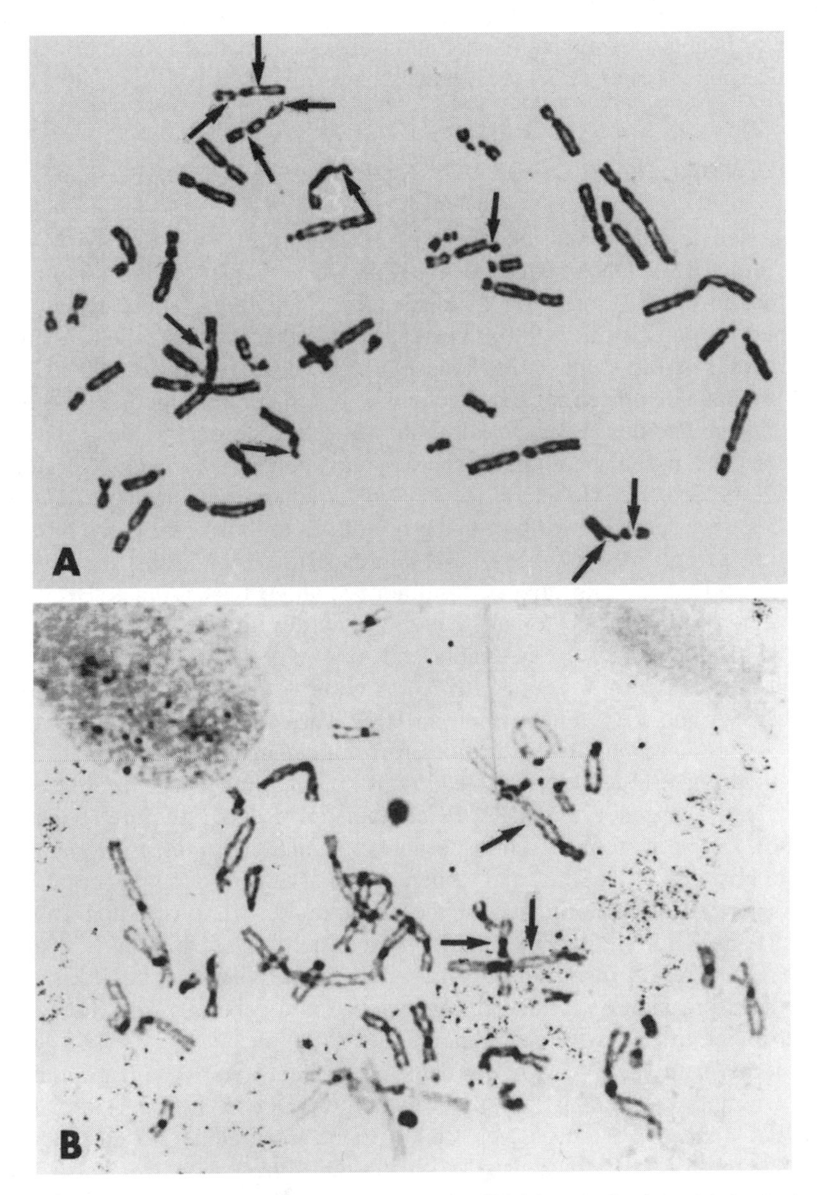

Figure 16.1. Metaphases of human embryonic fibroblast cell clones transfected by pRPneo-C, a plasmid containing BKV early region and expressing TAg (Trabanelli et al., 1998). Gaps and breaks (arrows) are evident in **A** and dicentric chromosomes (arrows) in **B**. Analysis of chromosome alterations was carried out in cell cultures before the appearance of the transformed phenotype. Giemsa staining. ×1000.

that the region of tAg binding to PP2A is not part of the TAg (Beck et al., 1998), suggesting that PP2A binding is a specific function of tAg. In addition, SV40 tAg is able to enhance transcription of E2F-activated promoters of early growth response genes (Loeken, 1992; Beck et al., 1998).

7. POTENTIAL ROLE OF BK VIRUS IN HUMAN TUMORS

The role of BKV in human neoplasia is still uncertain. Fiori and Di Mayorca (1976) found BKV DNA sequences by DNA–DNA reassociation kinetics in 5 of 12 human tumors and 3 of 4 human tumor cell lines. These results were confirmed by Pater et al. (1980), whereas three other reports failed to support these findings (Wold et al., 1978; Israel et al., 1978; Grossi et al., 1981) . These tumors contained full-length BKV genomes, but also rearranged and defective BKV DNA molecules. Because BKV shows a specific oncogenic tropism for the ependymal tissue, endocrine pancreas, and bones in rodents (Corallini et al., 1977, 1978, 1982; Uchida et al., 1979; Chenciner et al., 1980), BKV DNA sequences were searched for by Southern blotting in those rare types of human tumors most frequently induced by BKV in experimental animals such as ependymomas and other brain tumors, insulinomas, and osteosarcomas. BKV DNA was detected in a free, episomal state and generally in a low copy number in 19 of 74 (26%) human brain tumors and in 4 of 9 (44%) human tumors of pancreatic islets (Corallini et al., 1987b). A number of tumors expressed BKV-specific RNA and TAg. Furthermore, a BKV variant DNA, BKV-IR, was detected by Southern blot hybridization in a human insulinoma (Caputo et al., 1983). Virus was rescued by transfection of human embryonic fibroblasts with tumor DNA. The genome of BKV-IR contains an IS-like structure (Pagnani et al., 1986), a type of stem-loop transposable element able to integrate and excise from the host genome (Calos and Miller, 1980). The IS-like sequence of BKV-IR incorporates in its loop two of the early region transcriptional enhancer repeats (Pagnani et al., 1986). It may promote cell transformation by excision from viral DNA and insertion into the cell genome, thereby specifically activating the expression of cellular oncogenes or more generally as a mutagen by random integration into the cell genome.

In another study, BKV DNA was detected by Southern hybridization in 46% of brain tumors of the most common histotypes (Dörries et al., 1987). In this report BKV DNA sequences were found to be integrated into chromosomal DNA. Tumors typically associated with immunosuppression were also investigated by Southern hybridization, and BKV DNA was detected in Kaposi's sarcoma (KS) at a frequency of 20% (Barbanti-Brodano et al., 1987, 1988). Infectious BKV DNA was rescued from two KS tissues by transfection of cellular DNA into human embryonic fibloblasts. These rescued viruses, characterized by restriction endonucleases and nucleotide sequence analysis, were similar to BKV-IR, indicating that a specific BKV strain may be associated with certain types of human tumors (Negrini et al., 1990).

More recently, neoplastic and normal human tissues and tumor cell lines were investigated by PCR using specific primers for early region BKV DNA. The results of the studies conducted by Southern blot hybridization and PCR are summarized in Table 16.1, which reports positive and negative data relative to human tumors of different histotypes, obtained by various authors. Nucleotide sequence analysis of seven brain tumors, one osteosarcoma, two glioblastoma cell lines, one normal brain, and one normal bone tissue specimen confirmed that the amplified sequences correspond to the expected fragment of BKV early region (De Mattei et al., 1995). Expression of BKV early region was detected by Northern blot analysis or RT-PCR in several tumors, tumor cell lines, and normal tissues (De Mattei et al., 1995) (Table 16.1). In one study, SV40 and BKV sequences were searched for in human brain tumors. All the tumors harboring SV40 sequences were co-infected by BKV (Martini et al., 1996) (Table 16.1), suggesting a possible helper function of BKV to support SV40 infection of human cells. In another study, Flaegstad et al. (1999) detected BKV DNA by PCR in 18 neuroblastomas and in none of five normal adrenal gland samples (Table 16.1). The presence of BKV DNA was comfirmed by in situ hybridization in the tumor cells of 17 of the same neuroblastomas. BKV TAg was detected in tumor cells, but not in normal control samples. Finally, BKV TAg and p53 was co-localized in tumor cells by double immunostaining.

PCR amplification of DNA sequences from BKV early and regulatory regions was carried out in urinary tract tumors. Positive samples numbered 31 of 52 (60%) (Table 16.1), with a range of 50–67% in different tumor types (Monini et al., 1995a). In addition, BKV DNA sequences were amplified by PCR in carcinomas of the uterine cervix, vulva, lips, and tongue (Monini et al., 1995b, 1996). The percentage of positive samples in these neoplastic tissues of the urinary and genital tracts and of the oral cavity was similar to that detected in the corresponding normal tissues (61% and 59%, respectively). However, in tumors of the urinary bladder and prostate, two-dimensional gel electrophoresis and Southern blot hybridization analysis showed either a single integration of BKV DNA or integrated and episomal viral sequences (Monini et al., 1995a). In both the integrated and extrachromosomal viral sequences, the late region was disrupted. Viral episomes consisted of rearranged oligomers containing cellular DNA sequences whose size was apparently incompatible with encapsidation within a viral particle. Attempts to rescue these viral sequences by transfection of tumor DNA into permissive cells were unsuccessful, suggesting that in these tumors the process of integration and formation of episomal oligomers produced a rearrangement of viral sequences responsible for the elimination of viral infectivity and potentially leading to stable expression of BKV transforming functions. In other reports, however, the presence of BKV DNA sequences in human brain tumors, mostly malignant gliomas, bladder carcinomas, Kaposi's sarcomas, and lymphomas, analyzed by PCR, was not confirmed (Arthur et al., 1994; Völter et al., 1997) (Table 16.1). The negative results obtained by Völter et al. (1997) may be attributed to the use

Table 16.1. Presence and Expression of BKV DNA in Human Neoplastic and Normal Tissues and in Tumor Cell Lines

Tissues and Cell Lines	BKV DNA Positive Samples/ Samples Analyzed	(%)	Study	Method[a]	BVK RNA[b] Positive Samples/ Samples Analyzed
Tumors					
Brain	19/74	(26)	Corallini et al. (1987b)	SBH	
Brain	11/24	(46)	Dörries et al. (1987)	SBH	
Brain	0/75		Arthur et al. (1994)	PCR	
Brain	50/58	(86)	De Mattei et al. (1995)	PCR	
Brain	74/83	(89)	Martini et al. (1996)	PCR	
Brain	0/10		Völter et al. (1997)	PCR	
Brain	18/18	(100)	Flaegstad et al. (1999)	PCR	
Bone	11/25	(44)	De Mattei et al. (1995)	PCR	8/11
Insulinomas	4/9	(44)	Corallini et al. (1987b)	SBH	
Hodgkin's disease	0/5		Völter et al. (1997)	PCR	
Kaposi's sarcoma	4/20	(20)	Barbanti-Brodano et al. (1987)	SBH	
Kaposi's sarcoma	38/38	(100)	Monini et al. (1996)	PCR	
Kaposi's sarcoma	0/2		Völter et al. (1997)	PCR	
Urinary tract	31/52	(60)	Monini et al. (1995a,b)	PCR	
Urinary tract	0/15		Völter et al. (1997)	PCR	
Genital	32/42	(76)	Monini et al. (1995b, 1996)	PCR	

Cell lines from					
Brain tumors	8/10	(80)	De Mattei et al. (1995)	PCR	5/5
Brain tumors	21/26	(81)	Martini et al. (1996)	PCR	
Bone tumors	20/20	(100)	De Mattei et al. (1995)	PCR	6/8
Kaposi's sarcoma	6/8	(75)	Monini et al. (1996)	PCR	
Kaposi's sarcoma	0/14		Völter et al. (1997)	PCR	
Normal Tissues					
Brain	13/13	(100)	De Mattei et al. (1995)	PCR	
Bone	2/5	(40)	De Mattei et al. (1995)	PCR	2/2
Urinary tract	15/26	(58)	Monini et al. (1995a,b)	PCR	
Adrenal gland	0/5		Flaegstad et al. (1999)	PCR	
Cervix and vulva	15/21	(71)	Monini et al. (1995b, 1996)	PCR	
PBMC[c]	25/35	(71)	De Mattei et al. (1995)	PCR	8/8
PBMC[c]	53/70	(76)	Martini et al. (1996)	PCR	
Lymphnodes	4/4	(100)	Monini et al. (1996)	PCR	
Skin	25/33	(76)	Monini et al. (1996)	PCR	
Sperm	18/20	(90)	Martini et al. (1996)	PCR	
Sperm	18/19	(95)	Monini et al. (1996)	PCR	

[a]SBH = Southern blot hybridization; PCR = polymerase chain reaction.
[b]Detected by Northern blot hybrization or RT-PCR.
[c]Peripheral blood mononuclear cells from healthy donors.

of degenerate primers to simultaneously detect BKV, JCV, and SV40 sequences instead of primers specific for BKV DNA.

PCR analysis for the early and regulatory regions in 25 samples of KS (5 classic, 12 African, and 8 AIDS-associated) revealed 100% positivity for BKV DNA sequences. Analysis by PCR of eight KS cell lines disclosed BKV DNA sequences in six of them, whereas JCV and SV40 sequences were absent (Monini et al., 1996). In addition, BKV DNA sequences were detected in 15 of 26 (58%) prostatic tissues and in 18 of 19 (95%) seminal fluids (Monini et al., 1996), suggesting that BKV may be a candidate for the sexually transmitted infectious agent that was indicated by epidemiologic studies to be an important co-factor in KS (Peterman et al., 1993). A specific role of the sexual route for BKV transmission is suggested by the presence of JCV in only 4 of 19 (21%) prostatic tissues and in 4 of 19 (21%) seminal fluids (Monini et al., 1996). It is notable that polyomaviruses closely related to BKV induce angiogenic responses very similar to KS. Indeed, endothelial cells transformed by polyoma virus middle T antigen or by SV40 TAg induce hemangiomas or highly vascularized KS-like tumors in nude mice (Williams et al., 1989; O'Connel et al., 1991). Similar lesions are induced in nude mice by mouse brain and aortic endothelial cells transformed by BKV. These lesions later evolve toward development of a fibrotic wall around tumor cells, probably produced by infiltrating mouse fibroblasts and associated with increased expression of TGF-β (Corallini et al., manuscript submitted).

8. CONCLUSIONS

The role of BKV in human malignancy is still uncertain because of its ubiquity in the human population and the presence of BKV DNA sequences in normal human tissues during latency. The classic Koch's postulates do not seem to be applicable to ubiquitous viruses that produce a persistent, latent infection. New rules should be considered for these viruses in order to establish their oncogenic role in humans (zur Hausen, 1991): (1) presence and persistence of the virus or its nucleic acid in tumor cells; (2) cell immortalization or neoplastic transformation after transfection of the viral genome or its subgenomic fragments; (3) demonstration that the malignant phenotype of the primary tumor and the modifications induced by transfection of cultured cells depend on specific functions expressed by the viral genome; and (4) epidemiologic and clinical evidence that viral infection represents a risk factor for tumor development.

This chapter reviewed data showing that BKV may fulfill at least the first three criteria and suggesting that BKV may cooperate as a co-factor in the development or progression of human tumors. Indeed, BKV DNA is present and expressed in human tumors. The oncogenic risk related to BKV infection could be established by an epidemiologic survey of the incidence of tumors associated with BKV in segregated human populations of Brazil, Paraguay, and Malaysia, which are not infected by BKV (Brown et al., 1975). Several recent findings support the relationship of BKV to human neoplasia. In fact, BKV was demonstrated to be latent in many organs of infected individuals. Moreover, BKV was directly shown to produce numerical and structural chromosome aberrations in human cells. These observations, together with the ability of BKV TAg to bind p53 and pRB proteins, suggest that oncogenic mechanisms, activated by BKV, may participate in triggering neoplastic transformation of a BKV latently infected cell. Once chromosomal alterations are established in the host cell, viral sequences may be dispensable for the maintenance of the transformed phenotype and may therefore be lost by the neoplastic tissue.

ACKNOWLEDGMENTS

The work of the authors described in this chapter was supported by grants to G. Barbanti-Brodano, M. Negrini, and M. Tognon from Associazione Italiana per la Ricerca sul Cancro (A.I.R.C.) and from Ministero dell'Università e della Ricerca Scientifica e Tecnologica (M.U.R.S.T. ex 40% and ex 60%) and by grants to A. Corallini from Ministero dell'Università e della Ricerca Scientifica e Tecnologica (M.U.R.S.T. ex 60%).

REFERENCES

Alwine JC (1982): Evidence for simian virus 40 late transcriptional control: Mixed infections of wild-type simian virus 40 and a late leader deletion mutant exhibit trans effects on late viral RNA synthesis. J Virol 42:798–803.

Andrey G, Snoeck R, Vanputte M, De Clercq E (1997): Activities of various compounds against murine and primate polyomavirus. Antimicrob Agents Chemother 41(3):587–593.

Apperley JF, Rice SJ, Bishop JA, Chia YC, Krausz T, Gardner SD, Goldman JM (1987): Late-onset hemorrhagic cystitis associated with urinary excretion of polyomaviruses after bone marrow transplantation. Transplantation 43:108–112.

Arthur RR, Grossman SA, Ronnett BM, Bigner SH, Vogelstein B, Shah KV (1994): Lack of association of human polyomaviruses with human brain tumors. J Neuro-Oncol 20:55–58.

Arthur RR, Shah KV, Baust SJ, Santos GW, Saral R (1986): Association of BK viruria with hemorrhagic cystitis in recipients of bone marrow transplants. N Engl J Med 315:230–234.

Arthur RR, Shah KV, Charache P, Saral R (1988): BK and JC virus infections in recipients of bone marrow transplants. J Infect Dis 158:563–569.

Ault GS, Stoner GL (1992): Two major types of JC virus defined in progressive multifocal leukoencephalopathy brain by early and late coding region DNA sequences. J Gen Virol 73:2669–2678.

Ault GS, Stoner GL (1993): Human polyomavirus JC promoter/enhancer rearrangement patterns from progressive multifocal leukoencephalopathy brain are unique derivatives of a single archetypal structure. J Gen Virol 74:1499–1507.

Azzi A, Fanci R, Bosi A, Ciappi S, Zakrzewska K, Laszlo D, Guidi S, Saccardi R, Vannucchi AM (1994): Monitoring of polyomavirus BK in bone marrow transplantation patients by DNA hybridization assay and by polymerase chain reaction: An approach to assess the relationship between BK viruria and hemorrhagic cystitis. Bone Marrow Transplant 14:235–240.

Barbanti-Brodano G, Martini F, De Mattei M, Lazzarin L, Corallini A, Tognon M (1998): BK and JC human polyomavirus and simian virus 40: Natural history and infection in humans, experimental oncogenicity, and association with human tumors. Adv Virus Res 50:69–99.

Barbanti-Brodano G, Pagnani M, Balboni PG, Rotola A, Cassai E, Beth-Giraldo E, Giraldo G, Corallini A (1988): Studies on the association of Kaposi's sarcoma with ubiquitous viruses. In AIDS and Associated Cancers in Africa; Giraldo G, Beth-Giraldo E, Clumeck N, Gharbi Md-R, Kyalwazi SK, de Thè G, Eds; A. G. Karger: Basel, pp 175–181.

Barbanti-Brodano G, Pagnani M, Viadana P, Beth-Giraldo E, Giraldo E, Corallini A (1987): BK virus in Kaposi's sarcoma. Antibiot Chemother 38:113–120.

Barlow PW (1972): Differential division in human X chromosome mosaics. Humangenetik 14:122–127.

Baysal BE, Farr JE, Goss JR, Devlin B, Richard CW III (1998): Genomic organization and precise physical location of protein phosphatase 2A regulatory subunit A beta isoform gene on chromosome band 11q23. Gene 217:107–116.

Beck GR Jr, Zerler BR, Moran E (1998): Introduction to DNA tumor viruses: Adenovirus, simian virus 40, and polyomavirus. In Human Tumor Viruses; McCance DJ, Ed; ASM Press: Washington, DC, pp 51–86.

Beth E, Giraldo G, Schmidt-Ullrich R, Pater MM, Pater A, Di Mayorca G (1981): BK virus–transformed inbred hamster brain cells. I: Status of the viral DNA and the association of BK virus early antigens with purified plasma membranes. J Virol 40: 276–284.

Borgatti M, Costanzo F, Portolani M, Vullo C, Osti L, Masi M, Barbanti-Brodano G (1979): Evidence for reactivation of persistent infection during pregnancy and lack of congenital transmission of BK virus, a human papovavirus. Microbiologica 2: 173–178.

Bouck N, Beales N, Shenk T, Berg P, Di Mayorca G (1978): New region of the simian virus 40 genome required for efficient viral transformation. Proc Natl Acad Sci USA 75:2473–2477.

Bradley MK, Dougherty RM (1978): Transformation of African green monkey kidney cells with the RF strain of human papovavirus BKV. Virology 85:231–240.

Bratt G, Hammarin AL, Grandien M, Hedquist BG, Nennesmo I, Sundelin B, Seregard S (1999): BK virus as the cause of meningoencephalitis, retinitis and nephritis in a patient with AIDS. AIDS 13:1071–1075.

Brown P, Tsai T, Gajdusek DC (1975): Seroepidemiology of human papovaviruses: Discovery of virgin populations and some unusual patterns of antibody prevalence among remote peoples of the world. Am J Epidemiol 102:331–340.

Brunner M, Di Mayorca G, Goldman E (1989): Absence of BK virus sequences in transformed hamster cells transfected by human tumor DNA. Virus Res 12:315–330.

Calin G, Di Iasio MG, Caprini E, Vorechovsky I, Natali PG, Sozzi G, Croce CM, Barbanti-Brodano G, Russo G, Negrini M (2000): Low frequency of alterations of the α (PPP2R1A) and β (PPP2R1B) isoforms of the subunit A of the serine–threonine phosphatase in 2A in human neoplasms. Oncogene 19:1191–1195.

Calos MP, Miller JM (1980): Tansposable elements. Cell 20:579–595.

Caputo A, Corallini A, Grossi MP, Carrà L, Balboni PG, Negrini M, Milanesi G, Federspil G, Barbanti-Brodano G (1983): Episomal DNA of a BK virus variant in a human insulinoma. J Med Virol 12:37–49.

Chan PK, Ip KW, Shiu SY, Chiu EK, Wong MP, Yuen KY (1994): Association between polyomaviruria and microscopic haematuria in bone transplant recipients. J Infect 29:139–146.

Chenciner N, Meneguzzi G, Corallini A, Grossi MP, Grassi P, Barbanti-Brodano G, Milanesi G (1980): Integrated and free viral DNA in hamster tumors induced by BK virus. Proc Natl Acad Sci USA 77:975–979.

Chesters PM, Heritage J, McCance DJ (1983): Persistence of DNA sequences of BK virus and JC virus in normal human tissues and in diseases tissues. J Infect Dis 147: 676–684.

Coleman DV, Daniel RA, Gardner SD, Field AM, Gibson PE (1977): Polyomavirus in urine during pregnancy. Lancet 2:709–710.

Coleman DV, Gardner SD, Field AM (1973): Human polyomavirus infection in renal allograft recipients. BMJ 3:371–375.

Coleman DV, Gardner SD, Mulholland C, Fridiksdottir V, Portner AA, Lilford R, Valdimarsson H (1983): Human polyomavirus in pregnancy. A model for study of defense mechanisms to virus reactivation. Clin Exp Immunol 53:289–296.

Coleman DV, Wolfendale MR, Daniel RA, Dhanjal NK, Gardner SD, Gibson PE, Field AM (1980): A prospective study of human polyomavirus infection in pregnancy. J Infect Dis 142:1–8.

Corallini A, Altavilla G, Cecchetti MG, Fabris G, Grossi MP, Balboni PG, Lanza G, Barbanti-Brodano G (1978): Ependymomas, malignant tumors of pancreatic islets and osteosarcomas induced in hamsters by BK virus, a human papovavirus. J Natl Cancer Inst 61:875–883.

Corallini A, Altavilla G, Carrà L, Grossi MP, Federspil G, Caputo A, Negrini M, Barbanti-Brodano G (1982): Oncogenicity of BK virus for immunosuppressed hamsters. Arch Virol 73:243–253.

Corallini A, Barbanti-Brodano G, Bortoloni W, Nenci I, Cassai E, Tampieri M, Portolani M, Borgatti M (1977): High incidence of ependymomas induced by BK virus, a human papovavirus. J Natl Cancer Inst 59:1561–1563.

Corallini A, Giannì M, Mantovani C, Vandini A, Rimessi P, Negrini M, Giavazzi R, Bani MR, Milanesi G, Dal Cin P, van den Berghe H, Barbanti-Brodano G (1991): Transformation of human cells by recombinant DNA molecules containing BK virus early region and the human activated c-H-*ras* or c-*myc* oncogenes. Cancer J 4:24–34.

Corallini A, Pagnani M, Caputo A, Negrini M, Altavilla G, Catozzi L, Barbanti-Brodano G (1988): Cooperation in oncogenesis between BK virus early region gene and the activated human c-Harvey-*ras* oncogene. J Gen Virol 69:2671–2679.

Corallini A, Pagnani M, Viadana P, Camellin P, Caputo A, Reschiglian P, Rossi S, Altavilla G, Selvatici R, Barbanti-Brodano G (1987a): Induction of malignant subcutaneous sarcomas in hamsters by a recombinant DNA containing BK virus early region and the activated human c-Harvey-*ras* oncogene. Cancer Res 47:6671–6677.

Corallini A, Pagnani M, Viadana P, Silini E, Mottes M, Milanesi G, Gerna G, Vettor R, Trapella G, Silvani V, Gaist G, Barbanti-Brodano G (1987b): Association of BK virus with human brain tumors and tumors of pancreatic islets. Int J Cancer 39:60–67.

Costa T, Howley PM, Legallais F, Yee C, Young N, Rabson AS (1977): Oncogenicity of a nude mouse cell line transformed by a human papovavirus. J Natl Cancer Inst 58:1147–1151.

Costa T, Yee C, Tralka TS, Rabson AS (1976): Hamster ependymomas produced by intracerebral inoculation of human papovavirus (MMV): J Natl Cancer Inst 56:863–864.

Cubukcu-Dimopulo O, Greco A, Kumar A, Karluk D, Mittal K, Jagirdar J (2000): BK virus infection in AIDS. Am J Surg Pathol 248(1):145–149.

Dalrymple SA, Beemon KL (1990): BK virus T antigen induce kidney carcinomas and thymoproliferative disorders in transgenic mice. J Virol 64:1182–1191.

Daniel R, Shah K, Madden D, Stagno S (1981): Serological investigation of the possibility of congenital transmission of papovavirus JC. Infect Immun 33:319–321.

Dean FB, Bullock P, Murakami Y, Wobble R, Weissbach L, Hurwitz J (1987): Simian virus 40 (SV40) DNA replication: SV40 large T antigene unwinds DNA containing the SV40 origin of replication. Proc Natl Acad Sci USA 84:16–20.

De Mattei M, Martini F, Corallini A, Gerosa M, Scotlandi K, Carinci P, Barbanti-Brodano G, Tognon M (1995): High incidence of BK virus large T antigen coding sequences in normal human tissues and tumors of different histotypes. Int J Cancer 61:756–760.

De Mattei M, Martini F, Tognon M, Serra M, Baldini N, Barbanti-Brodano G (1994): Polyomavirus latency and human tumors. J Infect Dis 169:1175–1176.

de Silva LM, Bale P, de Courcy J, Brown D, Knowles W (1995): Renal failure due to BK virus infection in an immunodeficient child. J Med Virol 45:192–196.

Di Taranto C, Pietropaolo V, Orsi GB, Jin L, Sinibaldi L, Degener AM (1997): Detection of BK polyomavirus genotypes in healthy and HIV-positive children. Eur J Epidemiol 13:653–657.

Dörries K, Loeber G, Meixenberger J (1987): Association of polyomavirus JC, SV40 and BK with human brain tumors. Virology 160:268–270.

Dörries K, Vogel E, Günther S, Czub S (1994): Infection of human polyomavirus JC and BK in peripheral blood leukocytes from immunocompetent individuals. Virology 198:59–70.

Dougherty RM (1976): Induction of tumors in Syrian hamster by a human renal papovavirus, RF strain. J Natl Cancer Inst 57:395–400.

Dyson N, Bernards R, Friend SH, Gooding LR, Hassell JA, Major EO, Pipas JM, Van Dyke T, Harlow E (1990): Large T antigens of many polyomaviruses are able to form complexes with retinoblastoma protein. J Virol 64:1353–1356.

Dyson N, Buchkovich K, Whyte P, Harlow E (1989): The cellular 107K protein that binds to adenovirus E1A also associates with the large T antigens of SV40 and JC virus. Cell 58:249–255.

Elsner C, Dörries K (1992): Evidence of human polyomavirus BK and JC infection in normal brain tissue. Virology 191:72–80.

Fiori M, Di Mayorca G (1976): Occurrence of BK virus DNA in DNA obtained from certain human tumors. Proc Natl Acad Sci USA 73:4662–4666.

Flaegstad T, Andresen PA, Johnsen JI, Asomani SK, Jorgensen GE, Vignarjan S, Kjuul A, Kogner P, Traavik T (1999): A possible contributory role of BK virus infection in neuroblastoma. Cancer Res 59:1160–1163.

Flaegstad T, Permin H, Husebekk A, Husby G, Traavik T (1988): BK virus infection in patients with AIDS. Scand J Infect Dis 20:145–150.

Flaegstad T, Sundsfjord A, Arthur RR, Pedersen M, Traavik T, Subramani S (1991): Amplification and sequencing of the control regions of BK and JC virus from human urine by polymerase chain reaction. Virology 180:553–560.

Frisque RJ, Bream GL, Cannella MT (1984): Human polyomavirus JC virus genome. J Virol 51:458–469.

Frisque RJ, Rifkin DB, Walker DL (1980): Transformation of primary hamster brain cells with JC virus and its DNA. J Virol 35:265–269.

Galloway DA, McDougall JK (1983): The oncogenic potential of herpes simplex viruses: Evidence for a "hit and run" mechanism. Nature (London) 302:21–24.

Gardner SD (1973): Prevalence in England of antibody to human polyomavirus (BK): BMJ 1:77–78.

Gardner SD, Field AM, Coleman DV, Hulme B (1971): New human papovavirus (B.K.) isolated from urine after renal transplantation. Lancet 1:1253–1257.

Gardner SD, MacKenzie EFD, Smith C, Porter AA (1984): Prospective study of the human polyomaviruses BK and JC and cytomegalovirus in renal transplant recipients. J Clin Pathol 37:578–586.

Gillespie SM, Chang Y, Lemp G, Arthur R, Buchbinder S, Steimle A, Baumgartner J, Rando T, Neal D, Rutherford G, Schomberger L, Janssen R (1991): Progressive multifocal leukoencephalopathy in persons infected with human immunodeficiency virus, San Francisco, 1981–1989. Ann Neurol 30:597–604.

Goudsmit J, Baak ML, Slaterus KW, van der Noordaa J (1981): Human papovavirus isolated from the urine of a child with acute tonsillitis. BMJ 283:1363–1364.

Goudsmit J, Wetheim-van Dillen P, van Strien A, van der Noordaa J (1982): The role of BK virus in acute respiratory tract disease and the presence of BKV DNA in tonsils. J Med Virol 10:91–99.

Greenlee JE, Narayan O, Johnson RT, Hernodon RM (1977): Induction of brain tumors in hamsters with BK virus, a human papovavirus. Lab Invest 36:636–642.

Grinnel BW, Padgett BL, Walker DL (1983): Distribution of nonintegrated DNA from JC papovavirus in organs of patients with progressive multifocal leukoencephalopathy. J Infect Dis 147:669–675.

Grossi MP, Caputo A, Meneguzzi G, Corallini A, Carrà L, Portolani M, Borgatti M, Milanesi G, Barbanti-Brodano G (1982a): Transformation of human embryonic fibroblasts by BK virus, BK virus DNA and a subgenomic BK virus DNA fragment. J Gen Virol 63:369–403.

Grossi MP, Corallini A, Valieri A, Balboni PG, Poli F, Caputo A, Milanesi G, Barbanti-Brodano G (1982b): Transformation of hamster kidney cells by fragments of BK virus DNA. J Virol 41:319–325.

Grossi MP, Corallini A, Meneguzzi G, Chenciner N, Barbanti-Brodano G, Milanesi G (1982c): Tandem integration of complete viral genomes can occur in nonpermissive hamster cells transformed by linear BKV DNA with cohesive ends. Virology 120: 500–503.

Grossi MP, Meneguzzi G, Chenciner N, Corallini A, Poli F, Altavilla G, Alberti S, Milanesi G, Barbanti-Brodano G (1981): Lack of association between BK virus and ependymomas, malignant tumors of pancreatic islets, osteosarcomas and other human tumors. Intervirology 15:10–18.

Gualandi F, Morelli C, Pavan JV, Rimessi P, Sensi A, Bonfatti A, Gruppioni R, Possati L, Stanbridge EJ, Barbanti-Brodano G (1994): Induction of senescence and control of tumorigenicity in BK virus transformed mouse cells by human chromosome 6. Genes Chrom Cancer 10:77–85.

Harris KF, Christensen JB, Imperiale MJ (1996): BK virus large T antigen: Interactions with the retinoblastoma family of tumor suppressor proteins and effects on cellular growth control. J Virol 70:2378–2386.

Hashida J, Gaffney PC, Yunis EJ (1976): Acute hemorrhagic cystitis of childhood and papovavirus-like particles. J Pediatr 89:85–87.

Haukland HH, Vonen B, Traavik T (1992): Transformed rat pancreatic islet-cell lines established by BK virus infection in vitro. Int J Cancer 51:79–83.

Hay N, Skolnick-David H, Aloni Y (1982): Attenuation in the control of SV40 gene expression. Cell 29:183–193.

Hedquist BG, Bratt G, Hammarin AL, Grandien M, Nennesmo I, Sundelin B, Seregard S (1999): Identification of BK virus in a patient with acquired immnune deficiency syndrome and bilateral atypical retinitis. Ophthalmology 106:129–132.

Heritage J, Chesters PM, McCance DJ (1981): The persistence of papovavirus BK DNA sequences in normal human renal tissue. J Med Virol 8:143–150.

Hogan TF, Borden EC, McBain JA, Padgett BL, Walker DL (1980): Human polyomavirus infections with JC virus and BK virus in renal transplant recipients. Ann Intern Med 92:373–378.

Hogan TF, Padgett BL, Walker DL, Borden EC, Frias Z (1983): Survey of human polyomavirus (JCV, BKV) infections in 139 patients with lung cancer, breast cancer, melanoma, or lymphoma. Prog Clin Biol Res 105:311–324.

Howell DN, Smith SR, Butterly DW, Klassen PS, Krigman HR, Burchette JL Jr, Miller SE (1999): Diagnosis and management of BK polyomavirus interstitial nephritis in renal transplant recipients. Transplantation 68:1279–1288.

Howley PM, Martin MA (1977): Uniform representation of the human papovavirus BK genome in transformed hamster cells. J Virol 23:205–208.

Imperiale MJ (2000): The human polyomaviruses BKV and JCV: Molecular pathogenesis of acute disease and potential role in cancer. Virology 267:1–7.

Israel MA, Martin MA, Takemoto KK, Howley PM, Aaronson SA, Solomon D, Khoury G (1978): Evaluation of normal and neoplastic human tissue for BK virus. Virology 90:187–196.

Jin L, Pietropaolo V, Booth JC, Waard KH, Brown DWG (1995): Prevalence and subtypes distribution of BK virus in healthy people and immunocompromised patients detected by PCR-restriction enzyme analysis. Clin Diagn Virol 3:285–295.

Kang S, Folk WR (1992): Lymphotropic papovavirus transforms hamster cells without altering the amount or stability of p53. Virology 191:754–764.

Lecatsas G, Bernard MM (1982): BK virus excretion in Fanconi's anemia. S Afr Med J 62:467.

Lecatsas G, Pretorius F, Crewe-Brown H, Requadt E, Ackthun I (1977): Polyomavirus in urine in pernicious anemia. Lancet 2:147.

Lecatsas G, Prozesky OW, Van Wyk J, Els HJ (1973): Papovavirus in urine after renal transplantation. Nature (London) 241:343–344.

Lecatsas G, Schoub BD, Rabson AR, Joffe M (1976a): Papovavirus in human lymphocyte cultures. Lancet 2:907–908.

Lecatsas G, Schoub BD, Prozesky OW, Pretorius F, De Beer FC (1976b): Polyomavirus in urine in aplastic anemia. Lancet 1:259–260.

Lednicky JA, Butel JS (1999): Polyomaviruses and human tumors: A brief review of current concepts and interpretations. Front Biosci 4:153–164.

Loeber G, Dörries K (1988): DNA rearrangements in organ-specific variants of polyomavirus JC strain GS. J Virol 62:1730–1735.

Loeken MR (1992): Simian virus 40 small t antigen transactivates the adenovirus E2A promoter by using mechanisms distinct from those used by adenovirus E1A. J Virol 66:2551–2555.

Major EO, Di Mayorca G (1973): Malignant transformation of BHK21 clone 13 cells by BK virus, a human papovavirus. Proc Natl Acad Sci USA 70:3210–3212.

Mantyjarvi RA, Meurman OH, Vihma L, Berglund B (1973): A human papovavirus (BK), biological properties and seroepidemiology. Ann Clin Res 5:283–287.

Markowitz RB, Eaton BA, Kubik MF, Latorra D, McGregor JA, Dynan WS (1991): BK virus and JC virus shed during pregnancy have predominantly archetypal regulatory regions. J Virol 65:4515–4519.

Markowitz RB, Thompson HC, Mueller JF, Cohen JA, Dynan WS (1993): Incidence of BK virus and JC virus viruria in human immunodeficiency virus–infected and virus–uninfected subjects. J Infect Dis 167:13–20.

Martin JD, Padgett BL, Walker DL (1983): Characterization of tissue culture–induced heterogeneity in DNAs of independent isolates of JC virus. J Gen Virol 64:2271–2280.

Martin RG, Setlow VP, Edwards CAF, Vembu D (1979): The roles of simian virus 40 tumor antigens in transforming of chinese hamster lung cells. Cell 17:635–643.

Martini F, Iaccheri L, Lazzarin L, Carinci P, Corallini A, Gerosa M, Iuzzolino P, Barbanti-Brodano G, Tognon M (1996): SV40 early region and large T antigen in human brain tumors, peripheral blood cells, and sperm fluids from healthy individuals. Cancer Res 56:4820–4825.

Mason DHJ, Takemoto KK (1977): Transformation of rabbit kidney cells by BKV (MM) human papovavirus. Int J Cancer 19:391–395.

McCance DJ (1983): Persistence of animal and human papovaviruses in renal and nervous tissues. In Polyomaviruses and Human Neurological Disease; Sever JL, Madden DL, Eds; Alan R. Liss: New York, pp 479–481.

Meneguzzi G, Chenciner N, Corallini A, Grossi MP, Barbanti-Brodano G, Milanesi G (1981): The arrangement of viral integrated DNA is different in BK virus–transformed mouse and hamster cells. Virology 111:139–153.

Mininberg DT, Watson C, Desquitado M (1982): Viral cystitis with transient secondary vesicoureteral reflux. J Urol 127:983–985.

Monini P, Rotola A, Di Luca D, De Lellis L, Chiari E, Corallini A, Cassai E (1995a): DNA rearrangements impairing BK virus productive infection in urinary tract tumors. Virology 214:273–279.

Monini P, De Lellis L, Barbanti-Brodano G (1995b): Association of BK and JC human polyomaviruses and SV40 with human tumors. In DNA Tumor Viruses: Oncogenic mechanisms; Barbanti-Brodano G, Bendinelli M, Friedman H, Eds; Plenum Press: New York, pp 51–73.

Monini P, Rotola A, De Lellis L, Corallini A, Albini A, Benelli R, Parravicini C, Secchiero P, Barbanti-Brodano G, Cassai E (1996): Latent BK virus infection and Kaposi's sarcoma pathogenesis. Int J Cancer 66:717–722.

Nagao S, Iijima IS, Suzuki H, Yokota T, Shigeta S (1982): BK virus-like particles in the urine of a patient with nephrotic syndrome. An electron microscopic observation. Fukushima J Med Sci 29:45–49.

Nakashatri H, Pater MM, Pater A (1988): Functional role of BK virus tumor antigens in transformation. J Virol 62:4613–4621.

Nase LM, Karkkaiven M, Mantyjarvi RA (1975): Transplantable hamster tumors induced with the BK virus. Acta Pathol Microbiol Scand 83:347–352.

Negrini M, Castagnoli A, Pavan JV, Sabbioni S, Araujo D, Corallini A, Gualandi F, Rimessi P, Bonfatti A, Giunta C, Sensi A, Stanbridge EJ, Barbanti-Brodano G (1992): Suppression of tumorigenicity and anchorage-independent growth of BK virus–transformed mouse cells by human chromosome 11. Cancer Res 52:1297–1303.

Negrini M, Rimessi P, Mantovani C, Sabbioni S, Corallini A, Gerosa MA, Barbanti-Brodano G (1990): Characterization of BK virus variants rescued from human tumors and tumour cell lines. J Gen Virol 71:2731–2736.

Negrini M, Sabbioni S, Arthur RR, Castagnoli A, Barbanti-Brodano G (1991): Prevalence of the archetypal regulatory region and sequence polymorphism in nonpassaged BK virus variants. J Virol 65:5092–5095.

Ng S-C, Mertz JE, Sanden-Will S, Bina M (1985): Simian virus 40 maturation in cells harboring mutants deleted in the agnogene. J Biol Chem 260:1127–1132.

Noss G, Stauch G (1984): Oncogenic activity of the BK type of human papovavirus in inbred rat strains. Arch Virol 81:41–50.

Noss G, Stauch G, Mehraein P, Georgii A (1981): Oncogenic activity of the BK type of human papovavirus in newborn Wistar rats. Arch Virol 69:239–251.

O'Connel K, Landman G, Farmer E, Edidin M (1991): Endothelial cells transformed by SV40 T antigen cause Kaposi's sarcoma like tumors in nude mice. Am J Pathol 139:743–749.

O'Reilly RJ, Lee FK, Grossbard E, Kapoor N, Kirkpatrick D, Dinsmore R, Stutzer C, Shah KV, Nahmias AJ (1981): Papovavirus excretion following marrow transplantation. Incidence and association with hepatic disfunction. Transplant Proc 13:262–266.

Padgett BL, Walker DL (1976): New human papovaviruses. Progr Med Virol 22:1–35.

Padgett BL, Walker DL, Desquitado M, Kim DV (1983): BK virus and non-hemorrhagic cystitis in a child. Lancet 1:770.

Pagnani M, Corallini A, Caputo A, Altavilla G, Selvatici R, Cattozzi L, Possati L, Barbanti-Brodano G (1988): Co-operation in cell transformation between BK virus and the human c-Harvey-ras oncogene. Int J Cancer 42:405–413.

Pagnani M, Negrini M, Reschiglian P, Corallini A, Balboni PG, Scherneck S, Macino G, Milanesi G, Barbanti-Brodano G (1986): Molecular and biological properties of BK virus-IR, a BK virus variant rescued from a human tumor. J Virol 59:500–505.

Pallas DC, Shahrik LK, Martin BL, Jaspers S, Miller TB, Brautigan DL, Roberts TM (1990): Polyoma small and middle T antigens and SV40 small t antigen form a stable complexes with protein phosphatase 2A. Cell 60:167–176.

Pater A, Pater MM (1986): Transformation of primary human embryonic kidney cells to anchorage independence by a combination of BK virus DNA and the Harvey-ras oncogene. J Virol 58:680–683.

Pater MM, Pater A, Fiori M, Slota J, Di Mayorca G (1980): BK virus DNA sequences in human tumors and normal tissues and cell lines. In Viruses in Naturally Occurring Cancers. Book A. Cold Spring Harbor Conferences on Cell Proliferation; Essex M, Todaro G, zur Hausen H, Eds; Cold Spring Harbor Laboratory Press: New York, Vol 7, pp 329–341.

Peterman TA, Jaffe HW, Beral V (1993): Epidemiologic clues to the etiology of Kaposi's sarcoma. AIDS 7:605–611.

Pietropaolo V, Di Taranto C, Degener AM, Jin L, Sinibaldi L, Baiocchini A, Melis M, Orsi N (1998): Transplacental transmission of human polyomavirus BK. J Med Virol 56:372–376.

Portolani M, Barbanti-Brodano G, La Placa M (1975): Malignant transformation of hamster kidney cells by BK virus. J Virol 15:420–422.

Portolani M, Borgatti M, Corallini A, Cassai E, Grossi MP, Barbanti-Brodano G, Possati L (1978): Stable transformation of mouse, rabbit and monkey cells and abortive transformation of human cells by BK virus, a human papovavirus. J Gen Virol 38: 369–374.

Portolani M, Marzocchi A, Barbanti-Brodano G, La Placa M (1974): Prevalence in Italy of antibodies to a new human papovavirus (BK virus): J Med Microbiol 7:543–546.

Portolani M, Piani M, Gazzanelli G, Borgatti M, Bortoletti A, Grossi MP, Corallini A, Barbanti-Brodano G (1985): Restricted replication of BK virus in human lymphocytes. Microbiologica 8:59–66.

Purchio AF, Fareed GC (1979): Transformation of human embryonic kidney cells by human papovavirus BK. J Virol 29:763–769.

Quinlivan EB, Norris M, Bouldin TW, Suzuki K, Meeker R, Smith MS, Hall C, Kenney S (1992): Subclinical central nervous system infection with JC virus in patients with AIDS. J Infect Dis 166:80–85.

Ray FA, Peabody DS, Cooper JL, Cram LS, Kraemer PM (1990): SV40 T antigen alone drives karyotype instability that precedes neoplastic transformation of human diploid fibroblasts. J Cell Biochem 42:13–31.

Rhiza HJ, Belohradsky BH, Schneider V, Schwenk HU, Burkamm GW, zur Hausen H (1978a): BK virus. II. Serological studies in children with congenital disease and patients with malignant tumors and immunodeficiencies. Med Microbiol Immunol 165:83–92.

Rhiza HJ, Belohradsky BH, zur Hausen H (1978b): BK virus. I. Seroepidemiologic and serologic response to viral infection. Med Microbiol Immunol 165:73–81.

Rice SJ, Bishop JA, Apperley J, Gardner SD (1985): BK virus as a cause of haemorrhagic cystitis after bone marrow transplantation. Lancet 2:844–845.

Rubinstein R, Pare N, Harley EH (1987): Structure and function of the transcriptional control region of nonpassaged BK virus. J Virol 61:1747–1750.

Rubinstein R, Schoonakker BCA, Harley EH (1991): Recurring theme of changes in the transcriptional control region of BK virus during adaptation to cell culture. J Virol 65:1600–1604.

Rundell K, Major EO, Lanpert M (1981): Association of cellular 56,000- and 32,000-molecular-weight proteins with BK virus and polyoma virus t-antigens. J Virol 37: 1090–1093.

Sabbioni S, Negrini M, Possati L, Bonfatti A, Corallini A, Sensi A, Stanbridge EJ, Barbanti-Brodano G (1994): Multiple loci on human chromosome 11 control tumorigenicity of BK virus transformed cells. Int J Cancer 57:185–191.

Saitoh K, Sugae N, Koike N, Ahijama Y, Iwamura Y, Kimura H (1993): Diagnosis of childhood BK virus cystitis by electron microscopy and PCR. J Clin Pathol 46:773–775.

Sandberg AA (1983): The X chromosome in human neoplasia, including sex chromatin and congenital conditions with X-chromosome anomalies. In Cytogenetics of the Mammalian X Chromosome, Part B: X Chromosome Anomalies and Their Clinical Manifestations; Sandberg AA, Ed; Alan R. Liss: New York, pp 459–498.

Sandberg AA (1985): The Y chromosome in human neoplasia. In The Y Chromosome, Part B: Clinical Aspects of Y Chromosome Abnormalities; Sandber AA, Ed; Alan R. Liss: New York, pp 377–393.

Schneider EM, Dörries K (1993): High frequency of polyomavirus infection in lymphoid cell preparations after allogenic bone marrow transplantation. Transplant Proc 25:1271–1273.

Seehafer J, Downer DN, Salmi A, Colter JS (1979): Isolation and characterization of BK virus-transformed rat and mouse cells. J Gen Virol 42:567–578.

Seehafer J, Salmi A, Colter JS (1977): Isolation and characterization of BK virus transformed hamster cells. Virology 77:356–366.

Shah KV (2000): Polyoma viruses (JC virus, BK virus, and Simian virus 40) and human cancer. In Infectious Causes of Cancer: Targets for Intervention; Goedert JJ, Ed; Humana Press Inc.: Totowa, NJ, pp 461–474.

Shah KV, Daniel RW, Strandberg J (1975): Sarcoma in a hamster inoculated with BK virus, a human papovavirus. J Natl Cancer Inst 54:945–949.

Shah KV, Daniel R, Madden D, Stagno S(1980): Serological investigation of BK papovavirus infection in pregnant woman and their offspring. Infect Immun 30:29–35.

Shah KV, Daniel RW, Warszawski RM (1973): High prevalence of antibodies to BK virus, an SV40 related papovavirus, in residents of Maryland. J Infect Dis 128:784–787.

Shah KV, Hudson JC, Valis J, Strandberg D (1976): Experimental infection of human foreskin cultures with BK virus, a human papovavirus. Proc Soc Exp Biol Med 153: 180–186.

Shah KV, Ozer HL, Ghazey HN, Kelly TJ Jr (1977): Common structural antigen of papovaviruses of the simian virus 40-polyoma subgroup. J Virol 21:179–186.

Sheppard HM, Siska IC, Espiritu C, Gatti A, Liu X (1999): New insights into the mechanism of inhibition of p53 by simian virus 40 large T antigen. Mol Cell Biol 19:2746–2753.

Shinohara T, Matsuda M, Yasui K, Yoshike K (1989): Host range bias of the JC virus mutant enhancer with DNA rearrangement. Virology 170:261–263.

Shivakumar CV, Das GC (1996): Interaction of human polyomavirus BK with the tumor-suppressor protein p53. Oncogene 13:323–332.

Simmons DT, Melendy T, Usher D, Stillman B (1996): Simian virus 40 large T antigen binds to topoisomerase I. Virology 222:365–374.

Sleigh MJ, Topp WC, Hanich P, Sambrook JF (1978): Mutants of SV40 with an altered small t protein are reduced in their ability to transform cells. Cell 14:79–88.

Small JA, Khoruy G, Jay G, Howley PM, Scangos GA (1986): Early regions of JC virus and BK virus induce distinct and tissue-specific tumours in transgenic mice. Proc Natl Acad Sci USA 83:8288–8292.

Snider WD, Simpson DM, Nielsen S, Gold JW, Metroka CE, Posner JB (1983): Neurological complications of the acquired immune deficency syndrome: Analysis of 50 patients. Ann Neurol 14:403–418.

Stewart N, Bacchetti S (1991): Expression of SV40 large T antigen, but not small t antigen, is required for the induction of chromosomal aberrations in transformed human cells. Virology 180:49–57.

Sundsfjord A, Johansen T, Flaegstad T, Moens U, Villand P, Subramani S, Traavik T (1990): At least two types of control regions can be found among naturally occurring BK virus strains. J Virol 64:3864–3871.

Sundsfjord A, Osei A, Rosenquist H, Van Ghelue M, Silsand Y, Haga HJ, Rekvig OP, Moens U (1999): BK and JC viruses in patients with systemic lupus erythematosus: Prevalent and persistent BK viruria, sequence stability of the viral regulatory regions, and nondetectable viremia. J Infect Dis 180:1–9.

Taguchi F, Nagaki D (1978): BK papovavirus in urine of patients with systemic lupus erythematosus. Acta Virol 22:513.

Taguchi F, Nagaki D, Saito M, Haruyama C, Iwasaki K, Suzuki T (1975): Transplacental transmission of BK virus in humans. Jpn J Microbiol 19:395–398.

Takemoto KK, Martin MA (1976): Transformation of hamster kidney cells by BK papovavirus DNA. J Virol 17:247–253.

Takemoto KK, Mullarkey MF (1973): Human papovavirus, BK strain: Biological studies including antigenic relationship to simian virus 40. J Virol 12:625–631.

Takemoto KK, Linke H, Miyamura T, Fareed GC (1979): Persistent BK papovavirus infection of transformed human fetal brain cells. I. Episomal viral DNA in cloned lines deficient in T-antigen expression. J Virol 29:1177–1185.

Takemoto KK, Rabson AS, Mullarkey MR, Blaese MF, Garon CF, Nelson D (1974): Isolation of papovavirus from brain tumor and urine of a patient with Wiskott-Aldrich syndrome. J Natl Cancer Inst 53:1205–1207.

Tanaka R, Koprowski H, Iwasaki Y (1976): Malignant transformation of hamster brain cells in vitro by human papovavirus BK. J Natl Cancer Inst 56:671–673.

Theile M, Grabowski G (1990): Mutagenic activity of BKV and JCV in human and other mammalian cells. Arch Virol 113:221–233.

Tognon M, Casalone R, Martini F, De Mattei M, Granata P, Minelli E, Arcuri C, Collini P, Bocchini V (1996): Large T antigen coding sequences of two DNA tumor viruses, BK and SV40 and nonrandom chromosome chances in glioblastoma cell lines. Cancer Genet Cytogenet 90:17–23.

Tominaga T, Yogo Y, Kitamura T, Aso Y (1992): Persistence of archetypal JC virus DNA in normal renal tissue derived from tumor-bearing patients. Virology 186:736–741.

Trabanelli C, Corallini A, Gruppioni R, Sensi A, Bonfatti A, Campioni D, Merlin M, Calza N, Possati L, Barbanti-Brodano G (1998): Chromosomal aberrations induced by BK virus T antigen in human fibroblasts. Virology 243:492–496.

Traystman MD, Gupta PK, Shah KV, Reissig M, Cowles LT, Hillis WD, Frost JK (1980): Identification of viruses in the urine of renal transplant recipients by cytomorphology. Acta Cytol 24:501–510.

Uchida S, Watanabe S, Aizawa T, Kato F, Furuno A, Muto T (1976): Induction of papillary ependymomas and insulinomas in the Syrian golden hamster by BK virus, a human papovavirus. Gann 67:857–865.

Uchida S, Watanabe S, Aizawa T, Furuno A, Muto T (1979): Polioncogenicity and insulinoma-inducing ability of BK virus, a human papovavirus, in Syrian golden hamsters. J Natl Cancer Inst 63:119–126.

Vago L, Cinque P, Salam E, Nebuloni M, Caldarelli R, Racca S, Ferrante P, Trabottoni G, Costanzi G (1997): JCV-DNA and BKV-DNA in the CNS tissue and CSF of AIDS patients and normal subjects. Study of 41 cases and review of the literature. J Acquir Immune Defic Syndr Hum Retrovirol 12:139–146.

Vallbracht A, Löhler J, Gossmann J, Glück T, Pertersen D, Gerth H-J, Gencic M, Dörries K (1993): Disseminated BK type polyomavirus infection in an AIDS patient associated with CNS disease. Am J Pathol 143:1–11.

van der Noordaa J (1976): Infectivity, oncogenity and transforming ability of BK virus and BK virus DNA. J Gen Virol 30:371–373.

van der Noordaa J, De Jong W, Pauw W, Sol CJA, van Strien A (1979): Transformation and T antigen induction by linearized BK virus DNA. J Gen Viro! 44:843–847.

van der Noordaa J, van Strien A, Sol CJA (1986): Persistence of BK virus in human foetal pancreas cells. J Gen Virol 67:1485–1490.

van der Noordaa J, Wartheim-van Dillen P (1977): Rise in antibodies to human papovavirus BK and clinical disease. BMJ 1:1471.

Vasavada R, Eager KB, Barbanti-Brodano G, Caputo A, Ricciardi RP (1986): Adenovirus type 12 early region 1A proteins repress class I HLA expression in transformed human cells. Proc Natl Acad Sci USA 83:5257–5261.

Vazeux R, Cumont M, Girard PM, Nassif X, Trotot P, Marche C, Matthiessen L, Vedrenne C, Mikol J, Henin D, Katlama C, Bolgert F (1990): Severe encephalitis resulting from coinfections with HIV and JC virus. Neurology 40:944–948.

Völter C, Zur Hausen H, Alber D, De Villiers E (1997): Screening human tumor samples with a broad-spectrum polymerase chain reaction method for the detection of polyomaviruses. Virology 237:389–396.

Voltz R, Jager G, Seelos K, Fuhry L, Hohlfeld R (1997): BK virus encephalitis in an immunocompetent patient. Arch Neurol 53:101–103.

Walker DL, Frisque RJ (1986): The biology and molecular biology of JC virus. In The Papoviridae. The Polyomaviruses; Salzman NP, Ed; Plenum Press: New York, Vol 1, pp 327–377.

Walker DL, Padgett BL, Zu Rhein GM, Albert AE, Marsh RF (1973): Current study of an opportunistic papovavirus. In Slow Virus Diseases; Zeman W, Lennette EH, Eds; Williams & Wilkins: Baltimore, pp 49–58.

Walter G, Ruediger R, Slaughter C, Mumby M (1990): Association of protein phosphatase 2A with polyoma virus medium tumor antigen. Proc Natl Acad Sci USA 87:2521–2525.

Wang SS, Esplin ED, Li JL, Huang L, Gazdar A, Minna J, Evans GA (1998): Alterations of the PPP2R1B gene in human lung and colon cancer. Science 282:284–287.

Watanabe S, Yoshiike K (1982a): Change of DNA near the origin of replication enhances the transforming capacity of human papovavirus BK. J Virol 42:978–985.

Watanabe S, Yoshiike K, Nozawa A, Yuasa Y, Uchida S (1979): Viable deletion mutant of human papovavirus BK that induces insulinomas in hamsters. J Virol 32:934–942.

Watanabe S, Kotake S, Nozawa A, Muto T, Uchida S (1982b): Tumorigenicity of human BK papovavirus plaque isolates, wild-type and plaque morphology mutants, in hamsters. Int J Cancer 29:583–589.

Wiley CA, Grafe M, Kenney C, Nelson JA (1988): Human immunodeficiency virus (HIV) and JC virus in acquired immune deficiency syndrome (AIDS) patients with progressive multifocal leukoencephalopathy. Acta Neuropathol 76:338–346.

Williams RL, Risau W, Zerwes H-G, Drexler H, Aguzzi A, Wagner EF (1989): Endothelioma cells expressing the polyoma middle T oncogene induce hemangiomas by host cell recruitment. Cell 57:1053–1063.

Wold WSM, Mackey JK, Brackmann KH, Takemori N, Rigden P, Green M (1978): Analysis of human tumors and human cell lines for BK virus-specific DNA sequences. Proc Natl Acad Sci USA 75:454–458.

Wolman SR, Cannuto PM, Golimbu M, Schinella R (1988): Cytogenetic, flow cytometric, and ultrastructural studies of twenty-nine nonfamilial human renal carcinomas. Cancer Res 48:2890–2897.

Yang RCA, Wu R (1979): BK virus DNA: Complete nucleotide sequence of a human tumor virus. Science 206:456–462.

Yogo Y, Furuno A, Nozawa A, Uchida S (1981): Organization of viral genome in a T antigen–negative hamster tumor induced by human papovavirus BK. J Virol 38: 556–563.

Yogo Y, Furuno A, Watanabe S, Yoshiike K (1980): Occurrence of free, defective viral DNA in a hamster tumor induced by human papovavirus BK. Virology 103:241–244.

Yogo Y, Hara K, Guo J, Taguchi F, Nagashima K, Akatani K, Ikegami N (1993): DNA-sequence rearrangement required for the adaptation of JC polyomavirus to growth in a human neuroblastoma cell line (IMR-32). Virology 197:793–795.

Yogo Y, Kitamura T, Sugimoto C, Ueki T, Aso Y, Hara K, Taguchi F (1990): Isolation of a possible archetypal JC virus DNA sequence from nonimmunocompromised individuals. J Virol 64:3139–3143.

Yogo Y, Kitamura T, Sugimoto C, Hara K, Lida T, Taguchi F, Tajima A, Kawabe K, Aso Y (1991): Sequence rearrangement in JC virus DNAs molecularly cloned from immunosuppressed renal transplant patients. J Virol 65:2422–2428.

Yoshiike K, Takemoto KK (1986): Studies with BK virus and monkey lymphotropic papovavirus. In The Papovaviridae. The Polyomaviruses; Salzman NP, Ed; Plenum Press: New York, Vol 1, pp 295–326.

zur Hausen H (1991): Papillomavirus/host cell interactions in the pathogenesis of anogenital cancer. In Origins of Human Cancer. A Comprehensive Review; Brugge J, Curran T, Harlow E, McCormick F, Eds; Cold Spring Harbor Laboratory Press: New York, pp 695–705.

17

SV40 AND HUMAN TUMORS

AMY S. ARRINGTON, B.A., and JANET S. BUTEL, PH.D.

1. INTRODUCTION INTO THE HUMAN POPULATION

The discovery of SV40, as well as its widespread introduction into the human population, was tied to the development and distribution of poliovaccines. Early poliovaccines that were inadvertently contaminated with SV40 were distributed worldwide, exposing hundreds of millions of persons to the virus. This exposure involved predominantly the formalin-inactivated Salk vaccine and, to a lesser extent, the live attenuated Sabin vaccine. The viral contamination occurred because these early vaccines were prepared in primary cultures of kidney cells derived from rhesus monkeys, which are often naturally infected with SV40. It is estimated that over 98 million people, or over 62% of the U.S. population, were exposed to poliovaccines potentially contaminated with SV40 during the years of 1955 to 1963 (Shah and Nathanson, 1976). In addition to these vaccines, some adenovirus types 3 and 7 vaccines, distributed on a much more limited scale to military and civilian personnel from 1961 to 1965, also contained SV40 viral sequences (Lewis et al., 1973). It is possible that human infections by SV40 predate the vaccine era, perhaps as a result of exposures to infected primates or primate tissues, but the widespread use of the contaminated poliovaccines surely provided a significant exposure mechanism (Butel and Lednicky, 1999). The characteristic SV40-induced cytopathic effect of cell vacuolization in African green monkey kidney cells led to the recognition and isolation of the virus in 1960 by Sweet and Hilleman. Soon thereafter, measures

Human Polyomaviruses: Molecular and Clinical Perspectives, Edited by Kamel Khalili and Gerald L. Stoner.
ISBN 0-471-39009-7

were implemented to ensure that the vaccines were free from SV40 contamination.

Shortly after its discovery, SV40 was found to induce tumors in animals, particularly in Syrian hamsters (Eddy et al., 1962; Girardi et al., 1962; Butel et al., 1972; Diamandopoulos, 1972), and to possess the ability to transform a variety of cell types from different species in tissue culture. These experiments that established the potent oncogenic potential of SV40 raised concerns of whether SV40 might have caused infections in humans exposed to the virus via the contaminated vaccines and whether the vaccinees might be at a higher risk for disease. To ascertain whether humans could be infected by SV40, studies focused first on the immune response to the virus. A study was conducted in 1961 of 35 adult volunteers who were accidentally inoculated by the respiratory route with SV40 present in a candidate vaccine to respiratory syncytial virus. These volunteers, given 10^4 $TCID_{50}$ of virus, developed subclinical infections and detectable SV40-specific neutralizing antibody. Infectious SV40 was isolated from throat swabs of three of the test subjects 7 or 11 days after virus exposure (Kravetz et al., 1961; Morris et al., 1961; Shah and Nathanson, 1976). In 1962, a group of children who were 3–6 months of age at the time of virus feeding through the oral poliovaccine were reported to harbor subclinical infections, excreting SV40 in their stools for up to 5 weeks (Melnick and Stinebaugh, 1962). Although these studies suggested that SV40 was capable of establishing infections in human beings, the proposition that the virus might be involved in human carcinogenesis was virtually dismissed despite the evidence that it was highly tumorigenic in rodents.

The prevalence and distribution of SV40 infections in the human population today are not known, and the epidemiologic data that are available are subject to interpretational difficulties. It cannot be known with certainty which individuals actually received contaminated vaccines, the dosage of infectious SV40 present in a particular inoculum is indeterminable for those who were exposed (Shah, 1998), and it is extremely difficult to follow a cohort of subjects for decades after virus exposure to monitor for delayed adverse health effects. One long-term follow-up study included 1073 children born between 1960 and 1962 who received either oral or inactivated poliovaccine. No evidence of increased cancer risk was detected in those subjects 17–19 years later (Mortimer et al., 1981). Recent retrospective studies of the general vaccinated population (Geissler, 1990; Olin and Giesecke, 1998; Strickler et al., 1998) found no definitive evidence linking SV40 to the occurrence of specific types of human neoplasms. Some studies (Olin and Giesecke, 1998; Strickler et al., 1998) based their conclusions on the assumption that after vaccines were cleared of SV40 in 1963, no further SV40 exposure occurred in the human population. This assumption is arguable, considering the recent detection of SV40 DNA in pediatric tumors and tissues (Bergsagel et al., 1992; Lednicky et al., 1995; Suzuki et al., 1997; Butel et al., 1999a,b; Zhen et al., 1999). In contrast, one recent age-controlled epidemiologic study suggested a higher incidence of certain tumors in populations exposed to potentially contaminated vaccines (Fisher et al., 1999). The significance of these observations remains uncertain; epidemiologic studies that include SV40 serology are needed to address the prevalence and geographic distribution of SV40 infections in humans.

The advent of the polymerase chain reaction (PCR) technique has resulted in an increased ability to detect viral sequences in human tissue and has reopened the question of a possible role for SV40 in the etiopathogenesis of human tumors. Reports linking the presence of SV40 viral DNA with specific human tumor types have accumulated in recent years, making the potential oncogenicity of SV40 in humans a medically pertinent and scientifically pressing issue. The remainder of this chapter is devoted to a review of SV40 in human tumors, a discussion of the possible mechanisms by which this small DNA tumor virus may be contributing to human oncogenesis, and a consideration of criteria that must be met to confirm the role of SV40 in human tumor development.

2. SV40 ONCOPROTEINS AND CELLULAR TRANSFORMATION

An appreciation of the early gene products of SV40 and their interactions with cellular proteins is fundamental to an understanding of the possible role of SV40 in the etiology of human tumors. SV40 is classified as a member of the *Polyomavirus* genus of the family Papovaviridae. It encodes both early and late gene products, but for the purposes of this review only the early genes are discussed. More in-depth reviews of the structure and function of the viral genome are available (Butel et al., 1972; Butel, 1986; Pipas, 1992; Butel and Lednicky, 1999). The early proteins of SV40 consist of two nonstructural (replication) proteins, large T antigen (TAg) and small t antigen (tAg), which are produced soon after infection by alternative splicing of viral transcripts.

The most well-characterized protein of SV40 is TAg, a highly multifunctional oncoprotein that consists of 708 amino acids and several defined functional domains (Pipas, 1992; Conzen and Cole, 1994; Butel and Lednicky, 1999) (Fig. 17.1). TAg is required for initiation of viral DNA replication, stimulation of cell entry into S phase, and interaction with cellular proteins, the latter two roles of which are crucial in both viral replication and transformation of cells.

Functional domains of TAg that are indispensable in the subversion of cell cycle control include binding sites for cellular proteins p53, pRb, and Rb family members p107 and p130 (Conzen and Cole, 1994; Butel and Lednicky, 1999). The amino terminus of TAg contains two distinct domains important in transformation. The extreme amino terminus of TAg includes the J domain, which is involved in the formation and dissolution of protein complexes and in the prevention of apoptosis (Campbell et al., 1997; Kelley and Georgopoulos, 1997; Srinivasan et al., 1997; Stubdal et al., 1997; Slinskey et al., 1999). The second region of the amino terminus of TAg mediates the binding of pRb and

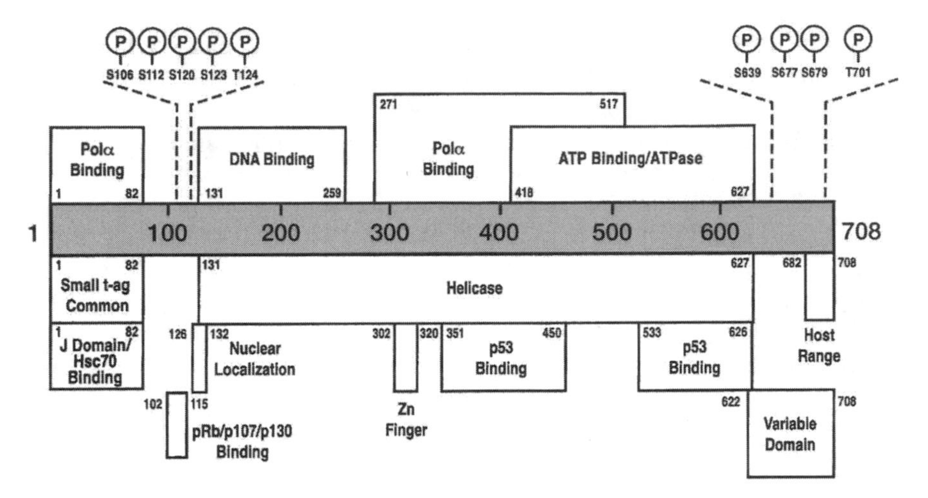

Figure 17.1. Functional domains of SV40 TAg. The numbers given are the amino acid residues using the numbering system for SV40-776. Regions are indicated as follows. tAg Common: region of TAg encoded in the first exon; the amino acid sequence in this region is common to both TAg and tAg. Polα Binding: regions required for binding to polymerase α–primase. J Domain/Hsc70 Binding: region required for binding the heat shock protein hsc70. pRb/p107/p130 Binding: region required for binding of the Rb tumor suppressor protein and the Rb-related proteins p107 and p130. Nuclear Localization: contains the nuclear localization signal. DNA Binding: minimal region required for binding to SV40 *Ori* DNA. Helicase: region required for full helicase activity. Zn Finger: region that binds zinc ions. p53 Binding: regions required for binding the p53 tumor suppressor protein. ATP Binding/ATPase: region containing the ATP binding site and ATPase catalytic activity. Host Range: region defined as containing the host range and adenovirus helper functions. Variable Domain: region containing amino acid differences among viral strains. The circles containing a P indicate sites of phosphorylation found on TAg expressed in mammalian cells. S indicates a serine and T indicates a threonine residue. Taken with permission from Stewart et al., 1996.

the pRb family members p107 and p130 (Fanning, 1992; Slinskey et al., 1999; Butel, 2000).

The mechanisms of action of p107 and p130 remain enigmatic, but the function of tumor suppressor protein pRb as a G1 checkpoint control is well understood. E2F transcription factors are functionally inactivated when bound to hypophosphorylated pRb, thereby holding cells in the G1 phase of the cell cycle. When cell cycle–regulated kinases phosphorylate pRb, E2F factors are released to initiate transcription of S-phase genes, and entry into S phase follows. In SV40-expressing cells, the amino terminus of SV40 TAg binds hypophosphorylated pRb and active E2F transcription factors are released; this unscheduled release results in a disruption of cell cycle regulation, ultimately leading to more favorable conditions for cellular transformation (Butel, 2000).

Another important cell cycle regulator targeted by SV40 TAg is p53. The p53 tumor suppressor protein, recognized as a guardian of the genome (Levine,

1997), was first identified as a cellular protein bound to SV40 TAg in SV40-transformed cells (Lane and Crawford, 1979; Linzer and Levine, 1979). p53 has been shown to play a critical role in mammalian cell cycle G1 checkpoint control and to induce apoptosis when overexpressed in cultured cells (Shaw et al., 1992; Amundson et al., 1998; Giaccia and Kastan, 1998; Oren, 1999). It is hypothesized that a possible mechanism of cell regulation by this "molecular policeman" is the induction of cell death to prevent potentially mutagenic events from being stabilized by the division of cells containing damaged, unrepaired DNA (Vogelstein et al., 2000; Vousden, 2000). p53 is found mutated or lost in up to 50% of all human cancers, emphasizing the importance of its functional loss in carcinogenesis (Harris, 1996; Hollstein et al., 1996; Levine, 1997; Hainaut and Hollstein, 2000). Two p53 binding domains that encompass amino acids 351–450 and 533–626 lie near the carboxy-terminal end of SV40 TAg; TAg functions by binding to and inhibiting p53 activity, thereby allowing unchecked progression through the cell cycle. With time, this atypical proliferation coupled with the absence of p53-induced death of damaged cells could result in an accumulation of cellular mutations and increased genomic instability.

The interaction of TAg with members of the CBP family of transcriptional coactivators, CBP (CREB-binding protein), p300 and p400, is less well-understood. Data suggest that TAg interacts with these proteins through multiple regions of TAg, including interactions at both the carboxy terminus (Lill et al., 1997) and the amino terminus near the cr2 domain (Eckner et al., 1996; Srinivasan et al., 1997). The binding details remain to be elucidated, but most likely a combination of direct and indirect interactions through multiprotein complexes allow for the association of CBP, p300, and p400 with SV40 TAg.

tAg shares 82 amino acids with the amino-terminal domain of TAg and inhibits protein phosphatase 2A (PP2A) activity by forming complexes with the catalytic subunit and one regulatory subunit of PP2A (Pallas et al., 1990; Yang et al., 1991). tAg is also believed to serve an inhibitory role to cellular apoptosis induced by TAg, leading to more efficient transformation of rat embryo fibroblasts (Kolzau et al., 1999). A recent study suggested that tAg may be involved in cell cycle progression into the G2 phase in SV40-infected CV-1 cells, possibly by affecting the regulation of cell cycle proteins such as cyclin D1 and $p27^{kip1}$ (Whalen et al., 1999). Transgenic animals carrying SV40 sequences with a deleted tAg gene consistently develop tumors in highly mitotic tissues, but show reduced tumor development in nondividing tissues relative to wild-type virus (Choi et al., 1988; Carbone et al., 1989), suggesting that tAg contributes to transformation of resting cells. Unlike TAg, tAg alone is incapable of cellular transformation. The major functions of tAg in transformation seem to be stimulation of growth and prevention of death of target cells, thereby increasing transformation driven by TAg (Porrás et al., 1999).

When considering the possible role of SV40 in human cancer etiology, it is useful to compare its transforming mechanisms to those of a well-established human cancer virus. Human papillomavirus (HPV) is recognized as the causative agent of cervical cancers, accounting for at least 90% of cases worldwide (zur Hausen, 1996; Herrero and Muñoz, 1999; Butel, 2000). HPV is similar to SV40 in that it targets both p53 and pRb for inactivation, but does so using two different viral proteins: E7, which binds the pRb family; and E6, which targets p53 for ubiquitin-mediated degradation. Like SV40 TAg, HPV E7 and E6 functionally inactivate those cellular tumor suppressors, abrogating cell cycle checkpoints, allowing for increased cellular mitoses and proliferation and, in some cases, progression to a cancerous state. Because SV40 modulates the same cellular control pathways as HPV, by analogy it possesses the genetic capability of causing malignant transformation of human cells. In fact, as described below, SV40 transformation of human cells has been demonstrated directly. Due to the accumulating reports associating SV40 DNA with human tumors, the possibility that this DNA tumor virus may be a factor in the etiopathogenesis of human neoplasms warrants serious scientific scrutiny.

3. SV40 TRANSFORMATION OF HUMAN CELLS

The transforming properties of SV40 have been characterized in a variety of cell types from many species, including humans, hamsters, mice, rats, cattle, and guinea pigs (Butel et al., 1972; Butel and Jarvis, 1986; Levine, 1988; Butel and Lednicky, 1999). Infection of cells that are nonpermissive for replication by SV40 sometimes results in the random integration of viral DNA into host chromosomal DNA. The integration of SV40 DNA is not site specific, but the early region of the viral genome is always retained in cells expressing transformed phenotypes. However, episomal viral genomes have also been detected in transformed cells and tumors (Hwang and Kucherlapati, 1980; Lednicky and Butel, 1999), and it remains to be determined whether persistent episomal viral DNA in the absence of integrated viral copies can maintain the transformed state.

SV40 genes function primarily for the production of progeny infectious virions, allowing perpetuation of viral life cycles. Cellular transformation and immortalization are secondary results of nonlytic infections of host cells, whereby viral proteins, particularly TAg, are expressed continuously, usually due to the integration of viral DNA into the host genome. Different cell types vary in permissivity to SV40 infections. Monkey cells, such as rhesus monkey kidney cells from which SV40 was first isolated, are considered to be permissive to SV40 infection, as the virus infects most cells and replicates to relatively high levels, resulting in large numbers of infectious progeny virions and lytic cell death. In contrast, mouse cells are termed nonpermissive, as SV40 infection typically leads to an abortive cycle that fails to produce infectious progeny; in very rare instances, the viral genome may be incorporated into host chromosomal DNA (Grodzicker and Hopkins, 1980). Human cells are considered to be "semipermissive" to SV40 infection: Viral replication occurs in most human cells, but relatively less efficiently as only low levels of infectious progeny are usually produced.

Many types of human cells have been immortalized by SV40 genes, ranging from fibroblasts to melanocytes to mesothelial cells (Bryan and Reddel, 1994). Diploid human fibroblasts, used as a model for the evaluation of SV40-mediated transforming effects on primary human cells, have revealed that morphologic transformation and immortalization are two distinct and separable events in human cells. The lifespan of primary fibroblasts in culture is approximately 50–60 generations before cells reach senescence, a phenomenon characterized by irreversible growth arrest without actual cell death (Jha et al., 1998). Infection by SV40 results in a characteristic transformed phenotype, including extension of lifespan up to 70–80 generations. Eventually, cells enter a "crisis" period, in which a significant proportion of cells die via apoptotic and other less well-characterized pathways (Jha et al., 1998). The few cells that survive this period are typically capable of continuous cell growth and, by convention, those reaching 100 population doublings are termed immortal. Immortalization is a nonspontaneous and rare event distinct from transformation, occurring at a frequency of 1×10^{-5} to less than 1×10^{-8} (Jha et al., 1998). This pattern in human cells is in contrast to the response of rodent cells, in which SV40-mediated transformation and immortalization occur simultaneously and in a relatively high proportion of infected cells (Tevethia et al., 1998). Assays that have been developed for the evaluation of cellular transformation and immortalization include anchorage-independent growth in a semisolid media (such as agarose), multilayer growth (loss of contact inhibition) with an increase in saturation density, focus formation (Brugge and Butel, 1975), and a reduction in the requirement for growth factors found in serum.

Transformation and immortalization of primary human fibroblasts by SV40 seem to occur through pRb and p53 pathways, as TAg mutants defective for binding those proteins were unable to complement an SV40 mutant temperature-sensitive for immortalization of cells at the nonpermissive temperature (Shay and Wright, 1989). Three domains of SV40 TAg are involved in immortalization of human fibroblasts: the extreme amino terminus, encompassing the J domain of TAg; a more central region of TAg, containing the Rb-binding domain; and the carboxy-terminus of TAg, containing the p53 binding regions (Conzen and Cole, 1994; Cole, 1996). Deletions within the carboxy-terminal region prevented the immortalization, but not transformation, of fibroblasts (O'Neill et al., 1995).

Animal models have played an important role in defining SV40 as a DNA tumor virus. Dr. Bernice Eddy first demonstrated that SV40-infected rhesus monkey kidney cells induced sarcomas in newborn hamsters at the site of inoculation (Eddy et al., 1962). Subsequently, intracranial injection of SV40 produced ependymomas in both newborn hamsters and *Mastomys* (Gerber and Kirschstein, 1962; Rabson et al., 1962), whereas intravenous injection of SV40 into weanling hamsters produced lymphocytic leukemia, lymphoma, soft tissue sarcoma, and osteosarcomas (Diamandopoulos, 1972). Intrapleural injection of

SV40 resulted in mesotheliomas in 100% of weanling hamsters, whereas intracardial and intraperitoneal inoculation caused more than 50% of animals to develop mesotheliomas (Cicala et al., 1993).

The consensus of these model systems is that SV40 is a strong oncogenic agent. Although rodent cells are typically more susceptible to SV40 transformation than normal human cells, common patterns between human and rodent systems suggest that SV40-mediated transformation is a plausible biologic possibility with respect to the association of SV40 DNA with human tumors. The expression of SV40 TAg and the association of TAg with cellular tumor suppressor proteins p53 and pRb in tumor cells are important factors to be considered in deciphering the possible role of SV40 in human tumorigenesis.

4. DETECTION OF SV40 IN HUMAN TUMORS

The earliest observation of an association between SV40 and human cancer was by Soriano et al. (1974), who reported the detection of SV40 in a patient with metastatic malignant melanoma. Virus was isolated from a lung metastasis, and both SV40 TAg and capsid (V) antigen were detected in lung, liver, and muscle metastases, but not in normal tissue; in addition, the patient was seropositive for SV40 neutralizing antibodies. Since then, reports linking SV40 and human tumors have accumulated dramatically, and the possible role of SV40 in cancer etiology now requires serious consideration. The advent of PCR technology has increased greatly the sensitivity of detection of viral genomes in either fresh-frozen or archival paraffin-embedded samples of human tumor tissue; sequencing of viral products, Southern blotting analysis, and other confirmatory techniques have enhanced the reliability of PCR-based studies. In the following sections, the detection and characterization of SV40 viral sequences in different types of human tumors are summarized. However, the mere detection of viral sequences in tumor tissue does not prove a direct role for the virus in the etiopathogenesis of the cancer. Approaches to prove causation are considered in the final section.

SV40 and Human Mesotheliomas

A large number of reports have described the association of SV40 with malignant mesothelioma, an aggressive, rare tumor of the lung pleura, pericardium, or peritoneum that is invariably fatal (Table 17.1). The predominant cause of this malignancy is believed to be the environmental carcinogen asbestos, although up to 20% of cases involve patients with no known asbestos exposure, suggesting the existence of another causative agent or co-factor (Carbone et al., 1994; Testa et al., 1998). It is estimated that only 10% of individuals exposed to high levels of asbestos develop malignant mesothelioma, and it is not known what determines the susceptibility of this small fraction of the exposed population to tumor development.

An emerging possibility of a viral co-factor is linked to the frequent association of SV40 with mesothelioma (Table 17.1). However, despite the many positive reports, several studies have failed to find viral sequences in similar samples, raising both technical and biologic issues to be considered (Strickler et al., 1996; Hirvonen et al., 1999; Mulatero et al., 1999; Emri et al., 2000). A multi-laboratory study by Testa et al. (1998) addressed several fundamental questions, including whether authentic SV40 DNA was detected in human mesotheliomas, whether viral sequences could be detected consistently in positive samples by different laboratories, and whether viral genes were expressed in tumor cells. SV40 viral DNA was detected by PCR using primers directed at two conserved regions of the SV40 TAg gene, and positive PCR products were verified by Southern blot hybridization and/or DNA sequencing. Each participating laboratory consistently demonstrated the presence of SV40 viral DNA in frozen mesothelioma samples. Immunostaining using two different anti-TAg monoclonal antibodies was performed on selected samples, and TAg expression was detected. Significantly, one of the participating laboratories had previously published negative findings from an examination of Finnish mesotheliomas. One possible explanation for the failure of that group to detect SV40 sequences in their previous study is based on geographical differences in the use of contaminated poliovaccines. SV40-tainted vaccines were not administered in Finland, whereas the multi-institutional study examined mesotheliomas from patients in the United States, where an estimated 98 million people were exposed to potentially contaminated vaccines (Shah and Nathanson, 1976; Butel and Lednicky, 1999).

Although a strong association between SV40 DNA and mesotheliomas has been established, PCR analysis alone does not establish expression or functionality of TAg, the viral transforming protein. Mesotheliomas are unique neoplasms in that they rarely, if ever, exhibit p53 or pRb mutations (Carbone et al., 1997; De Luca et al., 1997; Waheed et al., 1999). Based on this phenomenon, one pathway in the development of these tumors could involve the inactivation of wild-type tumor suppressors by some way other than mutation, such as by complex formation with SV40 TAg. Carbone et al. (1997) found using immunohistochemical assays that p53 and TAg expression were significantly correlated within the same mesothelioma cells and by using Western blot analysis that TAg co-precipitated with p53. These data suggest that in some human malignant mesothelioma cells TAg is capable of binding, stabilizing, and presumably inactivating wild-type p53. In a related study, De Luca et al. (1997) showed that TAg extracted from mesothelioma cells was able to bind each of the retinoblastoma family proteins, including pRb, p107, and pRb2/p130, further implicating TAg in the possible inactivation of tumor suppressor proteins within the tumor cells.

Considering that SV40 may promote oncogenesis, at least in part through a p53-mediated pathway, a prediction would be that the transformation phenotype of SV40 TAg-positive cells would be reversed upon disruption of p53/TAg complex formation. This hypothesis was tested in vitro by Waheed et al. (1999).

Table 17.1. Detection of SV40 in Mesotheliomas

Source of Tissue	No. Samples Pos./No. Tested	Method(s) of Analysis	Paraffin (P) or Frozen (F) Samples	Assay Conditions (primers; PCR cycles)	Study
Meso	29/48	PCR, sequencing, IHC	F	PYV.for/PYV.rev; SV.for3/SV.rev; 40 cycles	Carbone et al. (1994)
Meso	8/11	PCR	Both	SV.for3/SV.rev; 45 cycles	Cristaudo et al. (1995)
Meso	0/48	PCR	P	SV.for3/SV.rev; PYV.for/PYV.rev; 40 cycles	Strickler et al. (1996)
Meso	6/9	PCR	Both	SV.for3/SV.rev; PYV.for/PYV.rev	Pepper et al. (1996)
Meso	14/34	IHC, Western blot	F	N/A	Carbone et al. (1997)
Meso	30/35	PCR, sequencing, IHC	F	SV.for3/SV.rev; 40 cycles	De Luca et al. (1997)
Meso	10/12	PCR, sequencing, Southern blot, IHC	F	SV5/SV6; 2573/2902; 45 cycles	Testa et al. (1998)
Meso	6/9	PCR	Both	PYV.for/PYV.rev; SV.for3/SV.rev; 45 cycles	Gibbs et al. (1998)
Meso	26/26	PCR, sequencing, Southern blot	P	SV.for2/SV.rev; SV.for3/SV.rev; PYV.for/PYV.rev; LA1-LA2	Griffiths et al. (1998)
Meso	9/25	PCR, sequencing		SV.for3/SV.rev; 40 cycles	Mutti et al. (1998)
Meso	38/42	PCR, Southern blot	F	SV.for3/SV.rev; SV.for2/SV.rev; R1/R2; T7/T8; 45 cycles	Pass et al. (1998)
Meso	7/18	PCR, Southern blot	P	SV.for2/SV.rev; PYV.for/PYV.rev; LA1/LA2; 44 cycles	Procopio et al. (1998)
Meso	29/48	PCR, IHC, Western blot	F	SV.for/SV.rev; PYV.for/PYV.rev	Rizzo et al. (1998)

Meso	57/118	PCR, sequencing	F, P[a]	SV.for3/SV.rev	Shivapurkar et al. (1999)
Meso	13/28	PCR, IHC	Both	SV.for3/SV.rev; 45 cycles	Dhaene et al. (1999)
Meso (Finnish)	0/49	PCR	F	SV.for2/SV.rev; 40 cycles	Hirvonen et al. (1999)
Meso	5/11	PCR, sequencing	P	SV.for3/SV.rev; 45 cycles	Mayall et al. (1999)
Meso	0/12	PCR	P	PYV.for/PYV.rev; 40 cycles	Mulatero et al. (1999)
Meso	14/25	"PRINS," IHC	P	N/A	Ramael et al. (1999)
Meso	8/12	PCR, sequencing	F	SV5/SV6; R1/R2; 45 cycles	Rizzo et al. (1999)
Meso	1/1	PCR, sequencing	P	RA1/RA2; RA3/RA4; TA1/TA2; 70 cycles	Arrington et al. (2000)
Meso (Turkish, Italian)	0/29 Turkish 1/1 Italian	PCR	P	SV1/SV2; T3/T4	Emri et al. (2000)
Meso	10/18	PCR, Southern blot	P	RA3/RA4; 40 cycles	Cristaudo et al. (2000)
Meso	50/83	PCR, Southern blot	P	PYV.for/PYV.rev; 44 cycles	Procopio et al. (2000)
Meso	9/23	PCR, Southern blot	P	SV.for2/SV.rev; PYV.for/PYV.rev; 44 cycles	Strizzi et al. (2000)
Meso	44/78	PCR, sequencing	P	SV.for3/SV.rev; 45 cycles	Shivapurkar et al. (2000)
Patients	7/7	PCR	F	SV.for2/SV.for3; SV.rev; SV8/SV9	McLaren et al. (2000)

Meso = mesothelioma; PCR = polymerase chain reaction; IHC = immunohistochemistry; PRINS = primed in situ method; N/A = not applicable.
[a]Microdissected.

An adenovirus vector was used to express antisense SV40 early region transcripts in human pleural mesothelioma cell lines, some of which contained SV40 DNA, and the effect on cellular proliferation and apoptosis was measured. The antisense SV40 early region construct abrogated TAg, and this coincided with an upregulation of p21/WAF-1 expression, growth inhibition, and apoptosis. This antisense construct had no effect on cell lines that contained no SV40 DNA. These results suggest that TAg was acting through a p53-mediated pathway and was required for the maintenance of the transformed phenotype in the virus-positive cells (Waheed et al., 1999).

It is estimated that less than 10% of primary human fibroblasts infected by SV40 become transformed and immortalized (O'Neill et al., 1995; Ozer et al., 1996). The high percentages of mesotheliomas found to contain SV40 DNA suggest that mesothelial cells may be relatively more susceptible to SV40 infection and transformation than other human cells. To test this hypothesis, Bocchetta et al. (2000) examined the effect of SV40 infection on primary mesothelial cells in culture. Infected mesothelial cells expressed TAg in 100% of cells, contained high levels of p53/TAg complexes, and underwent limited cell lysis. The rate of transformation of infected mesothelial cells was 1000 times higher than that of human fibroblasts, perhaps because abrogation of p53 function by TAg resulted in reduced cell lysis and allowed prolonged exposure to the proliferative effects of SV40 TAg. Although this study did not address why mesothelial cells are targeted by virus infection in vivo rather than adjacent stromal cells (Shivapurkar et al., 1999; Bocchetta et al., 2000; Carbone, 2000; Jasani et al., 2000), it did suggest an explanation for a relationship between SV40 and human mesotheliomas.

Factors Affecting PCR Analyses of Human Tumors for Viral Sequences

The PCR technique permits a rapid evaluation of large numbers of samples, and the sensitivity of the assay permits detection of very low levels of viral sequences in specimens. The presence of viral sequences in mesotheliomas has been documented in numerous independent studies worldwide. However, several studies failed to find tumor-associated SV40 (Strickler et al., 1996; Hirvonen et al., 1999; Mulatero et al., 1999; Emri et al., 2000) (Table 17.1), raising questions regarding technical issues that should be considered.

The DNA extraction step determines whether tissues yield adequate and suitable DNA for PCR analysis. Unfortunately, with archival formalin-fixed, paraffin-embedded tissue, degradation of tissue is a common, but variable, problem, and the quality of recovered DNA may be poor. If only minimal amounts of paraffin-embedded tissue are available, the yield of DNA may be inadequate for analysis. The use of appropriate controls is very important in the proper analysis of tissue for viral DNA. Primers directed to a cellular gene should be used to establish the suitability of a DNA sample for PCR analysis. Because of the sensitivity of the PCR assay, inadvertent laboratory contamination of samples during processing or testing must be rigorously guarded against. It is best to process tissue samples and set up PCR reactions in containment facilities from which positive controls (plasmids, viruses) are excluded. Negative tissue controls, extracted and analyzed in parallel, as well as water controls (lacking template) should be included in each experiment to monitor for reagent contamination. It is necessary to confirm the authenticity of viral PCR products by specific techniques, such as direct DNA sequencing or Southern blot analysis with SV40-specific probes. Additionally, PCR analysis using sets of primers directed to different conserved regions of the SV40 genome, such as the amino terminus of TAg and the SV40 regulatory region, further confirms the presence of authentic SV40 viral genomes in human tumor tissue.

The possibility that SV40 sequences reportedly present in human tumors were due to laboratory contamination was ruled out by extensive sequence analyses. Studies showed that tumor-associated sequences could be distinguished from known laboratory strains of SV40, based on a newly recognized variable domain at the carboxy terminus of the TAg gene and on polymorphisms and complex duplications or rearrangements in the viral regulatory region (Lednicky et al., 1995, 1997; Stewart et al., 1998; Butel and Lednicky, 1999; Arrington et al., 2000).

Results of PCR-based studies are highly dependent on PCR conditions (i.e., primer selection, annealing temperature, and number of amplification cycles). When less than 50 cycles of PCR are used, it is possible that false-negative results may occur due to lack of sensitivity (i.e., inability to detect a low copy number of viral genomes per sample) (Griffiths et al., 1998; Jasani, 1999). This may have been a factor in some studies that reported negative results, such as one in which only 40 cycles of PCR were used to evaluate paraffin-embedded mesothelioma samples (Mulatero et al., 1999). Griffiths et al. (1998) also evaluated the specificity of particular primer pairs directed at different regions of the SV40 genome. Primer pairs giving smaller products ($\leq$250 bp) tended to be the most sensitive for viral DNA detection, presumably because fixative-induced damage and degradation of template DNA may result in a decreased ability to generate longer products (Griffiths et al., 1998; A. S. Arrington et al., unpublished data, 2001). Many of these problems can be avoided if fresh or frozen tissues are available for analysis (Lednicky and Butel, 1998).

Tumor samples may be heterogeneous mixtures containing both normal and cancerous cells. It is expected that if SV40 is involved in the etiology of a neoplasm, the virus should be detected in the tumor cells rather than in surrounding stromal cells. Shivapurkar and colleagues (1999) used the microdissection technique to determine which cells in malignant mesotheliomas were associated with SV40. Both normal stromal and malignant cells from 118 mesotheliomas, as well as from 20 lung cancers and 14 normal reactive pleural effusions, were tested, and SV40 was present in malignant and reactive mesothelial cells but not in surrounding normal tissue or in histologically distinct tumors of the lung. This study also concluded that both frozen and paraffin-embedded samples were suitable for analysis (Shivapurkar et al., 1999).

Finally, geographic differences may play a role, as suggested by the negative studies of Finnish and Turkish mesothelioma samples (Hirvonen et al., 1999; Emri et al., 2000). It was noted that SV40-contaminated poliovaccines were not used in these areas. As the distribution of SV40 in different human populations is not known, it is possible that tumors from distinct geographic regions may have different etiologies and may vary in their association with SV40.

SV40 and Human Brain Tumors

Brain tumors are uncommon, accounting for only 2% of tumors in general, and are typically poor in clinical prognosis. Very little is known about the etiology of human brain tumors and the involvement of possible co-factors. SV40 has been associated with human tumors of neural origin (Table 17.2). Although the human polyomaviruses JCV and BKV are neurotropic and can be found in human brain tissue (Corallini et al., 1987; Dörries et al., 1987; Elsner and Dörries, 1992; De Mattei et al., 1995), the tumor-associated SV40-related sequences have been distinguished from those of JCV and BKV. JCV infects oligodendrocytes in the brain and is recognized as a cause of progressive multifocal leukoencephalopathy (PML), which is more prevalent in immunocompromised subjects, such as AIDS patients, than in healthy individuals (Newman and Frisque, 1999; Hou and Major, 2000). PML and meningoencephalitis also occur in SV40-infected monkeys immunocompromised due to concurrent simian immunodeficiency virus infections, showing that SV40 is similarly neurotropic (Horvath et al., 1992; Lednicky et al., 1998; Newman et al., 1998; Simon et al., 1999).

The history of SV40 and its association with human brain tumors began soon after the discovery of the virus in contaminated poliovaccines. It was demonstrated that SV40 can induce neural tumors when injected into newborn hamsters (Eddy et al., 1962; Girardi et al., 1962) and that SV40 was capable of transforming primary human astrocytes in culture (Shein, 1967). Early studies identified several SV40-positive brain tumors using indirect immunofluorescence and DNA hybridization techniques (Table 17.2) (Weiss et al., 1975; Meinke et al., 1979; Krieg et al., 1981; Krieg and Scherer, 1984). In 1977, SV40 was isolated from the cerebrospinal fluid of a newborn child (Brandner et al., 1977).

Recent technological advances, such as PCR and Western blot analysis, have yielded numerous reports of association of SV40 with human brain neoplasms. Virus-positive tumor types include astrocytomas, ependymomas, choroid plexus papillomas, glioblastomas, gliomas, gliosarcomas, medulloblastomas, meningioma, oligodendromas, and pituitary adenomas (Table 17.2). It is noteworthy that SV40 has been associated with several pediatric brain tumors, particularly ependymomas, and choroid plexus tumors. The mean age for diagnosis of these tumors is 1–2 years and 10 months, respectively (Bergsagel et al., 1992). Berg-

Table 17.2. Detection of SV40 in Human Brain Tumors

Source of Tissue	No. Samples Pos./No. Tested	Method(s) of Analysis	Paraffin (P) or Frozen (F) Samples	Assay Conditions (primers; PCR cycles)	Study
Mening	2/7	IF	F	N/A	Weiss et al. (1975)
Mixed	2/39	IF	F	N/A	Tabuchi et al. (1978)
Mixed	1/7	DNA-DNA hybridization	F	N/A	Meinke et al. (1979)
Mixed	8/35	Southern blot	F	N/A	Krieg et al. (1981)
Mixed	11/32	In situ hybridization	F	N/A	Ibelgaufts and Jones (1982)
Mixed	0/29	DNA hybridization		N/A	Dörries et al. (1987)
CP, E	10/20 CP 10/11 E	PCR, sequencing, Southern blot, IHC	F	PYV.for/PYV.rev; SV.for2/SV.rev; SV.for3/SV.rev; 45–60 cycles	Bergsagel et al. (1992)
CP, E	14/17	PCR, sequencing	F	RA1/RA2; RA3/RA4; LA1/LA2; TA1/TA2; 60 cycles	Lednicky et al. (1995)
Mixed	32/83	PCR, sequencing	Both	PYV.for/PYV.rev; up to 105 cycles	Martini et al. (1996)
Mixed gliomas	7/33	PCR, Southern blot	Both	SVT-F/SVT-R; SVP-R; 45 cycles	Suzuki et al. (1997)
Mixed	71/199	PCR, Southern blot	P	SVTAGP1/SVTAGP2; SVTAGP1/SVTAGP3; 45 cycles	Huang et al. (1999)
Mixed	43/65	IP, silver staining, Western blot	F	N/A	Zhen et al. (1999)
CP, E, G (Finnish)	0/30	PCR, Southern blot			Ohgaki et al. (2000)
Mixed	5/274	PCR, sequencing, IHC	Both	PYV.for/PYV.rev; LA1/LA2; SV.for3/SV.rev; 40 cycles	Weggen et al. (2000)

E = ependymoma; G = glioblastoma; CP = choroid plexus; Mening = meningioma; Mixed = multiple types of brain tumors; IF = immunofluorescence; IP = immunoprecipitation; IHC = immunohistochemistry; PCR = polymerase chain reaction.

sagel et al. (1992) detected SV40 sequences in 10 of 20 choroid plexus tumors and in 10 of 11 ependymomas from pediatric patients, none of whom was old enough to have received contaminated poliovaccines. Subsequently, infectious SV40 was isolated from a choroid plexus carcinoma by lipofection of tumor DNA into monkey kidney cells (Lednicky et al., 1995). Sequence analysis of this virus revealed specific nucleotide changes in both the regulatory region and the extreme carboxy terminus of the TAg gene, which distinguished the tumor isolate from known laboratory strains. This was the first report of virus isolation in support of PCR detection of viral sequences in human tumors.

These SV40-positive findings in pediatric tumors raise the possibility of transplacental or perinatal infection and suggest that the virus may be present in the human population. A recent study determined the frequency of SV40-neutralizing antibody titers in a population of unselected hospitalized children who were born from 1980 to 1995; antibodies were detected in approximately 6% of children (Butel et al., 1999b). To confirm the presence of SV40 infections, archival biopsy tissues from several antibody-positive patients were analyzed. SV40 DNA was detected by PCR and confirmed by sequencing in biopsy specimens from four children (Butel et al., 1999a). These data suggest that SV40 is present in the human population in those not exposed to potentially contaminated poliovaccines, although presumably at a lower prevalence than the human polyomaviruses JCV and BKV. The mechanisms of person-to-person spread of SV40 and in vivo dissemination to the nervous system are yet to be elucidated.

It is noteworthy that a group who detected SV40 DNA in a collection of Swiss brain tumors (Huang et al., 1999) failed to find SV40 sequences in brain tumors from Finland (Ohgaki et al., 2000). This observation is reminiscent of that reported for Finnish mesotheliomas (described above).

To investigate whether SV40 TAg is bound to p53 in brain tumors, Zhen et al. (1999) utilized immunoprecipitation and Western blot analysis on various samples, including astrocytomas, pituitary adenomas, meningiomas, ependymomas, choroid plexus papillomas, glioblastomas, and medulloblastomas. Specific TAg/p53 complexes were detected in many of the brain tumor samples analyzed. To screen for the presence of TAg/pRb-specific complexes, 15 TAg-positive brain tumor samples were analyzed by the same methodology, and the findings indicated that SV40 TAg can form specific complexes with pRb in brain tumors, as well.

There are large differences among studies of brain tumors with respect to the frequency of detection of SV40 DNA (Table 17.2), differences that might be explained by some of the same parameters considered above for mesotheliomas. However, the association of SV40 genomic DNA with human brain tumors is remarkable, and, considering the lack of knowledge of the etiology of such tumors, the possible involvement of SV40 holds promise of yielding new insights into the tumor development processes.

SV40 and Other Human Tumors and Nontumor Tissues

In addition to mesotheliomas and brain tumors, SV40 has been associated with other human tumors and nontumor tissues, including osteosarcomas, non-Hodgkin's lymphomas, AIDS-related lymphomas, peripheral blood cells, sperm fluids, and kidney tissue from pediatric renal transplant patients (Table 17.3). The relationship between SV40 and these tumors and tissues is presently unclear, as there are fewer independent reports and generally lower positivity rates than for mesotheliomas and brain tumors. However, these combined data provide sufficient molecular and cellular evidence to link SV40 as a possible cofactor in the etiology of some human neoplasms.

The greater task is now at hand: the differentiation between causal and incidental connections of SV40 with human tumors. The remainder of this chapter is dedicated to a discussion of criteria that can be used to establish causal associations.

5. SV40 AND HUMAN TUMORS: RELATING ASSOCIATION TO CAUSATION

The reproducible detection of viral DNA associated with tumor tissue is a first step in a long process needed to establish a causal role for a virus in the etiology of human cancer. The fact that SV40 DNA is consistently found in certain human tumors cannot be dismissed; its presence is highly suggestive that this well-known DNA tumor virus has the potential to play a role in human disease.

The general problems involved in establishing causal relationships between viruses and cancer were summarized by Evans and Mueller (1990), and criteria were defined that should be met to prove an etiologic link. The authors identified several problems to be addressed: (1) A long incubation period between initial infection with the virus and appearance of the cancer; (2) the common and ubiquitous nature of most viruses and the rarity of the cancer with which they are associated; (3) the initial infection with the candidate virus is often subclinical so that the time of infection cannot be established; (4) the need for co-factors in most viral-related cancers; (5) the causes of cancer may vary in different geographic areas or by age; (6) different viral strains may have different oncogenic potential; (7) human host factors are important in susceptibility to cancer, especially age at the time of infection, genetic characteristics, and status of the immune system; (8) cancers result from a complex, multistage process in which a virus may play a role at different points in pathogenesis together with alterations in the immune system, oncogenes, chromosomal translocations, and other events at the molecular level; (9) the inability to reproduce human cancer in experimental animals with the putative cancer virus; and (10) the recognition that a virus, toxin, chemical, altered gene, or other inciting factor may all be capable of causing cancers with the same histologic features.

Table 17.3. Detection of SV40 in Other Human Tumors and Nontumor Tissues

Source of Tissue	No. Samples Pos./No. Tested	Method(s) of Analysis	Paraffin (P) or Frozen (F) Samples	Assay Conditions (primers; PCR cycles)	Study
1 patient: melanoma plus other tissues	Pos. tumor and lung, liver and muscle metastases	Radioimmunoassays (TAg and VAg)	F	N/A	Soriano et al. (1974)
Osteo; other bone tumors	40/126 osteo; 14/34 other bone tumors	PCR, sequencing, Southern blot	F	SV.for2/SV.rev; LA1/LA2; T1/T2; R1/R2; 45 cycles	Carbone et al. (1996)
PBC; SF	16/70 PBC; 9/20 SF	PCR, sequencing, IF	F	PYV.for/PYV.rev; 105 cycles	Martini et al. (1996)
Osteo	5/10 osteo	PCR, sequencing	Both	RA1/RA2; RA3/RA4; LA1/LA2/LA3; TA1/TA2; SV.for3/SV.rev; 62 cycles	Lednicky et al. (1997)
Osteo	9/35 osteo	PCR, Southern blot	F	PYV.for/PYV.rev	Mendoza et al. (1998)

Osteo	48/145	PCR, sequencing, Southern blot	P	PYV.for/PYV.rev; SV.for2/SVrev; 45 cycles	Rizzo et al. (1998)
NHL; LPD; AIDS lymph	3/29 NHL; 6/25 LPD; 2/25 AIDS lymph	PCR, sequencing, Southern blot	F	SV5/SV6; R1/R2; 45 cycles	Rizzo et al. (1999)
Pediatric patients	4/13 patients	PCR, sequencing	P	SV.for3/SV.rev; RA1/RA2; 65 cycles	Butel et al. (1999a)
Osteo; PBC	25/54 osteo; 13/30 PBC	PCR, Southern blot	F	R1/R2; SVfor2/SVrev; T1/T2; 60 cycles	Yamamoto et al. (2000)
GCTs	30/107	PCR, IHC	P	PYV.for/PYV.rev; SV.for2/SV.rev; 45 cycles	Gamberi et al. (2000)
PBC; non-AIDS lymph; AIDS lymph	18/115 PBC; 8/58 non-AIDS lymph; 7/21 AIDS lymph	PCR		SVrev/Svfor/Svprobe; VPF/VPR/Vprobe; RA1/RA2/RA3/5097; TA1/CTF-0/CTF/CTR	David et al. (2001)

Osteo = osteosarcomas; PBC = peripheral blood cells; SF = sperm fluid; NHL = non-Hodgkin's lymphoma; LPD = lymphoproliferative disorders; AIDS lymph = AIDS lymphoma; IF = immunofluorescence; PCR = polymerase chain reaction; GCTs = giant cell tumors.

The history involving now-accepted human cancer viruses (hepatitis B virus, Epstein-Barr virus, HPV) illustrates the confounding role of co-factors in efforts to link viruses to cancer (Butel, 2000). In the case of SV40 and mesothelioma, asbestos has been suggested as a possible co-factor (Testa et al., 1998; Carbone, 2000; Carbone et al., 2000). With other tumor types, potential co-factors have not been identified.

Epidemiologic data are currently lacking to relate SV40 to the etiopathogenesis of a human cancer. Guidelines suggested by Evans and Mueller (1990) to define a causal link include (1) The geographical distribution of infection with the virus should be similar to that of the tumor with which it is associated when adjusted for the age of infection and the presence of known cofactors; (2) the presence of viral markers should be higher in cases than in matched controls; (3) viral markers should precede the tumor, with a higher incidence of tumors in persons with the marker than in those without; and (4) prevention of virus infection should decrease tumor incidence. These guidelines are difficult to fulfill, but they are helpful to evaluate experimental observations as they become available.

Sir A.B. Hill also proposed criteria to differentiate between causation or mere association of an environmental factor and a disease (see Hill and Hill, 1991), including a correlation between the strength of the association between the factor and the disease, the consistency of the association, specificity, demonstration of a temporal relationship (i.e., does exposure to the factor precede the disease?), a biologic gradient as well as biologic plausibility, demonstrated coherence with known facts regarding the natural history of the disease, experimental evidence, and analogy to related biologic systems. Several of these criteria have been met for SV40 and human cancer (Butel and Lednicky, 1999), including strength of association, consistency, and specificity by the independent confirmatory reports associating bona fide SV40 sequences with certain types of human tumors. Association of SV40 with human neoplasms is both biologically plausible and coherent, based on the tumorigenicity of the virus in animal models, the evidence that SV40 causes human infections, and the widespread use of contaminated poliovaccines as a mechanism for introduction of the virus into the human population. Experimental evidence supports a potential causal role for SV40 in human cancer in that the TAg gene is detected consistently in tumors positive for SV40 DNA and TAg has been shown to interact with cellular proteins p53 and pRb in some tumors. Other criteria to prove causation await validation through epidemiologic studies, including evidence of a temporal relationship. However, the criteria that have been met are sufficient to suggest that SV40 may indeed be playing a role in the etiology of some human cancers.

6. CONCLUSIONS

The preponderance of data indicates that SV40 is associated with certain types of human tumors, and the number of positive reports continues to mount.

Therefore, the basic question of whether SV40 is present in human tumors has been answered in the affirmative. The follow-up step, that of proving whether a causal link exists between SV40 and human tumors, is a challenging task, both virologically and epidemiologically. It remains a possibility that SV40 is merely a passenger in the tumors and plays no active role in the etiology of any neoplasm. This theory seems unlikely in light of the vast amount of knowledge regarding SV40 as a tumor virus and TAg as a potent oncoprotein. As SV40 and TAg were studied in the laboratory for years as models for the molecular basis of carcinogenesis, it seems highly unlikely that such an oncoprotein would be functionally dormant in human cells in vivo when expressed and interacting with tumor suppressors p53 and pRb. Demonstration of the expression and function of SV40 gene products in different types of tumors should be a priority. Other questions that need to be addressed are how the virus is being transmitted from person to person, how it is distributed throughout the infected host, how it interacts with different cells and tissues, how the host responds, and whether different virus strains vary in oncogenic potential in humans. Information such as this will reveal the extent to which this viral agent is involved in the etiology of human tumors.

There would be significant benefits if a role for SV40 in the etiology of human cancer were established. Viral markers could be used to develop improved diagnosis and treatment approaches and might even prove to be of prognostic value. New means of cancer prevention could be developed aimed at preventing viral infection or disease. Ultimately, this small DNA tumor virus may provide significant insights into the etiology of poorly understood malignancies, such as mesotheliomas and tumors of the central nervous system. It would be ironic, indeed, if the virus first discovered as a contaminant of viral vaccines became a candidate for a vaccine itself.

REFERENCES

Amundson SA, Myers TG, Fornace AJJ (1998): Roles for p53 in growth arrest and apoptosis: Putting on the brakes after genotoxic stress. Oncogene 17:3287–3299.

Arrington AS, Lednicky JA, Butel JS (2000): Molecular characterization of SV40 DNA in multiple samples from a human mesothelioma. Anticancer Res 20:879–884.

Bergsagel DJ, Finegold MJ, Butel JS, Kupsky WJ, Garcea RL (1992): DNA sequences similar to those of simian virus 40 in ependymomas and choroid plexus tumors of childhood. N Engl J Med 326:988–993.

Bocchetta M, Di Resta I, Powers A, Fresco R, Tosolini A, Testa JR, Pass HI, Rizzo P, Carbone M (2000): Human mesothelial cells are unusually susceptible to simian virus 40–mediated transformation and asbestos cocarcinogenicity. Proc Natl Acad Sci USA 97:10214–10219.

Brandner G, Burger A, Neumann-Haefelin D, Reinke C, Helwig H (1977): Isolation of simian virus 40 from a newborn child. J Clin Microbiol 5:250–252.

Brugge JS, Butel JS (1975): Role of simian virus 40 gene *A* function in maintenance of transformation. J Virol 15:619–635.

Bryan TM, Reddel RR (1994): SV40-induced immortalization of human cells. Crit Rev Oncog 5:331–357.

Butel JS (1986): SV40 large T-antigen: Dual oncogene. Cancer Surv 5:343–365.

Butel JS (2000): Viral carcinogenesis: Revelation of molecular mechanisms and etiology of human disease. Carcinogenesis 21:405–426.

Butel JS, Arrington AS, Wong C, Lednicky JA, Finegold MJ (1999a): Molecular evidence of simian virus 40 infections in children. J Infect Dis 180:884–887.

Butel JS, Jafar S, Wong C, Arrington AS, Opekun AR, Finegold MJ, Adam E (1999b): Evidence of SV40 infections in hospitalized children. Hum Pathol 30:1496–1502.

Butel JS, Jarvis DL (1986): The plasma-membrane–associated form of SV40 large tumor antigen: Biochemical and biological properties. Biochim Biophys Acta 865: 171–195.

Butel JS, Lednicky JA (1999): Cell and molecular biology of simian virus 40: Implications for human infections and disease. J Natl Cancer Inst 91:119–134.

Butel JS, Tevethia SS, Melnick JL (1972): Oncogenicity and cell transformation by papovavirus SV40: The role of the viral genome. Adv Cancer Res 15:1–55.

Campbell KS, Mullane KP, Aksoy IA, Stubdal H, Zalvide J, Pipas JM, Silver PA, Roberts TM, Schaffhausen BS, DeCaprio JA (1997): DnaJ/hsp40 chaperone domain of SV40 large T antigen promotes efficient viral DNA replication. Genes Dev 11: 1098–1110.

Carbone M (2000): Simian virus 40 and human tumors: It is time to study mechanisms. J Cell Biochem 76:189–193.

Carbone M, Lewis AM Jr, Matthews BJ, Levine AS, Dixon K (1989): Characterization of hamster tumors induced by simian virus 40 small t deletion mutants as true histiocytic lymphomas. Cancer Res 49:1565–1571.

Carbone M, Pass HI, Rizzo P, Marinetti M, Di Muzio M, Mew DJ, Levine AS, Procopio A (1994): Simian virus 40-like DNA sequences in human pleural mesothelioma. Oncogene 9:1781–1790.

Carbone M, Rizzo P, Grimley PM, Procopio A, Mew DJ, Shridhar V, de Bartolomeis A, Esposito V, Giuliano MT, Steinberg SM, Levine AS, Giordano A, Pass HI (1997): Simian virus-40 large-T antigen binds p53 in human mesotheliomas. Nat Med 3: 908–912.

Carbone M, Rizzo P, Pass H (2000): Simian virus 40: The link with human malignant mesothelioma is well established. Anticancer Res 20:875–877.

Carbone M, Rizzo P, Procopio A, Giuliano M, Pass HI, Gebhardt MC, Mangham C, Hansen M, Malkin DF, Bushart G, Pompetti F, Picci P, Levine AS, Bergsagel JD, Garcea RL (1996): SV40-like sequences in human bone tumors. Oncogene 13:527–535.

Choi YW, Lee IC, Ross SR (1988): Requirement for the simian virus 40 small tumor antigen in tumorigenesis in transgenic mice. Mol Cell Biol 8:3382–3390.

Cicala C, Pompetti F, Carbone M (1993): SV40 induces mesotheliomas in hamsters. Am J Pathol 142:1524–1533.

Cole CN (1996): Polyomavirinae: The viruses and their replication. In Fields BN, Knipe DM, Howley PM, Chanock RM, Melnick JL, Monath TP, Roizman B, Straus SE, Eds; Fields Virology, 3rd ed. Lippincott-Raven: Philadelphia, pp 1997–2025.

Conzen SD, Cole CN (1994): The transforming proteins of simian virus 40. Semin Virol 5:349–356.

Corallini A, Pagnani M, Viadana P, Silini E, Mottes M, Milanesi G, Gerna G, Vettor R, Trapella G, Silvani V (1987): Association of BK virus with human brain tumors and tumors of pancreatic islets. Int J Cancer 39:60–67.

Cristaudo A, Powers A, Vivaldi A, Foddis R, Guglielmi G, Gattini V, Buselli R, Sensales G, Ciancia E, Ottenga F (2000): SV40 can be reproducibly detected in paraffin-embedded mesothelioma samples. Anticancer Res 20:895–898.

Cristaudo A, Vivaldi A, Sensales G, Guglielmi G, Ciancia E, Elisei R, Ottenga F (1995): Molecular biology studies on mesothelioma tumor samples: Preliminary data on H-ras, p21, and SV40. J Environ Pathol Toxicol Oncol 14:29–34.

David H, Mendoza S, Konishi T, Miller CW (2001): Simian virus 40 is present in human lymphomas and normal blood. Cancer Lett 162:57–64.

De Luca A, Baldi A, Esposito V, Howard CM, Bagella L, Rizzo P, Caputi M, Pass HI, Giordano GG, Baldi F, Carbone M, Giordano A (1997): The retinoblastoma gene family pRb/p105, p107, pRb2/p130 and simian virus-40 large T-antigen in human mesotheliomas. Nat Med 3:913–916.

De Mattei M, Martini F, Corallini A, Gerosa M, Scotlandi K, Carinci P, Barbanti-Brodano G, Tognon M (1995): High incidence of BK virus large-T-antigen–coding sequences in normal human tissues and tumors of different histotypes. Int J Cancer 61:756–760.

Dhaene K, Verhulst A, Van Marck E (1999): SV40 large T-antigen and human pleural mesothelioma: Screening by polymerase chain reaction and tyramine-amplified immunohistochemistry. Virchows Arch Int J Pathol 435:1–7.

Diamandopoulos GT (1972): Leukemia, lymphoma, and osteosarcoma induced in the Syrian golden hamster by simian virus 40. Science 176:173–175.

Dörries K, Loeber G, Meixensberger J (1987): Association of polyomaviruses JC, SV40, and BK with human brain tumors. Virology 160:268–270.

Eckner R, Ludlow JW, Lill NL, Oldread E, Arany Z, Modjtahedi N, DeCaprio JA, Livingston DM, Morgan JA (1996): Association of p300 and CBP with simian virus 40 large T antigen. Mol Cell Biol 16:3454–3464.

Eddy BE, Borman GS, Grubbs GE, Young RD (1962): Identification of the oncogenic substance in rhesus monkey cell cultures as simian virus 40. Virology 17:65–75.

Elsner C, Dörries K (1992): Evidence of human polyomavirus BK and JC infection in normal brain tissue. Virology 191:72–80.

Emri S, Kocagoz T, Olut A, Güngen Y, Mutti L, Baris YI (2000): Simian virus 40 is not a cofactor in the pathogenesis of environmentally induced malignant pleural mesothelioma in Turkey. Anticancer Res 20:891–894.

Evans AS, Mueller NE (1990): Viruses and cancer: Causal associations. Ann Epidemiol 1:71–92.

Fanning E (1992): Simian virus 40 large T antigen: The puzzle, the pieces, and the emerging picture. J Virol 66:1289–1293.

Fisher SG, Weber L, Carbone M (1999): Cancer risk associated with simian virus 40 contaminated polio vaccine. Anticancer Res 19:2173–2180.

Gamberi G, Benassi MS, Pompetti F, Ferrari C, Ragazzini P, Sollazzo MR, Molendini L, Merli M, Magagnoli G, Chiesa F, Gobbi AG, Powers A, Picci P (2000): Presence and expression of the simian virus-40 genome in human giant cell tumors of bone. Genes Chromosomes Cancer 28:23–30.

Geissler E (1990): SV40 and human brain tumors. Prog Med Virol 37:211–222.

Gerber P, Kirschstein RL (1962): SV40-induced ependymomas in newborn hamster. I. Virus–tumor relationship. Virology 18:582–588.

Giaccia AJ, Kastan MB (1998): The complexity of p53 modulation: Emerging patterns from divergent signals. Genes Dev 12:2973–2983.

Gibbs AR, Jasani B, Pepper C, Navabi H, Wynford-Thomas D (1998): SV40 DNA sequences in mesotheliomas. Dev Biol Stand 94:41–45.

Girardi AJ, Sweet BH, Slotnick VB, Hilleman MR (1962): Development of tumors in hamsters inoculated in the neonatal period with vacuolating virus, SV40. Proc Soc Exp Biol Med 109:649–660.

Griffiths DJ, Nicholson AG, Weiss RA (1998): Detection of SV40 sequences in human mesothelioma. Dev Biol Stand 94:127–136.

Grodzicker T, Hopkins N (1980): Origins of contemporary DNA tumor virus research. In Tooze J, Ed; DNA Tumor Viruses, 2nd ed; Cold Spring Harbor Laboratory: Cold Spring Harbor, pp 1–59.

Hainaut P, Hollstein M (2000): p53 and human cancer: The first ten thousand mutations. Adv Cancer Res 77:81–137.

Harris CC (1996): Structure and function of the p53 tumor suppressor gene: Clues for rational cancer therapeutic strategies. J Natl Cancer Inst 88:1442–1455.

Herrero R, Muñoz N (1999): Human papillomavirus and cancer. Cancer Surv 33:75–98.

Hill AB, Hill ID (1991): Bradford Hill's Principles of Medical Statistics, 12th ed. Edward Arnold: London, England.

Hirvonen A, Mattson K, Karjalainen A, Ollikainen T, Tammilehto L, Hovi T, Vainio H, Pass HI, Di Resta I, Carbone M, Linnainmaa K (1999): Simian virus 40 (SV40)-like DNA sequences not detectable in Finnish mesothelioma patients not exposed to SV40-contaminated polio vaccines. Mol Carcinog 26:93–99.

Hollstein M, Shomer B, Greenblatt M, Soussi T, Hovig E, Montesano R, Harris CC (1996): Somatic point mutations in the p53 gene of human tumors and cell lines: Updated compilation. Nucleic Acids Res 24:141–146.

Horvath CJ, Simon MA, Bergsagel DJ, Pauley DR, King NW, Garcea RL, Ringler DJ (1992): Simian virus 40–induced disease in rhesus monkeys with simian acquired immunodeficiency syndrome. Am J Pathol 140:1431–1440.

Hou J, Major EO (2000): Progressive multifocal leukoencephalopathy: JC virus induced demyelination in the immune compromised host. J Neurovirol 6:S98–S100.

Huang H, Reis R, Yonekawa Y, Lopes JM, Kleihues P, Ohgaki H (1999): Identification in human brain tumors of DNA sequences specific for SV40 large T antigen. Brain Pathol 9:33–42.

Hwang SP, Kucherlapati R (1980): Localization and organization of integrated simian virus 40 sequences in a human cell line. Virology 105:196–204.

Ibelgaufts H, Jones KW (1982): Papovavirus-related RNA sequences in human neurogenic tumours. Acta Neuropathol (Berl) 56:118–122.

Jasani B (1999): Simian virus 40 and human pleural mesothelioma. Thorax 54:750–751.

Jasani B, Cristaudo A, Emri SA, Gazdar AF, Gibbs A, Krynska B, Miller C, Mutti L, Ohgaki H, Radu C, Tognon M, Procopio A (2000): Association of SV40 with human tumors. Semin Cancer Biol 11:49–61.

Jha KK, Banga S, Palejwala V, Ozer HL (1998): SV40-mediated immortalization. Exp Cell Res 245:1–7.

Kelley WL, Georgopoulos C (1997): The T/t common exon of simian virus 40, JC, and BK polyomavirus T antigens can functionally replace the J-domain of the *Escherichia coli* DnaJ molecular chaperone. Proc Natl Acad Sci USA 94:3679–3684.

Kolzau T, Hansen RS, Zahra D, Reddel RR, Braithwaite AW (1999): Inhibition of SV40 large T antigen induced apoptosis by small T antigen. Oncogene 18:5598–5603.

Kravetz HM, Knight V, Chanock RM, Morris JA, Johnson KM, Rifkind D, Utz JP (1961): III. Production of illness and clinical observations in adult volunteers. JAMA 176:107–113.

Krieg P, Amtmann E, Jonas D, Fischer H, Zang K, Sauer G (1981): Episomal simian virus 40 genomes in human brain tumors. Proc Natl Acad Sci USA 78:6446–6450.

Krieg P, Scherer G (1984): Cloning of SV40 genomes from human brain tumors. Virology 138:336–340.

Lane DP, Crawford LV (1979): T antigen is bound to a host protein in SV40-transformed cells. Nature 278:261–263.

Lednicky JA, Arrington AS, Stewart AR, Dai XM, Wong C, Jafar S, Murphey-Corb M, Butel JS (1998): Natural isolates of simian virus 40 from immunocompromised monkeys display extensive genetic heterogeneity: New implications for polyomavirus disease. J Virol 72:3980–3990.

Lednicky JA, Butel JS (1998): Consideration of PCR methods for the detection of SV40 in tissue and DNA specimens. Dev Biol Stand 94:155–164.

Lednicky JA, Butel JS (1999): Polyomaviruses and human tumors: A brief review of current concepts and interpretations. Frontiers Biosci 4:153–164.

Lednicky JA, Garcea RL, Bergsagel DJ, Butel JS (1995): Natural simian virus 40 strains are present in human choroid plexus and ependymoma tumors. Virology 212:710–717.

Lednicky JA, Stewart AR, Jenkins JJ III, Finegold MJ, Butel JS (1997): SV40 DNA in human osteosarcomas shows sequence variation among T-antigen genes. Int J Cancer 72:791–800.

Levine AJ (1988): Oncogenes of DNA tumor viruses. Cancer Res 48:493–496.

Levine AJ (1997): p53, the cellular gatekeeper for growth and division. Cell 88:323–331.

Lewis AM Jr, Levine AS, Crumpacker CS, Levin MJ, Samaha RJ, Henry PH (1973): Studies of nondefective adenovirus 2-simian virus 40 hybrid viruses. V. Isolation of additional hybrids which differ in their simian virus 40-specific biological properties. J Virol 11:655–664.

Lill NL, Tevethia MJ, Eckner R, Livingston DM, Modjtahedi N (1997): p300 family members associate with the carboxyl terminus of simian virus 40 large tumor antigen. J Virol 71:129–137.

Linzer DI, Levine AJ (1979): Characterization of a 54K dalton cellular SV40 tumor antigen present in SV40-transformed cells and uninfected embryonal carcinoma cells. Cell 17:43–52.

Martini F, Iaccheri L, Lazzarin L, Carinci P, Corallini A, Gerosa M, Iuzzolino P, Barbanti-Brodano G, Tognon M (1996): SV40 early region and large T antigen in human brain tumors, peripheral blood cells, and sperm fluids from healthy individuals. Cancer Res 56:4820–4825.

Mayall FG, Jacobson G, Wilkins R (1999): Mutations of p53 gene and SV40 sequences in asbestos associated and non-asbestos-associated mesotheliomas. J Clin Pathol 52: 291–293.

McLaren BR, Haenel T, Stevenson S, Mukherjee S, Robinson BWS, Lake RA (2000): Simian virus (SV) 40 like sequences in cell lines and tumour biopsies from Australian malignant mesotheliomas. Aust NZ J Med 30:450–456.

Meinke W, Goldstein DA, Smith RA (1979): Simian virus 40–related DNA sequences in a human brain tumor. Neurology 29:1590–1594.

Melnick JL, Stinebaugh S (1962): Excretion of vacuolating SV-40 virus (papova virus group) after ingestion as a contaminant of oral poliovaccine. Proc Soc Exp Biol Med 109:965–968.

Mendoza SM, Konishi T, Miller CW (1998): Integration of SV40 in human osteosarcoma DNA. Oncogene 17:2457–2462.

Morris JA, Johnson KM, Aulisio CG, Chanock RM, Knight V (1961): Clinical and serologic responses in volunteers given vacuolating virus (SV_{40}) by respiratory route. Proc Soc Exp Biol Med 108:56–59.

Mortimer EAJ, Lepow ML, Gold E, Robbins FC, Burton GJ, Fraumeni JFJ (1981): Long-term follow-up of persons inadvertently inoculated with SV40 as neonates. N Engl J Med 305:1517–1518.

Mulatero C, Surentheran T, Breuer J, Rudd RM (1999): Simian virus 40 and human pleural mesothelioma. Thorax 54:60–61.

Mutti L, De Luca A, Claudio PP, Convertino G, Carbone M, Giordano A (1998): Simian virus 40-like DNA sequences and large-T antigen–retinoblastoma family protein pRb2/p130 interaction in human mesothelioma. Dev Biol Stand 94:47–53.

Newman JS, Baskin GB, Frisque RJ (1998): Identification of SV40 in brain, kidney and urine of healthy and SIV-infected rhesus monkeys. J Neurovirol 4:394–406.

Newman JT, Frisque RJ (1999): Identification of JC virus variants in multiple tissues of pediatric and adult PML patients. J Med Virol 58:79–86.

Ohgaki H, Huang H, Haltia M, Vainio H, Kleihues P (2000): More about: Cell and molecular biology of simian virus 40: Implications for human infections and disease. J Natl Cancer Inst 92:495–497.

Olin P, Giesecke J (1998): Potential exposure to SV40 in polio vaccines used in Sweden during 1957: No impact on cancer incidence rates 1960 to 1993. Dev Biol Stand 94:227–233.

O'Neill FJ, Frisque RJ, Xu X, Hu YX, Carney H (1995): Immortalization of human cells by mutant and chimeric primate polyomavirus T-antigen genes. Oncogene 10: 1131–1139.

Oren M (1999): Regulation of the p53 tumor suppressor protein. J Biol Chem 274: 36031–36034.

Ozer HL, Banga SS, Dasgupta T, Houghton J, Hubbard K, Jha KK, Kim SH, Lenahan M, Pang Z, Pardinas JR, Patsalis PC (1996): SV40-mediated immortalization of human fibroblasts. Exp Gerontol 31:303–310.

Pallas DC, Shahrik LK, Martin BL, Jaspers S, Miller TB, Brautigan DL, Roberts TM (1990): Polyoma small and middle T antigens and SV40 small t antigen form stable complexes with protein phosphatase 2A. Cell 60:167–176.

Pass H, Rizzo P, Donington J, Wu P, Carbone M (1998): Further validation of SV40-like DNA in human pleural mesotheliomas. Dev Biol Stand 94:143–145.

Pepper C, Jasani B, Navabi H, Wynford-Thomas D, Gibbs AR (1996): Simian virus 40 large T antigen (SV40LTAg) primer specific DNA amplification in human pleural mesothelioma tissue. Thorax 51:1074–1076.

Pipas JM (1992): Common and unique features of T antigens encoded by the polyomavirus group. J Virol 66:3979–3985.

Porrás A, Gaillard S, Rundell K (1999): The simian virus 40 small-t and large-T antigens jointly regulate cell cycle reentry in human fibroblasts. J Virol 73:3102–3107.

Procopio A, Marinacci R, Marinetti MR, Strizzi L, Paludi D, Iezzi T, Tassi G, Casalini A, Modesti A (1998): SV40 expression in human neoplastic and non-neoplastic tissues: Perspectives on diagnosis, prognosis and therapy of human malignant mesothelioma. Dev Biol Stand 94:361–367.

Procopio A, Strizzi L, Vianale G, Betta P, Puntoni R, Fontana V, Tassi G, Gareri F, Mutti L (2000): Simian virus-40 sequences are a negative prognostic cofactor in patients with malignant pleural mesothelioma. Genes Chromosomes Cancer 29:173–179.

Rabson AS, O'Conor GT, Kirschstein RL, Branigan WJ (1962): Papillary ependymomas produced in *Rattus (Mastomys) natalensis* inoculated with vacuolating virus (SV40). J Natl Cancer Inst 29:765–787.

Ramael M, Nagels J, Heylen H, De Schepper S, Paulussen J, De Maeyer M, Van Haesendonck C (1999): Detection of SV40 like viral DNA and viral antigens in malignant pleural mesothelioma. Eur Respir J 14:1381–1386.

Rizzo P, Carbone M, Fisher SG, Matker C, Swinnen LJ, Powers A, Di Resta I, Alkan S, Pass HI, Fisher RI (1999): Simian virus 40 is present in most United States human mesotheliomas, but it is rarely present in non-Hodgkin's lymphoma. Chest 116: 470S–473S.

Rizzo P, Di Resta I, Stach R, Mutti L, Picci P, Kast WM, Pass HI, Carbone M (1998): Evidence for and implications of SV40-like sequences in human mesotheliomas and osteosarcomas. Dev Biol Stand 94:33–40.

Shah K, Nathanson N (1976): Human exposure to SV40: Review and comment. Am J Epidemiol 103:1–12.

Shah KV (1998): SV40 infections in simians and humans. Dev Biol Stand 94:9–12.

Shaw P, Bovey R, Tardy S, Sahli R, Sordat B, Costa J (1992): Induction of apoptosis by wild-type p53 in a human colon tumor-derived cell line. Proc Natl Acad Sci USA 89:4495–4499.

Shay JW, Wright WE (1989): Quantitation of the frequency of immortalization of normal human diploid fibroblasts by SV40 large T-antigen. Exp Cell Res 184:109–118.

Shein HM (1967): Transformation of astrocytes and destruction of spongioblasts induced by a simian tumor virus (SV40) in cultures of human fetal neuroglia. J Neuropathol Exp Neurol 26:60–76.

Shivapurkar N, Wiethege T, Wistuba II, Milchgrub S, Muller KM, Gazdar AF (2000): Presence of simian virus 40 sequences in malignant pleural, peritoneal and noninvasive mesotheliomas. Int J Cancer 85:743–745.

Shivapurkar N, Wiethege T, Wistuba II, Salomon E, Milchgrub S, Muller KM, Churg A, Pass H, Gazdar AF (1999): Presence of simian virus 40 sequences in malignant mesotheliomas and mesothelial cell proliferations. J Cell Biochem 76:181–188.

Simon MA, Ilyinskii PO, Baskin GB, Knight HY, Pauley DR, Lackner AA (1999): Association of simian virus 40 with a central nervous system lesion distinct from progressive multifocal leukoencephalopathy in macaques with AIDS. Am J Pathol 154:437–446.

Slinskey A, Barnes D, Pipas JM (1999): Simian virus 40 large T antigen J domain and Rb-binding motif are sufficient to block apoptosis induced by growth factor withdrawal in a neural stem cell line. J Virol 73:6791–6799.

Soriano F, Shelburne CE, Gokcen M (1974): Simian virus 40 in a human cancer. Nature 249:421–424.

Srinivasan A, McClellan AJ, Vartikar J, Marks I, Cantalupo P, Li Y, Whyte P, Rundell K, Brodsky JL, Pipas JM (1997): The amino-terminal transforming region of simian virus 40 large T and small t antigens functions as a J domain. Mol Cell Biol 17: 4761–4773.

Stewart AR, Lednicky JA, Benzick US, Tevethia MJ, Butel JS (1996): Identification of a variable region at the carboxy terminus of SV40 large T-antigen. Virology 221: 355–361.

Stewart AR, Lednicky JA, Butel JS (1998): Sequence analyses of human tumor-associated SV40 DNAs and SV40 viral isolates from monkeys and humans. J Neurovirol 4:182–193.

Strickler HD, Goedert JJ, Fleming M, Travis WD, Williams AE, Rabkin CS, Daniel RW, Shah KV (1996): Simian virus 40 and pleural mesothelioma in humans. Cancer Epidemiol Biomarkers Prev 5:473–475.

Strickler HD, Rosenberg PS, Devesa SS, Hertel J, Fraumeni JFJ, Goedert JJ (1998): Contamination of poliovirus vaccines with simian virus 40 (1955–1963) and subsequent cancer rates. JAMA 279:292–295.

Strizzi L, Vianale G, Giuliano M, Sacco R, Tassi F, Chiodera P, Casalini P, Procopio A (2000): SV40, JC and BK expression in tissue, urine and blood samples from patients with malignant and nonmalignant pleural disease. Anticancer Res 20:885–889.

Stubdal H, Zalvide J, Campbell KS, Schweitzer C, Roberts TM, DeCaprio JA (1997): Inactivation of pRB-related proteins p130 and p107 mediated by the J domain of simian virus 40 large T antigen. Mol Cell Biol 17:4979–4990.

Suzuki SO, Mizoguchi M, Iwaki T (1997): Detection of SV40 T antigen genome in human gliomas. Brain Tumor Pathol 14:125–129.

Sweet BH, Hilleman MR (1960): The vacuolating virus, S.V.$_{40}$. Proc Soc Exp Biol Med 105:420–427.

Tabuchi K, Kirsch WM, Low M, Gaskin D, Van Buskirk J, Maa S (1978): Screening of human brain tumors for SV40-related T antigen. Int J Cancer 21:12–17.

Testa JR, Carbone M, Hirvonen A, Khalili K, Krynska B, Linnainmaa K, Pooley FD, Rizzo P, Rusch V, Xiao GH (1998): A multi-institutional study confirms the presence and expression of simian virus 40 in human malignant mesotheliomas. Cancer Res 58:4505–4509.

Tevethia MJ, Lacko HA, Conn A (1998): Two regions of simian virus 40 large T-antigen independently extend the life span of primary C57BL/6 mouse embryo fibroblasts and cooperate in immortalization. Virology 243:303–312.

Vogelstein B, Lane D, Levine AJ (2000): Surfing the p53 network. Nature 408:307–310.

Vousden KH (2000): p53: Death star. Cell 103:691–694.

Waheed I, Guo ZS, Chen GA, Weiser TS, Nguyen DM, Schrump DS (1999): Antisense to SV40 early gene region induces growth arrest and apoptosis in T-antigen–positive human pleural mesothelioma cells. Cancer Res 59:6068–6073.

Weggen S, Bayer TA, Von Deimling A, Reifenberger G, Von Schweinitz D, Wiestler OD, Pietsch T (2000): Low frequency of SV40, JC and BK polyomavirus sequences in human medulloblastomas, meningiomas and ependymomas. Brain Pathol 10:85–92.

Weiss AF, Portmann R, Fischer H, Simon J, Zang KD (1975): Simian virus 40–related antigens in three human meningiomas with defined chromosome loss. Proc Natl Acad Sci USA 72:609–613.

Whalen B, Laffin J, Friedrich TD, Lehman JM (1999): SV40 small T antigen enhances progression to >G2 during lytic infection. Exp Cell Res 251:121–127.

Yamamoto H, Nakayama T, Murakami H, Hosaka T, Nakamata T, Tsuboyama T, Oka M, Nakamura T, Toguchida J (2000): High incidence of SV40-like sequences detection in tumour and peripheral blood cells of Japanese osteosarcoma patients. Br J Cancer 82:1677–1681.

Yang SI, Lickteig RL, Estes R, Rundell K, Walter G, Mumby MC (1991): Control of protein phosphatase 2A by simian virus 40 small-t antigen. Mol Cell Biol 11:1988–1995.

Zhen HN, Zhang X, Bu XY, Zhang ZW, Huang WJ, Zhang P, Liang JW, Wang XL (1999): Expression of the simian virus 40 large tumor antigen (Tag) and formation of Tag–p53 and Tag–pRb complexes in human brain tumors. Cancer 86:2124–2132.

zur Hausen H (1996): Papillomavirus infections—A major cause of human cancers. Biochim Biophys Acta 1288:F55–F78.

18

MOLECULAR EVOLUTION AND EPIDEMIOLOGY OF JC VIRUS

HANSJÜRGEN T. AGOSTINI, M.D., DAVID V. JOBES, PH.D., and GERALD L. STONER, PH.D.

1. NO CONTINUITY WITHOUT CHANGE

Mechanisms of Change

The viral replicative elements populating our world are based on either DNA or RNA. Both are exposed to constantly changing environments. To maintain a high level of fitness or improve adaptation, the genomes of all organisms have to change. The molecular mechanisms involved are the substitution of nucleotides due to the limitations of proofreading mechanisms, other errors involving duplications or deletions, and molecular recombination. The genetic code is degenerate. One can distinguish synonymous or silent mutations and nonsynonymous mutations that change an amino acid in the encoded protein. With regard to point mutations at third codon positions, the frequency of synonymous mutations at this position of a codon varies. This leads one to assume that additional functional properties of the mRNA, besides being the template for translation, determine the selection of a mutation (Britten, 1993).

DNA- and RNA-based biologic units differ dramatically in their mutation rates during replication. The misincorporation rate in RNA viruses is as high

Human Polyomaviruses: Molecular and Clinical Perspectives, Edited by Kamel Khalili and Gerald L. Stoner.
ISBN 0-471-39009-7

as 10^{-4} substitutions per site per year (Fitch et al., 1991; Li et al., 1988) compared with about 10^{-9} in humans (Britten, 1986). Some DNA viruses like the herpesviruses, adenoviruses, and poxviruses are large enough to code for their own DNA polymerase, whereas smaller viruses like the polyomaviruses JCV, BKV, and SV40 depend on the enzymatic machinery of the infected cell. Viruses that do not utilize the DNA polymerase of the host and the cellular tools for fidelity control do not necessarily have increased rates of misincorporation. The DNA polymerase of herpes simplex virus performs under certain conditions more accurately than purified eukaryotic DNA polymerases (Abbotts et al., 1987).

Different virus families utilize different molecular mechanisms to maintain genetic plasticity. Homologous recombination of the genome, for example, is found in positive-strand RNA viruses but not in unsegmented negative-strand RNA viruses. In DNA viruses both homologous recombination and nonhomologous recombination can be important. Nonhomologous recombination by which viral DNA is inserted into the genome of its host or vice versa was demonstrated in cell culture for SV40 or polyomavirus at a high multiplicity of infection and in transformed tumor cells (Dora et al., 1989; Brockman et al., 1975). In the case of JCV, there is no evidence for nonhomologous recombination so far. Although homologous recombination is certainly not a major feature in JCV evolution, JCV Type 5 is an example of a possible recombinant between a Type 2B and a Type 6 strain (see below).

JCV Rearrangements: A "Quasispecies" on a DNA Basis?

M. Eigen created the concept of quasispecies in the early 1970s (Eigen, 1971). He tried to create a model for evolution at the earliest stage with replicons on the level of macromolecules able to replicate with a certain error rate. The quasispecies describes a dynamic steady state in which mutant replicons reach equilibrium. This concept offers an interesting approach to understanding population dynamics of viral replication, especially of RNA viruses. The error-prone reproduction of viral genomes within a cell leads to a population of closely related but nonidentical variants or quasispecies. Within these quasispecies it is impossible to define the prototype sequence. In a specific environment the mutants differ in their fitness, and selection takes place. In contrast to DNA viruses, RNA viruses, which lack proofreading mechanisms, mutate to an extent close to their elimination threshold. Increasing the mutation rate will increase the portion of nonfunctional isoforms and slow down adaptation to a given environment. In a constant environment the quasispecies will be limited over time to those replicons with the best replicative adaptability (reviewed by Domingo and Holland, 1997). To replicate at a high rate with a high error rate and the ability to recombine facilitates adaptation. It also becomes clear why the vaccination and treatment against RNA virus infections will remain difficult and host jumping in changing environments is likely.

In contrast to RNA viruses the concept of quasispecies has to be adapted to a much longer time scale for DNA viruses. In the case of JCV, the genotypes evolved over a time period long enough to distinguish early migration movements of its human host. RNA viruses adapt fast enough to coevolve with constantly changing cell lines in vitro. However, there is one aspect of genomic variation in the biology of JCV that shows parallels to the RNA quasispecies —the rearrangement of the JCV regulatory region to the late side of the origin of replication (*ori*). Here the transmissible archetypal form of the virus found in the kidneys undergoes changes while replicating in susceptible cell types of its host. When, where, and how these variants arise, and whether they circulate in the population, remain matters of debate among virologists. Possibilities that have been suggested include the upper respiratory tract, the lower gastrointestinal tract, the urinary tract, or glial cells in the central nervous system in immunocompromised individuals. The infection of the latter leads to the fatal demyelinating disease progressive multifocal leukoencephalopathy (PML) (see Chapters 11 and 12).

Coevolution

One has to assume that species with a similar genomic structure developed from a hypothetical ancestral virus strain. The distribution of this prototype virus and its variants is not only a result of molecular evolution, but is also closely linked to the evolution and migration of the viral host. In the case of the polyomavirus family, species-specific viruses were isolated in birds and mammals, as summarized in Table 18.1, and there are very few indications that polyomaviruses can change hosts. One possible exception to this rule of species specificity is simian vacuolating virus 40 (SV40) that was detected in 1960 by Sweet and Hilleman in primary monkey kidney cell cultures used to grow the inactivated polio vaccine. In the late 1950s, a large number of vaccinated individuals were inadvertently exposed to SV40, and many were infected. The

Table 18.1. Polyomaviruses (Py) and Their Hosts

Virus	Host
BK virus (BKV)	Human
JC virus (JCV)	Human
Simian virus 40 (SV40)	Monkey
Lymphotropic papovavirus (LPV)	Monkey
K polyomavirus (Kilham strain) (KV)	Mouse
Mouse Py	Mouse
Bovine Py	Cattle
Hamster Py	Hamster
Budgerigar fledgling disease virus (BFDV)	Bird

oncogenic activity of SV40 in humans is currently under investigation and is discussed critically in Chapters 17 and 20.

JCV and BKV are the traditional human polyomaviruses. They are also classified as *Polyomavirus hominis* type 1 (BKV) and type 2 (JCV). Different genotypes have been described for both virus species. The molecular evolution of genotypes took place within geographically defined interbreeding populations and serves as a marker for following the migration of its host. Different viral genotypes may come into competition when populations meet, mingle, and intermarry in which case the better-adapted genotype will prevail. These principles have been developed with regard to JCV, and it is currently uncertain whether and to what extent they also apply to BKV.

2. VARIATIONS IN THE JCV GENOME

The Story of JCV Genotypes and the Archetype Hypothesis

In 1984 Frisque and colleagues, working at Penn State University, sequenced the complete JCV genome. The JCV strain was named after the city of Madison in Wisconsin (Mad-1), where it had been isolated from PML brain tissue (see Chapter 3). Four years later the sequence of another JCV strain was added by Loeber and Dörries (1988) from Würzburg, Germany. They isolated variants from brain (GS/B) and kidney (GS/K) of the same PML patient. The genomes of these two new isolates differed from the Mad-1 strain in two ways: first, in the regulatory region to the late side of the origin of replication, which includes binding sites for promoters of viral replication and transcriptional activators; and, second, within the coding region (Fig. 18.1). The regulatory region showed variations based on deletions and duplications, whereas point mutations of about 2% of the nucleotide sequence accounted for the coding region difference.

The first evidence for genetic differences between Asian and European strains of JCV was presented in studies of restriction fragments and by regulatory region sequence analysis in 1987 (Matsuda et al., 1987) and 1990 (Yogo et al., 1990). A few years later it became clear that the coding region sequences of JCV(Mad-1) and JCV(GS) represented different genotypes (Yogo et al., 1991a; Ault and Stoner, 1992), with the GS strain in the Asian group. Subsequently, two parallel classifications of JCV genotypes have been developed. The classification scheme of Yogo et al. utilizes an alphabetical means of genotype assignment (A–D) (Yogo et al., 1991a; Guo et al., 1996; Kato et al., 1997). These were initially assigned by unique RFLP markers and later verified by V–T intergenic region sequences. These initial types were expanded as additional geographic variants were discovered. Our numerical typing scheme, initially based on the V–T intergenic region (Ault and Stoner, 1992), is now defined in an upstream *VP*1 gene fragment, supplemented with additional downstream fragments, and, as needed, by phylogenetic analysis of the com-

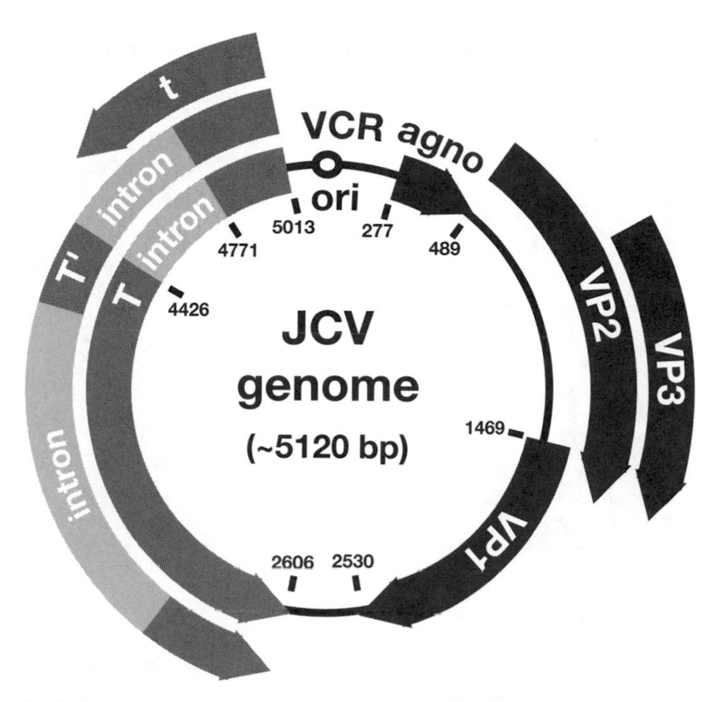

Figure 18.1. JCV genome. The circular double-strand DNA genome of ~5120 bp is divergently transcribed in early (counterclockwise) and late (clockwise) phases, as are all polyomaviruses. The origin of DNA replication (*ori*) is located within the ~384 bp viral control region (VCR). In addition to the large T antigen and small t antigen proteins, several other alternatively spliced forms have been described (Trowbridge and Frisque, 1995), known collectively as T′ (T prime) proteins. One of these is illustrated. Portions of *VP*1 and the VT-intergenic region have been mainly utilized for genotyping, but typing sites are located throughout the genome, and phylogenetic analysis of the entire genome provides the most reliable approach to genotyping.

plete genome (Jobes et al., 1998). The nomenclature based on these JCV genotypes and their phylogenetic relationship is described below. The differences between these two typing schemes reflect the fact that different genome sequences have been analyzed, different populations have been studied, and the fact that arbitrary choices must sometimes be made to "lump" or to "split" sequence variants. Even though both systems are "works in progress," there is fundamental agreement between them. A preliminary correlation of the two JCV typing systems is given in Table 18.2. (Additional types and subtypes of JCV have been described by Yogo and Sugimoto that are not listed in the table.)

A total of eight JCV genotypes (Types 1–8) and numerous subtypes have been identified. The genotype assignments are based on the *VP*1 gene fragment as well as on the analysis of the complete viral genome with the exception of the potentially rearranging enhancer and promoter region to the late side of *ori*.

Table 18.2. Nomenclature Comparison of JCV Genotypes

Agostini, Jobes, and Stoner	Yogo, Sugimoto, et al.
Type 1A, B	A (EU)
Type 4	A (EU)
Type 2A	B (MY)
Type 2B	B1-c
Type 2D	B1-b
Type 2E	—
Type 3A, B	B (AF2)
Type 5	—
Type 6	C (AF1)
Type 7A	B (SC)
Type 7B	B (CY)
Type 7C1	B1-a
Type 7C2	B2
Type 8A, B	—

The total number of JCV genotypes may still increase when more populations in remote areas of the world are tested for the JCV strains they carry.

The regulatory region changes led to the concept that JCV strains with a nonrearranged regulatory region are passed from individual to individual (Yogo et al., 1990) and that deletions and duplications of this archetypal sequence give rise to the unique sequences found in PML strains (Yogo et al., 1990; Agostini et al., 1997c; Elsner and Dörries, 1998; Ciappi et al., 1999; Vaz et al., 2000). The way in which the JCV regulatory region rearranges does not depend on the viral genotype defined by its coding region (Agostini et al., 1997c). The "quasispecies" that replicates in the brain might reflect the journey of JCV within its host from the site of primary infection until it reaches the organ where it persists or causes PML in immunocompromised individuals (Newman and Frisque, 1999).

How JCV Genotypes Are Defined

Dissimilar viral strains can be characterized by either their serotype or genotype. A serotype reflects the specific immune response of the host to viral antigenic epitopes of different structures. For BKV four serologically defined types have been described. The antigenic difference of BKV serotypes is based in part on variation in the central portion of VP1 (Jin et al., 1993a,b). In contrast, the JCV DNA does not show sufficient sequence variation to generate serologically distinguishable antigenic epitopes. JCV genotypes are therefore defined solely by the DNA sequence of the viral genome. The degree of variability of the DNA sequence among JCV genotypes differs in different genomic elements outside of the viral control region (VCR). Elements of high variability are the intergenic region between *VP*1 and large T-antigen genes, the T-antigen intron, the *VP*1 gene, and the 3′ end of exon 2 of the T-antigen gene (Agostini et al., 1997a). Elements with a very low variability are the intergenic region between the agnoprotein and the *VP*2 gene and the agnoprotein gene itself (Agostini et al., 1997a; Hatwell and Sharp, 2000).

For genotyping, sequence analysis following amplification of the 5′ portion of the *VP*1 gene by the polymerase chain reaction (PCR) proved to be very reliable. The 215 bp segment defined by positions 1710–1924 (Mad-1 numbering) with low GC-content can be detected with high sensitivity using primers JLP-15 and JLP-16, as shown in Table 18.3 and described in detail below. Within this fragment all known genotypes can be distinguished except some variants of Type 2A and Type 7B in China (X. Cui, et al., unpublished data). Supplemental methods are being developed by which additional sequence analyses of the *VP*1 gene or V–T intergenic region will be able to make these distinctions (Table 18.3). Otherwise, the single typing fragment results are highly congruent with methods based on the phylogenetic analysis of the V–T region or the complete genome (Ault and Stoner, 1992; Agostini et al., 1998b,d; Jobes et al., 1998) (Fig. 18.2).

The classification scheme we have proposed uses a combination of numbers followed by a single letter to designate types and subtypes. JCV strain 312A, for example, is a Type 3 strain (African) of the subtype A and the twelfth strain analyzed. Alternatively, strains are named after the town or region where they were first isolated (e.g., Mad-1, first strain isolated in Madison, WI). It should be noted that strains Tokyo-1 and Tky-1 are different Japanese strains with GenBank accession numbers of AF030085 and AB038254, respectively.

Besides the rearrangements by duplications and deletions as described above, the VCR of some genotypes show point mutations on both sides of *ori*. In general, these alterations are useful to confirm major clades of genotypes like Types 1 and 4 (European), Types 2 and 7 (Asian), and Types 3 and 6 (African). The precise distinction of JCV genotypes based only on the sequence of the VCR is not possible. However, positions 5017, 5026, and 5039 differ in Types 1 and 2 (Mad-1 numbering) (Yogo et al., 1991b; Ault and Stoner, 1992; Chang et al., 1996; Agostini et al., 1995) (Table 18.4). Types 3 and 6 can also be distinguished from Type 1, Type 2, and Type 7 at one or more of these positions. Mutations and small deletions in the promoter and enhancer region of the archetypal VCR can also help to define certain other genotypes. Some Type 2 strains can be distinguished at 217-G from Type 1 strains with 217-A (archetype numbering, Yogo et al., 1990). However, while 217-G is prominent in Japan, 217-A predominates in the Americas among Native-American Type 2A strains. In Africa, 133-C distinguishes Type 3 strains of the subtype 3A from those of all other genotypes. In strains from Taiwan and elsewhere in South China and Southeast Asia, several small deletions of 5 bp (most commonly nucleotides 218–222) distinguish the subtypes of strains termed Taiwan-1, 2 and 3, which we have classified as Type 7A (Fig. 18.3) (Jobes et al., 1998).

Table 18.3. JCV Typing and Complete Genome Primers

Primer	5′-Pos.[a]	Sequence (5′–3′)	Comment
JLP-15	1710	ACAGTGTGGCCAGAATTCCACTACC	Sensitive primer pair for initial typing in the *VP*1 gene
JLP-16	1924	TAAAGCCTCCCCCCCAACAGAAA	
JLP-19	2106	ATTTTGGGACACTAACAGGAGGAG	Distinguishes a subset of Type 7B strains (China); and Native American strains from Japanese Type 2A, etc.
JLP-20	2350	AATTGGGTAGGGGTTTTTAACCC	
JIG-17	2270	GGTTCCCAGCAGTGGAGAGGAC	Auxiliary typing and subtyping sites in the intergenic region for Asian strains
JIG-18	2773	GAAGATTCTGAAGCAGAAGACTCTGGAC	
JRR-25	4981	CATGGATTCCTCCCTATTCAGCA	Includes three typing sites and the rearranging part of the VCR
JRR-28	291	TCACAGAAGCCTTACGTGACAGC	
BAM-1	4306	GGGATCCTGTGTTTTCATCATCACTGGC	Overlapping at the *Bam*HI-site for complete genome PCR
BAM-2	4313	AGGATCCCAACACTCTACCCCACC	
ECO-7	1721	AGAATTCCACTACCCAATCTAAATGAGGATC	Overlapping at the *Eco*RI-site for complete genome PCR
ECO-12	1729	TGGAATTCTGGCCACACTGTAACAAG	

[a]Numbering of the 5′ position is based on JCV (Mad-1), GenBank No. J02227.

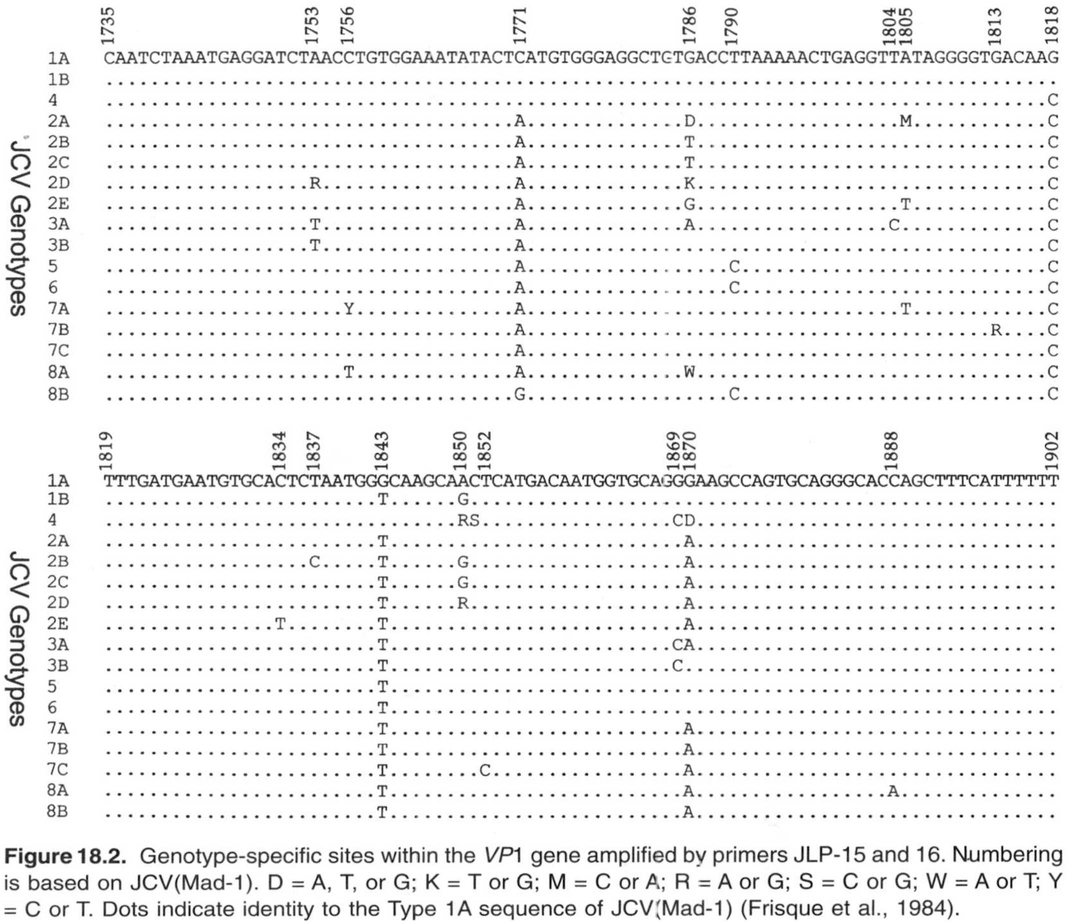

Figure 18.2. Genotype-specific sites within the *VP1* gene amplified by primers JLP-15 and 16. Numbering is based on JCV(Mad-1). D = A, T, or G; K = T or G; M = C or A; R = A or G; S = C or G; W = A or T; Y = C or T. Dots indicate identity to the Type 1A sequence of JCV(Mad-1) (Frisque et al., 1984).

Table 18.4. Typing Information in the Viral Control Region Left of *ori*

	Nucleotide Position[a]				
Type	5017	5025	5026	5039	5040
1, 4	A	G	C	C	T
2	T	G	T	G	T
2D	T	G	T	G	A
3	T	G	T	C	T
6	T	A	T	C	T
7	T	G	T	G	T
8	T	G	T	G	T

[a]Numbering based on JCV(Mad-1), GenBank No. J02227.

Describing JCV Regulatory Region Rearrangements

Each duplication or deletion in the VCR creates a breakpoint in the sequence. Several attempts were made to find blocks of preferentially rearranging sequences within the variable portion of the VCR or to classify the rearranged VCR in sequences similar or dissimilar to Mad-1 (Ault and Stoner, 1993; Elsner and Dörries, 1998; Iida et al., 1993; Vaz et al., 2000). However, the rearrangements seemed to be largely random, and none of the schemes proved to be universal. We therefore suggested a system that allows precise description of the deletions and duplications that an archetypal regulatory region has undergone (Agostini et al., 1997c). The numbering is based on the 267 bp sequence of the archetypal configuration of strain CY (GenBank accession number M35834; Yogo et al., 1990) with rearranged elements of the archetype in brackets. The deletion of a segment starting at nucleotide 115 and ending at 183 would be noted as [1–114][184–267]. In this example, the DNA sequence at both position 115 and position 184 shows an "A." The proximal breakpoint could therefore be between 114/115 or 115/116 and the distal breakpoint

→

Figure 18.3. Phylogenetic analysis of representative JCV strains. Fifty JCV sequences (~4854 bp each) were aligned with the PILEUP program. Trees were generated by heuristic search using the PAUPSEARCH program, and midpoint rooting was obtained with PAUPDISPLAY. Due to its possible recombinant nature, JCV Type 5 was not included. PNG = Papua New Guinea. (**A**) Neighbor-joining distance method. Bootstrap values were obtained with 500 replicates using PAUPSEARCH, and these confidence levels are indicated at the nodes. (**B**) Maximum parsimony algorithm, 50% majority rule tree. The percentage of most parsimonious trees showing each grouping is indicated at the node. The basic clade organization is identical to that shown in A. Both methods show a trifurcation at the base (Europe, Types 1 and 4; Africa, Type 6; and Asia and the Pacific, Types 2, 7, and 8). Type 3 is Afro-Asiatic. The details of the topology regarding Types 7 and 8 and the subtypes of Type 2 differ with the method used. These largely congruent analyses indicate the robustness of the data and the methods.

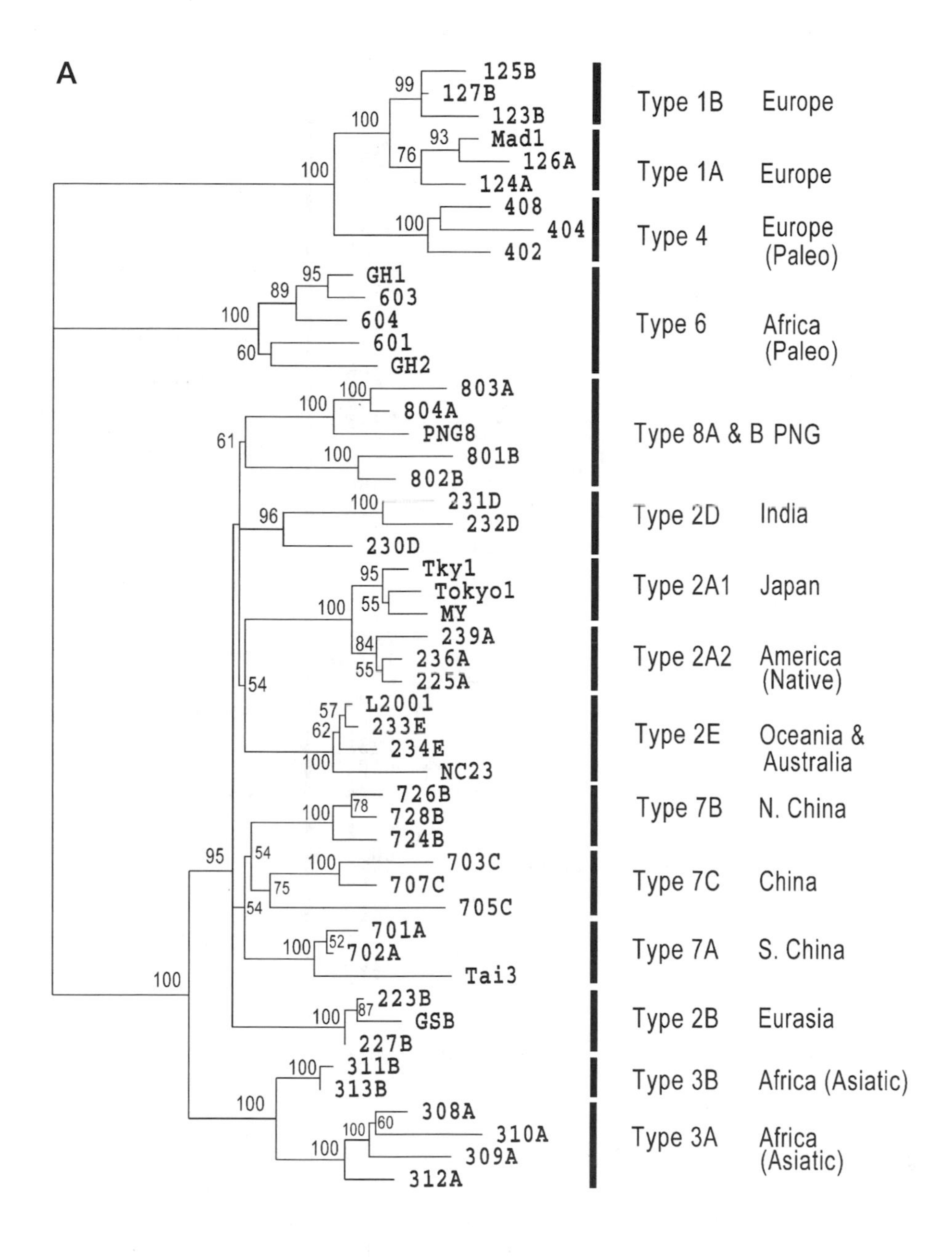
A
125B
127B
123B
Mad1
126A
124A
408
404
402
GH1
603
604
601
GH2
803A
804A
PNG8
801B
802B
231D
232D
230D
Tky1
Tokyo1
MY
239A
236A
225A
L2001
233E
234E
NC23
726B
728B
724B
703C
707C
705C
701A
702A
Tai3
223B
GSB
227B
311B
313B
308A
310A
309A
312A
Type 1B Europe
Type 1A Europe
Type 4 Europe (Paleo)
Type 6 Africa (Paleo)
Type 8A & B PNG
Type 2D India
Type 2A1 Japan
Type 2A2 America (Native)
Type 2E Oceania & Australia
Type 7B N. China
Type 7C China
Type 7A S. China
Type 2B Eurasia
Type 3B Africa (Asiatic)
Type 3A Africa (Asiatic)

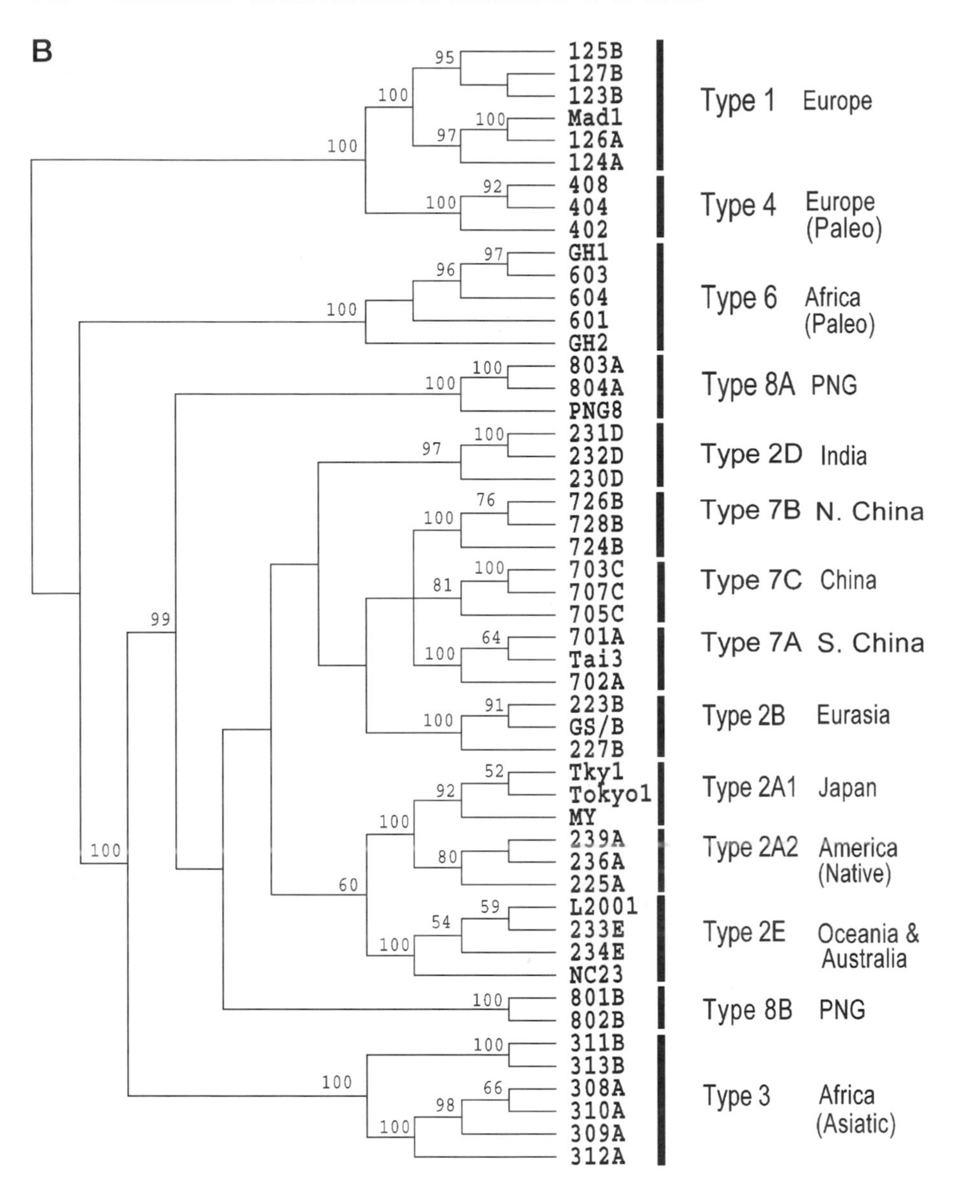

Figure 18.3. (*Continued*)

between 183/184 or 184/185. To allow for this ambiguity, the notation can be modified to [1–114] A [185–end]. The position of the "A" indicates that it could be placed to either side of the deletion, representing position 115 or position 184. This standard notation incorporates the ambiguities of breakpoints and will be useful for future studies of the molecular mechanisms of rearrangement. Thus far no convincing hypothesis exists.

PCR Amplification of JCV DNA from Urine Samples

JCV DNA is detectable in multiple organs (Newman and Frisque, 1999). The clinical diagnosis of PML can be confirmed by PCR detection of JCV in cerebrospinal fluid (Weber et al., 1997) or biopsy tissue stereotactically removed from the brain (Dörries et al., 1998). Lymphatic tissue from the tonsils, peripheral blood lymphocytes, or bone marrow is also a target for JCV infection (Gallia et al., 1997; Dörries et al., 1994; Tornatore et al., 1992) and may play a role as carrier for JCV within the human organism. Recent studies suggest that while B lymphocytes do not replicate the virus well, they may possess surface receptors that could passively transport the virus (Wei et al., 2000). However, for epidemiologic studies readily accessible samples are desirable. JCV complies with this criterion: It persists in the kidneys (Aoki et al., 1999; Tominaga et al., 1992), and viral DNA is frequently shed in the urine at easily detectable levels.

DNA Extraction from Urine Samples. Urine samples can be stored after collection for up to 3 weeks at 4°C before further processing. For our studies we used the cell pellet from 15–50 ml urine obtained by centrifugation at 4300 *g* and washed once in sterile phosphate-buffered saline (PBS; standard formulation without magnesium) to lower the content of urea, a potent PCR inhibitor. For extraction of total DNA the urinary cell pellet was digested for 2–14 hours at 56°C with proteinase K at a final concentration of 0.2 mg/ml lysis buffer (50 mM Tris, 1 mM EDTA, 0.45% NP40, 0.45% Tween 20, pH 8.0). To inactivate the proteinase, the sample was boiled for 10 minutes. Alternatively, a commercially available RNA/DNA extraction kit such as the QIAamp Viral RNA/DNA kit (Qiagen) can be used. It is possible to freeze the DNA extract at this point, but in our experience the rate of successful complete genome amplification by PCR is higher if the amplification is done before freezing.

PCR: Partial* VP*1 Gene. Primers JLP-15 and JLP-16 are highly sensitive for amplification of a target in the 5′ end of the *VP*1 gene that allows one to differentiate all major genotypes (except two subtypes in Japan and China) (Fig. 18.2). This short (215 bp) target is also suitable for amplification from formalin-fixed, paraffin-embedded tissue. The reaction mix contains the primers at a concentration of 300 nM with 1.5 mM $MgCl_2$ in a standard PCR buffer with a final volume of 100 μl. DNA extract (1–5 μl) is added. The amplification with primers JLP-15 and JLP-16 works with a variety of thermostable DNA polymerases with or without proofreading activity. Hot start or cold start is recommended to minimize mispriming. Following an initial denaturation at 95°C for 5 minutes, the 50 cycle, three-step PCR program includes 1 minute for denaturation at 95°C, followed by 1.5 minutes at 63°C for annealing and 1 minute at 72°C for elongation, with a final extension time of 10 minutes at 72°C. To confirm this typing and to distinguish a subset of Type 7B from Type

2A in China, another fragment can be amplified from the 3′ end of the *VP*1 gene using primers JLP-19 and JLP-20 (X. Cui, et al., unpublished data). Also useful for typing and for assigning subtypes are primers amplifying a 504 bp fragment from the intergenic region between the *VP*1 gene and exon 2 of the large T-antigen gene—JIG-17 and JIG-18 (Table 18.3).

***PCR: Regulatory Region Including the Typing Sites Left of* ori.** As the JCV regulatory region is GC rich compared with the rest of the viral genome, PCR amplification of this region is less often successful than with *VP*1 target sequence. Using primers JRR-25 and JRR-28 (Table 18.3) a 463 bp fragment that includes the three typing sites left of *ori* and the enhancer and promoter of the regulatory region could be amplified in about 60% of the JLP-15 and JLP-16 positive urinary samples. The 50 cycle program was run as described for JLP-15 and -16, including the hot start, but with the annealing temperature at 62°C.

PCR: Complete Genome (~5120 bp). To amplify the complete JCV genome in a single PCR reaction one takes advantage of the fact that JCV has a single recognition site for the restriction enzymes *Bam*HI and *Eco*RI. Primers were designed to overlap at their 5′ ends at one of these restriction sites as shown for primer pairs BAM-1 and BAM-2 (GGATCC) or ECO-7 and ECO-12 (GAATTC) (Table 18.3). One JCV strain has been identified (126A) that has a second *Eco*RI site in the genome beginning at position 2186 (Mad-1 numbering) due to a T to C mutation at 2191. For this reason, and due to the fact that the BAM-1 and -2 primers are more sensitive than the ECO-7 and -12 primer pair, amplification from the *Bam*HI site is preferred. However, cloning at this site interrupts the 5′ end of T-antigen exon 2 and prevents functional T-antigen expression, whereas cloning at the *Eco*RI site interrupts the *VP*1 gene.

If the PCR product is to be used for sticky-end cloning at the restriction enzyme site, extension of the 5′ end of the primer increases the cutting efficiency of the restriction enzyme. In some clinical samples it is helpful to predigest 5–15 μl of the DNA extract with 2–8 U of the appropriate restriction enzyme in a total volume of 30 μl for 30–120 minutes followed by partial heat inactivation at 65°C for 20 minutes. This pretreatment relieves the supercoiled structure of the JCV genome and allows full strand separation during the denaturation step of the PCR (Agostini and Stoner, 1995). The PCR reaction mix using a combination of proofreading and nonproofreading DNA polymerases (e.g., GeneAmp XL PCR, Applied Biosystems) contains 200 nM of each primer and 0.7 mM magnesium acetate. The hot start technique is recommended. The initial denaturation step at 94°C for 1 minute is followed by 15 cycles of 94°C for 40 seconds and 6 minutes at 64°C (BAM-1 and -2) or 65°C (ECO-7 and -12). This step is prolonged every six cycles by 2 minutes to give a total of 39 cycles. After a final elongation step at 72°C for 10 minutes, the reaction is stopped at 4°C.

Sequencing of PCR Products. There are two principal methods for sequencing PCR products. The product can be used either directly or after cloning in a suitable vector. For both methods the PCR products should be purified. Preparative gel electrophoresis purification avoids reduction of the cloning efficiency by unlabeled primers or side products. The direct use of PCR products in a cycle sequencing reaction helps to eliminate random errors that may be introduced during DNA polymerization. It has the disadvantage that the material for the sequencing reaction is limited, which is an important factor for automated sequencing of the complete viral genome. Cloning of the initial PCR product provides large amounts of template, and usually the sequence obtained is of better quality. However, if only a single PCR product clone is used, the chances of picking up PCR-introduced errors is high, especially if the original copy number of the template in the clinical sample is low and the DNA polymerase in the PCR does not feature a $3'-5'$ exonuclease (proofreading) activity. Therefore, one should use multiple clones (5–10) if possible from different PCR experiments and combine them in equimolar amounts after the plasmid isolation and before sequencing.

3. JCV EPIDEMIOLOGY

Why Use JCV as a Marker for Human Migration?

A viral marker for ancient and modern population migrations should fulfill certain conditions. First, the virus should be transmitted only by close contact. Although the exact mode of transmission is still unclear, JCV appears to meet this condition. JCV strains are acquired early in life, likely from exposure within the family or the immediate community (Kitamura et al., 1994; Kunitake et al., 1995).

Second, it should infect a high proportion of the population without putting the survival of its host at risk. JCV is a ubiquitous virus, and the primary infection is asymptomatic. In serologic studies more than 75% of adults tested positive for antibodies (Padgett and Walker, 1978), and there are no reports of JCV-free populations, although some remote tribes have been reported to have low antibody levels to JCV (Brown et al., 1975; Major and Neel, 1998).

Third, samples containing viral DNA should be easily collected and analyzed. JCV meets this requirement, as DNA of JCV is excreted in a single urine sample of 20–80% of adults over the age of 30 years (Agostini et al., 1996b; Chima et al., 1998). Studies of serial samples from patients with multiple sclerosis showed that over a period of observation of up to 18 months individuals were either chronic excreters, intermittent excreters, or nonexcreters (Agostini et al., 2000). Similar results have been obtained by Sundsfjord et al. (1999) in studies of systemic lupus erythematosus (SLE). Individuals younger than 30 years of age have a lower prevalence of JCV viruria (Agostini et al., 1996b). In contrast to BKV, the excretion rate of JCV does not appear to depend on the immunologic status of the host (Markowitz et al., 1993; Sundsfjord et al., 1994).

A final advantage, as noted above, is that the JCV genome is small enough that the complete JCV genome can be amplified from urine in a single PCR reaction (Agostini and Stoner, 1995). The product is suitable for cloning and/or sequencing (Agostini et al., 1997a, 1999; Jobes et al., 1998). In this way all of the information contained in the viral genome can be exploited.

Worldwide Distribution of JCV Genotypes

A total of eight major genotypes have been described thus far that differ by 1–2.6% of their DNA sequence (for convenience, the rearranging portion of the VCR from *ori* to the agnoprotein start site is excluded). These genotypes are linked to populations in specific geographic regions of the world. This distribution is best explained if the virus was tied to the dispersal of modern humans from Africa (Fig. 18.4).

Asia. Japan, China, and Southeast Asia are dominated by strains of the Type 2/Type 7 clade. Regional differences allow five major subtypes of Type 2 and three of Type 7 to be distinguished. Type 8 is a related JCV clade in Papua New Guinea and elsewhere in the South Pacific (see below). Type 2A (and a closely related variant Type 2C) are similar to the MY group found in Japan (Kitamura et al., 1998), but have not been found in China (Guo et al., 1998; X. Cui et al., unpublished data). Type 2B is likely to be widely distributed in

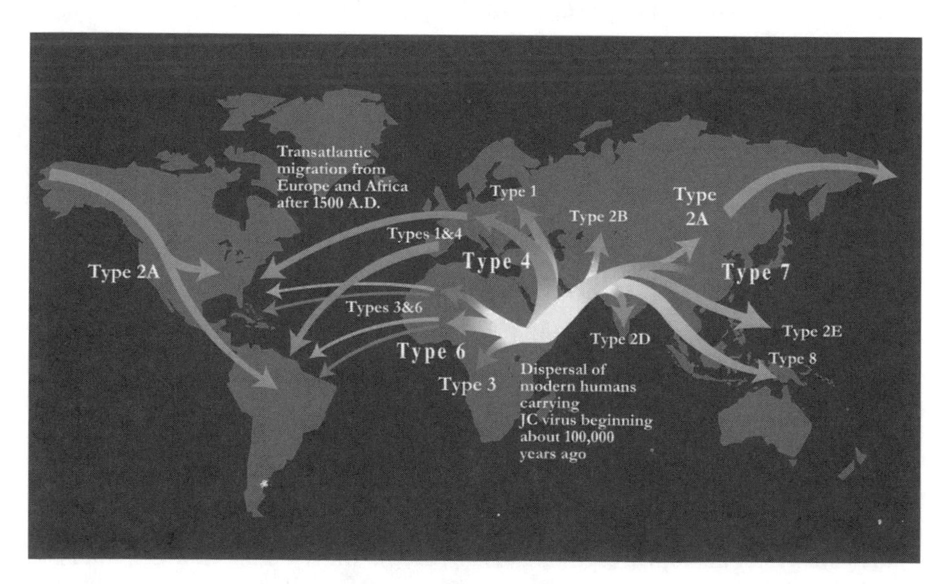

Figure 18.4. Proposed dispersal of JCV genotypes with early human migrations originating from Africa about 100,000 years ago.

Eurasia, but its boundaries are not yet well defined. Type 2D is found in India, but is also represented in Mongolian populations of China (X. Cui et al., unpublished data). Type 2E is a Pacific Island group (see below).

Type 7 is a large South China/Southeast Asia clade. Type 7A includes the prototype strain Tai-3 (Ou et al., 1997) and predominates in South China (X. Cui et al., unpublished data). Type 7B includes the CY group described by Yogo and colleagues (Guo et al., 1998) and predominates in North China. A third group, Type 7C, is found in both North and South China. The Bunun Aborigines in Taiwan mainly carry subtypes of Type 2 with the majority resembling strains from North China, as well as some Philippine isolates of Chinese origin (Chang et al., 1999). This is in contrast to the modern Taiwanese population characterized by South Chinese Type 7 and suggests that the ancient Taiwanese population migrated in two waves from the mainland to the island.

Africa. Two major genotypes have been identified in Africa thus far—Types 3 and 6 (Agostini et al., 1995, 1998c; Guo et al., 1996; Chima et al., 1998). From Type 3 two variants exist of which Type 3A seems to be more widely distributed and located in both East and West Africa, whereas Type 3B strains are more frequent in the west of the continent. Type 6 strains were found in Ghana (Guo et al., 1996) and were the dominant genotype in the Bantu from the Central African Republic (Chima et al., 1998). In the same study a group of Biaka Pygmies from the Central African Republic was tested for JCV excretion. In contrast to the Bantu, Biaka Pygmies carried Type 3, Type 6, and, in one individual, a Type 1 strain, which reflects the extent of the interactions this tribe maintained with their African neighbors as part of their itinerant lifestyle.

With the worldwide spread of HIV and the high prevalence of AIDS in African countries, the African continent is of special interest for JCV research, especially because the incidence of PML is much lower than in other regions of the world (Lucas et al., 1993; Chima et al., 1999). It soon became clear that this under-representation of PML in AIDS patients was not explained by a low prevalence of JCV in the African population (Agostini et al., 1995; Guo et al., 1996). Whether African genotypes of JCV are less likely to cause PML or the African host may be more resistant to PML is uncertain. Alternatively, it could be that lower standards of health care do not permit the long-term survival of AIDS patients, which would favor the eventual development of PML.

Europe. The most common JCV genotypes in East, Central, and Southwest Europe are Type 1 and Type 4. For both types variants exist, and European populations differ with respect to the JCV genotypes and subtypes secreted in their urine. Type 1 has two subtypes, Types 1A and 1B. Type 1A is the predominant JCV genotype in East Europe (Poland and Hungary), whereas Type 1B is more frequent in the Southwest (Spain) (Agostini et al., 2001). Type 4 is closely related to Type 1 in that it differs in only about 1% of its coding region, introns, and the constant portion of the early side of the regulatory region (Agostini et al., 1998b). In our classification scheme, different genotypes are generally characterized by a DNA sequence difference of 1% or more and are phylogenetically distinct, although the dividing line between types and subtypes is not absolute. Designation of a new clade as a new type or as a subtype of an existing type can sometimes be a matter of convenience. Compared with the general population, the prevalence of Type 4 is increased in Basques, Sinti, and Roma (Agostini et al., 2001).

We suggest that the geographic distributions of Type 4 and Type 1 are consistent with the Paleolithic and Neolithic migratory episodes that established the modern European gene pool (Semino et al., 2000). These genotypes are phylogenetically distinct (see below). If Type 4 is a remnant of Paleolithic settlement and Type 1 represents the demic diffusion of agriculturalists westward from the Middle East beginning about 10,000 years ago, the identification of these European strains as distinct types with several subtypes will be fully justified. Further detailed studies of the distribution of Types 1 and 4 in relation to Y chromosome and mitochondrial DNA lineages will be required to confirm this hypothesis. The eastern boundary where JCV of the European Type 1/Type 4 clade is replaced by Asian Type 2/Type 7 strains remains to be identified. The few Type 2 strains found in Central Europe are predominantly of the 2B subtype.

North and South America. In the United States, the dominant genotypes of two-thirds of the European Americans are Types 1B and 1A, followed by Type 4 and, with lower numbers, Types 2A, 2B, and 2C (Agostini et al., 1996b). However, Native Americans such as the Flathead in Montana or the Navaho in New Mexico carry predominantly JCV Type 2A strains of Asian origin. In the culturally more isolated Navaho people, Type 2A sequences accounted for 89% of the strains excreted in single urine samples. The number fell to 59% among the Flathead people where forced resettlement accelerated the admixture of Native Americans and European settlers and increased the proportion of strains of European origin (Agostini et al., 1997d). Type 2A (also known as Type B, MY) is characteristic of the people of Northeast Asia, specifically Japan (Sugimoto et al., 1997; Kato et al., 1997; Kitamura et al., 1998). It is therefore very likely that the ancestors of Native Americans brought these genotypes with them when they first arrived in North America over the Bering land bridge or by a coastal route 12,000 or more years ago. The Native-American strains (Type 2A2) can be distinguished from the Japanese strains (Type 2A1) in the JLP-19 and -20 amplified fragment (see below).

In African Americans the situation is different. Although the African Type 3 accounted for 18% of the strains found in an African-American population around Washington, DC, JCV Type 1 (32%) and Type 4 (44%) were the major JCV genotypes in this cohort (Chima et al., 2000). The West and Central African genotype, Type 6, was absent among this group of African Americans, suggesting a selective advantage of the European strains over Type 6 strains in African Americans following more than 400 years of admixture in the New World.

The Hispanic population in the Americas frequently represents the admixture of the indigenous people of a region, Europeans of mainly Spanish background, and Africans brought from various parts of Africa. This has been referred to as the trihybrid genetic model (Sans, 2000). The viral epidemiologic findings for JCV in a Puerto Rican cohort from San Juan and Ponce confirmed the hypothesis implied in this model. The genotypes found were Types 1 and 4 representing European influence, Types 3 and 6 contributed by Africans, and Type 2A from Asia representing the indigenous Taino population (Fernandez Cobo et al., 2001). Surprisingly, the majority of strains in this study group were of the Asian genotype found in Native Americans, Type 2A, although the Taino Indians became extinct as a recognizable ethnic entity some 200 years ago. Their enduring genetic legacy has included this previously unrecognized viral contribution to the present population of Puerto Rico (reviewed by Stoner et al., 2000). The situation will likely be similar throughout much of South America.

Pacific Islands. Studies from Guam, New Britain, and Papua New Guinea led to the discovery of JCV Type 8 in the Highlands of Papua New Guinea and Type 2E in other South Pacific islands. These Type 8 strains are of special interest because some of them (termed PNG-2B) show a deletion in the agnoprotein gene of transmissible virus (Jobes et al., 1999). Other strains spread in the population have an unusual rearrangement in the VCR that is reminiscent of PML-type rearrangements, but less extensive, being limited to the *ori*-proximal region (Ryschkewitsch et al., 2000). Other populations in the region such as the Chamorros in Guam are infected with Type 7 from the mainland or Type 2E, the South Pacific subtype (Agostini et al., 1997d; Ryschkewitsch et al., 2000; D. V. Jobes, et al., submitted). The Austronesian-speaking Tolai people of Papua New Guinea on the coastal island of New Britain are infected with variants of the major genotype of Papua New Guinea and the South Pacific, Type 8, and Type 2E (Jobes et al., 1999; Ryschkewitsch et al., 2000; D. V. Jobes, et al., in press). This constellation of JCV genotypes, Type 8 and Type 2E, correlates very nicely with the fact that the colonizing Highland population in Papua New Guinea with Type 8 evolved much longer (more than 40,000 years) and in a more isolated fashion than the sea-traveling coastal populations of the New Britain and Guam representing peoples of the Austronesian expansion occurring only 3500–5000 years ago and carrying Type 2E.

The successful assignment of genotypes in the Pacific Islands and the New World, including Puerto Rico, in relation to the history of the Old World strains distinctly infecting peoples in the major continental groups (Asia, Europe, and Africa) strongly suggests coevolution of the virus with modern humans during the dispersal from Africa (Out of Africa 2). To pin down that hypothesis, it will be necessary to show that JCV phylogenetics accommodates most of our knowledge of human dispersal. However, just as mtDNA and nrY chromosome data are contrasting but complementary in situations where there exists a sex bias in migration and intermarriage, we can suggest two things for JCV. First, its mode of inheritance (i.e., transmission) is more likely akin to the maternal inheritance of mtDNA. If so, it will provide insights in basic agreement with what we already know, but with a peculiar perspective due to its infectious nature and additional element of natural selection for more prolific, if not more aggressive, forms of the virus in an interbreeding population. Second, it appears that, as in the case of mtDNA and the non-recombining Y chromosome, recombination between JCV genotypes is sufficiently rare that it does not detract from the usefulness of viral variants to trace populations and to phylogenetically reconstruct their antecedents.

4. JCV PHYLOGENETICS

A Methodologic Overview

Either the short (215 bp) 5′ fragment within the *VP*1 gene, a longer (610 bp) part of the V–T intergenic region, or the complete genome (~5120 bp) can define JCV genotypes. The latter two can also define their phylogenetic relationships. The VCR left and right of *ori* contain additional information for genotyping, but the region right of *ori* (late side) tends to uniquely rearrange in PML brain and is therefore unsuitable for phylogenetics. In principle, the size of the sequence analyzed adds to the reliability of the phylogenetic analysis, subject to the limitations of different computational methods. All programs used in our analysis are available in the latest version of the GCG Wisconsin package (GCG, Oxford Molecular) under the names of PILEUP, DISTANCES, PAUPSEARCH, and PAUPDISPLAY.

The first step to analyze the phylogenetic relation of different viral sequences is to align them accurately using the PILEUP program in order to compare site-specific differences. The neighbor-joining method (Saitou and Nei, 1987) and the UPGMA clustering (Kim et al., 1993) are very fast approaches able to handle large data sets. Here the strategy includes searching for pairs of sequences with the highest similarity. UPGMA clustering unites the two most similar sequences (taxa) and forms an operational taxonomic unit (OTU) that is then clustered with the next most similar OTU and so on until the UPGMA tree with all sequences included in the analysis is created. The neighbor-joining method constructs a single tree from distance data using a step-by-step algorithm. It starts from a completely degenerate phylogenetic tree in which all the branches originate from the same point, also described as star phylogeny. Pairs of branches with minimal distance are then successively grouped together until the complete tree is formed. This process is much faster than any of the searching methods, but only one tree is calculated even if other trees with an equally good score exist. As in other algorithms based on the distance of aligned sequences, the best tree is the one with the minimum sum of branch lengths based on a corrected distance matrix calculated from the aligned sequences. Neighbor joining uses the same distance correction methods that are used by a search employing the distance criterion (Swofford, 2002). Another distance method available in PAUPSEARCH is the minimum evolution method. Here a distance matrix from the aligned sequences is calculated. The minimum evolution algorithm uses these values to compute the sum of the branch lengths for each tree. The optimal tree is the one with the minimum sum of branch lengths.

Other optimality criteria are maximum parsimony and maximum likelihood. With maximum parsimony, the optimal tree calculated explains the data with the least amount of evolutionary change. This tree is also referred to as the most parsimonious tree. Maximum likelihood is a statistics-based method that takes a model for evolutionary change into account. If one or more trees and the data set these trees result from are known, it is possible to calculate the likelihood that the data set resulted from each tree. The tree with the highest likelihood is considered to be the optimal tree. If more than one tree tie for the optimal tree for any of these criteria (distance, parsimony, or likelihood), all of the optimal trees are reported (Swofford, 2002).

Using minimum evolution distance, maximum parsimony, or maximum likelihood as search criteria, the optimal phylogenetic tree(s) will be obtained. Using the PAUPDISPLAY program, these can be used to construct a strict consensus tree, or a 50% majority rule tree in which each grouping must appear in more than half the trees in order to be retained in the consensus tree. The confidence level of a specific clade or grouping in this tree can be determined by an additional bootstrap analysis. It is the principle of the bootstrap analysis to run a resampled version of the original data, equal in size but with replacements, the "bootstrap replicate." The replacement of the data is done randomly. Some columns of the original data can be included once, more than once, or not at all. After a number of these "bootstrap replicates" have been analyzed the same way as the original data, a consensus bootstrap tree is created from the trees obtained from each of the bootstrap replicates. Those groupings are shown that represent the majority of the replicate trees and characterize the confidence level of this particular group (Swofford, 2002).

All phylogenetic trees resulting from the search with different optimality criteria should explain the sequence data with a minimum of necessary evolutionary changes between neighboring strains. The fewer steps to build a tree or the shorter the evolutional distance, the more likely that a tree reflects the true phylogenetic relationship between JCV isolates, assuming roughly equal rates of change.

The size of the viral DNA used for analysis can influence the result. This is the case for JCV when comparing phylogenetic trees based on the 610 bp V–T intergenic fragment with those based on the complete genome sequences (Jobes et al., 1998). This is especially important for methods like maximum parsimony that trace the evolution of individual characters within a sequence.

In the complete genome the number of informative sites is four times higher than in the V–T region alone.

Relationship and Possible Spread of JCV Genotypes

Representative phylogenetic trees with 50 complete genome sequences (without the VCR right of *ori*) are shown in Figure 18.3. The approach is based on either neighbor-joining (NJ) (Fig. 18.3A) or the maximum parsimony algorithm (Fig. 18.3B). The maximum parsimony tree is a 50% majority rule consensus tree rooted at the midpoint. Both neighbor joining and maximum parsimony algorithms support clade formation with high values for the NJ bootstrap replicates by Types 1A and 1B and 4, Type 2A–E, Type 3, Type 6, Type 7A–C, and Type 8A and 8B. (For a discussion of the status of Type 5, see below.)

Subtypes of Type 2A can be further divided into 2A1 (Japanese) and 2A2 (Native American). All major genotypes can be distinguished within the short fragment in the 5′ end of the *VP*1 gene defined by primers JLP-15 and -16. The only exception is the separation of a subset of Type 7B strains lacking the characteristic 1813G > A mutation from Type 2A strains in China. To characterize these strains a 245 bp fragment in the 3′ end of the *VP*1 gene can be amplified with primers JLP-19 and -20 (see Table 18.3). At position 2228, a T → C mutation defines the subset of Type 7B strains, which are not distinguished from Type 2A in the JLP-15 and -16 amplified fragment. In addition, this fragment provides a site that distinguishes Japanese and Native-American Type 2A strains. At position 2317 the nucleotide G characterizes Type 2A2 variants (American strains), whereas C is found in Type 2A1 variants (Japanese strains) (Fernandez Cobo et al., 2001). The consensus sequence at position 2317 of other strains is the nucleotide A. A third primer pair, JIG-17 and -18, amplifies a fragment including the intergenic region and allows confirmation of genotyping and assignment of additional subtypes (X. Cui, et al., manuscript in preparation).

The phylogenetics of JCV is still a work in progress. JCV genotypes characteristic of many ethnic groups of human genetic interest remain to be identified (Cavalli-Sforza et al., 1994). However, taking into account the findings of anthropologic disciplines like archaeology, linguistics, genetics of the Y chromosome, or mtDNA, a picture emerges of a virus that was carried by its human host all over the world, slowly evolving into distinguishable genotypes in each continental population. An outline of this picture is shown in Figure 18.4. It is assumed that the dispersal of JCV started with the migration of modern humans (*Homo sapiens sapiens*) from Africa beginning about 100,000 years ago, also referred to as "Out of Africa 2" or the "Garden of Eden" model.

Although the phylogenetic trees generated by NJ or maximum parsimony differ with regard to the order of branching of individual clades, both assign an essentially equivalent basal trifurcation to groups representing Europe (Types 1 and 4), Africa (Types 3 and 6), and Asia (Types 2 and 7). Asia, including the Indian subcontinent, offers the most complex diversity of JCV types and subtypes, especially of Type 2 strains, which were initially subdivided as subtypes Type 2A and 2B in European Americans (Agostini et al., 1996b) or Type B (CY and MY) in East Asia (Guo et al., 1996). Detailed studies in China, Taiwan, and Southeast Asia made clear that additional types and subtypes reflect distinguishable populations in this region. One example is JCV Type 7B strains (also known as Type B, CY). The initial short fragment analysis in the 5′ end of the *VP*1 gene suggested that some of these strains lacking the 1813 G > A mutation were members of the Type 2A group. However, based on phylogenetic analysis of the complete genome data, it became clear that they fall into the Type 7B clade rather than the Type 2A clade. Typing sites in the JLP-19 and -20 fragment or the JIG-17 and -18 fragment can also make this distinction.

Types 3 and 6 are indigenous to Africa and were carried over to North and South America by forced migration in the slave trade. Phylogenetically, JCV Type 3A, mainly found in East and West Africa, and Type 3B (West Africa) are more closely related to Asian strains than they are to Type 6 (Central and West Africa), which points toward an earlier separation of Type 6 in Africa. Furthermore, Type 3 strains could have been carried from Asia back to Africa, slowly replacing Type 6 strains by a westward advance. This picture of the emergence of JCV in sub-Saharan Africa with dispersal to other continents needs to be supported by additional studies in the other parts of Africa.

Time Frame of JCV Evolution

Based on the complete sequences of BKV and SV40 it was calculated that the primate polyomaviruses evolve at a rate of approximately 3.8×10^{-8} synonymous substitutions per site per year (Yasunaga and Miyata, 1982). This matches roughly with the mutation rate in the DNA of their hosts. However, it cannot be assumed that these viruses separated with the lineages of their hosts. At this rate of evolution it would have taken roughly a million years for the actual major JCV genotypes to develop, which is about 5–10 times longer than expected with "Out of Africa 2" as the starting point. Using this date (100,000 years ago) as a point of reference, and based on the sequence of the JCV V–T intergenic region, it is likely that JCV evolved about 10 times faster than originally assumed for BKV and SV40, that is, about $1–3 \times 10^{-7}$ per site per year (Sugimoto et al., 1997). Another calculation based on similar assumptions arrived at a figure of 4×10^{-7} synonymous substitutions per site per year (Hatwell and Sharp, 2000). The calculation of this rate will depend on the elapsed time assumed since divergence. The rate of mutation differs among viral genes of JCV. Forbidden synonymous substitutions could explain the conservation of third bases with low values for substitutions of synonymous sites in conserved viral genes like agnoprotein. This indicates additional functional requirements of these gene products or their transcripts (Britten, 1993).

JCV Type 5: A Homologous Recombinant?

Recombination events in the evolution of a species complicate phylogenetic studies. This is one reason why mitochondrial DNA, the Y chromosome, non-recombining parts of chromosome 21, or viruses like JCV are good candidates for anthropologic studies (Jin et al., 1999; Jorde et al., 2000). One requirement for homologous recombination of a virus is the double infection of a host. For JCV it was shown that individuals can be infected with different genotypes (Agostini et al., 1996a; Newman and Frisque, 1999). Type 4 strains, originally suggested to be a possible New World recombinant of Types 1 and 3 (Agostini et al., 1996a), proved to be a phylogenetically older European genotype closely related to Type 1. However, there is still one exception to the rule: JCV Type 5, represented by a single strain, 501 (Agostini et al., 1998d), was identified as a possible recombinant of Type 2B and Type 6, with the breakpoints being located in the late *VP*1 gene between sites 2087 and 2111 and the agnoprotein–*VP*2 region between sites 182 and 670 (Hatwell and Sharp, 2000). The geographic origin of the single Type 5 remains to be clarified. Clearly, recombination in JCV is a rare event that will not complicate studies of human evolution and migration. Similarly, many kidneys are doubly infected with JCV and BKV, but no hybrid viruses have ever been observed.

5. JCV GENOTYPES AND VIRAL BIOLOGY

Prevalence of JCV Excretion

The prevalence of JCV excretion in the urine varies with different ethnic groups and possibly with general health status as well. In healthy Africans of the Central African Republic, only about 20% were found to be excreting the virus (Chima et al., 1998). Among Central Europeans (Agostini et al., 2001; Stoner et al., 1998a) and European Americans, 30–40% were excreters (Agostini et al., 1996b; Shah et al., 1997). However, hospitalized Hungarian Roma showed a JCV excretion rate of 84% (Agostini et al., 2001). In Asians, Pacific Island populations, and the Native Americans of North America, 60–80% or more of the individuals tested were excreting the virus (Agostini et al., 1997d; Ryschkewitsch et al., 2000). It remains unclear whether differences in the viral excretion rates reflect the genetics of the host or those of the viral genotype that dominates in each population as a result of longtime coevolution.

JCV Genotypes and Disease

Progressive Multifocal Leukoencephalopathy. In general, it seems likely that all JCV genotypes can cause PML. The prototype strains for JCV Type 1 (Mad-1), for JCV Types 2A (Tokyo-1) and 2B (GS/B), and for Type 6 (601) are all isolates from PML tissue. In African PML patients from Abidjan, Côte d'Ivoire (Lucas et al., 1993; Chima et al., 1999), as well as a Gambian patient who died in Germany (Stoner et al., 1998b), several Type 3 and Type 6 strains could be identified. Type 6 in the United States was originally found in an African-American woman suffering from this central demyelinating disease (Ault and Stoner, 1992; Agostini et al., 1998c). When comparing the prevalence of JCV genotypes in PML tissue with the urinary excretion of a control group matched for age and ethnicity without PML, one finds a significant overrepresentation of Type 2, notably Type 2B strains (Agostini et al., 1997b, 1998a). Type 4 strains accounted for a number of PML cases in the United States. However, the incidence of Type 4 in PML was somewhat lower than expected from its prevalence in control urine samples, although the difference was not statistically significantly (Agostini et al., 1997b; Dubois et al., 2001). When using this epidemiologic approach to identify differences in the biologic behavior of JCV genotypes it is, of course, crucial that the ethnic backgrounds of test and control groups be comparable.

Multiple Sclerosis. Multiple sclerosis (MS) is a demyelinating disease of unknown etiology. While there are reports of latent JCV infection in the brain (White et al., 1992; Ferrante et al., 1995; Elsner and Dörries, 1992; Elsner and Dörries, 1998), as well as detection of JCV DNA in cerebrospinal fluid of 9% of MS patients (Ferrante et al., 1998), other studies failed to detect JCV DNA in brain lesions of MS patients (Buckle et al., 1992) or to find JCV protein in pathologic tissue samples (Stoner et al., 1986). In contrast to PML, the risk of disease or the clinical course of MS was not associated with the JCV genotypes that persist in these patients and were excreted in the urine (Agostini et al., 2000). Taking all results into account, we conclude that there is no obvious pathogenetic relationship between JCV or its genotypes and the onset or clinical course of MS. This does not rule out a role for other human polyomaviruses.

Genotyping of JCV Associated with Human Neurologic Tumors and HIV-Associated Retinopathy. It is possible to induce brain tumors in experimental animals, especially in hamsters, by JCV (Zu Rhein and Varakis, 1979). Similar to the more recent research on SV40 in human tumors (see Chapter 17), the role of JCV and BKV in brain tumors is under renewed consideration (see Chapters 15 and 16). In these kinds of studies, genotyping of virus strains can find real clinical applications. Genotyping of the JCV strains found in clinical specimens could be used to determine if tumor-associated strains are identical with those residing latently in the kidneys and being excreted in the urine of an individual. Genotyping will help not only to determine if tumor association of the virus is based on an endogenous infection but also to validate experimental data based on very sensitive methods like DNA detection by PCR that are prone to problems with laboratory contamination.

With regard to infection of the human retina by HIV or cytomegalovirus (CMV) it is of interest that both viruses were shown to transactivate JCV in vitro (Krachmarov et al., 1996; Heilbronn et al., 1993). We are currently investigating whether JCV contributes to the clinical retinopathy induced by HIV and CMV.

Possible Molecular Basis for Altered Neurovirulence of JCV Genotypes

JCV is very difficult to grow in cell cultures. For the first isolation human fetal glial cells were used (Padgett et al., 1971). Since then several additional cell culture systems were established to replicate the virus (Beckmann and Shah, 1983; Major et al., 1985; Mandl et al., 1987; Aksamit and Proper, 1988; Nukuzuma et al., 1995). Thus far, the behavior of different JCV genotype isolates has not been studied in these cell models. The differences may be minimized because a cell model might not include aspects of the genetic diversity that the host introduces into the virus–host relationship. However, based on the epidemiologic data one can define regions of potential functional importance in the viral genome where JCV genotypes differ.

T-Antigen Splicing. Alternative splicing of an early mRNA transcript leads to the expression of large T and small t antigen (Frisque, 1983). Small t antigen, which shares the amino-terminal domain of large T antigen, is not essential for viral replication. In PML tissue large T antigen is always expressed in the early phase of viral replication, but the level of small t antigen expression varies dramatically, indicating a regulation of the splicing process (Ishaq and Stoner, 1994). The basis for this is not known. Meanwhile, the picture has become more complicated. Additional regulatory proteins, designated T′ antigens, are now known to result from the use of additional downstream donor and acceptor splice sites of JCV mRNA (Trowbridge and Frisque, 1995). Splicing sites are well conserved during evolution. A consensus sequence of more than 130 sequences from different species is shown in Figure 18.5 (Mount, 1982) in comparison with the major splice sites of different JCV genotypes. It is evident that at the donor site positions +2 to −2 are invariant AG/GU, as are positions −2 and −1 at the acceptor site. However, conservation of the splice sites extends for up to five nucleotides, in line with the fact that, during initiation of the spliceosome assembly, the small nuclear RNA U1 binds to the first six nucleotides of the donor-intron site. In JCV Type 3A mutations change the +4 position of the donor and the −4 position of the acceptor. A correspondence of these changes is possible, but the intronic site of the acceptor is in general very variable at the −4 site, and the identical exchange of nucleotides in Type 3B strains is lacking the donor site mutation. Interestingly, the donor mutation in Type 3A reestablishes a phylogenetically conserved nucleotide (A) seen in the consensus sequence (Fig. 18.5).

T Antigen Zinc Finger. Zinc-finger motifs are common in proteins interacting with DNA. They are usually defined by two cysteine residues, which are frequently separated from two histidine residues by 12 amino acids within

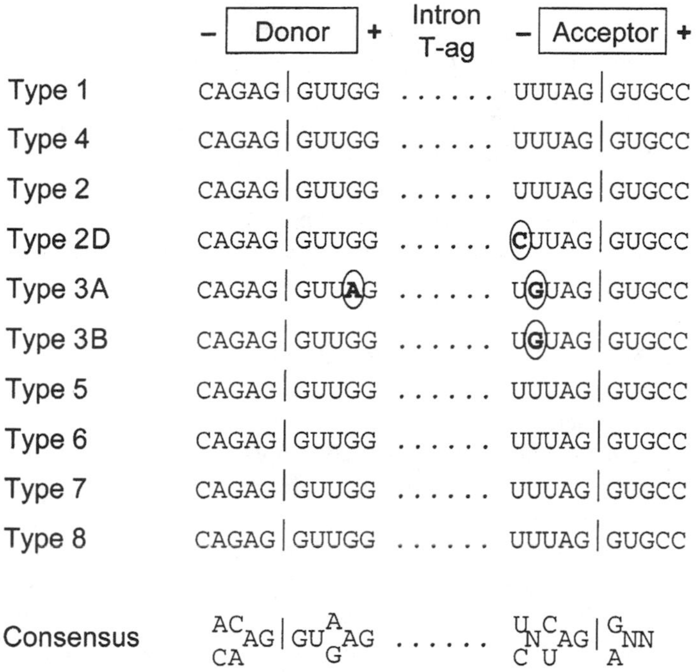

	– Donor +	Intron T-ag	– Acceptor +
Type 1	CAGAG | GUUGG	......	UUUAG | GUGCC
Type 4	CAGAG | GUUGG	......	UUUAG | GUGCC
Type 2	CAGAG | GUUGG	......	UUUAG | GUGCC
Type 2D	CAGAG | GUUGG	......	**C**UUAG | GUGCC
Type 3A	CAGAG | GUU**A**G	......	U**G**UAG | GUGCC
Type 3B	CAGAG | GUUGG	......	U**G**UAG | GUGCC
Type 5	CAGAG | GUUGG	......	UUUAG | GUGCC
Type 6	CAGAG | GUUGG	......	UUUAG | GUGCC
Type 7	CAGAG | GUUGG	......	UUUAG | GUGCC
Type 8	CAGAG | GUUGG	......	UUUAG | GUGCC
Consensus	$^{AC}_{CA}$AG | GU$^{A}_{G}$AG	......	$^{U}_{C}$N$^{C}_{U}$AG | $^{G}_{A}$NN

Figure 18.5. Splice sites within the large T antigen mRNA of different JCV genotypes. The consensus sequence is based on the analysis of more than 130 sequences of different species (Mount, 1982). N = random nucleotide; the upper nucleotides represent the more frequent alternative at this position. Circles indicate mutations found in Types 3A, 3B, and 2D close to the donor site (genomic position 4771|4770) and the acceptor site (4426|4427) based on Mad-1 numbering (Frisque et al., 1984).

a β-sheet. Folding of the protein at this element depends not only on the interaction of a zinc ion with the cysteine and histidine residues but also on the immediate sequence close to and within the protein loop as indicated by a conservation of residues in different species (Harrison, 1991). In its function as a regulatory, early protein of viral replication, the large T antigen includes such a site starting at Cys-303. Mutation experiments have shown that the motif is essential for viral replication (Swenson et al., 1996). Just two residues on the amino-terminal side of 303, that is, at amino acid 301, JCV genotypes 2A, 3, 6, 7, and 8 replace the hydrophilic glutamine residue found in Types 1, 4, 2B, and 5 with a hydrophobic leucine residue (Q301L).

Capsid Protein VP1. VP1 is the major capsid protein of JCV. Upon virus assembly it forms the outer surface of the virion and becomes responsible for the interaction with the putative cellular receptor, an N-linked glycoprotein containing terminal $\alpha(2,6)$-linked sialic acids (Liu et al., 1998a,b). It is unlikely that different JCV genotypes use different surface receptors for the cell entry, but it seems plausible that changes in the protein sequence and structure could modulate its binding energy. In Type 4 strains a DNA sequence change in the *VP*1 gene at nucleotide position 1959 changes the predicted amino acid from the basic Lys^{+} to neutral Thr^{0} as in Type 3, but downstream at positions 2462 and 2464, where Type 3 codes for neutral Gln^{0}, Type 4 retains the acidic Glu^{-1}. The result is a change in the net charge in this part of *VP*1 protein of Type 4 by −1 (Agostini et al., 1996a). Such a change has the potential to influence protein folding and virion assembly as well as the kinetics or stability of virus–receptor interaction.

Essential Parts of the JCV Genome with Low Variability. The higher the density of information essential for viral replication in a genomic region, the lower is the tolerance for evolutional changes. This hypothesis can be used to identify elements within the JCV genome essential for the viral replicative cycle. With the exception of the intergenic region between agnoprotein and the *VP*2 gene that is highly conserved among different JCV genotypes, introns and intergenic regions show a higher degree of variation than do coding regions (Agostini et al., 1997a). Within the VCR to the late side of *ori* there are only a few parts that are never deleted in the rearranged regulatory region found in PML tissue (see above). In the coding region the agnoprotein, *VP*2 and *VP*3 seem to be more highly conserved than *VP*1, which could be linked to their function in protein localization and viral assembly.

6. CONCLUSIONS

The phylogenetic approach to JCV research opens not only a new angle for understanding the natural history, molecular biology, and rare pathogenicity of this human polyomavirus, but it also gives insights into the migration of the human populations with which different JCV genotypes evolved. The characterization of this human–viral coevolution is a work in progress, as additional genotypes in remote populations may yet be found, and the full biologic implications of these adaptations are not yet understood. Furthermore, it is not yet clear how characteristic types arose and became fixed in early migrating human populations. Bottleneck phenomena, as well as genotypes that enjoy a selective advantage in a population, could influence the picture we get of JCV coevolution. However, the broad agreement among our work and that of other groups indicates that the major genotypes have been defined and form a coherent picture. Emerging from these studies is a remarkably stable, successful, and very highly adapted virus that has coevolved with its human host over tens of thousands, and perhaps hundreds of thousands, of years. Specific viral genotypes characterize certain continental populations. The fact that JCV is a ubiquitous but regionally diverse virus species that rarely recombines makes it a valuable supplement to the contributions of mtDNA and the nonrecombining portion of the Y chromosome in reconstructing the story of human evolution and dispersal.

Using readily available PCR techniques to detect JCV genotypes from urine or other clinical samples, it becomes possible to include large-scale epidemiologic data when exploring the pathogenesis of diseases caused by human polyomaviruses. One example is the increased incidence of JCV Type 2B in PML. The use of direct cycle sequencing of the PCR products minimizes mutation artifacts due to PCR-induced errors frequently found in cloned viral sequences. Thus far, the eight genotypes of JCV and their subtypes are well distinguishable within the variable part of the 5′ end of the *VP*1 gene amplified by primers JLP-15 and -16, except some Type 7 strains in China that are closely related to JCV Type 2A. For phylogenetic analysis, the V–T intergenic region of 610 bp is useful, but we recommend including the complete viral genome minus the rearranging part of the viral control region (~4854 bp). To date, over 100 JCV strains from all over the world have been completely sequenced and are available via international DNA sequence databases.

In addition to the sequence variation that evolved over tens of thousands of years, the JCV genome varies by rearranging parts of the archetypal control region after primary infection of an individual. While the underlying mechanisms of the rearrangement or the site where it occurs within an individual are not yet understood, the archetype hypothesis is well accepted (see Chapter 7). Analyzing the diversity of different viral genes not only allows the potential elucidation of the phylogenetic relationship of JCV genotypes and the evolution of its human host, but it may also hint toward functional aspects of genomic elements.

ACKNOWLEDGMENTS

We thank colleagues in our labs at the University of Freiburg and the National Institutes of Health, including Günther Schlunck, Annegret Mattes, Beatrix Flügel, Sylvester C. Chima, Mariana Fernandez Cobo, Xiaohong Cui, Alison Deckhut, Grace Ault, Jian C. Wang, Thomas Ng, Christopher Cubitt and Caroline Ryschkewitsch for their contributions to the work reviewed here. We also wish to acknowledge the contributions of Vivek Nerurkar, Richard Yanagihara, and Jonathan Friedlaender, as well as colleagues in many countries including Carolina Frias, Vicente Ausina, E. Pérez-Trallero, Rosina Girones, Charles S. Mgone, Marcin Prost, Samuel Komoly and Yasuhiro Yamamura.

REFERENCES

Abbotts J, Nishiyama Y, Yoshida S, Loeb LA (1987): On the fidelity of DNA replication: Herpes DNA polymerase and its associated exonuclease. Nucleic Acids Res 15: 1185–1198.

Agostini HT, Brubaker GR, Shao J, Levin A, Ryschkewitsch CF, Blattner WA, Stoner GL (1995): BK virus and a new type of JC virus excreted by HIV-1 positive patients in rural Tanzania. Arch Virol 140:1919–1934.

Agostini HT, Deckhut AM, Jobes DV, Girones R, Schlunck G, Prost M, Frias C, Pérez-Trallero E, Ryschkewitsch CF, Stoner GL (2001): Genotypes of JC virus in East, Central and Southwest Europe. J Gen Virol 82:1221–1331.

Agostini HT, Jobes DV, Chima SC, Ryschkewitsch CF, Stoner GL (1999): Natural and pathogenic variation in the JC virus genome. In Pandalai SG, Ed; Recent Research Developments in Virology; Transworld Research Network: Trivandrum, India, pp 683–701.

Agostini HT, Ryschkewitsch CF, Singer EJ, Baumhefner RW, Stoner GL (1998a): JC Virus Type 2B is found more frequently in brain tissue of progressive multifocal leukoencephalopathy (PML) patients than in control urines. J Hum Virol 1:200–206.

Agostini HT, Ryschkewitsch CF, Baumhefner RW, Tourtellotte WW, Singer EJ, Komoly S, Stoner GL (2000): Influence of JC virus coding region genotype on risk of multiple sclerosis and progressive multifocal leukoencephalopathy. J Neurovirol 6(Suppl 2):S101–S108.

Agostini HT, Ryschkewitsch CF, Brubaker GR, Shao J, Stoner GL (1997a): Five complete genomes of JC virus Type 3 from Africans and African Americans. Arch Virol 142:637–655.

Agostini HT, Ryschkewitsch CF, Mory R, Singer EJ, Stoner GL (1997b): JC virus (JCV) genotypes in brain tissue from patients with progressive multifocal leukoencephalopathy (PML) and in urine from controls without PML: Increased frequency of JCV Type 2 in PML. J Infect Dis 176:1–8.

Agostini HT, Ryschkewitsch CF, Singer EJ, Stoner GL (1996a): Co-infection with two JC virus genotypes in brain, cerebrospinal fluid or urinary tract detected by direct cycle sequencing of PCR products. J Neurovirol 2:259–267.

Agostini HT, Ryschkewitsch CF, Singer EJ, Stoner GL (1997c): JC virus regulatory region rearrangements and genotypes in progressive multifocal leukoencephalopathy: Two independent aspects of virus variation. J Gen Virol 78:659–664.

Agostini HT, Ryschkewitsch CF, Singer EJ, Stoner GL (1998b): JC virus Type 1 has multiple subtypes: Three new complete genomes. J Gen Virol 79:801–805.

Agostini HT, Ryschkewitsch CF, Stoner GL (1996b): Genotype profile of human polyomavirus JC excreted in urine of immunocompetent individuals. J Clin Microbiol 34:159–164.

Agostini HT, Ryschkewitsch CF, Stoner GL (1998c): Complete genome of a JC virus genotype Type 6 from the brain of an African American with progressive multifocal leukoencephalopathy. J Hum Virol 1:267–272.

Agostini HT, Shishido-Hara Y, Baumhefner RW, Singer EJ, Ryschkewitsch CF, Stoner GL (1998d): JC virus Type 2: Definition of subtypes based on DNA sequence analysis of ten complete genomes. J Gen Virol 79:1143–1151.

Agostini HT, Stoner GL (1995): Amplification of the complete polyomavirus JC genome from brain, cerebrospinal fluid and urine using pre-PCR restriction enzyme digestion. J Neurovirol 1:316–320.

Agostini HT, Yanagihara R, Davis V, Ryschkewitsch CF, Stoner GL (1997d): Asian genotypes of JC virus in Native Americans and in a Pacific Island population: Markers of viral evolution and human migration. Proc Natl Acad Sci USA 94:14542–14546.

Aksamit A, Proper J (1988): JC virus replicates in primary adult astrocytes in culture. Ann Neurol 24:471.

Aoki N, Kitamura T, Tominaga T, Fukumori N, Sakamoto Y, Kato K, Mori M (1999): Immunohistochemical detection of JC virus in nontumorous renal tissue of a patient with renal cancer but without progressive multifocal leukoencephalopathy. J Clin Microbiol 37:1165–1167.

Ault GS, Stoner GL (1992): Two major types of JC virus defined in progressive multifocal leukoencephalopathy brain by early and late coding region DNA sequences. J Gen Virol 73:2669–2678.

Ault GS, Stoner GL (1993): Human polyomavirus JC promoter/enhancer rearrangement patterns from progressive multifocal leukoencephalopathy brain are unique derivatives of a single archetypal structure. J Gen Virol 74:1499–1507.

Beckmann AM, Shah KV (1983): Propagation and primary isolation of JCV and BKV in urinary epithelial cell cultures. Prog Clin Biol Res 105:3–14.

Britten RJ (1986): Rates of DNA sequence evolution differ between taxonomic groups. Science 231:1393–1398.

Britten RJ (1993): Forbidden synonymous substitutions in coding regions. Mol Biol Evol 10:205–220.

Brockman WW, Lee TN, Nathans D (1975): Characterization of cloned evolutionary variants of simian virus 40. Cold Spring Harbor Symp Quant Biol 39(Pt 1):119–127.

Brown P, Tsai T, Gajdusek DC (1975): Seroepidemiology of human papovaviruses: Discovery of virgin populations and some unusual patterns of antibody prevalence among remote people of the world. Am J Epidemiol 102:331–340.

Buckle GJ, Godec MS, Rubi JU, Tornatore C, Major EO, Gajdusek DC, Asher DM (1992): Lack of JC viral genomic sequence in multiple sclerosis brain tissue by polymerase chain reaction. Ann Neurol 32:829–831.

Cavalli-Sforza LL, Menozzi P, Piazza A (1994): The History and Geography of Human Genes; Princeton University Press: Princeton, NJ.

Chang DC, Sugimoto C, Wang M, Tsai RT, Yogo Y (1999): JC virus genotypes in a Taiwan aboriginal tribe (Bunun): Implications for its population history. Arch Virol 144:1081–1090.

Chang DC, Wang ML, Ou WC, Lee MS, Ho HN, Tsai RT (1996): Genotypes of human polyomaviruses in urine samples of pregnant women in Taiwan. J Med Virol 48: 95–101.

Chima SC, Agostini HT, Ryschkewitsch CF, Lucas SB, Stoner GL (1999): Progressive multifocal leukoencephalopathy and JC virus genotypes in west African patients with acquired immunodeficiency syndrome—A pathologic and DNA sequence analysis of 4 cases. Arch Pathol Lab Med 123:395–403.

Chima SC, Ryschkewitsch CF, Fan KJ, Stoner GL (2000): Polyomavirus JC genotypes in an urban United States population reflect the history of African origin and genetic admixture in modern African Americans. Hum Biol 72:837–850.

Chima SC, Ryschkewitsch CF, Stoner GL (1998): Molecular epidemiology of human polyomavirus JC in the Biaka pygmies and Bantu of Central Africa. Mem Inst Oswaldo Cruz 93:615–623.

Ciappi S, Azzi A, De Santis R, Leoncini F, Sterrantino G, Mazzotta F, Mecocci L (1999): Archetypal and rearranged sequences of human polyomavirus JC transcription control region in peripheral blood leukocytes and in cerebrospinal fluid. J Gen Virol 80:1017–1023.

Domingo E, Holland JJ (1997): RNA virus mutations and fitness for survival. Annu Rev Microbiol 51:151–178.

Dora S, Schwarz C, Knippers R (1989): Excision of integrated simian virus 40 DNA involving homologous recombination between viral DNA sequences. J Mol Biol 206: 81–90.

Dörries K, Arendt G, Eggers C, Roggendorf W, Dörries R (1998): Nucleic acid detection as a diagnostic tool in polyomavirus JC induced progressive multifocal leukoencephalopathy. J Med Virol 54:196–203.

Dörries K, Vogel E, Günther S, Czub S (1994): Infection of human polyomaviruses JC and BK in peripheral blood leukocytes from immunocompetent individuals. Virology 198:59–70.

Dubois V, Moret H, Lafon ME, Brodard V, Icart J, Ruffault A, Guist'hau O, Buffet-Janvresse C, Abbed K, Dussaix E, Ingrand D (2001): JC virus genotypes in France: Molecular epidemiology and potential significance for progressive multifocal leukoencephalopathy. J Infect Dis 183:213–217.

Eigen M (1971): Selforganization of matter and the evolution of biological macromolecules. Naturwissenschaften 58:465–523.

Elsner C, Dörries K (1992): Evidence of human polyomavirus BK and JC infection in normal brain tissue. Virology 191:72–80.

Elsner C, Dörries K (1998): Human polyomavirus JC control region variants in persistently infected CNS and kidney tissue. J Gen Virol 79:789–799.

Fernandez Cobo M, Jobes DV, Yanagihara RT, Nerurkar VR, Yamamura Y, Ryschkewitsch CF, Stoner GL (2001): Reconstructing population history using JC virus: Amerinds, Spanish, and Africans in the ancestry of modern Puerto Ricans. Hum Biol 73:385–402.

Ferrante P, Caldarelli-Stefano R, Omodeo-Zorini E, Vago L, Boldorini R, Costanzi G (1995): PCR detection of JC virus DNA in brain tissue from patients with and without progressive multifocal leukoencephalopathy. J Med Virol 47:219–225.

Ferrante P, Omodeo-Zorini E, Caldarelli-Stefano R, Mediati M, Fainardi E, Granieri E, Caputo D (1998): Detection of JC virus DNA in cerebrospinal fluid from multiple sclerosis patients. Mult Scler 4:49–54.

Fitch WM, Leiter JM, Li XQ, Palese P (1991): Positive Darwinian evolution in human influenza A viruses. Proc Natl Acad Sci USA 88:4270–4274.

Frisque RJ (1983): Regulatory sequences and virus-cell interactions of JC virus. Prog Clin Biol Res 105:41–59.

Frisque RJ, Bream GL, Cannella MT (1984): Human polyomavirus JC virus genome. J Virol 51:458–469.

Gallia GL, Houff SA, Major EO, Khalili K (1997): JC virus infection of lymphocytes —Revisited. J Infect Dis 176:1603–1609.

Guo J, Kitamura T, Ebihara H, Sugimoto C, Kunitake T, Takehisa J, Na YQ, Al-Ahdal MN, Hallin A, Kawabe K, Taguchi F, Yogo Y (1996): Geographical distribution of the human polyomavirus JC virus type A and B and isolation of a new type from Ghana. J Gen Virol 77:919–927.

Guo J, Sugimoto C, Kitamura T, Ebihara H, Kato A, Guo Z, Liu J, Zheng SP, Wang YL, Na YQ, Suzuki M, Taguchi F, Yogo Y (1998): Four geographically distinct genotypes of JC virus are prevalent in China and Mongolia: Implications for the racial composition of modern China. J Gen Virol 79:2499–2505.

Harrison SC (1991): A structural taxonomy of DNA-binding domains. Nature 353:715–719.

Hatwell JN, Sharp PM (2000): Evolution of human polyomavirus JC. J Gen Virol 81: 1191–1200.

Heilbronn R, Albrecht I, Stephan S, Bürkle A, zur Hausen H (1993): Human cytomegalovirus induces JC virus DNA replication in human fibroblasts. Proc Natl Acad Sci USA 90:11406–11410.

Iida T, Kitamura T, Guo J, Taguchi F, Aso Y, Nagashima K, Yogo Y (1993): Origin of JC polyomavirus variants associated with progressive multifocal leukoencephalopathy. Proc Natl Acad Sci USA 90:5062–5065.

Ishaq M, Stoner GL (1994): Differential expression of mRNAs for JC virus large and small tumor antigens in brain tissues from progressive multifocal leukoencephalopathy patients with and without AIDS. Proc Natl Acad Sci USA 91:8283–8287.

Jin L, Gibson PE, Booth JC, Clewley JP (1993a): Genomic typing of BK virus in clinical specimens by direct sequencing of polymerase chain reaction products. J Med Virol 41:11–17.

Jin L, Gibson PE, Knowles WA, Clewley JP (1993b): BK virus antigenic variants: Sequence analysis within the capsid VP1 epitope. J Med Virol 39:50–56.

Jin L, Underhill PA, Doctor V, Davis RW, Shen PD, Cavalli-Sforza LL, Oefner PJ (1999): Distribution of haplotypes from a chromosome 21 region distinguishes multiple prehistoric human migrations. Proc Natl Acad Sci USA 96:3796–3800.

Jobes DV, Chima SC, Ryschkewitsch CF, Stoner GL (1998): Phylogenetic analysis of 22 complete genomes of the human polyomavirus JC virus. J Gen Virol 79:2491–2498.

Jobes DV, Friedlaender JS, Mgone CS, Koki G, Alpers MP, Ryschkewitsch CF, Stoner GL (1999): A novel JC virus variant found in the Highlands of Papua New Guinea has a 21-base pair deletion in the agnoprotein gene. J Hum Virol 2:350–358.

Jorde LB, Watkins WS, Bamshad MJ, Dixon ME, Ricker CE, Seielstad MT, Batzer M (2000): The distribution of human genetic diversity: A comparison of mitochondrial, autosomal, and Y-chromosome data. Am J Hum Genet 66:979–988.

Kato A, Kitamura T, Sugimoto C, Ogawa Y, Nakazato K, Nagashima K, Hall WW, Kawabe K, Yogo Y (1997): Lack of evidence for the transmission of JC polyomavirus between human populations. Arch Virol 142:875–882.

Kim J, Rohlf FJ, Sokal RR (1993): The accuracy of phylogenetic estimation using the neighbor-joining method. Evolution 47:471–486.

Kitamura T, Kunitake T, Guo J, Tominaga T, Kawabe K, Yogo Y (1994): Transmission of the human polyomavirus JC virus occurs both within the family and outside the family. J Clin Microbiol 32:2359–2363.

Kitamura T, Sugimoto C, Ebihara H, Kato A, Guo J, Taguchi F, Tominaga T, Ogawa Y, Ohta N, Kizu N, Imamura K, Funaki H, Kurosawa T, Ichikawa S, Suzuki T,

Chiba K, Nagashima K, Yasumoto S, Yogo Y (1998): Peopling of Japan as revealed by genotyping of urinary JC Virus DNA. Anthropol Sci 106:311–325.

Krachmarov CP, Chepenik LG, Barr-Vagell S, Khalili K, Johnson EM (1996): Activation of the JC virus Tat-responsive transcriptional control element by association of the Tat protein of human immunodeficiency virus 1 with cellular protein Purα. Proc Natl Acad Sci USA 93:14112–14117.

Kunitake T, Kitamura T, Guo J, Taguchi F, Kawabe K, Yogo Y (1995): Parent-to-child transmission is relatively common in the spread of the human polyomavirus JC virus. J Clin Microbiol 33:1448–1451.

Li WH, Tanimura M, Sharp PM (1988): Rates and dates of divergence between AIDS virus nucleotide sequences. Mol Biol Evol 5:313–330.

Liu CK, Hope AP, Atwood WJ (1998a): The human polyomavirus, JCV, does not share receptor specificity with SV40 on human glial cells. J Neurovirol 4:49–58.

Liu CK, Wei G, Atwood WJ (1998b): Infection of glial cells by the human polyomavirus JC is mediated by an N-linked glycoprotein containing terminal α(2-6)- linked sialic acids. J Virol 72:4643–4649.

Loeber G, Dörries K (1988): DNA rearrangements in organ-specific variants of polyomavirus JC strain GS. J Virol 62:1730–1735.

Lucas SB, Hounnou A, Peacock C, Beaumel A, Djomand G, N'Gbichi J-M, Yeboue K, Honde M, Diomande M, Giordano C, Doorly R, Brattegaard K, Kestens L, Smithwick R, Kadio A, Ezani N, Yapi A, De Cock KM (1993): The mortality and pathology of HIV infection in a West African city. AIDS 7:1569–1579.

Major EO, Miller AE, Mourrain P, Traub RG, de Widt E, Sever J (1985): Establishment of a line of human fetal glial cells that supports JC virus multiplication. Proc Natl Acad Sci USA 82:1257–1261.

Major EO, Neel JV (1998): The JC and BK human polyoma viruses appear to be recent introductions to some South American Indian tribes: There is no serological evidence of cross-reactivity with the simian polyoma virus SV40. Proc Natl Acad Sci USA 95:15525–15530.

Mandl C, Walker DL, Frisque RJ (1987): Derivation and characterization of POJ cells, transformed human fetal glial cells that retain their permissivity for JC virus. J Virol 61:755–763.

Markowitz RB, Thompson HC, Mueller JF, Cohen JA, Dynan WS (1993): Incidence of BK virus and JC virus viruria in human immunodeficiency virus–infected and virus–uninfected subjects. J Infect Dis 167:13–20.

Matsuda M, Jona M, Yasui K, Nagashima K (1987): Genetic characterization of JC virus Tokyo-1 strain, a variant oncogenic in rodents. Virus Res 7:159–168.

Mount SM (1982): A catalogue of splice junction sequences. Nucleic Acids Res 10: 459–472.

Newman JT, Frisque RJ (1999): Identification of JC virus variants in multiple tissues of pediatric and adult PML patients. J Med Virol 58:79–86.

Nukuzuma S, Yogo Y, Guo J, Nukuzuma C, Itoh S, Shinohara T, Nagashima K (1995): Establishment and characterization of a carrier cell culture producing high titres of polyoma JC virus. J Med Virol 47:370–377.

Ou W-C, Tsai R-T, Wang M, Fung C-Y, Hseu T-H, Chang D (1997): Genomic cloning and sequence analysis of Taiwan-3 human polyomavirus JC virus. J Formos Med Assoc 96:511–516.

Padgett BL, Walker DL (1978): Natural history of human polyomavirus infections. In Stevens JG, Ed; Persistent Viruses. Proceedings of the 1978 ICN-UCLA Symposia on Molecular and Cellular Biology held in Keystone, Colorado, February, 1978; Academic Press: New York, pp 751–758.

Padgett BL, Walker DL, Zu Rhein GM, Eckroade RJ, Dessel BH (1971): Cultivation of papova-like virus from human brain with progressive multifocal leucoencephalopathy. Lancet 1:1257–1260.

Ryschkewitsch CF, Friedlaender JS, Mgone CS, Jobes DV, Agostini HT, Chima SC, Alpers MP, Koki G, Yanagihara R, Stoner GL (2000): Human polyomavirus JC variants in Papua New Guinea and Guam reflect ancient population settlement and viral evolution. Microbe Infect 2:987–996.

Saitou N, Nei M (1987): The neighbor-joining method: A new method for reconstructing phylogenetic trees. Mol Biol Evol 4:406–425.

Sans M (2000): Admixture studies in Latin America: From the 20th to the 21st century. Hum Biol 72:155–177.

Semino O, Passarino G, Quintana-Murci L, Liu A, Beres J, Czeizel A, Santachiara-Benerecetti AS (2000): MtDNA and Y chromosome polymorphisms in Hungary: Inferences from the palaeolithic, neolithic and Uralic influences on the modern Hungarian gene pool. Eur J Hum Genet 8:339–346.

Shah KV, Daniel RW, Strickler HD, Goedert JJ (1997): Investigation of human urine for genomic sequences of the primate polyomaviruses simian virus 40, BK virus, and JC virus. J Infect Dis 176:1618–1621.

Stoner GL, Agostini HT, Ryschkewitsch CF, Komoly S (1998a): JC virus excreted by multiple sclerosis patients and paired controls from Hungary. Mult Scler 4:45–48.

Stoner GL, Agostini HT, Ryschkewitsch CF, Mazló M, Gullotta F, Wamukota W, Lucas S (1998b): Detection of JC virus in two African cases of progressive multifocal leukoencephalopathy including identification of JCV Type 3 in a Gambian AIDS patient. J Med Microbiol 47:733–742.

Stoner GL, Jobes DV, Fernandez Cobo M, Agostini HT, Chima SC, Ryschkewitsch CF (2000): JC virus as a marker of human migration to the Americas. Microbe Infect 2:1905–1911.

Stoner GL, Ryschkewitsch CF, Walker DL, Soffer D, Webster HD (1986): Immunocytochemical search for JC papovavirus large T-antigen in multiple sclerosis brain tissue. Acta Neuropathol (Berl) 70:345–347.

Sugimoto C, Kitamura T, Guo J, Al-Ahdal MN, Shchelkunov SN, Otova B, Ondrejka P, Chollet JY, El-Safi S, Ettayebi M, Grésenguet G, Kocagöz T, Chaiyarasamee S, Thant KZ, Thein S, Moe K, Kobayashi N, Taguchi F, Yogo Y (1997): Typing of urinary JC virus DNA offers a novel means of tracing human migrations. Proc Natl Acad Sci USA 94:9191–9196.

Sundsfjord A, Flaegstad T, Flo R, Spein AR, Pedersen M, Permin H, Julsrud J, Traavik T (1994): BK and JC viruses in human immunodeficiency virus type 1–infected persons: Prevalence, excretion, viremia, and viral regulatory region. J Infect Dis 169: 485–490.

Sundsfjord A, Osei A, Rosenqvist H, Van Ghelue M, Silsand Y, Haga HJ, Relvig OP, Moens U (1999): BK and JC viruses in patients with systemic lupus erythematosus: Prevalent and persistent BK viruria, sequence stability of the viral regulatory regions, and nondetectable viremia. J Infect Dis 180:1–9.

Swenson JJ, Trowbridge PW, Frisque RJ (1996): Replication activity of JC virus large T antigen phosphorylation and zinc finger domain mutants. J Neurovirol 2:78–86.

Swofford DL (2002): PAUP 4.0 User's Manual: Phylogenetic Analysis Using Parsimony; Sinauer Associates, Inc: Sunderland, MA.

Tominaga T, Yogo Y, Kitamura T, Aso Y (1992): Persistence of archetypal JC virus DNA in normal renal tissue derived from tumor-bearing patients. Virology 186:736–741.

Tornatore C, Berger JR, Houff SA, Curfman B, Meyers K, Winfield D, Major EO (1992): Detection of JC virus DNA in peripheral lymphocytes from patients with and without progressive multifocal leukoencephalopathy. Ann Neurol 31:454–462.

Trowbridge PW, Frisque RJ (1995): Identification of three new JC virus proteins generated by alternative splicing of the early viral mRNA. J Neurovirol 1:195–206.

Vaz B, Cinque P, Pickhardt M, Weber T (2000): Analysis of the transcriptional control region in progressive multifocal leukoencephalopathy. J Neurovirol 6:398–409.

Weber T, Klapper PE, Cleator GM, Bodemer M, Lüke W, Knowles W, Cinque P, Van Loon AM, Grandien M, Hammarin AL, Ciardi M, Bogdanovic G, and European Union Concerted Action Viral (1997): Polymerase chain reaction for detection of JC virus DNA in cerebrospinal fluid: A quality control study. J Virol Methods 69:231–237.

Wei G, Liu CK, Atwood WJ (2000): JC virus binds to primary human glial cells, tonsillar stromal cells, and B-lymphocytes, but not to T lymphocytes. J Neurovirol 6:127–136.

White FA, Ishaq M, Stoner GL, Frisque RJ (1992): JC virus DNA is present in many human brain samples from patients without progressive multifocal leukoencephalopathy. J Virol 66:5726–5734.

Yasunaga T, Miyata T (1982): Evolutionary changes of nucleotide sequences of papova viruses BKV and SV40: They are possibly hybrids. J Mol Evol 19:72–79.

Yogo Y, Iida T, Taguchi F, Kitamura T, Aso Y (1991a): Typing of human polyomavirus JC virus on the basis of restriction fragment length polymorphisms. J Clin Microbiol 29:2130–2138.

Yogo Y, Kitamura T, Sugimoto C, Hara K, Iida T, Taguchi F, Tajima A, Kawabe K, Aso Y (1991b): Sequence rearrangement in JC virus DNAs molecularly cloned from immunosuppressed renal transplant patients. J Virol 65:2422–2428.

Yogo Y, Kitamura T, Sugimoto C, Ueki T, Aso Y, Hara K, Taguchi F (1990): Isolation of a possible archetypal JC virus DNA sequence from nonimmunocompromised individuals. J Virol 64:3139–3143.

Zu Rhein GM, Varakis JN (1979): Perinatal induction of medulloblastomas in Syrian golden hamsters by a human polyoma virus (JC). In Rice JM, Ed; Perinatal Carcinogenesis, National Cancer Institute Monographs, No. 51; National Cancer Institute: Bethesda, MD, pp 205–208.

19

THE EPIDEMIOLOGY OF BK VIRUS AND THE OCCURRENCE OF ANTIGENIC AND GENOMIC SUBTYPES

WENDY A. KNOWLES, PH.D.

1. INTRODUCTION

Serologic data have shown that infection with BK virus (BKV) occurs commonly as a childhood infection throughout the world and is widely established in all developed and underdeveloped countries studied. BKV circulates independently from the other human polyomavirus, JC virus (JCV) (Brown et al., 1975). Thirty years after its discovery (Gardner et al., 1971), however, remarkably little is known about the natural history of this ubiquitous virus in the healthy population. BKV DNA persists in the kidneys (Heritage et al., 1981; Grinnell et al., 1983; Chesters et al., 1983), peripheral blood leukocytes (Dörries et al., 1994; Azzi et al., 1996; Degener et al., 1997), and possibly the brain (Elsner and Dörries, 1992) following primary infection and is excreted in the urine when immunity is compromised. However, the mode of transmission and maintenance within a population (e.g., the route of entry into the body, the site of initial replication, and the site and duration of possible virus excre-

Human Polyomaviruses: Molecular and Clinical Perspectives, Edited by Kamel Khalili and Gerald L. Stoner.
ISBN 0-471-39009-7

tion in normal healthy individuals at the time of primary infection or later) are still not understood. Presumably BKV transmission occurs within the human population from a common source of infectious virus, as no animal reservoir has been found. BKV antibody was not detected in chimpanzees, rhesus monkeys, or owl monkeys (Brown et al., 1975; Padgett and Walker, 1976), and the anti-BKV hemagglutination-inhibition (HI) antibody reported in rabbit and swine sera (Iwasaki et al., 1974) was probably due to nonantibody inhibitors (Seganti et al., 1981).

BKV is the only primate polyomavirus in which antigenic subtypes have been shown to occur, although there was one early report of a serologically distinct JCV isolate (Mad-11) from the brain tissue of a patient with progressive multifocal leukoencephalopathy (Padgett and Walker, 1983). The molecular basis for the antigenic variation in BKV has been determined within the late VP1 coding region of the genome (Tavis et al., 1989; Jin et al., 1993b), and a comparison of the reported nucleotide sequences of several JCV genomes at this location suggested that similar antigenic variation would not be expected in JCV (Chang et al., 1996a). While genomic changes in the noncoding control region and the concept of archetypal sequences are related to the evolution of the virus within the body and pathogenicity (see Chapter 7), any clinical significance of variations in the late coding region of the genome or association of various subtypes with different population groups remains to be established.

2. BKV EPIDEMIOLOGY

The Seroprevalence of BKV in Various Populations

Many assays have been developed for the detection of total or IgG antibody to BKV: hemagglutination-inhibition (HI) (Gardner et al., 1971; Gardner, 1973), complement fixation (CF) (Gardner et al.,1971), immune electron microscopy (IEM) (Gardner et al., 1971), virus neutralization (VN) (Shah et al., 1973; Flaegstad et al., 1986a), indirect fluorescent antibody (IFA) (Shah et al., 1973; Portolani et al., 1974), indirect enzyme-linked immunosorbent assay (ELISA) (Burguière et al., 1980; Iltis et al., 1983; Flaegstad and Traavik, 1985; Hamilton et al., 2000), immunoelectroosmophoresis (IEOP) (Dei and Urbano, 1982), and indirect radioimmunoassay (RIA) (Zapata et al., 1984). However, the HI assay, which measures both IgG and IgM antibodies to BKV capsid antigen (Mäntyjärvi et al., 1972; Takemoto and Mullarkey, 1973; Jung et al., 1975; Rziha et al., 1978b), has been the most widely used in BKV seroprevalence studies, the results being comparable with those obtained using the other assays and consistently more sensitive than CF (Gardner, 1973; Krech et al., 1975; Panà et al., 1976; Rziha et al., 1978b). Nonspecific glycoprotein inhibitors of BKV hemagglutination exist in human sera, and the methods used for their removal vary in effectiveness (de Stasio et al., 1979, 1980). Neuraminidase or KIO_4 has been used most widely to remove nonantibody inhibitors, but low levels may still remain, and serum dilutions ranging from 1 in 10 to 1 in 128 have been taken as the lowest dilution in which HI antibody can be regarded as being BKV specific from comparisons with other assays.

Despite these differences in test procedure, the overall seroprevalence of BKV in many populations of healthy individuals or unselected patient groups was found to be between 50% and 75% (Table 19.1). This moderately high rate was reflected in the high titers of HI antibody to BKV (strain RFV) reported in six batches of commercial human immune globulin prepared between 1953 and 1963 (Dougherty and DiStefano, 1974). However, all these figures may represent an underestimate of the numbers of people infected with BKV; in one study over 90% of pregnant women had neutralizing antibody to BKV compared with only 75% as measured by HI (Shah et al., 1980), and some molecular studies have indicated that nearly all adults have been exposed to BKV infection (Dörries et al., 1994). Brown et al. (1975), in an extensive study of 1544 sera from 28 diverse populations, many of which were isolated, from around the world, found that the prevalence of HI antibody to BKV ranged between 0% and 90%. In only a few small, extremely remote Indian or aboriginal tribes in Brazil, Paraguay, and Malaysia was anti-BKV antibody absent or at a very low prevalence (<13%). Similarly, Candeias et al. (1977) detected antibody to BKV in only 2.5–6.1% of the members of three isolated indigenous Brazilian tribes. The heterogeneity in BKV seroprevalence suggested that BKV may have been introduced to some groups only recently (Major and Neel, 1998). Generally, BKV antibody prevalence was the same in males and females, although Portolani et al. (1974) reported a slightly higher seroprevalence in adult males.

In very few populations has the effect of factors such as socioeconomic status, family size, and nursery groups on the transmission of BKV been investigated. Markoulatos et al. (1981) reported a higher prevalence and higher titers of anti-BKV antibody in residents of Athens than in rural areas of Greece, although only a very small number of samples were tested. However, Flaegstad et al. (1989) could find no significant difference in the prevalence or level of IgG antibody to BKV between urban and rural groups in Portugal, and the seroprevalence of BKV in children from a rural highland community in central Mexico with low socioeconomic status, large family size, and a low level of sanitation was reported to be similar to figures from countries with high standards of hygiene (Golubjatnikov et al., 1978). Mäntyjärvi et al. (1973), testing serial serum samples from young children, documented seroconversion or a fourfold rise in anti-BKV HI antibody titer in 6 of 32 children living in a residential nursery and a similar proportion (5 of 34) living at home, although the average ages of the children in the two groups were 1 1/2 and 3 1/2 years, respectively. Burguière et al. (1980) found no significant difference in either prevalence or level of antibody to BKV as measured by indirect ELISA, between staff working for 3 or 4 years in a chronic hemodialysis center or a renal transplant center in Paris, and matched control subjects and concluded that BKV was not spread in either setting.

Table 19.1. BKV Seroprevalence in Various Populations

Country	Study Subjects	Age Range	Assay Used (Serum Dilution)		No. of Sera Tested	No. of Sera Positive (%)		Study
Finland	Trauma patients, controls for SSPE cases	8–20 (mean, 13) years	HI	(1/10)	30	21	(70)	Meurman et al. (1972)
Finland	Controls for MS patients	Mean, 38 years	HI	(1/10)	50	23	(46)	Meurman et al. (1972)
England	Healthy children and adults, and patients tested for CMV or ASO titers	0 to >50 years	CF	(1/2)	508	276	(54)	Gardner (1973)
			HI	(1/40)	409	254	(62)	
Finland	Routine hospital patients with nonviral etiology, mostly trauma	2 to >60 years	HI	(1/10)	203	117	(58)	Mäntyjärvi et al. (1973)
United States	Healthy subjects in rubella vaccine study	1 to >36 years	HI	(1/20)	334	248	(74)	Shah et al. (1973)
United States	Patients for Wasserman testing	Young adults	HI[a]	(1/10)	400	375	(94)	Dougherty and DiStefano (1974)
			HI[a]	(1/20)	400	325	(81)	
Italy	Healthy children, adolescents, and blood donors	0.5–65 years	HI	(1/128)	453	305	(67)	Portolani et al. (1974)
			IFA	(1/2)	311	197	(63)	
Switzerland	Healthy blood donors	Adult	HI	(1/10)	158	134	(85)	Krech et al. (1975)

Japan	Healthy pregnant and nonpregnant women	20–29 years	HI	(1/40)	214	158	(74)	Taguchi et al. (1975)
Italy	Healthy blood donors, controls for tumor and transplant patients	Adult	HI	(1/128)	501	201	(40)	Corallini et al. (1976)
Italy	Healthy subjects and pediatric trauma patients	0 to >50 years	HI	(1/10)	582	380	(65)	Panà et al. (1976)
England	Healthy employees, controls for patients with malignant disease	Adults	HI	(1/40)	66	44	(67)	Flower et al. (1977)
Brazil	Cosmopolitan group	N.S.	HI	(1/40)	N.S.	N.S.	(70)	Candeias et al. (1977)
Hungary	Selected surgical patients and healthy subjects	0 to >60 years	HI	(1/10)	949	604	(64)	Szűcs et al. (1979)
Italy	Consecutive patients with various diagnoses at a childrens' hospital	0 to >12 years	IEOP	(N)	984	486	(49)	Dei et al. (1982)
Norway	Healthy adults and pediatric patients	0–82 years	IgG ELISA	(1/160)	461	340	(74)	Flaegstad et al. (1986b)
Portugal	Healthy subjects	1–82 years	IgG ELISA	(1/160)	320	218	(68)	Flaegstad et al. (1989)

HI = hemagglutination-inhibition; CF = complement fixation; IFA = indirect fluorescent antibody; IEOP = immunoelectroosmophoresis; ELISA = enzyme-linked immunosorbent assay; N = neat; SSPE = subacute sclerosing panencephalitis; MS = multiple sclerosis; CMV = cytomegalovirus; ASO = anti-streptolysin O; N.S. = not stated.

[a]Antigen used was BKV strain RF.

Age Prevalence of BKV Infection

Age prevalence studies in 10 countries have shown BKV to be an infection of childhood (Fig. 19.1), as is also the case in isolated communities (Brown et al., 1975). Maternal antibody is present at birth but is lost during the first few months of life. Gardner (1973) found a very low prevalence of anti-BKV antibody (5%) in infants between 4 and 11 months of age in England, but later studies in most other populations have indicated that up to 20–25% of infants may become infected with BKV before 1 year of age. While variations in test sensitivity and specificity may contribute to these differences, the effect of socioeconomic factors or breast feeding on the early acquisition of BKV antibody has not been investigated. Thereafter there is a rapid increase in the percentage of children infected with BKV during early childhood, with adult levels of seroprevalence, 65–90%, reached in most studies between 5 and 10 years of age. Szűcs et al. (1979) in Hungary reported a further small rise in early adulthood, and this trend can also be seen to a lesser extent in some other populations (Fig. 19.1). A slight to moderate fall in seroprevalence to BKV after the age of 40–50 years was noted in several studies (Shah et al., 1973; Mäntyjärvi et al., 1973; Portolani et al., 1974; Szűcs et al., 1979), probably reflecting low antibody levels in the older age groups that are undetected by the less sensitive assays (Gardner, 1973). Brown et al. (1975) were unable to explain a fall in seroprevalence from 50% in adolescents to 10% in later life in a study group in Iceland. In most studies the highest titers of antibody occurred in childhood when the incidence of infection is highest.

The presence of BKV-specific IgM antibody in different age groups has been investigated using techniques of varying sensitivity and specificity. Flaegstad et al. (1986b) were unable to detect BKV-specific IgM in any of 60 infants under 1 year of age by antibody capture ELISA, in contrast to 11% of healthy children between 1 and 15 years of age. Brown et al. (1984) found that 21.1% and 17.8% of unwell children aged, respectively, 1–11 years and 2–5 years had BKV-specific IgM as measured by a sensitive MACRIA. These results may be expected from the high incidence of BKV infection in childhood. Rziha et al. (1978a), however, were unable to detect BKV-specific IgM in the sera of 164 asymptomatic children less than 18 months of age and found only 3 of 99 (3%) children positive between 5 and 15 years of age using a less sensitive indirect fluorescent antibody test. Wide variation has occurred in the reported prevalence of BKV-specific IgM in healthy adults or blood donors: 0 of 107 (Jung et al., 1975), 0 of 63 (Taguchi et al., 1975), and 13 of 66 (Flower et al., 1977). However, very similar figures of 3.6% (Brown et al., 1984) and 5.7%

→

Figure 19.1. Age seroprevalence studies of BKV in 15 populations from 10 different countries. n = number of sera tested; HI = hemagglutination-inhibition; CF = complement fixation; IFA = indirect fluorescent antibody; IEOP = immunoelectroosmophoresis; ELISA = enzyme-linked immunosorbent assay.

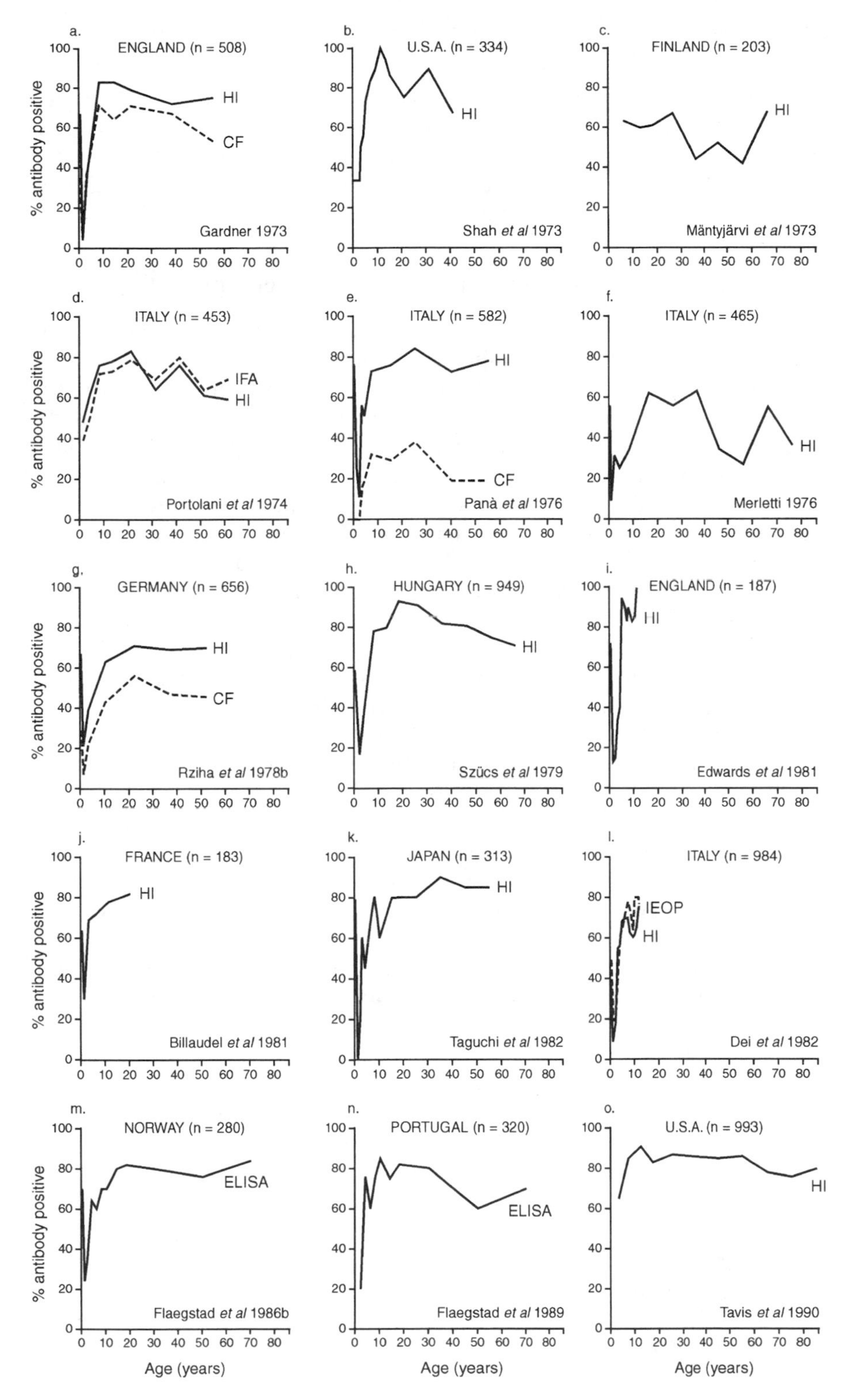

a.
ENGLAND (n = 508)
HI
CF
Gardner 1973
b.
U.S.A. (n = 334)
HI
Shah et al 1973
c.
FINLAND (n = 203)
HI
Mäntyjärvi et al 1973
d.
ITALY (n = 453)
IFA
HI
Portolani et al 1974
e.
ITALY (n = 582)
HI
CF
Panà et al 1976
f.
ITALY (n = 465)
HI
Merletti 1976
g.
GERMANY (n = 656)
HI
CF
Rziha et al 1978b
h.
HUNGARY (n = 949)
HI
Szücs et al 1979
i.
ENGLAND (n = 187)
HI
Edwards et al 1981
j.
FRANCE (n = 183)
HI
Billaudel et al 1981
k.
JAPAN (n = 313)
HI
Taguchi et al 1982
l.
ITALY (n = 984)
IEOP
HI
Dei et al 1982
m.
NORWAY (n = 280)
ELISA
Flaegstad et al 1986b
n.
PORTUGAL (n = 320)
ELISA
Flaegstad et al 1989
o.
U.S.A. (n = 993)
HI
Tavis et al 1990
% antibody positive
Age (years)

(Flaegstad et al., 1986b) were found when two different M-antibody capture methods were used. Whether these results indicate primary BKV infection or reinfection in adulthood or represent persistent BKV-specific IgM production in a few subjects is unknown, although it appears that BKV-specific IgM can persist for several years in healthy adults (Flower et al., 1977).

The Association of BKV with Illness in Immunocompetent Subjects

Several studies have been carried out on serial or paired serum samples from pediatric groups to try to determine any clinical features associated with primary BKV infection. Mäntyjärvi et al. (1973) retrospectively tested 476 serial serum samples from 66 children by HI and found seroconversion to BKV in 10 children with an antibody rise in one further patient. Seven of these children experienced mild upper respiratory tract infection and fever during the period in which the BKV antibody rise occurred, but respiratory viruses were also identified in three of these seven, and respiratory infections were common in children without a rise in BKV antibody. Similarly, Goudsmit et al. (1982) documented an antibody rise to BKV in 7 of 86 (8%) children admitted to the hospital over a 3-year period with acute upper respiratory tract disease; four of the seven also had febrile convulsions. BKV antibody rises were not detected in 91 children with lower respiratory tract symptoms. The mean age of the children with evidence of recent BKV infection in this study was 4.8 years.

In contrast, only 1 child of 82 with acute respiratory or intestinal disease studied by Corallini et al. (1976) had a significant antibody rise to BKV. Gardner (1973) and Edwards et al. (1981) failed to detect any significant BKV antibody titer rises, respectively, in paired sera from 31 children with illnesses of possible viral etiology or 43 children admitted to a pediatric department with a variety of symptoms, and Coleman et al. (1980) were unable to detect inclusion-bearing cells in the urine of 46 children under 12 years of age with acute febrile illness.

van der Noordaa and Wertheim-van Dillen (1977), studying serial serum samples from adults as well as children, detected primary BKV infection in 3 patients, aged 3, 33, and 45 years, respectively, out of 77. All three patients had acute upper respiratory infections in association with neurologic symptoms, convulsions in the child and Guillain-Barré syndrome in the adults; one of the adult patients also had serologic evidence of psittacosis infection. It is, therefore, difficult to identify primary BKV infection even in children unless serial samples are studied over a period of time.

A few single case reports have been published of immunocompetent children in whom primary or active BKV infection was documented. Goudsmit et al. (1981) reported a BKV antibody rise associated with dyspnea, high fever, cervical lymphadenopathy, conjunctival irritation, and tonsillitis in a 2-year-old mentally retarded boy; BKV was isolated from the urine, but not the throat, on day 3 of the illness. Polyomavirus particles were seen in the urine of a 5-year-old child with headache, malaise, and vomiting (Gardner and Knowles, 1995).

In four further cases urinary tract symptoms were present. An early report by Hashida et al. (1976) described a 5-year-old boy with acute hemorrhagic cystitis and inclusion-bearing cells in the urine. These inclusions contained papovavirus particles likely to be BK virus, and the altered cells were no longer detectable when the symptoms subsided. Similar cells were seen in the urine of a 3.5-year-old boy with transient nonhemorrhagic cystitis; BKV was isolated from the urine during the acute phase of illness (day 2) but not later (day 7 and 2 months). A high titer of anti-BKV antibody and BKV-specific IgM were present in a convalescent serum sample from this child, and a rise in antibody to adenovirus was also found (Mininberg et al., 1982; Padgett et al., 1983). BKV was documented by cytology, electron microscopy (EM), and polymerase chain reaction (PCR) in the urine of a further afebrile 5-year-old boy with cystitis and mild hematuria (Saitoh et al., 1993), and BKV was isolated from the urine of a 2-year-old boy with polyuria (Gardner and Knowles, 1995).

There is a single report of BKV encephalitis in an immunocompetent adult, detected by serology and the presence of BKV DNA in an acute phase cerebrospinal fluid (CSF) sample (Voltz et al., 1996).

It is concluded that most BKV infection in immunocompetent subjects is subclinical or associated with mild nonspecific symptoms, although occasionally upper respiratory tract, urinary, or neurologic involvement may be apparent.

Excretion of BKV by Immunocompetent Individuals

The high incidence of BKV infection in childhood would indicate a universally distributed source of virus and probable child-to-child transmission between, as well as within, families (Brown et al., 1975). However, the site of BKV excretion in the healthy population, associated either with primary infection or possibly later with viral persistence or reactivation, has not been established. BKV DNA is known to persist in the kidney (Heritage et al., 1981; Grinnell et al., 1983; Chesters et al., 1983), and BKV viruria can frequently be demonstrated when immunity is lowered (Arthur and Shah, 1989). It was, therefore, suggested that urine is the means by which BKV is transmitted (Brown et al., 1975), and urine has been the sample most often examined. Very few studies have been done specifically to investigate urinary excretion of BKV in immunocompetent children at an age when most virus transmission is occurring (Table 19.2). In only one report on a small number of samples (Jin et al., 1995) was the rate of BKV detection greater than 5%.

Further information on urinary BKV excretion in immunocompetent children and adults has come largely from control groups and healthy donors (Table 19.3). Although BKV was occasionally isolated in cell culture from asymptomatic or immunocompetent subjects (Gibson and Gardner, 1983; Beckmann et al., 1985), the prevalence of polyomavirus excretion was uniformly low in the early studies employing techniques for the detection of virus-infected cells or virus particles, and it was assumed that polyomaviruses were only rarely

Table 19.2. Urinary Excretion of BKV in Childhood

Age Range (Years)	Study Subjects	Detection Method	What Detected	No. of Subjects Tested	No. of Subjects Positive (%)	Study
3–16	Controls for patients with malignant disease	Electron microscopy on concentrated urine	Papovavirus particles[a]	12	0	Reese et al. (1975)
<12	Children with acute febrile illness	Cytology on urine deposit	Intranuclear inclusion-bearing cells[a]	46	0	Coleman et al. (1980)
Children	Unwell children, controls for immunosuppressed patients	Virus isolation from urine supernatant	Infectious BKV	21	0	Borgatti et al. (1981)
5–9	Pediatric patients	Virus isolation	Infectious BKV	100	1 (1)	Jin et al. (1993a)
		Single-round PCR on urine deposit	BKV DNA, VP1 region	100	5 (5)	
2–5	Pediatric patients	Single-round PCR on urine deposit	BKV DNA, VP1 region	15	4 (26.7)	Jin et al. (1995)
≤5	Pediatric patients	Nested PCR and DNA hybridization on urine deposit	BKV DNA, TAg region	100	4 (4)	Gibson (1995)
3–5	Healthy school children	Single-round PCR and DNA hybridization on urine deposit	BKV DNA, VP1 region	134	7 (5.2)	Di Taranto et al. (1997)
6–7	Healthy school children	Single-round PCR and DNA hybridization on urine deposit	BKV DNA, VP1 region	77	1 (1.3)	Di Taranto et al. (1997)
6–12	HIV-negative children in ENT ward	Single-round PCR and DNA hybridization on urine deposit	BKV DNA, VP1 region	56	0	Di Taranto et al. (1997)

PCR = polymerase chain reaction; HIV = human immunodeficiency virus; ENT = ear, nose, and throat.
[a]Method does not distinguish between BKV and JCV.

Table 19.3. Urinary Excretion of BKV in Nonimmunosuppressed Adults and Children

Age Range (Years)	Study Subjects	Detection Method	What Detected	No. of Subjects Tested	No. of Subjects Positive (%)	Study
20–69	Controls for patients with malignant disease	Electron microscopy on concentrated urine	Papovavirus particles[a]	9	0	Reese et al. (1975)
7–27	Healthy donors	Electron microscopy on urine deposit	Papovavirus particles[a]	15	0	Rziha et al. (1978a)
N.S.	Patients for routine urine cytology	Cytology on urine deposit	Intranuclear inclusion-bearing cells[a]	494	6[b] (1.2)	Coleman et al. (1980)
	Enlarged prostate			126	5	
N.S.	Adults attending family-planning clinic	Cytology on urine deposit	Intranuclear inclusion-bearing cells[a]	170	0	Coleman et al. (1980)
N.S.	Routine cytology on patients with vague urinary tract symptoms	Cytology on urine deposit	Intranuclear inclusion-bearing cells[a]	3648	12 (0.3)[c]	Kahan et al. (1980)
22–37	Healthy laboratory workers, controls for RT patients; serial samples	Cytology and indirect immunofluorescence on urine deposit	BKV infected urothelial cells	16	0	Hogan et al. (1980a)
N.S.	Normal persons	Electron microscopy on high-speed urine pellet	Papovavirus particles[a]	31	1 (3.2)	Hogan et al. (1980b)
0–39	Patients and healthy children and adults, controls for BMT patients	Dot hybridization on urine deposit	BKV DNA	125	0	Arthur et al. (1985)
		Double antibody indirect ELISA on urine supernatant	BKV antigen	125	1 (0.8)[d]	
N.S.	Immunocompetent patients	Hybridot on whole urine	BKV or JCV DNA	10	0	Gibson et al. (1985)

(*Table continues on following page*)

Table 19.3. (*Continued*)

Age Range (Years)	Study Subjects	Detection Method	What Detected	No. of Subjects Tested	No. of Subjects Positive (%)	Study
19–87	Genitourinary clinic patients	Cytology on urine deposit	Intranuclear inclusion-bearing cells[a]	33	0	Cobb et al. (1987)
		Hybridot on whole urine	BKV DNA	50	10 (20)	
Adult	Healthy individuals (urines negative by in situ filter hybridization)	Single-round PCR and DNA hybridization on urine pellet	BKV DNA, T antigen region	30	2 (6.7)	Arthur et al. (1989)
Adult	Healthy laboratory staff	Indirect immunofluorescence with anti-BKV MAb on urine deposit	BKV-infected urothelial cells	40	0	Marrero et al. (1990)
		Dot ELISA with anti-BKV MAb on PEG-treated urine supernatant	BKV antigen	40	5 (12.5)	
0–29	Urology clinic outpatients	Restriction enzyme digestion and blot hybridization on DNA extracted from urine pellet	BKV DNA	38	1 (2.6)	Kitamura et al. (1990)
30–59	Urology clinic outpatients	Restriction enzyme digestion and blot hybridization on DNA extracted from urine pellet	BKV DNA	38	0	Kitamura et al. (1990)

60–89	Urology clinic outpatients	Restriction enzyme digestion and blot hybridization on DNA extracted from urine pellet	BKV DNA	44	4	(9.1)[e]	Kitamura et al. (1990)
60–90	Patients without urologic symptoms	Restriction enzyme digestion and blot hybridization on DNA extracted from urine pellet	BKV DNA	23	2	(8.7)	Kitamura et al. (1990)
N.S.	Normal subjects	Single-round PCR and DNA hybridization on urine deposit	BKV DNA, T antigen region	12	0		Marshall et al. (1991)
18–69	Healthy HIV-seronegative homosexual males, controls for HIV-positive patients; multiple samples (urines negative by in situ filter hybridization)	Single-round PCR and DNA hybridization on urine pellet	BKV DNA, T antigen region	30	2	(6.7)	Arthur (1992)
Adult	Laboratory staff	Single-round PCR on urine deposit	BKV DNA, VP1 region	29	0		Jin et al. (1993a)
N.S.	Male HIV-negative patients at infectious disease clinic	Single-round PCR with radiolabeled primer on high-speed urine pellet	BKV DNA, control region	34	6	(17.6)	Markowitz et al. (1993)
16–64	Hospital patients, controls for HIV-positive group	Single-round PCR and DNA hybridization on urine pellet	BKV DNA, control region	56	0		Sundsfjord et al. (1994a)

(*Table continues on following page*)

Table 19.3. (*Continued*)

Age Range (Years)	Study Subjects	Detection Method	What Detected	No. of Subjects Tested	No. of Subjects Positive (%)	Study
13–65	Healthy volunteers and urology clinic patients, controls for renal transplant group	Restriction enzyme digestion and blot hybridization on DNA extracted from urine pellet	BKV DNA	65	1 (1.5)	Kitamura et al. (1994)
N.S.	Healthy adults	Single-round PCR on urine deposit	BKV DNA, VP1 region	18	0	Jin et al. (1995)
Adults	Immunocompetent donors	Nested PCR and DNA hybridization on PEG-treated urine supernatant	BKV DNA, control region	30	5 (16.7)	Azzi et al. (1996)
28–69	HIV-negative homosexual men	Single-round PCR and DNA hybridization on high-speed urine pellet	BKV DNA, T antigen region	78	4 (5.1)	Shah et al. (1997)

Mean age, 33.8	HIV-negative heterosexual patients at sexually transmitted diseases clinic, controls for HIV-positive patients	Single-round and nested PCRs on urine deposit	BKV DNA, VP1 and control regions	26	0		Degener et al. (1997)
20–26	Immunocompetent students	Single-round PCR, hybridization and sequencing on urine high-speed pellet	BKV DNA, control region	75	0		Tsai et al. (1997)
23–75	Healthy controls for SLE patients	Single-round PCR and DNA hybridization on urine pellet	BKV DNA, early region	88	0		Sundsfjord et al. (1999)
N.S.	Immunocompetent subjects	Nested PCR and DNA hybridization on PEG-treated urine supernatant	BKV DNA, control region	62	25	(40.3)	Azzi et al. (1999)

N.S. = not stated; RT = renal transplant; BMT = bone marrow transplant; HIV = human immunodeficiency virus; SLE = systemic lupus erythematosis; ELISA = enzyme-linked immunosorbent assay; Mab = monoclonal antibody; PEG = polyethylene glycol.

[a]Method does not distinguish between BKV and JCV.

[b]One patient had sarcoma of the bladder.

[c]One patient had a renal cell carcinoma; 6 of the other 11, males aged 56–79 years, had benign prostatic enlargement.

[d]A 19-year-old pregnant woman with diabetes.

[e]Two patients had benign prostatic hypertrophy; two had urinary tract tumors.

excreted by healthy subjects. Furthermore, urinary cytology and EM did not distinguish between BKV and JCV, although it has recently been suggested that BKV- and JCV-induced inclusion-bearing cells can be distinguished cytologically (Itoh et al., 1998). However, positives detected by either test could have been identified as BKV by, respectively, indirect immunofluorescence on urinary epithelial cells (Hogan et al., 1980b) and immune EM (Gardner et al., 1971; Gardner, 1977b; Penney and Narayan, 1973; Albert and ZuRhein, 1974). Assays for viral antigen and DNA later became available, and since 1989 a range of PCR protocols has been used leading to reports of widely differing rates of viruria of between 0% and 40% in various groups of immunocompetent adults. Excretion of BKV was found to increase with age, although not to the extent recorded for JCV (Kitamura et al., 1990).

It could be suggested that failure to isolate BKV in cell culture from non-immunosuppressed subjects is due to the fact that naturally occurring BKV strains in the urinary tract, which have an archetypal regulatory region, are known to grow poorly, if at all, in vitro in the cells commonly used (Mew et al., 1981; Sugimoto et al., 1989). BKV was isolated from the urine (positive by cytology) of a healthy male in urine-derived epithelial cells (Beckmann and Shah, 1983), and a human endothelial cell line, HUC-EC-C, may also be susceptible for the growth of archetypal BKV (Sundsfjord et al., 1994b). Furthermore, anti-BKV antibody may be present in the urine (Reese et al., 1975) and coat the virus particles (Gardner, 1977b; Hogan et al., 1980b; Gibson et al., 1985), thus neutralizing the virus in the convalescent stage.

The persistence of BKV viruria in immunocompetent subjects is unknown and may be different in primary infection in children and reactivated or persistent infection in adults. Inclusion-bearing urothelial cells and infectious BKV were detected only transiently in children with urinary tract symptoms (Hashida et al., 1976; Padgett et al., 1983; Saitoh et al., 1993), and Di Taranto et al. (1997) were unable to detect BKV DNA in follow-up urine samples taken after 8 months from eight positive healthy schoolchildren, whereas Kitamura et al. (1990) detected BKV DNA in urine samples taken 42 and 158 days apart, respectively, from two of three elderly males tested. A further possible reason for failure to detect BKV in the urine of immunocompetent patients until recently may be the excretion of very low amounts of virus, below the level of detection of all but the most sensitive molecular techniques. Although in some patients with urinary tract symptoms between 1 and 5 pg/ml of BKV DNA was detected (Kitamura et al., 1990), Azzi et al. (1999) and Markowitz et al. (1993) detected <4 fg of BKV DNA and <3 fg of BKV DNA/ml, respectively, in immunocompetent subjects and HIV-negative patients. Furthermore, the sample volume and method of preparation used varied considerably between different PCR studies. The region of DNA amplified in the PCR reaction may also have an affect on the sensitivity of the test. It should be noted, however, that, unlike cytology, EM, and virus isolation, the PCR studies reported here may simply be detecting a latent BKV genome in the urinary cells and do not necessarily indicate excretion of BK virus.

However, it is possible that the urinary tract is not the main route of BKV excretion in normal subjects, at least in primary infection, and, although it was postulated early that BKV may be a respiratory infection (Mäntyjärvi et al., 1973; van der Noordaa and Wertheim-van Dillen, 1977; Goudsmit et al., 1981, 1982), there are only two reports of attempts to isolate BKV from throat washings or swabs of immunocompetent children or those with a variety of diseases (Coleman et al., 1977; Possati and Bartolotta, 1981). BKV DNA was, however, detected in a throat washing (Jin, 1993). Sundsfjord et al. (1994b) failed to detect BKV DNA in saliva from 60 immunocompromised (HIV-infected) patients or 10 healthy adults at a dental clinic. However, Goudsmit et al. (1982) were able to show the presence of BKV DNA in tonsil tissue from 5 of 12 children with recurrent respiratory disease. Multiple copies of nonintegrated genome-length BKV DNA were detected per cell, but the authors were unable to demonstrate infectious virus by either transfection of tonsillar DNA or cocultivation with susceptible cells. Five of six children in whose tonsils BKV DNA was not detected had been infected with BKV as shown by the presence of anti-BKV HI antibody.

More recently, Sundsfjord et al. (1994b) were able to detect BKV DNA in nasopharyngeal aspirates from only 2 of 201 hospitalized children with serious respiratory infections; the positive children were aged 7 and 8 months, respectively, and one also had an infection with respiratory syncytial virus. As with the earlier findings in tonsillar tissue (Goudsmit et al., 1982), infectious BK virus was not demonstrated in either aspirate, and the noncoding control region of the BKV DNA was rearranged, thus suggesting that the BKV may be latent, possibly in lymphoid cells. Sundsfjord et al. (1994b) concluded that the respiratory tract was not an important site for either primary replication of BKV or persistence and suggested a fecal route of transmission should be investigated. However, Coleman et al. (1977) and Possati and Bartolotta (1981) had been unable to isolate BKV from the feces of immunocompetent children or those with a variety of diseases under 15 years of age, although the latter authors used an insensitive technique for BKV isolation. Interestingly, it has very recently been reported that BKV DNA can be widely detected in urban sewage at a concentration equivalent to 10^1–10^3 virus particles per 4 ml of sewage (Bofill-Mas et al., 2000).

BKV Infection During Pregnancy

Active BKV infection has been shown to occur during pregnancy, possibly mediated by either immunologic (Wolfendale et al., 1982; Coleman et al., 1983) or hormonal (Moens et al., 1994, 1999) changes. Several serologic studies on paired sera from pregnant women have indicated comparable low levels of active BKV infection. Taguchi et al. (1975) detected a fourfold or greater rise in HI antibody to BKV during pregnancy in 6 of 80 (7.5%) women, 2 of whom had BKV-specific IgM at delivery as shown by 2-mercaptoethanol (2-ME) reduction, and suggested that viral reactivation was occurring. Five of 100 (5%)

women studied by Shah et al. (1980) had a significant rise in anti-BKV HI antibody titer, and in each of the five women could BKV-specific IgM be demonstrated by IFA at delivery. Borgatti et al. (1979), studying single sera from 253 pregnant women, reported that the percentage of sera with high-titered anti-BKV antibody was significantly greater in the study group than in a control group, but was unable to detect BKV-specific IgM using the 2-ME method; the stage of pregnancy was not stated. However, in the same study, fourfold or greater anti-BKV titer rises were found in 3 of 33 (9%) women from whom paired serum samples were obtained.

In a larger study on paired sera from 430 pregnant women, Coleman et al. (1980) found high (>640) HI titers to BKV in 27 women (6.3%) and an antibody rise in a further 18 women (4.2%). Twenty-eight of these 45 women were found to have BKV-specific IgM by an IFA technique (Gibson et al., 1981), of which 20 were confirmed by HI following serum fractionation on a sucrose density gradient; 10 of these 20 were further confirmed by IEM on the IgM positive serum fractions. In these latter 10 patients the production of BKV-specific IgM increased as pregnancy progressed. Flaegstad et al. (1986b), using a sensitive antibody capture ELISA, detected BKV-specific IgM in 3 of 107 (3%) pregnant women at about 20 weeks of gestation. In none of the studies were primary BKV infections documented, indicating that the presence of active BKV infection during pregnancy is usually due to virus reactivation.

Following an early report of JCV in the urine of a pregnant woman (Coleman et al., 1977), Lecatsas et al. (1978) detected polyomavirus particles by negative stain EM in the urine pellet from 10 of 300 (3.3%) pregnant women at 7 to 41 weeks of gestation; the viruses, however, were not typed as BKV or JCV. A similar percentage of polyomavirus excretors was detected by cytology in the very large study of Coleman et al. (1980). Urine samples (n = 6380) from 1235 pregnant women were tested, and inclusion-bearing cells were found in 40 (3.2%) women. Excretion was intermittent and related to gestation as for IgM antibody, with the abnormal cells becoming detectable during the second or third trimester in many cases and continuing postpartum in a few women. Virus isolation in cell culture was attempted from the urine of the 40 excretors, but BKV could be identified in only two women; one of these isolates was a BKV variant, AS (see later). Isolates were obtained from 10 other women, at least four of whom were shown to be excreting JC virus; all isolates grew poorly, and by EM it was found that many of the urine samples examined contained only small numbers of partially disintegrated or antibody-coated polyomavirus particles. It is possible that JCV is selectively excreted during pregnancy, although it may be that the cytological screening method used favored the detection of JCV rather than BKV excretors, as the serologic results would suggest that BKV reactivation during pregnancy is more prevalent than it appeared from the virus isolation results. Subsequent tests on the urine samples from the cytologically positive women in this study, using two different PCR protocols, found 10 of 39 (Jin et al., 1993a) and 6 of 38 (Gibson, 1995) women to have BKV DNA in their urine.

The presence of BKV DNA in the urine of pregnant women was also investigated in three other studies using the PCR technique. Both Markowitz et al. (1991) and Chang et al. (1996b) amplified DNA from the noncoding control region of the genome and found, respectively, 18 of 117 (15%) and 3 of 31 (10%) patients to be positive. In each study the BKV DNA detected was archetypal, which may have contributed to the difficulty experienced by Coleman et al. (1980) in obtaining viral isolates in cell culture. In a third PCR study, Jin et al. (1995) reported BKV VP1 region DNA in 19 of 40 (47.5%) pregnant women. Chang et al. (1996b) studied women in the second and third trimesters, but the stage of pregnancy was not stated in the other studies. Attempts to detect BKV excretion from sites other than the urinary tract (e.g., throat or cervix) during pregnancy or in breast milk postpartum have not been reported, and the role of the mother in perinatal or early postnatal transmission of BKV has not been studied.

Congenital BKV Infection

Although the first BKV age seroprevalence study of Gardner (1973) showed a very low prevalence of anti-BKV antibody between 4 and 11 months of age, suggesting very little congenital, perinatal, or early postnatal infection, many subsequent studies in different populations indicated that early BKV infection was more common. The possibility of congenital BKV infection was investigated by several groups. In two early studies BKV-specific IgM antibody was reported in cord blood samples. Taguchi et al. (1975) found BKV-specific IgM by the 2-ME reduction method in cord blood samples from three of six infants whose mothers experienced a fourfold or greater rise in anti-BKV antibody during pregnancy, and Rziha et al. (1978b) reported BKV-specific IgM detected by IFA in cord blood of 77 of 846 (9.1%) infants born to healthy mothers. However, Shah et al. (1980) indicated that the IFA test may give nonspecific results, and hence the specificity of the above results has been questioned.

Borgatti et al. (1979) were unable to detect BKV-specific IgM by 2-ME reduction in cord blood sera containing high titers of anti-BKV HI antibody. Furthermore, Shah et al. (1980) failed to detect BKV-specific IgM by IFA in the cord blood of 387 infants, three of whose mothers had BKV-specific IgM at term. Similarly, BKV-specific IgM could not be demonstrated by either IFA or HI on IgM positive serum fractions of cord or neonatal blood samples from 309 infants, which included 39 infants whose mothers excreted inclusion-bearing cells in the urine during pregnancy and 45 infants whose mothers had high or rising anti-BKV antibody during pregnancy (Coleman et al., 1980; Gibson et al., 1981). The negative results in this study were later confirmed using a sensitive MACRIA, although 4 of 404 cord sera gave equivocal results (Brown et al., 1984). Further evidence against the transplacental transmission of BKV was provided by Coleman et al. (1980), who were unable to find inclusion-bearing cells in the urine of any of 29 babies born to cytology positive mothers and were also unable to detect polyomavirus particles or to isolate BK virus from 25 and 23 of these babies, respectively. There was also no evidence of BKV infection in 10 placentas or amniotic fluid samples from two patients in this study. Furthermore, Dörries et al. (1994) were unable to detect BKV DNA in cord blood leukocytes of 10 neonates.

These findings, however, do not exclude the possibility of BKV crossing the placenta and causing gross fetal pathology in rare cases of primary BKV infection during pregnancy, although no such cases have thus far been documented serologically. McCance and Mims (1977, 1979) described transplacental infection of polyoma virus in mice following primary, but not reactivated, infection in the mother during pregnancy. Pietropaolo et al. (1998) studied tissue from 15 abnormal aborted fetuses and found BKV DNA in brain and kidney of 12 and 9 fetuses, respectively; placental tissue was also positive in 12 cases and in 6 of 12 placentas with a normal pregnancy outcome. BKV control region sequences were found more frequently than those of the late VP1 region. BKV serology was not done in this retrospective study. Rziha et al. (1978a) reported finding BKV-specific IgM by IFA in the sera of 17 of 402 (4.2%) infants under 6 months of age with congenital disease compared with 0 of 68 controls; 13 of the positive children were 2 weeks of age or younger. In this study BKV was isolated from the urine of a 1-month-old boy with multiple dysplasias and hepatosplenomegaly.

3. BKV SUBTYPES

Antigenic Subtypes of BKV

As BK-like polyomaviruses were isolated in cell culture and identified by antisera raised against the prototype, it became clear that antigenic variants exist. Lecatsas et al. (1976) in South Africa reported an isolate, MG, from the urine of a renal transplant patient, which was serologically distinct from BKV by HI using hyperimmune anti-BKV and anti-MG rabbit antisera. Wright et al. (1976) compared several urine-derived polyomavirus isolates and also found MG to be distinct by HI from the prototype, GS, RF, and DW; small differences in the tryptic peptides of VP1 and VP3 were also noted. Further antigenic variants were subsequently reported: SB from a patient in England with lymphoma (Gardner, 1977a; Gibson and Gardner, 1983), AS from a pregnant woman (Coleman et al., 1980; Gibson and Gardner, 1983), BO from immunosuppressed children in Italy (Borgatti et al., 1981), and DB from a bone marrow transplant patient in North America (Tavis et al., 1990).

Tavis et al. (1989) confirmed that the BKV prototype and AS were distinct by HI using polyclonal rabbit antisera obtained 10 days after a single intravenous injection of virus; while the BKV antiserum reacted to high titer only with the homologous strain, the AS antiserum reacted equally with both prototype BKV and AS. The authors subsequently found a further isolate, DB, to be intermediate between prototype BKV and AS (Tavis et al., 1990). Knowles et al. (1989) produced a panel of rabbit antisera against isolates shown to differ from the BKV prototype either antigenically or with a restriction enzyme digest and were able to distinguish two distinct antigenic subtypes by both HI and virus neutralization, with two further isolates, SB and AS, broadly cross reactive but distinct from one another (Table 19.4); these results were confirmed by kinetic neutralization tests (Knowles, 1986). The antigenic subtypes were subsequently designated as serogroups I to IV (Jin et al., 1993b). The rabbit antisera used in this study had been raised by a single intravenous injection of high-titered, purified virus, the animals being bled at 11 days and at 25–27 days postinoculation. Cross reactions seen with the early antisera were often much less pronounced with the later bleeds. It was suggested by Tavis et al. (1990) that using antisera raised to prototype BKV as opposed to a more broadly reactive subtype, for the detection of polyomavirus antigens may lead to some strains being missed.

Gardner (1977a) found that the distribution of antibody to SB in the general population was similar to that of the BKV prototype, although titers to SB were lower. Limited age prevalence studies on the antigenic subtypes of BKV have been reported from two laboratories, and the results are shown in Figure 19.2a,b. In each population the adult seroprevalence of AS was lower than that of prototype BKV: about 50% in England (Gibson and Gardner, 1983) and 30–40% in the United States (Tavis et al., 1989, 1990). However, in the United States infection with AS appeared to occur in early childhood, whereas in England antibody was acquired during late childhood and adolescence. The age seroprevalence of isolate DB, later shown to belong to serogroup IV (Jin et al., 1995), was intermediate between prototype BKV and AS (Tavis et al., 1990). A small study by W. A. Knowles (unpublished data, 1991) found that the

Table 19.4. HI Titers of Rabbit Antisera Raised to BKV Subtypes

Isolate		Antiserum[a]						
		BK	GS	PG	SB	AS	MG	IV
BK	(I)[b]	*320*	640	640	160	≤20	<20	<20
GS	(I)	320	*640*	640	80	20	<20	<20
PG	(I)	640	640	*640*	80	20	<20	<20
SB	(II)	80	160	80	*1280*	40	40	80
AS	(III)	<20	20	<20	80	*320*	≤20	40
MG	(IV)	20	40	<20	160	80	*640*	1280
IV	(IV)	<20	40	<20	160	80	640	*1280*

HI = hemagglutination-inhibition.

[a]Produced in rabbits by a single intravenous injection of purified virus; 25–27 day bleed.

[b]Serogroups I to IV (Jin et al., 1993b).

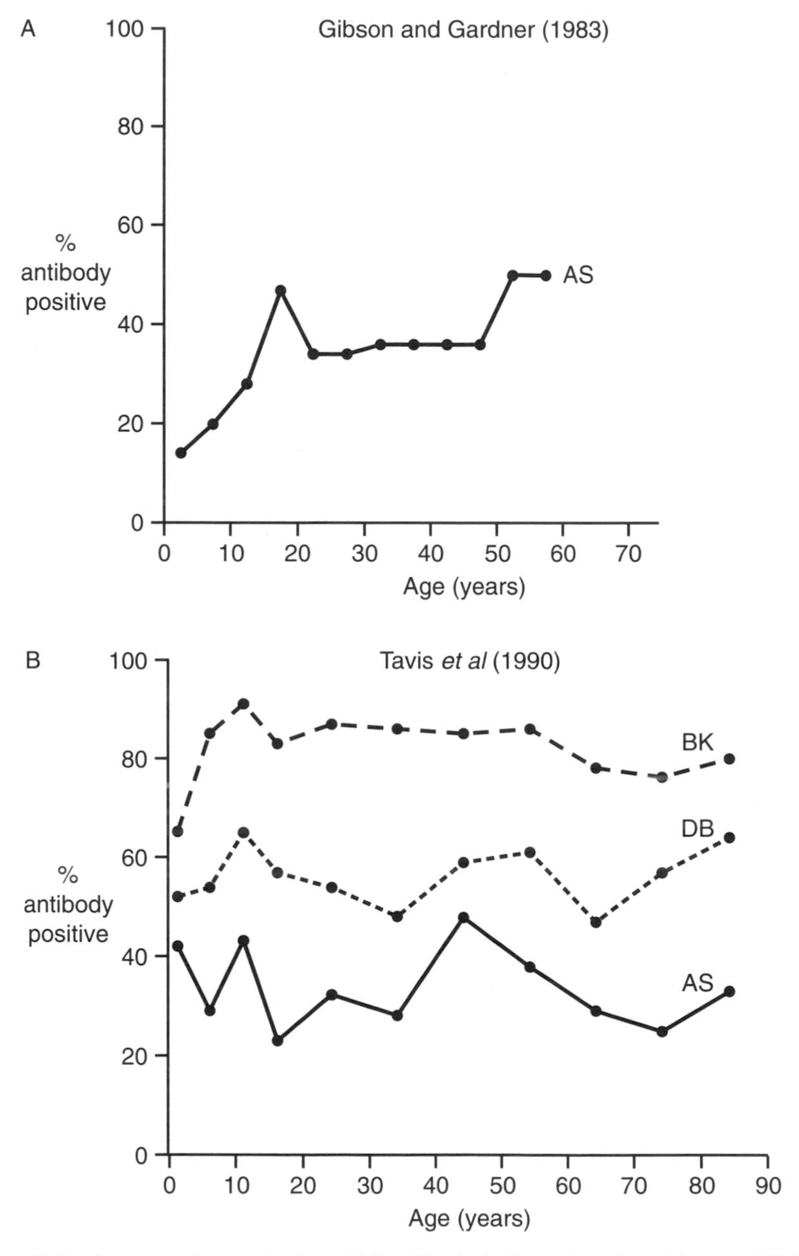

Figure 19.2. Age prevalence studies of HI antibody to the antigenic subtypes of BKV. (a) Antibody to isolate AS (subtype III) in individuals in England. (b) Antibody to isolates BK (prototype, subtype I), DB (subtype IV) and AS (subtype III) in individuals in the U.S.A. (c) Antibody to isolates of subtypes I to IV in children in England. HI = hemagglutination-inhibition.

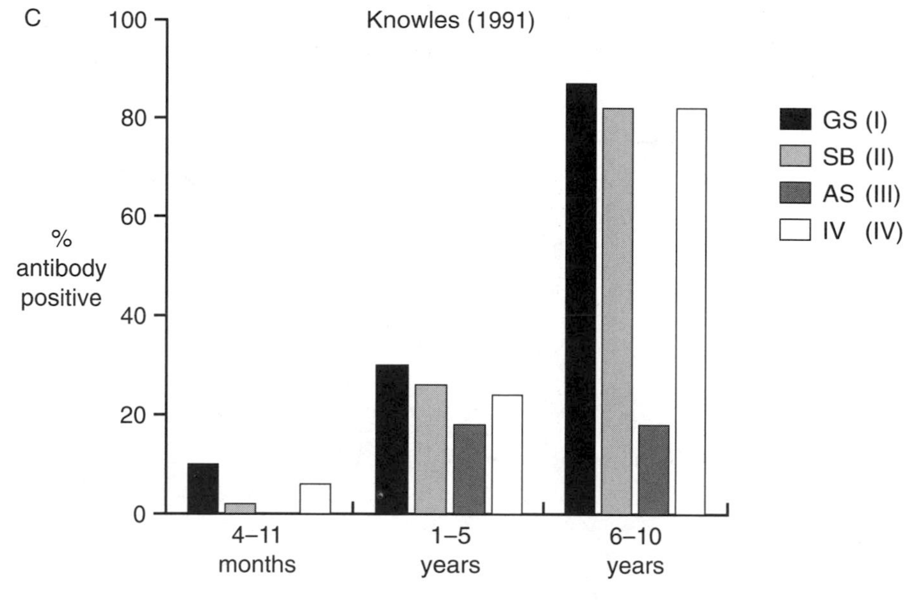

Figure 19.2. (*Continued*)

antibody prevalence to isolates of serogroups I, II, and IV was almost identical in children under 10 years of age, whereas the seroprevalence to serogroup III (AS) was much lower (Fig. 19.2c); the antibody titers to AS were also lower than to the other subtypes. In this study most sera were cross reactive with several subtypes (except AS), although the titers to different subtypes varied, and only a small proportion of sera showed a reaction to just one subtype, as has also been reported by others (Borgatti et al., 1981; Tavis et al., 1990). Thus it is not possible to determine the distribution of subtypes in a population using serologic data.

Genomic Subtypes of BKV

The molecular basis for the existence of antigenic subtypes in BKV has been identified and is localized to a region of 23 amino acids in the amino-terminal quarter of VP1 (Fig. 19.3). Whereas the nucleotide sequence homology between the subtypes and the BKV prototype is >95% over VP1 as a whole, between nucleotides 1744 and 1812 (numbering according to BKV[DUN]; Seif et al., 1979), corresponding to amino acids 61–83, the homology is less than 70% (Jin et al.,1993b). Seven of 20 amino acid differences between VP1 of BKV(DUN) and AS are clustered in this region (Tavis et al., 1989), as are all the five amino acid differences between VP1 of SB and AS (Jin et al., 1993b). This region, corresponding to the BC loop of SV40 (Liddington et al.,1991), is hydrophilic and as such is likely to be externally situated on the virion and to form an important antigenic site (Tavis et al., 1989; Jin et al., 1993b). A

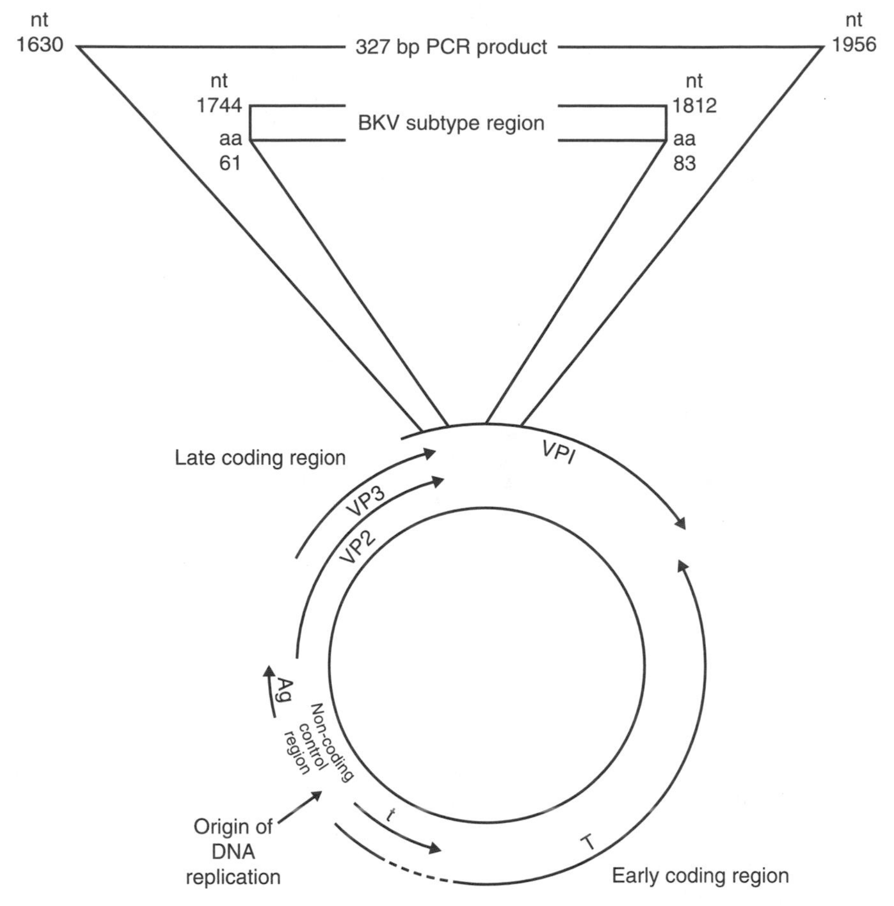

Figure 19.3. Location of the subtype region in the genome of BKV. nt = nucleotide; bp = base pair; aa = amino acid; PCR = polymerase chain reaction; VP1 = virion polypeptide 1; VP2 = virion polypeptide 2; VP3 = virion polypeptide 3; Ag = agnoprotein; T = large T antigen; t = small t antigen. Nucleotide numbering as BKV(DUN) (Seif et al., 1979).

scheme for determining the genetic subtype of BKV by nucleotide sequencing or restriction enzyme analysis of a PCR-amplified 327 base pair product (nucleotides 1630 to 1956) was developed (Jin, 1993; Jin et al., 1993a). In all isolates tested there was complete agreement between the genetic and antigenic subtypes determined (Jin et al., 1993a), and all strains belonged to one of the four subtypes; no further subtypes or intermediate subtypes were found.

Very recently a yeast cell system (Sasnauskas et al., 1999) has been used to express the VP1 of SB and AS (Hale et al., 1999), and the virus-like particles (VLPs) produced have been cross-tested by HI using the rabbit antisera of Knowles et al. (1989). In each case the results matched those obtained using crude cell culture antigen, thus confirming that the site of subtype variation resides within VP1.

Distribution of BKV Subtypes in Various Patient Groups

As yet there are no data with which to determine a possible association of BKV subtype with particular population groups as can be done for JCV (see Chapter 18), as in most cases the ethnic origin of the excretors is unknown; for example, prototype BK was isolated in London from a Sudanese patient (Gardner et al., 1971). A few attempts have been made, however, to associate the BKV subtypes with various clinical conditions. A summary of the antigenic subtyping results on all BKV strains isolated in the laboratory of Sylvia Gardner between 1971 and 1992 was given by Gardner and Knowles (1995). Subtype I (BKV prototype) was by far the most prevalent subtype in all patient groups, with subtypes II (SB) and IV (IV) being isolated from a few renal (RT) and bone marrow (BMT) transplant patients and patients with malignant disease; subtype III (AS) was associated with pregnancy.

As serologic typing by HI can only be done on virus isolates, it could be possible that these results were influenced by differences between the subtypes in the ease of isolation in cell culture. However, genotyping directly from clinical material gave a similar distribution of subtypes (Jin et al., 1993a). Subsequent PCR studies showed that all four subtypes were present in the urine of HIV-infected patients, but the association of only subtypes I and III with pregnancy remained, and subtypes I, III, and IV were all detected in healthy young children (Agostini et al., 1995; Jin et al., 1995; Degener et al., 1997; Di Taranto et al., 1997). Dual infection with more than one subtype, but always including subtype I, was documented in HIV-infected patients and pregnant women but not in BMT recipients. Furthermore, in the latter group the subtype detected in one third of patients remained the same for up to 3 months (Jin et al., 1995).

It seems, therefore, that four natural, stable subtypes of BKV are circulating independently and that passage in cell culture (as shown by the varied passage history of the isolates used for antiserum preparation (Knowles et al., 1989)) does not alter the subtype. The origin and significance of the minor subtypes are unknown.

REFERENCES

Agostini HT, Brubaker GR, Shao J, Levin A, Ryschkewitsch CF, Blattner WA, Stoner GL (1995): BK virus and a new type of JC virus excreted by HIV-1 positive patients in rural Tanzania. Arch Virol 140:1919–1934.

Albert AE, Zu Rhein GM (1974): Application of immune electron microscopy to the study of the antigenic relationships between three new human papovaviruses. Int Arch Allergy Appl Immunol 46:405–416.

Arthur RR (1992): Detection of JC and BK viruses in pathological specimens by polymerase chain reaction. In Becker Y, Daria G, Eds; Frontiers of Virology, I, Diagnosis of Human Viruses by Polymerase Chain Reaction Technology, Springer-Verlag: Berlin, pp 219–227.

Arthur RR, Beckmann AM, Li CC, Saral R, Shah KV (1985): Direct detection of the human papovavirus BK in urine of bone marrow transplant recipients: Comparison of DNA hybridization with ELISA. J Med Virol 16:29–36.

Arthur RR, Dagostin S, Shah KV (1989): Detection of BK virus and JC virus in urine and brain tissue by the polymerase chain reaction. J Clin Microbiol 27:1174–1179.

Arthur RR, Shah KV (1989): Occurrence and significance of papovaviruses BK and JC in the urine. Prog Med Virol 36:42–61.

Azzi A, Cesaro S, Laszlo D, Zakrzewska K, Ciappi S, De Santis R, Fanci R, Pesavento G, Calore E, Bosi A (1999): Human polyomavirus BK (BKV) load and haemorrhagic cystitis in bone marrow transplantation patients. J Clin Virol 14:79–86.

Azzi A, De Santis R, Ciappi S, Leoncini F, Sterrantino G, Marino N, Mazzotta F, Laszlo D, Fanci R, Bosi A (1996): Human polyomaviruses DNA detection in peripheral blood leukocytes from immunocompetent and immunocompromised individuals. J Neurovirol 2:411–416.

Beckmann AM, Shah, KV (1983): Propagation and primary isolation of JCV and BKV in urinary epithelial cell cultures. Prog Clin Biol Res 105:3–14.

Beckmann AM, Shah KV, Mounts P (1985): Genetic heterogeneity of the human papovaviruses BK and JC. J Med Virol 15:239–250.

Billaudel S, Le Bris JM, Soulillou JP, Chippaux-Hyppolite C, Courtieu AL (1981): Anticorps inhibant l'hémagglutination du virus BK: brève surveillance de 52 transplantés rénaux et prévalence dans différents groupes d'âge de l'ouest de la France. Ann Virol (Inst. Pasteur) 132E:337–345.

Bofill-Mas S, Pina S, Girones R (2000): Documenting the epidemiologic patterns of polyomaviruses in human populations by studying their presence in urban sewage. Appl Environ Microbiol 66:238–245.

Borgatti MA, Baserga M, Nicoli A, Portolani M, Rosito P, Paolucci G (1981): Antigenic variant of the BK virus isolated from immunosuppressed children. Microbiologica 4:327–331.

Borgatti M, Costanzo F, Portolani M, Vullo C, Osti L, Masi M, Barbanti-Brodano G (1979): Evidence for reactivation of persistent infection during pregnancy and lack of congenital transmission of BK virus, a human papovavirus. Microbiologica 2: 173–178.

Brown DWG, Gardner SD, Gibson PE, Field AM (1984): BK virus specific IgM responses in cord sera, young children and healthy adults detected by RIA. Arch Virol 82:149–160.

Brown P, Tsai T, Gajdusek DC (1975): Seroepidemiology of human papovaviruses. Am J Epidemiol 102:331–340.

Burguière AM, Fortier B, Bricout F, Huraux JM (1980): Control of BK virus antibodies in contacts of patients under chronic hemodialysis or after renal transplantation (by an enzyme linked immunosorbent assay). Pathol Biol 28:541–544.

Candeias JAN, Baruzzi RG, Pripas S, Iunes M (1977): Prevalence of antibodies to the BK and JC papovaviruses in isolated populations. Rev Saúde Públ S Paulo 11:510–514.

Chang D, Liou Z-M, Ou W-C, Wang K-Z, Wang M, Fung C-Y, Tsai R-T (1996a): Production of the antigen and the antibody of the JC virus major capsid protein VP1. J Virol Methods 59:177–187.

Chang D, Wang M, Ou W-C, Lee M-S, Ho H-N, Tsai R-T (1996b): Genotypes of human polyomaviruses in urine samples of pregnant women in Taiwan. J Med Virol 48:95–101.

Chesters PM, Heritage J, McCance DJ (1983): Persistence of DNA sequences of BK virus and JC virus in normal human tissues and in diseased tissues. J Infect Dis 147: 676–684.

Cobb JJ, Wickenden C, Snell ME, Hulme B, Malcolm ADB, Coleman DV (1987): Use of hybridot assay to screen for BK and JC polyomaviruses in non-immunosuppressed patients. J Clin Pathol 40:777–781.

Coleman DV, Daniel RA, Gardner SD, Field AM, Gibson PE (1977): Polyoma virus in urine during pregnancy. Lancet ii:709–710.

Coleman DV, Gardner SD, Mulholland C, Fridiksdottir V, Porter AA, Lilford R, Valdimarsson H (1983): Human polyomavirus in pregnancy. A model for the study of defence mechanisms to virus reactivation. Clin Exp Immunol 53:289–296.

Coleman DV, Wolfendale MR, Daniel RA, Dhanjal NK, Gardner SD, Gibson PE, Field AM (1980): A prospective study of human polyomavirus infection in pregnancy. J Infect Dis 142:1–8.

Corallini A, Barbanti-Brodano G, Portolani M, Balboni PG, Grossi MP, Possati L, Honorati C, La Placa M, Mazzoni A, Caputo A, Veronesi U, Orefice S, Cardinali G (1976): Antibodies to BK virus structural and tumor antigens in human sera from normal persons and from patients with various diseases, including neoplasia. Infect Immun 13:1684–1691.

Degener AM, Pietropaolo V, Di Taranto C, Rizzuti V, Ameglio F, Cordialifei P, Caprilli F, Capitanio B, Sinibaldi L, Orsi N (1997): Detection of JC and BK viral genome in specimens of HIV-1 infected subjects. Microbiologica 20:115–122.

Dei R, Marmo F, Corte D, Sampietro MG, Franceschini E, Urbano P (1982): Age-related changes in the prevalence of precipitating antibodies to BK virus in infants and children. J Med Microbiol 15:285–291.

Dei R, Urbano P (1982): Immunoelectroosmophoresis for the human papovavirus BK. J Virol Methods 3:311–318.

De Stasio A, Mastromarino P, Panà A, Seganti L, Sinibaldi L, Valenti P, Orsi N (1980): Characterization of a glycoprotein inhibitor present in human serum and active towards BK virus hemagglutination. Microbiologica 3:293–305.

De Stasio A, Orsi N, Panà A, Seganti L, Sinibaldi L, Valenti P (1979): The importance of pre-treatment of human sera for the titration of hemagglutination-inhibiting antibodies towards BK virus. Ann Sclavo 21:249–257.

Di Taranto C, Pietropaolo V, Orsi GB, Jin L, Sinibaldi L, Degener AM (1997): Detection of BK polyomavirus genotypes in healthy and HIV-positive children. Eur J Epidemiol 13:653–657.

Dörries K, Vogel E, Günther S, Czub S (1994): Infection of human polyomaviruses JC and BK in peripheral blood leukocytes from immunocompetent individuals. Virology 198:59–70.

Dougherty RM, DiStefano HS (1974): Isolation and characterization of a papovavirus from human urine. Proc Soc Exp Biol Med 146:481–487.

Edwards JMB, Kessel I, Gardner SD, Eaton BR, Pollock TM, Fleck DG, Gibson P, Woodroof M, Porter AD (1981): A search for a characteristic illness in children with serological evidence of viral or toxoplasma infection. J Infect 3:316–323.

Elsner C, Dörries K (1992): Evidence of human polyomavirus BK and JC infection in normal brain tissue. Virology 191:72–80.

Flaegstad T, Rönne K, Filipe AR, Traavik T (1989): Prevalence of anti BK virus antibody in Portugal and Norway. Scand J Infect Dis 21:145–147.

Flaegstad T, Traavik T (1985): Detection of BK virus antibodies measured by enzyme-linked immunosorbent assay (ELISA) and two haemagglutination inhibition methods: A comparative study. J Med Virol 16:351–356.

Flaegstad T, Traavik T, Christie KE, Joergensen J (1986a): Neutralization test for BK virus: Plaque reduction detected by immunoperoxidase staining. J Med Virol 19: 287–296.

Flaegstad T, Traavik T, Kristiansen B-E (1986b): Age-dependent prevalence of BK virus IgG and IgM antibodies measured by enzyme-linked immunosorbent assays (ELISA). J Hyg Camb 96:523–528.

Flower AJE, Banatvala JE, Chrystie IL (1977): BK antibody and virus-specific IgM responses in renal transplant recipients, patients with malignant disease, and healthy people. BMJ ii:220–223.

Gardner SD (1973): Prevalence in England of antibody to human polyomavirus (B.K.). BMJ i:77–78.

Gardner SD (1977a): Implication of papovaviruses in human diseases. In Kurstak E, Kurstak C, Eds; Comparative Diagnosis of Viral Diseases 1, Human and Related Viruses, Part A; Academic Press: New York pp 41–84.

Gardner SD (1977b): The new human papovaviruses: Their nature and significance. In Waterson AP, Ed; Recent Advances in Clinical Virology, No. I; Churchill Livingstone: Edinburgh, pp 93–115.

Gardner SD, Field AM, Coleman DV, Hulme B (1971): New human papovavirus (B.K.) isolated from urine after renal transplantation. Lancet i:1253–1257.

Gardner SD, Knowles WA (1995): Human polyomaviruses. In Zuckerman AJ, Banatvala JE, Pattison JR, Eds; Principles and Practice of Clinical Virology, 3rd ed; John Wiley: Chichester, pp 635–651.

Gibson PE (1995): Detection of human polyomaviruses by PCR. In Clewley JP, Ed; The Polymerase Chain Reaction (PCR) for Human Viral Diagnosis; CRC Press: Boca Raton, FL, pp 197–203.

Gibson PE, Field AM, Gardner SD, Coleman DV (1981): Occurrence of IgM antibodies against BK and JC polyomaviruses during pregnancy. J Clin Pathol 34:674–679.

Gibson PE, Gardner SD (1983): Strain differences and some serological observations on several isolates of human polyomaviruses. Prog Clin Biol Res 105:119–132.

Gibson PE, Gardner SD, Porter AA (1985): Detection of human polyomavirus DNA in urine specimens by hybridot assay. Arch Virol 84:233–240.

Golubjatnikov et al. (1978): Cited in Padgett BL, Walker DL (1978): Natural history of human polyomavirus infections. In Stevens JG, Todaro GJ, Fox CF, Eds; Persistent Viruses; ICN-UCLA Symposia on Molecular and Cellular Biology XI; Academic Press: New York, pp 751–758.

Goudsmit J, Baak ML, Slaterus KW, van der Noordaa J (1981): Human papovavirus isolated from urine of a child with acute tonsillitis. BMJ 283:1363–1364.

Goudsmit J, Wertheim-van Dillen P, van Strien A, van der Noordaa J (1982): The role of BK virus in acute respiratory tract disease and the presence of BKV DNA in tonsils. J Med Virol 10:91–99.

Grinnell BW, Padgett BL, Walker DL (1983): Distribution of nonintegrated DNA from JC papovavirus in organs of patients with progressive multifocal leukoencephalopathy. J Infect Dis 147:669–675.

Hale A, Dargevieiute A, Jin L, Knowles W, Brown D, Sasnauskas K (1999): Mapping antigenic determinants of BK virus. Abstracts of the XIth International Congress of Virology, August 9–13, Sydney, Australia, pp 304–305.

Hamilton RS, Gravell M, Major EO (2000): Comparison of antibody titers determined by hemagglutination inhibition and enzyme immunoassay for JC virus and BK virus. J Clin Microbiol 38:105–109.

Hashida Y, Gaffney PC, Yunis EJ (1976): Acute hemorrhagic cystitis of childhood and papovavirus-like particles. J Pediatr 89:85–87.

Heritage J, Chesters PM, McCance DJ (1981): The persistence of papovavirus BK DNA sequences in normal human renal tissue. J Med Virol 8:143–150.

Hogan TF, Borden EC, McBain JA, Padgett BL, Walker DL (1980a): Human polyomavirus infections with JC virus and BK virus in renal transplant patients. Ann Intern Med 92:373–378.

Hogan TF, Padgett BL, Walker DL, Borden EC, McBain JA (1980b): Rapid detection and identification of JC virus and BK virus in human urine by using immunofluorescence microscopy. J Clin Microbiol 11:178–183.

Iltis JP, Cleghorn CS, Madden DL, Sever JL (1983): Detection of antibody to BK virus by enzyme-linked immunosorbent assay compared to hemagglutination inhibition and immunofluorescent antibody staining. Prog Clin Biol Res 105:157–168.

Itoh S, Irie K, Nakamura Y, Ohta Y, Haratake A, Morimatsu M (1998): Cytologic and genetic study of polyomavirus-infected or polyomavirus-activated cells in human urine. Arch Pathol Lab Med 122:333–337.

Iwasaki K, Yano K, Yanagisawa Y, Yamazaki K, Taguchi F (1974): Incidence of antibody against BK virus among Tokyoites and animals in Tokyo. Ann Rep Tokyo Metr Res Lab 25:69–72.

Jin L (1993): Rapid genomic typing of BK virus directly from clinical specimens. Mol Cell Probes 7:331–334.

Jin L, Gibson PE, Booth JC, Clewley JP (1993a): Genomic typing of BK virus in clinical specimens by direct sequencing of polymerase chain reaction products. J Med Virol 41:11–17.

Jin L, Gibson PE, Knowles WA, Clewley JP (1993b): BK virus antigenic variants: Sequence analysis within the capsid VP1 epitope. J Med Virol 39:50–56.

Jin L, Pietropaolo V, Booth JC, Ward KH, Brown DW (1995): Prevalence and distribution of BK virus subtypes in healthy people and immuncompromised patients detected by PCR-restriction enzyme analysis. Clin Diagn Virol 3:285–295.

Jung M, Krech U, Price PC, Pyndiah MN (1975): Evidence of chronic persistent infections with polyomaviruses (BK type) in renal transplant recipients. Arch Virol 47:39–46.

Kahan AV, Coleman DV, Koss LG (1980): Activation of human polyomavirus infection —Detection by cytologic technics. Am J Clin Pathol 74:326–332.

Kitamura T, Aso Y, Kuniyoshi N, Hara K, Yogo Y (1990): High incidence of urinary JC virus excretion in nonimmunosuppressed older patients. J Infect Dis 161:1128–1133.

Kitamura T, Yogo Y, Kunitake T, Suzuki K, Tajima A, Kawabe K (1994): Effect of immunosuppression on the urinary excretion of BK and JC polyomaviruses in renal allograft recipients. Int J Urol 1:28–32.

Knowles WA (1986): Biological and Antigenic Comparison of Six Selected Isolates of Human Polyomavirus with the Strains BK and JC. Ph.D. Thesis, University of London.

Knowles WA, Gibson PE, Gardner SD (1989): Serological typing scheme for BK-like isolates of human polyomavirus. J Med Virol 28:118–123.

Krech U, Jung M, Price PC, Thiel G, Sege D, Reutter F (1975): Virus infections in renal transplant recipients. Z Immun-Forsch Bd 148:S341–355.

Lecatsas G, Crew-Brown H, Boes E, Ackthun I, Pienaar J (1978): Virus-like particles in pregnancy urine. Lancet ii:433–434.

Lecatsas G, Schoub BD, Prozesky OW (1976): Development of a new human polyoma virus strain (MG). Arch Virol 51:327–333.

Liddington RC, Yan Y, Moulai J, Sahli R, Benjamin TL, Harrison SC (1991): Structure of simian virus 40 at 3.8-A resolution. Nature 354:278–284.

Major EO, Neel JV (1998): The JC and BK human polyoma viruses appear to be recent introductions to some South American Indian tribes: there is no serological evidence of cross-reactivity with the simian polyoma virus SV40. Proc Natl Acad Sci USA 95:15525–15530.

Mäntyjärvi RA, Arstila PP, Meurman OH (1972): Hemagglutination by BK virus, a tentative new member of the papovavirus group. Infect Immun 6:824–828.

Mäntyjärvi RA, Meurman OH, Vihma L, Berglund B (1973): A human papovavirus (B.K.), biological properties and seroepidemiology. Ann Clin Res 5:283–287.

Markoulatos P, Spirou N, Ghubril V, Vincent J (1981): Données préliminaires sur la présence d'anticorps antivirus BK (groupe papova) dans deux populations différentes de la Grèce. Arch Inst Past Hellén 27:45–51.

Markowitz RB, Eaton BA, Kubik MF, Lattora D, McGregor JA, Dynan WS (1991): BK virus and JC virus shed during pregnancy have predominantly archetypal regulatory regions. J Virol 65:4515–4519.

Markowitz RB, Thompson HC, Mueller JF, Cohen JA, Dynan WS (1993): Incidence of BK virus and JC virus viruria in human immunodeficiency virus–infected and virus–uninfected subjects. J Infect Dis 167:13–20.

Marrero M, Pascale F, Alvarez M, De Saint Maur G, Rosseau E, Nicolas JC, Garbarg-Chenon A, Bricout F (1990): Dot ELISA for direct detection of BK virus in urine samples. Acta Virol 34:563–567.

Marshall WF, Telenti A, Proper J, Aksamit AJ, Smith TF (1991): Survey of urine from transplant recipients for polyomaviruses JC and BK using the polymerase chain reaction. Mol Cell Probes 5:125–128.

McCance DJ, Mims CA (1977): Transplacental transmission of polyoma virus in mice. Infect Immun 18:196–202.

McCance DJ, Mims CA (1979): Reactivation of polyoma virus in kidneys of persistently infected mice during pregnancy. Infect Immun 25:998–1002.

Merletti L (1976): Sieroepidemiologia umana del polyoma virus BK in Umbria. Boll Ist Sieroter Milan 55:573–576.

Meurman OH, Mäntyjärvi RA, Salmi AA, Panelius M (1972): Prevalence of antibodies to a human papova virus (BK virus) in subacute sclerosing panencephalitis and multiple sclerosis patients. Z Neurol 203:191–194.

Mew RT, Lecatsas G, Prozesky OW, Harley EH (1981): Characteristics of BK papovavirus DNA prepared directly from human urine. Intervirology 16:14–19.

Mininberg DT, Watson C, Desquitado M (1982): Viral cystitis with transient secondary vesicoureteral reflux. J Urol 127:983–985.

Moens U, van Ghelue M, Johansen B, Seternes OM (1999): Concerted expression of BK virus large T- and small t-antigens strongly enhances oestrogen receptor-mediated transcription. J Gen Virol 80:585–594.

Moens U, Subramaniam N, Johansen B, Johansen T, Traavik T (1994): A steroid hormone response unit in the late leader of the noncoding control region of the human polyomavirus BK confers enhanced host cell permissivity. J Virol 68:2398–2408.

Padgett BL, Walker DL (1976): New human papovaviruses. Prog Med Virol 22:1–35.

Padgett BL, Walker DL (1983): Virologic and serologic studies of progressive multifocal leukoencephalopathy. Prog Clin Biol Res 105:107–117.

Padgett BL, Walker DL, Desquitado MM, Kim DU (1983): BK virus and non-haemorrhagic cystitis in a child. Lancet i:770.

Panà A, Di Arca SU, Castello C, Maldarizzi B (1976): Prevalence in the Rome healthy population of antibodies to a human polyoma-virus (BK-strain). Boll Ist Sieroter Milan 55:18–22.

Penney JB, Narayan O (1973): Studies of the antigenic relationships of the new human papovaviruses by electron microscopy agglutination. Infect Immun 8:299–300.

Pietropaolo V, Di Taranto C, Degener AM, Jin L, Sinibaldi L, Baiocchini A, Melis M, Orsi N (1998): Transplacental transmission of human polyomavirus BK. J Med Virol 56:372–376.

Portolani M, Marzocchi A, Barbanti-Brodano G, La Placa M (1974): Prevalence in Italy of antibodies to a new human papovavirus (BK virus). J Med Microbiol 7:543–546.

Possati L, Bartolotta E (1981): Attempts to isolate BK virus from children affected by various diseases. Acta Virol 25:254–255.

Reese JM, Reissig M, Daniel RW, Shah KV (1975): Occurrence of BK virus and BK virus–specific antibodies in the urine of patients receiving chemotherapy for malignancy. Infect Immun 11:1375–1381.

Rziha H-J, Belohradsky BH, Schneider U, Schwenk HU, Bornkamm GW, zur Hausen H (1978a): BK virus: II. Serologic studies in children with congenital disease and patients with malignant tumors and immunodeficiencies. Med Microbiol Immunol 165:83–92.

Rziha H-J, Bornkamm GW, zur Hausen H (1978b): BK virus. I. Seroepidemiologic studies and serologic response to viral infection. Med Microbiol Immunol 165:73–81.

Saitoh K, Sugae N, Koike N, Akiyama Y, Iwamura Y, Kimura H (1993): Diagnosis of childhood BK virus cystitis by electron microscopy and PCR. J Clin Pathol 46:773–775.

Sasnauskas K, Buzaite O, Vogel F, Jandrig B, Razanskas R, Staniulis J, Scherneck S, Krüger DH, Ulrich R (1999): Yeast cells allow high-level expression and formation of polyomavirus-like particles. Biol Chem 380:381–386.

Seganti L, Mastromarino P, Superti F, Panà A, Orsi N (1981): Comparative study of non-antibody inhibitors present in plasma of different animal species and active towards BK virus, a human papovavirus. Microbiologica 4:395–402.

Seif L, Khoury G, Dhar R. (1979): The genome of human papovavirus BKV. Cell 18: 963–977.

Shah K, Daniel R, Madden D, Stagno S (1980): Serological investigation of BK papovavirus infection in pregnant women and their offspring. Infect Immun 30:29–35.

Shah KV, Daniel RW, Strickler HD, Goedert JJ (1997): Investigation of human urine for genomic sequences of the primate polyomaviruses simian virus 40, BK virus, and JC virus. J Infect Dis 176:1618–1621.

Shah KV, Daniel RW, Warszawski RM (1973): High prevalence of antibodies to BK virus, an SV40-related papovavirus, in residents of Maryland. J Infect Dis 128:784–787.

Sugimoto C, Hara K, Taguchi F, Yogo Y (1989): Growth efficiency of naturally occurring BK virus variants in vivo and in vitro. J Virol 63:3195–3199.

Sundsfjord A, Flaegstad T, Flø R, Spein AR, Pedersen M, Permin H, Julsrud J, Traavik T (1994a): BK and JC viruses in human immunodeficiency virus type 1–infected persons: Prevalence, excretion, viremia, and viral regulatory regions. J Infect Dis 169:485–490.

Sundsfjord A, Spein A-R, Lucht E, Flaegstad T, Seternes OM, Traavik T (1994b): Detection of BK virus DNA in nasopharyngeal aspirates from children with respiratory infections but not in saliva from immunodeficient and immunocompetent adult patients. J Clin Microbiol 32:1390–1394.

Sundsfjord A, Osei A, Rosenqvist H, Ghelue MV, Silsand Y, Haga H-J, Rekvig OP, Moens U (1999): BK and JC viruses in patients with systemic lupus erythematosus: Prevalent and persistent BK viruria, sequence stability of the viral regulatory regions, and nondetectable viremia. J Infect Dis 180:1–9.

Szűcs G, Kende M, Új M (1979): Haemagglutination-inhibiting antibodies to B.K. virus in Hungary. Acta Microbiol Acad Sci Hung 26:173–178.

Taguchi F, Kajioka J, Miyamura T (1982): Prevalence rate and age of acquisition of antibodies against JC virus and BK virus in human sera. Microbiol Immunol 26: 1057–1064.

Taguchi F, Kajioka J, Shimada N (1985): Presence of interferon and antibodies to BK virus in amniotic fluid of normal pregnant women. Acta Virol 29:299–304.

Taguchi F, Nagaki D, Saito M, Haruyama C, Iwasaki K, Suzuki T (1975): Transplacental transmission of BK virus in human. Jpn J Microbiol 19:395–398.

Takemoto KK, Mullarkey MF (1973): Human papovavirus, BK strain: Biological studies including antigenic relationship to simian virus 40. J Virol 12:625–631.

Tavis JE, Frisque RJ, Walker DL, White FA (1990): Antigenic and transforming properties of the DB strain of the human polyomavirus BK virus. Virology 178:568–572.

Tavis JE, Walker DL, Gardner SD, Frisque RJ (1989): Nucleotide sequence of the human polyomavirus AS virus, an antigenic variant of BK virus. J Virol 63:901–911.

Tsai R-T, Wang M, Ou W-C, Lee Y-L, Li S-Y, Fung C-Y, Huang Y-L, Tzeng T-Y, Chen Y, Chang D (1997): Incidence of JC viruria is higher than that of BK viruria in Taiwan. J Med Virol 52:253–257.

van der Noordaa J, Wertheim-van Dillen P (1977): Rise in antibodies to human papova virus BK and clinical disease. BMJ 1:1471.

Voltz R, Jäger G, Seelos K, Fuhry L, Hohlfeld R (1996): BK virus encephalitis in an immunocompetent patient. Arch Neurol 53:101–103.

Wolfendale MR, Williams MJH, Coleman DV, Gardner SD, Gibson PE, Field AM (1982): Clinical significance of human polyomavirus infection in pregnancy. J Obstet Gynecol 2:215–218.

Wright PJ, Bernhardt G, Major EO, di Mayorca G (1976): Comparison of the serology, transforming ability and polypeptide composition of human papovaviruses isolated from urine. J Virol 17:762–775.

Zapata M, Mahony JB, Chernesky MA (1984): Measurement of BK papovavirus IgG and IgM by radioimmunoassay (RIA). J Med Virol 14:101–114.

20

THE EPIDEMIOLOGY OF SV40 INFECTION DUE TO CONTAMINATED POLIO VACCINES: RELATION OF THE VIRUS TO HUMAN CANCER

DANA E.M ROLLISON, SC.M., and KEERTI V. SHAH, M.D., Dr.P.H.

1. INTRODUCTION

Simian virus 40 (SV40) is a polyomavirus that has recently received much attention because viral sequences have reportedly been identified from several human cancers such as brain tumors and mesotheliomas. SV40, a natural infection of the rhesus macaque (*Macaca mulatta*) in North India (Fig. 20.1), was discovered to be a contaminant of inactivated Salk polio vaccines that were distributed between 1955 and 1963, potentially exposing an unknown fraction of 98.4 million U.S. residents to the virus (Shah and Nathanson, 1976). Virus lots for the vaccine were prepared in primary simian cultures, which were inadvertently contaminated with the indigenous SV40.

SV40 has been shown to be oncogenic in laboratory animals. Experiments in hamsters revealed that susceptibility to SV40-induced tumors was highest in

Human Polyomaviruses: Molecular and Clinical Perspectives, Edited by Kamel Khalili and Gerald L. Stoner.
ISBN 0-471-39009-7

Figure 20.1. Rhesus macaques in North India. (**A**) Animals on the roadside. (**B**) Female with infant. (Courtesy of Dr. C. H. Southwick.)

newborn animals and associated only with high doses of virus. Most often, tumors occurred at the site of inoculation. Ependymomas were produced through intracerebral injection of SV40, while intravenous inoculation produced lymphomas and osteogenic sarcomas (Diamandopoulos, 1973).

Numerous epidemiologic studies have been carried out to investigate if exposure to SV0 has resulted in human cancer. This chapter provides a detailed summary and critique of the epidemiologic evidence, incorporating studies in a previous review (Strickler and Goedert, 1998) and evidence published since their review (Strickler et al., 1998b; Olin and Giesecke, 1998; Fisher et al., 1999). We also provide suggestions for future research.

2. BACKGROUND

The extent of population exposure to SV40 has been previously reviewed (Shah and Nathanson, 1976; Strickler and Goedert, 1998). Information that is particularly relevant to epidemiologic studies of SV40 exposure and cancer is summarized here. In addition to the 98.4 million Americans, some of whom were potentially exposed to contaminated Salk polio vaccine, residents of several countries that imported American vaccine (i.e., Sweden, Germany, USSR) also may have been exposed. Although the number of individuals potentially exposed to SV40 through polio vaccines is very high, the actual number of people who were infected with SV40 is not known and may have been smaller. Not all inactivated vaccine lots were contaminated with SV40, and, in the contaminated lots, formalin inactivation can be expected to have reduced the titer of live SV40. While SV40 was present in experimental live poliovirus vaccines, it was required to be excluded from licensed Sabin vaccines (Shah and Nathanson, 1976).

Several factors would have influenced the probability of infection with SV40 from administration of contaminated Salk polio vaccine.

Proportion of Vaccine Lots Contaminated with SV40

Fraumeni et al. (1963) attempted to measure the extent of vaccine contamination with live SV40 by testing stored vaccine lots that were administered to children across the United States in May and June 1955. When the lots were linked to individual states in which they had been administered, it was reported that 71% of states and 80% of the children received vaccine lots that contained some residual live SV40 (Table 20.1) (Fraumeni et al., 1963). These lots were administered in May and June 1955 and may not be representative of all Salk polio vaccines administered in the United States. However, these are the only published data regarding the extent of contamination of the national vaccine supply with SV40. Shah et al. (1972) examined serum specimens of Maryland residents for SV40 neutralizing antibodies in an attempt to estimate the proportion of Maryland residents who may have received SV40-contaminated vac-

Table 20.1. Per Capita Dose of SV40 in Polio Vaccines Administered to Children 6–8 Years Old, United States, May and July 1955[a]

Level of SV40	Quantity of Vaccine Containing SV40 Per Child	States[b]		Children	
		No.	%	No.	%
High	0.75–0.97 ml	19	38.78	3,938,400	41.50
Low	0.01–0.74 ml	16	32.65	3,670,800	38.68
None	—	14	28.57	1,879,900	19.81
Total		49		9,489,100	

[a]Adapted from Fraumeni et al. (1963).
[b]Includes the 48 continental states and the District of Columbia.

cine. Significant titers of neutralizing antibodies to SV40 were detected in the sera of about 20% of Maryland children who were born between 1955 and 1957 and bled in 1969 when they were 12–14 years old. Almost all of these children were probably vaccinated with potentially contaminated Salk vaccine. This 20% antibody prevalence may reflect the extent of the exposure of Maryland residents to SV40-contaminated vaccine. It was not possible to determine what proportion of the vaccines that produced the antibodies contained live residual SV40 in addition to inactivated SV40. Interestingly, Maryland was one of the states classified as having a "high level" of SV40 contamination in the study conducted by Fraumeni et al. (1963).

Amount of SV40 in Contaminated Vaccine

Table 20.1 provides the per capita dose of vaccine that contained live SV40 in lots tested by Fraumeni et al. (1963). Lots received by about 39% of the states contained 0.75–0.97 ml of contaminated vaccine per child, while the lots from about 33% of states contained 0.01–0.74 ml of contaminated vaccine per child. About 29% of states received lots that were free of SV40. Fraumeni et al. (1963) categorized the states as receiving high level (0.75–0.97 ml), low level (0.01–0.074 ml), or no SV40, but they did not provide any data on the quantities of live SV40 in the contaminated vaccines.

Measurements made by other investigators on stored batches of vaccine have shown that both formalin-treated inactivated polio vaccines (IPV) and live virus oral polio vaccines (OPV) may contain SV40, but the concentrations of the virus in the two preparations were different (Shah and Nathanson, 1976). OPV contained SV40 in concentrations up to 10^4–10^6 infectious doses/ml (Hull, 1968; Magrath et al., 1961; Shah and Nathanson, 1976; Sweet and Hilleman, 1969), while IPV, when it contained residual live SV40, had the virus in lower concentrations ($10^{2.3}$–$10^{3.3}$ infectious units/ml). The lower amounts of the virus in IPV were due to inactivation of SV40 by the formalin that was employed to inactivate the polioviruses (Gerber, 1961).

Route of Administration

Data on antibody response to SV40 administered by different routes illustrate that immune response was not the same for OPV and IPV. SV40 administered orally produced no detectable antibody response, while a minimal antibody response was detected after intranasal inoculation, and a moderate to high antibody response persisting for years was detected after subcutaneous inoculation (Sweet and Hilleman, 1960; Morris et al., 1961; Goffe et al., 1961; Melnick and Stinebaugh, 1962; Horvath and Fornosi, 1964; Shah, 1972; Sabin, 1960; Gerber et al., 1967).

Given the lack of data on the amounts of live SV40 in vaccine lots that were administered and the variability in factors that influence host infection with SV40, it is difficult to estimate how many individuals were actually infected with SV40 through exposure to contaminated polio vaccine.

In several case series, SV40 sequences have been identified in different human tumors. The results from these studies are discussed in Chapter 17. Briefly, SV40 sequences have been recovered from many types of brain tumors, bone tumors, and mesotheliomas, but the percentages of tumors that tested positive for SV40 DNA span a very wide range. Epidemiologic evidence is needed to evaluate if the virus infection contributes to the development of the cancer. We review here the evidence available to date.

3. REVIEW OF EPIDEMIOLOGICAL STUDIES

Eleven epidemiologic studies published between 1963 and 1999 are reviewed in this chapter. In addition to these, there was an unpublished study, sponsored by the American Cancer Society, in which over 700,000 participants (32–91 years old) provided their histories of polio vaccinations (Hammond, 1966; Shah and Nathanson, 1976). Data on deaths in this group were obtained for the years 1962–1964. No differences in cancer deaths were observed between participants who gave histories of receiving polio vaccine versus those who did not. The results of this study have not been formally published.

Three of the 11 published studies reviewed in this chapter are matched case–control studies, 1 is a prospective cohort study, and 7 are retrospective cohort studies. Table 20.2, adapted from Strickler and Goedert (1998), summarizes information from the studies reviewed here. There are significant limitations to all of the studies, most of which are acknowledged by the authors themselves.

Fraumeni et al. (1963) (*JAMA*)

The first epidemiologic study to investigate the association between poliovirus vaccination and subsequent cancer risk was published by Fraumeni et al.

Table 20.2. Summary of Epidemiologic Studies Addressing the Association Between Exposure to SV40-Contaminated Polio Vaccine and Cancer

Study	Design	Study Pop.	Exposure/Measure	Outcome	Follow-Up	Results
Fraumeni et al. (1963)	Retrospective cohort	Over 9 million children, 6–8 years old, vaccinated in the U.S. in May–July, 1995	IPV at ages 6–8 years; exposure assigned by birth cohort and SV40 measurements of stored vaccine lots	Leukemia, all cancers other than leukemia	4 years	No association
Stewart and Hewitt (1965)	Matched case–control	2107 British childhood cancer cases (identified by the Oxford survey) and matched controls	IPV; vaccine history obtained from Oxford survey	Leukemia, other cancers	9 years	No association
Innis (1968)	Matched case–control	816 childhood cancer cases treated in two Australian hospitals, 1958–67; hospital-based controls matched on sex, age, time of admission	IPV in early childhood or infancy; exposure assessed through medical record review	All childhood malignancies	Case ages not given	Statistically significant difference in polio vaccine history between cases and controls in children >1 year
Heinonen et al. (1973)	Prospective cohort	50,897 children born to women who were pregnant in 1959–65 from 12 U.S. hospitals	Maternal OPV and IPV; exposure assessed by interview and medical record review	All malignant neoplasms	4 years	Increased risks observed for all cancers, neural tumors, and leukemias
Farwell et al. (1979)	Matched case–control	40 children with CNS tumors, born 1956–62 in Connecticut—80 controls selected from birth certificates and matched to cases on gender, date of birth, and town of birth	Maternal IPV; exposure assessed by questionnaire mailed to delivering obstetricians	All childhood neoplasms	Case ages: 0–19 years	Statistically significant difference in maternal history of polio vaccine between cases and controls

Mortimer et al. (1981)	Retrospective cohort	1073 children born in Cleveland in 1960–62 who received contaminated polio vaccine at 0–3 days old as part of an efficacy study	Mostly OPV, some IPV at 0–3 days old; exposure recorded as part of study	Childhood cancer	17–19 years	No association
Geissler (1990)	Retrospective cohort	885,783 children born in 1959–61 and 891,231 children born in 1962–64 in the German Democratic Republic	OPV at birth; exposure assigned by birth cohort	All malignancies	22 years (minimum)	Increased numbers of gliomas, medulloblastomas, spongioblastomas, oligodendroglioma
Olin and Giesecke (1998)	Retrospective cohort	Swedish cancer registry rates for exposed (born in 1946–53) and nonexposed (born in 1941–55 and 1954–58)	IPV in childhood; exposure assigned by birth cohort	Brain cancer, ependymoma, osteosarcoma, mesothelioma	30–39 years; used rates for ages 35–39 years	No association
Strickler et al. (1998)	Retrospective cohort	SEER cancer registry rates for exposed (born in 1947–52 and 1956–62) and nonexposed (born in 1964–69) birth cohorts	IPV in infancy or childhood; exposure assigned by birth cohort	Ependymoma, osteosarcoma, mesothelioma, brain tumors	24–46 years; used rates through 1993	Significant protective effect for brain tumors
Strickler et al. (1999)	Retrospective cohort	Calculated incidence rates using Connecticut registry cases and census pop. estimates; also conducted SEER analysis (see above)	IPV in infancy or childhood; exposure assigned by birth cohort	Medulloblastoma	19 years	No significant increase in Connecticut or SEER data
Fisher et al. (1999)	Retrospective cohort	SEER cancer registry rates for exposed (born in 1955–59) and nonexposed (born in 1963–67) birth cohorts	IPV in infancy or preschool; exposure assigned by birth cohort	All cancers, ependymoma, other brain, osteosarcoma, other bone, mesothelioma	18–26 years; used rates for ages 18–26 years	Increases in all rates observed after adjustment

(1963). A cohort analysis was conducted among over 9 million children ages 6–8 who received Salk poliomyelitis vaccine in May and June of 1955, a time during which vaccine supply was limited. Lot numbers for vaccines distributed to 49 states (including Washington, DC) during this time were obtained from the Communicable Diseases Center of the U.S. Public Health Service (USPHS), and stored specimens from these lots were tested for SV40 contamination. Based on specimen testing and the geographic distribution of the lots, states were grouped into three SV40 exposure categories: high, low, and none (Table 20.1).

The number of deaths from leukemia and all cancers other than leukemia were obtained for each state from the National Vital Statistics division of the USPHS. Annual state population estimates were obtained from linear interpolation of the 1950 and 1960 census estimates. Annual age-specific rates of leukemia and all other cancers between 1950 and 1959 were calculated for each state. Overall, there were no increases in leukemia or other cancer rates over time for individuals under 25 years of age. The authors noted that no increases were observed for brain, kidney, or connective tissue cancers either. Among the birth cohort that was 6–8 years old in 1955, there was an increase in the rate of cancers other than leukemia in the high and low exposure groups of states in 1955, the year of vaccination, compared with 1954. From 1954 to 1955, the rates for cancers other than leukemia per 100,000 children increased from approximately 4.0 to 5.4 in the high exposure group and from 4.0 to 4.7 in the low exposure group. This increase was not observed for the group of states with no exposure. However, rates in 1951–1953 were higher than those in 1954 and 1955 for both the low and high exposure groups, indicating that the apparent rate increase from 1954 to 1955 may have been a product of the aberrantly low rates in 1954 and not the influence of polio vaccination in 1955. No tests for statistical significance were presented.

A strength of this early study was that exposure to SV40 was approximated using actual laboratory measurements of live SV40 in stored specimens linked to lots distributed to particular states. Although actual measurements were available, they were still for groups of people rather than for individuals. It could not be ascertained if the individuals who developed cancer had, in fact, received contaminated vaccine.

A potential limitation of this study was the identification of cases. There may have been underreporting of the number of cancer deaths in some states, and the underreporting could have varied by state. The analysis that incorporated the state-specific exposure estimates was restricted to the cohort that received the vaccine at ages 6–8 years. Newborn hamsters are most susceptible for the oncogenic effect of SV40. Therefore, children aged 6–8 may not be the most susceptible group. Follow-up of the cohort was limited to 4 years, prohibiting the investigation of tumors occurring later in childhood or in adulthood. Finally, data were not presented separately for tumors of different sites and histologic types.

Stewart and Hewitt (1965) (*Lancet*)

Stewart and Hewitt (1965) reported on the immunization histories of 999 childhood leukemia cases and 1108 other childhood cancer cases and of their matched controls. Selection of participants and collection of exposure data were obtained through the Oxford survey, which identified deaths from cancer up to 9 years of age in England, Wales, and Scotland from 1956 to 1960. Among the 999 leukemia cases, 270 had a history of polio vaccination compared with 259 matched controls. Two hundred fifty-nine of the 1108 other childhood cancers and 265 of their matched controls also had histories of polio vaccination. The authors concluded that their data did not provide evidence for an association between polio vaccination and childhood cancer.

The authors did not present a statistical analysis of the data, and only the unmatched data are presented in the article, so we calculated unmatched odds ratios to roughly estimate the associations between a history of polio immunization and childhood leukemia and other childhood cancers. The unmatched odds ratio was 1.06 (95% CI: 0.86–1.26) for childhood leukemia, and 0.97 (95% CI: 0.77–1.17) for other cancers, supporting the authors' conclusion that there was no association between history of polio vaccination and childhood cancers in this data set.

There is minimal explanation of what the Oxford survey is and how information was collected for vaccination histories and health outcomes. It is impossible to know what types of biases could have affected the data. Although there is an interesting discussion of birth season and leukemia risk, there are no data presented for ages at vaccination. There was no breakdown of the "other cancers" category into specific cancers.

Innis (1968) (*Nature*)

Innis (1968) conducted a matched case–control study among 600 childhood cancer cases treated in two Australian hospitals from 1958 to 1967. Hospital-based controls were matched to the cases on age, gender, and time of admission. Exposures to vaccines for a variety of diseases including polio were assessed through chart review. Twenty-nine cancer cases and 28 controls out of 86 case–control pairs under the age of 1 year were identified as having a history of polio vaccination. For children over 1 year of age, 618 cancer cases and 569 controls out of 706 case–control pairs had a history of polio vaccination ($\chi^2 = 12.182$, $p < 0.0005$).

Although the chi-squared statistic is presented, there are no odds ratios calculated from the data. Because the data are presented in unmatched form, we calculated unmatched odds ratios to obtain a rough estimate of the magnitude of association between history of polio immunization and malignant disease at greater than 1 year of age. An odds ratio of 1.69 (95% CI: 1.40–1.98) was obtained, supporting the author's observation of an association in this age group.

A strength of the study was the discussion of socioeconomic status as a potential confounder. After observing no differences in rates of vaccination for other diseases among cases and controls, it was concluded that potential differences in access to medical care did not affect the findings.

The study had several limitations. It is unclear whether the authors used a matched analysis, which was called for by the study design. Chi-square results are discussed, but individual data are not presented, making it impossible to calculate matched odds ratios from the raw data. There is no discussion of the potential for contamination of vaccines with SV40 in this Australian population. It is unclear whether they were using American vaccines or vaccines manufactured within Australia. The reasons for hospitalization in the controls are not defined, nor are the catchment areas of the two hospitals. Selection bias could have affected the results. Ages at immunization were not considered, so the relevance of the window of exposure addressed in the study cannot be discussed. Information on specific types of cancer was not presented, and the ages of the study participants were not summarized, so it is unclear which types of childhood tumors are included in the study. The author mentions that in many cases the actual dose of vaccine administered was noted in the medical record. It would have been interesting to look at case–control status by categories of vaccine dose.

Heinonen et al. (1973) (*International Journal of Epidemiology*)

In 1973, Heinonen et al. reported on a cohort of 50,897 children who were born to women between the years 1959 and 1965. These women were recruited from 12 U.S. hospitals as part of a National Institute of Neurological Disorders and Stroke (NINDS) study on birth defects (Heinonen et al., 1973). Women were interviewed every 4 weeks throughout their pregnancies to obtain information on vaccinations for polio and influenza, as well as for viral infections. Interview and chart review information was confirmed by attending physicians. Children were followed up for all malignant neoplasms through age 4 years. Cancer incidence data and vital statistics were available for the first year of life, whereas only mortality and autopsy data were available for ages 1–4 years.

Fourteen malignancies were observed in the 18,342 children whose mothers received IPV during pregnancy compared with 10 malignancies observed in the 32,555 children whose mothers did not receive IPV. This difference was reported by the authors to be statistically significant ($p < 0.05$) and can be expressed as a relative risk of 2.48. When microscopic tumors were excluded, the numbers of malignancies dropped to 12 and 9 in the exposed and unexposed groups, respectively. This exclusion reduced the relative risk estimate to 2.32. No malignancies were observed in the children born to the 3056 women who received OPV.

There were two astrocytic neural tumors (a glioma and an astrocytoma) observed in this study population, one in each exposure group. However, because the number of children born to nonexposed women is almost twice the number of children born to exposed women, the relative risk estimate is 1.8. One medulloblastoma was identified in the exposed group compared with none in the nonexposed group. Four leukemia cases were identified in each exposure group, also yielding a relative risk estimate of 1.8. Finally, after exclusion of a microscopic nephroblastoma, there were three and four other tumors observed in the exposed and nonexposed groups, respectively, generating a relative risk estimate of 1.33. These relative risks are based on very small numbers and are not statistically significant, which may be why the authors presented the raw data rather than the relative risks.

The prospective design and large sample size are strengths of this study. Additionally, confounding and bias were thoroughly addressed in the discussion. There was no recall bias because the study was prospective. There was also a unique investigation into clustering by season of diagnosis and the interval between the two vaccine injections in the mother. However, the children were only followed up for 4 years. There was no direct measurement of SV40 infection in the mothers or in the children. History of vaccination was obtained through interviews and may have been incomplete. It is unclear if the physicians' review of patients' reporting was comprehensive.

Farwell et al. (1979) (*Transactions of the American Neurological Association*)

During the 1950s, the Connecticut Tumor Registry was one of the most advanced state registries in the United States. Farwell et al. (1979) conducted a matched case–control study of 40 childhood (defined as ages 0–19 years) CNS tumor cases randomly selected from the total 120 cases identified through the registry from 1956 to 1962. Controls were selected from birth certificates, and two controls were matched to each case on date of birth, gender, and town of birth. Questionnaires were sent to the obstetricians who delivered the cases and controls, obtaining information on whether or not the mothers were vaccinated for polio during pregnancy.

The authors reported that mothers of 37% of 52 cases of CNS tumors and of 21% of 38 controls were vaccinated for polio during pregnancy. A p value of 0.15 is reported for this comparison, indicating that the association between "exposure to SV40" and CNS tumors was not statistically significant. When CNS tumors were broken down into specific tumor types, there was a statistically significant association observed between polio vaccination and medulloblastoma: 10 of 15 mothers of cases were vaccinated during pregnancy compared with 8 of 38 mothers of controls ($p < 0.01$). A "suggestive" association was also reported for glioma patients, although a statistical significance test was not presented.

Like the other matched case–control studies reviewed here, the authors did not present matched data, prohibiting the calculation of matched odds ratios from the raw data. Unmatched odds ratios were calculated to obtain rough estimates of the associations of polio vaccine exposure with different brain cancers. Statistically significant associations were observed for medulloblastomas (OR = 7.50, 95% CI: 6.17–8.83) and overall CNS tumors (OR = 2.16, 95% CI: 1.20–3.12). The association for gliomas was not statistically significant (OR = 2.00, 95% CI: 0.84–3.16).

The population-based approach within the context of a superior tumor registry is a strength of this study. Additionally, the study provided data on different types of brain tumors that were investigated. There were also several limitations of this study. First, it is unclear why a sample of one-third of cases was selected because the tumors are rare and all available information would have been helpful. There is a discrepancy in the number of cases stated. A one-third sample of 120 cases would yield 40 cases, while 52 cases are reported in the text. The use of the term "exposed to SV40" is misleading because there is no evidence showing that the vaccines the mothers were exposed to actually contained SV40. The statistical tests used to compare cases and controls were not described, and it is possible that matched analysis was not used for the matched data. The response rate was poor in the controls, raising the possibility of selection bias.

Strickler et al. (1998a) (*Medical and Pediatric Oncology*)

Strickler et al. (1998a) conducted another retrospective cohort study to address the above medulloblastoma findings presented by Farwell et al. (1979). They returned to the Connecticut Tumor Registry data and calculated childhood medulloblastoma incidence rates for four 5 year time periods: 1950–1954 (before distribution of contaminated polio vaccine), 1955–1959 and 1960–1964 (during administration), and 1965–1969 (after administration). Age-specific incidence rates per 100,000 person-years were calculated for children 0–4, 5–9, and 10–14 years of age using U.S. Census data for the denominators. The medulloblastoma incidence rate for 1955–1959 was approximately 1.4 per 100,000 person-years, a nonsignificant increase from 0.7 per 100,000, the approximate rate for 1950–1954. The rates for 1960–1964 and 1965–1969 were similar to the rate for 1950–1954. No trends in incidence rates corresponding to polio vaccine distribution were observed for children 5–9 and 10–14 years of age.

One might hypothesize that the nonsignificant increase in the medulloblastoma rate for 1955–1959 is associated with the distribution of contaminated polio vaccine in Connecticut, which occurred between 1955 and 1963. However, Strickler et al. (1998a) argue that if exposure to contaminated vaccine was, in fact, associated with medulloblastoma incidence, then the increase in incidence should be observed at a time later than 1955–1959 for two reasons: First, the cumulative exposure to contaminated vaccine peaks in the 1960–1964 time period, and, second, a latency period needs to be considered. Because the rates for 1960–1964 and 1965–1969 were similar to the rate for 1950–1954, the period before vaccination, the increase observed for 1955–1959 is not related to exposure to contaminated polio vaccine. Strickler et al. (1999)

suggest that the difference in interpretations of data between themselves and Farwell et al. may lie in Farwell's failure to stratify the cases by age.

Additionally, Strickler et al. (1998a) investigated the time trends for medulloblastoma incidence rates using SEER data in an analysis similar to their previous analyses (described later in this section) of ependymomas, osteosarcomas, mesotheliomas, and all brain tumors in 1998 (Strickler et al., 1998b). Using the same birth cohorts to approximate different exposures to contaminated polio vaccines in the 1998 analysis, no increased risks of medulloblastomas were observed for children exposed as infants (RR = 0.742, 95% CI: 0.55–1.00) or for children exposed in childhood (RR = 0.565, 95% CI: 0.34–0.94) compared with children who were never exposed.

Mortimer et al. (1981) (*New England Journal of Medicine*)

In 1981, Mortimer et al. presented findings for the only population for which there are data on actual exposures to SV40. As part of a vaccine efficacy study conducted in Cleveland, 1073 newborn infants born between 1960 and 1962 received polio vaccines that were later determined to be contaminated with SV40. One hundred sixty-three of the children received intramuscular injections, while the remaining children received an oral vaccine. Children were followed up for cancer through 1979. Information on vital status and history of cancer was obtained by personal correspondence with the study participants. A variety of sources were used to track down the participants, including school registration records, drivers' licenses, Ohio death certificates, and visits to former neighbors. Although 87% of participants were located in 1979, responses regarding health status were received from less than half of the participants. The expected mortality from cancer in this cohort was obtained by applying annual U.S. age-specific rates of cancer mortality to the cohort.

Only one child reported a history of cancer: A 15-year-old girl had a mixed tumor of the salivary gland. There were no deaths due to cancer reported from this cohort. The expected number of deaths due to cancer was one.

This study population is unique in that exposure was ascertained by measurement of SV40, and the exposure occurred in the neonatal period. However, the follow up for occurrence of cancer was incomplete. Information for vital status and cause of death was obtained annually throughout the study period. It was unclear how cancer incidence was ascertained. Responses from inquiries into health status were returned by less than 50% of participants. It is possible that some incident cancer cases went unreported. Even if follow up had been complete, the study would have had limited statistical power due to small sample size.

The use of U.S. rates as a comparison group may dilute the association because the U.S. population was also exposed to contaminated polio vaccine. Most of the children in this study were exposed to oral vaccine, while parenteral exposure may be of greater etiologic importance. Although one group received the vaccine intramuscularly, the authors point out that the dose of SV40 was still considerably lower than the dose given to animals in experimental studies.

Geissler (1990) (*Progress in Medical Virology*)

Geissler (1990) conducted a cohort study in the German Democratic Republic (GDR). Over 86% of the 885,783 children born between 1959 and 1961 were vaccinated against polio with the Sabin-Chumakov live virus vaccine in the first year of life. In 1963, GDR began using SV40-free polio vaccines. Cancer incidence for a minimum of 22 years of follow up for the 1959–1961 cohort was obtained from the National Cancer Registry within the GDR and compared with the incidence for the 891,321 people born between 1962 and 1964 who were presumably not exposed to the SV40-contaminated polio vaccines.

The numbers of cases of many different kinds of brain tumors and other selected cancers were presented for the exposed and nonexposed cohorts. Rates were only presented for overall cancer, while numbers of cases were reported for specific tumor types. From these numbers, we calculated the relative risks of various brain tumors associated with membership to the exposed cohort. Significantly increased risks were observed for gliomas (RR = 1.31, 95% CI: 1.22–1.39), medulloblastomas (RR = 1.30, 95% CI: 1.25–1.36), and oligodendrogliomas (RR = 1.38, 95% CI: 1.38–0.96). Significant protective effects were observed for ependymomas (RR = 0.70, 95% CI: 0.61–0.78) and meningiomas (RR = 0.65, 95% CI: 0.41–0.90). These hand calculations are based on the original number of births in these cohorts and do not take into account migration and death due to other causes.

The ability to break down different types of brain tumors was a strength of this study, although the statistical analysis could have been more thorough. The calendar year through which follow up was assessed was not presented, and it is unclear how many years each of these cohorts was followed. Because the German cohort received OPV and Americans received IPV, generalizability of the German findings to the U.S. population is questionable.

Olin and Giesecke (1998) (*Developments in Biological Standardization*)

In a cohort analysis similar to the one conducted in the GDR (Geissler, 1990), Olin and Giesecke (1998) reported on birth cohorts from Sweden. The Swedish national polio vaccine program was implemented in 1957. Before this program, American Salk vaccine was administered to over 70% of Swedish children born from 1946 to 1949 and to 59% of children born from 1950 to 1953. These cohorts were presumably exposed to SV40-contaminated inactivated vaccine. Cancer incidence rates in the exposed group were defined as the incidence rates for 35–39 year olds in 1985–1987, the years in which the greatest proportion of this age group was comprised of people who were potentially exposed to contaminated vaccine. Rates in the unexposed group were defined as cancer incidence rates for 35–39 year olds in 1980 and 1993, the 2 years that straddle the time period in which the 35–39 year age group contained anyone who was potentially exposed. Incidence rates for all brain cancers, ependymomas, osteosarcomas, and mesotheliomas were obtained from the Swedish Cancer Registry.

There was an overall increase in brain cancer incidence from 1960 to 1990 for both males (9.0–13.1 per 100,000) and females (8.4–12.7 per 100,000). No other increases were observed for the other cancer types except for pleural mesothelioma in males. Relative risks were calculated by comparing age-specific incidence rates in the exposed cohort to those in the nonexposed cohort. No increased relative risks were observed for any of the cancer types studied for 35–39-year-old males or females. Several relative risks were below one, although confidence limits were not presented, so the significance of these findings remains in question.

It is unclear how the relative risks were calculated. It is stated that the average incidence rate for 1985–1987 was used as the rate in the exposed, and this rate was compared with the average incidence rate for 1980 and the average rate for 1993. However, only one relative risk is presented. If the 1980 and 1993 years were averaged, this average would be an artificial value assigned to the unexposed group because the brain cancer incidence rates increased from 1980 to 1993. Limitations of this study include the failure to adjust for the increase in brain tumor incidence over time and the inability to stratify by different brain tumor types. Also, the ages studied do not capture the ages at which brain tumor and mesothelioma incidence rates are at their highest.

Strickler et al. (1998b) (*JAMA*)

In 1998, Strickler et al. conducted a cohort analysis using cancer incidence data from the Surveillance, Epidemiology, and End Results (SEER) program and the Connecticut Tumor Registry (for ependymomas). Three birth cohorts were selected to reflect different polio vaccine exposure groups. Individuals born between 1956 and 1962 were considered to be at high risk of exposure in infancy. Individuals born between 1947 and 1952 were classified as exposed in childhood, and those born in 1964–1969 were considered to be unexposed to the SV40-contaminated polio vaccine. Age-specific incidence rates per 100,000 person-years from 1973 through 1993 were calculated for brain ependymomas, all primary brain tumors, osteosarcomas, and mesotheliomas and compared among the three birth cohorts.

There were no significant increases in ependymomas, osteosarcomas, or mesotheliomas reported for either of the exposed birth cohorts compared with the nonexposed cohort. There were protective effects of borderline significance against all primary brain tumors observed in both the cohort exposed to vaccine as infants (RR = 0.90, 95% CI: 0.82–0.99) and as children (RR = 0.82, 95% CI: 0.73–0.92).

The brain tumor category was not stratified by specific types except for ependymomas. It would have been of value to see stratification of 4000 brain tumor cases by specific tumor types. Although one focus of the paper was ependymomas, a primarily pediatric tumor, follow-up time was limiting for the investigation of other brain tumors and mesotheliomas for which incidence peaks later in life. The authors mention that the protective effect observed for brain tumors may be an artifact of the increase in brain tumor incidence over time.

Fisher et al. (1999) (*Anticancer Research*)

Soon after the paper of Strickler et al. (1998b) was published, Fisher et al. (1999) published one, also using SEER data, and criticized Strickler et al. for employing data and statistical techniques that "may not have provided an optimal assessment of risk." Although this criticism was not explained, a major difference in the analyses was the inclusion by Fisher et al. of a formal adjustment for trends in cancer incidence over time.

Fisher et al. (1999) limited their two exposure groups by age and birth year. Individuals born between 1955 and 1959 were considered to be exposed to the contaminated polio vaccine in infancy or preschool years, while individuals born between 1963 and 1967 were considered to be not exposed. Data analysis was restricted to ages 18 through 26 years, including all the ages for which there is SEER data for both birth cohorts. The years 1979 and 1987 were chosen as midpoint years for the follow-up time periods for the exposed and nonexposed birth cohorts, respectively. These 2 years reflect ages 20–24 years for both cohorts and are the years that the population estimates were obtained. The number of tumor occurrences over 9 years of follow up for each cohort was divided by the number of 20–24 year olds in the midpoint year (1979 or 1987) to obtain an incidence rate for each of the two cohorts.

The exposed and unexposed birth cohorts were comprised of approximately 2 million people each. Overall, there were 5512 and 5764 cases of cancer identified in the exposed and unexposed groups, respectively. Over 300 cases of brain tumors other than ependymoma were identified in each group. The frequencies of the other tumors were somewhat lower: 32 cases of ependymoma and choroid plexus tumors, 98 osteogenic sarcomas, 160 other bone tumors, and 8 mesotheliomas.

Based on these numbers, incidence rates were calculated for the two exposure groups, and percent increases in incidence were determined. When comparing the exposed cohort to the nonexposed cohort, there was an 11% decrease in all cancers, 19.6% increase in ependymomas and choroid plexus tumors, 8.4% decrease in other brain tumors, 9.8% increase in osteogenic sarcomas, 16.6% increase in other bone tumors, and a 178% increase in mesotheliomas. After obtaining these percent changes, Fisher et al. (1999) attempted to adjust for the increase in cancer incidence over time by subtracting the expected change from the observed change and dividing by the expected change, where the expected change was the overall 13% increase in cancer rates from 1979 to 1987.

After adjusting for the secular increase in cancer incidence rates over time, there were positive percent increases from 1979 to 1987 for all cancers (2.3%), ependymomas and choroid plexus tumors (37%), other brain tumors (5%), osteogenic sarcomas (26%), other bone cancers (34%), and mesotheliomas (230%). Tests for statistical significance were not presented.

Increasing cancer incidence rates over time may be due to several factors, including advances in diagnostic technology and increased access to medical care, improved reporting to cancer registries, recent screening efforts, and increases in exposures to etiologic agents that contribute to cancer incidence. These factors must be considered when drawing conclusions from comparisons of birth cohorts. Fisher et al. (1999) attempted to account for these factors by choosing the 13% increase in all cancers to adjust the cancer-specific percent increases. Perhaps it would have been more appropriate to use cancer-specific increases to adjust the cancer-specific estimates. Not all cancers increased at the same rate as cancers overall. For example, part of the increases in breast and prostate cancers can be attributed to screening efforts in the past decade. If a large part of the 13% increase in all cancers is attributed to increases in cancers other than the ones looked at by Fisher et al. (1999), it may have been inappropriate to use the increase in all cancers to adjust the increases in individual cancers.

Some of the results are based on very small numbers. For example, the adjusted 230% increase in mesotheliomas was based on eight cases. Secular trends in exposure such as exposure to asbestos also need to be considered.

4. CRITIQUE OF EPIDEMIOLOGIC STUDIES

A careful review of the published literature on exposure to SV40-contaminated polio vaccine and subsequent cancer development reveals several epidemiologic challenges in the design, conduct, and interpretation of findings.

Exposure Measurement

The greatest challenge is measurement of SV40 exposure. Studies varied widely in this respect, from actual measurements of SV40 titers in stored vaccine lots, to review of medical records and interview of physicians for polio immunization histories, to the assignment of exposure based on birth cohort.

Only one of the retrospective cohort studies obtained exposure measurements on the individual level. Mortimer et al. (1981) measured actual amounts of SV40 in the context of a vaccine efficacy study. Fraumeni et al. (1963) measured the amounts of vaccine that contained residual live SV40, and these levels were extrapolated to state child populations. The other retrospective cohort studies assigned exposures according to birth cohorts (Geissler, 1990; Olin and Giesecke, 1998; Strickler et al., 1998a,b, 1999; Fisher et al., 1999). The prospective cohort study (Heinonen et al., 1973) and the three matched case–control studies (Stewart and Hewitt, 1965; Innis, 1968; Farwell et al., 1979) obtained surrogate measures of exposure for individuals. Most of these studies used history of vaccination as a surrogate measure for SV40 exposure. All of the other studies have measurement errors that arise from using surrogate markers of exposure, as well as from extrapolating group exposure data to individuals. These misclassifications of the exposure would bias findings toward the null. Thus, one must exercise caution in accepting the conclusion that they were negative studies.

Timing and Route of Exposure

Other concerns related to exposure measurement include the timing of exposure and route of exposure. Both OPV and IPV were investigated in various studies. By the time OPV was licensed for administration to the general population, vaccines in the United States were free of SV40. Therefore, in the United States, almost all potential exposure to live SV40 was through IPV administration.

Also, the time windows of exposure to vaccines used in these studies included in utero, neonatal, infancy, preschool, and school aged. In animal studies, the period of highest susceptibility to SV40-induced cancers is the neonatal period.

Length of Follow Up

Timing is also an issue when considering length of follow up. For example, observation through early childhood may be enough to evaluate the association between SV40 and childhood brain tumors. However, if SV40 exposure was associated with tumors that occur later in life, such as gliomas and mesotheliomas, a longer follow-up period would be necessary to observe the association. Neither of the two studies based on SEER data included a follow-up time long enough to detect these associations (Strickler et al., 1998b; Fisher et al., 1999). This lengthy follow up would require monitoring of the exposed population as it continues to age.

Sample Size

Small sample size limited the ability to detect small differences in rates of childhood cancers and rare adult tumors, especially in the cohort studies. The study with the best exposure measurement, the cohort study of 1073 children (Mortimer et al., 1981), was limited by its low statistical power. The expected number of cancer deaths in this cohort was one. Small sample sizes also limited stratification by more than a couple of classes of tumors. In addition to etiologic differences by tumor site, there may be differences by tumor cell type. Although stratification is not always feasible, failure to stratify can lead to misclassification of the outcome, which can also bias findings toward the null.

Trends Over Time in Cancer Incidence and Exposures

A final challenge in the interpretation of results was highlighted in the discrepancy of findings between the two retrospective cohort studies published in 1998 and 1999 by Strickler et al. and Fisher et al. (1999), respectively. Both groups used SEER data, but they defined their exposed birth cohorts differently. Fisher et al. adjusted for the background increase in cancer incidence. Over the past few decades, improvements in diagnostic technologies, access to care, and reporting to registries have contributed to the observed increase in cancer incidence over time. Changes in the prevalence of cancer risk factors may also account for some of the increase. Whatever the increase is attributed to, it must be adjusted for in the analysis of birth cohorts. Fisher et al. (1999) adjusted for this increase over time, and they reported a marked increase in incidence in the SV40-exposed birth cohorts compared with the nonexposed cohorts for every type of cancer studied.

In addition to trends in cancer incidence, exposures have also changed over time. For example, the time period of exposure to asbestos overlapped with the time period of distribution of SV40-contaminated polio vaccine. In the case of SV40 and mesothelioma, analyses by birth cohort may not be adequate to assess whether there is an association because SV40 exposure would have occurred concurrently with exposure to asbestos, a known risk factor for mesothelioma.

All of the cohort studies that assigned exposure by birth cohort made an important assumption, that SV40 exposure was limited to a time period between 1955 and 1963 and there was no exposure either before or after that period of time. If the assumption is not valid, then the use of birth cohort analysis to investigate an association between SV40 exposure and human cancer is questionable. Numerous recent studies have reported the presence of SV40 sequences in tumors of individuals who were born many years after 1963, for example, in pediatric brain tumors and osteosarcomas. This implies that SV40 infection is continuing to occur at the present time.

5. IS SV40 CIRCULATING IN THE GENERAL POPULATION?

Mother-to-Child Transmission

This mode of transmission has been proposed for pediatric brain tumors that are positive for SV40 (Bergsagel et al., 1992; Lednicky et al., 1995). No studies have formally investigated mother-to-child transmission of SV40. There is no serologic evidence that human polyomaviruses BKV and JCV are transmissible from mother to child (Andrews et al., 1983). However, a recent study reported

BKV sequences in 12 of 15 aborted fetus placentas and in 6 of 12 placentas from normal deliveries. The authors suggested that the data are evidence for transplacental transmission of BKV (Pietropaolo et al., 1998).

Person-to-Person Transmission

It has been suggested that SV40 may have been seeded into the human communities as a result of administration of contaminated vaccines and that is has since become adapted to humans and is now circulating by person-to-person transmission. Evidence regarding person-to-person transmission of SV40 in the general population is limited. Investigations of SV40 circulation in the general population include one study that searched for SV40 in urine specimens and another that examined sewage samples. Shah et al. (1997) reported that none of 166 urine samples from homosexual men, 88 of whom were HIV positive, contained SV40, whereas JCV and BKV were detected in 34% and 14% of specimens, respectively. Bofill-Mas et al. (2000) conducted an investigation of polyomaviruses in urban sewage sampled from Spain, France, South Africa, and Sweden. Although JCV and BKV were detected in 93% and 79% of 28 analyzed sewage samples, respectively, no SV40 could be isolated.

In contrast to these negative findings, there have been some reports of SV40 antibodies identified from human sera collected before the introduction of contaminated vaccine (Geissler et al., 1985) or from individuals who very likely did not receive contaminated polio vaccines (Shah et al., 1971). Additionally, Butel et al. (1999) recently published an estimate of SV40 antibody prevalence of 6% for children born between 1980 and 1995. However, serologic evidence by itself is not definitive proof of transmission because the possibility that SV40 antibodies may represent immunologic cross reactivity with human polyomaviruses BKV and JCV has not been excluded.

6. IS SV40 AN ETIOLOGIC AGENT IN THE DEVELOPMENT OF HUMAN CANCER?

As described in this chapter, the epidemiologic evidence for the contribution of SV40 to the development of human cancers is inconclusive. The studies could not have detected a modest increase in rare cancers. A large number of recent case series (reviewed in Chapter 17) report the identification of SV40 sequences from a variety of human cancers. There is a marked divergence of views with regard to the significance of these findings. Some have concluded that these reports provide incontrovertible evidence for the presence of SV40 in some human cancers and have suggested that the virus may contribute to the development of some of these cancers (Butel and Lednicky, 1999; Carbone et al., 2000, see also Chapter 17). Others have expressed skepticism (Shah, 2000; Strickler, 2001). This controversy needs to be resolved.

Resolution may be sought through the evaluation of data within the framework of causal criteria (Hill, 1965). Tumor case series provide limited evidence for this evaluation. For example, *strength of association* cannot be measured in a case series. *Specificity* of SV40 to cancer tissue has been demonstrated by Shivapurkar et al. (2000), who isolated viral sequences from malignant mesothelial cells, but not from adjacent normal tissues. However, the diversity of tumors in which SV40 has been found is evidence against specificity. Tumor case series have added to the *plausibility* by looking for mechanistic signatures, such as TAg–p53 binding (Zhen et al., 1999), and by demonstration that the human mesothelium is more susceptible to transformation by SV40 than human fibroblasts (Bocchetta et al., 2000). Case series of tumor tissue alone cannot address the question *of temporal sequence* (i.e., if SV40 exposure preceded the development of tumors) or of a *dose–response relationship*. *Consistency* of the findings is weak because SV40 has been found in a plethora of tumors by some and in none by others. In the absence of knowledge about if and how SV40 is circulating in the community, the *coherence* of the proposed association is very weak.

Further studies of tumor tissue need to be complemented with epidemiologic studies in order to provide adequate evidence with which to evaluate the causal criteria for SV40 and human cancer.

7. SUGGESTIONS FOR FUTURE RESEARCH

Whether or not SV40 infection contributes to human cancer, it is important to know if a simian polyomavirus has crossed species and has become a human infection circulating in U.S. communities by person-to-person transmission. Also, if SV40 is found to be present in tumor cells, for example, in mesothelioma cells, it could provide a potential target for SV40-based immunotherapy to prevent or treat these cancers (Carbone et al., 2000). It is anticipated that thousands of cases of mesothelioma will occur in future years.

We suggest that future studies consider incorporating the following features in their design:

1. Studies should look for the complete "signature" of SV40 infection. As described earlier, the greatest problem in the epidemiologic studies to date has been the uncertainty about SV40 exposure. Most recent case series have searched only for SV40 DNA sequences in cancer tissues. In addition, one could compare cases and controls for serum antibodies to capsid proteins and to T proteins, virus shedding in urine, evidence of infection in family members, and so forth. An SV40 infection that has resulted in a cancer should be distinguishable from a transient infection. A virus-caused tumor is likely to have the viral genome in every tumor cell, integration of the viral genome into cellular DNA, and consistent transcription of the viral oncogenes.

2. Studies should examine SV40 infection in the context of other known risk factors for the specific cancers. As an example, it will be impossible to know if and how SV40 infection contributes to the development of mesothelioma unless SV40 exposure is examined in the context of asbestos exposure in the same population. The effect of SV40 may be independent of asbestos, or additive to asbestos, or SV40 infection may be found not to influence the risk of development of mesothelioma.
3. Studies should incorporate rigorous quality-control procedures. All specimens being tested, as well as negative and positive control samples, should be masked. Each run should include specimens that would allow for estimation of sensitivity, specificity, and reproducibility of the assay. This would go a long way toward avoiding the contentious debates that have sometimes characterized this field.

In conclusion, the role of SV40 in human cancer will be clarified when the results from epidemiologic and virologic studies are mutually corroborative. Excellent tools are at hand to conduct these studies.

REFERENCES

Andrews CA, Daniel RW, Shah KV (1983): Serologic studies of papovavirus infections in pregnant women and renal transplant recipients. In Sever JL, Madden DL, Eds., Polyomaviruses and Human Neurological Diseases; Alan R. Liss, Inc: New York, pp 133–141.

Bergsagel DJ, Finegold MJ, Butel JS, Kupsky WJ, Garcea RL (1992): DNA sequences similar to those of simian virus 40 in ependymomas and choroid plexus tumors of childhood. N Engl J Med 326:988–993.

Bocchetta M, Di Resta I, Powers A, Fresco R, Tosolini A, Testa JR, Pass HI, Rizzo P, Carbone M (2000): Human mesothelial cells are unusually susceptible to simian virus 40–mediated transformation and asbestos cocarcinogenicity. Proc Natl Acad Sci USA 97(18):10214–10219.

Bofill-Mas S, Pina S, Girones R (2000): Documenting the epidemiologic patterns of polyomaviruses in human populations by studying their presence in urban sewage. Appl Environ Microbiol 66(1):238–245.

Butel JS, Jafar S, Wong C, Arrington AS, Opekun AR, Finegold MJ, Adam E (1999): Evidence of SV40 infections in hospitalized children. Hum Pathol 30:1496–1502.

Butel JS, Lednicky JA (1999): Cell and molecular biology of simian virus 40: Implications for human infection and disease [review]. J Natl Cancer Inst 91:119–134.

Carbone M, Rizzo P, Pass H (2000): Simian virus 40: The link with human malignant mesothelioma is well established. Anticancer Res 20:875–878.

Diamandopoulos G (1973): Induction of lymphocytic leukemia, lymphosarcoma, reticulum cell sarcoma and osteogenic sarcoma in the Syrian golden hamster by oncogenic DNA simian virus 40. J Natl Cancer Inst 50:1347–1359.

Farwell JR, Dohrmann GJ, Marrett LD, Meigs JW (1979): Effect of SV40 virus-contaminated polio vaccine on the incidence and type of CNS neoplasms in children: A population-based study. Trans Am Neurol Assoc 104:261–264.

Fisher SG, Weber L, Carbone M (1999): Cancer risk associated with simian virus 40 contaminated polio vaccine. Anticancer Res 19:2173–2180.

Fraumeni JF, Ederer F, Miller RW (1963): An evaluation of the carcinogenicity of simian virus 40 in man. JAMA 185(9):85–90.

Geissler E (1990): SV40 and human brain tumors. Prog Med Virol 37:211–222.

Geissler E, Konzer P, Scherneck S, Zimmerman W (1985): Sera collected before introduction of contaminated polio vaccine contain antibodies against SV40. Acta Virol 29(5):420–423.

Gerber P (1961): Patterns of antibodies to SV40 in children following the last booster with inactivated poliomyelitis vaccines. Proc Soc Exp Biol Med 108:205–209.

Gerber P, Hottle GA, Grubbs RE (1967): Inactivation of vacuolating virus (SV40) by formaldehyde. Proc Soc Exp Biol Med 125:1284–1287.

Goffe AP, Hale J, Gardner PS (1961): Poliomyelitis vaccines. Lancet 1:612.

Hammond EC (1966): Cancer Mortality in Relation to SV40 in Polio Vaccine. Presented at the American Cancer Society Science Writers' Seminar, Washington, D.C.

Heinonen OP, Shapiro S, Monson RR, Hartz SC, Rosenberg L, Slone D (1973): Immunization during pregnancy against poliomyelitis and influenza in relation to childhood malignancy. Int J Epidemiol 2:229–235.

Hill AB (1965): The environment and disease: Association or causation. Proc R Soc Med 58:295–300.

Horvath BL, Fornosi F (1964): Excretion of SV40 after oral administration of contaminated polio vaccine. Acta Microbiol Acad Sci (Hungary) 11:271–275.

Hull RN (1968): The Simian Viruses. Virology Monograph No. 2; Springer-Verlag: New York.

Innis MD (1968): Oncogenesis and poliomyelitis vaccine. Nature 219:972–973.

Lednicky JA, Garcea RL, Bergsagel DJ, Butel JS (1995): Natural simian virus 40 strains are present in human choroid plexus and ependymoma tumors. Virology 212:710–717.

Magrath DI, Russell K, Tobin JOH (1961): Vacuolating agent. BMJ 2:287–288.

Melnick JL, Stinebaugh S (1962): Excretion of vacuolating SV40 virus (papovavirus group) after ingestion as a contaminant of oral poliovaccine. Proc Soc Exp Biol Med 109:965–968.

Morris JA, Johnson KM, Aulisio CG, Chanock RM, Knight V (1961): Clinical and serologic responses in volunteers given vacuolating virus (SV40) by the respiratory route. Proc Soc Exp Biol Med 108:56–59.

Mortimer EA, Lepow ML, Gold E, Robbins FC, Burton GJ, Fraumeni JF (1981): Long-term follow-up of persons inadvertently inoculated with SV40 as neonates. N Engl J Med 305(25):1517–1518.

Olin P, Giesecke J (1998): Potential exposure to SV40 in polio vaccines used in Sweden during 1957: No impact on cancer incidence rates 1960 to 1993. Dev Biol Stand 94:227–233.

Pietropaolo V, Di Taranto C, Degener AM, Jin L, Sinibaldi L, Baiocchini A, Melis M, Orsi N (1998): Transplacental transmission of human polyomavirus BK. J Med Virol 56:372–376.

Sabin AB (1960): Live poliovirus vaccines. Presented at the Second International Conference, Pan American Health Organization, Washington, DC, pp 87–88.

Shah KV (1972): Evidence for an SV40-related papovavirus infection of man. Am J Epidemiol 95:199–206.

Shah KV (2000): Does SV40 infection contribute to the development of human cancers? Rev Med Virol 10(1):31–43.

Shah KV, Daniel RW, Strickler HD, Goedert JJ (1997): Investigation of human urine for genomic sequences of the primate polyomaviruses simian virus 40, BK virus, and JC virus. J Infect Dis 176:1618–1621.

Shah KV, McCrumb FR, Daniel RW, Ozer HL (1972): Serologic evidence for a simian-virus-40–like infection of man. J Natl Cancer Inst 48:557–561.

Shah K, Nathanson N (1976): Human exposure to SV40: Review and comment. Am J Epidemiol 103(1):1–12.

Shah KV, Ozer HL, Pond HS, Palma LD, Murphy GP (1971): SV40 neutralizing antibodies in sera of US residents without history of polio immunization. Nature 231(5303):448–449.

Shivapurkar N, Wiethege T, Wistuba II, Salomon E, Milchgrub S, Muller KM, Churg A, Pass H, Gazdar AF (2000): Presence of simian virus 40 sequences in malignant mesotheliomas and mesothelial cell proliferations. J Cell Biochem 76:181–188.

Stewart AM, Hewitt D (1965): Aetiology of childhood leukaemia. Lancet 2:789–790.

Strickler HD (2001): Simian virus 40 (SV40) and human cancers. Einstein Q J Biol Med 18:14–21.

Strickler HD, Goedert JJ (1998): Exposure to SV40-contaminated poliovirus vaccine and the risk of cancer—A review of the epidemiological evidence. Dev Biol Stand 94:235–244.

Strickler HD, Rosenberg PS, Devesa SS, Fraumeni JF, Goedert JJ (1998a): Contamination of poliovirus vaccine with SV40 and the incidence of medulloblastoma. Med Pediatr Oncol 32:77–78.

Strickler HD, Rosenberg PS, Devesa SS, Hertel J, Fraumeni JF, Goedert JJ (1998b): Contamination of poliovirus vaccines with simian virus 40 (1955–1963) and subsequent cancer rates. JAMA 279:292–295.

Sweet BJ, Hilleman MR (1960): The vacuolating virus, SV40. Proc Soc Exp Biol Med 105:420–427.

Zhen H, Zhang X, Bu X, Zhang ZW, Huang WJ, Zhang P, Liang JW, Wang XL (1999): Expression of the simian virus 40 large tumor antigen (Tag) and formation of Tag–p53 and Tag–pRb complexes in human brain tumors. Cancer 86:2124–2132.

21

THE IMMUNE RESPONSE TO SV40, JCV, AND BKV

SATVIR S. TEVETHIA, PH.D., and TODD D. SCHELL, PH.D.

1. INTRODUCTION

The simian papovavirus SV40 was originally identified as a contaminant of poliovirus vaccine grown in rhesus monkey kidney cell cultures (Sweet and Hilleman, 1960) and was found to cause persistent and latent infection in the kidneys of rhesus monkeys (Shah et al., 1968; Shah and Nathanson, 1976). The importance of SV40 was realized by its ability to induce tumors in hamsters (Eddy et al., 1961, 1962; Girardi et al., 1962) and its ability to transform a wide variety of mammalian cells in culture that are invariably tumorigenic in the immune compromised host (Butel et al., 1972; Tevethia, 1980). Two sister human viruses, JCV (Padgett et al., 1971) and BKV (Gardner et al., 1971), share extensive homology with SV40. All three viruses have the ability to establish latent infections and to induce tumors in the heterologous hosts (reviewed by Padgett in Tooze, 1980).

These observations raised several important issues: (1) their prevalence in the human population, (2) their oncogenic potential in human hosts, and (3) the mechanism of persistence. It was soon established that these three viruses share the capacity for reactivation from latency (SV40 and BKV in the kidneys and JCV in the brain) in their natural hosts upon immunosuppression. JCV establishes latent infection mainly in the oligodendroglia in the brain; and its activation results in the development of multifocal leukoencephalopathy (PML)

Human Polyomaviruses: Molecular and Clinical Perspectives, Edited by Kamel Khalili and Gerald L. Stoner.
ISBN 0-471-39009-7

in humans (Padgett et al., 1971; Narayan et al., 1973). BKV establishes latent infections mainly in the kidneys (Gardner et al., 1971); and its activation causes a number of disorders of kidneys, including hemorrhagic cystitis (Arthur et al., 1986). These viruses are reactivated in transplant patients undergoing immunosuppressive therapy (Gardner et al., 1971) and in AIDS patients with low CD4 counts (Degener et al., 1997). SV40 is likewise activated in rhesus monkeys suffering from SIV-induced immunosuppression (Lednicky et al., 1998; Newman et al., 1998).

These observations led to the obvious conclusion that the immune response plays a major role in maintenance of latency. Although a large body of evidence is available to support that conclusion, the mechanism by which the immune response maintains virus latency is not known. Both monkeys and humans develop antibody responses to the respective viruses. The continued presence of virus neutralizing antibodies in patients at the time of reactivation upon immunosuppression has pointed to the T-cell-mediated immune response as the main effector mechanism for maintaining latency. Whether the cell-mediated immune response is required during establishment of latent infections has not been addressed.

The recent flood of publications demonstrating the existence of the SV40 T antigen (transforming protein) DNA in a variety of human tumors, and in some cases the protein itself, has raised the possibility that SV40 may circulate in the human population, and the expression of T antigen may be involved in the etiology of human tumors (reviewed in Chapter 17). We do not know if SV40 is pathogenic in humans and causes disease similar to JCV or BKV. The main goal of this chapter is to analyze the immune response (humoral and cellular) to these viruses during infection of their natural hosts and during tumorigenesis in the heterologous hosts.

2. ROLE OF CELLULAR IMMUNE RESPONSE DURING VIRUS INFECTION AND VIRAL CARCINOGENESIS

Current View of T-Cell-Meditated Responses

The main effector in the cellular immune response to viruses is the $CD8^+$ cytotoxic T lymphocyte (CTL), which lyses virus-infected cells following recognition of virus-encoded epitopes presented on the cell surface by major histocompatibility complex (MHC) class I molecules (Yewdell and Bennink, 1992). Multiple factors contribute to successful $CD8^+$ T-cell recognition of virus epitopes, including the proper generation of the virus epitopes, successful loading and stable binding of virus epitopes to MHC class I molecules, proper trafficking of the complex to the cell surface, and the presence of T cells in the host repertoire capable of recognizing the virus epitopes (Yewdell and Bennink, 1999).

MHC class I molecules are composed of a transmembrane heavy chain that associates noncovalently with β_2-microglobulin and a bound peptide representing the T-cell-recognized epitope. The $\alpha 1$ and $\alpha 2$ extacellular domains of the heavy chain form a groove at the distal surface of the molecule, which accommodates the peptide in an extended conformation (Madden, 1995). The epitopes recognized by $CD8^+$ T cells are typically peptides of 8–10 amino acids in length that can be derived from a variety of virus proteins (Rammensee et al., 1995). This length requirement is imposed by the structure of the peptide binding groove in which the closed ends of the groove accommodate the amino and caboxy termini of the peptide (Madden, 1995). The interactions between conserved residues of the MHC molecule and the free amino and carboxy termini of the peptide serve to stabilize peptide binding. The specificity of peptide binding to MHC molecules, however, is determined by the peptide sequence. Peptides that bind to a given MHC molecule typically contain two or three "anchor" residues that contribute to successful binding within the peptide binding groove (Rammensee et al., 1995). The side chains of these anchor residues are oriented into the peptide binding groove, where they are positioned into chemically and structurally compatible regions of the groove (Madden, 1995). Thus, peptides that bind to a particular MHC molecule generally share a sequence motif of anchor residues.

Peptides are liberated from virus proteins in the cytosol, most likely through the activity of a large cytosolic protease called the *proteasome* (Nandi et al., 1998). This multisubunit, multicatalytic protease has been implicated in the generation of $CD8^+$ T-cell-recognized epitopes through the use of specific inhibitors as well as by the finding that some proteasome subunits are encoded within the MHC region of the chromosome, and their association with the proteasome can modify its activity. In particular, the activity of the proteasome leads to the liberation of peptides containing hydrophobic residues at the carboxy terminus, which typically serve as one of the anchor residues for binding to MHC class I molecules.

Peptides are actively transported from the cytosol to the endoplasmic reticulum (ER) by the transporter associated with antigen processing (TAP), which also has a preference for peptides containing a hydrophobic residue at the carboxy terminus (Pamer and Cresswell, 1998). Once inside the ER, peptides are loaded onto partially folded MHC class I molecules. If peptides are not of optimal length for stable binding to MHC class I, some trimming can occur in the ER (Yewdell and Bennink, 2001). Once a stable complex is formed, the folded complex is released from the ER and transits to the cell surface where it can be recognized by T cells specific for the particular epitope bound. Although a large number of peptides might be liberated from virus proteins, only those that form stable complexes with MHC class I are presented at the cell surface (Yewdell and Bennink, 1999). The stability of peptide/MHC class I complexes also contributes to the number of specific complexes found at the cell surface. Differences in stability can influence the hierarchy of $CD8^+$ T-cell responses in instances where multiple virus epitopes are presented for T-cell recognition (Chen et al., 2000). Finally, the $CD8^+$ T-cell response of the host will be limited by the T-cell receptors available in the host's repertoire. Thus, even though multiple virus-derived peptides might be presented at the cell surface for T-cell recognition, the host might contain T cells that can respond to only a few of these epitopes (Yewdell and Bennink, 1999).

Helper, or $CD4^+$, T cells recognize peptide antigens presented by MHC class II molecules expressed on specialized cells of the immune system, such as B lymphocytes and professional antigen-presenting cells (APCs). In contrast to antigens that enter the MHC class I processing pathway, exogenous antigens are generally processed for presentation by MHC class II molecules in endosomal vesicles (Germain and Margulies, 1993; Pieters, 2000). MHC class II molecules are composed of two transmembrane proteins, the α and β chains, which adopt a three-dimensional structure similar to that of MHC class I molecules (Stern and Wiley, 1994). Importantly, the ends of the peptide binding groove of MHC class II molecules are more open than those observed for MHC class I molecules. This usurps the length restriction imposed on MHC class I binding peptides such that peptides of varying lengths can bind in the groove of MHC class II molecules. Consequently, the stability of peptide binding to MHC class II molecules is maintained through interactions with the main chain atoms of the peptide. Although less well defined than for MHC class I binding peptides, MHC class II binding peptides also contain anchor residues for binding to a particular MHC class II molecule (Rammensee, 1995).

MHC class II molecules are co-translationally inserted into the ER but do not bind to peptides in the ER due to interaction with a chaperone molecule called the *invariant chain* (Ii) that blocks access to the peptide binding groove of MHC class II molecules (Pieters, 2000). In addition, Ii targets MHC class II molecules to the endosomal/lysosomal pathway in which limited proteolysis degrades the major portion of Ii, leaving a peptide fragment, called *CLIP*, which remains bound in the peptide binding groove. Another molecule encoded within the MHC region, called *HLA-DM*, mediates the exchange of CLIP for peptides generated from proteins that have been shuttled into the endosomal/lysosomal pathway. Once stable peptide/MHC class II complexes are formed, they traffic to the cell surface for presentation to helper T lymphocytes.

Activated T cells specific for a particular epitope may recognize and destroy any virus-infected cell that expresses an appropriate peptide/MHC complex, but the initial activation of T lymphocytes requires an encounter with professional APCs. Professional APCs, the most prominent of which is the dendritic cell, can capture, process, and present antigens as well as provide the necessary co-stimulation needed to activate naive T cells (Banchereau et al., 2000). Entry of virus proteins into the MHC class I antigen presentation pathway might occur through either direct infection of APCs or by "cross-presentation" in which virus-infected cells or cellular debris are phagocytosed, and the antigens are shuttled into the MHC class I processing pathway (Pamer and Cresswell, 1998). Recently, $CD4^+$ T cells have been shown to prime APCs via signaling through the CD40/CD40 ligand receptor pair (Heath and Carbone, 1999). These activated APCs are then capable of delivering co-stimulatory signals via B7–CD28 or similar engagements for the activation of naive $CD8^+$ T cells. Studies addressing the requirement for $CD4^+$ T-cell help for successful priming of virus-specific $CD8^+$ T-cell responses have suggested that some virus-specific responses are independent of $CD4^+$ T-cell help, likely due to alternate mechanisms for activation of APCs (Ruedl et al., 1999; Whitmire et al., 1999; Andreasen et al., 2000).

Cellular Immune Response During Viral Lytic Cycle

It is now clear that both adaptive and innate immunity are involved in defense against viral infections. However, the participation of various arms of the immune response depends on the nature of virus–host interaction. Virus neutralizing antibodies play a definitive role in interrupting viral spread, whereas T-lymphocyte-mediated mechanisms are involved in the elimination of virus-infected target cells. The main player in this arm of effector mechanism is the $CD8^+$ T lymphocytes (CTL), which lyse virus-infected cells by recognizing an viral epitope presented by the MHC class I antigens. The effectiveness of this system depends on the nature of the epitopes that are processed from viral gene products, the presence of anchor residues that enables the MHC class I antigen to selectively bind the peptide, and transport to the cell surface for recognition by CTL. A major consideration is the presence of T-cell precursors in the repertoire. In a typical immune response to viral infection, natural killer cells that recognize virus-infected cells are recruited followed by a burst of CTL after about 4–7 days (Borrow and Oldstone, 1997). As the virus load declines, the levels of CTL fall to a baseline that then persists as memory T cells and can be recalled upon re-exposure to the viral gene products. The CTL effector mechanism is obligatory for the elimination of virus-infected cells, especially for those viruses that cause persistent infections (Borrow and Oldstone, 1997). The CD4-mediated T-cell responses are involved in helping B cells produce antibodies and produce cytokines involved in the proliferation of $CD8^+$ T cells and activation of APC.

There have been only a handful of attempts made to study the involvement of T-cell-mediated immunity during infection of natural hosts by the three viruses, JCV, BKV, and SV40. In one report, Shah's group (Drummond et al., 1985) studied $CD4^+$ T-cell responses to BKV in seropositive individuals by the lymphocyte proliferation assay. The results showed that all of 15 seropositive individuals were positive for BKV antigens as measured by $CD4^+$ T-cell responses. Seronegative subjects were nonreactive. Unfortunately, this interesting lead has not been followed up. There may be several reasons for this. First, for human polyomaviruses there are no available, reliable animal models in which virus infection leads to its replication in vivo. However, cellular immune response to SV40 could be studied in its natural host, the rhesus monkey, which is being utilized for the study of SIV. Second, the tools needed to explore this problem such as purified viruses and their components are not yet available.

Third, most of the attention has been focused on T antigens encoded by these viruses as inducers of immunity against virus-induced tumors (Tevethia, 1990). In that arena considerable progress has been made. Tumors induced in mice by T antigen expressed as a transgene have served as valuable models toward understanding a T-cell response during tumorigenesis (Schell et al., 1999, 2000). Another consideration that may complicate the interpretation of data regarding T-cell-mediated immunity during infection of natural hosts is the cross reactivity between both the nonvirion and virion proteins encoded by these three viruses. Despite these difficulties, the recent report by Koralnik et al. (2001) of JCV-specific CTL in PML patients suggests a promising new approach to T-cell-mediated antiviral immunity in humans.

There is a report (Bates et al., 1988) that has addressed the role of $CD8^+$ T cells in interrupting the SV40 infectious cycle in vitro. The investigators utilized murine CTL clones specific for $H2\text{-}K^b$ and $H2\text{-}D^b$ restricted SV40 T antigen epitopes as probes for abrogating the SV40 infectious cycle in permissive monkey cells. To utilize this strategy, a continuous line of monkey kidney cells, TC-7, was tranfected with murine $H2\text{-}K^b$ and $H2\text{-}D^b$ class I antigens, thus allowing the presentation of T-antigen epitopes to SV40 T-antigen-specific CTL clones generated in B6 mice. The results showed that the interaction of T-antigen-specific CTL clones with SV40-infected $TC7/H\text{-}2D^b$ or $TC\text{-}7/H\text{-}2K^b$ cells for 5 hours reduced the SV40 yield by 70–90% as measured by the infectious center assay. The results of this study lead to two important conclusions: (1) The target for CTL is a nonvirion protein, T antigen, which is synthesized before viral DNA replication; and (2) the CTL-mediated events may also take place in the natural host undergoing SV40 infection. It is also tempting to speculate that a loss of this kind of CTL-mediated surveillance may be directly involved in the reactivation of BKV, JC, and SV40 from latency.

Cellular Immune Response During Viral Carcinogenesis

It is well established that SV40, BKV, and JCV induce tumors in newborn hamsters and in mice when T antigen is expressed as a transgene. In the latter case, T antigen can induce tumors in a variety of tissues depending on the promoter used to express T antigen. In addition, cells transformed by SV40 T antigen are almost always transplantable in nude mice that lack T cells (Choi et al., 1983; Tevethia et al., 2001; Thompson et al., 1990). In this section, we focus on the role of the immune response in carcinogenesis by SV40, as more information has been gathered with this system. However, information about BKV and JCV is included whenever relevant.

Tumor induction by SV40 in hamsters was found to be age dependent (Diamandopoulos and McLane, 1975). This age-related resistance to tumor induction is immunologically mediated as adult hamsters that have been immunocomprised by X-iradiation before SV40 inoculation become susceptible to tumor induction (Allison et al., 1967). In elegant experiments, Diamadopoulos (1973) has demonstrated that the age-related resistance could be by-passed if the hamsters were inoculated with SV40 intravenously. A wide range of tumors was observed, from leukemias to lymphosarcomas, in these animals. Hamsters inoculated with SV40 in the cheek pouch, which is considered an immunologically privileged site, also developed tumors (Allison et al., 1967). The SV40 tumors induced in hamsters have been shown generally to be transplantable in syngeneic hamsters (Tevethia et al., 2001). The first evidence of the specific antigenicity of these SV40-induced tumors was provided by a number of investigators (Tevethia, 1980), who demonstrated that prior immunization of hamsters with either SV40 or SV40-transformed cells lead to the development of immunologic resistance against a tumor challenge. This immunity, demonstrated by the transplantation rejection test, was specific for SV40 and mediated by T lymphocytes as treatment of SV40-immunized hamster with anti-T-lymphocyte serum abrogated the immunity against tumor challenge (Tevethia et al., 1968). Similar approaches later on were utilized to demonstrate the antigenicity of BKV- and JCV-transformed cells. Although studies carried out in inbred hamsters provided significant insights into the induction of the cellular immune response against virus specified antigenicity on SV40 transformed cells, detailed analysis of the cellular immune response and the nature of antigen could not be undertaken due to the limited knowledge about hamster immunology.

In inbred mice, inoculation of SV40 in newborns did not result in the induction of tumors in a number of mouse strains (Abramczuk et al., 1984). Only in Balb/C mice, which have been classified as low responders, did a small percentage of newborn mice that received SV40 develop tumors. In contrast, mice expressing SV40, BKV, or JCV T antigen as a transgene developed progressively growing tumors in the tissue targeted for T-antigen expression (Brinster et al., 1984; Small et al., 1986). In transgenic mice, T-cell response is compromised by either central or peripheral tolerance. Mouse cells transformed by SV40 as a rule are not transplantable in the immunocompetent adult host as is observed in hamsters (Tevethia et al., 2001). Studies have shown that mouse cells transformed by SV40 must be cultured in vitro for prolonged time periods before becoming transplantable in syngeneic mice, and even then immunosuppression of the host is required. For example, Balb/C cells transformed by SV40 required over 50 generations of in vitro passage followed by passage in immunosuppressed hosts before inducing progressively growing tumors in adult Balb/C mice (Tevethia and MacMillan, 1974). These tumor cells maintained the expression of T antigen and the MHC class I antigens. In the case of C57BL/6 mice, which have been classified as high responders for the induction of SV40 T-antigen-specific CTL, SV40-transformed cells only become transplantable in syngeneic mice at the expense of H-2^b class I antigen and only after prolonged in vitro culture and passage in newborn mice (Flyer et al., 1983). These observations indicated that SV40, as well as BKV and JCV, transformed cells contain strong antigenicity resulting in the induction of a highly effective cellular immune response in the host.

3. T ANTIGEN AS A TARGET FOR T-CELL-MEDIATED CELLULAR RESPONSE

SV40, JCV, and BKV T antigens are largely homologous in structure and function (Pipas, 1992) and are the first virus-encoded nonstructural protein synthesized by the infected cells during permissive and nonpermissive infection (Tooze, 1980). The T antigen is a highly immunogenic protein that induces both a humoral and cellular immune response in the host (Tevethia, 1980, 1990). Because T and small t antigen, also encoded by the SV40 early region, are the only proteins synthesized in the transformed or tumor cells, it is logical to assume that either of these two proteins plays a dominant role in inducing the immune response and thus are the targets for cellular immune response mediated by T lymphocytes.

A large amount of data published during the last 25 years have definitively shown that the T antigen is the target for immune response and participates in inducing tumor rejection in the experimental host preimmunized with SV40 (Rapp et al., 1966; Chang et al., 1979; Tevethia et al., 1980; Flyer and Tevethia, 1982). BKV and JCV are expected to behave similarly. The earlier evidence for the role of T antigen can be summarized as follows. (1) Cells from different species transformed by SV40 and expressing T antigen share the ability to immunize against the transplantation of tumors induced by SV40 in either hamsters or mice. Interestingly, a number of studies carried out with BKV, JCV, and SV40 transformed cells observed cross reactivity between the CTL epitopes encoded by these viruses (Campbell et al., 1983; Deckhut et al., 1991; Tevethia et al., 1998). (2) Adenovirus–SV40 hybrid viruses, which express either full-length T antigen or a fragment of T antigen, were able to confer immunity to hamsters and mice against tumor transplantation (Rapp et al., 1966; Lewis and Rowe, 1973). This finding suggested that the T antigen is responsible for inducing the rejection response. In addition, T antigen synthesized during the permissive cycle also induced tumor immunity (Tevethia and Tevethia, 1976). (3) Partially purified T antigen induced an effective response in syngeneic mice against SV40 tumors in vivo (Chang et al., 1979), followed by the demonstration that the T antigen purified to homogeneity not only induced SV40-specific tumor immunity in Balb/C mice, but in high responder C57BL/6 mice induced a cytotoxic T-cell response specific for SV40 T antigen (Tevethia et al., 1980).

This later observation identified T antigen as responsible for inducing both a T-cell response and tumor immunity. Our laboratory had previously demonstrated that spleen cells from SV40-immunized mice admixed with the tumor cells inhibited tumor progression (Zarling and Tevethia, 1973). The candidate for the effective tumor inhibitory capacity was identified as a T lymphocyte (Tevethia et al., 1974). Further evidence that T antigen directly participates in priming the cellular immune response came from the finding by Flyer et al. (1983) that a loss of SV40 T-antigen expression in polyomavirus-induced tumor cells expressing SV40 T antigen when subjected to immunologic pressure in vivo led to the loss of recognition by the SV40 T-antigen-specific CTL as a result of the loss of T antigen itself.

The generation of CTL specific for SV40-transformed cells has been demonstrated in a number of studies in which high responder mice of C57BL/6 origin (H-2^b) upon immunization with either syngeneic or allogeneic transformed cells develop CTL specific for SV40 T antigen (Trinchieri et al., 1976; Knowles et al., 1979; Pretell et al., 1979) in an MHC class I–restricted manner. These observations combined with the direct demonstration of T antigen as the target for CTL have led to the understanding that CTL are the major player in inducing immune resistance against tumor development. The understanding of MHC restriction (Zinkernagel and Doherty, 1979), the development of CTL clones specific for T antigen, and the demonstration that short synthetic peptides of 8–10 amino acids are presented by the MHC class I molecules to the CTL led to the identification of CTL epitopes in SV40 T antigen (Campbell et al., 1983; Tanaka et al., 1988; Tevethia et al., 1990; Deckhut et al., 1992).

Initial isolation of two distinct CTL clones from H-2^b mice immunized with syngeneic SV40 T-antigen-transformed cells that differed in their specificities for the recognition of SV40 and BKV T-antigen epitopes and their close proximity in the amino terminal half of the T-antigen provided the first evidence that the T-antigen posseses multiple epitopes and that CTL specificity may be dictated by the amino acid sequence (Campbell et al., 1983). Using a large collection of CTL clones generated in B6 mice immunized with SV40-transformed B6 cells, four distinct epitopes were identified that were specific for SV40 T antigen. By using the T-antigen deletion mutants truncated at either carboxy or amino terminal ends or carrying internal deletions and using sythetic peptides, the four CTL epitopes were mapped to residues 206–215 (epitope I), recognized by CTL clones Y-1 and K11; residues 223–231 (epitope II/III), recognized by CTL clones Y-2 and Y-3; residues 404–411 (epitope IV), recognized by CTL clone Y-4; and residues 489–497 (epitope V), recognized by CTL clones Y-5 and H-1. The epitopes I, II/III, and V are H-D^b restricted, whereas epitope IV is H-K^b restricted (Deckhut et al., 1992; Mylin et al., 1995b).

The multiplicity of CTL epitopes on T antigen raises the question of the immunodominant and immunorecessive nature of these CTL epitopes within T antigen and their relative roles in inducing a class I–restricted CTL response. An immunologic hierarchy has been demonstrated among these four T-antigen epitopes. Immunization of B6 mice with full-length T antigen expressed either in transformed cells or cloned in a vaccinia virus vector leads to the induction of CTL specific for epitopes I, II/III, and IV but not V as judged by the frequency analysis of pre-CTL specific for each of the epitopes (Mylin et al., 1995a, 2000; Fu et al., 1998). However, B6 mice immunized with T antigen from which epitopes I, II/III, and IV have been inactivated by site-directed mutagenesis or by deletion did induce CTL specific for epitope V (Tanaka et al., 1989; Mylin et al., 2000). This example of immunodomination has important implications for the use of immunotherapuetic approaches. Interestingly,

CTL epitopes preceeded by a signal sequence that allows direct access of the epitope peptide into the ER and expressed in a recombinant vaccinia virus induced CTL to epitope V extremely efficiently (Mylin et al., 2000).

4. CELLULAR IMMUNE RESPONSE IN SV40 T ANTIGEN TRANSGENIC MICE

Expression of the T antigens from SV40, JCV, and BKV as a transgene in mice can induce tumors at distinct anatomic sites depending on the promoter used to drive expression of the transgene. These mice provide powerful tools to examine the effect of endogenous T-antigen expression and tumor progression on the immune response. Expression of SV40 T antigen in vivo can have varied effects on $CD8^+$ T-cell immunity, ranging from the spontaneous perturbation of autoimmune disease to the induction of profound tolerance, depending on the site and timing of T-antigen expression. Because detailed studies on the immune responsiveness of BKV and JCV T antigen transgenic mice remain to be performed, this section focuses on results obtained with SV40 T antigen transgenic mice.

Expression of T antigen in the thymus during T-cell development most often results in the loss of T antigen epitope-specific T cells by clonal deletion (Hanahan, 1985; Faas et al., 1987; Geiger et al., 1993; Schell et al., 1999). Potentially autoreactive T cells do survive negative selection in some T antigen transgenic mouse models where T antigen expression occurs after T-cell development. Such models have been used to study the effect of T-antigen-induced peripheral tolerance for both $CD4^+$ and $CD8^+$ T-antigen-specific T lymphocytes. Flavell and co-workers described a system in which the fate of $H2\text{-}K^k$–restricted SV40 T-antigen-specific T-cell receptor (TCR) transgenic T cells was determined in line 177-5 ($H2^k$) mice expressing SV40 T antigen from the rat elastase I promoter (Ornitz et al., 1985; Geiger et al., 1992; Antonia et al., 1995). Although expression of the transgene was detected in the thymus as well as the pancreas, TCR transgenic $CD8^+$ T cells were positively selected and entered the periphery where they were involved in the establishment of autoreactivity in the pancreas at an early age (Geiger et al., 1992). These T-antigen-specific $CD8^+$ T cells, however, became anergic and disappeared from the periphery by 5 months of age (Antonia et al., 1995), after which pancreatic tumors became apparent. This onset of anergy was linked to events that occurred in the thymus before exit of the T cells into the periphery, resulting in a gradual loss of T-cell responsiveness to T antigen.

A different scenario was observed for RT3 mice, which express T antigen from the insulin promoter (Hanahan, 1985; Geiger et al., 1993). In RT3 mice crossed with $H2\text{-}K^k$–restricted TCR transgenic mice, the $CD8^+$ T cells develop normally and show no apparent signs of tolerance in the periphery even after 5 months. This might be attributed to the delayed expression of T antigen in RT3 mice, which is not detected until 10–12 weeks of age (Adams et al., 1987).

These double transgenic mice show signs of autoimmunity in the pancreas similar to that of young line 177-5 mice (Geiger et al., 1993), suggesting that the mechanisms responsible for the maintenance of peripheral tolerance cannot be maintained in the presence of a large population of potentially autoreactive T cells.

In contrast, TCR transgenic $CD8^{+}$ T lymphocytes ignored expression of the $H2\text{-}K^{k}$–restricted CTL epitope in the islet cells of mice expressing a nontransforming T-antigen fragment from the insulin promoter (Soldevila et al., 1995). Lack of spontaneous autoimmunity in this model was attributed, in part, to the absence of signals that could occur if transformed cells were present. Co-expression of the B7.1 co-stimulatory molecule on the pancreatic islet cells with the T antigen fragment resulted in destruction of the pancreatic islets by activated TCR transgenic T cells, indicating an inherent lack of co-stimulation for the activation of potentially autoreactive T cells in this model. These results indicate that SV40 T-antigen expression in mice can result in autoimmunity, immunologic ignorance, or antigen-induced tolerance.

The role of T-antigen-specific immunity in the control of spontaneous tumor progression in T antigen transgenic mice was addressed (Ye et al., 1994) using the RIP1-Tag4 ($H2^{b}$) line of mice, which express full-length T antigen from the insulin promoter. T antigen is detected in the pancreas at 10–12 weeks of age in RIP1-Tag4 mice (Adams et al., 1987). This expression results in the progressive growth of insulinomas that become life threatening by 8–9 months of age (Ye et al., 1994). RIP1-Tag4 mice develop functional T-antigen-specific $CD8^{+}$ T cells that can be activated by specific immunization with SV40 to control the progressive growth of tumors if the mice are immunized before 9 weeks of age. Immunization after 9 weeks of age failed to induce a significant delay in tumor progression. Whether this failure is due to the gradual onset of tolerance with increasing age similar to that which occurs in line 177-5 mice was not directly determined. The insulinomas in RIP1-Tag4 mice, however, were shown to down-regulate MHC class I expression, which is a common mechanism of escape from CTL-mediated immunity for some tumors (Melief et al., 2000).

The ability of individual CTL epitopes to induce the control of spontaneous tumor progression was investigated in SV11 mice (Schell et al., 1999) that express T antigen as a transgene from the SV40 enhancer/promotor and develop choroid plexus papillomas (Brinster et al., 1984; Van Dyke et al., 1987). SV11 ($H2^{b}$) mice are tolerant to the immunodominant T-antigen epitopes I, II/III, and IV due to expression of T antigen in the thymus during T-cell ontogeny (Schell et al., 1999). CTL specific for the immunodominant $H2\text{-}K^{b}$–restricted epitope IV were established in SV11 mice following adoptive transfer of naive C57BL/6 splenocytes into SV11 recipients and immunization with recombinant vaccinia viruses (rVVs) expressing full-length T antigen or epitope IV minigenes. This resulted in a significant increase in the life span of SV11 mice. Additionally, reconstitution of SV11 mice with naive C57BL/6 splenocytes following a nonlethal dose of γ-irradiation resulted in priming of epitope IV–specific CTL against the endogenous T antigen and a highly significant increase in the life span of SV11 mice due to inhibition of tumor progression. In similar studies by others, adoptive transfer of T antigen immune lymphocytes or a CTL line specific for T-antigen epitope II/III has been shown to moderate the progressive growth of T-antigen-induced prostate (Granziero et al., 1999) and liver (Romieu et al., 1998) tumors, repectively, in T antigen transgenic mice.

The fate of endogenous CTL specific for individual $H2^b$ epitopes was examined in the 501 lineage of mice that express T antigen from the late liver α-amylase promoter and is first detected in the salivary glands at 3 months of age. After 8 months of age 501 mice develop T-antigen-expressing osteosarcomas that can metastasize to the liver and lungs (Knowles et al., 1990; Marton et al., 2000). Spontaneous autoimmunity was not detected in 501 mice, in contrast to some of the T antigen transgenic mouse models discussed above. Immunization of 501 mice with rVVs expressing individual T-antigen epitopes revealed a sequential loss of CTL responses against the T-antigen epitopes (Schell et al., 2000). Loss of epitope I–specific CTL occurred by 6 months of age, corresponding with increased levels of T-antigen expression in the salivary glands. Importantly, loss of epitope IV–specific CTL responses correlated with the appearance of T-antigen-expressing osteosarcomas. The use of MHC class I/epitope IV tetramers revealed that there was a progressive decrease in the number of epitope IV–specific $CD8^+$ T cells that could be recruited by immunization, culminating in complete loss of responsiveness with the appearance of tumors. These results indicate that autoreactive and potentially tumor-reactive T lymphocytes are tolerized over time due to the endogenous expression of T antigen. This tolerance likely occurs through cross-presentation of the antigen, which leads to either deletion or the induction of T-cell anergy as has been described for other transgenic systems (Heath et al., 1998).

Fewer studies have addressed the role of T-antigen-specific $CD4^+$ T cells in the control of T-antigen-induced tumors in transgenic mice. Two studies (Forster et al., 1995; Ganss and Hanahan, 1998) directly addressed the role of T-antigen epitope-specific $CD4^+$ T cells in the control of insulinomas using the RIP1-Tag5 (RT5) line of mice, which express full-length T antigen from the insulin promoter at 10–12 weeks of age in the pancreas. RT5 mice succumb to tumors at approximately 4–5 months of age. RT5 mice were crossed with TagTCR2 mice, which express a transgenic TCR specific for an H-2^k–presented $CD4^+$ T-cell epitope in order to increase the frequency of antigen-specific $CD4^+$ T cells (Forster et al., 1995). While there was an increase in the degree of infiltration of T-antigen-expressing islet cells in these double transgenic mice, no effect on tumor progression was observed. These endogenous T-antigen-specific $CD4^+$ T cells could be activated in vivo if B7.1 also was expressed on pancreatic islet cells, resulting in insulitis and diabetes before tumor formation (Ganss and Hanahan, 1998). In contrast, adoptive transfer of activated lymphocytes from normal or TagTCR2 transgenic mice into tumor-bearing RT5 mice resulted in a significant infiltration of early-stage tumors if the recipient mice were first irradiated. The authors suggest that irradiation might contribute to increased permeability of the tumors. These studies support the idea that presentation of T-antigen epitopes in the absence of proper co-stimulation leads to tolerance instead of T-cell activation.

The effect of the endogenous expression of BKV or JCV T antigen on T-antigen-specific immunity remains to be determined. Several lines of BKV and JCV transgenic mice have been developed and await investigation of immune function and the role of immunity in control of T-antigen-induced tumors. Mice expressing the BKV T antigen from the viral promoter developed T-antigen-expressing primary renal and hepatocellular tumors as well as thymic hyperplasia and thymomas (Small et al., 1986; Dalrymple and Beemon, 1990). Mice expressing JCV T antigen develop neuroectodermal tumors and metastatic adrenal neuroblastomas (Franks et al., 1996; Krynska et al., 1999). The status of the immune response awaits elucidation in these models.

5. CROSS REACTIVITY AMONG SV40 T ANTIGEN WITH BKV AND JCV T-ANTIGEN CTL EPITOPES

Understanding CTL epitope specificity of SV40, BKV, and JCV T antigens is important, as the T antigens of these viruses show considerable amino acid homology, and all three viruses have now been isolated from humans. The epitope specificity also provides a measure of selective cellular immune response to each of these viruses. An understanding of this immune response will allow the elucidation of the level of immunosurveillance not only against virus infections but also against neoplastic development. The selection of CTL epitopes is dictated by the rules of determinant selection, which include the presence of anchor residues specific for each of the MHC class I haplotypes (Falk et al., 1991). In the case of SV40 T antigen, epitopes have been more thoroughly defined for the $H\text{-}2^b$ haplotype. Thus, for a peptide to be presented by the $H\text{-}2D^b$ molecule, it needs an asparagine residue at position five and a hydrophobic residue at position 9 at the carboxy-terminal end, whereas for $H\text{-}2K^b$, peptides contain tyrosine or phenylalanine at position 5 with a hydrophobic residue at position eight at the carboxy terminus (Falk et al., 1991). In addition to having the proper MHC class I binding motif, there are other requirements that affect the generation of effector T cells (Yewdell and Bennink, 1999).

Our approach to discriminating among T antigens encoded by SV40, JCV, and BKV is based on the fine specificity of SV40 T-antigen CTL clones. Only those epitopes in BKV and JCV T antigens that show identity in critical amino acids with the corresponding SV40 T antigen will be recognized by the SV40 T-antigen-specific CTL clones. We therefore compared the sequences of SV40 T-antigen epitopes I, II/III, V, and IV with the corresponding sequences in T antigens encoded by JCV and BKV and tested the reactivity of SV40 T-antigen-specific CTL clones directed to epitopes I, II/III, V, and IV with cells transformed by SV40, JCV, or BKV T antigens (Tevethia et al., 1998). This panel of CTL clones allowed us to distinguish between the T antigens expressed by SV40, JCV, and BKV. In the same study, we determined that three viruses isolated from humans are indeed authentic SV40. One important step was to transform primary mouse cells of C57BL/6 origin (H-2^b) with the human isolates. The use of C57BL/6 cells was necessary because the CTL clones that recognize SV40 T-antigen epitopes are H-$2D^b$ and H-$2K^b$ restricted, and for this reason the CTL epitopes processed from T antigen must be presented by mouse (H-2^b) MHC class I molecules. It is possible, nonetheless, to test cells derived from human tumors suspected of expressing a papovavirus T antigen for the presence of CTL epitopes detected by the CTL clones by expressing H-$2K^b$ and/or H-$2D^b$ genes into the tumor cells. Alternatively, the H-$2K^b$ and/or H-$2D^b$ gene products could be introduced by infecting the tumor cells with vaccinia virus vectors that express these genes. Thus, the approach we have used to document the identity of the papovavirus T antigen in human tumors need not rely on isolating infectious virus or amplifying T-antigen-coding sequences from the tumors.

6. ANTIBODY RESPONSES DURING VIRUS INFECTIONS IN THE PERMISSIVE HOST

The natural host for SV40 is rhesus monkey, and the virus appears to be latent in the kidneys. The antibody response to SV40 in these monkeys and virus transmission in the monkey colonies have been nicely summarized by Shah and Nathanson (1976). Based on the antibody profiles, the rhesus and a few other macaques are susceptible to SV40 infection. As adults, a large proportion of rhesus monkeys become seropositive for SV40 (Shah and Nathanson, 1976). Seronegative rhesus monkeys can be infected with SV40 via a variety of routes and demonstrate viremia (Shah et al., 1968). An earlier study (Rapp et al., 1967) showed that African green monkeys inoculated with autologous cells infected with SV40 developed antibodies to the virus as well as T antigen. No sign of neoplasia was noticed at the site of inoculation in these monkeys, again showing that the development of antibodies to T antigen is not a sign of neoplasia.

The antibody responses to BKV in humans has been described elsewhere in this book (see Chapter 19). Infection of humans by BKV and JCV is widespread, and most of the young population develop antibodies to these two viruses.

7. ANTIBODY RESPONSES IN SEMIPERMISSIVE HOST

In an attempt to understand antiviral immune response in a permissive host, rabbits, which are considered a semipermissive host, were inoculated intravenously with either a high dose (1×10^{12}) or a low dose (1×10^{9}) of purified

SV40 and were studied for the development of antibodies to T antigen, virion antigen, and virus neutralizing antibodies for various periods of time. Rabbits inoculated with a high dose of SV40 synthesized high levels of virus neutralizing antibodies as well as antibodies to T antigen. This pattern of antibody response lasted for up to 200 days of the observation period. Rabbits inoculated with a low dose of virus synthesized antibodies to the virus only after a rechallenge with a high dose of virus (Tevethia, 1970). These results suggested that SV40 establishes a persistent infection and continues to produce virus and T antigen. Any virus released from the cells is neutralized by antiviral antibodies. These results further suggested at that time that the presence of antibodies to T antigen does not necessarily indicate the presence of T-antigen-induced tumor. It merely indicates that a virus infection has occurred that may or may not lead to transformation of cells.

With a number of reports documenting the detection of SV40 DNA sequences in a variety of human tumors and in some cases the synthesis of T antigen, it is quite likely that a select population of humans may become exposed to SV40 (see Chapter 17). However, SV40 administered to humans as a contaminant of polio and RSV vaccines is capable of inducing SV40 neutralizing antibodies that persist in some cases for prolonged periods, suggesting that virus may have established a persistent infection. Viral neutralizing antibodies were induced by either the subcutaneous or intranasal route but not by the oral route. Whether SV40 induces a clinical disease similar to BKV or JCV remains to be established (Melnick and Stinebaugh, 1962; Gerber, 1967).

8. ANTIBODY RESPONSES TO T ANTIGEN DURING TUMORIGENESIS IN NONPERMISSIVE HOSTS

SV40 T antigen was identified in SV40-transformed or tumor cells by the complement fixation test using sera from hamsters bearing SV40 tumors (Black et al., 1963). The nuclear location of T antigen was documented by the immunofluorescence test (Pope and Rowe, 1964; Rapp et al., 1964). The almost universal development of high titer antibodies to T antigen in hosts bearing tumors induced by virus-free transformed or tumor cells demonstrated that the T antigen is a highly antigenic protein and is capable of inducing the antibody response during tumorigenesis either by the virus or virus-induced tumor or transformed cells (Tevethia, 1980). Hamsters and mice bearing tumors expressing SV40 T antigen respond by making antibodies not only to T antigen but also to small t antigen. Furthermore, multiple epitopes in T antigen, mostly clustered in the amino- and carboxy-terminal ends of T antigen, are recognized by antibodies from the tumor-bearing host (Greenfield et al., 1980; Ransom et al., 1982). The multiplicity of epitopes in T antigen is supported by the isolation of a large group of monoclonal antibodies directed to different specificities by a large group of investigators, and the data have been summarized previously (Mole et al., 1987).

The immunogenicities of BKV and JCV T antigens are essentially very similar to that of SV40 T antigen. There is extensive cross reactivity between epitopes of BKV, JCV, and SV40 T antigen epitopes. Only some of the monoclonal antibodies made against SV40 T antigen are specific for SV40 T antigen. JCV T-antigen-specific monoclonal antibodies were isolated by immunizing SV11 T antigen transgenic mice that are tolerant to epitopes on SV40 T antigen with JCV-transformed mouse cells followed by selection of JCV T-antigen-specific antibodies (Tevethia et al., 1992). The JCV T-antigen-specific epitopes were localized in the amino terminus of T antigen (Tevethia et al., 1992). Isolation of other JCV-specific antibodies also has been reported (Bollag and Frisque, 1992).

9. IMMUNOLOGIC APPROACHES FOR ASSOCIATING SV40 WITH NEOPLASIA IN HUMANS

SV40 is being associated with the development of a number of human neoplasias (mesothelioma, osteogenic sarcomas, and others) based on the presence of T-antigen sequences amplified by PCR (Bergsagel et al., 1992; Carbone et al., 1994; Arrington and Butel, 2001). In only a few cases has the expression of T-antigen protein been demonstrated. However, the role of SV40 as an etiologic agent of human neoplasia remains to be firmly established. A large number of studies using experimental models have shown that adult rodents exposed to SV40 are resistant to tumor development largely due to a strong specific immunosurveillance mediated by the cell-meditated immune responses against any emerging potentially transformed cells in vivo. However, differences in the mode of exposure might influence the outcome. In certain cases where the virus is administered by alternative routes, adult hamsters develop a variety of tumors (Diamandopoulos, 1973; Diamandopoulos and McLane, 1975). Studies in T antigen transgenic systems have shown that T-cell responses specific for T antigen are silenced by tolerance before tumor progression (Schell et al., 1999, 2000). Applying the immunologic findings from the experimental SV40 tumor system to the involvement of SV40 T antigen in tumors in humans will provide convincing supporting data either for or against SV40 as an etiologic agent.

There are three host–SV40 interactions that may lead to the development of immune responses to either viral or T antigens. The first is the transient infection of SV40 in humans without virus persistence or establishment of transient foci of persistently infected or transformed cells. This abortive infection may not provide a sufficient antigenic load to trigger an immune response either to virus proteins or to T antigen. It is possible that multiple abortive exposures by SV40 in the same individual may lead to the induction of virus neutralizing antibodies that are sufficient to control any future infection by SV40. In that case, the viral antibody titers are likely to remain low or undetectable. However, the studies reported in the literature have detected significant titers of virus neutralizing antibodies in childern (Butel et al., 1999). An intriguing possibility is that the virus may target dendritic cells, the professional APCs. These cells, after processing the antigen, viral or T antigen, will trigger an efficient immune response by sensitizing both CD8 and CD4 T cells. Viral or T antigen produced from cells that are not APC must undergo cross-priming by the APC, a process that may compromise the efficiency of antigen presentation (Heath and Carbone, 1999). Because persistence of memory T cells does not require a continuous antigenic stimulus, their presence in the circulation of these T cells can nevertheless be detected and quantitated by the sensitive, highly specific in vitro techniques currently available. If the virus targets a tissue such as kidney, there is the likelihood that a massive replication of the virus can occur, such as is seen in infections in humans by BKV, until the virus undergoes latency. In that case, a strong immune response will result in the synthesis of antiviral antibodies and a T-cell response to T antigen as well. T-antigen synthesis in the kidneys has been demonstrated in patients under immunosuppression (Randhawa and Demetris, 2000). The problem of defining humans at risk of SV40 infection is that this virus has not been linked to a clinical syndrome, whereas BKV and JCV have been implicated directly and routinely with nephropathy and PML, respectively (Padgett et al., 1971; Narayan et al., 1973; Arthur et al., 1986).

The second type of virus–host interaction is the establishment of persistent infection in which cells will continue to synthesize T antigen as well as viral proteins with some production of virus that could infect other susceptible cells. In this case, a strong antibody immune response to both the virus and T antigen will occur. In addition, a vigorous cellular response will occur. These responses will be maintained at high levels due to continuous antigenic stimulation. Neutralizing antibodies will eventually terminate the persistent state, and the virus may undergo latency that may lead to down-regulation of immune responses. Reactivation of the virus under immunosuppression may restart the cycle. In this context, it is interesting to note that transplant patients have been shown to develop high neutralizing antibody titers to SV40 (Butel et al., 1999).

The third type of SV40–host interaction may be establishment of fully transformed cells that are capable of forming a foci that may develop into a progressively growing tumor accompanied by other cellular changes. In this case, the T antigen in the transformed cells will induce a vigorous antibody response. Experimental evidence suggests that the antibody titers correlate with the size of the tumor and, in later stages, the T-antibody titers may drop due to the formation of immune complexes (Ransom et al., 1982). In addition, the antibody responses may spread to epitopes other than the dominant epitopes in the carboxy and amino termini of T antigen (Greenfield et al., 1980; Ransom et al., 1982). The development of antibody to T antigens in this type of interaction will depend on the synthesis of adequate levels of the proteins by the transformed or tumor cells. Furthermore, antibodies to small t antigen specific epitopes may also develop in the host.

For the detection of T antigen in potential tumor cells, it is essential to use well-defined monoclonal antibodies directed to specific epitopes. In addition, it is desirable to use monoclonal antibodies that recognize both ends of T antigen. Efforts should be made to utilize monoclonal antibodies that are specific for SV40 T antigen. A large number of monoclonal antibodies that are made against SV40 T antigen have been shown to cross react with BKV and JCV T antigens. It would also be desirable to use antibodies that work well in all types of assays. For the detection of antibodies to T antigen in human sera, care should be taken to include adequate controls to eliminate nonspecific reactions.

For the detection of antiviral antibodies in a patient's serum, strict specificity controls are needed to differentiate reactivities between SV40, JCV, and BKV. It would also be desirable for all investigators to utilize a standard strain of SV40 and to utilize purified viruses. The last point is important for the measurement of antiviral antibodies by ELISA.

Tracking cellular immune responses in humans suspected of being infected with SV40 or patients with cancer in whom SV40 is suspected as the etiologic agent would provide substantial supporting and convincing evidence for a role of SV40 in human neoplasia. The presence of $CD8^+$ T cells specific for epitopes for T antigen in humans will suggest that infection by SV40 has occurred and that the expression of T antigen would induce a cellular response. The magnitude of CD8 T cells responding to T antigen might indicate a continued antigenic stimulation. To explore the CD8 T-cell responses in humans, it is essential to seek an experimental approach that would identify HLA-restricted T-antigen epitopes.

Our laboratory (Schell et al., 2001) utilized HLA-A2.1 transgenic mice that have been shown to respond to HLA-A2.1–restricted T-cell epitopes in influenza virus and human p53 (Le et al., 1989; Theobald et al., 1995). HLA-A2.1 transgenic mice were immunized with B6 cells transformed by T antigen from which the known $H\text{-}2^b$–restricted epitopes were inactivated by mutagenizing the anchor residues. $CD8^+$ T cells from these mice were shown to be T-antigen specific but only in association with the HLA-2.1 class I antigen. Thus, any cell expressing HLA-A.2.1 and T antigen would be lysed by the CTL clone that is specific for T antigen in association with HLA-A.2.1. The HLA-A.2.1–restricted epitope was mapped to T antigen residues 281–289. This epitope represents a potential specific CTL recognition epitope for humans. It was interesting to note that the 281–289 T-antigen epitope did not cross react with either BKV or JCV. By synthesizing the HLA-A.2.1/T281 tetramers, it was possible to quantitate the number of $CD8^+$ T cells in HLA-A.2.1 transgenic mice immunized with T antigen.

This powerful approach will allow investigators to track SV40 T-antigen-specific T-cell responses at least in HLA-A.2.1–positive individuals. This approach will also allow the identification of HLA-A.2.1–restricted epitopes specific for BKV and JCV T antigens. This may be an additional way to distinguish between SV40, BKV, and JCV.

ACKNOWLEDGMENTS

This work was supported by grant CA25000 from the National Cancer Institute, National Institutes of Health, Bethesda, MD. We thank Dr. M. Judith Tevethia for critically reading the manuscript.

REFERENCES

Abramczuk J, Pan S, Maul G, Knowles BB (1984): Tumor induction by simian virus 40 in mice is controlled by long-term persistence of the viral genome and the immune response of the host. J Virol 49:540–548.

Adams TE, Alpert S, Hanahan D (1987): Non-tolerance and autoantibodies to a transgenic self antigen expressed in pancreatic beta cells. Nature 325:223–228.

Allison AC, Chesterman FC, Baron S (1967): Induction of tumors in adult hamsters with simian virus 40. J Natl Cancer Inst 38:567–572.

Andreasen SO, Christensen JE, Marker O, Thomsen AR (2000): Role of CD40 ligand and CD28 in induction and maintenance of antiviral CD8+ effector T cell responses. J Immunol 164:3689–3697.

Antonia SJ, Geiger T, Miller J, Flavell RA (1995): Mechanisms of immune tolerance induction through the thymic expression of a peripheral tissue-specific protein. Int Immunol 7:715–725.

Arthur RR, Shah KV, Baust SJ, Santos GW, Saral R (1986): Association of BK viruria with hemorrhagic cystitis in recipients of bone marrow transplants. N Engl J Med 315:230–234.

Banchereau J, Briere F, Caux C, Davoust J, Lebecque S, Liu YJ, Pulendran B, Palucka K (2000): Immunobiology of dendritic cells. Annu Rev Immunol 18:767–811.

Bates MP, Jennings SR, Tanaka Y, Tevethia MJ, Tevethia SS (1988): Recognition of simian virus 40 T antigen synthesized during viral lytic cycle in monkey kidney cells expressing mouse H-2Kb– and H-2Db–transfected genes by SV40-specific cytotoxic T lymphocytes leads to the abrogation of virus lytic cycle. Virology 162: 197–205.

Bergsagel DJ, Finegold MJ, Butel JS, Kupsky WJ, Garcea RL (1992): DNA sequences similar to those of simian virus 40 in ependymomas and choroid plexus tumors of childhood. N Engl J Med 326:988–993.

Black PH, Rowe WP, Turner HC, Huebner RJ (1963): A specific complement-fixing antigen present in SV40 tumor and transformed cells. Proc Natl Acad Sci USA 50: 1148–1156.

Bollag B, Frisque RJ (1992): PAb 2000 specifically recognizes the large T and small t proteins of JC virus. Virus Res 25:223–239.

Borrow P, Oldstone MBA (1997): Lymphocytic choriomeningitis virus. In Viral Pathogenesis; Nathanson N, Rafi A, Eds. Lippincott-Raven: Philadelpha, pp 593–627.

Brinster RL, Chen HY, Messing A, van Dyke T, Levine AJ, Palmiter RD (1984): Transgenic mice harboring SV40 T-antigen genes develop characteristic brain tumors. Cell 37:367–379.

Butel JS, Jafar S, Wong C, Arrington AS, Opekun AR, Finegold MJ, Adam E (1999): Evidence of SV40 infections in hospitalized children. Hum Pathol 30:1496–1502.

Butel JS, Tevethia SS, Melnick JL (1972): Oncogenicity and cell transformation by papovavirus SV40: The role of the viral genome. Adv Cancer Res 15:1–55.

Campbell AE, Foley FL, Tevethia SS (1983): Demonstration of multiple antigenic sites of the SV40 transplantation rejection antigen by using cytotoxic T lymphocyte clones. J Immunol 130:490–492.

Carbone M, Pass HI, Rizzo P, Marinetti M, Di Muzio M, Mew DJ, Levine AS, Procopio A (1994): Simian virus 40-like DNA sequences in human pleural mesothelioma. Oncogene 9:1781–1790.

Chang C, Martin RG, Livingston DM, Luborsky SW, Hu CP, Mora PT (1979): Relationship between T-antigen and tumor-specific transplantation antigen in simian virus 40–transformed cells. J Virol 29:69–75.

Chen W, Anton LC, Bennink JR, Yewdell JW (2000): Dissecting the multifactorial causes of immunodominance in class I-restricted T cell responses to viruses. Immunity 12:83–93.

Choi KH, Tevethia SS, Shin S (1983): Tumor formation by SV40-transformed human cells in nude mice: The role of SV40 T antigens. Cytogenet Cell Genet 36:633–640.

Dalrymple SA, Beemon KL (1990): BK virus T antigens induce kidney carcinomas and thymoproliferative disorders in transgenic mice. J Virol 64:1182–1191.

Deckhut AM, Lippolis JD, Tevethia SS (1992): Comparative analysis of core amino acid residues of H-2D(b)-restricted cytotoxic T-lymphocyte recognition epitopes in simian virus 40 T antigen. J Virol 66:440–447.

Deckhut AM, Tevethia MJ, Haggerty S, Frisque RJ, Tevethia SS (1991): Localization of common cytotoxic T lymphocyte recognition epitopes on simian papovavirus SV40 and human papovavirus JC virus T antigens. Virology 183:122–132.

Degener AM, Pietropaolo V, Di Taranto C, Rizzuti V, Ameglio F, Cordiali Fei P, Caprilli F, Capitanio B, Sinibaldi L, Orsi N (1997): Detection of JC and BK viral genome in specimens of HIV-1 infected subjects. New Microbiol 20:115–122.

Diamandopoulos GT (1973): Induction of lymphocytic leukemia, lymphosarcoma, reticulum cell sarcoma, and osteogenic sarcoma in the Syrian golden hamster by oncogenic DNA simian virus 40. J Natl Cancer Inst 50:1347–1365.

Diamandopoulos GT, McLane MF (1975): Effect of host age, virus dose, and route of inoculation on tumor incidence, latency, and morphology in Syrian hamsters inoculated intravenously with oncogenic DNA simian virus 40. J Natl Cancer Inst 55: 479–482.

Drummond JE, Shah KV, Donnenberg AD (1985): Cell-mediated immune responses to BK virus in normal individuals. J Med Virol 17:237–247.

Eddy BE, Borman GS, Berkley WH (1961): Tumors induced in hamsters by injection of rhesus monkey kidney cell extracts. Proc Soc Exp Biol Med 107:191–197.

Eddy BE, Borman GS, Grubbs GE, Young RD (1962): Identification of the oncogenic substance in rhesus monkey cell cultures as simian virus 40. Virology 17:65–75.

Faas SJ, Pan S, Pinkert CA, Brinster RL, Knowles BB (1987): Simian virus 40 (SV40)–transgenic mice that develop tumors are specifically tolerant to SV40 T antigen. J Exp Med 165:417–427.

Falk K, Rotzschke O, Stevanovic S, Jung G, Rammensee H-G (1991): Allele-specific motifs revealed by sequencing of self-peptides eluted from MHC molecules. Nature (Lond) 351:290–296.

Flyer DC, Pretell J, Campbell AE, Liao WS, Tevethia MJ, Taylor JM, Tevethia SS (1983): Biology of simian virus 40 (SV40) transplantation antigen (TrAg). X. Tumorigenic potential of mouse cells transformed by SV40 in high responder C57BL/6 mice and correlation with the persistence of SV40 TrAg, early proteins and sequences. Virology 131:207–220.

Flyer DC, Tevethia SS (1982): Biology of simian virus 40 (SV40) transplantation antigen (TrAg). VIII. Retention of SV40 TrAg sites on purified SV40 large T antigen following denaturation with sodium dodecyl sulfate. Virology 117:267–270.

Forster I, Hirose R, Arbeit JM, Clausen BE, Hanahan D (1995): Limited capacity for tolerization of $CD4^+$ T cells specific for a pancreatic beta cell neo-antigen. Immunity 2:573–585.

Franks RR, Rencic A, Gordon J, Zoltick PW, Curtis M, Knobler RL, Khalili K (1996): Formation of undifferentiated mesenteric tumors in transgenic mice expressing human neurotropic polymavirus early protein. Oncogene 12:2573–2578.

Fu TM, Mylin LM, Schell TD, Bacik I, Russ G, Yewdell JW, Bennink JR, Tevethia SS (1998): An endoplasmic reticulum–targeting signal sequence enhances the immunogenicity of an immunorecessive simian virus 40 large T antigen cytotoxic T-lymphocyte epitope. J Virol 72:1469–1481.

Ganss R, Hanahan D (1998): Tumor microenvironment can restrict the effectiveness of activated antitumor lymphocytes. Cancer Res 58:4673–4681.

Gardner SD, Field AM, Coleman DV, Hulme B (1971): New human papovavirus (B.K.) isolated from urine after renal transplantation. Lancet 1:1253–1257.

Geiger T, Gooding LR, Flavell RA (1992): T-cell responsiveness to an oncogenic peripheral protein and spontaneous autoimmunity in transgenic mice. Proc Natl Acad Sci USA 89:2985–2989.

Geiger T, Soldevila G, Flavell RA (1993): T cells are responsive to the simian virus 40 large tumor antigen transgenically expressed in pancreatic islets. J Immunol 151: 7030–7036.

Gerber P (1967): Patterns of antibodies to SV40 in children following the last booster with inactivated poliomyelitis vaccines. Proc Soc Exp Biol Med 125:1284–1287.

Germain RN, Margulies DH (1993): The biochemistry and cell biology of antigen processing and presentation. Annu Rev Immunol 11:403–450.

Girardi AJ, Sweet BH, Slotnick VB, Hilleman MR (1962): Development of tumors in hamsters inoculated in the neonatal period with vacuolating virus, SV40. Proc Soc Exp Biol Med 109:649–660.

Granziero L, Krajewski S, Farness P, Yuan L, Courtney MK, Jackson MR, Peterson PA, Vitiello A (1999): Adoptive immunotherapy prevents prostate cancer in a transgenic animal model. Eur J Immunol 29:1127–1138.

Greenfield RS, Flyer DC, Tevethia SS (1980): Demonstration of unique and common antigenic sites located on the SV40 large T and small t antigens. Virology 104:312–322.

Hanahan D (1985): Heritable formation of pancreatic β-cell tumours in transgenic mice expressing recombinant insulin/simian virus 40 oncogenes. Nature 315:115–122.

Heath WR, Carbone FR (1999): Cytotoxic T lymphocyte activation by cross-priming. Curr Opin Immunol 11:314–318.

Heath WR, Kurts C, Miller JF, Carbone FR (1998): Cross-tolerance: A pathway for inducing tolerance to peripheral tissue antigens. J Exp Med 187:1549–1553.

Knowles BB, Koncar M, Pfizenmayer K, Solter D, Aden DP, Trinchieri G (1979): Genetic control of the cytotoxic T cell response to SV40 tumor-associated specific antigen. J Immunol 122:1798–1806.

Knowles BB, McCarrick J, Fox N, Solter D, Damjanov I (1990): Osteosarcomas in transgenic mice expressing an α-amylase–SV40 T-antigen hybrid gene. Am J Pathol 137:259–262.

Koralnik IJ, Du Pasquier RA, Letvin NL (2001): JC virus-specific cytotoxic T lymphocytes in individuals with progressive multifocal leukoencephalopathy. J Virol 75: 3483–3487.

Krynska B, Otte J, Franks R, Khalili K, Croul S (1999): Human ubiquitous JCV(CY) T-antigen gene induces brain tumors in experimental animals. Oncogene 18:39–46.

Le AX, Bernhard EJ, Holterman MJ, Strub S, Parham P, Lacy E, Engelhard VH (1989): Cytotoxic T cell responses in HLA-A2.1 transgenic mice. Recognition of HLA alloantigens and utilization of HLA-A2.1 as a restriction element. J Immunol 142: 1366–1371.

Lednicky JA, Arrington AS, Stewart AR, Dai XM, Wong C, Jafar S, Murphey-Corb M, Butel JS (1998): Natural isolates of simian virus 40 from immunocompromised monkeys display extensive genetic heterogeneity: new implications for polyomavirus disease. J Virol 72:3980–3990.

Lewis AM, Rowe WP (1973): Studies of nondefective adenovirus 2–simian virus 40 hybrid viruses. 8. Association of simian virus 40 transplantation antigen with a specific region of the early viral genome. J Virol 12:836–840.

Madden DR (1995): The three-dimensional structure of peptide–MHC complexes. Annu Rev Immunol 13:587–622.

Marton I, Johnson SE, Damjanov I, Currier KS, Sundberg JP, Knowles BB (2000): Expression and immune recognition of SV40 Tag in transgenic mice that develop metastatic osteosarcomas. Transgenic Res 9:115–125.

Melief CJ, Toes RE, Medema JP, van der Burg SH, Ossendorp F, Offringa R (2000): Strategies for immunotherapy of cancer. Adv Immunol 75:235–282.

Melnick JL, Stinebaugh S (1962): Excretion of vacuolating SV40 virus (papovavirus group) after ingestion as a contaminant of oral poliovaccine. Proc Soc Exp Biol Med 109:965–968.

Mole SE, Gannon JV, Ford MJ, Lane DP (1987): Structure and function of SV40 large-T antigen. Philos Trans R Soc Lond B Biol Sci 317:455–469.

Mylin LM, Bonneau RH, Lippolis JD, Tevethia SS (1995a): Hierarchy among multiple H-2b–restricted cytotoxic T-lymphocyte epitopes within simian virus 40 T antigen. J Virol 69:6665–6677.

Mylin LM, Deckhut AM, Bonneau RH, Kierstead TD, Tevethia MJ, Simmons DT, Tevethia SS (1995b): Cytotoxic T lymphocyte escape variants, induced mutations, and synthetic peptides define a dominant H-2K^b-restricted determinant in simian virus 40 tumor antigen. Virology 208:159–172.

Mylin LM, Schell TD, Roberts D, Epler M, Boesteanu A, Collins EJ, Frelinger JA, Joyce S, Tevethia SS (2000): Quantitation of CD8(+) T-lymphocyte responses to multiple epitopes from simian virus 40 (SV40) large T antigen in C57BL/6 mice immunized with SV40, SV40 T-antigen–transformed cells, or vaccinia virus recombinants expressing full-length T antigen or epitope minigenes. J Virol 74:6922–6934.

Nandi D, Marusina K, Monaco JJ (1998): How do endogenous proteins become peptides and reach the endoplasmic reticulum. Curr Top Microbiol Immunol 232:15–47.

Narayan O, Penney JB, Johnson RT, Herndon RM, Weiner LP (1973): Etiology of progressive multifocal leukoencephalopathy. Identification of papovavirus. N Engl J Med 289:1278–1282.

Newman JS, Baskin GB, Frisque RJ (1998): Identification of SV40 in brain, kidney and urine of healthy and SIV-infected rhesus monkeys. J Neurovirol 4:394–406.

Ornitz DM, Palmiter RD, Messing A, Hammer RE, Pinkert CA, Brinster RL (1985): Elastase I promoter directs expression of human growth hormone and SV40 T antigen genes to pancreatic acinar cells in transgenic mice. Cold Spring Harbor Symp Quant Biol 50:399–409.

Padgett BL, Walker DL, Zu Rhein GM, Eckroade RJ, Dessel BH (1971): Cultivation of papova-like virus from human brain with progressive multifocal leukoencephalop athy. Lancet 1:1257–1260.

Pamer E, Cresswell P (1998): Mechanisms of MHC class I–restricted antigen processing. Annu Rev Immunol 16:323–358.

Pieters J (2000): MHC class II-restricted antigen processing and presentation. Adv Immunol 75:159–208.

Pipas JM (1992): Common and unique features of T antigens encoded by the polyomavirus group. J Virol 66:3979–3985.

Pope JH, Rowe WP (1964): Detection of a specific antigen in SV40 transformed cells by immunofluorescence. J Exp Med 120:121–128.

Pretell J, Greenfield RS, Tevethia SS (1979): Biology of simian virus 40 (SV40) transplantation antigen (TrAg). V In vitro demonstration of SV40 TrAg in SV40 infected nonpermissive mouse cells by the lymphocyte mediated cytotoxicity assay. Virology 97:32–41.

Rammensee HG (1995): Chemistry of peptides associated with MHC class I and class II molecules. Curr Opin Immunol 7:85–96.

Rammensee H-G, Friede T, Stevanovic S (1995): MHC ligands and peptide motifs: First listing. Immunogenetics 41:178–228.

Randhawa PS, Demetris AJ (2000): Nephropathy due to polyomavirus type BK. N Engl J Med 342:1361–1363.

Ransom JH, Thompson DL, Tevethia SS (1982): Kinetics of the immune response of tumor-bearing hamsters to two simian virus 40 coded non-structural polypeptides present in simian virus 40 tumor cells. Int J Cancer 29:217–222.

Rapp F, Butel JS, Melnick JL (1964): Virus induced intranuclear antigen in cells transformed by papovavirus. Proc Soc Exp Biol Med 116:1131–1135.

Rapp F, Tevethia SS, Melnick JL (1966): Papovavirus SV40 transplantation immunity conferred by an adenovirus–SV40 hybrid. J Natl Cancer Inst 36:703–708.

Rapp F, Tevethia SS, Rawls WE, Melnick JL (1967): Production of antibodies to papovavirus SV40 Tumor antigen in African green monkeys. Proc Soc Exp Biol Med 125:794–798.

Romieu R, Baratin M, Kayibanda M, Lacabanne V, Ziol M, Guillet J-G, Viguier M (1998): Passive but not active $CD8^+$ T cell-based immunotherapy interferes with liver tumor progression in a transgenic mouse model. J Immunol 161:5133–5137.

Ruedl C, Kopf M, Bachmann MF (1999): CD8(+) T cells mediate CD40-independent maturation of dendritic cells in vivo. J Exp Med 189:1875–1884.

Schell TD, Knowles BB, Tevethia SS (2000): Sequential loss of cytotoxic T lymphocyte responses to simian virus 40 large T antigen epitopes in T antigen transgenic mice developing osteosarcomas. Cancer Res 60:3002–3012.

Schell TD, Lippolis JD, Tevethia SS (2001): Cytotoxic T lymphocytes from HLA-A2.1 transgenic mice define a potential human epitope from simian virus 40 large T antigen. Cancer Res 61:873–879.

Schell TD, Mylin LM, Georgoff I, Teresky AK, Levine AJ, Tevethia SS (1999): Cytotoxic T-lymphocyte epitope immunodominance in the control of choroid plexus tumors in simian virus 40 large T antigen transgenic mice. J Virol 73:5981–5993.

Shah K, Nathanson N (1976): Human exposure to SV40: Review and comment. Am J Epidemiol 103:1–12.

Shah KV, Willard S, Myers RE, Hess DM, DiGiacomo R (1968): Experimental infection of rhesus with simian virus 40 (SV40). Proc Soc Exp Biol Med 130:196–203.

Small JA, Khoury G, Jay G, Howley PM, Scangos GA (1986): Early regions of JC virus and Bk virus induce distinct and tissue specific tumors in transgenic mice. Proc Natl Acad Sci USA 83:8288–8292.

Soldevila G, Geiger T, Flavell RA (1995): Breaking immunologic ignorance to an antigenic peptide of simian virus 40 large T antigen. J Immunol 155:5590–5600.

Stern LJ, Wiley DC (1994): Antigenic peptide binding by class I and class II histocompatibility proteins. Structure 2:245–251.

Sweet BH, Hilleman MR (1960): The vacuolating virus, SV-40. Proc Soc Exp Biol Med 105:420–427.

Tanaka Y, Anderson RW, Maloy WL, Tevethia SS (1989): Localization of an immunorecessive epitope on SV40 T antigen by H-2Db–restricted cytotoxic T-lymphocyte clones and a synthetic peptide. Virology 171:205–213.

Tanaka Y, Tevethia MJ, Kalderon D, Smith AE, Tevethia SS (1988): Clustering of antigenic sites recognized by cytotoxic T lymphocyte clones in the amino terminal half of SV40 T antigen. Virology 162:427–436.

Tevethia MJ, Tevethia SS (1976): Biology of SV40 transplantation antigen (TrAg). I. Demonstration of SV40 TrAg on glutaraldehyde-fixed SV40-infected African green monkey kidney cells. Virology 69:474–489.

Tevethia SS (1970): Immune response of rabbits to purified papovavirus SV40. J Immunol 104:72–78.

Tevethia SS (1980): Immunology of simian virus 40; In Klein G, Ed., Viral Oncology; Raven Press: New York, pp 581–601.

Tevethia SS (1990): Recognition of simian virus 40 T antigen by cytotoxic T lymphocytes. Mol Biol Med 7:83–96.

Tevethia S, Beachy T, Schell T, Lippolis J, Newmaster R, Mylin L, Tevethia M (2001): Role of CTL host responses and their implication for tumorigenicity testing and the use of tumor cells as vaccine substrate. Dev Biol Stand (in press).

Tevethia SS, Blasecki JW, Vaneck G, Goldstein AL (1974): Requirement of thymus-derived theta-positive lymphocytes for rejection of DNA virus (SV 40) tumors in mice. J Immunol 113:1417–1423.

Tevethia S, Dreesman GR, Lausch RN, Rapp F (1968): Effect of anti-hamster thymocyte serum on papovavirus SV40-induced transplantation immunity. J Immunol 101: 1105–1110.

Tevethia SS, Epler M, Georgoff I, Teresky A, Marlow M, Levine AJ (1992): Antibody response to human papovavirus JC (JCV) and simian virus 40 (SV40) T antigens in SV40 T antigen-transgenic mice. Virology 190:459–464.

Tevethia SS, Flyer DC, Tjian R (1980): Biology of simian virus 40 (SV40) transplantation antigen (TrAg). VI. Mechanism of induction of SV40 transplantation immunity in mice by purified SV40 T antigen (D2 protein). Virology 107:13–23.

Tevethia SS, Lewis M, Tanaka Y, Milici J, Knowles B, Maloy WL, Anderson R (1990): Dissection of H-2Db–restricted cytotoxic T-lymphocyte epitopes on simian virus 40 T antigen by the use of synthetic peptides and H-2Dbm mutants. J Virol 64:1192–1200.

Tevethia SS, MacMillan VL (1974): Acquisition of malignant properties by SV40-transformed mouse cells: Relationship to type-C viral antigen expression. Intervirology 3:269–276.

Tevethia SS, Mylin L, Newmaster R, Epler M, Lednicky JA, Butel JS, Tevethia MJ (1998): Cytotoxic T lymphocyte recognition sequences as markers for distinguishing among tumour antigens encoded by SV40, BKV and JCV. Dev Biol Stand 94:329–339.

Theobald M, Biggs J, Dittmer D, Levine AJ, Sherman LA (1995): Targeting p53 as a general tumor antigen. Proc Natl Acad Sci USA 92:11993–11997.

Thompson DL, Kalderon D, Smith AE, Tevethia MJ (1990): Dissociation of Rb-binding and anchorage-independent growth from immortalization and tumorigenicity using SV40 mutants producing N-terminally truncated large T antigens. Virology 178:15–34.

Tooze J, Ed (1980): DNA Tumor Viruses. Molecular Biology of Tumor Viruses, Cold Spring Harbor Laboratory: Cold Spring Harbor.

Trinchieri G, Aden DP, Knowles BB (1976): Cell-mediated cytotoxicity to SV40-specific tumour-associated antigens. Nature (Lond) 261:312–314.

Van Dyke TA, Finlay C, Miller D, Marks J, Lozano G, Levine AJ (1987): Relationship between simian virus 40 large tumor antigen expression and tumor formation in transgenic mice. J Virol 61:2029–2032.

Whitmire JK, Flavell RA, Grewal IS, Larsen CP, Pearson TC, Ahmed R (1999): CD40–CD40 ligand costimulation is required for generating antiviral CD4 T cell responses but is dispensable for CD8 T cell responses. J Immunol 163:3194–3201.

Ye X, McCarrick J, Jewett L, Knowles BB (1994): Timely immunization subverts the development of peripheral nonresponsiveness and suppresses tumor development in simian virus 40 tumor antigen-transgenic mice. Proc Natl Acad Sci USA 91:3916–3920.

Yewdell JW, Bennink JR (1992): Cell biology of antigen processing and presentation to major histocompatability complex molecule–restricted T lymphocytes. Adv Immunol 52:1–123.

Yewdell JW, Bennink JR (1999): Immunodominance in major histocompatibility complex class I–restricted T lymphocyte responses. Annu Rev Immunol 17:51–88.

Yewdell JW, Bennink JR (2001): Cut and trim: generating MHC class I peptide ligands. Curr Opin Immunol 13:13–18.

Zarling JM, Tevethia SS (1973): Transplantation immunity to simian virus 40–transformed cells in tumor-bearing mice. I. Development of cellular immunity to simian virus 40 tumor–specific transplantation antigens during tumorigenesis by transplanted cells. J Natl Cancer Inst 50:137–147.

Zinkernagel RM, Doherty PC (1979): MHC-restricted cytotoxic T cells: Studies on the biological role of polymorphic major transplantation antigens determining T-cell restriction-specificity, function, and responsiveness. Adv Immunol 27:51–177.

22

THE HUMAN POLYOMAVIRUSES: PAST, PRESENT, AND FUTURE

GERALD L. STONER, PH.D., and ROLAND HÜBNER, PH.D.

1. THE PAST: ORIGIN AND EVOLUTIONARY HISTORY OF THE POLYOMAVIRUSES

Introduction

To truly understand the present and future course of viruses in the human population, we must put them into an evolutionary context (Ewald, 1994). Therefore, to conclude this survey of the human polyomaviruses we examine the evolutionary past of these viruses and assess our scientific progress since the discovery of SV40 in rhesus monkey kidney cells in 1960 and the discovery in 1971 of the two human polyomaviruses BKV and JCV in urine and brain, respectively. After taking a sweeping look at where polyomavirus investigations have been and where they are going, we suggest some directions for future research. Although wide-ranging, this will not be a comprehensive discussion of the large and growing field of the molecular biology, genetics, and pathogenesis of the human polyomaviruses. The details have been provided in the preceding chapters. In this summation we take the editorial liberty to be selec-

Human Polyomaviruses: Molecular and Clinical Perspectives, Edited by Kamel Khalili and Gerald L. Stoner.
ISBN 0-471-39009-7

tive in our coverage. Many studies that are as important as those we cite must go unmentioned due to limitations of space.

We do, however, take an extensive critical look at the major controversy currently embroiling this field, that is, the origins and oncogenicity of SV40 in humans. The polyomaviruses, a genus named for the prototype species, mouse polyoma virus, have long been placed among the DNA tumor viruses, as their very name implies (Friedlander and Patarca, 1999; Tooze, 1981; Villarreal, 1989). The question of their actual role in human oncogenesis, a major concern when SV40 was first discovered, was overshadowed for many years by the discovery of retroviruses, but is front and center once again (Butel and Lednicky, 1999; Carbone et al., 1995). SV40 has been a major laboratory molecular biologic workhorse since the beginnings of recombinant DNA research nearly 30 years ago (Levine, 1994). During this time it has been viewed as an important experimental tool to elucidate eukaryotic molecular biology and to probe mechanisms of oncogenesis (Simmons, 2000).

The mouse polyoma virus has also been an important model for studies of mechanisms of cellular transformation and viral oncogenesis, but it has not generally been included in the scope of this book (but see Chapters 5 and 9). Some aspects of polyoma virus tumorigenesis in the mouse may be unique to the murine system. Susceptibility to tumor induction by polyoma virus varies with the inbred mouse strain, and a variety of mechanisms may be involved (Freund et al., 1992). One of the dominant susceptibility genes in C3H mice, named PyvS, does not encode cell receptors or alter productive infection, viral dissemination, or intracellular events essential for cell transformation, but rather encodes an endogenous mouse mammary tumor virus (MMTV)–associated transmembrane glycoprotein with superantigen activity (Lukacher et al., 1995).

Each polyomavirus will contribute its own unique insights. For example, the African green monkey B-lymphotropic papovavirus (LPV) is a primate polyomavirus for which cellular specificity of productive virus infection is clearly dependent on the nature of the cell surface receptor. Its infectivity for human cells is strictly limited to a few Burkitt's lymphoma cell lines (see Chapter 9). The mechanism regulating this susceptibility is thought to be a highly specific cellular receptor (Haun et al., 1993).

As presented in the preceding chapters, the polyomaviruses are all characterized by circular genomes of ~5000 bp transcribed bidirectionally from overlapping transcriptional controls in a single regulatory region. They have a virally encoded, alternatively spliced, multifunctional regulatory protein, T antigen, that serves as an initiator of viral DNA synthesis as well as an upregulator of late region transcription and a repressor of its own expression. In addition, T antigen modulates cellular transcriptional activity (see Chapter 6). The genomic DNAs of polyomaviruses show varying levels of similarity, but this pattern of transcription and the functions of the encoded proteins mark them as members of the same virus genus.

Viruses such as the polyomaviruses that have a very small coding capacity are of necessity highly dependent on the enzymatic functions of the host cell and are closely tied to the evolution of their hosts (Shadan and Villarreal, 1995; Villarreal, 1999). Therefore, it is not surprising that emphasis has been placed on the probable coevolution of polyomaviruses and their hosts ever since the homology of sequence and genomic organization among these viruses first became clear (Soeda et al., 1980).

Like mouse polyomavirus, the bovine polyomavirus is potentially tumorigenic in rodents (Schuurman et al., 1992). Possible contamination of batches of calf serum used in cell culture with this bovine polyomavirus is a cause for concern in experimental studies (Kappeler et al., 1996).

The first avian polyomavirus identified, budgerigar fledgling disease virus (BFDV), causes acute hepatic and splenic necrosis in its natural avian hosts (Phalen et al., 1999). There are no reports that avian polyomaviruses are oncogenic in rodent species.

Nine of the polyomavirus complete genome sequences currently available are listed in Table 22.1. While all are comparable in their genetic organization, many of these viruses are highly divergent in sequence. However, scattered conserved regions are similar enough that degenerate primers can be designed to detect multiple viruses and, possibly, to amplify as yet undiscovered species (Völter et al., 1998). However, it is not at all clear that the picture is as simple as having species-specific polyomavirus sequences whose level of divergence reflects the evolutionary distance of their hosts. Table 22.2 shows the percent similarity between these DNA sequences calculated by the GAP program of GCG.

There is no simple correlation between the DNA sequences of these viruses and the phylogenetic relationships of their host. While one of the primate polyomaviruses—SV40, whose natural hosts are species of the genus *Macaca*—is closely related to the human viruses, the other—African green monkey B-lymphotropic papovavirus (LPV)—is not. Species of *Macaca* and *Chlorocebus* (vervet, grivet, and African green monkeys) may last have shared a common ancestor with humans (catarrhine primates) about 30 million years ago, while species of *Macaca* diverged from other Old World monkeys about 8 million years ago, with spread to Asia around 3 million years ago (Jones et al., 1992). It seems highly unlikely that JCV and BKV diverged before the departure of *Homo erectus* to Asia about 1.7 million years ago. How then could SV40 be so similar to JCV and BKV if it split with their last common ancestor 30 million years ago? Furthermore, the macaque is no more closely related to humans than is the African green monkey, but SV40 is much closer phylogenetically to JCV and BKV than is LPV. It is notable also that the DNA sequence of the hamster polyomavirus is no more closely related to the polyoma virus of the mouse than it is to primate viruses such as SV40 (Table 22.2). Yet their affinity is indicated by the fact that these two rodent polyomaviruses are the only ones in which alternative splicing of the early region yields a middle-sized T antigen (Cole, 1996).

There are two implications of these observations. First, although the evolutionary relatedness of the polyomaviruses is clear from their genomic orga-

Table 22.1. Polyomaviruses for Which Complete Genome Sequences Are Available

Virus	Natural Host Species	Year Originally Isolated	Disease in Natural Host	Genome Length (bp)	References	GenBank Accession No.
JC virus (JCV, Mad-1)	Human	1971	PML in immunosuppressed patients	5130	Frisque et al. (1984), Padgett et al. (1971)	J02226
BK virus (BKV)	Human	1971	BKV nephritis, hemorrhagic cystitis	5153	Gardner et al. (1971), Seif et al. (1979)	J02038
BK virus (AS)	Human	1980	BKV nephritis, hemorrhagic cystitis	5098	Coleman et al. (1980), Tavis et al. (1989)	M23122
Simian virus 40 (SV40)	Asian macaque	1960	PML in SAIDS with interstitial pneumonia, renal tubular necrosis	5243	Fiers et al. (1978), Sweet and Hilleman (1960)	J02400
Monkey B-lymphotropic papovavirus (LPV)	African green monkey	1979	Unknown	5270	Pawlita et al. (1985)	K02562
Mouse polyomavirus (PyV)[a]	Mouse	1953	Variety of tumors from epithelial and mesenchymal cell lineages	5297	Griffin et al. (1981)	J02288
Hamster polyomavirus (HaPyV)	Hamster	1967	Epithelioma, lymphoma	5366	Bastien and Feunteun (1988), Delmas et al. (1985), Prokoph et al. (1996)	X02449
Bovine polyomavirus (BovPyV)	Cattle	1974	Unknown	4697	Parry et al. (1983); Schuurman et al. (1990)	D13942
Avian polyomavirus (BFDV)[b]	Psittacine birds	1981	Disseminated infection with liver and splenic necrosis, as well as encephalopathy	4981	Bozeman (1981), Latimer et al. (1996), Rott et al. (1988)	AF118150

[a]Another mouse polyoma virus has been sequenced known as K (Kilham) virus (GenBank Acc. No. M57473) (Mayer and Dörries, 1991).
[b]First isolated from the budgerigar and named budgerigar fledgling disease virus (BFDV).

Table 22.2. Similarity Among Nine Polyomavirus Genomic DNA Sequences, Exclusive of the Rearranging Regulatory Region, Including JCV (4854 bp), BKV (4823 bp), and SV40 (4909 bp), Using the GAP Program (Genetics Computer Group, Oxford Molecular)[a]

	% Similarity								
Virus	JCV	BKV	BKV(AS)	SV40	LPV	PyV	HaPyV	BovPyV	BFDV (Avian)
JCV	—	77.3	78.2	72.5	55.3	55.0	53.7	54.4	51.4
BKV		—	95.1	73.7	56.1	55.9	54.2	54.8	51.0
BKV(AS)			—	74.2	55.7	55.9	54.5	55.2	50.8
SV40				—	56.3	55.5	53.8	55.8	51.2
LPV					—	54.5	55.2	54.0	49.4
PyV						—	53.8	57.4	51.4
HaPyV								54.0	51.0
BovPyV								—	50.8

[a]Gap creation penalty set at 30 with a penalty of 2 for gap extension.

nization and the function of their protein products, separate evolution over many millions of years can efface the DNA sequence relationships of these viruses so that they do not reflect the true evolutionary relatedness of their hosts. The extent of divergence is so great and so many nucleotide substitutions have accumulated that most of the sites capable of changing have already changed before, that is, the sequence has become saturated (Page and Holmes, 1998). The protein sequences, however, will be more highly conserved than the DNA sequences. Most highly conserved will be the three-dimensional structure and function of the protein products. Second, there have likely been cross-species transfers during this long evolutionary history (see below).

The three closely related primate polyomaviruses are considered here in an order (JCV, BKV, and SV40) that is the reverse of their reported discovery, but perhaps follows their order of appearance as persistent viruses of the human host species *Homo sapiens sapiens*.

The Origin and Evolutionary History of JCV

JCV was discovered as the cause of progressive multifocal leukoencephalopathy (PML) in 1971 (see Chapter 3). PML itself had been described in 1958 (Åström et al., 1958) (see Chapter 1) and was identified as a viral disease by electron microscopy in 1964 (Zu Rhein and Chou, 1965) (see Chapter 2). Dr. Gabriele M. Zu Rhein presented her controversial findings to the Association for Research in Nervous and Mental Diseases in December 1964 (Zu Rhein and Chou, 1968). "With the risk of heresy" (Zu Rhein, 1969), she used autopsy tissue that had been kept in 10% formalin for 2 years and took some criticism for her elegant results and far-reaching conclusions. The published transcript of this discussion attests that, despite the negative remarks by none other than Dr. Albert B. Sabin, who did not accept ultrastructural results from the new electron microscopy science alone as adequate evidence for a virus, she did not recant (Zu Rhein and Chou, 1968). Dr. Zu Rhein's fundamental proposal was that "in PML the oligodendroglial cells succumb to the cytocidal effect of a pathogenic virus replicating massively within their nuclei and that demyelination results chiefly from the abolition of a myelin maintaining morphological and functional oligoglial–myelin relationship" (Zu Rhein and Chou, 1968). This was at a time when the oligodendroglial–myelin relationship was itself considered to be a hypothesis rather than a fact. Needless to say, her findings in PML have been amply confirmed. The same discovery was independently made in 1964 (Silverman and Rubinstein, 1965), and within a few years similar virus particles had been seen in 27 cases of this rare disorder (Zu Rhein, 1969).

JCV is the first polyomavirus for which evidence has been presented of worldwide distribution, suggesting coevolution with modern humans (Agostini et al., 1997b; Stoner et al., 2000; Sugimoto et al., 1997). The origins and evolutionary history of JCV have been dealt with in detail in Chapter 18. There it is suggested that JCV has evolved into a limited number of well-defined genotypes in various populations of the world (Agostini et al., 1997b; Sugimoto et al., 1997). This may have begun within a small founding population of *Homo sapiens sapiens*, numbering perhaps as few as 10,000 individuals (Zietkiewicz et al., 1998).

It is possible that all individuals in this small group were infected by a single viral prototype strain that then evolved within small departing bands by drift and selection into distinct genotypes that now differ from each other by no more than 1–3%. Unlike true cellular genetic markers that are lineally inherited from one parent, such as mitochondrial DNA (mtDNA) or the nonrecombining portion of the Y chromosome (nrY), or biparentally inherited markers in nuclear genes, JCV is an infectious agent and as such is likely to have some ability to spread horizontally among a mating population.

Alternatively, the initial outmigrations to Asia and Europe may have involved small population groups in which a founder effect fixed genotypes already present in these initial human clans with little subsequent change in the dispersing populations. This possibility would predict a larger number of genotypes remaining behind in Africa than in the presumably smaller outmigrating populations. In fact, the diversity among JCV strains in Asia, Australia, Papua New Guinea, the South Pacific, and pre-Columbian America (Types 2A–E, Types 7A–C, and Types 8A and 8B) is much greater than that presently evident in Africa. In contrast, although description of JCV genotypes in Africa is currently limited, there seems to be one major paleo-African JCV genotype in West and Central Africa, that is, Type 6 (Chima et al., 1998; Guo et al., 1996). The other major African type, Type 3 in East and West Africa, is more closely related to the diverse Asian strains than it is to the other African genotype, Type 6 (see Chapter 18).

Whatever the exact origins of the JCV variants, the existence of closely related, geographically based genotypes, with no distinct serotypes, indicates a very stable DNA sequence and virus structure over a long period of evolutionary time. JCV is a virus utilizing cellular repair mechanisms in DNA synthesis that is well adapted to its evolutionary niche with apparently little tolerance for mutation. Its inherent genetic stability can also be attributed to its ability to establish and maintain a benign persistent state in the host (Shadan and Villarreal, 1995; Villarreal et al., 2000). It has a low mutation rate of perhaps $1–4 \times 10^{-7}$ per site per year (Hatwell and Sharp, 2000; Sugimoto et al., 1997). The advantage of such a low mutation rate for phylogenetic studies is that the number of mutations is a truer reflection of the evolutionary distance, as there are fewer multiple substitutions at a single site that can obscure the actual evolutionary history of the sequences being compared. Thus, there is an excellent correspondence between fundamentally different methods of phylogenetic reconstruction. Whether a distance method measuring overall sequence similarity is applied (neighbor joining or minimum evolution) or a discrete method (maximum parsimony) that reflects nucleotide mutations at each individual site, the results are remarkably similar. Another indication of the stability of the JCV genome is the fact that no type-specific deletions or insertions have been identified in any coding region of the DNA, except in a subtype of Type 8 strains from Papua New Guinea (designated PNG-2B) (Jobes et al., 1999). These reveal a deletion of 21 bp in the agnoprotein gene (seven surface amino acids; residues 57–63 of this 71 residue protein). This inherent stability of the genomic sequence suggests that observations of mutations or nucleotide insertion or deletion following polymerase chain reaction (PCR) amplification that predict translational frameshifts and/or truncated proteins should be further evaluated to eliminate the possibility that a PCR error has been introduced in the amplification and cloning process.

The universal distribution of JCV among humans, with ancestral Native-American populations infected with Asian strains of the virus (Stoner et al., 2000) and remote Pacific Island populations infected with a distinct Asian genotype or subtype (Ryschkewitsch et al., 2000), is consistent with the existence of this persistent infection in the human population ever since the "Out of Africa 2" dispersal began about 100,000 years ago. Unlike acute epidemic infections that necessarily die out in small populations due to the induction of a protective immune response and virus clearance, small migrating bands with a persistent kidney infection would have carried the virus with them.

The Origin and Evolutionary History of BKV

BKV, while closely related to JCV, differs in several key respects. Transmission occurs earlier, judging by age of seroconverion (see Chapter 19), but, paradoxically, the number of excreters in the healthy population may be much lower based on the experience of most investigators (Shah et al., 1997; Sundsfjord et al., 1994a). Furthermore, the immunologic control of the kidney infection is very different. BKV excretion is clearly controlled by the immune response as HIV-infected individuals excrete BKV more frequently and more heavily, unlike JCV for which excretion is largely independent of the state of immunocompetence of the host (Markowitz et al., 1993; Sundsfjord et al., 1994a).

The range of cell types in which BKV can be cultured is much broader than for JCV (Hogan et al., 1991). Thus, while these two viruses are closely related and serologically cross-reactive (although not cross-protective) and both persistently infect the kidney (presumably after a viremic phase accompanying primary infection), they differ in many details and are clearly different virus species. Interestingly, while BKV has a broader host range in vitro than does JCV, and grows much better than JCV in primary human fetal glial (PHFG) cells, BKV does not usually infect CNS glial cells in vivo. For unknown reasons, BKV, despite its broader tissue specificity, does not cause PML. The very rare cases of BKV brain infection have been a meningoencephalitis, which is pathologically quite distinct from PML (Bratt et al., 1999; Vallbracht et al., 1993). This suggests that the barrier to BKV glial cell infection is at another level involving some other aspect of the biology of the infection, for example, the trafficking of the virus in the body, or the nature of the cellular receptors, or another mechanism allowing viral entry into the brain parenchyma. Conversely, BKV, but not JCV, is a cause of the polyomavirus nephritis associated with end-stage renal disease in AIDS (Smith et al., 1998) and renal allograft nephritis (Drachenberg et al., 1999; Nickeleit et al., 2000; Randhawa et al., 1999; Rosen et al., 1983), even though detectable amounts of JCV are frequently excreted along with BKV (Smith et al., 1998). In rare cases, disseminated BKV infections have been described (Bratt et al., 1999; Vallbracht et al., 1993) as have disseminated JCV infections (Newman and Frisque, 1997, 1999). Human cases with clinical polyomavirus disease or overt pathology in multiple organs are notable for their rarity.

Unlike JCV, the variations in DNA sequence within BKV genotypes are extensive enough that they have given rise to distinct serotypes. These were in fact first recognized immunologically before methods of DNA sequencing were developed (see Chapter 19). In the case of the serotype AS, the serologic difference was so striking that this virus strain was first considered to be a distinct new polyomavirus species, the AS virus (ASV). Subsequently, DNA sequencing established that ASV shares 95% sequence similarity to BKV and can be considered to be a subtype of BKV (Tavis et al., 1989).

It is clear that the evolutionary histories of JCV and BKV are quite different, but why these two closely related but distinct human viruses have taken separate evolutionary pathways is unknown. We shall assume that JCV has coevolved with humans over the past 100,000–200,000 years and that the JCV genotypes found in various geographic regions and population groups evolved over this time frame. We shall further assume that BKV had a quite different history that has involved adaptation to a wider group of hosts as also reflected in its ability to infect a wider variety of tissues in vitro. There are several possible explanations for these data. On the one hand, the greater diversity of BKV strains could be consistent with a longer evolutionary history in humans for BKV than for JCV. On the other hand, given the well-known propensity of avian influenza virus strains to evolve rapidly when newly acquired by a human host (Webster, 1999), BKV might be evolving more rapidly than JCV because it represents a "newly" introduced virus adapting to its host. Presumably the donor species would have been another primate from which the virus might have been acquired in the not too distant past (e.g., on the order of 5000 years ago). Alternatively, the several distinct BKV genotypes might represent multiple jumps from another primate species. These possibilities will be noted in more detail below. There has as yet been no clear indication of geographic or ethnic associations for BKV genotypes or serotypes (see Chapter 19), although this possibility deserves further study.

In line with the first scenario above, it is reasonable to assume that the common ancestor of BKV and JCV left Africa with the migration of *Homo erectus* or *Homo ergaster* to Europe and Asia, a dispersal known as "Out of Africa 1" that archaeologists have dated to about 1.7 million years ago (Gabunia et al., 2000).

According to this hypothesis, JCV evolved from this ancestral virus within the immediate African ancestors of *Homo sapiens sapiens* who remained behind. Thus, JCV became the polyomavirus most highly adapted to modern humans. During the time of the separation of these evolving human populations, BKV would have evolved independently in *Homo erectus* populations in Asia and Europe and in their descendants, including, perhaps, the Neandertal lineage. This long, separate, and diverse history of BKV evolution in primitive humans outside of Africa would explain the greater diversity of this virus as it exists today in the human population. After modern humans dispersed from Africa beginning around 100,000 years ago ("Out of Africa 2"), sufficient contact, and perhaps interbreeding, ensued between the moderns and the more primitive *Homo erectus* descendants in various parts of the world that BKV may then have spread back to modern humans, having mutated to the point that immunity to JCV was no longer cross-protective. Modern humans then had two closely related species of polyomaviruses: their original JCV as well as the BKV that returned to the modern human population in several regional variants. We presume that the regional features of BKV may be less readily associated with specific modern populations because of its greater transmissibility, as well as the varied contacts between primitive and modern humans during the 30,000–40,000 years in which they may have shared habitats in parts of Asia and Europe. However, we do not rule out the possibility that some discernible patterns may persist.

If the second possibility obtains (i.e., recent cross-species transfer), the donor primate species may be one harboring a virus such as the baboon papovavirus SA12 that appears to be closely related to BKV (Cunningham and Pipas, 1985). Such a virus may have jumped across species to humans and evolved as BKV. If it was transferred only once at an earlier time, BKV serotypes may have evolved within the human population. If it jumped across species more recently and multiple times, BKV serotypes may reflect evolution that occurred largely in its primate host(s) before its introduction to humans.

These two scenarios are quite different and should be distinguishable. If the first obtains, then the explanation for a BKV-like virus in the baboon may be human-to-baboon transfer, rather than the reverse.

The Origin and Evolutionary History of SV40

The history of SV40 in macaques is even more difficult to decipher than the history of JCV and BKV in humans. Does each primate species have its own polyomavirus, many of them waiting to be discovered? A new polyomavirus has recently been described in the cynomolgus monkey (Van Gorder et al., 1999), although the complete sequence of this new virus has not yet been reported. There may be other primate polyomaviruses that are rarely pathogenic that remain to be detected.

As noted above, it seems unlikely that a virus related as closely as is SV40 to BKV and JCV would have split from the homologous human viruses when their primate ancestors diverged about 30 million years ago. Why the African green monkey B-lymphotropic papovavirus (LPV) is much more distantly related to the highly homologous group (JCV, BKV, SV40; Table 22.2) remains to be explained. Presumably, the LPV sequence represents host–virus coevolution stemming from a much earlier division in the primate lineage. In light of the proposal regarding BKV evolution above, an early transfer of the BKV/JCV progenitor from *Homo erectus* to macaques in Asia seems the most plausible alternative. According to this scenario, SV40 could then have evolved as a separate virus species for the past 1.5 million years (or whatever length of time has elapsed since the virus was transferred from *Homo erectus* to macaques in Asia). We refer to this hypothetical progenitor virus ancestral to *J*CV, *B*KV, and *S*V40 as the JBS* virus (the asterisk indicating a hypothetical ancestral virus of presently undefined DNA sequence). The sequence relationships of the JBS* virus and its progeny (JCV, BKV, and SV40) with the cynomolgus polyomavirus (Van Gorder et al., 1999), the baboon papovavirus SA12 (Cunningham and Pipas, 1985), LPV (Pawlita et al., 1985), and baboon polyomavirus 2 (Gardner et al., 1989) will help to sort out polyomavirus evolution in relation to primate evolutionary descent and cross-species transfer.

While the prehistory of SV40 in primates is still obscure, the modern history of exposure of humans to SV40 is not in dispute. Between 1954, when human polio vaccine trials began, and 1963, SV40 was inadvertently introduced into millions of humans around the world by way of vaccination with both killed and live oral polio vaccines grown in rhesus monkey kidney cells (Sweet and Hilleman, 1960; see also Chapters 17 and 20).

This became a cause of concern as soon as SV40 was found to infect humans (Sweet and Hilleman, 1960) and was demonstrated to cause tumors in experimental animals (Eddy et al., 1962). In the intervening years, the mechanisms by which malignant cellular transformation occurs in animal models and, presumably, could also occur in humans, have become increasingly clear (see Chapters 15, 16, and 17). The question has become: Is SV40 now endemic in the human population, and, if so, what was the source of that infection and in which human tumors, if any, does SV40 play an etiologic role? Certainly, human contact with monkeys did not begin with polio vaccination in the 1950s. These two primate species have interacted since the dawn of human existence in territories they share.

2. THE PRESENT: THE STATE OF RESEARCH IN THE HUMAN POLYOMAVIRUSES

Introduction

Tremendous progress has been made since the discovery of these viruses 30–40 years ago, and much of the recent work has been detailed in this book. We now know that JCV and BKV, as well as SV40, may be implicated in tumorigenesis in humans (see Chapters 15, 16, and 17). Whether these associations reflect actual causation or are a matter of "guilt by association" is a problem remaining to be adequately addressed (Blaho and Aaronson, 1999).

As noted above, interaction with cellular transcription factors is the modus operandi of the early proteins of these "bare-bones" viruses. These interactions activate the permissive cell for virus replication, and in nonpermissive cells viral early functions may alter cellular growth control and contribute to malignant transformation.

The interactions of polyomavirus T antigens with tumor suppressor p53 (Villarreal, 1989) and the pRb family of proteins (Dyson et al., 1990) are the best understood. An increasingly detailed picture of these interactions has been emerging (Simmons, 2000; Sullivan et al., 2000).

In addition to detailed studies of the molecular biologic mechanisms of oncogenesis of these viruses, there have been notable achievements in the last 10 years in structural biology. These have included the X-ray structure of the capsid protein VP1 of SV40 at 3.1 Å resolution (Stehle et al., 1996) and the structure of the *ori* DNA binding domain of SV40 T antigen by nuclear magnetic resonance imaging (NMR) (Luo et al., 1996). In the mouse polyoma virus the structure of the T antigen DnaJ-like domain has been determined by NMR (Berjanskii et al., 2000). Increasingly, molecular biologic studies and structural studies will be integrated as molecular biologic processes and regulatory interactions are understood at the structural level (see Chapter 8).

JCV: The Present

Transmission of Polyomaviruses. Is the site of primary infection the tonsil or the colon? The transmission of BKV has been thought to involve exit from the body by urinary excretion, with primary infection in the upper respiratory tract, possibly the tonsil. BKV has been isolated from cultures of tonsillar tissue during acute respiratory disease (Goudsmit et al., 1982) and from the urine of a child with acute tonsillitis (Goudsmit et al., 1981). Others have found few samples positive for BKV or JCV in nasopharyngeal aspirates or saliva of children with acute respiratory disease and propose the alimentary tract as a more likely portal of polyomavirus entry (Sundsfjord et al., 1994b).

For JCV both possibilities have been supported, with some evidence favoring the upper respiratory tract (Monaco et al., 1998) and other investigators promoting the lower gastrointestinal tract (Ricciardiello et al., 2000). Thus, while JCV is excreted from the kidney through the urinary tract, the likely site(s) of primary infection remain in doubt. Because little is known about the symptoms of primary infection by either virus, this issue will not be easily pinned down. However, given the extensive evidence for the archetypal configuration of the regulatory region as the excreted, and therefore transmissible, form of the virus (see Chapter 7), it is this stable form of the virus rather than rearranged PML types that would be expected to characterize the primary infection. It can be suggested that the important consequence of infection, whatever the route of entry, is that viremia frequently results, followed by kidney infection and the potential for latent infection of the hematopoietic system and possibly the brain. The latter may depend on timing of the primary infection and on as yet unknown co-factors. Variation in the mutable regulatory region of JCV has been minimal for the past 100,000 years. This stabilization is best explained if it represents a very narrow range of configurations required for optimal replication in *two* tissues. Either the kidney maintains a tight control on the replicating viral regulatory region, or the range of configurations thrown off by this tissue is screened by the specificity of a different tissue that is the target of the primary infection.

A lower gastrointestinal tract primary infection has been suggested based on the presence of the Mad-1 strain of JCV in both neoplastic colon tissue and adjacent normal colon tissue (Ricciardiello et al., 2001). As the original Mad-1 strain, which is characterized by a Type 1A coding region and a 98 bp tandem repeat in the regulatory region (see Chapter 18), has not been amplified by PCR from urine and has not been detected in PML brain since its isolation in 1971, the significance of this apparent colon infection is not clear.

The possibility of transplacental transmission of human polyomaviruses is raised by the evidence for viral reactivation during pregnancy (Coleman et al., 1980) and evidence from animal models (McCance and Mims, 1979) (see Chapter 19). Whether and to what extent such prenatal transmission occurs is currently unclear, but it is clear that the child is heavily exposed to the mother after birth as well, and the mother may be persistently infected and may continue to excrete one or both of these viruses. Although such early exposure seems inconsistent with seroconversion between ages 5 and 10 years for BKV, or between ages 10 and 15 years for JCV (Hogan et al., 1991), it has been speculated that polyomaviruses may have evolved a mechanism to circumvent the host acute phase reaction (Shadan and Villarreal, 1995). If so, seroconversion may reflect a host response to virus reactivation accompanying hormonal changes occurring at puberty rather than to primary infection.

Distribution of JCV in Body Tissues and Fluids. Twelve studies with data on detection of JCV in tissues of PML patients, as well as HIV-positive and HIV-negative controls are summarized in Table 22.3. These studies of JCV DNA in urine, plasma, peripheral blood mononuclear cells (PBMC), and cerebrospinal fluid (CSF) represent a variety of patient groups and a range of PCR methodologies with varying levels of sensitivity. Not surprisingly, the results vary. One matter of disagreement concerns the question of latent JCV infection in non-PML tissue (see Chapter 10), especially in lymphocytes (Dörries, 1994; Tornatore et al., 1992) and the CNS (Elsner and Dörries, 1998; White et al., 1992). One group has reported that some HIV-positive individuals without PML harbor JCV in PBMC and in the CSF, while HIV-negative controls do not (Koralnik et al., 1999). Additional studies with large numbers of both HIV-positive and HIV-negative individuals are needed.

Table 22.3. JCV Detected by PCR in PML Patients and HIV-Positive and HIV-Negative Controls

	Patient Groups											
	PML				HIV-Positive				HIV-Negative			
Study	Urine	Plasma	PBMC	CSF	Urine	Plasma	PBMC	CSF	Urine	Plasma	PBMC	CSF
Quinlivan et al. (1992)							0/10	4/13[a] (31%)				1/11 (9%)
Tornatore et al. (1992)			17/19[b] (89%)				10/26 (38%)				0/30	
Markowitz et al. (1993)					20/88 (23%)				29/122 (24%)			
Dörries (1994)											24/29[c] (83%)	
Sundsfjord et al. (1994a)					13/82 (16%)		0/42		12/56 (21%)			
McGuire et al. (1995)				24/26 (92%)				10/114 (9%)				1/16 (6%)
Azzi et al. (1996)					11/30 (37%)		15/32 (47%)		7/30 (23%)		14/36 (39%)	
Perrons et al. (1996)	3/8 (38%)		0/7	19/23 (83%)	3/29 (10%)		0/10	0/62				
Ferrante et al. (1997)	9/12 (75%)		9/12 (75%)	11/12 (92%)		15/52 (29%)	5/52 (10%)	0/52				
Dubois et al. (1998)			9/19 (47%)				13/96 (14%)					
Koralnik et al. (1999)	2/6 (33%)	3/7 (43%)	5/8 (63%)	10/11 (91%)	7/22 (32%)	7/32 (22%)	13/103 (13%)	1/65 (2%)	5/15 (33%)	0/13	0/18	0/27
Andréoletti et al. (1999)			6/10 (60%)	10/12 (83%)			31/151 (21%)					

[a]Samples taken from brain rather than CSF.
[b]Two regulatory regions were Mad-1-like, one was Mad-4-like (one from a non-PML control patient).
[c]6/6 regulatory regions similar to Mad-1. Cord blood was negative (0/10).

However, there is agreement that JCV is detected in the CSF of most PML patients, but few, if any, individuals without PML. This makes JCV detection by PCR in the CSF a useful replacement for brain biopsy in clinical PML diagnosis. However, to support published conclusions on the viral etiology of demyelinating disease, the standard should include pathologic proof of the clinical diagnosis.

There is also general agreement that JCV is excreted in the urine of many individuals, irrespective of their immune status. As this does not change appreciably between PML patients and controls, this test has no direct clinical relevance. As noted above, there is less agreement regarding the presence of JCV in PBMC, the nature of the JCV regulatory region in strains found there, and the role of this infection in the pathogenesis of disease. While there is general agreement that JCV regulatory regions in PML brains are usually rearranged (PML type) and those in urinary strains are archetypal (see Chapter 7) (Agostini et al., 1999), there is no consensus on the question of whether lymphocytes (PBMC) or brain are sometimes infected in HIV-negative controls (see Chapter 10).

The significance of the PBMC infection in HIV-positive patients is uncertain. It may be that these patients are at higher risk for PML (Tornatore et al., 1992). However, as there are cases of PML that are negative for JCV in PBMC, and cases of HIV-1–positive patients who are positive for JCV in PBMC but do not develop PML (Andréoletti et al., 1999), it appears that while JCV may be latent in PBMC, these cells may not necessarily be the immediate and necessary carrier of JCV to the brain. The failure to demonstrate early or late JCV mRNA in PBMC (Andréoletti et al., 1999; Dubois et al., 1997; Lafon et al., 1998; Tornatore et al., 1992) supports viral B-cell latency rather than reactivation. Alternatively, the existence of B-cell surface receptors for JCV could make them passive carriers of JCV into the brain (see Chapter 9) (Wei et al., 2000). Reports of the presence of JCV-infected PBMC cells in the PML brain are limited to the finding of a few cells in the Virchow-Robin spaces in an AIDS-PML patient at autopsy (Houff et al., 1988).

However, even the demonstration of actively infected B cells in the brain parenchyma of fully developed PML cases would give no clue as to where those cells became infected and when they entered the brain in relation to onset of disease. The failure of other active neurologic diseases that damage the blood–brain barrier and initiate inflammatory reactions in the CNS to promote JCV reactivation and the onset of PML (Dubois et al., 1998) raises further doubts as to whether mere access of circulating B cells to the brain is the proximal cause of PML. Convincing evidence on this point might be obtained in a prospective study in which the circulating viral strains in an HIV-positive patient are characterized as to genotype and regulatory region rearrangement before the virus replicates in the brain and causes PML in that patient.

To summarize, the significance of the PBMC infection is currently unknown. Further studies of HIV-1–positive patients with follow up are essential. If it can be demonstrated that PBMC containing rearranged regulatory regions (other than Mad-1 and Mad-4, which are commercially available laboratory strains) circulate before the onset of PML, the case for an etiologic role of this B-cell infection, as suggested by others (Jensen and Major, 1999), would be strengthened. A representative of JCV Type 8 from Papua New Guinea in the South Pacific, which has an agnogene deletion and a unique regulatory region rearrangement (Jobes et al., 1999; Ryschkewitsch et al., 2000), is now available from the American Type Culture Collection (VRMC-24). Use of this distinctive strain in the laboratory as a positive control will not compromise the integrity of PCR searches for JCV DNA in clinical samples with low level infection by European, African, or Asian strains.

An alternative to JCV infection of the brain by circulating PBMC immediately preceding onset of PML would involve reactivation of latent JCV infection within the brain (see Chapter 12). To make that simple statement says nothing, however, about when and how the brain became latently infected or why the virus reactivates in only about 5% of AIDS patients. Are those 5% the only ones in the population latently infected in the CNS by JCV? The likelihood of brain latency may reflect the age at primary infection or the level of primary viremia. Alternatively, most individuals may be latently infected in the CNS, as they are in the kidney, but are lacking some additional co-factor conferring susceptibility to PML induction. Study of rare early cases of PML may be instructive for understanding the pathogenesis of the disease. In one of two subclinical PML cases diagnosed at autopsy, very small focal lesions were characterized by collections of reactive astrocytes, but with little or no demyelination and no infected oligodendrocytes (see Chapter 12).

Beyond the diagnostic use of PCR analysis for JCV in the CSF, where do molecular studies presently stand? It is to be hoped that the quantitative PCR assay of the CSF (see Chapter 13) (Koralnik et al., 1999; McGuire et al., 1995) will allow rapid assessment of the therapeutic effect of experimental treatments, although a case has been noted in which virus was cleared from the CSF, but the disease progressed and the patient died (Guillaume et al., 2000). If JCV in the circulation of HIV-1–positive individuals proves to be a useful prognostic indicator, as has been suggested (Dubois et al., 1998), it may also be useful to add JCV genotyping of the positive samples. We have found in retrospective studies that JCV Type 2B is present at a higher frequency in PML brain relative to controls (relative risk ~3) (Agostini et al., 1998) (see Chapter 18). However, this point needs to be demonstrated in a prospective study. Results have been reported recently from similar studies of JCV genotypes in PML brain in France (Dubois et al., 2001).

Archetypal Regulatory Region versus Rearranged PML Type. A decade of work detailed in Chapter 7 has shown that the form of JCV excreted in the urine bears a noniterated regulatory region (termed *archetypal*), whereas the transcriptional control region from PML brain is apparently randomly reconfigured, having sustained deletions and duplications of the archetypal configuration. Frequently, as in Mad-1 and Mad-4, a deletion (23 and 66 bp in this example) is present in both copies of the duplication (98 bp in the Mad-1 strain, 79 bp in Mad-4) so that it is clear that the deletion preceded the duplication.

The adaptation of JCV to the brain involves enhancement of replication in glial cells, but it is likely a dead-end infection as these neurotropic and neurovirulent strains are unlikely to replicate in the kidney and circulate in the population. This has clearly been true over evolutionary time, as the archetype has persisted in populations around the world and is by far the dominant configuration excreted in urine. Mad-1 and Mad-4 have not yet been isolated from urine, suggesting that these strains do not generally circulate in the population. The alternative requires that the archetypal strains of the Type 1A genotype sustain a 23 bp deletion followed by a 98 or 79 bp duplication to generate these strains repeatedly de novo. This is possible, but quite unlikely to re-occur repeatedly in precisely the same way.

As noted above, it is unlikely that the brain-adapted (PML-type) virus strains would have enhanced transmissibility due to a replicative advantage in the kidney or in the primary site(s) of infection. The little evidence available suggests that archetypal JCV dominates in the kidneys of PML patients even though rearranged regulatory regions can be found there if archetypal structures are first digested away (Ault and Stoner, 1994). This tissue specificity of excreted virus may be all that stands between the healthy population and an epidemic of JCV with enhanced neurotropism and/or neurovirulence. Although unlikely, the appearance in the future of circulating neurotropic strains cannot be excluded, and it is possible that some brain-adapted JCV strains with enhanced infectivity will begin to circulate. The serious implications of this possibility suggest that continued study and monitoring is warranted.

The Role of Other Viruses in the Transactivation of JCV and the Pathogenesis of PML. PML brains may be co-infected with HIV-1 itself or any of the other opportunistic agents that commonly infect PML brains, including *Toxoplasma gondii* (toxoplasmosis), *Cryptococcus neoformans* (cryptococcal meningitis), and cytomegalovirus (CMV). There is little evidence that HIV or these agents regularly and characteristically alters the pathology of PML, although the extent and distribution of JCV-induced lesions may be affected (see Chapter 12). Among the viruses for which molecular evidence has been provided for a role in JCV reactivation are HIV-1 itself (Chowdhury et al., 1992) and cytomegalovirus (CMV) (Heilbronn et al., 1993; Winklhofer et al., 2000). The interaction of polyomaviruses and CMV could also be a two-way street. Evidence has been presented that BKV T antigen can induce CMV immediate–early and early proteins in doubly infected semipermissive cells (Kristoffersen et al., 1997). Additional pathologic study is required to determine which of these interactions may be clinically relevant.

A virus for which a pathologic association has recently been proposed in PML is the new human herpesvirus 6 (HHV-6) (Mock et al., 1999). However, HHV-6 infections have been associated with a variety of neurologic pathologies, including meningoencephalitis with PML (Ito et al., 2000) and without

PML (Torre et al., 1998), as well as in both normal and neoplastic brain tissue (Cuomo et al., 2001). Moreover, these frequent infections were never described in electron microscopic studies and seem unlikely to be associated with a cytolytic pathology in which productive infection by this enveloped herpesvirus is identifiable ultrastructurally (see Chapter 12). It is thus difficult to assign an etiologic role to this agent. As HHV-6 DNA is also a frequent inhabitant of multiple sclerosis (MS) lesions (Challoner et al., 1995), the possibility remains that this ubiquitous virus superinfects regions of altered blood–brain barrier function with an uncertain contribution to the pathogenesis of the demyelinated lesion in either condition. Clearly, the potential interactions of these herpesviruses in PML, and as well as others such as Epstein-Barr virus (Farge et al., 1994), present areas for future investigation, and some of the enigmas of JCV pathogenesis may be solved by an understanding of their role.

The Immunology of the Human Polyomaviruses. Other than the pioneering work of Tevethia and colleagues described in Chapter 21, there has been relatively little experimental effort devoted to the immunology of the polyomaviruses since the early work demonstrating that SV40 and mouse polyoma viruses expressed an antigen generating antitumor immunity (Tooze, 1981). That early work led to the designation of the polyomavirus early proteins as large and small T (tumor) antigens. In the case of JCV in humans, immunologic study has been largely represented by the diagnostic hemagglutination-inhibition (HI) antibody assays developed by Walker and colleagues (see Hogan et al., 1991). Similar assays have been applied in studies of the serology of BKV (see Chapter 19). The HI method requires large amounts of virions, which are not readily available for JCV given the difficulty of virus cultivation. ELISA methods have also been developed (see Chapters 13 and 19). Studies of JCV antibodies in the serum are of no diagnostic value unless combined with intrathecal antibody detection (see Chapter 13).

At present, the role of virus-specific immunity in controlling JCV (and BKV) infection is not clear. The level of excretion of BKV in the urine of AIDS patients is related to the degree of immunosuppression, but the same is not true for JCV excretion (Knowles et al., 1999; Markowitz et al., 1993; Sundsfjord et al., 1994a). In a recent study of AIDS patients, PML patients, and controls, JCV excretion was found in about 33% of all three groups (Koralnik et al., 1999). For many acute viral infections, the clearance of virus may lead to positive selection of mutants to escape the specific immune response of the host. In the case of polyomaviruses, however, an asymptomatic primary infection frequently persists as a chronic infection of the kidney (see Chapter 10). The immune response usually does not clear the virus and is unlikely to exert any selective pressure for mutation of the persistent virus. Although the immune response does not clear JCV from the kidney, there is still hope that the immune response can control viral replication in the permissive (oligodendrocyte) and semipermissive (astrocyte) cells in the brain. It is therefore encouraging that CD8-positive cytotoxic JCV-specific T lymphocytes have recently been demonstrated in the circulation of long-term surviving PML patients (Koralnik et al., 2001).

PML can occur in the early stages of AIDS when the immune system is still relatively intact. In these patients a specific boost of antiviral immunity early in the disease might be crucial to controlling it. The MHC class I and class II antigens required to modulate the immune response are present within PML lesions (Achim and Wiley, 1992). It is thus conceivable that appropriate virus-specific vaccination utilizing residual T-cell function might provide immunotherapy by boosting immunity in AIDS patients to contain the threat of pathogenic JCV replication in the brain. A vaccinia virus construct of the SV40 T antigen was designed that retains immunogenic domains but deletes those with oncogenic potential, including the Rb protein binding site, the p53 binding site, and the oncogenic CR1 and J domains (Xie et al., 1999). A similar immunogen could be developed for therapeutic use in controlling JCV replication in the PML brain. It is interesting in this context to note that an inactivated BFDV vaccine has been demonstrated to be effective for control of acute avian polyomavirus infection in flocks of parrots and parakeets (Ritchie et al., 1998).

The Present: BKV

The state of our knowledge of BKV, including the possible role of BKV in tumorigenesis, has been presented in Chapters 14, 16, and 19. The clinical picture of BKV emphasized in most textbooks is the association of BKV with ureteral stenosis in renal allograft recipients and hemorrhagic cystitis in bone marrow transplant patients (Azzi et al., 1999; Peinemann et al., 2000). Whether the latter association reflects the etiology of this condition has been discussed by others (Hogan et al., 1991).

Currently, a pressing clinical question is the role of BKV in the interstitial nephritis associated with both renal allograft failure and end-stage renal disease in AIDS (Boubenider et al., 1999; Drachenberg et al., 1999; Howell et al., 1999; Mouratoff et al., 2000; Nickeleit et al., 2000; Randhawa et al., 1999; Smith et al., 1998). Early diagnosis is essential as control of virus replication requires reduction of the immunosuppressive drug regimen (Binet et al., 1999).

The possibility that mutant BKV strains with greater propensity for cytolytic replication in the kidney are circulating in the population should be considered (Smith et al., 1998). A BKV strain identified in Cincinnati in an AIDS patient with end-stage renal disease had a Q169L mutation in T antigen in the *ori* DNA binding region (Simmons et al., 1990). This change restores the residue in BKV to the nonpolar character of the Ala residue present in SV40 T antigen. NMR studies of the structure of the *ori* binding domain of SV40 T antigen (Luo et al., 1996) suggest that this residue is in an alpha helix that stabilizes the protein structure through interaction with the face of a β sheet. Whether this mutation also influences the DNA binding activity of the protein remains to be investigated, but the question of whether particular genotypes of BKV (see Chapter

19) or particular regulatory region rearrangements (see Chapter 14) are more likely than others to be associated with BKV nephritis should be considered.

The Present: SV40

There is no more controversial topic in this book than the question of SV40 infection of humans and its role, if any, in tumorigenesis of several uncommon cancers. In this book, a strong case has been presented for a role for SV40 in certain human tumors (see Chapter 17), while others have cast a critical eye over these claims from an epidemiological point of view (see Chapter 20). This adversarial approach not only makes entertaining reading, it is also good science and speeds scientific progress just as the adversarial approach makes good law. If we can find the possible flaws of logic or method in two sides carefully argued, it may be possible to improve the data or the controls or to devise the decisive experiment that settles the question.

Before reviewing some of the points at issue, it is worth noting that despite the impression given by some press coverage of this subject (Fig. 22.1), scientists have never been engaged in a cover-up of either the discovery of SV40 as a contaminant of some polio vaccine lots or the fact that SV40 is tumorigenic in animals and poses an unknown risk to humans. The discovery of the virus (Sweet and Hilleman, 1960) and its oncogenicity in hamsters (Eddy et al., 1962) were promptly reported in the scientific literature, and these reports initiated a long series of investigations as rigorous as the methodology of the day permitted (Hilleman, 1998). While no one wished to ignore the risk to humans, it seemed equally important that the public not be needlessly panicked into rejecting a vaccine that was reversing what had been a rising summer epidemic, particularly because the vaccine could be satisfactorily cleaned up once the problem was recognized and production was begun in alternative cell types free of endogenous virus contamination (Hilleman, 1998; Levine, 1994). Accordingly, the search for the tumorigenic mechanisms of SV40 and the sequelae of this "accident," if any, has been pursued diligently, but without fanfare. The recent media "frenzy" that ensued from misleading reporting in the press has been documented in the *Journal of the National Cancer Institute* (Kuska, 1999).

SV40, along with adenovirus, papillomaviruses, and others, has provided an invaluable tool as a model system for cell biology and has played a key role in the development of molecular biology over the last 30 years (Hilleman, 1998; Levine, 1994). These findings include the Nobel Prize–winning discovery of "split genes" in which the splicing of mRNA removes extraneous sequence (introns) before its translation into protein utilizing the coding sequences known as exons (Sharp, 1994).

As an experimental model of oncogenesis, SV40 has helped to elucidate several mechanisms by which malignant cell transformation occurs, including interaction with the tumor suppressors p53 and the Rb family of proteins (Barbanti-Brodano et al., 1995; Villarreal, 1989). Has this work now borne unexpected fruit in that the virus is not merely a model for delineating hypo-

Figure 22.1. "Polio Vaccine Scare" shares the cover of *New York Magazine* with Rudy Giuliani and the Yanks on November 11, 1996. Inside, the story implicating SV40 contamination of early polio vaccines in human tumors was entitled: "*A Shot in the Dark*" (Wechsler, 1996). A blurb declares: "***Monkey business***: *According to Maurice Hilleman of Merck, who discovered SV40, the government kept the contamination of the vaccine secret to avoid public hysteria.*" In fact, Hilleman's findings were reported in June 1960 to the Second International Live Poliomyelitis Vaccine conference held under the sponsorship of the Sister Elizabeth Kinney foundation at the Pan American Health Organization headquarters in Washington, DC, and published that same year in the *Proceedings of the Society for Experimental Biology and Medicine* (Hilleman, 1998). Shortly thereafter, a report appeared from another group published in the leading medical journal *Lancet* (Goffe et al., 1961). The government could hardly have suppressed this published information had it wished to do so. Studies continued and continued to be published (see Chapters 18 and 21). This open approach has maintained public confidence in the medical research establishment. (Taken with permission from Wechsler, 1996).

thetical mechanisms, but is itself the very culprit? Or is the virus merely an innocent bystander, caught up in a sweep of the usual suspects, but eventually to be exonerated? If these findings of SV40 involvement in tumorigenesis in humans are confirmed, it will be as though the detective on the case is revealed to be guilty of the unsolved crime himself. Understandably, there was a lot of skepticism on the part of many workers in this field, and, at the same time, a willingness on the part of erstwhile friends and defenders of the virus to pursue the case against it as a potential infectious risk factor for cancers with regularly lethal outcomes. Scientists could not know whether they were risking their own lives to study the mechanisms of cellular transformation by this monkey virus, yet the experiments went on. There was enough concern about the potential harmful effects of repeated exposure in the laboratory to SV40 that when a prominent SV40 researcher at NIH died of lymphoma in his early 40s in 1987, his internal organs were examined by PCR for SV40 DNA at autopsy, and the negative results were reported by five prominent virologists in the *New England Journal of Medicine* (Howley et al., 1991).

SV40 and Human PML. The first attempt to link SV40 to human disease came only a decade after discovery of the virus. In 1971, Padgett, Walker and colleagues in Madison, WI, reported the isolation of an apparently new human polyomavirus from PML brain in primary human fetal glial (PHFG) cells. They named the new virus JC, strain Mad-1, from the initials of the deceased patient (see Chapters 2 and 3). This work was published in *The Lancet* (Padgett et al., 1971). (Interestingly, this report was immediately preceded in the same number of *The Lancet* by the serendipitous discovery of the other human polyomavirus, BKV [see Chapter 4].) About the same time a group in Baltimore decided to test PML tissue for the presence of SV40 utilizing a system in which cultured cells from PML brain were fused after several passages with primary African green monkey kidney (PAGMK) cells, or homogenates were inoculated directly onto PAGMK cells. The first two PML cases examined were found to be positive for SV40, and these were reported in the *New England Journal of Medicine* in 1972. A guest editorial comment in the same issue pointed out that although the donor monkeys had tested seronegative for SV40, PAGMK cells can harbor up to 20 endogenous viral agents. The editorial suggested that BSC-1 cells, a continuous cell line derived from AGMK cells that was free of endogenous agents, would have been a more judicious choice. The two new SV40 strains were characterized in a series of papers as authentic SV40 and were designated EK (obtained from a PML brain biopsy specimen) and DAR (from a PML autopsy case). However, when the new culture system of Padgett, Walker, and colleagues using PHFG cells was adopted, the next nine PML cases from the same laboratory proved to be associated with JCV infection (Narayan et al., 1973). It is now known that JCV would not have grown in the PAGMK cells, although SV40 could have grown in the culture system of Padgett et al., which was adapted from the one Dr. Harvey Shein first used for growth of SV40 (see Chapters 2 and 3). No other PML cases in this country were ever attributed to SV40. However, in 1985, a case of PML associated with SV40 infection was reported in Japan, but subsequently retracted (Eizuru et al., 1993).

The cause of confusion in the latter case seemed clear. Immune serum to SV40 used for immunocytochemistry was high-titered and sufficiently cross reactive to react with JCV antigens in the tissue, whereas the antibody specific for JCV was relatively weak and failed to react with JCV antigens in the paraffin-embedded tissue. A case of PML due to SV40 was also reported from Europe in which the virus was isolated by culture of a brain homogenate on a monkey kidney cell line (CV-1 cells) (Scherneck et al., 1981). We obtained the paraffin-embedded tissue some 10 years later and found, using both in situ hybridization with specific DNA probes and immunocytochemistry, as well as PCR, that this PML case represented a fulminant JCV Type 1A infection (Stoner and Ryschkewitsch, 1998). Type 1A is the dominant JCV genotype in Eastern Europe. No SV40 could be detected in these tissues by virus-specific in situ methods. Studies on the DAR case using tissue provided by Dr. L.J. Rubinstein identified JCV target antigen and viral DNA by immunocytochemistry and in situ hybridization, respectively (Stoner and Ryschkewitsch, 1998). The PCR-amplified DNA was identified as JCV Type 4. There are no reports of virus-specific in situ or PCR methods exploring the antigen or DNA in the biopsy tissue known as EK. We conclude that there are currently no credible reports linking SV40 to cases of human PML. A recent report of SV40 in the CSF of an AIDS patient with PML was based on a clinical, not a pathologic, diagnosis of PML (Tognon et al., 2001).

The problem of explaining the original isolates of SV40 from PML brain remains. A clue may come from the editorial comments that accompanied the original publication. Use of cells capable of harboring latent SV40 is a concern, but why would the primary monkey cells fused with cultured cells grown from human PML brain produce SV40, while uninoculated control cultures manipulated and maintained in parallel did not? The answer may come from studies with hybrid viruses showing that T antigens of each species are capable of transactivating the heterologous virus (Frisque and White, 1992). In this case, since the primary African green monkey kidney cells would not have supported active JCV replication, the alternative was for JCV T antigen in human PML glial cells introduced into the cultures to transactivate the SV40 promoter of latent SV40 virus.

Clinicians need to exercise caution when inferring the etiology of PML, as data on SV40 such as those just summarized are in the textbooks and are still being cited in the literature. Before the SV40–PML association was discredited, a doctor wrote in his memoir of neurologic practice about the first case of PML that he diagnosed, which happened to be in a boyhood friend:

> I know exactly how and when Marv Rotblatt (not his real name) got his SV-40. My father gave it to him, in the kitchen of our apartment as soon as the first shipment of Salk vaccine arrived in Chicago. My mother insisted on it. My father pulled some strings, and Marv and I and a few other selected lucky friends became the first Chi-

> cagoans to get a series of Salk vaccine shots. . . . The SV-40 virus also entered us, injected in the same syringe, through the same needle. The SV-40 viruses had been growing away, unbeknownst to anyone, in the same monkey kidneys that grew the polioviruses. . . . (Klawans, 1988)

It makes a good story, but it is extremely unlikely, without some supporting evidence, that SV40, though it may have been present in that lot of polio vaccine, caused any cases of PML.

The Role of SV40 in Human Tumors. With the advent of the PCR technique, an ever-increasing number of reports became available that associated the monkey polyomavirus SV40 with human disease (reviewed in Chapter 17). Because of its tumorigenicity in animal models and the ability to transform human cells in vitro through the oncogenic functions of its large T antigen, it is biologically quite plausible that SV40 might contribute to the development of some human cancers. In this context, however, it is good to remember that the high frequency of SV40-induced transformations in both laboratory animals and cell cultures are only observed under very special conditions, such as direct delivery of high viral doses into susceptible cell types. The susceptible laboratory animals may be highly inbred, and the age and immune status of the host seem to be crucial (Tevethia, 1980). For instance, adult hamsters appear to take care of SV40 even in very high doses.

Because SV40 is known to have entered the human population 45 years ago through exposure to contaminated polio vaccines (Shah and Nathanson, 1976), numerous follow-up studies were conducted in order to detect the consequences of this major period of exposure (reviewed in Chapter 20). These epidemiologic studies fail thus far to provide any clear support for a relationship between the exposure to polio vaccines, of which some were contaminated, and the subsequent development of human cancers. Despite some difficulties in defining exactly the exposed study population, a major impact of exposure to adventitious virus for the human population can nevertheless be safely excluded with regard to common malignancies.

The prevalence of SV40 infection in the human population based on seropositivity has been estimated at 5–10% (Butel and Lednicky, 1999). Given this low prevalence, the chance of becoming infected early in life with this monkey virus from other persons appears to be limited, unless a parent or other close contact were to be infected. Considering the experimental data about tumor induction in animals, one would expect any detrimental effects from an SV40 infection to occur in early childhood, while a healthy adult person might be naturally protected. Indeed, infection of laboratory workers could be easily documented by immunologic approaches (reviewed by Butel and Lednicky, 1999), but a higher frequency of tumors has thus far never been reported from this seemingly high-risk population.

Among the reports that suggest a link between SV40 and human tumors are high frequency associations with malignant mesothelioma of the pleura (see

Chapter 17). Nevertheless, SV40 is mostly described to be associated only sporadically or at low frequency with defined human diseases. Examples are individual types of rare childhood brain tumors, including ependymomas and choroid plexus papillomas, and osteosarcomas (Carbone et al., 1996; Heinsohn et al., 2000; Stewart et al., 1998; Weggen et al., 2000). This overall paucity of cases can therefore hardly be seen as sufficient evidence for a causal relationship between SV40 and the individual human diseases. Even for mesothelioma entire case series could not demonstrate a consistent association with SV40 (e.g., Emri et al., 2000; Hirvonen et al., 1999). In other words, many patients with a given type of tumor suspected to be triggered by SV40 do not harbor viral sequences, which casts serious doubt on whether SV40 infection is a necessary condition for development of these tumors.

In stark contrast, in those cancers known to have a viral etiology, such as cervical carcinoma and human papillomavirus (HPV), the virus appears to be necessary for tumor development, and viral strains of a specific genotype can be detected consistently (Schiller, 1999). If one considers additionally that the viral load of most tumor specimens carrying SV40 DNA is rather low, then this pattern may simply indicate that SV40 settles in and remains associated with some tumor cells because of a favorable environment in rapidly dividing cells. Indeed, robust growth of SV40 is common in most cultured human tumors (O'Neill et al., 1998), and the in vitro experiments by Bocchetta et al. (2000) show that fast turnover cells, such as the mesothelium, are far more easily infected than other human cells. The question that needs to be addressed becomes, therefore, how can we distinguish a human cell that got infected by SV40 and was subsequently converted into a malignant cell or cell clone from an already malignant human cell into which SV40 has adventitiously entered? If the same type of tumors that are believed to be induced by SV40 can develop in the absence of the virus, then SV40 should at least contribute something significant to the malignant transformation to be a plausible etiologic agent or a significant co-factor.

Malignant mesothelioma of the pleura has become the emblematic human neoplasm where SV40 is thought to have contributed something significant to the pathogenesis of the disease (reviewed by Carbone et al., 1997b). The most important reasons can be summarized as follows: First, numerous PCR investigations reported frequencies of SV40 DNA detection ranging from 50% to 90% (Carbone et al., 2000). Furthermore, large T antigen expression was documented in parallel by immunohistochemistry in some series (e.g., Testa et al., 1998). Second, mesothelioma patients were alive during the early polio vaccination period in contrast to osteosarcoma or pediatric tumor patients, who were born decades later and for which the route of infection remains therefore more problematic (Butel and Lednicky, 1999). Third, mesothelioma incidence experienced a sharp increase in frequency over the past four decades (Price, 1997). Fourth, mesothelioma cells do not feature the common mutational inactivation of the two major tumor-suppressor proteins (i.e., p53 and pRb), which are detected in other malignancies (see Lechner et al., 1997). The findings of binding of large T antigen isolated from mesotheliomas with p53 and the pRb family proteins (Carbone et al., 1997a; De Luca et al., 1997) provided a plausible mechanism for SV40 to play a role in the etiology of this tumor (Mayall et al., 1999).

In contrast to other human tumor types linked to SV40, mesothelioma development is clearly associated with a well-known risk factor: Occupational and environmental exposure to mineral fibers, such as asbestos (McDonald and McDonald, 1996; Zeren et al., 2000). Hence, the vast majority of mesotheliomas are found associated with past asbestos exposure in industrialized countries. Although Carbone et al. (1997b) pointed out several important lacunas regarding the etiologic role of asbestos, the observed relationship with SV40 has also not yet provided a better explanation for these observations. However, the recent tissue culture studies by Bocchetta and colleagues (2000) could nicely explain the greater susceptibility of mesothelial cells to transformation by asbestos exposure and SV40 infection acting synergistically. It becomes even more puzzling why the contribution of SV40 after infection four decades ago does not shorten the disease's characteristic long latency period of 25-40 years.

More detailed studies are needed to rigorously examine the role of SV40 exposure in relation to asbestos exposure and p53 mutation. Regarding the PCR results, it is apparent that in a majority of these studies only small fragments of the viral genome were detected. In an extensive analysis of several human tumors, not including mesothelioma, it was possible to isolate full-length viral DNA from several specimens and identify authentic SV40 strains (Stewart et al., 1998). However, numerous studies on mesothelioma show a clear difficulty in the capacity to amplify extended genome fragments along the important viral early region. Specifically, the primer combination SV2/SV1, which amplifies around the intron of the tumor antigen, appears to be limiting. The net result is a drop in detection frequency to 25% instead of the 50–90% obtained with primers targeting shorter segments (McLaren et al., 2000; Pass et al., 1998; Rizzo et al., 1998). The available preliminary sequence data for mesotheliomas (Pass et al., 1998) point to a sequence identical to the most common laboratory strain of SV40, strain 776 (but see Arrington et al., 2000), and hence the priming sites should be present.

An additional concern is the reported inability of a recent multicenter study to find any of 25 mesothelioma specimens obtained from archival samples to be reproducibly positive for SV40 DNA (Strickler et al., 2001). While overall the assays were sensitive, specific, and reproducible, they could not reproducibly demonstrate the presence of SV40 DNA in the 25 selected human mesothelioma or control tissue specimens. The high prevalence of SV40 DNA detected in some earlier studies of mesothelioma makes the controlled and blind multicenter observations difficult to reconcile with previous studies. Overall, the results suggest that SV40 DNA was not present in the mesothelioma tumor DNA samples analyzed, or was present at a level below the sensitivity of the assays in use by most of the collaborating laboratories. There are thought to be geographic differences in the prevalence of SV40 DNA in tumor tissue, and it cannot be excluded that such differences or chance sampling bias might have contributed to these negative results (Strickler et al., 2001).

Furthermore, specific staining with antibodies directed against the large T antigen can be virtually absent in entire series associated with a high frequency for short DNA segments (e.g., Dhaene et al., 1999; Galateau-Salle et al., 1998). The immunostaining by Galateau-Salle et al. (1998) was performed on formalin-fixed, paraffin-embedded tissues compared with frozen sections in the study by Dhaene et al. (1999). These mesotheliomas from France and Belgium stained predominantly in the cytoplasm with only occasional diffuse nuclear staining. The authors were therefore quite puzzled by their failure to demonstrate the expected nuclear staining with the virus-specific monoclonal antibodies PAb419 and PAb101. The PAb419 reagent was also found by Testa et al. (1998) to stain the cytoplasm of malignant mesothelioma cells in addition to the nuclear staining in their specimens.

Several interpretations to try to link this immunoreactivity with SV40 have been put forward. For instance, because PAb419 recognizes an epitope shared with the small t antigen that resides in the cytoplasm, Dhaene et al. (1999) as well as Ramael and Nagels (1999) attribute this cytoplasmic staining to the expression of small t antigen by SV40. However, this interpretation is questionable because other studies have shown that PAb419 does not recognize the cytoplasmic form of small t antigen (Montano and Lane, 1984, 1989). Second, Ramael and Nagels (1999) speculate that PAb419 might not work on formalin-fixed, paraffin-embedded tissue sections, which could explain the failure to detect nuclear signals in the French mesothelioma specimens. Nevertheless, both PAb419 and PAb101 have been shown to stain the SV40 tumor antigen in the nuclei of formalin-fixed, paraffin-embedded animal material in which the T antigen is highly expressed (Ressetar et al., 1993; Simon et al., 1999). Because SV40 DNA was amplified from samples that show only nonspecific immunostaining with PAb419, and no nuclear staining with PAb101, Dhaene and colleagues (1999) invoked the "hit and run" model. This mechanism was proposed by Carbone et al. (1997b) to account for the generally low numbers of tumor cells showing immunostaining for the SV40 oncoprotein in a given specimen. However, both the frozen Belgian tissue specimens as well as the cell lines of the French study were reexamined and found devoid of SV40 protein and DNA by PCR with multiple primer sets (Pilatte et al., 2000; Hübner and Van Marck, unpublished). Therefore, the cytoplasmic staining is likely to stem from a cross-reacting cellular protein (Darmon and Jat, 2000). This interpretation suggests that other studies failing to demonstrate specific immunostaining despite the detection of SV40 DNA by PCR (e.g., Procopio et al., 2000) should be further scrutinized. One might note that Shivapurkar et al. (2000) could not link SV40 to the sarcomatous and mixed types of malignant mesothelioma, whereas Procopio et al. (2000) found a more elevated association with these types than with the common epithelial form.

An additional example of a reevaluation of mesothelioma cell cultures originally shown to carry SV40 DNA and expressing variable levels of large T antigen (Waheed et al., 1999) is given by Modi et al. (2000). Western blotting could not confirm that SV40 T antigen was produced in any of their cell lines.

There has been considerable discussion about whether the early polio vaccines could have been the source of SV40 in the actual subset of mesothelioma patients with detectable viral DNA (Carbone et al., 1997b; see also Chapter 20). The case for involvement of the early polio vaccines was strengthened when SV40 DNA was not found to be associated with mesotheliomas from Finland or Turkey, both countries that reportedly never received contaminated vaccines (Emri et al., 2000; Hirvonen et al., 1999). In this context it might be interesting to mention that the predominant inactivated polio vaccine (IPV) used in Finland was manufactured in Belgium and was also the sole polio vaccine distributed in Belgium before 1963 (Andre, 1979). In 1963, the Belgians introduced an oral polio vaccine (OPV) produced and controlled by the same manufacturer, whereas the Finns continued to rely exclusively on the inactivated vaccine. Therefore, one would have expected to find no SV40 in Belgian specimens. Nevertheless, there are two reports of studies finding traces of SV40 in Belgian mesotheliomas (Dhaene et al., 1999; Ramael et al., 1999; see also Chapter 17). It turns out that even the situation for the polio vaccine exposure in Finland is far from clear because the Belgian batches of IPV that were distributed in Finland were not the only vaccines used in that country. Indeed, Finland took part in the Francis field trial of the Salk vaccine (Francis et al., 1957), and further small-scale vaccinations took place in that country between 1954 and 1956 (Parvala, 1958). Possible contamination of some of these vaccine lots might, however, not have reached potential high-risk populations. Mass vaccinations started in Finland in 1957. The manufacturer of the IPV subsequently used was unable to indicate when their first batches were delivered to Finland. Hence, it might be possible that another polio vaccine was used during the first mass vaccinations. The origin of batches alone is not sufficient to evaluate possible exposure to SV40. The vaccination strategies varied from region to region, and the vaccines were intended for schoolchildren, with adults participating with varying frequency in the programs (and most mesothelioma patients were aged above 20–30 years when these vaccines became available). If one-third of the early polio vaccines administered in the United States were contaminated (Shah and Nathanson, 1976), then about 20% of the concerned age group might have become directly infected (not taking virus dose or the effect of the host immune response into account).

For Belgium, adult participation started in 1963, and, considering their age, most SV40-positive patients reported by Dhaene et al. (1999) could have only been directly infected by OPV (oral polio vaccine), that is, when more rigorous biosafety standards were in place. If, however, infection happened through indirect exposure, stemming from SV40 excretion observed in some children exposed to batches of OPV (Melnick and Stinebaugh, 1962), then it is again mysterious why SV40 did not spread into most members of high-risk groups. Comparison of cohorts between well-selected countries might perhaps ease some of the design questions for epidemiologic studies (discussed in Chapter 20).

More recently, it has been proposed that SV40 could still be present in later batches of polio vaccine because slower growing archetypal variants (i.e., with a single copy of the 72 bp enhancer) of SV40 might have passed through biosafety screens (Rizzo et al., 1999). This claim needs more thorough experimental assessment in realistic vaccine production settings. It should be noted that monkey kidney donors for polio virus culture have been bred in isolation facilities free of SV40 for more than three decades, and PCR has failed to detect SV40 in 190 batches of OPV used in the United Kingdom from 1971 to 1996 (Sangar et al., 1998). Nevertheless, questions raised about production practices by United States' OPV suppliers need to be addressed (Kops, 2000).

With respect to the suggested role for SV40 in mesothelioma to inactivate the function of both major tumor suppressor proteins (p53 and pRb) through binding by T antigen, two recent studies (Modi et al., 2000; Pilatte et al., 2000) cast doubt on the significance of that proposal. First, the amount of large T antigen in most infected mesothelioma cells was found to be very low (Waheed et al., 1999), and hence one needs to consider whether it could completely inactivate these tumor suppressor proteins. Given the low amount of T-antigen oncoprotein, it seems important to consider that both above-mentioned investigations identified a protein contaminant in the commercial antibody preparations with approximately the same molecular weight as the SV40 large T antigen. Both studies showed that this contaminant, which could be easily confounded with the viral oncoprotein, does not bind either p53 (Pilatte et al., 2000) or the pRb family proteins (Modi et al., 2000).

Most importantly, Waheed et al. (1999) and Modi et al. (2000) highlight another pertinent fact: In nearly all primary mesothelioma tumors and cell lines the cell cycle regulatory protein cyclin-dependent protein kinase (CDK) inhibitor p16 was already inactivated through chromosomal mutation. The p16 (INK4a) protein is a specific inhibitor of kinases CDK4 and CDK6, which mediate pRb phosphorylation and thereby inactivate its growth suppressive properties (Shapiro et al., 2000). Loss of p16 expression allows pRb to be phosphorylated and thus cell growth to be unregulated. This, together with p14/ARF silencing, appears to obviate the need for additional interference in cell cycle control and leaves the role of SV40 in mesothelioma induction presently undefined. The inhibition of these pathways formed the molecular basis to account for both the high proliferation potential and tumor resistance to chemotherapy so characteristic for malignant mesothelioma from all geographic regions of the world (Lechner et al., 1997). Given that a significant proportion of mesotheliomas are associated with SV40 DNA, and a portion of these tumor cells express the oncoprotein (perhaps the fastest growing ones), it is possible that anti-SV40 therapy may be successful. However, in view of the fact that the regulatory pathways appear to be inactivated by other means than through

SV40 T antigen, these expectations must be tempered with realism. Nevertheless, all available avenues for therapeutic attack should be explored.

3. THE FUTURE: FRONTIERS FOR POLYOMAVIRUS RESEARCH IN THE NEW MILLENNIUM

Introduction

Predicting the future is always a perilous enterprise. While the directions that extend current research can be easily discerned, the utility of the effort can be questioned. There was a revealing exchange in the 1964 meeting alluded to above in which Dr. Lucien J. Rubinstein, coming to Dr. Zu Rhein's defense, comments: "Nevertheless, we have here what we believe to be morphological evidence at the ultrastructural level which is strong enough to impel us to draw attention to it. These particles, as seen in electron micrographs, are sufficiently suggestive to stir up the interest of the virologist." Dr. Sabin's acerbic comment in reply, "That is a good way *not* to get a virologist interested." No scientist, least of all a virologist, wishes to be seen as merely providing evidence for someone else's pet theory. Nevertheless, here we go.

Ahead to the Past. As noted above, one of the prime questions for future analysis of polyomaviruses is the question of their evolutionary history as a genus. Ironically, a major area of research in the new millennium will be to deduce the history of these viruses in the previous 2 million years. Genomic DNA sequencing has made it possible.

As noted above, these are viruses with strictly limited coding capacity of only about 5000 bp and around 1500 amino acids, divided equally into regulatory proteins (T antigens), structural proteins (VP1–3), and the small agnoprotein, which has been assigned a variety of ancillary functions. These viruses are highly dependent on the host cell, and their regulatory proteins interact intimately with the host's DNA replication and transcriptional machinery. For this reason, and because they remain persistent in the host, they change very slowly. Polyomaviruses are living fossils, the coelacanth of the viral world.

Many important questions about the primate polyomaviruses remain. First, how many primate polyomaviruses exist, and what are their host species? Which of these viruses have coevolved with their vertebrate hosts, and which ones have jumped across species and when? As more sequence data become available, more of these questions will be answered. An active search should be made in the urine of immunocompromised primates with degenerate PCR primers. Application of a primer set capable of detecting a broad spectrum of known and unknown viruses has been suggested for screening tumor samples (Völter et al., 1998) and could be applied to urine samples as well. A recent study of partial genomic sequences of 20 avian polyomaviruses showed three subgroups (Phalen et al., 1999) capable of infecting four orders of birds, but failed to find evidence for species-specific genotypes.

Apparently, the mammalian polyomaviruses may have evolved differently and with greater host specificity than the majority of avian polyomaviruses. However, a distinct goose polyomavirus, goose hemorrhagic polyomavirus, has recently been described (Guerin et al., 2000). This question of species specificity of the polyomaviruses remains to be fully explored. The serologic evidence for bovine polyomavirus infection in humans (Parry and Gardner, 1986) calls into question the strict species specificity of infection by mammalian viruses as well.

What's in a Name? Is the term *polyomaviruses*, as applied to JCV, BKV, and SV40, a misnomer? The viruses are oncogenic in hamsters, but are they the cause of multiple tumors in humans? The possibilities are numerous: brain tumors including medulloblastomas and gliomas in the case of JCV (see Chapter 15); neuroblastomas, insulinomas, and brain tumors in the case of BKV (see Chapter 16); and osteosarcomas, mesotheliomas, ependymomas, and choroids plexus papillomas in the case of SV40 (see Chapter 17). Or are these viruses, whose lifestyle is the very essence of persistence and which show exquisite tissue specificity, perhaps merely passengers in these tumorous tissues? If they are etiologic agents, exactly how does the virus perturb and subvert cell growth control pathways? This interference leads to tumorigenesis in association with what co-factors (apparently asbestos in the case of mesotheliomas)? How can interactions in the oncogenic pathways be targeted therapeutically to restore normal growth control? What other pathways in addition to p53 and the Rb family of proteins are targeted? How does each virus differ, and how do these differences foreordain the unique spectrum of tumor types induced? How do mechanisms in humans differ from those in rodent model systems (Sheppard et al., 1999)? These questions and related ones will be a major focus of research in the human polyomavirus field in the coming decades.

JCV: Future Research

Role of Genotypes in JCV Biology. If genotypes were spread within small founding human populations (i.e., did not arise simply by accumulation of mutations in lineal descendants as occurs in mtDNA), then some of the type-defining mutations may be adaptive rather than neutral. Amino acid mutations in proteins affecting viral regulation or assembly are the obvious possibilities, but mutations in noncoding control regions and even those synonymous mutations in the third codon position that do not alter the encoded amino acid could be involved. There are various mechanisms by which these mutations might promote virus propagation in the population. Some have been discussed in Chapter 18. These include effects of VP1 mutations on interaction with cellular receptors and the possible modulation of alternate splicing by sequence differences near splice donor (acceptor) sites, as in the case of African genotype, Type 3. Additional points of interest are effects of type-specific amino acid mutations on zinc-finger function, and T antigen mutations in the origin-binding domain that may modulate its replication activity. The latter is suggested by a Q169L mutation in the BKV strain obtained from an AIDS patient with end-stage renal disease (see above). This illustrates the point that the potential for mutations with a selective advantage also exists for BKV, and findings from studies on each virus will illuminate the other as well.

In addition to the interactions of viral regulatory proteins with each other and with viral DNA, these proteins interact with numerous cellular replication factors, transcription factors, and enzymes. These range from DNA polymerase α-primase to the protein kinases and phosphatases that regulate the function of T antigen by phosphorylation and dephosphorylation (Moarefi et al., 1993; Simmons, 2000; Swenson et al., 1996; Virshup et al., 1992). Of particular interest are the interactions with those cellular proteins regulating cellular proliferation such as the tumor suppressor protein p53. In addition, mutations in p53 have been examined for a role in PML where this protein is expressed in JCV-infected glial cells (Bruner and Hair, 1993; Power et al., 2000). The role of interactions of JCV T antigen with a variety of other cellular factors showing single nucleotide polymorphisms (SNPs) will be an area for future study. As more is learned about the human genome, the potential influence of its polymorphisms on interactions with viral variants will become apparent.

JCV Phylogenetics, Human Dispersal, and Human Genetics. JCV phylogenetic trees show a trifurcation at the base (see Chapter 18)—Europe (Types 1 and 4), Africa (Type 6), and Asia (Types 2, 7, and 8)—consistent with predictions based on the genetics of human mtDNA, nrY, and other nuclear genes such as a nonrecombining region of chromosome 21 (Jin et al., 1999), as well as the fossil evidence. The African genotype known as Type 3 (also Type B, subtype AF2; Sugimoto et al., 1997) is an Asian-related genotype (see Chapter 18) that may have migrated back to Africa or may represent the African group from which all Asian strains were derived. However, contrary to expectations from many human genetics studies that find the widest diversity among Africans (consistent with human origins there), the diversity among paleo-African JCV strains (Type 6) appears to be much less than the diversity in the Asian group (Types 2, 7, and 8). Additional sampling in Africa is required before firm conclusions can be drawn. If the effect is real, however, it could be argued that the nature of the African landscape may have provided fewer physical barriers (mountains, seas, and so forth) to gene flow (in this case "virus flow") than experienced by populations in Asia and Oceania. It is also conceivable that this coevolving virus may be revealing aspects of human evolution and dispersal not evident from other perspectives.

In the future, patterns of JCV genotype dispersal will be compared with the human genetic markers of mtDNA and nrY as a means to follow human migration and human genetics. Of particular interest will be admixed populations such as those in Central and South America and the Caribbean (Sans, 2000). In some of these populations directional mating means that studies of maternally inherited mtDNA and paternally inherited nrY provide very different views of population history (Merriwether et al., 1997). Will the information from JCV genetics follow the lead of the mtDNA data or the nrY data, or will it represent a middle ground? Preliminary results from Puerto Rico (Fernandez Cobo et al., 2001) suggest that it may mimic the preferential maternal inheritance of Native-American mtDNA haplotypes observed in South America (Merriwether et al., 1997). This question of whether maternal genotypes are favored in a population carrying mixed genotypes relates directly to the modes of JCV transmission.

Changing Neurotropism and the Role of JCV Mad-1 Strain. It appears nearly certain that JCV is one of those persistent viruses that can never be eradicated from the human population. It has always been with us and always will be. Hopefully, it will remain relatively benign so that, despite the adverse influence of HIV-1, it will not increase in neurotropism or neurovirulence. The maintenance of the virus in a benign form over the next millennium requires that PML-type brain-adapted virus not be excreted from kidneys in a transmissible form. Standing between us and disaster is only the tissue specificity of this exquisitely adapted virus. If a viral variant with broader tissue specificity emerges, these protective barriers could be breached.

The Mad-1 (and the closely related Mad-4) strain of JCV continues to challenge scientists 30 years after its isolation. Is it transmissible and repeatedly isolable, or do subsequent isolations represent laboratory contaminants in laboratories where the virus has been grown or cloned? It is, of course, theoretically possible that the signature regulatory region rearrangement with its 23 and 66 bp deletions followed by a 98 bp tandem repeat has arisen independently multiple times. In the future, the issue of the origins of the regulatory region of this Type 1A genotype and its transmissibility will be settled. To summarize, there are three possibilities: (1) Mad-1 was a one-time laboratory isolate from PML tissue in 1971, and subsequent "sightings" of it represent detection of laboratory contaminants; (2) this strain is one of a few "PML-type" strains circulating in the population as proposed by others (Elsner and Dörries, 1998); or (3) this rearrangement (23 and 66 bp del/98 bp dup) arises independently in the host and does so much more frequently than any other rearrangement found in PML brain. Our experience is that Mad-1/Mad-4 strains have not been found in studies of hundreds of urine samples from all over the world, nor have these strains been found in the numerous PML brains we have studied (Agostini et al., 1997a). Moreover, while deletion of the 66 bp region is relatively common, the rare 23 bp deletion relative to archetype (see Chapter 7) makes independent occurrence of this complex rearrangement unlikely. Strains identified as Mad-1–like should be examined further to identify their origins. To confirm that a strain of this regulatory region configuration is authentic, original Mad-1, two sites at which the Mad-1 Type 1A sequence differs from the other three whole genome Type 1A sequences available (124A, 126A, and G2) should be examined. At position 2311 in the *VP*1 gene, Mad-1 has G rather than T, while at position 3134 in the T antigen gene it has T rather than

C. If, on the other hand, the Mad-1 type regulatory region is being generated anew by a consistent cellular mechanism, then the characteristic viral regulatory region need not associate exclusively with these unusual coding region polymorphisms at positions 2311 and 3134 nor, indeed, with only Type 1A coding region genotypes. It should be noted that random point mutations in molecularly cloned PCR-amplified regulatory region products or coding regions may reflect DNA polymerase-induced PCR error and should not be considered evidence for unique strains. In order to reduce the possiblity of contamination from the DNA of laboratory JCV Mad-1 or Mad-4 strains (Type 1), laboratories analyzing clinical samples should obtain other genotypes for positive controls. These are available from the American Type Culture Collection. A Type 2 clone from Japan is available (CY/cl1, ATCC No. VRMC-1), as is a Type 8 clone from Papua New Guinea (pJCPNG-Ag, ATCC No. VRMC-24). Type 2 (also known as CY) and Type 8 will be equally well amplified with most JCV primers in use, although the latter has a 21-bp deletion in the agnogene (Jobes et al., 1999).

In this same context future work needs to address the mechanisms of rearrangement in the brain-adapted regulatory region. Is it a random collection of replication errors a few of which result in a virus better adapted to replication in the brain, or does the virus subvert a cellular DNA recombining system (e.g., in the B cell) to its own advantage (or disadvantage)? Where and how does the rearrangement occur, and how does the rearranged virus infect the brain and spread within?

Prediction, Prevention, Diagnosis, Prognosis, and Treatment of CNS Disease. While JCV is a ubiquitous, persistent infectious agent that cannot be eradicated, hopefully PML as a disease can be controlled. Prediction of those individuals at greatest risk, prevention of disease in the at-risk group, early diagnosis in those who develop disease, assessment of their prognosis, and design of effective treatment are the clear goals.

While JCV infection is unlikely to be preventable, prophylactic immunizations should be considered. Recent demonstration of cytotoxic T cells in long-term surviving PML patients (Koralnik et al., 2001) is an important step in ascertaining what immunologic defect exists in PML patients and at what stage it can be repaired by immunization.

Hopefully, in the future treatment modalities can be more rapidly evaluated by quantitative PCR detection of JCV in the CSF (Drews et al., 2000; Koralnik et al., 1999; Matsiota-Bernard et al., 1997). Patients in whom the CSF clears of JCV but who die anyway (Guillaume et al., 2000) must be studied pathologically for clues to the underlying pathogenetic mechanisms.

Role of JCV in Human Brain Tumors. JCV is a potent inducer of brain tumors in hamsters and provides one of the few models for viral induction of brain tumors (astrocytomas) in primates (see Chapter 12). In the hamster a prominent tumor is the primitive neuroectodermal tumor (PNET) or medulloblastoma. In humans these medulloblastomas are among the more common brain tumors of childhood, and efforts exploring the possible etiologic role of JCV are underway (see Chapter 15). The roles of JCV, BKV, and SV40 in human tumors will continue to be an area of intense investigation.

BKV: Future Research

Much of the agenda outlined above for JCV can be adapted in studies of BKV. Further studies of the geographic distribution of BKV genotypes are warranted (see Chapter 19), although no parallel with the recent (100,000 years ago) dispersal from Africa postulated for JCV genotypes is expected. Nevertheless, if BKV evolved in one or more of the primitive *Homo erectus* populations as postulated above, attempts could be made to correlate BKV genotypes with regional descendants of these early people who overlapped with early modern humans.

In the case of BKV, the scene shifts from the brain to the kidney. The role of BKV nephritis in kidney transplant failure and in AIDS-associated end-stage renal disease are important areas of current and future exploration. What is the role, if any, of virulent BKV mutants, and what is the contribution of new, potent immunosuppressives? What is the role of regulatory region rearrangement (see Chapter 14)? Does it parallel JCV in that the transmissible virus is ordinarily "archetypal," but the rearranged forms are associated with increased virulence, now in the kidney? How does BKV rearrangement occur, and what exactly is its role in the pathogenesis of tubulointerstitial nephritis? In what ways is it analogous to that proposed for adaptation of JCV in the brain? Are PBMC involved, as suggested by Chatterjee et al. (2000). In light of the fact that BKV nephritis need not be a dead-end infection, as JCV is likely to be in the PML brain, will altered and more pathogenic forms of BKV begin to circulate in the population? Similarities and differences between JCV and BKV with regard to regulatory region rearrangements will be important to delineate.

As noted above, questions to be addressed include the etiologic role of BKV in human tumorigenesis (see Chapter 16). BKV is highly tumorigenic in hamsters, as is JCV, but its role in human tumors remains to be established. Both JCV and BKV cause a different spectrum and wider variety of tumors in hamsters than does SV40 (see Chapter 12). The possible interaction of BKV T antigen with mutant human p53 has been considered in relation to BKV tumorigenesis (Martini et al., 1995). We need to know more about the role of T antigen interactions of both BKV and JCV with tumor suppressors in tumorigenesis. In what respects do they mimic SV40 T antigen, and in what respects are they unique? Their mechanisms of tumorigenesis should be explored in the same detail as for SV40. How do BKV and JCV contribute to the etiologies of human cancer. If they are major contributors, what are the essential cofactors?

SV40: Future Research

The future of SV40 research will include investigation of its relation to other primate polyomavirus species and a better understanding of the natural infection in species of *Macaca*. This will in turn provide further insights into PML due to SV40 in macaques with naturally or experimentally acquired simian immunodeficiency virus (SIV) or simian HIV (SHIV). These SV40 brain infections will present with a variety of pathologies in addition to PML, including meningoencephalitis and SV40-induced interstitial nephritis and pneumonitis (Simon et al., 1999). These latter may represent sequelae of primary infection, while PML (in macaques and humans) may represent the reactivation of a latent infection as argued in Chapter 12. This hypothesis requires further exploration.

The Evolution of SV40. The evolutionary relationship of SV40 to JCV and BKV seems unlikely to be the result of the division in primate ancestry going back 30 million years. The DNA sequence similarity (72–74%) is too high to be consistent with such a long separation. That means SV40 has jumped species one or more times during its history. In our anthropocentric world we think of animals transmitting diseases to humans and fail to consider that human viruses might equally well be transferred to primates. The history of SV40 virus in humans and macaques should be investigated, and the number and timing of such jumps should be deduced. If the original, ancestral sequence of each virus (SV40, BKV, and JCV) can be recreated, it may eventually be possible to reconstruct the sequence of the hypothetical ancestor of all three viruses, the JBS* virus.

Search for SV40 in Humans. The search for SV40 in neoplastic and normal human tissues by serologic studies, PCR studies, and in situ studies with virus-specific probes and antibodies will continue to be important. In addition, in situ PCR could provide a very sensitive detection system that can reveal the nature and distribution of the infected cells. The DNA sequence of viral strains detected should be determined and these genomes characterized as to both the genotype of the coding region sequence (e.g., in the COOH-terminal region of T antigen) and, where possible, the arrangement of the regulatory region (number of 21 bp GC-rich domains and number of 72 bp enhancer elements) (Ilyinskii et al., 1992; Lednicky et al., 1998; Newman et al., 1998; Stewart et al., 1996). Also important to explore is the presence of DNA sequences representing exon 1 of T antigen (small t antigen) and the intron sequence. Small t antigen is important for mesothelioma induction in the hamster model (Cicala et al., 1993) and is the site of binding of the monoclonal antibody PAb419. Judging by findings in JCV, the intron sequence as well as the intergenic region may be expected to be regions of maximal variability between SV40 strains representing different genotypes. Given the number of different genotypes of JCV (see Chapter 18) and BKV (see Chapter 19), it seems likely that more wild-type genotypes of SV40 will be discovered. It is, however, possible that

SV40 remained sequestered in just a few interbreeding monkey populations for much of its evolutionary history.

SV40-specific serologic studies have been designed that do not correlate with responses to BKV (Jafar et al., 1998), but the uncertainty due to possible cross reaction with antibodies to JCV, BKV, and other polyomaviruses remains. The suggestion that bovine polyomavirus can infect humans (Parry and Gardner, 1986), as well as serologic evidence for an LPV-related virus in humans that has not yet been discovered (Yoshiike and Takemoto, 1986), complicates interpretation of SV40 serology. Questions will always be raised about cross reactivity with known and unknown polyomaviruses. In the future it may be helpful to focus immunologic assays on reactivity to particular major viral epitopes, some of which may be virus specific. Monoclonal antibodies obtained from mice immunized with SV40 capsid protein VP1 suggested the existence of a major epitope between Arg_{313} and Pro_{321} (Babé et al., 1989). This epitope is identical in BKV (Dun) but changed in both JCV and BKV (AS) (Fig. 22.2).

Role of Immunity to T Antigens in Prevention and Treatment of SV40-Associated Infections and Disease. The immunology of SV40 is far ahead of similar studies with BKV or JCV and will undoubtedly continue to lead the way. Nevertheless, the evolutionary distance between these viruses requires that they be studied independently to determine their similarities and differences. Understanding the role of T-antigen-specific CD8-positive cytotoxic T lymphocytes in diagnosis and treatment of SV40-associated diseases will be a major scientific goal in the coming years, as indeed it will be in all other polyomavirus diseases. If SV40 T antigen is involved in the etiology of human tumors, it should leave its mark on the immune system. At the same time, boosting that immune response with T-antigen immunization might provide a means of immune-mediated attack on the tumor (see Chapter 21) (Xie et al., 1999).

Better understanding of the evolution of SV40 in relation to JCV and BKV should reveal whether its history in humans predates its introduction through contaminated polio vaccines in 1954–1963. This is of more than theoretical

```
                           315
SV40         Arg Thr Gln Arg Val Asp Gly Gln Pro
BKV(Dun)     --- --- --- --- --- --- --- --- ---
BKV(AS)      --- --- --- Lys --- --- --- --- ---
JCV          --- --- Pro --- --- --- --- --- ---
                           307
```

Figure 22.2. VP1 amino acid sequence in a major epitope defined by monoclonal antibodies raised to SV40 VP1 (Babé et al., 1989). Dashes indicate identity to the SV40 amino acid sequence. JCV and BKV(AS) are mutated at positions 315 (position 307 in JCV VP1) and 316, respectively (Walker and Frisque, 1986). JCV numbering is altered by an eight-residue deletion near the amino-terminus of the protein.

interest. If endemic SV40 in humans predates the polio vaccination campaign, hope of eradicating it may be nil. On the other hand, if the introduction occurred solely through polio vaccination, it may be that SV40 virus-specific vaccination could be used to eradicate the newly endemic infection within the next century. Future areas for epidemiologic research have been laid out by Rollison and Shah in Chapter 20.

Molecular Biology. SV40 T antigen is one of the most complex, multifunctional regulatory proteins known (Simmons, 2000). Details of its interactions with various cell cycle regulators and their role in tumorigenesis will continue to be explored. If SV40 is not merely an experimental laboratory model of cell transformation and viral tumorigenesis, but an actual human pathogen, this search will take on a new urgency. Elucidation of these points of interaction with cellular proteins will suggest new ways in which the oncogenic properties of SV40 can be blocked.

4. CONCLUSIONS

The human polyomaviruses represent an increasingly important public health concern (see Chapter 5). In renal allograft nephritis (see Chapter 14), in AIDS neurology and neuropathology (see Chapter 11), and in tumorigenesis in a variety of tissues (see Chapters 15, 16, 17, and 20), the polyomaviruses must be taken seriously. No longer merely disease curiosities or molecular biologic tools, the questions they raise also impinge on the fundamental processes of life. These range from issues surrounding human evolution, to the evolution and molecular epidemiology of persistent viruses, to basic questions of cellular growth control involving the pathways mediated by p53 and the pRb family of proteins. Fortunately, the recognition of the importance of human polyomaviruses is growing in the scientific community and with it the commitment to them of research resources and scientific talent. There are many lessons to be drawn from the presentations in this book, and we hope the next 30 years will be as productive as the last 30 years have been since the discovery of the BK and JC human polyomaviruses in 1971.

To conclude, we would like to make two observations from the discussions in this chapter. First, each of these viruses, though closely related to each other, has a unique evolutionary history in humans and other primates, and their biology will therefore differ significantly. While there are many similarities, there are also many differences, and each needs to be investigated independently, contributing its own insights into the persistent life strategy of polyomaviruses. Encoded in each in ways not yet known are the tissue specificities of their primary infections, their persistent, subacute, and life-threatening infections of brain and kidney under immunosuppressive conditions, and their transformation of specific cells as co-factors for tumorigenesis. If the speculations in this chapter are near the mark, each has evolved independently for perhaps the past 1.5–2 million years. An appreciation of that evolutionary history since divergence from their last common viral ancestor will be essential to understanding the life history of these viruses in their natural hosts. This information is essential for predicting and, possibly, controlling, the future course of these endemic and persistent infections. Second, the misconception that SV40 could be tied etiologically to some cases of PML took 25 years to clarify. The much more complex question of the etiologic role of SV40 (and BKV and JCV) in tumorigenesis in humans may also take many years to settle, but with a sharp focus, critical thinking, and a willingness to consider all the evidence, we believe the truth that we all seek will prevail. We hope that the multifaceted presentations in this volume will serve that purpose.

ACKNOWLEDGMENTS

We thank our colleagues for many helpful discussions, and Hansjürgen T. Agostini and David V. Jobes for critically reading the manuscript.

REFERENCES

Achim CL, Wiley CA (1992): Expression of major histocompatibility complex antigens in the brains of patients with progressive multifocal leukoencephalopathy. J Neuropathol Exp Neurol 51:257–263.

Agostini HT, Jobes DV, Chima SC, Ryschkewitsch CF, Stoner GL (1999): Natural and pathogenic variation in the JC virus genome. In Pandalai SG, Ed; Recent Research Developments in Virology; Transworld Research Network: Trivandrum, India, pp 683–701.

Agostini HT, Ryschkewitsch CF, Singer EJ, Baumhefner RW, Stoner GL (1998): JC virus Type 2B is found more frequently in brain tissue of progressive multifocal leukoencephalopathy (PML) patients than in control urines. J Hum Virol 1:200–206.

Agostini HT, Ryschkewitsch CF, Singer EJ, Stoner GL (1997a): JC virus regulatory region rearrangements and genotypes in progressive multifocal leukoencephalopathy: Two independent aspects of virus variation. J Gen Virol 78:659–664.

Agostini HT, Yanagihara R, Davis V, Ryschkewitsch CF, Stoner GL (1997b): Asian genotypes of JC virus in Native Americans and in a Pacific Island population: Markers of viral evolution and human migration. Proc Natl Acad Sci USA 94:14542–14546.

Andre FE (1979): Poliomyelitis vaccines in Belgium: 20 years of experience. Dev Biol Stand 43:187–193.

Andréoletti L, Dubois V, Lescieux A, Dewilde A, Bocket L, Fleury HJ, Wattre P (1999): Human polyomavirus JC latency and reactivation status in blood of HIV-1–positive immunocompromised patients with and without progressive multifocal leukoencephalopathy. AIDS 13:1469–1475.

Arrington AS, Lednicky JA, Butel JS (2000): Molecular characterization of SV40 DNA in multiple samples from a human mesothelioma. Anticancer Res 20:879–884.

Åström KE, Mancall EL, Richardson EP Jr (1958): Progressive multifocal leukoencephalopathy: A hitherto unrecognized complication of chronic lymphocytic leukemia and Hodgkin's disease. Brain 81:93–111.

Ault GS, Stoner GL (1994): Brain and kidney of progressive multifocal leukoencephalopathy patients contain identical rearrangements of the JC virus promoter/enhancer. J Med Virol 44:298–304.

Azzi A, Cesaro S, Laszlo D, Zakrzewska K, Ciappi S, De Santis R, Fanci R, Pesavento G, Calore E, Bosi A (1999): Human polyomavirus BK (BKV) load and haemorrhagic cystitis in bone marrow transplantation patients. J Clin Virol 14:79–86.

Azzi A, De Santis R, Ciappi S, Leoncini F, Sterrantino G, Marino N, Mazzotta F, Laszlo D, Fanci R, Bosi A (1996): Human polyomaviruses DNA detection in peripheral blood leukocytes from immunocompetent and immunocompromised individuals. J Neurovirol 2:411–416.

Babé LM, Brew K, Matsuura SE, Scott WA (1989): Epitopes on the major capsid protein of simian virus 40. J Biol Chem 264:2665–2671.

Barbanti-Brodano G, Bendinelli M, Friedman H (1995): DNA Tumor Viruses Oncogenic Mechanisms; Plenum Press: New York.

Bastien C, Feunteun J (1988): The hamster polyomavirus transforming properties. Oncogene 2:129–135.

Berjanskii MV, Riley MI, Xie AY, Semenchenko V, Folk WR, Van Doren SR (2000): NMR structure of the N-terminal J domain of murine polyomavirus T antigens—Implications for DnaJ-like domains and for mutations of T antigens. J Biol Chem 275:36094–36103.

Binet I, Nickeleit V, Hirsch HH, Prince O, Dalquen P, Gudat F, Mihatsch MJ, Thiel G (1999): Polyomavirus disease under new immunosuppressive drugs—A cause of renal graft dysfunction and graft loss. Transplantation 67:918–922.

Blaho JA, Aaronson SA (1999): Convicting a human tumor virus: Guilt by association? Proc Natl Acad Sci USA 96:7619–7621.

Bocchetta M, DiResta I, Powers A, Fresco R, Tosolini A, Testa JR, Pass HI, Rizzo P, Carbone M (2000): Human mesothelial cells are unusually susceptible to simian virus 40–mediated transformation and asbestos cocarcinogenicity. Proc Natl Acad Sci USA 97:10214–10219.

Boubenider S, Hiesse C, Marchand S, Hafi A, Kriaa F, Charpentier B (1999): Post-transplantation polyomavirus infections. J Nephrol 12:24–29.

Bozeman LH (1981): Characterization of a papovavirus isolated from fledging budgerigars. Avain Dis 25:972–980.

Bratt G, Hammarin AL, Grandien M, Hedquist BG, Nennesmo I, Sundelin B, Seregard S (1999): BK virus as the cause of meningoencephalitis, retinitis and nephritis in a patient with AIDS. AIDS 13:1071–1075.

Bruner JM, Hair LS (1993): Detection of abnormal p53 in progressive multifocal leukoencephalopathy. J Neuropathol Exp Neurol 52:271.

Butel JS, Lednicky JA (1999): Cell and molecular biology of simian virus 40: Implications for human infections and disease. J Natl Cancer Inst 91:119–134.

Carbone M, Rizzo P, Grimley PM, Procopio A, Mew DJY, Shridhar V, De Bartolomeis A, Esposito V, Giuliano MT, Steinberg SM, Levine AS, Giordano A, Pass HI

(1997a): Simian virus–40 large-T antigen binds p53 in human mesotheliomas. Nature Med 3:908–912.

Carbone M, Rizzo P, Pass H (2000): Simian virus 40: The link with human malignant mesothelioma is well established. Anticancer Res 20:875–877.

Carbone M, Rizzo P, Pass HI (1995): Association of simian virus 40 with rodent and human mesotheliomas. In Barbanti-Brodano G, Bendinelli M, Friedman H, Eds; DNA Tumor Viruses Oncogenic Mechanisms; Plenum Press: New York, pp 75–90.

Carbone M, Rizzo P, Pass HI (1997b): Simian virus 40, poliovaccines and human tumors: A review of recent developments. Oncogene 15:1877–1888.

Carbone M, Rizzo P, Procopio A, Giuliano M, Pass HI, Gebhardt MC, Mangham C, Hansen M, Malkin DF, Bushart G, Pompetti F, Picci P, Levine AS, Bergsagel JD, Garcea RL (1996): SV40-like sequences in human bone tumors. Oncogene 13:527–535.

Challoner PB, Smith KT, Parker JD, MacLeod DL, Coulter SN, Rose TM, Schultz ER, Bennett JL, Garber RL, Chang M, Schad PA, Stewart PM, Nowinski RC, Brown JP, Burmer GC (1995): Plaque-associated expression of human herpesvirus 6 in multiple sclerosis. Proc Natl Acad Sci USA 92:7440–7444.

Chatterjee M, Weyandt TB, Frisque RJ (2000): Identification of archetype and rearranged forms of BK virus in leukocytes from healthy individuals. J Med Virol 60: 353–362.

Chima SC, Ryschkewitsch CF, Stoner GL (1998): Molecular epidemiology of human polyomavirus JC in the Biaka pygmies and Bantu of Central Africa. Mem Inst Oswaldo Cruz 93:615–623.

Chowdhury M, Taylor JP, Chang C-F, Rappaport J, Khalili K (1992): Evidence that a sequence similar to TAR is important for induction of the JC virus late promoter by human immunodeficiency virus type 1 Tat. J Virol 66:7355–7361.

Cicala C, Pompetti F, Carbone M (1993): SV40 induces mesotheliomas in hamsters. Am J Pathol 142:1524–1533.

Cole CN (1996): Polyomavirinae: The viruses and their replication. In Fields BN, Knipe DM, Howley PM, Eds; Fields Virology; Lippincott-Raven Publishers: Philadelphia, pp 1997–2025.

Coleman DV, Wolfendale MR, Daniel RA, Dhanjal NK, Gardner SD, Gibson PE, Field AM (1980): A prospective study of human polyomavirus infection in pregnancy. J Infect Dis 142:1–8.

Cunningham TP, Pipas JM (1985): Simian agent 12 is a BK virus-like papovavirus which replicates in monkey cells. J Virol 54:483–492.

Cuomo L, Trivedi P, Cardillo MR, Gagliardi FM, Vecchione A, Caruso R, Calogero A, Frati L, Faggioni A, Ragona G (2001): Human herpesvirus 6 infection in neoplastic and normal brain tissue. J Med Virol 63:45–51.

Darmon AJ, Jat PS (2000): BAP37 and prohibitin are specifically recognized by an SV40 T antigen antibody. Mol Cell Biol Res Commun 4:219–223.

Delmas V, Bastien C, Scherneck S, Feunteun J (1985): A new member of the polyomavirus family: The hamster papovavirus; complete nucleotide sequence and transformation properties. EMBO J 4:1279–1286.

De Luca A, Baldi A, Esposito V, Howard CM, Bagella L, Rizzo P, Caputi M, Pass HI, Giordano GG, Baldi F, Carbone M, Giordano A (1997): The retinoblastoma gene family pRb/p105, p107, pRb2/p130 and simian virus-40 large T-antigen in human mesotheliomas. Nature Med 3:913–916.

Dhaene K, Verhulst A, Van Marck E (1999): SV40 large T-antigen and human pleural mesothelioma—Screening by polymerase chain reaction and tyramine-amplified immunohistochemistry. Virchows Arch Int J Pathol 435:1–7.

Dörries K (1994): Infection of human polyomavirus JC and BK peripheral blood leukocytes from immunocompetent individuals. Virology 198:59–70.

Drachenberg CB, Beskow CO, Cangro CB, Bourquin PM, Simsir A, Fink J, Weir MR, Klassen DK, Bartlett ST, Papadimitriou JC (1999): Human polyoma virus in renal allograft biopsies: Morphological findings and correlation with urine cytology. Hum Pathol 30:970–977.

Drews K, Bashir T, Dörries K (2000): Quantification of human polyomavirus JC in brain tissue and cerebrospinal fluid of patients with progressive multifocal leukoencephalopathy by competitive PCR. J Virol Methods 84:23–36.

Dubois V, Dutronc H, Lafon ME, Poinsot V, Pellegrin JL, Ragnaud JM, Ferrer AM, Fleury HJA (1997): Latency and reactivation of JC virus in peripheral blood of human immunodeficiency virus type 1–infected patients. J Clin Microbiol 35:2288–2292.

Dubois V, Moret H, Lafon ME, Brodard V, Icart J, Ruffault A, Guist'hau O, Buffet-Janvresse C, Abbed K, Dussaix E, Ingrand D (2001): JC virus genotypes in France: Molecular epidemiology and potential significance for progressive multifocal leukoencephalopathy. J Infect Dis 183:213–217.

Dubois V, Moret H, Lafon ME, Janvresse CB, Dussaix E, Icart J, Karaterki A, Ruffault A, Taoufik Y, Vignoli C, Ingrand D (1998): Prevalence of JC virus viraemia in HIV-infected patients with or without neurological disorders: A prospective study. J Neurovirol 4:539–544.

Dyson N, Bernards R, Friend SH, Gooding LR, Hassell JA, Major EO, Pipas JM, Vandyke T, Harlow E (1990): Large T antigens of many polyomaviruses are able to form complexes with the retinoblastoma protein. J Virol 64:1353–1356.

Eddy BE, Borman GS, Grubbs GE, Young RD (1962): Identification of the oncogenic substance in rhesus monkey kidney cell cultures as SV40. Virology 17:65–75.

Eizuru Y, Sakihama K, Minamishima Y, Hayashi T, Sumiyoshi A (1993): Reevaluation of a case of progressive multifocal leukoencephalopathy previously diagnosed as simian virus 40 (SV40) etiology. Acta Pathol Jpn 43:327–332.

Elsner C, Dörries K (1998): Human polyomavirus JC control region variants in persistently infected CNS and kidney tissue. J Gen Virol 79:789–799.

Emri S, Kocagoz T, Olut A, Güngen Y, Mutti L, Baris YI (2000): Simian virus 40 is not a cofactor in the pathogenesis of environmentally induced malignant pleural mesothelioma in Turkey. Anticancer Res 20:891–894.

Ewald PW (1994): Evolution of Infectious Disease. Oxford University Press: Oxford.

Farge D, Herve R, Mikol J, Sauvaget F, Ingrand D, Singer B, Ferchal F, Auperin I, Gray F, Sudaka A, Degos L, Rouffy J (1994): Simultaneous progressive multifocal leukoencephalopathy, Epstein- Barr virus (EBV) latent infection and cerebral parenchymal infiltration during chronic lymphocytic leukemia. Leuk 8:318–321.

Fernandez Cobo M, Jobes DV, Yanagihara RT, Nerurkar VR, Yamamura Y, Ryschkewitsch CF, Stoner GL (2001): Reconstructing population history using JC virus:

Amerinds, Spanish, and Africans in the ancestry of modern Puerto Ricans. Hum Biol 73:385–402.

Ferrante P, Caldarelli-Stefano R, Omodeo-Zorini E, Cagni AE, Cocchi L, Suter F, Maserati R (1997): Comprehensive investigation of the presence of JC virus in AIDS patients with and without progressive multifocal leukoencephalopathy. J Med Virol 52:235–242.

Fiers W, Contreras R, Haegeman G, Rogiers R, van de Voorde A, van Heyverswyn H, van Herreweghe J, Volckaert G, Ysebaert M (1978): Complete nucleotide sequence of SV40 DNA. Nature 273:113–120.

Francis TJ, Napier JA, Voight RB, Hemphill FM, Werner HA, Korns RF, Boisen M, Tolchinsky E, Diamond EL (1957): Evaluation of the 1954 Field Trial of Poliomyelitis Vaccine: Final Report; Ann Arbor: Edwards Brothers, 327–331.

Freund R, Dubensky T, Bronson R, Sotnikov A, Carroll J, Benjamin T (1992): Polyoma tumorigenesis in mice: Evidence for dominant resistance and dominant susceptibility genes of the host. Virology 191:724–731.

Friedlander A, Patarca R (1999): DNA viruses and oncogenesis. Crit Rev Oncol 10: 161–238.

Frisque RJ, Bream GL, Cannella MT (1984): Human polyomavirus JC virus genome. J Virol 51:458–469.

Frisque RJ, White FA (1992): The molecular biology of JC virus, causative agent of progressive multifocal leukoencephalopathy. In Roos RP, Ed; Neurovirology—Pathogenesis of Viral CNS Infections. Humana Press: Totowa, NJ, pp 25–158.

Gabunia L, Vekua A, Lordkipanidze D, Swisher CC, Ferring R, Justus A, Nioradze M, Tvalchrelidze M, Antón SC, Bosinski G, Jöris O, De Lumley MA, Majsuradze G, Mouskhelishvili A (2000): Earliest Pleistocene hominid cranial remains from Dmanisi, Republic of Georgia: Taxonomy, geological setting, and age. Science 288: 1019–1025.

Galateau-Salle F, Bidet P, Iwatsubo Y, Gennetay E, Renier A, Letourneux M, Pairon JC, Moritz S, Brochard P, Jaurand MC, Freymuth F (1998): SV40-like DNA sequences in pleural mesothelioma, bronchopulmonary carcinoma, and nonmalignant pulmonary diseases. J Pathol 184:252–257.

Gardner SD, Field AM, Coleman DV, Hulme B (1971): New human papovavirus (B.K.) isolated from urine after renal transplantation. Lancet 1:1253–1257.

Gardner SD, Knowles WA, Hand JF, Porter AA (1989): Characterization of a new polyomavirus (*Polyomavirus papionis*-2) isolated from baboon kidney cell cultures. Arch Virol 105:223–233.

Goffe AP, Hale J, Gardner PS (1961): Poliomyelitis vaccine. Lancet 1:612.

Goudsmit J, Baak ML, Slataerus KW, Vandernoordaa J (1981): Human papovavirus isolated from the urine of a child with acute tonsilitis. BMJ 283:1363–1364.

Goudsmit J, Wertheim-van Dillen P, Van Strien A, van der Noordaa J (1982): The role of BK virus in acute respiratory tract disease and the presence of BKV DNA in tonsils. J Med Virol 10:91–99.

Griffin BE, Soeda E, Barrell BG, Staden R (1981): Appendix B: Sequence and analysis of polyoma virus DNA. In Tooze J, Ed; DNA Tumor Viruses; Cold Spring Harbor: Cold Spring Harbor Laboratory, pp 843–910.

Guerin JL, Gelfi J, Dubois L, Vuillaume A, Boucraut-Baralon C, Pingret JL (2000): A novel polyomavirus (goose hemorrhagic polyomavirus) is the agent of hemorrhagic nephritis enteritis of geese. J Virol 74:4523–4529.

Guillaume B, Sindic CJ, Weber T (2000): Progressive multifocal leukoencephalopathy: Simultaneous detection of JCV DNA and anti-JCV antibodies in the cerebrospinal fluid. Eur J Neurol 7:101–106.

Guo J, Kitamura T, Ebihara H, Sugimoto C, Kunitake T, Takehisa J, Na YQ, Al-Ahdal MN, Hallin A, Kawabe K, Taguchi F, Yogo Y (1996): Geographical distribution of the human polyomavirus JC virus type A and B and isolation of a new type from Ghana. J Gen Virol 77:919–927.

Hatwell JN, Sharp PM (2000): Evolution of human polyomavirus JC. J Gen Virol 81: 1191–1200.

Haun G, Keppler OT, Bock CT, Herrmann M, Zentgraf H, Pawlita M (1993): The cell surface receptor is a major determinant restricting the host range of the B-lymphotropic papovavirus. J Virol 67:7482–7492.

Heilbronn R, Albrecht I, Stephan S, Bürkle A, zur Hausen H (1993): Human cytomegalovirus induces JC virus DNA replication in human fibroblasts. Proc Natl Acad Sci USA 90:11406–11410.

Heinsohn S, Scholz RB, Weber B, Wittenstein B, Werner M, Delling G, Kempf-Bielack B, Setlak P, Bielack S, Kabisch H (2000): SV40 sequences in human osteosarcoma of German origin. Anticancer Res 20:4539–4545.

Hilleman MR (1998): Discovery of simian virus 40 (SV_{40}) and its relationship to poliomyelitis virus vaccines. Dev Biol Stand 94:183–190.

Hirvonen A, Mattson K, Karjalainen A, Ollikainen T, Tammilehto L, Hovi T, Vainio H, Pass HI, DiResta I, Carbone M, Linnainmaa K (1999): Simian virus 40 (SV40)–like DNA sequences not detectable in Finnish mesothelioma patients not exposed to SV40-contaminated polio vaccines. Mol Carcinog 26:93–99.

Hogan TF, Padgett BL, Walker DL (1991): Human polyomaviruses. In Belshe RB, Ed; Textbook of Human Virology; St. Louis: Mosby, pp 970–1000.

Houff SA, Major EO, Katz DA, Kufta CV, Sever JL, Pittaluga S, Roberts JR, Gitt J, Saini N, Lux W (1988): Involvement of JC virus–infected mononuclear cells from the bone marrow and spleen in the pathogenesis of progressive multifocal leukoencephalopathy. N Engl J Med 318:301–305.

Howell DN, Smith SR, Butterly DW, Klassen PS, Krigman HR, Burchette JL Jr, Miller SE (1999): Diagnosis and management of BK polyomavirus interstitial nephritis in renal transplant recipients. Transplantation 68:1279–1288.

Howley PM, Levine AJ, Li FP, Livingston DM, Rabson AS (1991): Lack of SV40 DNA in tumors from scientists working with SV40 virus [letter]. N Engl J Med 324:494.

Ilyinskii PO, Daniel MD, Horvath CJ, Desrosiers RC (1992): Genetic analysis of simian virus 40 from brains and kidneys of macaque monkeys. J Virol 66:6353–6360.

Ito M, Baker JV, Mock DJ, Goodman AD, Blumberg BM, Shrier DA, Powers JM (2000): Human herpesvirus 6–meningoencephalitis in an HIV patient with progressive multifocal leukoencephalopathy. Acta Neuropathol (Berl) 100:337–341.

Jafar S, Rodriguez-Barradas M, Graham DY, Butel JS (1998): Serological evidence of SV40 infections in HIV-infected and HIV-negative adults. J Med Virol 54:276–284.

Jensen PN, Major EO (1999): Viral variant nucleotide sequences help expose leukocytic positioning in the JC virus pathway to the CNS. J Leukocyte Biol 65:428–438.

Jin L, Underhill PA, Doctor V, Davis RW, Shen PD, Cavalli-Sforza LL, Oefner PJ (1999): Distribution of haplotypes from a chromosome 21 region distinguishes multiple prehistoric human migrations. Proc Natl Acad Sci USA 96:3796–3800.

Jobes DV, Friedlaender JS, Mgone CS, Koki G, Alpers MP, Ryschkewitsch CF, Stoner GL (1999): A novel JC virus variant found in the Highlands of Papua New Guinea has a 21-base pair deletion in the agnoprotein gene. J Hum Virol 2:350–358.

Jones S, Pilbeam D, Bunney S, Jones SR, Martin RD (1992): Cambridge Encyclopedia of Human Evolution; Cambridge University Press: Cambridge.

Kappeler A, Lutz-Wallace C, Sapp T, Sidhu M (1996): Detection of bovine polyomavirus contamination in fetal bovine sera and modified live viral vaccines using polymerase chain reaction. Biologicals 24:131–135.

Klawans HL (1988): Toscanini's Fumble and Other Tales of Clinical Neurology; Contemporary Books: Chicago, pp 124–125.

Knowles WA, Pillay D, Johnson MA, Hand J, Brown DWG (1999): Prevalence of long-term BK and JC excretion in HIV-infected adults and lack of correlation with serological markers. J Med Virol 59:474–479.

Kops SP (2000): Oral polio vaccine and human cancer: A reassessment of SV40 as a contaminant based upon legal documents. Anticancer Res 20:4745–4749.

Koralnik IJ, Boden D, Mai VX, Lord CI, Letvin NL (1999): JC virus DNA load in patients with and without progressive multifocal leukoencephalopathy. Neurology 52:253–260.

Koralnik IJ, Du Pasquier RA, Letvin NL (2001): JC virus–specific cytotoxic T lymphocytes in individuals with progressive multifocal leukoencephalopathy. J Virol 75: 3483–3487.

Kristoffersen AK, Johnsen JI, Seternes OM, Rollag H, Degré M, Traavik T (1997): The human polyomavirus BK T antigen induces gene expression in human cytomegalovirus. Virus Res 52:61–71.

Kuska R (1999): SV40 bugaboo: Spinning the news. J Natl Cancer Inst 91:662–664.

Lafon ME, Dutronc H, Dubois V, Pellegrin I, Barbeau P, Ragnaud JM, Pellegrin JL, Fleury HJA (1998): JC virus remains latent in peripheral blood B lymphocytes but replicates actively in urine from AIDS patients. J Infect Dis 177:1502–1505.

Latimer KS, Niagro FD, Steffens WL, Ritchie BW, Campagnoli RP (1996): Polyomavirus encephalopathy in a Ducorps' cockatoo (*Cacatua ducorpsii*) with psittacine beak and feather disease. J Vet Diagn Invest 8:291–295.

Lechner JF, Tesfaigzi J, Gerwin BI (1997): Oncogenes and tumor-suppressor genes in mesothelioma—A synopsis. Environ Health Perspect 105:1061–1067.

Lednicky JA, Arrington AS, Stewart AR, Dai XM, Wong C, Jafar S, Murphey-Corb M, Butel JS (1998): Natural isolates of simian virus 40 from immunocompromised monkeys display extensive genetic heterogeneity: New implications for polyomavirus disease. J Virol 72:3980–3990.

Levine AJ (1994): The origins of the small DNA tumor viruses. Adv Cancer Res 65: 141–168.

Lukacher AE, Ma Y, Carroll JP, Abromson-Leeman SR, Laning JC, Dorf ME, Benjamin TL (1995): Susceptibility to tumors induced by polyoma virus is conferred by an endogenous mouse mammary tumor virus superantigen. J Exp Med 181:1683–1692.

Luo XL, Sanford DG, Bullock PA, Bachovchin WW (1996): Solution structure of the origin DNA-binding domain of SV40 T-antigen. Nature Struct Biol 3:1034–1039.

Markowitz RB, Thompson HC, Mueller JF, Cohen JA, Dynan WS (1993): Incidence of BK virus and JC virus viruria in human immunodeficiency virus–infected and virus–uninfected subjects. J Infect Dis 167:13–20.

Martini F, De Mattei M, Iaccheri L, Corallini A, Barbanti-Brodano G, Tognon M (1995): TAg coding sequences and mutations of p53 tumor suppressor gene in human brain tumors and normal brain tissues. Min Biotecnol 7:129–132.

Matsiota-Bernard P, De Truchis P, Gray F, Flament-Saillour M, Voyatzakis E, Nauciel C (1997): JC virus detection in the cerebrospinal fluid of AIDS patients with progressive multifocal leucoencephalopathy and monitoring of the antiviral treatment by a PCR method. J Med Microbiol 46:256–259.

Mayall FG, Jacobson G, Wilkins R (1999): Mutations of p53 gene and SV40 sequences in asbestos associated and nonasbestos-associated mesotheliomas. J Clin Pathol 52: 291–293.

Mayer M, Dörries K (1991): Nucleotide sequence and genome organization of the murine polyomavirus, Kilham strain. Virology 181:469–480.

McCance DJ, Mims CA (1979): Reactivation of polyoma virus in kidneys of persistently infected mice during pregnancy. Infect Immun 25:998–1002.

McDonald JC, McDonald AD (1996): The epidemiology of mesothelioma in historical context. Eur Respir J 9:1932–1942.

McGuire D, Barhite S, Hollander H, Miles M (1995): JC virus DNA in cerebrospinal fluid of human immunodeficiency virus–infected patients: Predictive value for progressive multifocal leukoencephalopathy. Ann Neurol 37:395–399.

McLaren BR, Haenel T, Stevenson S, Mukherjee S, Robinson BWS, Lake RA (2000): Simian virus (SV) 40 like sequences in cell lines and tumour biopsies from Australian malignant mesotheliomas. Aust NZ J Med 30:450–456.

Melnick JL, Stinebaugh S (1962): Excretion of vacuolating SV40 virus (papova virus group) after ingestion as a contaminant of oral poliovaccine. Proc Soc Exp Biol Med 109:965–968.

Merriwether DA, Huston S, Iyengar S, Hamman R, Norris JM, Shetterly SM, Kamboh MI, Ferrell RE (1997): Mitochondrial versus nuclear admixture estimates demonstrate a past history of directional mating. Am J Phys Anthropol 102:153–159.

Moarefi IF, Small D, Gilbert I, Höpfner M, Randall SK, Schneider C, Russo AAR, Ramsperger U, Arthur AK, Stahl H, Kelly TJ, Fanning E (1993): Mutation of the cyclin-dependent kinase phosphorylation site in simian virus 40 (SV40) large T antigen specifically blocks SV40 origin DNA unwinding. J Virol 67:4992–5002.

Mock DJ, Powers JM, Goodman AD, Blumenthal SR, Ergin N, Baker JV, Mattson DH, Assouline JG, Bergey EJ, Chen BJ, Epstein LG, Blumberg BM (1999): Association of human herpesvirus 6 with the demyelinative lesions of progressive multifocal leukoencephalopathy. J Neurovirol 5:363–373.

Modi S, Kubo A, Oie H, Coxon AB, Rehmatulla A, Kaye FJ (2000): Protein expression of the RB-related gene family and SV40 large T antigen in mesothelioma and lung cancer. Oncogene 19:4632–4639.

Monaco MCG, Jensen PN, Hou J, Durham LC, Major EO (1998): Detection of JC virus DNA in human tonsil tissue: Evidence for site of initial viral infection. J Virol 72:9918–9923.

Montano X, Lane DP (1984): Monoclonal antibody to simian virus 40 small t. J Virol 51:760–767.

Montano X, Lane DP (1989): Monoclonal antibody analysis of simian virus 40 small t-antigen expression in infected and transformed cells. J Virol 63:3128–3134.

Mouratoff JG, Tokumoto J, Olson JL, Chertow GM (2000): Acute renal failure with interstitial nephritis in a patient with AIDS. Am J Kidney Dis 35:557–561.

Narayan O, Penney JB, Johnson RT, Herndon RM, Weiner LP (1973): Etiology of progressive multifocal leukoencephalopathy. Identification of papovavirus. N Engl J Med 289:1278–1282.

Newman JS, Baskin GB, Frisque RJ (1998): Identification of SV40 in brain, kidney and urine of healthy and SIV-infected rhesus monkeys. J Neurovirol 4:394–406.

Newman JT, Frisque RJ (1997): Detection of archetype and rearranged variants of JC virus in multiple tissues from a pediatric PML patient. J Med Virol 52:243–252.

Newman JT, Frisque RJ (1999): Identification of JC virus variants in multiple tissues of pediatric and adult PML patients. J Med Virol 58:79–86.

Nickeleit V, Klimkait T, Binet IF, Dalquen P, Del Zenero V, Thiel G, Mihatsch MJ, Hirsch HH (2000): Testing for polyomavirus type BK DNA in plasma to identify renal-allograft recipients with viral nephropathy. N Engl J Med 342:1309–1315.

O'Neill FJ, Carney H, Hu Y (1998): Host range analysis of simian virus 40, BK virus and chimaeric SV40/BKV: Relative expression of large T-antigen and Vp1 in infected and transformed cells. Dev Biol Stand 94:191–205.

Padgett BL, Walker DL, Zu Rhein GM, Eckroade RJ, Dessel BH (1971): Cultivation of papova-like virus from human brain with progressive multifocal leucoencephalopathy. Lancet 1:1257–1260.

Page RDM, Holmes EC (1998): Molecular Evolution—A Phylogenetic Approach; Blackwell Science: Oxford, pp. 147.

Parry JV, Gardner SD (1986): Human exposure to bovine polyomavirus: A zoonosis? Arch Virol 87:287–296.

Parry JV, Lucas MH, Richmond JE, Gardner SD (1983): Evidence for a bovine origin of the polyomavirus detected in foetal rhesus monkey kidney cells, FRhK-4 and -6. Arch Virol 78:151–165.

Parvala ME (1958): Reports of Official Delegates. In O'Connor B, Ed; Poliomyelitis. Papers and Discussions Presented at the Fourth International Poliomyelitis Conference; J.B. Lippincott Co: Philadelphia, pp 32–33.

Pass HI, Donington JS, Wu P, Rizzo P, Nishimura M, Kennedy R, Carbone M (1998): Human mesotheliomas contain the simian virus-40 regulatory region and large tumor antigen DNA sequences. J Thorac Cardiovasc Surg 116:854–859.

Pawlita M, Clad A, zur Hausen H (1985): Complete DNA sequence of lymphotropic papovavirus (LPV): Prototype of a new species of the polyomavirus genus. Virology 143:196–211.

Peinemann F, De Villiers EM, Dörries K, Adams O, Vögeli TA, Burdach S (2000): Clinical course and treatment of haemorrhagic cystitis associated with BK type of human polyomavirus in nine paediatric recipients of allogeneic bone marrow transplants. Eur J Pediatr 159:182–188.

Perrons CJ, Fox JD, Lucas SB, Brink NS, Tedder RS, Miller RF (1996): Detection of polyomaviral DNA in clinical samples from immunocompromised patients: Correlation with clinical disease. J Infect 32:205–209.

Phalen DN, Wilson VG, Gaskin JM, Derr JN, Graham DL (1999): Genetic diversity in twenty variants of the avian polyomavirus. Avian Dis 43:207–218.

Pilatte Y, Vivo C, Renier A, Kheuang L, Greffard A, Jaurand MC (2000): Absence of SV40 large T-antigen expression in human mesothelioma cell lines. Am J Respir Cell Mol Biol 23:788–793.

Power C, Gladden JGB, Halliday W, Del Bigio MR, Nath A, Ni W, Major EO, Blanchard J, Mowat M (2000): AIDS- and non-AIDS–related PML association with distinct p53 polymorphism. Neurology 54:743–746.

Price B (1997): Analysis of current trends in United States mesothelioma incidence. Am J Epidemiol 145:211–218.

Procopio A, Strizzi L, Vianale G, Betta P, Puntoni R, Fontana V, Tassi G, Gareri F, Mutti L (2000): Simian virus–40 sequences are a negative prognostic cofactor in patients with malignant pleural mesothelioma. Genes Chromosomes Cancer 29:173–179.

Prokoph H, Arnold W, Schwartz A, Scherneck S (1996): In vivo replication of hamster polyomavirus DNA displays lymphotropism in hamsters susceptible to lymphoma induction. J Gen Virol 77:2165–2172.

Quinlivan EB, Norris M, Bouldin TW, Suzuki K, Meeker R, Smith MS, Hall C, Kenney S (1992): Subclinical central nervous system infection with JC virus in patients with AIDS. J Infect Dis 166:80–85.

Ramael M, Nagels J (1999): Re. SV40-like DNA sequences in pleural mesothelioma, bronchopulmonary carcinoma and non-malignant pulmonary disease. J Pathol 189: 628–629.

Ramael M, Nagels J, Heylen H, De Schepper S, Paulussen J, De Maeyer M, Van Haesendonck C (1999): Detection of SV40 like viral DNA and viral antigens in malignant pleural mesothelioma. Eur Respir J 14:1381–1386.

Randhawa PS, Finkelstein S, Scantlebury V, Shapiro R, Vivas C, Jordan M, Pickin MM, Demetris AJ (1999): Human polyoma virus–associated interstitial nephritis in the allograft kidney. Transplantation 67:103–109.

Ressetar HG, Prakash O, Frisque RJ, Webster HD, Re RN, Stoner GL (1993): Expression of viral T-antigen in pathological tissues from transgenic mice carrying JC-SV40 chimeric DNAs. Mol Chem Neuropathol 20:59–79.

Ricciardiello L, Chang DK, Laghi L, Goel A, Chang CL, Boland CR (2001): Mad-1 is the exclusive JC virus strain present in the human colon, and its transcriptional control region has a deleted 98-base-pair sequence in colon cancer tissues. J Virol 75:1996–2001.

Ricciardiello L, Laghi L, Ramamirtham P, Chang CL, Chang DK, Randolph AE, Boland CR (2000): JC virus DNA sequences are frequently present in the human upper and lower gastrointestinal tract. Gastroenterology 119:1228–1235.

Ritchie BW, Vaughn SB, St Leger J, Rich GA, Rupiper DJ, Forgey G, Greenacre CB, Latimer KS, Pesti D, Campagnoli R, Lukert PD (1998): Use of an inactivated virus vaccine to control polyomavirus outbreaks in nine flocks of psittacine birds. J Am Vet Med Assoc 212:685–690.

Rizzo P, DiResta I, Powers A, Ratner H, Carbone M (1999): Unique strains of SV40 in commercial poliovaccines from 1955 not readily identifiable with current testing for SV40 infection. Cancer Res 59:6103–6108.

Rizzo P, Di Resta I, Powers A, Matker CM, Zhang A, Mutti L, Kast WM, Pass H, Carbone M (1998): The detection of simian virus 40 in human tumors by polymerase chain reaction. Monaldi Arch Chest Dis 53:202–210.

Rosen S, Harmon W, Krensky AM, Edelson PJ, Padgett BL, Grinnell BW, Rubino MJ, Walker DL (1983): Tubulo-interstitial nephritis associated with polyomavirus (BK type) infection. N Engl J Med 308:1192–1196.

Rott O, Kroger M, Muller H, Hobom G (1988): The genome of budgerigar fledgling disease virus, an avian polyomavirus. Virology 165:74–86.

Ryschkewitsch CF, Friedlaender JS, Mgone CS, Jobes DV, Agostini HT, Chima SC, Alpers MP, Koki G, Yanagihara R, Stoner GL (2000): Human polyomavirus JC variants in Papua New Guinea and Guam reflect ancient population settlement and viral evolution. Microbes Infect 2:987–996.

Sangar DV, Wood DJ, Minor PD (1998): Examination of poliovaccines for the presence of SV40 sequences. Dev Biol Stand 94:221–225.

Sans M (2000): Admixture studies in Latin America: From the 20th to the 21st century. Hum Biol 72:155–177.

Scherneck S, Geissler E, Jänisch W, Rudolph M, Vogel F, Zimmermann W (1981): Isolation of a SV40-like virus from a patient with progressive multifocal leukoencephalopathy. Acta Virol 25:191–198.

Schiller JT (1999): Papillomavirus-like particle vaccines for cervical cancer. Mol Med Today 5:209–215.

Schuurman R, Sol C, van der Noordaa J (1990): The complete nucleotide sequence of bovine polyomavirus. J Gen Virol 71:1723–1735.

Schuurman R, Van Strien A, Van Steenis B, van der Noordaa J, Sol C (1992): Bovine polyomavirus, a cell-transforming virus with tumorigenic potential. J Gen Virol 73: 2871–2878.

Seif I, Khoury G, Dhar R (1979): The genome of human papovavirus BKV. Cell 18: 963–977.

Shadan FF, Villarreal LP (1995): The evolution of small DNA viruses of eukaryotes: Past and present considerations. Virus Genes 11:239–257.

Shah KV, Daniel RW, Strickler HD, Goedert JJ (1997): Investigation of human urine for genomic sequences of the primate polyomaviruses simian virus 40, BK virus, and JC virus. J Infect Dis 176:1618–1621.

Shah K, Nathanson N (1976): Human exposure to SV40: Review and comment. Am J Epidemiol 103:1–12.

Shapiro GI, Edwards CD, Rollins BJ (2000): The physiology of p16 (INK4A)–mediated G1 proliferative arrest. Cell Biochem Biophys 33:189–197.

Sharp PA (1994): Split genes and RNA splicing. Cell 77:805–815.

Sheppard HM, Corneillie SI, Espiritu C, Gatti A, Liu XA (1999): New insights into the mechanism of inhibition of p53 by simian virus 40 large T antigen. Mol Cell Biol 19:2746–2753.

Shivapurkar N, Wiethege T, Wistuba II, Salomon E, Milchgrub S, Muller KM, Churg A, Pass H, Gazdar AF (2000): Presence of simian virus 40 sequences in malignant mesotheliomas and mesothelial cell proliferations. J Cell Biochem 76:181–188.

Silverman L, Rubinstein LJ (1965): Electron microscopic observations on a case of progressive multifocal leukoencephalopathy. Acta Neuropathol (Berl) 5:215–224.

Simmons DT (2000): SV40 large T antigen functions in DNA replication and transformation. Adv Virus Res 55:75–134.

Simmons DT, Loeber G, Tegtmeyer P (1990): Four major sequence elements of simian virus 40 large T antigen coordinate its specific and nonspecific DNA binding. J Virol 64:1973–1983.

Simon MA, Ilyinskii PO, Baskin GB, Knight HY, Pauley DR, Lackner AA (1999): Association of simian virus 40 with a central nervous system lesion distinct from progressive multifocal leukoencephalopathy in macaques with AIDS. Am J Pathol 154:437–446.

Smith RD, Galla JH, Skahan K, Anderson P, Linnemann C, Ault GS, Ryschkewitsch CF, Stoner GL (1998): Tubulointerstitial nephritis due to a mutant polyomavirus BK strain, BKV (Cin), causing end-stage renal disease. J Clin Microbiol 36:1660–1665.

Soeda E, Maruyama T, Arrand JR, Griffin BE (1980): Host-dependent evolution of three papova viruses. Nature 285:165–167.

Stehle T, Gamblin SJ, Yan YW, Harrison SC (1996): The structure of simian virus 40 refined at 3.1 Å resolution. Structure 4:165–182.

Stewart AR, Lednicky JA, Benzick US, Tevethia MJ, Butel JS (1996): Identification of a variable region at the carboxy terminus of SV40 large T-antigen. Virology 221: 355–361.

Stewart AR, Lednicky TA, Butel JS (1998): Sequence analyses of human tumor-associated SV40 DNAs and SV40 viral isolates from monkeys and humans. J Neurovirol 4:182–193.

Stoner GL, Jobes DV, Fernandez Cobo M, Agostini HT, Chima SC, Ryschkewitsch CF (2000): JC virus as a marker of human migration to the Americas. Microbe Infect 2:1905–1911.

Stoner GL, Ryschkewitsch CF (1998): Reappraisal of progressive multifocal leukoencephalopathy due to simian virus 40. Acta Neuropathol 96:271–278.

Strickler HD, Goedert JJ, Butel JS, Daniel R, Freymuth F, Gibbs A, Griffiths DJ, Jablons D, Jasani B, Jones C, Lednicky JA, Miller CW, Radu C, Richards WG, Shah KV, You L, Corson JM, Gerwin B, Harris C, Sugarbaker DJ, Egan W, Lewis AM, Krause PR, Peden K, Levine AS, Melnick S, Cosentino M, da Costa M, Devairrakam V, Ji J, Palefsky J, Rasmussen L, Shea K, Wacholder S, Waters D, Wright T (2001): A multicenter evaluation of assays for detection of SV40 DNA and results in masked mesothelioma specimens. Cancer Epidemiology Biomarkers & Prevention 10:523–532.

Sugimoto C, Kitamura T, Guo J, Al-Ahdal MN, Shchelkunov SN, Otova B, Ondrejka P, Chollet JY, El-Safi S, Ettayebi M, Grésenguet G, Kocagöz T, Chaiyarasamee S, Thant KZ, Thein S, Moe K, Kobayashi N, Taguchi F, Yogo Y (1997): Typing of urinary JC virus DNA offers a novel means of tracing human migrations. Proc Natl Acad Sci USA 94:9191–9196.

Sullivan CS, Cantalupo P, Pipas JM (2000): The molecular chaperone activity of simian virus 40 large T antigen is required to disrupt Rb-E2F family complexes by an ATP-dependent mechanism. Mol Cell Biol 20:6233–6243.

Sundsfjord A, Flaegstad T, Flo R, Spein AR, Pedersen M, Permin H, Julsrud J, Traavik T (1994a): BK and JC viruses in human immunodeficiency virus type 1–infected persons: Prevalence, excretion, viremia, and viral regulatory region. J Infect Dis 169: 485–490.

Sundsfjord A, Spein AR, Lucht E, Flaegstad T, Morten Seternes O, Traavik T (1994b): Detection of BK virus DNA in nasopharyngeal aspirates from children with respiratory infections but not in saliva from immunodeficient and immunocompetent adult patients. J Clin Microbiol 32:1390–1394.

Sweet BH, Hilleman MR (1960): The vacuolating virus, SV40. Proc Soc Exp Biol Med 105:420–427.

Swenson JJ, Trowbridge PW, Frisque RJ (1996): Replication activity of JC virus large T antigen phosphorylation and zinc finger domain mutants. J Neurovirol 2:78–86.

Tavis JE, Walker DL, Gardner SD, Frisque RJ (1989): Nucleotide sequence of the human polyomavirus AS virus, an antigenic variant of BK virus. J Virol 63:901–911.

Testa JR, Carbone M, Hirvonen A, Khalili K, Krynska B, Linnainmaa K, Pooley FD, Rizzo P, Rusch V, Xiao GH (1998): A multi-institutional study confirms the presence and expression of simian virus 40 in human malignant mesotheliomas. Cancer Res 58:4505–4509.

Tevethia SS (1980): Immunology of simian virus 40. In Klein G, Ed; Viral Oncology; Raven Press: New York, pp 581–601.

Tognon M, Martini F, Iaccheri L, Cultrera R, Contini C (2001): Investigation of the simian polyomavirus SV40 as a potential causative agent of human neurological disorders in AIDS patients. J Med Microbiol 50:165–172.

Tooze J (1981): DNA Tumor Viruses: Molecular Biology of Tumor Viruses; Cold Spring Harbor, NY: Cold Spring Harbor Laboratory.

Tornatore C, Berger JR, Houff SA, Curfman B, Meyers K, Winfield D, Major EO (1992): Detection of JC virus DNA in peripheral lymphocytes from patients with and without progressive multifocal leukoencephalopathy. Ann Neurol 31:454–462.

Torre D, Speranza F, Martegani R, Ferrante P, Omodeo-Zorini E, Mancuso R, Fiori GP (1998): Meningoencephalitis caused by human herpesvirus-6 in an immunocompetent adult patient: Case report and review of the literature. Infection 26:402–404.

Vallbracht A, Löhler J, Gossmann J, Glück T, Petersen D, Gerth H-J, Gencic M, Dörries K (1993): Disseminated BK type polyomavirus infection in an AIDS patient associated with central nervous system disease. Am J Pathol 143:29–39.

Van Gorder MA, Della Pelle P, Henson JW, Sachs DH, Cosimi AB, Colvin RB (1999): Cynomolgus polyoma virus infection—A new member of the polyoma virus family causes interstitial nephritis, ureteritis, and enteritis in immunosuppressed cynomolgus monkeys. Am J Pathol 154:1273–1284.

Villarreal LP (1989): Common Mechanisms of Transformation by Small DNA Tumor Viruses; American Society for Microbiology: Washington, DC.

Villarreal LP (1999): DNA virus contribution to host evolution. In Domingo E, Webster R, Holland J, Eds; Origin and Evolution of Viruses; Academic Press: San Diego, pp 391–420.

Villarreal LP, DeFilippis VR, Gottlieb KA (2000): Acute and persistent viral life strategies and their relationship to emerging diseases. Virology 272:1–6.

Virshup DM, Russo AAR, Kelly TJ (1992): Mechanism of activation of simian virus 40 DNA replication by protein phosphatase 2A. Mol Cell Biol 12:4883–4895.

Völter C, zur Hausen H, Alber D, De Villiers EM (1998): A broad spectrum PCR method for the detection of polyomaviruses and avoidance of contamination by cloning vectors. Dev Biol Stand 94:137–142.

Waheed I, Guo ZS, Chen GA, Weiser TS, Nguyen DM, Schrump DS (1999): Antisense to SV40 early gene region induces growth arrest and apoptosis in T-antigen–positive human pleural mesothelioma cells. Cancer Res 59:6068–6073.

Walker DL, Frisque RJ (1986): The biology and molecular biology of JC virus. In Salzman NP, Ed; The Papovaviridae, Vol 1, The Polyomaviruses; Plenum Press: New York, pp 327–377.

Webster RG (1999): Antigenic variation in influenza viruses. In Domingo E, Webster R, Holland J, Eds; Origin and Evolution of Viruses; Academic Press: San Diego, pp 377–390.

Wechsler P (1996): A Shot in the Dark. *New York Magazine* 29:38–43, 85.

Weggen S, Bayer TA, Von Deimling A, Reifenberger G, Von Schweinitz D, Wiestler OD, Pietsch T (2000): Low frequency of SV40, JC and BK polyomavirus sequences in human medulloblastomas, meningiomas and ependymomas. Brain Pathol 10:85–92.

Wei G, Liu CK, Atwood WJ (2000): JC virus binds to primary human glial cells, tonsillar stromal cells, and B-lymphocytes, but not to T lymphocytes. J Neurovirol 6:127–136.

White FA, Ishaq M, Stoner GL, Frisque RJ (1992): JC virus DNA is present in many human brain samples from patients without progressive multifocal leukoencephalopathy. J Virol 66:5726–5734.

Winklhofer KF, Albrecht I, Wegner M, Heilbronn R (2000): Human cytomegalovirus immediate-early gene 2 expression leads to JCV replication in nonpermissive cells via transcriptional activation of JCV T antigen. Virology 275:323–334.

Xie YC, Hwang C, Overwijk W, Zeng Z, Eng MH, Mule JJ, Imperiale MJ, Restifo NP, Sanda MG (1999): Induction of tumor antigen–specific immunity in vivo by a novel vaccinia vector encoding safety-modified simian virus 40 T antigen. J Natl Cancer Inst 91:169–175.

Yoshiike K, Takemoto KK (1986): Studies with BK virus and monkey lymphotropic papovavirus. In Salzman NP, Ed; The Papovaviridae, Vol 1, The Polyomaviruses; Plenum Press: New York, pp 295–326.

Zeren EH, Gumurdulu D, Roggli VL, Zorludemir S, Erkisi M, Tuncer I (2000): Environmental malignant mesothelioma in Southern Anatolia: A study of fifty cases. Environ Health Perspect 108:1047–1050.

Zietkiewicz E, Yotova V, Jarnik M, Korab-Laskowska M, Kidd KK, Modiano D, Scozzari R, Stoneking M, Tishkoff S, Batzer M, Labuda D (1998): Genetic structure of the ancestral population of modern humans. J Mol Evol 47:146–155.

Zu Rhein GM (1969): Association of papova-virions with a human demyelinating disease (progressive multifocal leukoencephalopathy). Prog Med Virol 11:185–247.

Zu Rhein GM, Chou SM (1965): Particles resembling papova viruses in human cerebral demyelinating disease. Science 148:1477–1479.

Zu Rhein GM, Chou SM (1968): Papova virus in progressive multifocal leukoencephalopathy. Res Publ Assoc Res Nervous Ment Dis 44:307–362.

INDEX